Table of Atomic Weights

Element	Symbol	Atomic Number	Atomic Weight
Actinium	Ac	89	(227)[a]
Aluminum	Al	13	26.98154
Americium	Am	95	(243)
Antimony	Sb	51	121.75
Argon	Ar	18	39.948
Arsenic	As	33	74.9216
Astatine	At	85	(210)
Barium	Ba	56	137.34
Berkelium	Bk	97	(247)
Beryllium	Be	4	9.01218
Bismuth	Bi	83	208.9804
Boron	B	5	10.81
Bromine	Br	35	79.904
Cadmium	Cd	48	112.40
Calcium	Ca	20	40.08
Californium	Cf	98	(251)
Carbon	C	6	12.01115
Cerium	Ce	58	140.12
Cesium	Cs	55	132.9054
Chlorine	Cl	17	35.453
Chromium	Cr	24	51.996
Cobalt	Co	27	58.9332
Copper	Cu	29	63.546
Curium	Cm	96	(247)
Dysprosium	Dy	66	162.50
Einsteinium	Es	99	(254)
Erbium	Er	68	167.26
Europium	Eu	63	151.96
Fermium	Fm	100	(257)
Fluorime	F	9	18.99840
Francium	Fr	87	(223)
Gadolinium	Gd	64	157.25
Gallium	Ga	31	69.72
Germanium	Ge	32	72.59
Gold	Au	79	196.9665
Hafnium	Hf	72	178.49
Hahnium	Ha	105	(260)[b]
Helium	He	2	4.00260
Holmium	Ho	67	164.9304
Hydrogen	H	1	1.00797
Indium	In	49	114.82
Iodine	I	53	126.9045
Iridium	Ir	77	192.22
Iron	Fe	26	55.847
Krypton	Kr	36	83.80
Kurchatovium	Ku	104	(260)[c]
Lanthanum	La	57	138.9055
Lawrencium	Lr	103	(256)
Lead	Pb	82	207.19
Lithium	Li	3	6.941
Lutetium	Lu	71	174.97
Magnesium	Mg	12	24.305
			54.9380
			(258)
			200.59
Molybdenum	Mo	42	95.94
Neodymium	Nd	60	144.24
Neon	Ne	10	20.179
Neptunium	Np	93	237.0482
Nickel	Ni	28	58.71
Niobium	Nb	41	92.9064
Nitrogen	N	7	14.0067
Nobelium	No	102	(255)
Osmium	Os	76	190.2
Oxygen	O	8	15.9994
Palladium	Pd	46	106.4
Phosphorus	P	15	30.97376
Platinum	Pt	78	195.09
Plutonium	Pu	94	(244)
Polonium	Po	84	(210)
Potassium	K	19	39.098
Praseodymium	Pr	59	140.9077
Promethium	Pm	61	(147)
Protactinium	Pa	91	231.0359
Radium	Ra	88	226.0254
Radon	Rn	86	(222)
Rhenium	Re	75	186.2
Rhodium	Rh	45	102.9055
Rubidium	Rb	37	85.4678
Ruthenium	Ru	44	101.07
Samarium	Sm	62	150.4
Scandium	Sc	21	44.9559
Selenium	Se	34	78.96
Silicon	Si	14	28.086
Silver	Ag	47	107.868
Sodium	Na	11	22.98977
Strontium	Sr	38	87.62
Sulfur	S	16	32.06
Tantalum	Ta	73	180.9479
Technetium	Tc	43	98.9062
Tellurium	Te	52	127.60
Terbium	Tb	65	158.9254
Thallium	Tl	81	204.37
Thorium	Th	90	232.0381
Thulium	Tm	69	168.9342
Tin	Sn	50	118.69
Titanium	Ti	22	47.90
Tungsten	W	74	183.85
Uranium	U	92	238.029
Vanadium	V	23	50.9414
Xenon	Xe	54	131.30
Ytterbium	Yb	70	173.04
Yttrium	Y	39	88.9059
Zinc	Zn	30	65.38
Zirconium	Zr	40	91.22

[a] Value in parentheses is the mass number of the most stable or best-known isotope.
[b] Suggested by American workers but not yet accepted internationally.
[c] Suggested by Russian workers. American workers have suggested the name Rutherfordium.

General Chemistry: Principles and Structure

James E. Brady
ST. JOHN'S UNIVERSITY

Gerard E. Humiston
HARCUM JUNIOR COLLEGE

General Chemistry: Principles and Structure

John Wiley & Sons, Inc.
NEW YORK LONDON SYDNEY TORONTO

Library of Congress Cataloging in Publication Data

Brady, James E
  General chemistry.

  Includes index.
   1. Chemistry.   I. Humiston, Gerard E.,
joint author.   II. Title.
QD31.2.B7   540   74-22187
ISBN 0-471-09530-3

Printed in the United States of America

10 9 8 7 6 5 4 3 2

Preface

Why do many students find chemistry difficult? There is no simple answer, but it is probably because of the difficulty of conceptual communication between the instructor, the textbook, and the student. Our goal is to provide clear, complete, and understandable explanations and illustrations of the principles and concepts that will help the student to succeed in a general chemistry course.

In simple language we develop concepts and provide solutions to many example problems. We have assumed no prior learning in chemistry (even though the student may have had one year of high school chemistry), and we have carefully defined each new term before using it in the development of concepts. New terms are in boldface type and are indexed. Important definitions are italicized. Comprehensive problem sets are included at the end of each chapter, and the answers to the odd-numbered problems are given in Appendix E.

In terms of mathematical background, we have assumed only an ability to handle relatively simple algebra and an exposure to logarithms. A review of mathematical concepts is included in Appendix C. We have tried not to frighten the student with endless equations but, instead, to impress on him the importance of quantitative ideas in chemistry and the other sciences. Since most students of general chemistry have not been exposed to calculus, we avoid its use entirely.

Graphic illustrations are used liberally and, in this area, we feel that we have overcome a major deficiency in other textbooks, which rely on two-dimensional illustrations for the communication of the three-dimensional concepts. Although these illustrations are often well drawn and the objects that they depict are recognizable by chemists, students are not chemists. At the end of a general chemistry course, most students have only a vague appreciation for the three-dimensional aspects of chemistry. Furthermore, some two-dimensional representations are difficult to understand, even by chemists. (Consider, for instance, the closest packing of spheres.) If chemists cannot understand them, consider the poor freshman!

The solution to these difficulties is the use of stereoscopic illustrations, and we have included them liberally wherever it seemed pedagogically desirable. Our students' reactions to these illustrations have been favorable. They find them very useful instructional devices. A stereo viewer is included with the text.

We have retained many of the familiar metric system units, particularly the calorie (and kcal) for energy and the atmosphere for pressure. The complete changeover to SI units is several years away, and most practicing chemists and biologists still think and talk in metric units.

The sequence of topics follows a logical rather than purely historical development of concepts. The book begins with the development of the quantitative relationships involving atomic weights, formulas, and chemical equations to facilitate the early introduction of quantitative experiments in the laboratory.

The material on stoichiometry is followed by a discussion (in Chapter 3) of the structure of the atom and the periodic table. We have taken a historical approach, since the evolution of these concepts gives a historical perspective to our currently accepted notions about atomic structure.

The treatment of chemical bonding is divided into two chapters. Chapter 4 deals with the "classical" description of the chemical bond. The ideas developed will carry the student through about two thirds of the book. The modern theories of bonding and molecular structure are discussed in Chapter 16 so that they can be applied to the discussion of descriptive chemistry in the chapters that follow. Bonding theory has been introduced at this late stage for two reasons: (1) the more sophisticated theories are presented after the student has had an opportunity to mature somewhat in his or her understanding of chemistry; and (2) the student will not have forgotten the salient features of these theories by the time they are applied to descriptive chemistry. We believe that when the complexities of bonding are discussed in the first semester they will have to be reviewed extensively before they can be used in the second semester.

Our treatment of solutions is also divided between two chapters. Chapter 5 focuses on solutions (particularly aqueous solutions) as a medium for carrying out chemical reactions. Chapter 9, which follows a discussion of liquids (Chapter 8), deals with the physical properties of solutions as they are affected by the interactions between solute and solvent.

Chapter 5 introduces many important concepts that are developed in greater detail in later chapters (for example, chemical equilibrium and acid-base reactions). The stoichiometry of solutions, the concepts of ionic reactions, and acid-base and redox titrations are also discussed in Chapter 5. This chapter, at a relatively early stage, prepares students for a variety of quantitative and qualitative laboratory experiments that deal with reactions in solution. Furthermore, it reflects our attitude toward descriptive chemistry: students must known in detail certain aspects of the subject. This viewpoint is expressed in Chapter 5. We also feel that there are aspects of descriptive chemistry that a student should know about. We pursue the second approach in Chapters 17 to 19, where trends and similarities in properties (reactivity and structure) are stressed.

Students who have had a high-school chemistry course may be familiar with a good deal of the material in Chapter 5, and the instructor may assign portions of it for review. We think, however, that every student who has had a course in general chemistry should know this material thoroughly.

Thermodynamics (Chapter 10) and kinetics (Chapter 11) are included sequentially to relate the importance of these two factors in determining the outcome of a chemical reaction. The interplay between thermodynamics and kinetics is discussed later in connection with descriptive chemistry.

In Chapter 12, on equilibrium, the equilibrium law is discussed first as an experimental phenomenon, and then it is analyzed in terms of kinetics and thermodynamics. There follows a chapter on acids and bases, and the discussion of equilibrium is concluded with a chapter on ionic equilibrium.

Electrochemistry is considered in Chapter 15, which includes practical applications to electroplating, energy production, and the electrochemical measurement of concentrations.

As previously mentioned, the intention of the descriptive chemistry chapters (Chapters 17 to 19) is to display trends and similarities in the structure and reactivity of the elements and their compounds. These chapters are intended to illustrate points under discussion, and they should not be memorized by the student. Their dominant theme is structure, and the stereoscopic illustrations, particularly, fulfill an important role.

We include chapters on organic and biochemistry, since most students who take general chemistry also take a separate organic chemistry course. Therefore, we have illustrated the breadth of this subject instead of dealing with a particular topic in depth. In Chapter 21, on biochemistry, we show how complex biomolecules are composed of relatively simple building blocks and how their structures and biological functions are accounted for.

The final chapter is on nuclear chemistry. It includes, in addition to the usual topics, illustrations of how the chemist can take advantage of nuclear phenomena to aid him in his understanding of chemical processes.

For completeness, more information has been included here than can usually be presented in a two-semester course. What, then, can be cut away? This decision must be made by the instructor. Since it is often the descriptive chemistry that is pruned, we have made each section, as nearly as possible, a self-contained unit. Thus the instructor can stress the areas that he or she feels are important.

The order of chapters reflects our own bias about the sequence of topics in a general chemistry course. If an instructor prefers not to divide the discussions on bonding between two semesters, the topics in Chapter 16 easily can be presented after Chapter 4. Similarly, a great deal of the material in Chapter 5 can be included in Chapter 9.

Finally, we thank all of the people who helped us to produce this book. Among them are Professors John Burmeister, Jack Cummins, Luther Erickson, I. C. Hisatsune, John Holum, John Mason, Jack Powell, Steven Schullery, Michael Wartell, and Kenneth Whitten who offered criticism and helpful suggestions; our colleagues who served as sounding boards and gave constructive comments; Professors Ernest Birnbaum and Eugene Holleran who counseled us on several critical chapters; Mrs. Leatrice Groffman, William Moschini, and the staff of the St. John's Computer Center who helped to produce the stereoscopic illustrations; Drs. Don Cromer and Carroll Johnson who provided copies of some of these illustrations; Edward Brady for lending us the alum crystal shown in Figure 7.1; and Patricia Cunningham and Anna Rixinger who worked out the solutions to the problem sets. We express appreciation for the patience of our editor, Gary Carlson, of Wiley, who cajoled, threatened, and persuaded us through a host of deadlines. And, of course, we must thank our wives and children who put up with all this.

James E. Brady
Gerard E. Humiston

Additional stereopticon viewers may be ordered
in quantities of ten or more by writing to:

The Taylor-Merchant Corporation
25 West 45th Street
New York, New York 10036

To the Student

This textbook contains a substantial number of computer-drawn stereo (three-dimensional) illustrations that are intended to help you visualize some of the 3-D aspects of chemistry. Each stereo illustration, like the one below, consists of a pair of drawings that, at first glance, appear to be identical; actually they are slightly different. When viewed in such a way that the left eye focuses on the left drawing and the right eye focuses on the right drawing, your mind brings them together and creates a three-dimensional image.

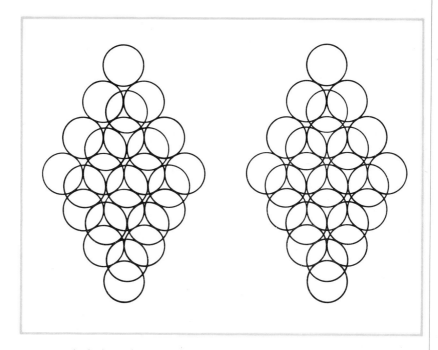

A viewer is included inside the back cover of the book to help you obtain a 3-D illusion. To get accustomed to using the viewer, assemble it according to the directions printed on the viewer and locate the bottom edge along the solid line under the drawing above. The viewer should be placed so that the folded support panel is placed between the two drawings. Now look through the lenses of the viewer, keeping both eyes open. Start with your eyes a few inches above the viewer. At first you may find that it takes a moment for the stereo image to fuse. You may have to move the viewer slightly if a double image persists. The drawing should appear to be two layers of tangent circles, one above the other.

In your study of chemistry you will encounter many new terms whose meaning you will need to know to understand discussions that follow. These are set in **boldface** type the first time that they appear in the text. Each of these is also included in the index for later reference. Important definitions have been set in *italics* to call your attention to them.

Contents

1
Introduction

Never before in history has man found himself able to influence his physical environment, for its good or detriment, as he can today. This situation is a result of his scientific discovery. This book deals with a branch of physical science called chemistry which concerns itself with the composition of substances, the ways in which their properties are related to their composition, and the interaction of these substances with one another to produce new materials.

The degree to which chemistry has changed civilization is evident everywhere. A good part of the clothing we wear, the automobiles we drive, and other products we encounter daily are composed of materials that simply did not exist at the turn of the century. In recent years the realization that a living organism is a complex chemical "factory" has generated a strong interest in the study of biochemistry and has brought great advances in our knowledge of the nature of life. Medicines created in the laboratory have made us healthier and, through the cure of disease, have prolonged our lives. It has been only recently, however, that we have also become painfully aware of a host or problems arising from this growth of technology. It is the solution of such problems that poses much of the challenge for chemistry in the future.

In this first chapter we consider how science operates, see the materials and concepts that chemists work with and are concerned about, and see how the concept of the atom became firmly established. We shall also introduce you to some of the jargon used by chemists. It is important to become familiar with chemical terminology (which will undoubtedly require some memorization), since many of the difficulties that students encounter in the study of chemistry can be traced to their inability to "speak the language."

1.1 The Scientific Method

Chemistry is called a science because its investigations of nature employ a specific procedure termed the **scientific method.** Any science, whether it be chemistry, biology, physics, or psychology, uses this method of investigation. It consists of a series of steps, the first of which is **observation.** A chemist starts out by collecting bits of information (called **data**) about nature. For example, he might observe that when hydrogen gas and oxygen gas are heated together, a violent explosion results and water vapor is produced. This type of observation, which is devoid of numerical information, is said to be **qualitative.** A different chemist might make some measurements and find that, under the same conditions of temperature and pressure, one cubic foot of hydrogen gas will completely consume only one-half cubic foot of oxygen gas to produce one cubic foot of water vapor. This is a **quantitative** observation since it results in numerical data. Quantitative measurements are generally more useful to a scientist than are qualitative observations because the former provide more information.

After a large amount of data has been collected, it is desirable to find a way to summarize the information in a concise way. Statements that accomplish this goal are called **laws** and, in a sense, simply serve as a means of convenient storage for vast quantities of experimental facts. They also provide a means of predicting the results of some, as yet, untried experiment. For instance, after a series of measurements regarding the relative quantities of hydrogen and oxygen that will react with one another, a chemist would conclude that when these two substances interact at the same temperature and pressure to form water, one volume of oxygen gas consumes two volumes of hydrogen gas. This simple statement is a law dealing with the reaction of hydrogen with oxygen. If we had five cubic feet of oxygen gas, we would predict that the optimum production of water would require 10 cubic feet of hydrogen.

A law may be a simple verbal statement, such as, "the force of attraction between oppositely charged particles decreases as their distance of separation increases," or it may be expressed in the generally more useful form of an equation, such as

$$F = \frac{q_1 q_2}{r^2}$$

in which F is the force of attraction between two oppositely charged particles, q_1 and q_2 are the charges on the particles, and r is their distance of separation. Laws quite commonly are expressed in equation form.

As we have noted, a law simply correlates large quantities of information. Laws in themselves do not explain why nature behaves as it does. Scientists, being human (despite what you may have heard to the contrary), are not satisfied with simple statements of fact and seek to explain their observations. Thus the second step in the scientific method is to propose tentative explanations, or **hypotheses,** that may be tested by experiment. If these prove sound, they develop into **theories.** Theories themselves always serve as guides to new experiments and are constantly being tested. When a theory is proven incorrect by experiment, it must either be discarded in favor of a new one or, as is often the case, modified so that all of the experimental observations may be accounted for. Science develops, then, by a constant interplay between theory and experiment.

No science can proceed very far without resorting to quantitative observations. This means that scientists, and more specifically chemists, must make measurements. The process of measurement usually involves the reading of numbers from some device; because of this, there is nearly always some limitation on the number of meaningful digits that may be obtained in an experimentally determined quantity. For example, let us consider the measurement of the length of a piece of wood with two different rulers, as shown in Figure 1.1.

Using the ruler in Figure 1.1a we might read the length of the piece of wood as 3.2 inches. To arrive at this number we are forced to estimate the second digit; that is, we must decide whether the length lies closer either to 3.2 or to 3.3 inches. Because we are making an estimate, some uncertainty exists in the second digit (the 2) and the third digit, for all practical purposes,

(a)   (b)

Figure 1.1

Measuring the length of a piece of wood with two different rulers. (a) Length = 3.2 in.
(b) Length = 3.24 in.

is completely unknown. Therefore, for measurements made with the ruler in Figure 1.1a, we are not justified in reporting numbers containing more than two figures.

Numbers (or digits) that arise as the result of measurement are called **significant figures.** When a number is written to represent the result of a measurement, it is always assumed that, unless stated otherwise, only the rightmost digit is uncertain. Thus, the measurement illustrated in Figure 1.1a yields a number with two significant figures.

In Figure 1.1b it is possible to estimate the length of the piece of wood to be 3.24 inches because of the greater number of calibration marks on the ruler. In this case there is no uncertainty about the 2, and the third digit is estimated to be equal to four. The number 3.24 contains three significant figures.

The importance of significant figures is that they indicate to us the reliability of our measurements. In the determination of the length of the piece of wood described above, we saw two different values, obtained with two different measuring devices; our intuition tells us that we can place more confidence in the value with the greater number of significant figures. Laws and theories are derived from measured quantities, and our confidence in them is directly related to the quality of the data on which they are based.

In discussing measured quantities, two words are often encountered. The term **precision** refers to the number of significant figures in a measurement and the term **accuracy** refers to how close an experimental observation lies to the true value. Generally a measurement is not accurate if it is not precise. In our example above, the value 3.24 in. has greater precision than 3.2 in. and also lies closer to the true length (whatever that may be).

There are instances where a number may be precise but not particularly accurate. The ruler in Figure 1.2, for instance, is improperly calibrated. Failure to notice this error in calibration (hence, failure to adjust appropriately the measurements made with this ruler) would lead to results that would all be wrong by one inch, even though three significant figures can be read from the scale.

In almost all cases, measured numbers are used to calculate other quantities. In these situations we must exercise care to assure ourselves of obtaining the proper number of significant figures in the computed result. To see how problems might arise, suppose we wish to calculate the area of a

Figure 1.2
An improperly calibrated
ruler. All measurements
made with this ruler will
be in error by one inch.

rectangle whose sides have been measured to be 6 and 7.0 inches long. We know that the area is the product of these two numbers, that is 6 in. × 7.0 in. = 42 in.² The question is, how many significant figures are justified in the answer? To determine this, let us consider the number of significant figures in each of the measured lengths.

The number 6 has one significant figure, which implies that there is some uncertainty in this digit (i.e., the length of the side could actually be either 5 or 7 inches). We thus are confident in the length of this side to, at best, one part in six, or about 17%. As for the other side of the rectangle, the minimum uncertainty implied by the number 7.0 is ± 0.1 in.; therefore the percent uncertainty is (0.1/7.0) × 100 = 1.4%. Our degree of confidence in the value computed for the area of the rectangle depends on how reliable are the measured lengths of the sides. If one side is uncertain to the extent of 17%, we cannot expect the uncertainty in the area to be less than 17%, which means that the number 42 in.², which implies an uncertainty of only (1/42) × 100 = 2.4%, must be rounded off to 40 in.² (which implies an uncertainty of 25%). To avoid this rather tedious analysis, every time we perform a computation, we can use the general rule that *for multiplication or division, the product or quotient should not possess any more significant figures than the least precisely known factor in the calculation.*

For addition and subtraction the procedure used to determine the number of significant figures in an answer is slightly different. Here the number we write as the result of a calculation is determined by the figure with the fewest number of decimal places. Thus, in the sum

$$\begin{array}{r} 4.371 \\ 302.5 \\ \hline 306.871 \end{array}$$

we must round off the answer to 306.9. The reason for this is that the two digits which follow the 5 in the number 302.5 are completely unknown, that is they could conceivably have any value from 0 to 9. As a result, the last two digits in the answer 306.871 must also be completely uncertain. If we follow our rules for writing numbers such that *only* these digits with real significance are included, we are not justified in writing these last two digits and must round off the answer to 306.9. For addition and subtraction, then, the rule is that *the absolute uncertainty in a sum or difference cannot be smaller than the largest absolute uncertainty in any of the terms in the calculation.* In the example above, the absolute uncertainty implied by the number 4.371 is ± 0.001 whereas the uncertainty in 302.5 is implied to be ±0.1. According to our rule, the sum of these two cannot have an uncertainty *less* than ±0.1; therefore the sum 306.871 must be rounded to 306.9 in order to suggest an uncertainty of ±0.1.

In some calculations we employ numbers that come from definitions (such as 3 feet = 1 yard) or that are the result of a direct count (such as the number of people in a room). These numbers are called exact numbers and contain no uncertainty (e.g., there are exactly 3 feet in 1 yard, or the number of people in a room must be a whole number). When such quantities are used in computation, they may be considered to possess an infinite number of significant figures. Thus, the conversion of a measured length of 4.27 yards into feet would be accomplished as

$$4.27 \text{ yd} \left(\frac{3 \text{ ft}}{1 \text{ yd}} \right) = 12.8 \text{ ft}$$

where the number of significant figures in the product is determined by the number of significant figures in the measured length. Note that in performing this calculation we have cancelled the units, yards.

1.3
The Metric System of Units

Units (sometimes called dimensions) form an integral part of any measurement. For instance, to say that the distance between two points is "three" is meaningless unless a specific unit or units (inches, feet, miles, etc.) are associated with the number. Chemists have traditionally used the **metric system** of units in all their measurements. Recently a slightly modified version of the metric system has been recommended for adoption. This revised set of units is known as the **International System of Units** (SI). The basic defined SI units are shown in Table 1.1.

The SI units are slowly being accepted; however, the older metric system is slow to fade away and is still used by most practicing chemists. Furthermore, its existence in the scientific literature demands that we be aware of metric system units. Some of the SI units are used in this book, but in many cases the more familiar metric system units are retained.

In chemistry it is necessary to measure, on a routine basis, mass, length, and volume. The units that chemists ordinarily use to express these quantities are based on the **gram** (abbreviated, **g**), the **meter** (**m**) and the **liter** (**l**), respectively. Other units within the metric system are derived from these by the use of an appropriate prefix and are related to them in a decimal fashion (see Table 1.2). Thus, for example, 1 meter = 100 centimeters = 1000 millimeters. A relationship also exists between the units of length and of volume since the

Table 1.1
Basic SI Units

Physical Quantity	Name of Unit	Symbol
Mass	Kilogram	kg
Length	Meter	m
Time	Second	s, sec
Electric current	Ampere	A
Temperature	Kelvin	K
Luminous intensity	Candela	cd

Table 1.2
Relationships between Units in the Metric System

Prefix	Multiplication Factor	Examples	Abbreviation
Kilo-	1000	1 kilometer = 1000 meter 1 kilogram = 1000 gram	km kg
Deci-	1/10	1 decimeter = 0.1 meter	dm
Centi-	1/100	1 centimeter = 0.01 meter	cm
Milli-	1/1,000	1 millimeter = 0.001 meter 1 milliliter = 0.001 liter 1 milligram = 0.001 gram	mm ml mg
Micro-	1/1,000,000	1 micrometer = 1 micron 1 micron = 0.000,001 meter 1 microliter = 0.000,001 liter	μ μl

liter is currently defined as exactly 1000 cm³, and therefore, 1 ml = 1 cm³. Table 1.3 gives the approximate English equivalents of some of these units. You will probably find it to your advantage to develop a feel for the magnitude of these quantities.

When we express measurements in some of these units, either very large or very small numbers are often encountered. To avoid having to write down a large number of zeros, it is convenient to express these quantities as the product of a number lying between 1 and 10 multiplied by 10 raised to some power. This type of representation is called **exponential notation** or **scientific notation**[1]. Using this notation we could, for example, write 1 kilogram = 1×10^3 grams rather than 1 kilogram = 1000 grams. If we write numbers in this fashion, the real usefulness of the metric system is apparent since converting from one unit to another merely involves changing the exponent on the 10. For example,

$$1 \text{ km} = 1 \times 10^3 \text{ m} = 1 \times 10^5 \text{ cm} = 1 \times 10^6 \text{ mm}$$

Table 1.3
Comparison of the English and Metric Systems

Length	1 meter = 39.37 inches 2.540 centimeter = 1 inch
Mass	1 kilogram = 2.204 pounds 453.6 grams = 1 pound
Volume	1 liter = 1.057 quarts 29.57 milliliters = 1 ounce 28.32 liters = 1 cubic foot

[1] A complete discussion of exponential notation and how it can be applied to performing numerical calculations will be found in Appendix C.

You might compare this to the conversions between miles, yards, feet, and inches in the English system.

There are occasions when the presence of zeros leads to difficulties in determining the number of significant figures in a number. The use of exponential notation allows us to eliminate any confusion that might arise. For example, suppose we had measured the length of an object with a ruler and had found it to be 1.2 m long. This number possesses two significant figures and implies an uncertainty of about one part in 12 ($0.1/1.2 = 1/12$). We could also express this length as 1200 millimeters and, because it still represents the same measurement, it must still only possess two significant figures, the 1 and the 2. The two zeros in the 1200 are used only to locate the decimal point. Unfortunately, someone unfamiliar with our experiment might think that all four digits are significant, implying an uncertainty of only one part in 1200, which is not at all what we intended to convey.

By writing numbers using exponential notation we can eliminate this ambiguity. Thus our 1200 mm could be written as 1.2×10^3 mm; here the first portion of the number expresses only two significant figures. Had we wanted to specify four significant figures, we could have done so by writing the number as 1.200×10^3 mm.

To summarize, then, the only time that zeros are considered as significant figures is when they are not present for the sole purpose of locating the decimal point. Thus, the quantity 0.0072 has but two significant figures while 0.007020 has four, since these are written as 7.2×10^{-3} and 7.020×10^{-3}, respectively. In the latter example, the zero between the 7 and the 2, as well as the rightmost zero are significant because they are not needed to position the decimal point.

1.4 Matter

In their investigations, chemists study the properties and transformations of **matter.** Matter is anything that occupies space and possesses **mass.** In setting down this definition a careful distinction is made between the terms mass and weight, even though we generally use them as if they were interchangeable. The **mass** of a body is a measure of its resistance to a change in velocity. A ping-pong ball moving at 20 mph, for example, is easily deflected by a soft breeze while a cement truck is not. Quite clearly, the mass of the cement truck is considerably greater than that of the ping-pong ball. The term, **weight,** refers to the force with which an object is attracted to the earth. Force and mass are related to each other by the equation

$$F = ma$$

where F = force, m = mass, and a = acceleration. In order to accelerate a body, a force must be applied to it. When an object is dropped, it accelerates because of the gravitational attraction of the earth while an object resting on the earth exerts a force (its weight, W) that is equal to its mass, m, multiplied by the acceleration due to gravity, g; that is,

$$W = mg$$

The weight of an object thus depends on g, a quantity that varies slightly from place to place on the earth's surface. Since the weight of an object depends on where the measurement is made, it is considered more desirable to express the amount of matter present in the object in terms of its mass.

Figure 1.3

A balance. Known masses are added to the right pan until the pointer is centered. The contents of each pan then have the same weight and therefore also possess the same mass.

The determination of mass (a process, oddly enough, called **weighing**) is actually performed by comparing the weights of two objects, one of known mass, the other of unknown mass. The apparatus used for this is called a **balance** (Figure 1.3). The object to be weighed is placed on the left pan of the balance and objects of known mass are added to the other until the two pans are balanced. At this point the contents of both pans weigh the same and, since they both experience the same gravitational acceleration, both pans contain equal masses. In chemistry, as mentioned previously, we generally measure mass in grams.

A modern single pan analytical balance is shown in Figure 1.4. To use this balance the object to be weighed is placed on the pan and the weights are adjusted internally, by turning the knobs on the face until balance is achieved. Although this modern apparatus looks quite different from its older brother, they both operate on exactly the same principle, that is, a balance beam that is adjusted to have the same weight on either side of the knife-edge pivot.

The properties that we use to describe matter may be classified into two broad categories; **extensive properties,** which depend on the size of a sample of matter, and **intensive properties,** which are independent of sample size. Of the two, intensive properties are the more useful since a substance[2] will exhibit the same intensive property regardless of how much of it we examine.

Examples of extensive properties are mass and volume, for as the quantity of a substance increases, its mass and volume also increase. Some examples of intensive properties are melting point, boiling point, and **density** (which is defined as the ratio of an object's mass to its volume). Water, for instance, has a density of 1 g/ml. This means that if we had one gram of water, it would occupy a volume of 1 ml. If instead we had 20 g of water, we would find that it occupied a volume of 20 ml, but the ratio of the water's mass to

---

[2] The word *substance* is used frequently by the chemist and is taken to mean the material of which an object is composed. For example, an ice cube is composed of the substance water.

Figure 1.4
A modern analytical balance.

volume, 20 g/20 ml, is still the same as 1 g/1 ml. Therefore we see that density is a property that does not depend on sample size.

Notice that the intensive property, density, is computed as the ratio of two extensive properties, mass and volume. Later in our discussion of chemistry, we shall encounter quite a few intensive properties defined in a similar fashion.

In speaking of the properties of substances, we also distinguish between **physical properties** and **chemical properties.** A physical property can be specified without reference to any other substance. All of the examples cited earlier in this section are physical properties. A chemical property, on the other hand, states some interaction between chemical substances. We say, for instance, that sodium is very reactive toward water. Reactivity is a chemical property that refers to the tendency of a substance to undergo a particular chemical reaction. However, to say simply that a substance is very reactive, without specifying "with what" or under what conditions, is not particularly helpful. Sodium, for example, is very reactive with water but quite unreactive toward the gas, helium.

<h3>1.6
Mixtures,
Compounds,
and Elements</h3>

The three words that form the title to this section lie very close to the heart of chemistry because it is with mixtures, compounds, and elements that the chemist in the laboratory works. He must therefore understand what they are and be able to distinguish between them before he can hope to make use of them.

Mixtures differ from elements and compounds in that they may be of variable composition. A solution of sodium chloride (table salt) in water is a mixture of two substances, and we know that by dissolving varying quantities of salt in water we can obtain solutions with a wide range of compositions. Most materials found in nature, or prepared in the laboratory, are not pure but, instead, are mixtures. One of the greatest problems faced by the chemist is the separation of mixtures into their components. This can usually be accomplished by some physical process (as opposed to a chemical reaction). Our solution of sodium chloride, for instance, if left to evaporate will leave the salt behind as a solid. Had we wished to recover the water as well, we could have boiled the solution in an apparatus similar to the one in Figure 1.5 and collected the water after it had condensed from the steam. This is one method, incidently, that is used to desalinate sea water.

Mixtures can be described as being either **homogeneous** or **heterogeneous.** A homogeneous mixture, such as a solution of sodium chloride in water, has uniform properties throughout. If we were to sample any portion of a sodium chloride solution, we would find that it has the same properties (e.g., composition) as any other portion of the solution; we say that is consists of a single **phase.** Thus, we define a **phase** as any part of a system that has uniform properties and composition.

A heterogeneous mixture, such as oil and water, is not uniform (Figure

Water out

Condenser

Water in

Figure 1.5

Distillation apparatus. Sodium chloride solution is boiled in the flask. The steam is converted to the liquid in the condenser.

Figure 1.6
Oil and water—a heterogeneous mixture. (a) Before mixing. (b) After mixing.

Before mixing

(a)

After mixing

(b)

1.6). If we were to sample one portion of the mixture, it would have the properties of water while some other part of the mixture would have the properties of oil. This mixture consists of two phases, the oil and the water. If we shook the mixture so that the oil was dispersed throughout the water as small droplets (as in a salad dressing), all of the oil droplets taken together would still constitute but a single phase, since the oil in one droplet has the same properties as the oil in another. If we added an ice cube to this "brew," we would then have three phases; the ice (a solid), the water (a liquid), and the oil (another liquid). In all of these examples, we can detect the presence of two or more phases because a **boundary** exists between them.

A **compound** differs from a mixture because its composition is not variable. For example, all samples of pure water, regardless of their source, contain the same relative quantities of hydrogen and oxygen. Also, when hydrogen is reacted with oxygen to produce water, the relative amounts of hydrogen and oxygen that combine are always the same. Thus, whenever 1.0 g of hydrogen reacts, it is always observed that 8.0 g of oxygen are consumed.

A useful feature of pure compounds is that they undergo phase changes (e.g., solid to liquid or liquid to gas) at constant temperature. Ice, for instance, melts at a temperature of 32° F, a temperature that remains constant while the water undergoes the change from solid to liquid. When mixtures exhibit phase changes, they generally do so over a range of temperatures. This phenomenon provides us with one experimental test to determine when we have obtained a pure compound.

Chemical compounds, as we have implied, are composed of still simpler substances called **elements.** As a rule, a compound can be decomposed to produce its elements only by chemical reaction, and processes that serve to separate the components of a mixture will not suffice to separate the elements of a compound. Thus the procedures that can be employed to separate the components of a solution of water and sodium chloride are not the same as those that must be used to break down the compound water into its elements, hydrogen and oxygen. Elements are also the simplest forms of matter that can exist under conditions that we encounter in a chemical laboratory; thus they are the simplest forms of matter with which the chemist deals directly.

In the early days of chemistry it was not at all easy to determine if a substance was an element. If it could be decomposed, it surely was not; however, failure to decompose a substance did not necessarily prove that it was an element. It could have been that the proper chemical reaction or set of conditions required to break the material down had not yet been found. Only in relatively recent times have instruments and techniques become available that have allowed us to make unambiguous decisions about whether some new substance is, in fact, a new element.

It is not surprising that the early history of chemistry was marked by incorrect theories about what occurred during chemical reactions. It had long been observed, for example, that when the combustion of wood took place, the resulting ash was very light and fluffy. Metals, too when heated and caused to react with air, changed their appearance. The resulting product was less dense than the original metal and thus appeared lighter. These observations led to the conclusion that "something," which the German chemists Becher and Stahl called **phlogiston,** was lost by substances when they burned. Even when it was pointed out that metals gained weight when they were heated in air, the theory was salvaged by concluding that phlogiston was simply lighter than air!

The reluctance to abandon the crumbling phlogiston theory demonstrates a general human phenomenon. New theories are difficult to come by, and old ones become so thoroughly entrenched that it is often tempting to try to shore up a sagging theory rather than to dream up a new one that will do a better job of explaining all of the observed facts.

It was Antoine Lavoisier, a French chemist, who finally laid the phlogiston theory to rest and set chemistry on the proper course again. He demonstrated by his experiments that the combustion process actually occurred by the reaction of substances with oxygen. He also showed, *through careful measurements,* that if a reaction is carried out in a closed container, so that none of the products of reaction escape, the total mass present after the reaction has occurred is the same as before the reaction began. These observations form the basis of the **law of conservation of mass,** which states that *mass is neither created nor destroyed in a chemical reaction.*[3]

The work of Lavoisier clearly demonstrated the importance of careful measurement. After his book, *Traité Elémentaire de Chemie,* appeared in 1789, many chemists were inspired to investigate the quantitative aspects of chemical reactions. These investigations led to another important chemical law, the law of definite proportions.

The **law of definite proportions** states that, *in a pure chemical substance, the elements are always present in definite proportions by mass.* In the substance water, for instance, the ratio of the mass of hydrogen to the mass of oxygen is always 1/8, regardless of the source of the water. Thus, if 9.0 g of water were decomposed, 1.0 g of hydrogen and 8.0 g of oxygen would be obtained, while if 18.0 g of water were broken down, 2.0 g of hydrogen and

---

[3] Einstein has shown that there is a relationship between mass and energy, $E = mc^2$, where c is the speed of light. Energy changes that take place during chemical reactions therefore are also accompanied by mass changes; however, the changes in mass are far too small to be detected experimentally. For example, the energy change associated with the reaction of 2 g of hydrogen with 16 g of oxygen happens to be equivalent to a mass change of approximately 10^{-16} g. Sensitive analytical balances can only detect mass differences of 10^{-6} to 10^{-7} g. Consequently, as far as the average chemist is concerned there are no observable mass increases or decreases accompanying chemical reactions.

16.0 g of oxygen would be produced. Furthermore, if 2.0 g of hydrogen are mixed with 8.0 g of oxygen and the mixture ignited, 9.0 g of water are formed and 1.0 g of hydrogen remains unreacted. Again, in the product the hydrogen to oxygen mass ratio is 1/8. Thus, varying the amounts of hydrogen and oxygen present during reaction does not alter the composition of the water produced.

1.8
The Atomic Theory of Dalton

The real father of modern chemistry could well be considered to be the Englishman, John Dalton, who proposed his atomic theory of matter around 1803. The concept of the atom (from the Greek *atomos* meaning indivisible) did not originate with Dalton. The Greek philosophers Leucippus and Democritus suggested, as early as 400 to 500 B.C., that matter cannot be forever divided into smaller and smaller parts and that ultimately particles would be encountered that would be indivisible. These early proposals, however, were not based on the results of experiments and were little more than exercises in thought. Dalton's theory was different because it was based on the laws of conservation of mass and definite proportions, laws that were derived from many direct observations.

The theory Dalton proposed can be expressed by the following postulates:

1. Matter is composed of indivisible particles called atoms.
2. All atoms of a given element have the same properties (e.g., size, shape, and mass) and differ from the properties of all other elements.
3. Chemical reaction consists merely of a reshuffling of atoms from one set of combinations to another.

The test of any theory, of course, is how well it explains already existing fact and whether it can predict as yet undiscovered laws. Dalton's theory proved successful on both counts.

First, it accounts for the law of conservation of mass. If a chemical reaction does nothing more than redistribute atoms and no atoms are lost from the system, it follows that the total mass must remain constant when the reaction occurs.

Second, it explains the law of definite proportions. To see this, we imagine a substance formed from two elements, say, *A* and *B*, in which each molecule of the substance is composed of one atom of *A* and one atom of *B*. We shall define a **molecule** as *a group of atoms bound tightly enough together that they behave as, and can be recognized as, a single particle* (just as a car is composed of many parts held together tightly enough so that we identify them as a car). Let us also suppose that the mass of an atom of *A* is twice the mass of an atom of *B*. Then in one molecule of this substance, twice as much mass is contributed by *A* as by *B* and the ratio of the mass of *A* to the mass of *B* in this molecule is 2/1. If we take a large collection of these molecules, we will always have equal numbers of *A* and *B* atoms; therefore, regardless of the size of the sample, we still always have a mass ratio (*A* to *B*) of 2/1. Also, if we were to react *A* and *B* together to form this compound, each atom of *A* would combine with only one atom of *B*. If we were to mix 100 atoms of *A* with 110 atoms of *B*, after the reaction was over, we would be left with 10 atoms of *B* unreacted.

Third, Dalton's theory predicted the **law of multiple proportions.** This law states that, *when two different compounds are formed from the same two elements, the masses of one element, which react with a fixed mass of the* other, are in a ratio of small whole numbers. Actually, this sounds more complicated than it really is. Let us consider the two compounds formed by carbon and oxygen. In one of them (carbon monoxide) we find 1.33 g of oxygen combined with 1.00 g of carbon while in the second (carbon dioxide) there are 2.66 g of oxygen combined with 1.00 g of carbon. If we examine the ratio of the masses of oxygen (1.33 g/2.66 g) that combines with a fixed mass of carbon (1.00 g), we observe a ratio of small whole numbers,

$$\frac{1.33}{2.66} = \frac{1.33/1.33}{2.66/1.33} = \frac{1}{2}$$

This is consistent with the atomic theory if we consider that carbon monoxide contains *one* atom of carbon and *one* of oxygen whereas carbon dioxide contains *one* atom of carbon and *two* atoms of oxygen. Since carbon dioxide has twice as many oxygen atoms bound to a carbon atom as does carbon monoxide, the weight of oxygen in a molecule of carbon dioxide *must* be twice the weight of oxygen in a molecule of carbon monoxide.

Example 1.1

Nitrogen forms several different compounds with oxygen. In one it is observed that 2.62 g of nitrogen are combined with 1.50 g of oxygen while in another, 0.656 g of nitrogen are combined with 1.50 g of oxygen. Show that these data demonstrate the law of multiple proportions.

Solution

In both cases we are dealing with a weight of nitrogen that combines with 1.50 g of oxygen. If these data do fit the law of multiple proportions, the ratio of the masses of nitrogen in the two compounds should be a ratio of small whole numbers. Let us take the ratio

$$\frac{2.62}{0.656}$$

Dividing the numerator and denominator by 0.656, we get

$$\frac{4.00}{1.00}$$

which is indeed a ratio of small whole numbers

Example 1.2

Sulfur forms two compounds with fluorine. In one of them it is observed that 0.447 g of sulfur is combined with 1.06 g of fluorine while in the other, 0.438 g of sulfur is combined with 1.56 g of fluorine. Show that these data illustrate the law of multiple proportions.

Solution

Let us first calculate the mass of sulfur that combines with the same weight of fluorine in the two compounds. For the first compound the equivalence between the weights of sulfur and fluorine is expressed as

$$1.06 \text{ g fluorine} \sim 0.447 \text{ g sulfur}$$

where the symbol $\sim$ is taken to mean "equivalent to." The equivalence here stems from the experimental fact that these weights of fluorine and sulfur are chemically combined with each other; in this case, we are speaking of a chemical equivalence.

For the second compound we write

$$1.56 \text{ g fluorine} \sim 0.438 \text{ g sulfur}$$

For the purposes of calculation the equivalence implied by the $\sim$ behaves the same as an equality so that this sort of equivalence relationship can be used to construct conversion factors useful in mathematical computations. For the second compound, then, we can find the weight of sulfur combined with 1.06 g of fluorine.

$$1.06 \; \cancel{\text{g fluorine}} \times \left(\frac{0.438 \text{ g sulfur}}{1.56 \; \cancel{\text{g fluorine}}} \right) \sim 0.298 \text{ g sulfur}$$

We now know the weights of sulfur that combine with the same weight of fluorine (1.06 g) in these two compounds. The next step is to look at the ratio of these weights; that is,

$$\frac{0.447}{0.298} = \frac{0.447/0.298}{0.298/0.298} = \frac{1.50}{1.00}$$

which is the same as 3/2, a ratio of small whole numbers.

1.9 Atomic Weights

Dalton's atomic theory proved so successful at explaining the laws of chemistry that it was accepted almost immediately. Since the key to the success of the theory had been the concept that each element had a characteristic atomic mass, chemists at once set out to measure them. It was at this point that a serious problem arose.

Because of their small size, there certainly was no way to determine the masses of individual atoms. All that scientists could hope to do was to arrive at a set of relative atomic weights. (Note that the terms atomic mass and atomic weight are used interchangeably.) We might determine, for instance, that in one compound formed between carbon and oxygen 3.0 g of carbon were combined with 4.0 g of oxygen. In other words, oxygen contributed 4/3 (or $1\frac{1}{3}$) times as much mass toward the compound as did the carbon. If this substance is made up of molecules, each containing one atom of carbon and one atom of oxygen, then it follows that each oxygen atom must weigh $1\frac{1}{3}$ times as much as one carbon atom; thus it appears that we have established the relative masses of these two elements.

The arguments just presented rest on a very critical assumption; that is, that the compound we were discussing is composed of one atom of carbon and one atom of oxygen. If this assumption about the number of carbon and oxygen atoms in a molecule is false, we have arrived at the wrong relative weights. Because of this kind of difficulty a self-consistent table of atomic weights was not developed until some 60 years after Dalton had introduced his theory. We take a closer look at the solution of this problem in Chapter 6.

A complete table of atomic weights appears on the inside back cover of the book. This is a table of relative atomic weights with the masses of atoms expressed in units called **atomic mass units** (amu). The size of this unit is chosen in a rather arbitrary way. To see this, let us return to our discussion of the compound between carbon and oxygen.

Let us assume, for the moment, that the substance is in fact composed of one atom of carbon and one of oxygen; then, as mentioned, one oxygen atom weighs 1.33 times as much as one carbon atom. If we were to define the amu as equal to the mass of one carbon atom, we would then assign carbon an atomic weight of 1.00 amu and oxygen an atomic weight of 1.33 amu. Since we might find it more convenient to have these numbers as whole numbers (or as near to whole numbers as possible), we might define the amu as equal to 1/3 of the mass of a carbon atom. If we did this, we would assign carbon an atomic weight of 3.00 amu and oxygen, which is 4/3 times heavier, would have an atomic weight equal to 4.00 amu. Thus the size of the amu, and hence the values that appear in a table of atomic weights, are really quite arbitrary.

As originally defined by chemists, the amu was taken to be 1/16 of the mass of naturally occurring oxygen. Oxygen was chosen as a standard because it formed compounds with nearly all of the elements. However, as we shall discuss in Chapter 3, not all atoms of an element have precisely the same mass, and naturally occurring oxygen was later found to be composed of a mixture of atoms of slightly different masses. Since the relative proportion of these different atoms (called **isotopes**) could conceivably change over a period of time, the entire atomic weight table could also change since the size of the amu was based on this mixture. To avoid this problem, the atomic mass unit is currently defined as 1/12 of the mass of one particular isotope of carbon.

In a certain sense, learning chemistry is like learning a language such as Greek (in fact, some students have even been heard to say, "Chemistry is Greek to me!") We might compare the chemical symbols for the elements with an alphabet, the chemical formulas that we construct from the symbols with words, and the chemical equations that we write with sentences. In learning any new language we must start at the beginning, with the alphabet.

At the present time a total of 105 different elements are known. Each element is identified by its name and can also be represented by its **chemical symbol.** Usually, the symbol bears a resemblance to the English name for the element. For instance, carbon = C, chlorine = Cl, nitrogen = N, and zinc = Zn. Some elements, however, have symbols that do not seem to correspond at all with their names. In nearly all of these cases, the elements have been known since the early history of chemistry when Latin was used as the universal language among scientists. For this reason the symbols are derived from their Latin names, for example potassium (L. *kalium*) = K, sodium (*natrium*) = Na, silver (*argentum*) = Ag, mercury (*hydrargyrum*) = Hg, and copper (*cuprum*) = Cu. Regardless of the origin of the symbol, the first letter is always capitalized. A complete list of the elements with their chemical symbols appears on the inside back cover of the book.

A chemical compound is represented symbolically by its **chemical formula.** Thus, water is represented by H_2O, carbon dioxide by CO_2, methane by CH_4, and aspirin by $C_9H_8O_4$. Chemical formulas show the composition of substances. The subscripts in a formula denote the relative number of atoms of each element that appear in the compound; when no subscript is written, a subscript of 1 is implied. The formula H_2O, then, describes a substance containing two hydrogen atoms for every one oxygen atom. Similarly, the compound CH_4 contains one atom of carbon for every four atoms of hydrogen.

1.10
Symbols,
Formulas,
and Equations

Some compounds are more complex, and their formulas are written containing parentheses. For example, $(NH_4)_2SO_4$ denotes the presence of two NH_4 units, for a total of two nitrogen atoms and eight hydrogen atoms, plus one sulfur atom and four oxygen atoms. As we shall see later, there are good reasons for writing this formula as $(NH_4)_2SO_4$ rather than $N_2H_8SO_4$ although both represent the same number of atoms.

Chemical equations are written to show the chemical changes that occur during chemical reactions. For example, the equation

$$Zn + S \longrightarrow ZnS$$

describes a reaction in which zinc (Zn) reacts with sulfur (S) to produce zinc sulfide (ZnS).[4] The substances on the left side of the arrow are known as the **reactants,** while the substances on the right of the arrow are called the **products.** (In the example above there is only one product). The arrow is read as "react to yield" or simply "yield." This equation is read as, "zinc plus sulfur react to yield zinc sulfide" or, "zinc plus sulfur yield zinc sulfide." Sometimes it is desirable (or necessary) to indicate also whether the reactants and products in a chemical reaction are solids, liquids, or gases. This is accomplished by placing the letters $s = $ solid, $l = $ liquid, or $g = $ gas in parentheses following for formulas of the substances in the equation. For instance, the equation

$$CaO\ (s) + H_2O\ (l) + 2CO_2\ (g) \longrightarrow Ca(HCO_3)_2\ (s)$$

describes a reaction between solid calcium oxide, CaO, liquid water, and gaseous carbon dioxide to yield solid calcium bicarbonate, $Ca(HCO_3)_2$.

Many of the equations that we write will contain coefficients preceding the chemical formulas. For example,

$$2H_2 + O_2 \longrightarrow 2H_2O$$

This equation is interpreted to mean that two hydrogen molecules plus one oxygen molecule (a coefficient of 1 is assumed when none is written) react to yield two molecules of H_2O. Such an equation is said to be **balanced** because it contains the same number of atoms of each element on both sides of the arrow. The techniques that we employ to help us balance equations will be discussed at a later time.

**1.11
Energy**

Whenever a chemical change occurs, it is accompanied by either an absorption or release of **energy.** These energy changes, as we shall see, tell us a great deal about the nature of the substances that are reacting and for this reason they are of deep interest to the chemist.

Energy, as you may be aware, is the capacity to perform work and, in fact, an identity exists between the two, that is, work $\equiv$ energy. There are two kinds of energy, **kinetic energy** and **potential energy.** Kinetic energy is associated with motion and is equal to one-half of an object's mass multiplied by its velocity squared.

$$K.E. = \tfrac{1}{2}mv^2$$

[4] The naming of chemical compounds is discussed in Chapter 4 and in Appendix B. For now, we use these names simply as labels.

Thus we see that the amount of work a moving body can do depends on both its mass and its velocity. For example, a truck moving at 20 mph can do more "work" on the rear end of a car than a bicycle moving at the same speed. We also know that a truck moving at 80 mph can do more "work" on a car than one traveling at only 5 mph.

Potential energy represents stored energy and is only present when there is either an attractive force or repulsive force between objects. For example, consider two balls connected by a coiled spring as shown in Figure 1.7. When the balls are pulled apart, the spring is stretched and the energy spent in separating the balls is stored in the stretched spring. We say therefore that the potential energy of the two balls has increased. This energy is released and converted to kinetic energy when the balls are allowed to move toward one another. In this case there was an attractive force (the stretched spring) between the two objects.

Using this same apparatus energy can also be stored by compressing the spring, thereby producing a repulsive force between the two balls. This stored energy is released as kinetic energy as the balls are permitted to move away from one another.

The changes in potential energy that accompany changes in the relative position of objects that either attract or repel one another are very important to remember, since they will help us analyze many physical and chemical changes in the chapters ahead.

As we have indicated, energy contained in chemical substances can be released through chemical reaction. Wood, for example, can react with oxygen, present in the air, in the process known as combustion. As the products of the reaction are formed, rather sizable quantities of energy are released. This chemical energy is initially present in the wood and oxygen and is released as the reaction proceeds. A process that results in the release of energy, such as the combustion of wood, is said to be **exothermic.** Processes that absorb energy, on the other hand, are termed **endothermic.**

The amount of energy released or absorbed in a chemical reaction depends on the quantity of materials that react. The burning of a match, for example, releases only a very small amount of energy while a large bonfire produces much more. Energy, therefore, is an extensive quantity.

The total amount of energy that a body possesses is equal to the sum of its kinetic energy and its potential energy. In a closed system, such as our universe, the total energy is constant and *energy is neither created nor destroyed*

Figure 1.7
Potential energy exists between objects that either attract or repel each other. When the spring is either stretched or compressed the P.E. of the two balls increases.

but, instead, can only be transformed from one kind of energy to another. This statement is called the **law of conservation of energy** and is another example of the many conservation laws that appear to govern our physical world.

Energy is transmitted from one body to another in a variety of ways; for example, as light, sound, electricity, and or heat. These various forms of energy can be converted from one to another and therefore are ultimately equivalent. The kind of units in which energy is measured generally depends on the form in which it appears. Thus kinetic energy, which has dimensions of mass times the velocity-squared, has the units $g \ (cm/sec)^2 = g \ cm^2/sec^2$, and the unit, the **erg,** is defined as $1 \ erg = 1 \ g \ cm^2/sec^2$. This represents the kinetic energy possessed by an object with a mass of two grams traveling at a velocity of one centimeter per second (in English units, an object weighing about 0.07 ounces moving at a velocity of about 2 feet/minute). The larger unit, the **joule,** is equal to 1×10^7 ergs (the kinetic energy of an object weighing 2.2 pounds traveling at a speed of about 195 feet/minute or 2.2 mph). Since all forms of energy usually end up as heat, it is often convenient to express quantities of energy in terms of heat.

The measurement of heat energy involves the concept of **temperature,** which is a measure of the intensity of heat or hotness. Another useful definition of temperature is that it is an intensive quantity that defines the direction and rate of heat flow. We know that heat always flows from a warm object to a cool one. We also know that the rate of heat transfer depends on the difference in temperature between two objects. If we want to warm something slowly, we place it in contact with an object just slightly warmer while if we wish to heat it quickly, we place it in contact with a very hot object.

The measurement of temperature is usually accomplished with a thermometer consisting of a narrow capillary tube connected to a thin-walled reservoir filled with some liquid (usually mercury). As the temperature of the liquid is raised, it expands and as its volume increases, the fluid rises in the capillary. The height of fluid in the capillary then becomes directly proportional to temperature.

A number of temperature scales have been devised. The **Fahrenheit** scale, in common use in the United States, is defined by the freezing point and boiling point of water.

At its freezing point (which is also its melting point) a pure substance such as water can exist as both solid (ice) and liquid in contact with each other *at the same temperature.* If heat is added to this mixture, some solid melts and more liquid is formed, but the temperature remains constant while the solid is melting. If heat is removed from the mixture, liquid freezes to produce more solid, once again without a temperature change. At its boiling point a pure liquid can exist in contact with its vapor at the same temperature. In this case, removal of heat causes some vapor to condense whereas adding heat causes more liquid to evaporate, all at a constant temperature. The constancy of temperature during these phase changes makes them ideal temperatures to serve as calibration points on a temperature scale.

On the Fahrenheit scale the freezing point of water is assigned a temperature of 32° F and the boiling point a value of 212° F. The difference between these two reference points, 180 Fahrenheit degrees, thus defines the size of the degree unit. In the sciences the temperature scale that is employed is called the **Celsius** scale (also called the **centigrade** scale) which defines 0°C as the freezing point of water and 100° C as the boiling point of water. Therefore, we see that 100 Celsius degrees are equal to 180 Fahrenheit degrees, so that the Celsius degree is nearly twice as large as that on the Fahrenheit scale. Temperatures in °C are related to temperatures in °F by the equation,

$$°C = \tfrac{5}{9} \, (°F - 32)$$

With this concept of temperature the unit of heat energy, the **calorie** (abbreviated, **cal**) is defined as the amount of energy required to raise the temperature of one gram of water at 15° C by one degree Celsius.[5] The larger unit, the **kilocalorie** (kcal), is another convenient quantity that we shall have occasion to use.

The calorie, a unit of heat energy, and the joule, a unit of mechanical kinetic energy, are related to one another by, 1 cal = 4.18 joules. In other words, if 4.18 joules of kinetic energy could be transferred to one gram of water, for example, via some stirring device, the temperature of the water would be raised by 1° C.

Review
Questions
and Problems

1.1 What is the difference between a theory and a law?

1.2 How many significant figures are there in the following numbers: 1.0370, 0.000417, 0.00309, 100.1, 9.0010?

1.3 Write each of the numbers in Question 1.2 in exponential form.

1.4 Identify the phases that exist in a copper pan containing two iron nails, a quart of water, and four glass marbles.

1.5 Perform the following conversions: 4.3 tons to kg; 1 cubic mile to m³; −40 °F to °C; 40 miles/hour to cm/sec; 1.0 × 10⁸ ml to dm³.

1.6 What is the difference between an extensive and an intensive property? Can you think of any examples that were not mentioned in the text?

1.7 There are many examples of homogeneous and heterogeneous mixtures in the world around us. How would you classify sea water, air (unpolluted), smog, smoke, homogenized milk, black coffee, and a penny.

1.8 Two compounds are formed between phosphorus and oxygen. 1.50 g of one compound was found to contain 0.845 g of phosphorus while a 2.50 g sample of the other contained 1.09 g of phosphorus. Show that these data are consistent with the law of multiple proportions.

1.9 Three samples of a solid substance composed of elements X and Y were prepared. The first was found to contain 4.31 g X and 7.69 g Y; the second was composed of 35.9% X and 64.1% Y; it was observed that 0.718 g X reacted with Y to form 2.00 g of the third sample. Show how these data demonstrate the law of definite composition.

1.10 In Example 1.1, the first compound (2.62 g of N, 1.50 g O) has the formula N_2O (the substance is nitrous oxide, laughing gas). Suggest a possible formula for the second compound.

1.11 Distinguish between the terms: atom and molecule; element and compound.

---

[5] It is necessary to specify the temperature of the water because the amount of heat required to raise the temperature of 1 gram of water 1° C varies slightly with the temperature of the water.

1.12 In a certain compound, 6.92 g of X were found combined with 0.584 g of carbon. If the atomic weight of carbon = 12.0 amu and if four atoms of X are combined with one atom of carbon, calculate the atomic weight of X.

1.13 If the atomic mass unit was defined such that a single fluorine atom weighed 1 amu, what would be the atomic weights of carbon and hydrogen?

1.14 Write the chemical symbols for these elements; iron, sodium, potassium, antimony, tin, lead, mercury, gold, silver, tungsten, copper.

1.15 How many elements of each kind are indicated in each of the following formulas: K_2S, Na_2CO_3, $(NH_4)_3PO_4$, $K_4Fe(CN)_6$, $Na_3Ag(S_2O_3)_2$?

1.16 What is meant by endothermic and exothermic?

1.17 How does potential energy differ from kinetic energy?

1.18 A car, whose mass is 1500 kg, skids to a halt from a speed of 60 m/sec. How many calories of heat are generated by friction between the car and the pavement?

1.19 A student has just returned from Germany with a car that he purchased while on vacation. The speedometer is calibrated in kilometers per hour (km/hr). As he drives away from the pier, he notices a sign that posts the speed limit at 35 mph. What is the maximum speed that he can reach, in km/hr, without having to worry about receiving a speeding ticket?

1.20 After receiving a speeding ticket, the student in the preceding question is informed by the policeman that the courthouse is located "4.3 miles straight down the road. You can't miss it!" The odometer in the student's car measures kilometers (km). How far must he travel, in km, before arriving at the courthouse.

1.21 Normal body temperature in man is 98.6° F. In Europe the centigrade temperature scale is used almost universally so that clinical thermometers are calibrated in °C. If you had one of these thermometers, what temperature would you expect, in °C, for a normal healthy individual? If your temperature registered 39° C, what would your temperature be in °F?

1.22 Naphthalene (used in moth balls) has a melting point of 80° C and a boiling point of 218° C. Suppose this substance were used to define a new temperature scale on which the melting point of naphthalene was 0° N and the boiling point of naphthalene was 100° N. What would be the freezing point and boiling point of water in °N? What general equation could we use to relate temperatures in °C and °N?

1.23 If the potential energy of an object decreases as it is moved away from another object, what kind of force (attractive or repulsive) must exist between the two?

2
Stoichiometry: Chemical Arithmetic

In Chapter 1 you were introduced to many of the important basic concepts in chemistry: the ideas of atoms and atomic weight, elements and compounds, and several major chemical laws. Chemists, physicists, and biologists have made extensive studies of these ideas so that today they routinely think of chemical, physical, and biological processes taking place between atoms and molecules on a microscopic, atomic scale. However, we cannot see individual atoms or molecules, and in the laboratory it is necessary to work with huge numbers of these small particles. Chemists have devised ways of carrying out chemical reactions so that observations on a large (macroscopic) scale can be readily translated into the language of the atomic world. That this can be done with relative ease is really quite fascinating.

In order to carry out these investigations it is necessary for a chemist (or even a chemistry student) to have a knowledge of the quantitative relationships that exist between the substances that enter into chemical reactions. Stoichiometry (derived from the Greek *stoicheion* = element and *metron* = measure) is the term used to refer to all of the quantitative aspects of chemical composition and reaction. We will now see how chemical formulas are determined and how chemical equations prove useful for predicting the proper amounts of reactants that must be mixed together to get a complete reaction.

Suppose we wish to prepare molecules of carbon monoxide, CO. In principle, we could do this by taking one atom of carbon and one atom of oxygen to form one molecule of CO. If we took two atoms of carbon and two atoms of oxygen, we could produce two molecules of CO. This process, of course, could be extended so that we could produce any desired number of CO molecules. The only requirement is that we must be sure to choose equal numbers of carbon and oxygen atoms.

Since it is impossible for us to actually count individual atoms, we must ask: How do we assure ourselves of taking atoms in equal numbers? The solution to this problem lies in the utilization of the concept of atomic weight. For simplicity, in the following discussion we shall deal with "average" atoms (i.e., ignore isotopic differences).

An average carbon atom has a mass of 12.01115 amu; an average oxygen atom, a mass of 15.9994 amu. When we take one atom of each element we have

$$\text{Atom ratio} = \frac{1}{1}$$

2.1
The Gram-Atom

$$\text{Mass ratio} = \frac{12.01115}{15.9994}$$

If we take two atoms each of carbon and oxygen, the mass ratio is

$$\frac{2 \times 12.01115}{2 \times 15.9994} = \frac{12.01115}{15.9994}$$

and the atom ratio is 2/2, which is the same, of course, as 1/1. We see then, that as long as an atom ratio of 1/1 is maintained, a mass ratio of 12.01115/15.9994 will exist. More important, however, when a mass ratio of 12.01115/15.9994 is maintained, an atom ratio of 1:1 exists. If we wished, instead, to have an atom ratio of 1 atom of carbon to 2 atoms of oxygen, we would require a mass ratio of 12.01115/(2 × 15.9994). It is clear, then, that *any desired atom ratio can be obtained merely by choosing the proper mass ratio.*

Example 2.1
What must the mass ratio of H to O be in order to have an atom ratio of 2H atoms/1O atom?

Solution

$$\text{The atom ratio desired} = \frac{2 \text{ atom H}}{1 \text{ atom O}}$$

1 atom H = 1.00797 amu
2 atom H = 2(1.00797 amu) = 2.01594 amu
Also,   1 atom O = 15.9994 amu

The mass ratio, then, is $\dfrac{2.01594}{15.9994}$.

In the laboratory we measure mass in the units, grams. If we weigh out 12.01115 g of carbon and 15.9994 g of oxygen, our mass ratio tells us that our atom ratio is 1/1. Regardless of the actual number of atoms of carbon or oxygen, we are assured that we have equal numbers of each. Thus, 12.01115 g of carbon contains the same number of atoms as 15.9994 g of oxygen.

The argument we have just presented leads us to define a quantity, called the **gram-atom** or **gram atomic weight** (abbreviated g-atom) as *a collection of atoms whose mass, in grams, is numerically equal to their atomic weight.* Thus,

12.01115 g carbon = 1 g-atom carbon
15.9994 g oxygen = 1 g-atom oxygen
207.19 g lead = 1 g-atom lead

To summarize: (1) **the weight of one gram-atom of any element may be obtained by simply writing its atomic weight followed by the unit, grams;** and (2) **one gram-atom of an element contains the same number of atoms as one gram-atom of any other element.** With this as background, let us look at some sample problems dealing with the gram-atom.[1]

[1] If you are unfamiliar with this method of solving problems, see Appendix C.

Example 2.2

How many gram-atoms of Si are in 30.5 grams of Si?

Solution

Our problem is one of converting the units gram Si to gram-atom Si, that is, 30.5 g Si = (?) g-atom Si.

We know from the table of atomic weights that

$$1 \text{ g-atom Si} = 28.1 \text{ g Si}$$

To convert from g Si to g-atom Si we must multiply 30.5 g Si by a factor that contains the units g Si in the denominator, that is,

$$\frac{1 \text{ g-atom Si}}{28.1 \text{ g Si}} = 1$$

When we set up this problem,

$$30.5 \text{ g Si} \times \left(\frac{1 \text{ g-atom Si}}{28.1 \text{ g Si}}\right) = 1.09 \text{ g-atom Si}$$

Thus 30.5 g Si = 1.09 g-atom Si.

Example 2.3

How many grams of Cu are there in 2.55 g-atom Cu?

Solution

From the table of atomic weights,

$$1 \text{ g-atom Cu} = 63.5 \text{ g Cu}$$

Our conversion factor must have the units, g-atom Cu in the denominator, that is,

$$\frac{63.5 \text{ g Cu}}{1 \text{ g-atom Cu}} = 1$$

Setting up the problem, we have

$$2.55 \text{ g-atom Cu} \times \left(\frac{63.5 \text{ g Cu}}{1 \text{ g-atom Cu}}\right) = 162 \text{ g Cu}$$

Example 2.4

How many g-atom Ca are required to react with 2.50 g-atom Cl to produce the compound $CaCl_2$?

Solution

Our problem: 2.50 g-atom Cl ~ (?) g-atom Ca.

We know from the formula that 1 atom of Ca combines with 2 atoms of Cl. We also know that 1 g-atom Ca and 1 g-atom Cl contain the same number of atoms, therefore, to maintain a ratio of 1 atom of Ca to 2 atoms of Cl, we state that 1 g-atom Ca combines with 2 g-atom Cl.

$$1 \text{ g-atom Ca} \sim 2 \text{ g-atom Cl}$$

We obtain our answer as follows:

$$2.50 \text{ g-atom } \cancel{Cl} \times \left(\frac{1 \text{ g-atom Ca}}{2 \text{ g-atom } \cancel{Cl}} \right) = 1.25 \text{ g-atom Ca}$$

Example 2.5

How many grams of Ca must react with 41.5 g Cl to produce $CaCl_2$?

Solution

Our problem: 41.5 g Cl ∼ (?) g Ca.
What quantities do we need and what do we know?
We know that

$\quad\quad\quad$ 1 g-atom Ca ∼ 2 g-atom Cl (Why?)

Also, $\quad\quad$ 1 g-atom Cl = 35.5 g Cl (atomic weight table)
$\quad\quad\quad$ 1 g-atom Ca = 40.1 g Ca (atomic weight table)

Here is the solution to the problem:

$$41.5 \text{ g } \cancel{Cl} \times \left(\frac{1 \text{ g-atom Cl}}{35.5 \text{ g } \cancel{Cl}} \right) = 1.17 \text{ g-atom Cl}$$

$$1.17 \text{ g-atom } \cancel{Cl} \times \left(\frac{1 \text{ g-atom Ca}}{2 \text{ g-atom } \cancel{Cl}} \right) \sim 0.584 \text{ g-atom Ca}$$

$$0.584 \text{ g-atom } \cancel{Ca} \times \left(\frac{40.1 \text{ g Ca}}{1 \text{ g-atom } \cancel{Ca}} \right) = 23.4 \text{ g Ca}$$

And here are all of these steps written together.

$$41.5 \text{ g } \cancel{Cl} \times \left(\frac{1 \text{ g-atom } \cancel{Cl}}{35.5 \text{ g } \cancel{Cl}} \right) \times \left(\frac{1 \text{ g-atom } \cancel{Ca}}{2 \text{ g-atom } \cancel{Cl}} \right) \times \left(\frac{40.1 \text{ g Ca}}{1 \text{ g-atom } \cancel{Ca}} \right) = 23.4 \text{ g Ca}$$

In this last example we have strung together a series of conversion factors. When we set up the problem, we link these together by considering what units must be eliminated by cancellation. Thus, the first factor had to have g Cl in the denominator; the second factor had to have g-atom Cl in the denominator; the third was required to have g-atom Ca in the denominator. At this point we stop since our units are now those of the answer, and we have only to perform the arithmetic to obtain the correct result.

2.2
Avogadro's Number, The Mole

We have seen that the gram-atom is a useful concept for dealing with laboratory-size quantities of atoms and that the utility of the gram-atom lies in the fact that it represents a fixed quantity of atoms. The number of atoms in one gram-atom, called **Avogadro's number** (named after Amadeo Avogadro) has been found to be 6.02×10^{23}. Thus,

$\quad\quad$ 1 g-atom O = 6.02×10^{23} atom O = 15.9994 g O
$\quad\quad$ 1 g-atom C = 6.02×10^{23} atom C = 12.01115 g C
$\quad\quad$ 1 g-atom Fe = 6.02×10^{23} atom Fe = 55.847 g Fe

The value of having a quantity that represents a fixed number of things is so useful in chemistry that we do not wish to restrict its use merely to atoms.

The more general term that is used to represent Avogadro's number of things is called the **mole.** We can think of this unit in the same way as we think of the dozen, since it represents a numerical quantity.

$$1 \text{ dozen} = 12 \text{ objects}$$
$$1 \text{ mole} = 6.02 \times 10^{23} \text{ objects}$$

Another useful way of thinking of the mole is as a weightless box with 6.02×10^{23} compartments. If the box were filled with Ca atoms, it would contain 1 g-atom Ca and have a mass of 40.1 g. If it were filled with Fe atoms, it would weigh 55.8 g. Thus, for atoms, the mole and the gram-atom represent the same thing:

$$1 \text{ mole Ca} = 1 \text{ g-atom Ca} = 40.1 \text{ g Ca}$$

Let us see how much one mole of CO_2 molecules weighs. One mole of CO_2 contains 6.02×10^{23} molecules of CO_2, each of which could be taken apart to give one carbon atom and two oxygen atoms. If one box full (1 mole) of CO_2 molecules were disassembled, we could fill one box with carbon atoms and two boxes with oxygen atoms, that is, 1 mole CO_2 contains 1 mole C and 2 mole O. The total mass of carbon dioxide is the sum of the masses of carbon plus oxygen. Since a box (mole) of carbon weighs 12.0 g and each box (mole) of oxygen weighs 16.0 g, the total mass of one mole of CO_2 is 44.0 g.

The simplest way of obtaining the weight of one mole of a substance is to merely add up the atomic weights of all of the elements present in the compound. If the substance is composed of molecules (e.g., CO_2, H_2O, or NH_3), the sum of the atomic weights is called the **molecular weight.** Thus the molecular weight of CO_2 is obtained as

C	1×12.0 amu	$= 12.0$ amu
O	2×16.0 amu	$= \underline{32.0 \text{ amu}}$
CO_2	Total	$\overline{44.0}$ amu

Similarly, the molecular weight of $H_2O = 18.0$ amu and that of $NH_3 = 17.0$ amu. Clearly the weight of one mole of a substance (also called the **gram molecular weight**) is obtained by writing its molecular weight followed by the units, grams. Thus,

$$1 \text{ mole } H_2O = 18.0 \text{ g}$$
$$1 \text{ mole } NH_3 = 17.0 \text{ g}$$

In later chapters we shall encounter many compounds that do not contain discrete molecules. Often, when certain atoms react, they gain or lose negatively charged particles called **electrons.** Sodium and chlorine happen to react in this way, so that when sodium chloride, NaCl, is formed from the elements, each Na atom loses one electron and each Cl atom gains one. Since Na and Cl are electrically neutral to start, these atoms acquire a charge when NaCl is formed. These are written as Na^+ (positive because Na has lost a negatively charged electron) and Cl^- (negative because Cl has gained an electron). Atoms or groups of atoms that have acquired an electrical charge are called ions; since solid NaCl is composed of Na^+ and Cl^- ions, this compound is said to be ionic.

This entire topic is explored further in Chapters 3 and 4. For now, it is only necessary for you to know that compounds that are ionic do not contain molecules. Their formulas simply state the ratio of the different atoms in the substance. In NaCl the atoms are in a one-to-one ratio. In the ionic compound $CaCl_2$ the ratio of Ca to Cl atoms is one to two (relax — at this point you were not expected to know that $CaCl_2$ is ionic). Rather than refer to molecules of NaCl or $CaCl_2$ we use the term **formula unit** to specify a NaCl pair, or a set of three ions in $CaCl_2$.

For ionic compounds the sum of the atomic weights of the elements present in an ionic solid is known as the **formula weight.** For NaCl this is $22.99 + 35.45 = 58.44$ and one mole of NaCl (one **gram formula weight**) would contain 58.44 g NaCl (6.02×10^{23} formula units of NaCl). It is also true that 6.02×10^{23} formula units (1 mole) of NaCl contains 6.02×10^{23} Na^+ ions and 6.02×10^{23} Cl^- ions. Use of the term formula weight, of course, is not restricted to ionic compounds and it can be applied also to molecular substances, in which case the terms formula weight and molecular weight mean the same thing.

2.4
Chemical Formulas

A chemist uses a formula to convey certain kinds of information, including elemental composition, relative numbers of each kind of atom present, the actual numbers of each kind of atom in a molecule of the substance, or the structure of the compound. We can classify formulas according to the amount of information they provide.

A formula that simply gives the relative number of atoms of each element present is called a **simplest formula.** It is also called an **empirical formula** because it is invariably obtained as the result of some experimental analysis. The formulas NaCl, H_2O, and CH_2 are empirical formulas.

A formula that states the actual number of each kind of atom found in a molecule is called a **molecular formula.** H_2O is a molecular formula (as well as an empirical formula) since a molecule of water contains two atoms of H and one atom of O. The formula C_2H_4 is a molecular formula for a substance (ethylene) containing two atoms of carbon and four atoms of hydrogen. Note that the simplest formula of this compound is CH_2 since the carbon to hydrogen ratio is $1:2$. A substance whose empirical formula is CH_2 could have a molecular formula CH_2, C_2H_4, C_3H_6, etc. Molecular formulas for ionic substances do not exist, of course, because such compounds do not contain molecules.

A third type of formula is a **structural formula;** for example,

acetic acid (present in vinegar)

In a structural formula the dashes between the different atomic symbols represent the chemical bonds that hold the atoms together in the molecule. We will see more of them in Chapter 4. A structural formula gives us information about the way in which the atoms in a molecule are linked together, and provides information that allows us to write the molecular and empirical formulas. Thus, for acetic acid shown above we can also write its molecular formula ($C_2H_4O_2$) and its empirical formula (CH_2O).

The most desirable kind of formula to have, of course, is the structural formula since it also contains all of the information provided by the other two types. However, in chemistry, as in the rest of life, we never get something for nothing. The more information a formula conveys the more difficult it is to arrive at experimentally. We shall see how empirical and molecular formulas are derived; however, most of the procedures involved for the determination of structural formulas are beyond the scope of this book.

Since the simplest formula gives the relative numbers of atoms present in a compound, it must also give the relative number of moles of each element. Here are some examples that show how we might obtain this information.

Example 2.6

A sample of a brown-colored gas that is a major air pollutant is found to contain 2.34 g of N and 5.34 g of O. What is the simplest formula of the compound?

Solution

We proceed by calculating the number of moles of each element present. We know that

$$1 \text{ g-atom N} = 1 \text{ mole N} = 14.0 \text{ g N (Why?)}$$
$$1 \text{ g-atom O} = 1 \text{ mole O} = 16.0 \text{ g O (Why?)}$$

Therefore,

$$2.34 \text{ g N} \times \left(\frac{1 \text{ mole N}}{14.0 \text{ g N}}\right) = 0.167 \text{ mole N}$$

$$5.34 \text{ g O} \times \left(\frac{1 \text{ mole O}}{16.0 \text{ g O}}\right) = 0.334 \text{ mole O}$$

We might write our formula $N_{0.167}O_{0.334}$. It does indeed tell us the relative number of moles of N and O; however, since the formula should have meaning on a molecular level, we prefer the subscripts to be integers. If we divide each subscript by the smallest one, we obtain

$$N_{\frac{0.167}{0.167}}O_{\frac{0.334}{0.167}} = NO_2$$

Example 2.7

What is the empirical formula of a compound composed of 43.7% P and 56.3% O by weight?

Solution

It is quite common to have a chemical analysis in the form of percent composition by weight. From the analysis we know that 100 g of the compound contains 43.7 g of P and 56.3 g of O; that is, 43.7 g P is combined with 56.3 g O. We convert these quantities to moles and proceed as before.

$$43.7 \text{ g P} \times \left(\frac{1 \text{ mole P}}{31.0 \text{ g P}}\right) = 1.41 \text{ mole P}$$

$$56.3 \text{ g O} \times \left(\frac{1 \text{ mole O}}{16.0 \text{ g O}}\right) = 3.52 \text{ mole O}$$

Our formula is

$$P_{1.41}O_{3.52} = P_{\frac{1.41}{1.41}}O_{\frac{3.52}{1.41}} = PO_{2.5}$$

Whole numbers may be obtained by doubling each of these values. Thus the empirical formula is P_2O_5.

2.6
Molecular Formulas

Not only does the molecular formula provide the information contained in the empirical formula, but it also tells us how many atoms of each element are present in a molecule of a substance. Remember that an empirical formula of CH_2 is found for any molecule that possesses twice as many hydrogen atoms as carbon atoms. To distinguish among all of the possible choices, we require the molecular weight of the compound. This is because the molecular weight is an integral multiple of the empirical formula weight (see Table 2.1). To find the number of times that the empirical formula is repeated in the molecular formula, we simply divide the experimentally determined molecular weight[2] by the empirical formula weight.

Table 2.1
Molecular Weights as Multiples of the Empirical Formula Weight

Formula	Molecular Weight
CH_2	$14.0 = 1 \times 14.0$
C_2H_4	$28.0 = 2 \times 14.0$
C_3H_6	$42.0 = 3 \times 14.0$
C_4H_8	$56.0 = 4 \times 14.0$
C_nH_{2n}	$n \times 14.0$

Example 2.8

A colorless liquid whose empirical formula also is NO_2 has a molecular weight of 92.0. What is its molecular formula?

Solution

The formula weight of NO_2 is 46.0.

---

[2] We shall see how molecular weights may be determined in Chapters 6 and 9.

The number of times the empirical formula, NO_2, occurs in the compound is

$$\frac{92.0}{46.0} = 2$$

The molecular formula is then $(NO_2)_2 = N_2O_4$ (dinitrogen tetroxide).

N_2O_4 is the preferred answer since $(NO_2)_2$ implies a knowledge of the structure of the molecule (i.e., that two NO_2 units are somehow joined together).

A very simple and also often very useful computation is the calculation of the percentage composition of a compound, that is, the percentage of the total mass contributed by each element. The procedure to determine the percent composition is illustrated by Example 2.9.

2.7
**Percentage
Composition**

Example 2.9
What is the percentage composition of CH_3Cl?

Solution
The total mass of one molecule of CH_3Cl is obtained from the molecular weight,

$$(12.01 + 3 \times 1.008 + 35.45) \text{ amu} = 50.48 \text{ amu}$$

$$\% \text{ C} = \frac{\text{weight carbon}}{\text{weight } CH_3Cl} \times 100$$

$$\% \text{ C} = \frac{12.01 \text{ amu}}{50.48 \text{ amu}} \times 100 = 23.79$$

$$\% \text{ H} = \frac{3.024 \text{ amu}}{50.48 \text{ amu}} \times 100 = 5.99$$

$$\% \text{ Cl} = \frac{35.45 \text{ amu}}{50.48 \text{ amu}} \times 100 = \underline{70.22}$$

$$\text{Total } \% = 100.00$$

To illustrate the usefulness of percentage composition, let us consider the chemical analysis of an organic compound containing the elements carbon, hydrogen, and oxygen.

Example 2.10
0.1000 g of ethyl alcohol (grain alcohol) known to contain only carbon, hydrogen, and oxygen was completely reacted with oxygen to produce the products CO_2 and H_2O. These products were trapped separately and weighed. 0.1910 g of CO_2 and 0.1172 g of H_2O were found. What is the empirical formula of the compound?

Solution
To many students this problem at first appears impossible; however, let us consider what we know and what we can calculate.

We can calculate the percentage composition of CO_2 and H_2O and, thus, compute the weight of carbon and the weight of hydrogen in the CO_2 and H_2O, respectively. Since the only source of C and H was the original compound, the difference between the weight of the compound taken (0.1000 g) and the total weight of carbon and hydrogen must be the weight of oxygen in the original 0.1000 g. We must obtain the weight of oxygen in this way, rather than from the total weight of oxygen in the H_2O and CO_2, since only a portion of the oxygen in the products came from the original compound. Once we know the weights of carbon, hydrogen, and oxygen, we can calculate how many moles of each are in the 0.1000 g and, hence, the empirical formula of the unknown compound. Thus, having planned our course of action we now proceed with the computations.

The formula weights of CO_2 and H_2O are 44.0 and 18.0, respectively. The fraction of the mass of CO_2 that is carbon is

$$\frac{12.0 \text{ g C}}{44.0 \text{ g } CO_2}$$

Likewise, the fraction of H_2O that is hydrogen is equal to

$$\frac{2.01 \text{ g H}}{18.0 \text{ g } H_2O}$$

The weight of carbon in the original compound is equal to the weight of CO_2 multiplied by the fraction of the weight that is due to carbon.

$$0.1910 \text{ g } CO_2 \times \left(\frac{12.0 \text{ g C}}{44.0 \text{ g } CO_2} \right) \sim 0.0521 \text{ g C}$$

Similarly, the weight of hydrogen in the original sample is

$$0.1172 \text{ g } H_2O \times \left(\frac{2.01 \text{ g H}}{18.0 \text{ g } H_2O} \right) \sim 0.0131 \text{ g H}$$

The total weight contributed by the carbon and hydrogen is

$$0.0521 \text{ g C} + 0.0131 \text{ g H} = 0.0652 \text{ g}$$

The weight of oxygen $= 0.1000 \text{ g} - 0.0652 \text{ g} = 0.0348 \text{ g O}$. Next, we calculate the number of moles of C, H, and O.

$$0.0521 \text{ g C} \times \left(\frac{1 \text{ mole C}}{12.0 \text{ g C}} \right) = 4.34 \times 10^{-3} \text{ mole C}$$

Similar calculations for hydrogen and oxygen give 1.31×10^{-2} mole H and 2.17×10^{-3} mole O. The empirical formula, then, is

$$C_{0.00434}H_{0.0131}O_{0.00217} = C_{\frac{0.00434}{0.00217}}H_{\frac{0.0131}{0.00217}}O_{\frac{0.00217}{0.00217}}$$

or

$$C_2H_6O$$

2.8
Balancing Chemical Equations

Recall that a chemical equation is a shorthand description of the changes that occur during a chemical reaction. One of the most useful properties of a chemical equation is that it allows us to determine the quantitative relationships that exist between reactants and products. To be helpful in this way, however, the equation must be **balanced,** that is, it must obey the law of conservation of mass by having the same number of atoms of each kind on both sides of the arrow.[3]

In order to minimize errors, writing a balanced chemical equation should always be considered a two-step process.

1. We first write an unbalanced equation with correct formulas for all of the reactants and products.
2. We balance the equation by adjusting the coefficients that precede the formulas.

There is never any excuse for having an improperly balanced equation, since it is always possible, by counting atoms on each side of the equation, to determine whether the equation is, in fact, balanced.

Most simple chemical equations can be easily balanced by inspection. This involves examining the equation and adjusting the coefficients until equal numbers of each element are present among both the reactants and products. For example, consider the reaction of sodium carbonate with hydrochloric acid (HCl) to produce sodium chloride, carbon dioxide, and water.

To obtain a properly balanced chemical equation we proceed as follows:

1. We write the unbalanced equation

$$Na_2CO_3 + HCl \longrightarrow NaCl + H_2O + CO_2$$

2. Coefficients are introduced to balance the equation.
   In this step, since there are two Na atoms on the left, we begin by placing a 2 in front of the formula NaCl. This gives us

$$Na_2CO_3 + HCl \longrightarrow 2NaCl + H_2O + CO_2$$

Since there are two Cl atoms on the right but only one on the left, we place a 2 before the HCl to get

$$Na_2CO_3 + 2HCl \longrightarrow 2NaCl + H_2O + CO_2$$

A fast inspection reveals that this equation is now balanced.

It was stated above that a balanced equation obeys the law of conservation of mass. For the equation that we have just balanced, and in fact for any equation, there are an infinite number of sets of coefficients that fulfill this requirement. Thus, the equations

$$\tfrac{1}{2}Na_2CO_3 + HCl \longrightarrow NaCl + \tfrac{1}{2}CO_2 + \tfrac{1}{2}H_2O$$
$$5Na_2CO_3 + 10HCl \longrightarrow 10NaCl + 5CO_2 + 5H_2O$$

are also balanced. The usual practice, however, is to use the smallest possible set of whole number coefficients, although there are occasions, as we shall see, where other choices are advantageous.

A chemical equation can be interpreted in several ways. For example, on a molecular level the balanced equation for the combustion of ethylene,

2.9

Calculations Based on Chemical Equations

[3] We shall see that chemical equations are sometimes written to show the reaction between ions. These equations must demonstrate the law of conservation of electrical charge by having the total charge on both sides of the equation the same.

$$C_2H_4 + 3O_2 \longrightarrow 2CO_2 + 2H_2O$$

can be read as,

1 molecule C_2H_4 + 3 molecules $O_2 \longrightarrow$

$\qquad\qquad\qquad$ 2 molecules CO_2 + 2 molecules H_2O

Alternatively, in terms of laboratory-sized quantities, we can express the reaction as

$\qquad$ 1 mole C_2H_4 + 3 moles $O_2 \longrightarrow$ 2 moles CO_2 + 2 moles H_2O

In this case, we have merely scaled everything up by a factor equal to Avogadro's number. The key point here is that *the coefficients in a chemical equation provide the* **ratio** *in which moles of one substance react with moles of another*. In this equation, the number of moles of O_2 consumed is always equal to three times the number of moles of C_2H_4 that react. For example, if 5 moles of C_2H_4 were available, 15 moles of O_2 would be needed for complete reaction. We could also use the coefficients and say that if 5 moles of C_2H_4 were reacted, 10 moles of CO_2 and 10 moles of H_2O would be formed. Here are some sample problems involving calculations based on chemical equations.

Example 2.11
How many moles of CO_2 are produced from the combustion of 3.30 mole C_2H_4?

Solution

$\qquad$ The question is: 3.30 moles $C_2H_4 \sim$ (?) mole CO_2.

$\qquad$ From the balanced equation we can say that two moles of CO_2 are formed every time one mole C_2H_4 is burned. Stated mathematically,

$$1 \text{ mole } C_2H_4 \sim 2 \text{ mole } CO_2$$

This equation is meant to merely express the relationship between the number of moles of C_2H_4 consumed and the number of moles of CO_2 produced; as such it may be used to construct a conversion factor. The solution to the problem, therefore, is

$$3.30 \text{ mole } C_2H_4 \times \left(\frac{2 \text{ mole } CO_2}{1 \text{ mole } C_2H_4} \right) \sim 6.60 \text{ mole } CO_2$$

Example 2.12
How many grams of O_2 are required for the combustion of 3.30 mole C_2H_4?

Solution

$\qquad$ Our problem is: 3.30 mole $C_2H_4 \sim$ (?) g O_2.

$\qquad$ According to the balanced equation

$$1 \text{ mole } C_2H_4 \sim 3 \text{ mole } O_2$$

Thus, the number of moles of O_2 required is

$$3.30 \text{ mole } C_2H_4 \times \left(\frac{3 \text{ mole } O_2}{1 \text{ mole } C_2H_4} \right) \sim 9.90 \text{ mole } O_2$$

To convert moles of O_2 to grams of O_2 we use the relationship

$$1 \text{ mole } O_2 = 32.0 \text{ g } O_2$$

$$9.90 \text{ mole } O_2 \times \left(\frac{32.0 \text{ g } O_2}{1 \text{ mole } O_2} \right) = 317 \text{ g } O_2$$

We could have combined these two steps:

$$3.30 \text{ mole } C_2H_4 \times \left(\frac{3 \text{ mole } O_2}{1 \text{ mole } C_2H_4} \right) \times \left(\frac{32.0 \text{ g } O_2}{1 \text{ mole } O_2} \right) \sim 317 \text{ g } O_2$$

Example 2.13
Calculate the number of grams of H_2O produced from 1.43 g O_2 during the combustion of C_2H_4.

Solution
Our problem: 1.43 g $O_2 \sim$ (?) g H_2O.
The relationships available to us are

$$32.0 \text{ g } O_2 = 1 \text{ mole } O_2$$
$$3 \text{ mole } O_2 \sim 2 \text{ mole } H_2O \text{ (from the chemical equation)}$$
$$1 \text{ mole } H_2O = 18.0 \text{ g } H_2O$$

Using these to construct conversion factors, we have

$$1.43 \text{ g } O_2 \times \left(\frac{1 \text{ mole } O_2}{32.0 \text{ g } O_2} \right) \times \left(\frac{2 \text{ mole } H_2O}{3 \text{ mole } O_2} \right)$$
$$\times \left(\frac{18.0 \text{ g } H_2O}{1 \text{ mole } H_2O} \right) \sim 0.536 \text{ g } H_2O$$

Example 2.14
How many grams of water will be formed when a mixture containing 1.93 g C_2H_4 and 5.92 g O_2 is ignited?

Solution
Unless C_2H_4 and O_2 are present in precisely the proper ratio, one of these reactants will be used up before the other when reaction occurs. If this takes place the reactant that disappears first is called the **limiting reactant,** because once this reactant is gone no more product can be formed. There-fore, our calculation of the amount of product formed must be based on the reactant that is completely consumed, that is, the quantity of limiting reactant. With this in mind, let us tackle the problem by computing the number of grams of O_2 necessary to react with all of the C_2H_4.

$$1.93 \text{ g } C_2H_4 \times \left(\frac{1 \text{ mole } C_2H_4}{28.0 \text{ g } C_2H_4} \right) \times \left(\frac{3 \text{ mole } O_2}{1 \text{ mole } C_2H_4} \right) \times \left(\frac{32 \text{ g } O_2}{1 \text{ mole } O_2} \right) \sim 6.62 \text{ g } O_2$$

Notice that the 1.93 g C_2H_4 requires more O_2 (6.62 g) than is available (5.92 g). Clearly, then, the reaction will cease when all of the O_2 has reacted; therefore the amount of H_2O formed depends on the quantity of O_2 initially present (O_2 is thus the limiting reactant). Once O_2 is estab-lished as the limiting reactant, we use the amount of O_2 in the original mixture to calculate the amount of H_2O formed.

$$5.92 \text{ g } O_2 \times \left(\frac{1 \text{ mole } O_2}{32.0 \text{ g } O_2}\right) \times \left(\frac{2 \text{ mole } H_2O}{3 \text{ mole } O_2}\right) \times \left(\frac{18 \text{ g } H_2O}{1 \text{ mole } H_2O}\right) \sim 2.22 \text{ g } H_2O$$

We can also calculate the amount of C_2H_4 that was in excess, that is, the amount of C_2H_4 left unreacted when the reaction ceases. The quantity of C_2H_4 consumed in the reaction is

$$5.92 \text{ g } O_2 \times \left(\frac{1 \text{ mole } O_2}{32.0 \text{ g } O_2}\right) \times \left(\frac{1 \text{ mole } C_2H_4}{3 \text{ mole } O_2}\right) \times \left(\frac{28.0 \text{ g } C_2H_4}{1 \text{ mole } C_2H_4}\right) \sim 1.73 \text{ g } C_2H_4$$

The weight of C_2H_4 left unreacted is simply the difference between the weight of C_2H_4 initially present and the weight of C_2H_4 that reacts with the O_2.

$$\text{Excess } C_2H_4 = 1.93 \text{ g} - 1.73 \text{ g} = 0.20 \text{ g}$$

There is an interesting postscript to the last problem. Although not stated explicitly, our assumption was that even without sufficient oxygen to consume all of the C_2H_4, any C_2H_4 that did react was converted completely to CO_2 and H_2O. If this were the case, one aspect of automotive air pollution would be removed. What actually occurs when the hydrocarbon (in this case C_2H_4) is present in excess is that some of it is converted to CO. In the internal combustion engine gasoline (which is composed of a mixture of hydro-carbons) is burned in a limited supply of oxygen and the incomplete combus-tion therefore produces a mixture of CO, CO_2, and H_2O.

Review Questions and Problems

2.1 Define stoichiometry, gram-atom, and mole.

2.2 What is the weight of (a) one g-atom U, (b) 0.375 g-atom nitrogen, (c) 2.57 g-atom Ni?

2.3 Which quantity weighs the most: (a) 0.52 g-atom Na, (b) 12.5 g nitrogen, or (c) 0.25 g-atom Ca? Which of the above contains the largest number of atoms?

2.4 Carbon atoms have a diameter of ap-proximately 1.5×10^{-8} cm. If carbon atoms were laid in a row 3 cm long, what would be the total mass of carbon?

2.5 Calculate the mass of Cu required to react with 5.00×10^{20} molecules of S_8 to form Cu_2S.

2.6 A philosophical question that had raged for some time was how many angels could stand on the head of a pin? If an angel is the size of an atom of iron (1.16×10^{-8} cm in diameter), calculate the number of angels that could stand on the head of a pin that is 2.00 mm in diameter.

2.7 What would be the weight of one mole-cule of sucrose ($C_{12}H_{22}O_{11}$)? How much would one mole of sucrose weigh?

2.8 Calculate the percentage composition of each of the following: (a) (benzene) C_6H_6; (b) (ethyl alcohol) C_2H_5OH; (c) (potassium dichromate) $K_2Cr_2O_7$; (d) (xenon tetrafluoride) XeF_4.

2.9 What is the empirical formula of each of the following?

$(NH_4)_2S_2O_8$ Al_2Cl_6 C_6H_6 $C_3H_8O_3$
$C_6H_{12}O_6$

2.10 Balance the following equations by in-spection.
(a) $ZnS + HCl \rightarrow ZnCl_2 + H_2S$
(b) $HCl + Cr \rightarrow CrCl_3 + H_2$
(c) $Al + Fe_3O_4 \rightarrow Al_2O_3 + Fe$
(d) $H_2 + Br_2 \rightarrow HBr$
(e) $Na_2S_2O_3 + I_2 \rightarrow NaI + Na_2S_4O_6$

2.11 Balance the following equations.
(a) $FeCl_3 + Na_2CO_3 \rightarrow$
$Fe_2(CO_3)_3 + NaCl$
(b) $NH_4Cl + Ba(OH)_2 \rightarrow$
$BaCl_2 + NH_3 + H_2O$
(c) $Ca(OH)_2 + H_3PO_4 \rightarrow$
$Ca_3(PO_4)_2 + H_2O$
(d) $Fe_2(CO_3)_3 + H_2SO_4 \rightarrow$
$Fe_2(SO_4)_3 + H_2O + CO_2$
(e) $Na_2O + (NH_4)_2SO_4 \rightarrow$
$Na_2SO_4 + H_2O + NH_3$

2.12 A 1.31 g sample of sulfur was reacted with an excess of chlorine to produce 4.22 g of a product that contains only sulfur and chlorine. What is the empirical formula of the compound?

2.13 A substance was found to be composed of 60.8% sodium, 28.5% boron, and 10.5% hydrogen. What is the empirical formula of the compound?

2.14 A 0.537 g sample of an organic compound containing only carbon, hydrogen, and oxygen was burned in air to produce 1.030 g of CO_2 and 0.632 g H_2O. What is the empirical formula of the compound?

2.15 A 1.35 g sample of a substance containing carbon, hydrogen, nitrogen, and oxygen was burned to produce 0.810 g H_2O and 1.32 g CO_2. In a separate reaction, 0.735 g of the substance yielded 0.284 g of NH_3. Determine the empirical formula of the substance.

2.16 Consider the following balanced equation,

$$6\ ClO_2 + 3\ H_2O \longrightarrow 5\ HClO_3 + HCl$$

(a) How many moles of $HClO_3$ are produced from 14.3 g ClO_2?
(b) How many grams of H_2O are needed to produce 5.74 g HCl?
(c) How many grams of $HClO_3$ are produced when 4.25 g of ClO_2 are added to 0.853 g of H_2O?

2.17 Acetylene, which is used as a fuel in welding torches, is produced in a reaction between calcium carbide and water.

$$CaC_2 + 2H_2O \longrightarrow Ca(OH)_2 + C_2H_2\ (g)$$
calcium carbide   **acetylene**

(a) How many moles of C_2H_2 would be produced from 2.50 moles of CaC_2?
(b) How many grams of C_2H_2 would be formed from 0.500 moles of CaC_2?
(c) How many moles of water would be consumed when 3.20 moles of C_2H_2 are formed?
(d) How many grams of $Ca(OH)_2$ are produced when 28.0 g of C_2H_2 are formed?

2.18 Aspirin (which many students take after working on chemistry problems) is prepared by the reaction of salicylic acid $(C_7H_6O_3)$ with acetic anhydride $(C_4H_6O_3)$ according to the reaction,

$$C_7H_6O_3 + C_4H_6O_3 \longrightarrow C_9H_8O_4 + C_2H_4O_2$$
(aspirin)

How many grams of salicylic acid must be used to prepare two 5 grain aspirin tablets (1 gram = 15.4 grains)?

2.19 The insecticide DDT (which ecologists now recognize as a serious environmental pollutant) is manufactured in a reaction between chlorobenzene and chloral

$$2C_6H_5Cl + C_2HCl_3O \longrightarrow C_{14}H_9Cl_5 + H_2O$$
chlorobenzene   chloral   **DDT**

How many kilograms of DDT can be produced from 1000 kilograms of chlorobenzene.

2.20 Polystyrene has a structure similar to that below.

Here the basic styrene unit (within brackets) is repeated a large number, n, times to give long chainlike molecules. A particular sample of this plastic was found to have an average molecular weight of 1 million. What is the average number of styrene units in a chain?

2.21 Often, when a chemist prepares a compound in a laboratory experiment products other than those he wishes are also formed in the reaction. In a typical experiment a student reacted benzene, C_6H_6, with bromine, Br_2, in an attempt to prepare bromobenzene, C_6H_5Br. This reaction also produced, as a by-product, dibromobenzene, $C_6H_4Br_2$. On the basis of the equation,

$$C_6H_6 + Br_2 \longrightarrow C_6H_5Br + HBr$$

(a) What is the maximum amount of C_6H_5Br that the student could have hoped to obtain from 15.0 g of benzene? (This is called the theoretical yield.)

(b) In his experiment the student obtained 2.50 g of $C_6H_4Br_2$. How much C_6H_6 was not converted to C_6H_5Br?

(c) What was the student's actual yield of C_6H_5Br?

(d) Calculate the percentage yield for this reaction.

$$\left(\text{Percentage yield} = \frac{\text{actual yield}}{\text{theoretical yield}} \times 100\right)$$

2.22 Dimethylhydrazine, $(CH_3)_2NNH_2$, has been used as a fuel in the Apollo lunar descent module, with liquid N_2O_4 as the oxidizer. The products of the reaction between these two in the rocket engine are H_2O, CO_2, and N_2.

(a) Write a balanced chemical equation for the reaction.

(b) Calculate the mass of N_2O_4 required to burn 50 kg of dimethylhydrazine.

2.23 The fermentation of sugar to produce ethyl alcohol follows the equation

$$C_6H_{12}O_6 \xrightarrow{\text{yeasts}} 2C_2H_5OH + 2CO_2$$

What is the maximum weight of alcohol that can be obtained from 500 g of sugar?

2.24 Freon 12, a gas used as a refrigerant, is prepared by the reaction

$$3CCl_4 + 2SbF_3 \longrightarrow 3CCl_2F_2 + 2SbCl_3$$
$$\text{freon 12}$$

If 150 g of CCl_4 is mixed with 100 g of SbF_3,

(a) How many grams of CCl_2F_2 can be formed?

(b) How many grams of which reactant will remain after reaction has ceased?

2.25 Vanillin is composed of carbon, 63.2%; hydrogen, 5.26%; and oxygen, 31.6%. What is the empirical formula of vanillin?

3
Atomic Structure And The Periodic Table

The atomic theory as proposed by Dalton represented a major breakthrough in the development of chemistry. All of the computations you learned to perform in the last chapter, in fact, are based on his idea that atoms of each element have a characteristic atomic mass. While Dalton's theory accounted for the mass relationships observed in chemical reactions, however, it was not able to explain why substances react the way they do. It could be determined, for example, that one oxygen atom was able to react with a maximum of only two atoms of hydrogen, but no one understood why. Furthermore, as additional evidence came to light, it became increasingly clear that the simple picture of an indivisible atom was just not sufficient to account for all of the facts, and it was by a rather fascinating piecing together of bits of information that our current picture of the atom developed. In this chapter we shall follow the evolution of modern atomic theory; we shall look at some of the results of the theory; and we shall see how the theory can help to explain and correlate chemical and physical properties of the elements.

In 1834, Michael Faraday reported the results of his experiments on electrolysis in which he showed that chemical change could be caused by the passage of electricity through water solutions of chemical compounds. These experiments demonstrated that matter was electrical in nature and led G. J. Stoney, 40 years later, to propose the existence of particles of electricity that he called **electrons.**

Around the same time, physicists began to investigate the flow of electric current in **gas discharge tubes.** These tubes were made of glass with metal electrodes at each end (Figure 3.1). It was found that when a high voltage was applied across the electrodes and air was partially evacuated from the tube, a flow of electricity was observed and the residual gas within the tube began to glow (a neon sign is a common example of a gas discharge tube). If virtually all of the gas was removed, the glow was no longer observed but the electrical discharge continued. A screen, coated with a zinc sulfide phosphor, placed between the electrodes was found to glow on the side facing the negative electrode (the **cathode**), demonstrating that the discharge originates at the cathode and flows toward the positive electrode (the **anode**). These "rays" were thus called **cathode rays.**

Further investigations showed that cathode rays

1. Are always the same regardless of the nature of the material composing the electrodes or the kind of residual gas within the tube.
2. Normally travel in straight lines.
3. Cast shadows.

3.1
The Electrical Nature of Matter

Figure 3.1
The gas discharge tube.

4. Can turn a pinwheel placed in their path, suggesting that cathode rays are composed of particles.
5. Heat a metal foil placed between the electrodes.
6. Can be bent by an electric or magnetic field oriented perpendicular to the length of the tube (Figure 3.2).

These observations showed that cathode rays are composed of highly energetic, negatively charged fundamental particles of matter. They were, in fact, the electrons described by Stoney.

J. J. Thomson, in 1897, used a cathode ray tube, quite similar to present-day television picture tubes, to measure the ratio of the charge to the mass of an electron. This device is shown schematically in Figure 3.3. Electrons generated at the cathode are accelerated toward the anode, which has a hole in it. Some electrons pass through the hole and continue on, striking the face of the tube, which is coated with a phosphor, at B. If oppositely charged plates are placed above and below the tube, the beam is deflected toward the positive plate and strikes the face at A. The amount of deflection that the particle undergoes will be directly proportional to its charge, since a highly negatively charged particle will be attracted to the positive plate more than one with a small charge. The amount of deflection will also be inversely propor-

(a)   (b)

Figure 3.2
(a) Cathode rays are deflected by an electric field and (b) by a magnetic field.

Figure 3.3
A cathode ray tube used to measure the charge-to-mass ratio of the electron.

tional to the mass of the particle, since a heavy particle will be less affected by the electrostatic attraction, as it passes between the plates, than a particle of smaller mass. Thus the deflection that is observed depends on the ratio of charge to mass (e/m) of the particle.

If a magnetic field is generated at right angles to the electric field, as shown in Figure 3.3, the electrons are deflected in a direction exactly opposite to that caused by the charged plates. In the absences of the electric field, the electron beam is bent by the magnetic field so that it collides with the surface of the tube at C.

In practice, Thomson applied a magnetic field of known strength across the tube and noted the deflection of the electron beam. Charge was then applied to the plates until the beam was brought back to its original point of impact, B. From the magnitude of the electric and magnetic fields, Thomson calculated the charge-to-mass ratio, e/m, for the electron to be -1.76×10^8 coulombs/gram.[1]

The charge on the electron was determined by means of a rather clever experiment performed in 1908 by R. A. Millikan at the University of Chicago. In his apparatus, illustrated in Figure 3.4, a fine mist of oil droplets was sprayed above a pair of parallel metal plates. As droplets settled through a hole in the upper plate, the air between the plates was briefly irradiated with X-rays. Electrons, knocked from gas atoms by the X-rays, were picked up by the oil droplets, thereby giving them a negative charge. Millikan found that by placing an electric charge on the plates (upper plate positive, lower plate negative) the downward motion of negatively charged drops could be stopped. A knowledge of the mass of a drop (measured by observing its rate of fall in the absence of the electric field) and the amount of charge on the plates required to keep that drop suspended permitted Millikan to calculate the amount of charge on the drop.

---

[1] A **coulomb** is a unit of electric charge and is equal to the amount of charge that passes through a wire in one second if the current is one ampere.

Figure 3.4
Millikan oil drop experiment.

After he performed this experiment many times, Millikan observed that the charge on the oil drops was always a multiple of -1.60×10^{-19} coulombs. He reasoned that since the oil drops could pick up one, two, three, or more electrons, the total charge on any drop must be a multiple of the charge on a single electron. This suggested, therefore, that the charge on the electron is -1.60×10^{-19} coulombs. Once the charge on the electron had been measured, its mass, 9.11×10^{-28} g, was obtained from the already known charge-to-mass ratio.

**3.3
Positive Particles, the Mass Spectrometer**

Since negative particles (electrons) are present in a gas discharge, and since ordinary matter is observed to be electrically neutral, it seems reasonable to suspect the existence of positively charged particles in a gas discharge as well. When, in fact, a perforated cathode was used in a discharge tube, streamers of light, which were called "canal rays," were observed emanating from the holes at the rear of the negative electrode (Figure 3.5).

During an electric discharge, electrons, emitted from the cathode, collide with neutral gas atoms, knocking electrons off of them. The atoms, by losing electrons, become positively charged particles (positive **ions**[2]) and are at-

Figure 3.5
Canal rays. Positive ions passing through holes in the cathode appear as canal rays at the rear of the electrode. Bursts of light are seen when they collide with a phosphor on the end of tube.

[2] Charged particles that result from the addition or removal of electrons from atoms or molecules are called **ions.** We shall come across many examples of ions during future discussions.

tracted toward the cathode. Most of them collide with the cathode; however, some travel through the perforations and emerge at the rear where they are observed as the "canal rays." If the rear wall of the discharge tube is coated with a phosphor, flashes of light can also be seen when these positive particles, which have passed through the cathode, hit the wall.

An instrument designed to determine the charge-to-mass ratio of positive ions is called a **mass spectrometer** (Figure 3.6). Gaseous material is introduced at A and ionized by an electric discharge across electrodes B and C. The positive ions thus produced are accelerated through the wire grid, E, and through the slits F and G which form a narrow beam that is fed between the poles of a powerful magnet. The magnetic field acts to deflect the particles into a circular path, with the degree of curvature determined by the charge to mass ratio of the ions. For ions with the same charge the extent to which their paths are bent depends on their masses, with lighter particles being deflected more than heavier ones. For ions with the same mass, the degree to which their paths are curved is directly proportional to their charge. By adjusting the strength of the magnetic field, ions with any desired e/m ratio may be focused on the detector at H. Ions with higher e/m ratios are deflected more (e.g., to P) while those of lower e/m ratios are deflected less (e.g., to Q).

The measurement of e/m for positive particles reveals the following information.

1. Positive ions always have e/m ratios that are *much smaller* than that of the electron. This means either that they are much more massive than the electron (i.e., m is very large) or that they carry very small positive charges (i.e., e is small). Since they are formed from neutral atoms by the loss of electrons, the charge that they carry is some integral (i.e., whole number) multiple of the charge on the electron. This means, in effect, that in order to have a much smaller e/m than the electron their masses must be much greater.

Atomic Structure and the Periodic Table

Figure 3.6

The mass spectrometer.

2. The *e/m* ratio is dependent on the nature of the gas introduced into the mass spectrometer, which shows that not all positive ions have the same *e/m*.

When hydrogen, the lightest of all gases, is placed in the mass spectrometer, the *e/m* ratio of the hydrogen ion is found to be $+9.63 \times 10^4$ coulombs/gram. This is the largest *e/m* observed for any positive ion, and it is thus assumed that the hydrogen ion represents a fundamental particle of positive charge, the **proton.** A neutral hydrogen atom, therefore, is composed of one electron and one proton. If we compare the charge-to-mass ratios of the proton and electron, we find the proton to be 1836 times as heavy as the electron; thus nearly all of the atom's mass is associated, somehow, with its positive charge.

Since ions are formed by the addition or loss of electrons, each of which either adds or removes 1.60×10^{-19} coulombs of charge, it is convenient to express the charges on particles in units of this size. Thus, for example, the electron would possess one unit of negative charge; on this scale, this constitutes a charge of $-1e$; likewise, two units of positive charge would be represented as $+2e$. We commonly indicate the charge on an ion formed from an atom by writing the number of units of positive or negative charge as a superscript on the right side of the chemical symbol. Thus the ion He^{2+} is an ion formed from the helium atom by the loss of two electrons while the ion, O^{2-}, is an ion formed from oxygen by the addition of two electrons.

3.4 Radioactivity

Still another bit of evidence that demonstrates the existence of subatomic particles was discovered by Henri Becquerel. He observed that some substances spontaneously emit radiation; we say these substances are **radioactive.** There are three important types of radiation.

1. Alpha radiation, composed of He^{2+} ions called alpha particles (α-particles).

2. Beta radiation, consisting of electrons, in this instance, called beta particles (β-particles).

3. Gamma radiation (γ-rays), highly energetic, very penetrating light waves similar to X-rays.

3.5 The Nuclear Atom

One of the most significant developments in our understanding of the structure of the atom was provided by Ernest Rutherford in 1911. Prior to this time it was thought that the atom had a nearly uniform density throughout, with the electron being buried in a glob of positive charge, much like raisins in a pudding. With this picture of a rather mushy atom in mind, one of Rutherford's students made an astonishing discovery when he set out to investigate the scattering of alpha particles by thin metal foils. He found that when a narrow beam of α-particles was directed at a thin gold foil, not all of them passed through the metal essentially undisturbed as had been expected. Some of the alpha particles were scattered at large angles and a small fraction, in fact, were even deflected back toward the source (Figure 3.7). Rutherford could only explain such observations by concluding that the atom contained a very small, extremely dense positive **nucleus** containing all of the protons and nearly all of the atom's mass. Since the nucleus contains the positive charge in the atom, it follows that the electrons must be distributed somewhere in the remaining volume of the atom.

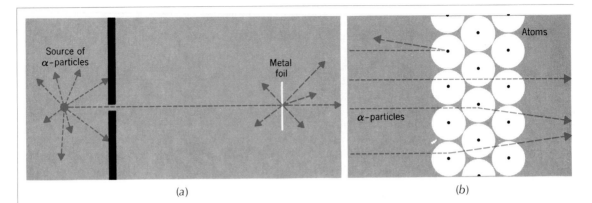

(a)

(b)

Figure 3.7

(a) Scattering of α-particles by a metal foil. (b) Deflection of α-particles by atomic nuclei.

Roentgen, in 1895, discovered that when high energy electrons in a discharge tube collide with the anode, penetrating radiation, which he called X-rays, was produced (see Figure 3.8). X-rays, visible light, infrared and ultraviolet radiation, and also radio waves are all forms of **electromagnetic radiation** that travel as waves through space at a constant speed c (called the speed of light; 3.0×10^{10} cm sec^{-1}). These waves are characterized by their **wavelength,** λ (see Figure 3.9), which is the distance between consecutive peaks (or troughs) in the wave, and by their **frequency,** ν, which is the number of peaks that pass by a given point per second. Wavelength and frequency are related to each other by the equation

$$\lambda \cdot \nu = c \qquad (3.1)$$

Moseley, in 1914, found that he could assign a number, called the **atomic number,** to an element based on the frequency of the X-rays that the element emits. Experiments by Rutherford and his students, in which they were able to measure the charge on the nucleus, enabled Moseley to conclude that this atomic number represented the number of protons in the nucleus.

Figure 3.8
Production of X-rays. High voltage electric discharge across the electrodes causes the anode to emit X-rays.

Figure 3.9

The electromagnetic spectrum.

Rutherford had observed that only about one-half of the nuclear mass could be accounted for by protons; he therefore suggested that particles of zero charge and of mass nearly the same as that of the proton are also present in the nucleus. The existence of these particles was confirmed in 1932 by an English scientist, J. Chadwick, who bombarded beryllium with alpha particles and found that highly energetic, uncharged particles were emitted. These particles, called **neutrons,** have a mass only slightly larger than that of the proton.

In summary, then, the atom is composed of a dense nucleus, containing protons and neutrons that provide nearly all of the atom's mass, surrounded in some fashion by electrons distributed throughout the remaining volume of the atom. The nucleus is extremely small, having a diameter of approximately

Table 3.1

Some Properties of Subatomic Particles

	Mass		Charge	
	Grams	Atomic Mass Units (amu)	Coulombs	Electronic Charge Units
Proton	1.67×10^{-24}	1.007274	$+1.602 \times 10^{-19}$	+1
Neutron	1.67×10^{-24}	1.008665	0	0
Electron	9.11×10^{-28}	0.000549	-1.602×10^{-19}	−1

10^{-13} cm, compared to the atom itself whose diameter is of the order of 10^{-8} cm, or 1 Å (the **angstrom** is a unit of length that is convenient for expressing atomic dimensions; 1 Å $= 1 \times 10^{-8}$ cm). The properties of the three major particles found in the atom are summarized in Table 3.1.

In Chapter 1 we noted that, contrary to Dalton's original hypothesis, not all atoms of the same element have identical masses. We referred to these different kinds of atoms as **isotopes.** The existence of isotopes is a common phenomenon, with most of the elements occurring naturally as mixtures of isotopes.

3.8

Isotopes

As we shall see, the properties of an element are almost entirely determined by the number and distribution of its electrons; therefore it is the atomic number (or number of protons) that serves, indirectly, to distinguish an atom of one element from an atom of another. Any mass differences that exist between atoms of the same element, then, must arise from different numbers of neutrons.

A particular isotope of an element is identified by specifying its atomic number, Z, and its **mass number,** A, which is the *sum* of the number of protons and neutrons in the atom. The number of neutrons present can be obtained from the difference, $A - Z$. We indicate the atom symbolically by writing the mass number as a superscript and the atomic number as a subscript; both precede the atomic symbol,

$$^A_Z X$$

Thus, the carbon atom (atomic number $= 6$) containing six neutrons would have a symbol of $^{12}_6 C$. It is this isotope of carbon, incidentally, that serves as the basis of the current scale of atomic weights; that is, the mass of one atom of $^{12}_6 C$ is defined as exactly 12 atomic mass units (amu).

The fact that atoms of a given element are not all alike has been put to practical use. One example is in archeological dating. Radioactive isotopes, such as ^{14}C, undergo nuclear transformations that cause them to be transformed into other elements. The rate of this decay is known, and from the relative abundance of ^{14}C and ^{12}C (a nonradioactive isotope of carbon) in both living and fossil material the age of the fossil can be estimated. The details of radioactive decay will be discussed in greater detail in Chapter 22.

Another application of radioactive isotopes is in chemotherapy. Treatment of thyroid cancer, for instance, is accomplished by administering carefully controlled doses of radioactive ^{131}I, which tends to concentrate in the thyroid glands where the radiation produced by the ^{131}I causes destruction of cancerous cells. In this case use is made of the body's natural tendency to concentrate iodine (either radioactive or nonradioactive) in the thyroid gland.

As we noted above, nearly all elements as found in nature occur as mixtures of isotopes. For example, the element copper is found to contain the naturally occurring isotopes $^{63}_{29}Cu$ and $^{65}_{29}Cu$ whose masses have been accurately determined to be 62.9298 and 64.9278 amu, respectively. Their relative abundances are 69.09% and 30.91%, and the observed atomic weight of copper, 63.54, is obtained as an average that is weighted according to the mass contributed by each isotope, as shown in Example 3.1.

Example 3.1

Using the data supplied in the paragraph above, calculate the average atomic mass of copper.

Solution

The amount of mass that each isotope contributes toward the average atomic mass is equal to the product of its mass multiplied by its fractional abundance.

$$^{63}Cu \quad 62.9298 \text{ amu} \times 0.6909 = 43.48 \text{ amu}$$
$$^{65}Cu \quad 64.9278 \text{ amu} \times 0.3091 = \underline{20.07 \text{ amu}}$$
$$\text{Total} \quad 63.55 \text{ amu}$$

In conclusion, observe the distinction between the mass number of an isotope and its actual mass. The mass number is simply the total count of protons plus neutrons and is not quite equal to the mass of the atom.

3.9
The Periodic Law and the Periodic Table

Scientists, even as early as 1800, had accumulated a significant amount of information concerning the physical and chemical properties of the known elements. This knowledge, however, existed for the most part as isolated and unrelated facts that needed to be correlated in some fashion before their total significance could be grasped. The first successful attempt at classifying elements according to their properties was made by J. W. Döbereiner in 1817. He noticed that, in some instances, when three elements of similar chemical properties were arranged in order of increasing atomic weight, the atomic weight of the middle member was very nearly the average of the other two. Furthermore, many other properties of the element in the center of the "triad" lay also between those of the other two. Table 3.2 shows some data that demonstrate this for the elements, lithium, sodium, and potassium.

Progress after 1817 was slow, mainly because accurate atomic weights for the elements had not yet been obtained. It was not until after 1860, when a reliable experimental method for determining atomic weights was developed, that progress toward the modern periodic classification of the elements was renewed.

In 1866 John Newlands, a British chemist and a lover of music, presented to the Chemical Society of London a table of the elements (Figure 3.10) arranged in order of increasing atomic weight. He pointed out that similarities among chemical and physical properties repeated themselves every eighth element. Thus, hydrogen, element number 1 in his table, bore certain similarities to fluorine (element number 8). Likewise, lithium (element number 2) and sodium (element number 9) have similar properties. Newlands likened this behavior to the repetition of octaves on the musical scale and termed the recurrence of properties the "law of octaves." This was a rather unfortunate choice of comparisons since it obscured the true significance of his work and subjected him to rather severe criticisms by the other members of the Chemical Society (one member even asked, rather sarcastically, if he had tried arranging the elements in alphabetical order).

Some of the criticisms directed at Newlands were justified since there were several drawbacks to the table he had proposed. In the years preceding

Table 3.2
Properties of Elements in the Dobereiner Triad;
Lithium, Sodium, Potassium

Properties That the Elements Have in Common

The free elements are metallic, soft, and react vigorously with water to produce compounds with the general formula MOH (M = Li, Na, or K). In all of their compounds these elements exist as ions with a charge of +1. Nearly all of their compounds are soluble in water, and they impart brilliant colors to flames (Li, red; Na, yellow; K, violet).

	Lithium	Potassium	Sodium
Atomic weights (amu)	6.94	39.10	22.99
	Average for Li and K: 23.02		
Melting Point (°0)	179	63.65	97.8
Boiling point (°0)	1317	774	892
Density (g/ml)	0.534	0.862	0.971

the introduction of his table many new elements had been discovered, and it may have been assumed that additional elements would be discovered in the future; yet his table left no room for them. Also, there were several instances where he was forced to place more than one element in the same position in the table. Furthermore, in many locations (particularly after calcium) elements occurred whose properties were clearly different from others in the same

1 H	8 F	15 Cl	22 Co, Ni	29 Br	36 Pd	42 I	50 Pt, Ir
2 Li	9 Na	16 K	23 Cu	30 Rb	37 Ag	44 Ca	53 Os
3 Be	10 Mg	17 Ca	25 Zn	31 Sr	38 Cd	45 Ba, V	54 Hg
4 B	11 Al	19 Cr	24 Y	33 Ce, La	40 U	46 Ta	56 Tl
5 C	12 Si	18 Ti	26 In	32 Zr	39 Sn	47 W	52 Pb
6 N	13 P	20 Mn	27 As	34 Di, Mo	41 Sb	48 Nb	55 Bi
7 O	14 S	21 Fe	28 Se	35 Ro, Ru	43 Te	49 Au	51 Th

Figure 3.10
Newlands' table of the elements (revised form).

column (e.g., chlorine, a gas, and cobalt, a metal). Nevertheless, Newlands had been on the right track, and 21 years later he was awarded the Davy medal in belated recognition of his contribution.

The modern periodic table is due mainly to the efforts of two chemists, a Russian, Dmitri Mendeleev, and a German, Julius Lothar Meyer, both of whom worked independently of the other and produced similar tables about the same time. Mendeleev presented the results of his work to the Russian Chemical Society in the early part of 1869, while Meyer's table did not appear until December of that same year.

Mendeleev's table, in the revised form that appeared in 1871 (Figure 3.11), is very similar to the one we use today and is essentially the "short form" of the current table except for the absence of the **noble gases,** helium, neon, argon, krypton, xenon, and radon, which had not yet been discovered. The periodic table of Mendeleev was an improvement over Newlands' in several respects. First, elements were placed in the table on the basis of similarities in properties among elements in a column, or **group.** For example, tellurium and iodine, whose atomic weights in 1869 were thought to be 128 and 127 amu respectively, were placed into the table in reverse order (according to atomic weight) because their properties dictated that tellurium be placed in Group VI and iodine be placed in Group VII. Because of this insistence upon placing elements of similar properties in the same group, Mende-

	Group I	Group II	Group III	Group IV	Group V	Group VI	Group VII	Group VIII
1	H 1							
2	Li 7	Be 9.4	B 11	C 12	N 14	O 16	F 19	
3	Na 23	Mg 24	Al 27.3	Si 28	P 31	S 32	Cl 35.5	
4	K 39	Ca 40	— 44	Tc 48	V 51	Cr 52	Mn 55	Fe 56, Co 59 Ni 59, Cu 63
5	(Cu 63)	Zn 65	— 68	— 72	As 75	Se 78	Br 80	
6	Rb 85	Sr 87	?Yt 88	Zr 90	Nb 94	Mo 96	— 100	Ru 104, Rh 104 Pd 105, Ag 100
7	(Ag 108)	Cd 112	In 113	Sn 118	Sb 122	Te 125	I 127	
8	Cs 133	Ba 137	?Di 138	?Ce 140	—	—	—	— — — —
9	—	—	—	—	—	—	—	
10	—	—	?Er 178	?La 180	Ta 182	W 184	—	Os 195, Ir 517 Pt 198, Au 199
11	(Au 199)	Hg 200	Tl 204	Pb 207	Bi 208	—		
12	—	—	—	Th 231	—	U 240	—	— — — —

Figure 3.11
Mendeleev's Periodic Table (1871)

leev was forced to leave empty spaces in the table which, presumably, would eventually be occupied by new elements after they had been discovered.

A second advantage of Mendeleev's table was that it was possible to predict the properties of the missing elements because elements in any particular column had to have similar properties. For example, germanium, which lies below silicon and above tin in Group IV, had not been discovered when Mendeleev constructed his table and therefore a blank space appears at this spot in the chart. On the basis of its position in the table, Mendeleev predicted that the properties of germanium, which he called "eka-silicon," should lie intermediate between those of silicon and tin. Table 3.3 shows how closely he predicted the properties that were found for germanium when it was discovered in 1886.

The necessity of reversing the order of atomic weights, which occurred when placing tellurium and iodine in the periodic table, was repeated after the noble gases were discovered. It was found that the atomic weight of argon (39.9 amu) was greater than that of potassium (39.1 amu); however, on the basis of physical and chemical properties, potassium clearly belonged in Group I (following argon) while argon had to be included in a separate group with the other noble gases. These reversals present no problems at all and the proper arrangement is obtained when the elements are placed in order of increasing atomic number (Section 3.6) rather than atomic weight. This leads us to the modern statement of the **periodic law:** *When the elements are arranged in order of increasing atomic number, there occurs a periodic repetition of physical and chemical properties.* Thus it is the charge on the nucleus,

Table 3.3
Predicted Properties of Eka-silicon and Observed Properties of Germanium

Property	Silicon	Tin	Predicted for Eka-silicon (Es)	Currently Accepted for Germanium
Atomic weight (amu)	28	118	72	72.59
Density (g/ml)	2.33	7.28	5.5	5.3
Melting point (°C)	1410	232	High	947
Physical form at room temperature	Gray nonmetal	White metal	Gray metal	Gray metal
Reaction with acids and alkalies	Acid — no reaction Alkalies — slow reaction	Slow attack by acids. No reaction with alkalies	Very slow attack by acids. No reaction with alkalies	No reaction with acids. Will react only with concentrated alkalies
Number of chemical bonds usually formed	4	4	4	4
Formula of chloride	$SiCl_4$	$SnCl_4$	$EsCl_4$	$GeCl_4$
Density of chloride (g/ml)	1.50	2.23	1.9	1.88
Boiling point of chloride (°C)	57.6	114	100	84

Figure 3.12

The modern periodic table of the elements

Periods	IA	IIA	IIIB	IVB	VB	VIB	VIIB	VIII	VIII	VIII	IB	IIB	IIIA	IVA	VA	VIA	VIIA	0 (Noble gases)
1	1 H 1.00797																	2 He 4.00260
2	3 Li 6.941	4 Be 9.01218											5 B 10.81	6 C 12.01115	7 N 14.0067	8 O 15.9994	9 F 18.99840	10 Ne 20.179
3	11 Na 22.98977	12 Mg 24.305											13 Al 26.98154	14 Si 28.086	15 P 30.97376	16 S 32.06	17 Cl 35.453	18 Ar 39.948
4	19 K 39.098	20 Ca 40.08	21 Sc 44.9559	22 Ti 47.90	23 V 50.9414	24 Cr 51.996	25 Mn 54.9380	26 Fe 55.847	27 Co 58.9332	28 Ni 58.71	29 Cu 63.546	30 Zn 65.38	31 Ga 69.72	32 Ge 72.59	33 As 74.9216	34 Se 78.96	35 Br 79.904	36 Kr 83.80
5	37 Rb 85.4678	38 Sr 87.62	39 Y 88.9059	40 Zr 91.22	41 Nb 92.9064	42 Mo 95.94	43 Tc 98.9062	44 Ru 101.07	45 Rh 102.9055	46 Pd 106.4	47 Ag 107.868	48 Cd 112.40	49 In 114.82	50 Sn 118.69	51 Sb 121.75	52 Te 127.60	53 I 126.9045	54 Xe 131.30
6	55 Cs 132.9054	56 Ba 137.34	57 *La 138.9055	72 Hf 178.49	73 Ta 180.9479	74 W 183.85	75 Re 186.2	76 Os 190.2	77 Ir 192.22	78 Pt 195.09	79 Au 196.9665	80 Hg 200.59	81 Ti 204.37	82 Pb 207.19	83 Bi 208.9804	84 Po (210)	85 At (210)	86 Rn (222)
7	87 Fr (223)	88 Ra 226.0254	89 †Ac (227)	104 Ku (261)	105 Ha (260)													

Atomic number → (1 H 1.00797) ← Atomic mass

*	58 Ce 140.12	59 Pr 140.9077	60 Nd 144.24	61 Pm (147)	62 Sm 150.4	63 Eu 151.96	64 Gd 157.25	65 Tb 158.9254	66 Dy 162.50	67 Ho 164.9304	68 Er 167.26	69 Tm 168.9342	70 Yb 173.04	71 Lu 174.97
†	90 Th 232.0381	91 Pa 231.0359	92 U 238.029	93 Np 237.0482	94 Pu (244)	95 Am (243)	96 Cm (247)	97 Bk (247)	98 Cf (251)	99 Es (254)	100 Fm (257)	101 Md (258)	102 No (255)	103 Lr (256)

which we associate with the atomic number, and the number of electrons in the neutral atom that are important in determining the sequence in which the elements occur and that are responsible for their properties.

The periodic table in use today (sometimes called the "long" form of the periodic table) is shown in Figure 3.12. We see that, like Mendeleev's table, it is constructed of a number of vertical columns, called **groups,** each containing a *family of elements.* These groups are identified by a Roman numeral and a letter, either A or B. Groups IA through VIIA and Group O are referred to collectively as the **representative elements,** while Groups IB through VIIB and Group VIII (actually composed of the three short columns in the center of the table) constitute the **transition elements.** Similarities between properties of the A- and B-group elements exist, although the similarities are often very weak.

Finally, we see that there are two long rows of elements lying just below the main part of the table. These elements, called the **inner transition elements,** actually belong in the body of the table but are placed where they are simply to conserve space. The first of these rows, elements 58 to 71, fit into the chart following lanthanum and is collectively called the **lanthanides** or the **rare earths.** The second row, elements 90 to 103, belongs between actinium ($Z = 89$) and Kurchatovium ($Z = 104$). The elements in this latter series are termed the **actinides.**

The horizontal rows in the periodic table are called **periods** and are designated by means of Arabic numerals. The elements hydrogen and helium are members of the first period; lithium through neon are known as second period elements, and so on.

Certain families of elements are characterized by names as well as by their group number. For example, the Group IA elements are frequently spoken of as the **alkali metals.** Similarly, the Group IIA elements are called the **alkaline earth metals;** the Group VIIA elements, the **halogens;** and the Group O elements, the **noble gases** (these are also sometimes called the **inert gases** because of their extremely limited ability to react chemically).

The elements can also be broadly classified either as **metals, nonmetals,** or **metalloids.** You are probably familiar with most of the physical properties that serve to identify metals: high electrical conductivity, luster, generally high melting points, **ductility** (ability to be drawn into wires), and **malleability** (ability to be hammered into thin sheets). Nonmetals, on the other hand, are uniformly very poor conductors of electricity, do not possess that characteristic luster found with metals and, as solids, are brittle. Metalloids have properties that lie intermediate between those of metals and nonmetals. They are, for example, useful as semiconductors of electricity.

In the periodic table, elements on the left are metals and those on the right are nonmetals. The heavy jagged line drawn from boron to astatine approximately represents the boundary between metallic and nonmetallic behavior, with elements lying immediately adjacent to the line generally having metalloid properties.

In conclusion, the periodic table is probably the most useful aid that the chemist has at his disposal. We will see how it can be used to correlate much of the theoretical and factual information that you should take with you when you leave this course.

When atoms combine during chemical reactions, it is the electrons surrounding the nucleus that interact. Therefore, the chemical properties of the elements are determined by the way in which the electrons are arranged. The nucleus serves mainly to determine the number of electrons that must be present to give a neutral atom. The key that has permitted the deduction of the electronic structure of the elements is atomic spectra.

If sunlight, or the light from an incandescent lamp, is formed into a narrow beam by a slit and is passed through a prism onto a screen, a rainbow of colors is observed (Figure 3.13). This spectrum is composed of visible light of all wavelengths and is called a **continuous spectrum.** However, if the source of light is a gas discharge tube, containing hydrogen, for example, the observed spectrum consists of a number of lines (the image of the slit) projected on the screen (Figure 3.14) and is called a **line spectrum.** Obviously the visible light emitted by hydrogen does not contain radiation of all wavelengths, as sunlight does, but rather only a relatively few wavelengths. Similar, yet distinctive line spectra are produced by all of the elements when they are caused to emit light. The wavelengths of the lines are characteristic of a particular element, and can be used to identify new elements. Atomic spectra can also be used to identify the compositions of mixtures. For example, as you have probably seen and heard on T.V. crime dramas, a sample of paint taken from the clothing of a hit-and-run victim can be analyzed for the elements it contains, as well as the relative amounts of each; this is done by using atomic spectra. The results can be compared with a similar analysis of a paint scraping taken from a suspect's car. If they match, there is fairly strong evidence that the suspect is indeed the hit-and-run driver.

Figure 3.13
Production of a continuous spectrum.

The hydrogen spectrum in Figure 3.14 shows only the lines that appear in the visible region of the spectrum. Light is also emitted by hydrogen in the infrared and ultraviolet regions (Figure 3.9a). The occurrence of line spectra baffled physicists for many years. In 1885 Balmer found that there was a relatively simple equation that was able to account for all of the lines in the visible spectrum of hydrogen.

$$\frac{1}{\lambda} = 109{,}678 \text{ cm}^{-1} \left(\frac{1}{2^2} - \frac{1}{n^2} \right) \tag{3.2}$$

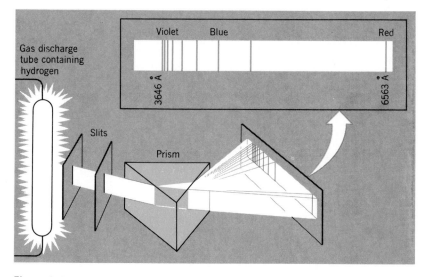

Figure 3.14
The line spectrum of hydrogen.

Table 3.4
Series of Lines in the Hydrogen Spectrum

	n_2	$\lambda(\overset{\circ}{A})$
Lyman Series ($n_1 = 1$)	2	1,215.7
	3	1,025.7
	4	972.5
	5	949.7
	6	937.8
	.	.
	.	.
	.	.
	∞	911.7
Balmer Series ($n_1 = 2$)	3	6,564.7
	4	4,862.7
	5	4,341.7
	6	4,102.9
	7	3,971.2
	.	.
	.	.
	.	.
	∞	3,647.1
Paschen Series ($n_1 = 3$)	4	18,756
	5	12,821
	6	10,941
	7	10,052
	8	9,548
	.	.
	.	.
	.	.
	∞	8,205
Brackett Series ($n_1 = 4$)	5	40,523
	6	26,259
	7	21,661
	8	19,451
	9	18,179
	.	.
	.	.
	.	.
	∞	14,588

	n_2	$\lambda(\text{Å})$
Pfund Series ($n_1 = 5$)	6	74,599
	7	46,538
	8	37,406
	9	32,970
	10	30,392
	.	.
	.	.
	.	.
	∞	22,794

where λ is the wavelength and n is an integer that can have values of 3, 4, 5, 6, . . . , ∞. By choosing a particular value of n the wavelength of a line in the spectrum can be calculated. Thus, when $n = 3$,

$$\frac{1}{\lambda} = 109,678 \text{ cm}^{-1} \left(\frac{1}{4} - \frac{1}{9}\right)$$

$$\frac{1}{\lambda} = 15233 \text{ cm}^{-1}$$

or

$$\lambda = 6.564 \times 10^{-5} \text{ cm} = 6564 \text{ Å}$$

Similarly, when $n = 4$, 5, and 6, we compute λ to be 4863, 4324, and 4103 A, respectively. These values, as we can see from Figure 3.14, are equal to the wavelengths of the lines in the visible portion of the hydrogen spectrum. If higher values of n are substituted into Equation 3.2, wavelengths of lines in the ultraviolet are obtained. All of the lines related by Equation 3.2 constitute what is called the **Balmer series.** Other series of lines are also observed for hydrogen and can be fit to the general equation (called the Rydberg equation)

$$\frac{1}{\lambda} = 109,678 \text{ cm}^{-1} \left(\frac{1}{n_1{}^2} - \frac{1}{n_2{}^2}\right) \tag{3.3}$$

where n_1 and n_2 are integers that may assume values of 1, 2, 3, . . . , ∞ with the requirement that n_2 is always greater than n_1. Thus, when $n_1 = 1$, the values of n_2 can be 2, 3, 4, . . . , ∞ and the lines in the **Lyman series** are obtained. When $n_1 = 2$ and $n_2 = 3$, 4, 5, . . . , ∞, we get the Balmer series. These and other series are summarized in Table 3.4.

3.11
The Bohr Theory of the Hydrogen Atom

Early attempts to account for the existence of line spectra on the basis of the motion of electrons in the atom met with complete failure. If an electron is moving about a nucleus, it must be following a curved path; otherwise it would simply leave the atom. A particle following a curved path undergoes an acceleration[3] and according to the accepted laws of physics at that time, a charged particle (such as the electron) undergoing an acceleration should continuously lose energy by radiating electromagnetic radiation. This means that the electron should gradually spiral in toward the nucleus, resulting in the collapse of the atom. Since atoms do not collapse, physics was faced with a paradox.

The way out of this paradox found its origin in the work of Max Planck (1900) and Albert Einstein (1905). They had demonstrated that, in addition to possessing wave properties, light also has particle properties. Thus, there are instances where light behaves as if it were composed of tiny packets, or **quanta,** of energy (later called **photons**). The energy, E, of the photon emitted or absorbed by a substance is proportional to the frequency of the light, ν. These two quantities are related by the equation,

$$E_{\text{photon}} = h\nu \tag{3.4}$$

where h, the proportionality constant, called **Planck's constant,** has a value of 6.63×10^{-27} erg sec.

Niels Bohr, in 1913, incorporated the ideas of Planck and Einstein into a theory that attempted to explain the line spectrum of hydrogen. His novel approach to the problem that had caused so much difficulty was, in effect, to simply ignore it! He postulated that since atoms obviously do not collapse and since light is emitted by an atom at only certain frequencies (which means that only certain specific energy changes occur), the electron in an atom can possess only certain, restricted quantities of energy. This is often phrased in a somewhat more esoteric way by saying that the energy of the electron is **quantized.** This means that the electron can have only certain discrete amounts of energy and none in between. We express this by saying that the electron is restricted to specific **energy levels** in the atom.

Bohr treated the electron in the hydrogen atom as if it traveled about the nucleus in circular orbits of fixed, or quantized energy; he derived an equation for the energy of the electron that had the form

$$E = -A\,\frac{1}{n^2}$$

---

[3] For example, consider what happens to an object, such as a coin, when it is placed on the edge of a rapidly spinning phonograph record. Anyone who has attempted this has found that the object is thrown outward, away from the center of the record (Figure 3.15). The coin obviously experiences a force (centrifugal force) that causes it to fly off the edge. Since the coin possessses mass, and since there is the relationship that force = mass × acceleration, the coin must also be experiencing an acceleration as it moves in its *circular* path at the edge of the record.

Figure 3.15
A coin is thrown outward from the center of a spinning record. The coin experiences an acceleration directed away from the center of the record.

Atomic Structure and the Periodic Table

in which the constant A could be evaluated from a knowledge of the mass and charge of the electron and Planck's constant. The quantity n is an integer, called a **quantum number,** that can have values of 1, 2, 3, etc., up to infinity. The energy of an electron in a particular orbit depends on the value of n (Figure 3.16), the lowest energy level being obtained when $n = 1$, since this yields the largest value for the fraction $1/n^2$ and thus the most negative (and therefore lowest) E. The idea of a negative energy seems rather odd at first glance. Actually, the minus sign occurs because of a rather arbitrary choice of the zero point on the energy scale. Since we can only measure differences in energy, the choice of where the zero point is placed is really unimportant.

Bohr's theory also yielded an equation that gives the radii r of the orbits

$$r = Bn^2$$

in which the constant, B, can also be calculated from the same quantities used to compute A, above. The radii of the orbits, like the energy, are determined by the quantum number n, with the smallest orbit occurring for $n = 1$ (Figure 3.17). This also corresponds to the lowest energy orbit; thus as the radius of the orbit increases, so does the energy.

The success of Bohr's theory, as with any theory, depended on how well it could account for experimental fact. According to Bohr, when energy is absorbed by an atom, for example in an electric discharge, the electron is raised in energy from one level to another, and when the electron returns to a lower energy level, a photon is emitted whose energy is equal to the difference between the two levels (see Figures 3.16 and 3.17). If we take n_2 to be the quantum number of the upper level and n_1 to be that of the lower (so that $n_2 > n_1$), the difference in energy, ΔE, between the two is

$$\Delta E = E_2 - E_1$$

$$\Delta E = \left(-A\,\frac{1}{n_2{}^2}\right) - \left(-A\,\frac{1}{n_1{}^2}\right)$$

Figure 3.16
Energies of Bohr orbits.

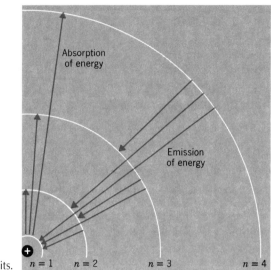

Absorption
of energy

Emission
of energy

Figure 3.17
The radii of Bohr orbits.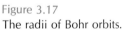

which can be written as

$$\Delta E = A \left(\frac{1}{n_1{}^2} - \frac{1}{n_2{}^2} \right) \tag{3.5}$$

If this energy difference appears as a photon, it would have a frequency, ν, that could be calculated from Equation 3.4,

$$\Delta E = h\nu$$

which, by incorporating Equation 3.1, can be expressed as

$$\Delta E = h \frac{c}{\lambda} = hc \frac{1}{\lambda}$$

Substituting this into Equation 3.5 we get

$$hc \frac{1}{\lambda} = A \left(\frac{1}{n_1{}^2} - \frac{1}{n_2{}^2} \right)$$

which, upon rearrangement yields

$$\frac{1}{\lambda} = \frac{A}{hc} \left(\frac{1}{n_1{}^2} - \frac{1}{n_2{}^2} \right)$$

The quantity A/hc has a value of 109.730 cm^{-1} so that our final equation is

$$\frac{1}{\lambda} = 109{,}730 \text{ cm}^{-1} \left(\frac{1}{n_1{}^2} - \frac{1}{n_2{}^2} \right) \tag{3.6}$$

Comparing Equations 3.3 and 3.6, we see that they are virtually identical. The Rydberg Equation (3.3), is obtained from experimental observation while Equation 3.6 is derived from theory. The fact that the theory agrees so well with experiment seems to indicate that Bohr was on the right track. Unfortunately, his approach was not at all successful with atoms more complex than hydrogen; however, his introduction of the notion of quantum numbers and quantized energy levels played a significant role in the development of our understanding of atomic structure.

3.12
Wave Mechanics

The currently accepted theory explaining the behavior of subatomic particles is called **wave mechanics,** which has its roots in a hypothesis put forward by Louis de Broglie in 1924. De Broglie suggested that if light can behave in some instances as if it were composed of particles, perhaps particles, at times, exhibit properties that we normally associate with waves.

Einstein had shown that the energy equivalent, E, of a particle of mass, m, is equal to

$$E = mc^2 \qquad (3.7)$$

where c is the speed of light. A photon whose energy is E could thus be said to have an effective mass equal to m. Max Planck had shown that the energy of a photon is given by Equation 3.4,

$$E = h\nu = \frac{hc}{\lambda}$$

Equating these two gives us

$$\frac{hc}{\lambda} = mc^2$$

When we solve for λ, the wavelength, we obtain

$$\lambda = \frac{h}{mc}$$

If this equation also applies to particles, such as the electron, the equation can be written as

$$\lambda = \frac{h}{mv} \qquad (3.8)$$

where we have replaced c, the speed of light, by v, the speed of the particle.

Experimental evidence for this dual wave-particle nature of matter exists. Some particles, including electrons, protons, and neutrons, have been observed to exhibit diffraction,[4] a property that can only be explained by wave

---

[4] *Diffraction.* If light is allowed to pass through a pinhole whose diameter is about the same as the wavelength of the light, the hole behaves as if it were a tiny light source, scattering light in all directions. This phenomenon is called **diffraction.** If two such pinholes are placed along side of one another, each hole behaves as a separate source of light, sending out light waves in all directions. When a screen is placed such that this light falls upon it, we observe a pattern, called a **diffraction pattern,** that consists of light and dark areas as shown in Figure 3.18. In the bright areas light waves that arrive from each hole are **in phase;** that is, the peaks and troughs of the two waves are lined up so that the amplitudes of the waves add together to produce a resultant wave of greater intensity. This is illustrated in Figure 3.19a. In the darkened areas, waves that arrive from the two pinholes are **out of phase** with each other, which means that the peaks of one wave coincide with the troughs of the other. When this happens, the amplitudes of the waves cancel each other so that zero intensity and hence darkness, is observed (Figure 3.19b). The factors that determine where the waves are in phase and where they are out of phase in a diffraction pattern will be considered in Chapter 7.

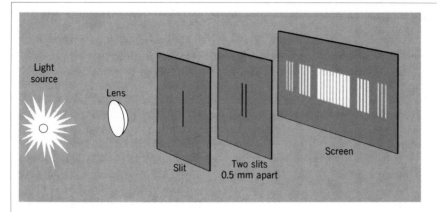

Figure 3.18
Production of a diffraction pattern.

motion. The reason that the wave nature of matter had not been discovered earlier is that objects large enough to see, either with the naked eye or with the aid of a microscope, possess so much mass that their wavelengths are much too short to be observed (see Example 3.2).

Example 3.2
What is the wavelength of a grain of sand that weighs 0.000010 g and is moving at a speed of 1.0 cm/sec (approximately 0.02 mph)?

Solution
    From Equation 3.7

$$\lambda = \frac{h}{mv}$$

Planck's constant has a value of

$$h = 6.63 \times 10^{-27} \text{ erg sec}$$

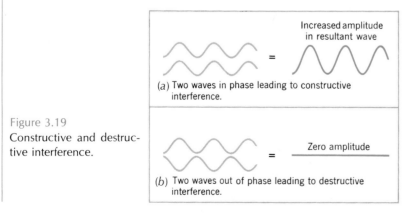

(a) Two waves in phase leading to constructive interference.

Figure 3.19
Constructive and destructive interference.

(b) Two waves out of phase leading to destructive interference.

Since 1 erg $= 1$ g cm²/sec², we can write h as

$$h = 6.63 \times 10^{-27} (\text{g cm}^2/\text{sec}^2) \text{ sec} = 6.63 \times 10^{-27} \text{ g cm}^2/\text{sec}$$

We are given that

$$m = 1.0 \times 10^{-5} \text{ g}$$
$$v = 1.0 \text{ cm/sec}$$

Substituting these quantities into our equation we have

$$\lambda = \frac{6.63 \times 10^{-27} \text{ g cm}^2/\text{sec}}{(1.0 \times 10^{-5} \text{ g})(1.0 \text{ cm/sec})} = 6.63 \times 10^{-22} \text{ cm}$$

This wavelength is far too small to be detected by any device existing at this time. Larger objects have even larger masses and therefore still smaller wavelengths.

We saw in Section 3.11 that one of the significant results of the Bohr theory of the atom was the introduction of integer quantum numbers. If we consider the electron to be moving as a wave about the nucleus, we find that the appearance of integers occurs in a very natural way. For example, let us imagine a wave moving about the nucleus along the circumference of a circle. Unless the wavelength of the wave is properly chosen, the wave will be out of phase with itself after it has completed one revolution (Figure 3.20a). When this occurs, cancellation of the wave will result, and the wave will simply disappear. Since the electron cannot just disappear, it follows that only waves that stay in phase can exist (Figure 3.20b). This requirement means that the wavelength must be repeated an integral number (whole number) of times along the circumference of the circle. Stated another way, the length of the circumference must be equal to an integral number of

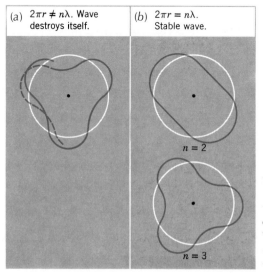

(a) $2\pi r \neq n\lambda$. Wave destroys itself.

(b) $2\pi r = n\lambda$. Stable wave.

$n = 2$

$n = 3$

Figure 3.20
Destructive and constructive interference produced by waves traveling along the circumference of the circle. (a) $2\pi r \neq n\lambda$. Wave destroys itself. (b) $2\pi r = n\lambda$. Stable wave.

wavelengths; that is,

$$2\pi r = n\lambda$$

where $2\pi r$ is the length of the circumference, λ is the wavelength, and n is an integer having values of $n = 1, 2, 3, \ldots, \infty$. Waves that satisfy this condition for $n = 2$ and 3 are illustrated in Figure 3.20*b*.

Because the electron has wavelike properties, it is not surprising that its motion in the atom can be described by means of a wave equation. Physicists had already developed equations to treat wave motion in general and Erwin Schrödinger, in 1926, employed the ideas of de Broglie to arrive at a wave equation that could be applied to the hydrogen atom. The branch of physics that deals with the solution of wave equations is called either **wave mechanics** or **quantum mechanics.**

The solution of the wave equation is very complicated and is certainly far beyond the scope of this book. We can, however, look at some of the results that come from the theory. To begin with, the wave equation can really only be solved for a very few species. Fortunately, one of them is the hydrogen atom, and the results obtained for hydrogen, as it turns out, can be extended quite successfully to the other elements in the periodic table.

In the Bohr theory we saw that the electron was restricted to certain energy levels and was thought to move in certain prescribed circular orbits about the nucleus. We were able to identify these energy levels and orbits by means of an integer that we called a quantum number. In the course of solving the wave equation, a series of **wave functions** emerge (usually designated by the Greek letter psi, ψ), each of which is characterized by a set of three integer quantum numbers. *Each wave function corresponds to a certain energy and describes a region about the nucleus* (called an **orbital,** to distinguish it from Bohr's orbits) *where an electron may move.*

According to wave mechanics, the energy levels in the atom are composed of one or more orbitals and in atoms that contain more than one electron, the distribution of the electrons about the nucleus is determined by the number and kind of energy levels that are occupied. Therefore, in order to investigate the way the electrons are arranged in space, we must first examine the energy levels in the atom. This is best accomplished through a discussion of the quantum numbers mentioned above.

1. The **principal quantum number,** n. The energy levels in the atom are arranged roughly into shells as determined by the principal quantum number, n. As in the Bohr theory, n may have values of 1, 2, 3, . . . , etc. up to infinity. Letters are also frequently associated with these shells as shown below.

Principal quantum number	1	2	3	4	. . .
Letter designation		K	L	M	N . . .

For example, we would refer to the shell with $n = 1$ as the K shell.

2. The **azimuthal quantum number,** l. Wave mechanics predicts that each main shell is composed of one or more **subshells,** each of which is specified by a secondary quantum number, l, that is called

the azimuthal quantum number. For any given shell this quantum number may have values of 0, 1, 2 etc. up to a maximum of $n - 1$. Thus when $n = 1$, the largest (and only) value of l that is allowed is $l = 0$; therefore the K shell consists of but one subshell. When $n = 2$, two values of l occur, $l = 0$ and $l = 1$; hence the L shell is made up of two subshells. The values of l that occur for each value of n are summarized below.

n	l
1	0
2	0, 1
3	0, 1, 2
4	0, 1, 2, 3
.	.
.	.
.	.
n	0, 1, 2, . . . , $n - 1$

We see that the number of subshells in any given shell is simply equal to its value of n.

For the purposes of discussing the distribution of electrons in the atom, it is common practice to associate letters with the various values of l:

l	0	1	2	3	4	5	6 . . .
Subshell designation	s	p	d	f	g	h	i . . .

The first four letters find their origin in the atomic spectra of the alkali metals (lithium through cesium). In these spectra four series of lines were observed and were termed the "sharp," "principal," "diffuse," and "fundamental" series, hence the letters s, p, d, and f. For $l = 4, 5, 6$ etc. we just continue on with the alphabet; however, for our purposes we will only be interested in s, p, d, and f subshells, since these are the only ones that are populated by electrons in atoms in their **ground state** (state of lowest energy).

To describe a subshell within a given shell, we write the value of n for the shell followed by the letter designation of the subshell. For example, the s subshell of the second shell ($n = 2$, $l = 0$) would be called the 2s subshell. In like manner, the p subshell of the second shell ($n = 2$, $l = 1$) would be the 2p subshell.

3. The **magnetic quantum number, m.** Each subshell is composed of one or more orbitals, where an orbital within a particular subshell is distinguished by its value of m, the magnetic quantum number. This quantum number, which derives its name from the fact that it can be used to explain the appearance of additional lines in atomic spectra produced when atoms emit light while in a magnetic field, has in-

Table 3-5

Summary of Quantum Numbers

Principal Quantum Number, n (shell)	Azimuthal Quantum Number, l (subshell)	Subshell Designation	Magnetic Quantum Number, m (orbital)	Number of Orbitals in Subshell
1	0	1s	0	1
2	0	2s	0	1
	1	2p	$-1\ 0 +1$	3
3	0	3s	0	1
	1	3p	$-1\ 0 +1$	3
	2	3d	$-2 -1\ 0 +1 +2$	5
4	0	4s	0	1
	1	4p	$-1\ 0 +1$	3
	2	4d	$-2 -1\ 0 +1 +2$	5
	3	4f	$-3 -2 -1\ 0 +1 +2 +3$	7

teger values that range between $-l$ and $+l$. When $l = 0$, only one value of m is permitted, $m = 0$; therefore an s subshell consists of only one orbital (we call it an s orbital). A p subshell ($l = 1$) contains three orbitals corresponding to m equal to -1, 0, and $+1$. In a similar fashion we find that a d subshell ($l = 2$) is composed of five orbitals and an f subshell ($l = 3$), seven. This is summarized in Table 3.5.

If we wished, we could identify a particular orbital in the atom by its set of three quantum numbers. Thus, an orbital for which $n = 3$, $l = 1$, and $m = 0$ corresponds to one of the three orbitals that belong to the p subshell of the third shell.

The energies of these shells, subshells, and orbitals are perhaps best illustrated by means of Figure 3.21. There are several points about this diagram that are worth noting. First, we see that the energy of the shells increase with increasing value of the principal quantum number, n. Thus the K shell, with $n = 1$, lies lowest in energy; above that there is the L shell with $n = 2$ (composed of the 2s and 2p subshells); higher still we find the M shell ($n = 3$), and so on.

Also note that as n becomes larger, the spacing between successive shells becomes less, as illustrated on the right side of Figure 3.21. Because of this narrowing energy separation, we begin to observe overlap among the subshells of the third and higher shells. The 4s subshell, for example, lies lower in energy than does the 3d subshell. This overlap is even more pronounced in higher shells where the 5s subshell lies below the 4d, the 6s and 4f below the 5d, and the 7s and 5f below the 6d.

Also in Figure 3.21 we have indicated each orbital by means of a dash. Each s subshell is shown as a single dash to stress that it is composed of but one orbital. Likewise, p subshells are shown as three dashes, d subshells as five dashes, and f subshells as seven dashes. Observe that each orbital of a given subshell is shown to have the same energy. This applies for isolated atoms but not always for atoms in chemical compounds.

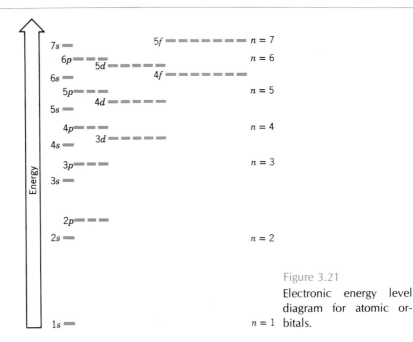

Figure 3.21

Electronic energy level diagram for atomic orbitals.

The sequence of energy levels described by Figure 3.21 turns out to be of critical importance in determining the arrangement of electrons in the atom. Before discussing this, however, we must look at yet another quantum number.

3.13 Electron Spin and the Pauli Exclusion Principle

In addition to the three quantum numbers, n, l, and m, which come directly from the solution of the wave equation, there is yet another number, the spin quantum number, s. This quantum number arises because the electron behaves as if it were spinning (in much the same manner as the earth spins about its axis). The circular motion of electric charge that results causes the electron to act as a tiny electromagnet, just as passing an electric current through a wire wrapped about a nail causes the nail to become magnetic (see Figure 3.22). Since the electron can only spin in either of two directions, s may have but two values. These turn out to be $+1/2$ and $-1/2$.

We find, then, that each electron in an atom can be assigned a set of values for its four quantum numbers, n, l, m, and s, which determine the orbital in which the electron will be found and the direction in which the electron will be spinning. There is a restriction, however, on the values that we may assign to these quantum numbers. This is expressed as the **Pauli exclusion principle,** which states that *no two electrons in any one atom may have all four quantum numbers the same.* This means that if we choose a particular set of values for n, l, and m corresponding to a particular orbital (e.g., $n = 1$, $l = 0$, $m = 0$; the 1s orbital), we are still able to have two electrons with different values of the spin quantum number, s (i.e., either $s = +1/2$ or $s = -1/2$). In effect, this limits the number of electrons in any given orbital to two, and it also requires that the spins of these two electrons be in opposite directions.

Figure 3.22
The spin of the electron. (a) The electron behaves as if it is spinning about an axis through its center. Since there are two directions of spin, there are two values of the spin quantum number. (b) Spinning charge produces a magnetic field just as circulation of charge through a wire wrapped about a nail causes the nail to become magnetic.

One electron has $s = +\frac{1}{2}$, the other has $s = -\frac{1}{2}$

Iron nail is magnetized when current flows through the wire

(a)

(b)

Since the Pauli exclusion principle leads to a restriction of a maximum of two electrons in any orbital, the maximum number of electrons that can be accommodated in s, p, d, and f subshells is summarized as below.

Subshell	Number of Obritals	Maximum Number of Electrons
s	1	2
p	3	6
d	5	10
f	7	14

The maximum number of electrons permitted in any shell is equal to $2n^2$. Thus for example, the K shell ($n = 1$) can hold up to two electrons and the L shell ($n = 2$) a maximum of eight.

The spin of the electron is also responsible for most of the magnetic properties that we find associated with atoms and molecules. Materials that are **diamagnetic** experience no attraction for another magnet.[5] In these substances there are the same number of electrons of each spin so that their magnetic effects cancel. **Paramagnetic** substances, on the other hand, are weakly attracted to a magnetic field. In these materials there are more electrons of one spin than the other (as will always be true when an atom or molecule has an odd number of electrons) and total cancellation does not occur. The extra electrons of one spin cause the atom or molecule, as a whole, to behave as if

[5] They are, in fact, repelled slightly by a magnetic field. This is a result of the motion of the electrons in the atom and is not associated with the electron's spin.

it were itself a tiny magnet. **Ferromagnetic** substances, of which iron is the most common example, owe their very strong magnetic behavior to interactions between paramagnetic atoms in the solid state. This phenomenon is discussed in more detail in Chapter 19.

3.14
The Electron Configurations of the Elements

As we have intimated earlier, the way that the electrons are arranged in an atom (i.e., its **electron configuration**) is determined by the order in which the subshells occur on the scale of increasing energy. This is so because, in an atom in its ground state, the electrons will be found in the lowest energy levels available. In hydrogen, for instance, the single electron will be located in the $1s$ subshell because it is this level that has the lowest energy. To indicate that the $1s$ subshell is populated by one electron, we use a superscript (in this case, 1), on the subshell designation. Thus we would denote the electron configuration of hydrogen as $1s^1$. As we proceed in this discussion, it will also be necessary to keep tabs on the electron spins. One method that is often employed is to symbolize an electron with its spin in one direction by an arrow pointing up, $\uparrow$, and an electron with opposite spin as an arrow pointing down $\downarrow$. To indicate the distribution of electrons among the orbitals of the atom, we then place the arrows over bars that symbolize orbitals. Hydrogen, for example, is represented as

$$H \quad \frac{\uparrow}{1s}$$

This kind of representation of the electron configuration is usually called an *orbital diagram.*

To obtain the electron configuration of the other elements in the periodic table, let us proceed from one atom to the next by adding a proton (plus assorted neutrons) to the nucleus followed by an electron, which we place into the lowest available energy level. As you follow this discussion, refer both to the periodic table (Figure 3.12) and the energy level diagram in Figure 3.21. A complete table of the electron configurations of the elements is contained in Table 3.6.

Hydrogen is the simplest element, consisting of but a single proton and one electron. The next element, atomic number, 2, is helium. Here there are two electrons to consider and, since the $1s$ orbital can accommodate both of them, the electronic structure of helium would be $1s^2$ and its orbital diagram is

$$He \quad \frac{\uparrow\downarrow}{1s}$$

Note that in placing the electrons in the same orbital we have indicated that their spins are in opposite directions as required by the Pauli exclusion principle. We refer to this by saying that their spins are *paired,* or simply that the electrons are paired.

The next two elements following He are Li and Be, which have three and four electrons, respectively. In each of these, the first two electrons will enter the $1s$ subshell and, since no more than two electrons can occupy an s subshell, the remaining electron(s) must occupy the $2s$ subshell. The electron configurations of Li and Be, then, are Li, $1s^2 2s^1$ and Be, $1s^2 2s^2$. We could also show this as

Table 3.6
The Electron Configurations of the Elements

Atomic Number			
1	H	$1s^1$	
2	He	$1s^2$	
3	Li	[He]	$2s^1$
4	Be	[He]	$2s^2$
5	B	[He]	$2s^22p^1$
6	C	[He]	$2s^22p^2$
7	N	[He]	$2s^22p^3$
8	O	[He]	$2s^22p^4$
9	F	[He]	$2s^22p^5$
10	Ne	[He]	$2s^22p^6$
11	Na	[Ne]	$3s^1$
12	Mg	[Ne]	$3s^2$
13	Al	[Ne]	$3s^23p^1$
14	Si	[Ne]	$3s^23p^2$
15	P	[Ne]	$3s^23p^3$
16	S	[Ne]	$3s^23p^4$
17	Cl	[Ne]	$3s^23p^5$
18	Ar	[Ne]	$3s^23p^6$
19	K	[Ar]	$4s^1$
20	Ca	[Ar]	$4s^2$
21	Sc	[Ar]	$4s^23d^1$
22	Ti	[Ar]	$4s^23d^2$
23	V	[Ar]	$4s^23d^3$
24	Cr	[Ar]	$4s^13d^5$
25	Mn	[Ar]	$4s^23d^5$
26	Fe	[Ar]	$4s^23d^6$
27	Co	[Ar]	$4s^23d^7$
28	Ni	[Ar]	$4s^23d^8$
29	Cu	[Ar]	$4s^13d^{10}$
30	Zn	[Ar]	$4s^23d^{10}$
31	Ga	[Ar]	$4s^23d^{10}4p^1$
32	Ge	[Ar]	$4s^23d^{10}4p^2$
33	As	[Ar]	$4s^23d^{10}4p^3$
34	Se	[Ar]	$4s^23d^{10}4p^4$
35	Br	[Ar]	$4s^23d^{10}4p^5$

Atomic Number			
36	Kr	[Ar]	$4s^23d^{10}4p^6$
37	Rb	[Kr]	$5s^1$
38	Sr	[Kr]	$5s^2$
39	Y	[Kr]	$5s^24d^1$
40	Zr	[Kr]	$5s^24d^2$
41	Nb	[Kr]	$5s^14d^4$
42	Mo	[Kr]	$5s^14d^5$
43	Tc	[Kr]	$5s^24d^5$
44	Ru	[Kr]	$5s^14d^7$
45	Rh	[Kr]	$5s^14d^8$
46	Pd	[Kr]	$4d^{10}$
47	Ag	[Kr]	$5s^14d^{10}$
48	Cd	[Kr]	$5s^24d^{10}$
49	In	[Kr]	$5s^24d^{10}5p^1$
50	Sn	[Kr]	$5s^24d^{10}5p^2$
51	Sb	[Kr]	$5s^24d^{10}5p^3$
52	Te	[Kr]	$5s^24d^{10}5p^4$
53	I	[Kr]	$5s^24d^{10}5p^5$
54	Xe	[Kr]	$5s^24d^{10}5p^6$
55	Cs	[Xe]	$6s^1$
56	Ba	[Xe]	$6s^2$
57	La	[Xe]	$6s^25d^1$
58	Ce	[Xe]	$6s^24f^15d^1$
59	Pr	[Xe]	$6s^24f^3$
60	Nd	[Xe]	$6s^24f^4$
61	Pm	[Xe]	$6s^24f^5$
62	Sm	[Xe]	$6s^24f^6$
63	Eu	[Xe]	$6s^24f^7$
64	Gd	[Xe]	$6s^24f^75d^1$
65	Tb	[Xe]	$6s^24f^9$
66	Dy	[Xe]	$6s^24f^{10}$
67	Ho	[Xe]	$6s^24f^{11}$
68	Er	[Xe]	$6s^24f^{12}$
69	Tm	[Xe]	$6s^24f^{13}$
70	Yb	[Xe]	$6s^24f^{14}$

71	Lu	[Xe]	$6s^24f^{14}5d^1$
72	Hf	[Xe]	$6s^24f^{14}5d^2$
73	Ta	[Xe]	$6s^24f^{14}5d^3$
74	W	[Xe]	$6s^24f^{14}5d^4$
75	Re	[Xe]	$6s^24f^{14}5d^5$
76	Os	[Xe]	$6s^24f^{14}5d^6$
77	Ir	[Xe]	$6s^24f^{14}5d^7$
78	Pt	[Xe]	$6s^14f^{14}5d^9$
79	Au	[Xe]	$6s^14f^{14}5d^{10}$
80	Hg	[Xe]	$6s^24f^{14}5d^{10}$
81	Tl	[Xe]	$6s^24f^{14}5d^{10}6p^1$
82	Pb	[Xe]	$6s^24f^{14}5d^{10}6p^2$
83	Bi	[Xe]	$6s^24f^{14}5d^{10}6p^3$
84	Po	[Xe]	$6s^24f^{14}5d^{10}6p^4$
85	At	[Xe]	$6s^24f^{14}5d^{10}6p^5$
86	Rn	[Xe]	$6s^24f^{14}5d^{10}6p^6$
87	Fr	[Rn]	$7s^1$
88	Ra	[Rn]	$7s^2$
89	Ac	[Rn]	$7s^26d^1$
90	Th	[Rn]	$7s^26d^2$
91	Pa	[Rn]	$7s^25f^26d^1$
92	U	[Rn]	$7s^25f^36d^1$
93	Np	[Rn]	$7s^25f^46d^1$
94	Pu	[Rn]	$7s^25f^6$
95	Am	[Rn]	$7s^25f^7$
96	Cm	[Rn]	$7s^25f^76d^1$
97	Bk	[Rn]	$7s^25f^9$
98	Cf	[Rn]	$7s^25f^{10}$
99	Es	[Rn]	$7s^25f^{11}$
100	Fm	[Rn]	$7s^25f^{12}$
101	Md	[Rn]	$7s^25f^{13}$
102	No	[Rn]	$7s^25f^{14}$
103	Lw	[Rn]	$7s^25f^{14}6d^1$

Li   ⇅   ↑
Be   ⇅   ⇅
$1s$ $2s$

Since both Li and Be have a completed $1s$ subshell, which corresponds to the electron configuration of He, they can also be written as

Li   [He] ↑
Be   [He] ⇅
$2s$

Here we focus our attention on the electronic structure of the outermost shell which, in chemical reactions, is responsible for chemical changes. In this example, the inner filled $1s$ subshell is called the helium **core.** We shall frequently find it useful to consider only those electrons that occur outside a core of electrons corresponding to one of the noble gases.

At beryllium, which has four electrons, the $2s$ subshell is completed. The fifth electron of boron ($Z = 5$), then, must enter the next lowest available subshell, which is the $2p$, to give boron the configuration $1s^22s^22p^1$. Likewise, the fifth and sixth electrons of carbon must enter the $2p$ subshell; thus, we represent carbon as $1s^22s^22p^2$. However, if we examine the distribution of the electrons over the various orbitals, we must face a choice: the electrons could be arranged in the following three ways;[6]

[He] ⇅   ⇅ __ __
[He] ⇅   ↑ ↓ __
[He] ⇅   ↑ ↑ __
$2s$ $2p$

The last two electrons can be paired in the same orbital, paired in different orbitals, or arranged so that their spins are in the same direction (unpaired).

As it turns out, the last diagram is correct. In these situations, where a choice is available, we apply **Hund's rule,** which states that *electrons, entering a subshell containing more than one orbital, will be spread out over the available orbitals with their spins in the same direction.* For nitrogen ($Z = 7$), then, the electron configuration would be written as $1s^22s^22p^3$, and it would have the orbital diagram

N   [He]   ⇅   ↑ ↑ ↑
$2s$ $2p$

---

[6] These are the only three possibilities that we have to consider since in an isolated atom, each of the p orbitals are equivalent in energy. Thus the arrangements,

⇅ __ __   __ ⇅ __   __ __ ⇅
$2p$ $2p$ $2p$

are indistinguishable from one another experimentally.

Finally, the elements oxygen, fluorine, and neon ($Z = 8$, 9 and 10, respectively) lead to the completion of the $2p$ subshell.

O [He] ⇅ ⇅ ↑ ↑

F [He] ⇅ ⇅ ⇅ ↑

Ne [He] ⇅ ⇅ ⇅ ⇅
              $2s$ $2p$

After the $2p$ subshell is filled at Ne, the next lowest available energy level is the $3s$. This becomes populated with Na and Mg ($Z = 11$ and 12). After this the $3p$ subshell is gradually filled by the next six electrons as we complete the atoms Al through Ar ($Z = 13$ to 18). Then, since the $4s$ subshell lies at lower energy than the $3d$, it is occupied next by the 19th and 20th electrons of K and Ca ($Z = 19$ and 20).

Examination of Figure 3.21 reveals that after the $4s$ subshell is completed additional electrons begin to populate the $3d$ subshell. Scandium, therefore, will have the electron configuration $1s^2 2s^2 2p^6 3s^2 3p^6 4s^2 3d^1$ or

Sc [Ar] ⇅ ↑ _ _ _ _
           $4s$ $3d$

As we proceed through Ti and V ($Z = 22$ and 23), two more electrons are added to the $3d$ subshell; however, when we get to Cr ($Z = 24$), we find the structure

Cr [Ar] ↑ ↑ ↑ ↑ ↑ ↑
           $4s$ $3d$

instead of

Cr [Ar] ⇅ ↑ ↑ ↑ ↑ _
           $4s$ $3d$

What we observe occurs because a half-filled or completely filled subshell has an added stability associated with it. This phenomenon is quite significant and is important to keep in mind. We see it again in period 4, for example, when we get to copper. On the basis of our energy level diagram in Figure 3.21, we would predict copper to have the electron configuration

Cu [Ar] ⇅ ⇅ ⇅ ⇅ ⇅ ↑
           $4s$ $3d$

The actual structure is given by

Cu [Ar] ↑ ⇅ ⇅ ⇅ ⇅ ⇅
           $4s$ $3d$

By promoting an electron from the $4s$ to the $3d$ subshell, one filled and one half-filled subshell is produced, instead of the filled $4s$ and the neither filled nor half-filled $3d$ subshell that we initially would predict.

After the $3d$ subshell is completed at atomic number 30 (zinc), the $4p$ subshell is filled as we proceed from Ga to Kr ($Z = 31$ to 36). This is followed by the completion of the $5s$ subshell from Rb to Sr ($Z = 37$, 38); the $4d$ subshell as we progress across the second row of transition elements ($Z = 39 - 48$); the $5p$ from In to Xe ($Z = 49 - 54$); and the $6s$ with Cs and Ba ($Z = 55$, 56).

Based on the energy level sequence in Figure 3.21, we would expect that, after the $6s$ subshell had been filled, we would begin to populate the $4f$ subshell next. Actually at La ($Z = 57$) the last electron enters the $5d$ subshell instead. The $4f$ subshell is filled afterward, with only one other irregularity at Gd where the $5d$ subshell is again the recipient of an electron so as to provide Gd with a half-filled subshell. As we go to higher and higher shells, these irregularities become more frequent because the spacing between subshells becomes less and less. As we proceed from atom to atom, the energy of the various subshells shift about somewhat as the nuclear charge increases. The result is that it is difficult to predict accurately the electron configuration of elements of very high atomic number. Nevertheless, we can account for the occurrence of the lanthanide elements by the filling of the $4f$ subshell (an f subshell can accommodate 14 electrons and the lanthanide series consists of 14 elements). Likewise, we can account for the actinide elements as the result of the filling of the $5f$ subshell.

In the last section we saw that the results of wave mechanics could be used to predict the electron configurations of the elements. These electron configurations are based on theory and, to be considered useful and valid, they must somehow manifest themselves in obvious ways. One of the strongest supports for the assignment of electron configurations is the periodic table itself. Recall that, in constructing the current periodic table, elements were arranged under each other in groups because of their similar chemical properties. For example, all of the elements in Group IA are metals which, when they react, form ions with a charge of $+1$. If we examine the electron configurations of these elements, we see that the outer shell (shell of highest n) for each has but one electron in an s subshell. Similarly, all of the elements in Group IIA have an outer-shell electron configuration that we might generalize as ns^2. In fact, by examining any group within the periodic table, we see that all of the elements in the group possess essentially identical outer-shell electronic structures. It is not really surprising that similar electronic structures lead to similar chemical and physical properties.

Because the properties of the elements depend on their electron configurations, it is important that you develop the ability to write them down. There are a variety of ways to remember the sequence in which the various levels are filled; however, the best aid is the periodic table itself. As we have just seen, the order of filling the energy levels can be used to account for the structure of the periodic table. We can also work in the other direction and use the periodic table to deduce electronic structure.

If we look back over the procedure for determining the electronic structure of the elements, we find that for any element in Groups IA and IIA the final electron was added to an s subshell, and that the principal quantum number of that subshell was the same as the period number. Sodium, for instance, a period 3 element, has its outer electron in the $3s$ subshell. For elements in Groups IIIA to O the last electron is added to a p subshell whose

3.15
The Periodic Table and Electron Configurations

value of n is also the same as the period number. In the case of the transition elements, the final electron that we add is placed into a d subshell with n equal to one less than the period number. For example, with iron (a fourth period element), the last electron enters a $3d$ subshell. Finally, the electronic structure of an inner transition element (i.e., one from the lanthanide or actinide series) is completed by an electron in an f subshell whose principal quantum number is two less than the period number.

In arriving at the electron configuration of an element, let us again imagine that we start at hydrogen and proceed through the periodic table, in order of increasing atomic number, until we arrive at the element we are interested in. We soon realize that an s subshell is filled when we cross Groups IA and IIA; a p subshell when we go through Groups IIIA to O; a d subshell as we pass across a row of transition elements; and an f subshell when we pass through a row of inner transition elements. We also realize that for the s and p subshells their value of n is equal to the period number; for the d subshells, n is equal to the period number minus one; and for the f subshells n is equal to the period number minus two.

We can see how this knowledge allows us to derive electron configurations by choosing lead ($Z = 82$) as an example. Figure 3.23 shows how we determine which subshells become filled. These are obtained from the regions in the table occupied by all of the elements up to and including the atom, lead. This means that in building up the lead atom we cross through periods 1 to 5 and part of 6, and that upon proceeding from left to right across one period after another, we fill, in order, the subshells $1s$, $2s$, $2p$, $3s$, $3p$, $4s$, $3d$, $4p$, $5s$, $4d$, $5p$, $6s$, $4f$, $5d$, and finally end by placing two electrons into the $6p$ subshell. Taking into account the maximum population of each subshell, then, we obtain as the electron configuration of lead,

$$1s^2 2s^2 2p^6 3s^2 3p^6 4s^2 3d^{10} 4p^6 5s^2 4d^{10} 5p^6 6s^2 4f^{14} 5d^{10} 6p^2$$

Sometimes it is preferred to write all subshells of a given shell together. Thus for lead we would have

$$1s^2 2s^2 2p^6 3s^2 3p^6 3d^{10} 4s^2 4p^6 4d^{10} 4f^{14} 5s^2 5p^6 5d^{10} 6s^2 6p^2$$

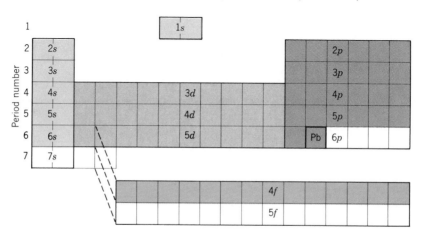

Figure 3.23
The use of the periodic table for predicting electron configurations.

In wave mechanics our view of where the electron may be found about the nucleus is far different from the idea of circular orbits imagined by Bohr. This is a consequence of the **uncertainty principle** of Heisenberg, which states that if we attempt to measure, at the same time, both the position and momentum of a particle, our measurements will be subject to errors that are related to one another by the equation

$$\Delta x \cdot \Delta(mv) \geqslant \frac{h}{4\pi}$$

This equation states that the product of the uncertainty in the position of the particle, Δx, times the uncertainty in its momentum, $\Delta(mv)$, must be greater than or equal to Planck's constant divided by 4π. This statement actually says that we are limited in our ability to know simultaneously where the electron is and where it is going. It leads us, instead, to refer to the probability of finding the electron in some small element of volume at various places around the nucleus. More specifically, it is the square of the wave function, ψ^2, which is taken to specify this probability.

On the basis of this concept, let us look at the **probability distribution** (the way the probability varies throughout the volume of an atom) for the single electron in the 1s orbital of the hydrogen atom. A graph of ψ^2 as a function of distance from the nucleus, r, is shown in Figure 3.24. It can be seen that those regions in which the probability of observing the electron is greatest lies close to the nucleus and that, as we might expect, the probability decreases as we move away from the nucleus, gradually approaching zero as r approaches infinity. In other words, we might expect to find the electron in the hydrogen atom almost anywhere, but most of the time it stays fairly close to the nucleus. Because the electron moves so fast, it effectively spreads itself out to surround the entire nucleus in a cloud of electronic charge. The electron will spend most of its time in those regions where the electron probability is high; thus the concentration of charge, which we might call **electron density,** will be large. In other areas the charge is thinly spread and the electron density is small.

There are several ways to indicate the distribution of charge in an orbital. One way is to plot ψ^2 as we have already done. Another, in two dimensions, is to illustrate the charge cloud as shown in Figure 3.25, where the darker shaded areas represent regions of high electron density. In three dimensions (Figure 3.26) we see this as a scatter of dots in which regions of high electron density are illustrated by the occurrence of a large number of points. Still another way to describe the shape of the orbital is to draw a surface on which the probability of observing the electron is constant. In the case of an electron

Figure 3.24
Probability of finding the electron as a function of distance from the nucleus for the 1s orbital of hydrogen.

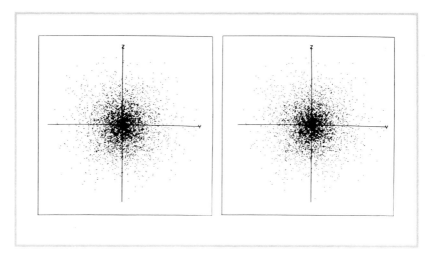

Figure 3.25
A representation of the charge cloud in hydrogen.

in a 1s orbital, or any s orbital for that matter, this surface is a sphere and it is often said that an s orbital has a spherical "shape." This simply means that the probability of finding the electron at some given distance from the nucleus will be the same regardless of the direction in which we move away from the center of the atom.

All s orbitals, as we have said, are spherical in shape. However, if we draw the sphere so that the probability of finding the electron within the sphere is, say, 90%, the size of that sphere increases from 1s to 2s to 3s, etc. In other words, the size of the charge cloud gets larger with increasing principal quantum number, not only for s orbitals, but also for p, d, and f orbitals. This means that electrons in orbitals of higher n will be at a greater average distance from the nucleus and that the atom gets larger as its higher energy subshells become populated.

The "shape" of the electron cloud characteristic of a p orbital is illustrated in Figure 3.27. We see that for a p orbital the electron density is not distributed symmetrically about the nucleus as in an s orbital, but rather is concentrated in particular regions along a straight line passing through the

Figure 3.26
Three-dimensional representation of charge cloud. From D. T. Cromer, *Journal of Chemical Education*, Vol. 45, p. 626, October, 1968, used by permission.

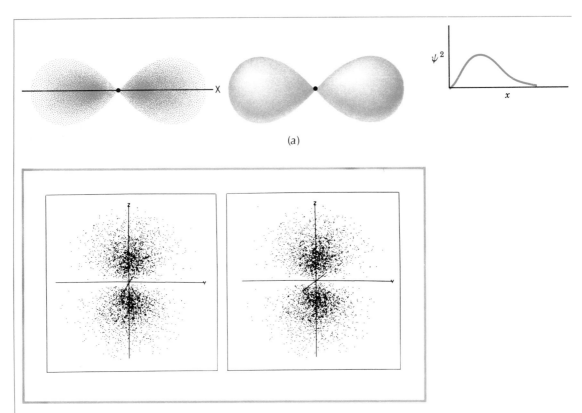

(a)

(b)

Figure 3.27
The shape of the p orbital. (a) Two-dimensional representations of a $2p$ orbital. (b) A three-dimensional representation. From D. T. Cromer, *Journal of Chemical Education* Vol. 45, p. 626, October, 1968. Used by permission.

nucleus. Electron density occurs on both sides of the nucleus so that an electron in a p orbital spends part of its time on each side of the atom. The concentration of the electron density that occurs along certain specific directions causes a p orbital to have definite directional properties which, as we will come to see, allow us to understand why molecules have the shapes they do.

A p subshell is composed of three p orbitals, each one having the same "shape." They differ from one another only in the directions in which their electron density is concentrated. These directions lie at right angles to one another as shown in Figure 3.28. Here a simple loop is used to represent a p orbital so that the directional character is more easily seen. Remember that this is a much simplified representation of a p orbital and that they are not really flat. The loop is used only to depict the directional nature of the orbital, and we shall use this kind of description again in later chapters to illustrate certain three-dimensional aspects of chemical bonding involving orbitals. Since the p orbitals can be drawn on a set of XYZ axes, we identify the orbitals by the notation, p_x, p_y, p_z.

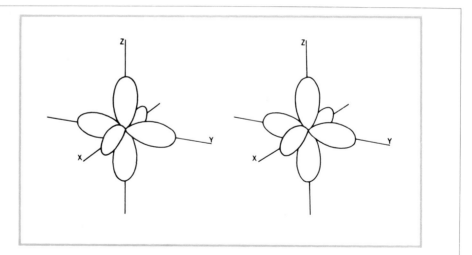

Figure 3.28
Stereo drawing of the three *p* orbitals.

We have seen that *s* and *p* orbitals differ in the "shapes" of their electron clouds. Orbitals in *d* and *f* subshells also have characteristic "shapes"; however, they are considerably more complicated than *p* orbitals. The *d* orbitals play an important role in the chemistry of the transition elements and we shall discuss them in some detail when we get to Chapter 19. The shapes of *f* orbitals are very complex and, since they are only required to discuss the chemistry of the inner transition elements, we will not attempt to describe them further.

3.17
The Variation of Properties with Atomic Structure

Many of the properties of the elements vary in a more or less regular fashion as we proceed from left to right within a period or from top to bottom within a group in the periodic table. Most of these variations can be accounted for directly in terms of variations in the electronic structures of the elements. Let us briefly look at three of these properties: atomic size, ionization energy, and electron affinity.

ATOMIC SIZE. We have seen that the electron density in an atom does not end abruptly at some particular distance from the nucleus but, instead, trails off gradually, approaching zero at very large distances from the center of the atom. Because of this, it is difficult to define precisely what is meant by the size of the atom. Since atoms never occur all by themselves in chemical systems, but are always in the neighborhood of other atoms, the radius of an atom could be taken to be half the distance between neighboring atoms when the element is present in its most dense form (i.e., most highly compacted form, which is usually the solid). Even this definition, however, is complicated because when atoms enter into a bond, as they do in molecules like H_2 or Cl_2, they approach each other more closely than do nonbonded atoms (e.g., the noble gases when they are frozen). Also, the atomic radius that we measure for atoms of a pure element will not necessarily be the same in compounds. For example, carbon atoms in diamond (pure carbon) are separated by a distance of 1.54 Å[7] and we would thus assign carbon a radius of 0.77 Å. In the ethane molecule, C_2H_6, the carbon-carbon distance is also 1.54 Å; however, in ethylene, C_2H_4, and acetylene, C_2H_2, we find carbon-carbon distances equal to 1.37 and 1.20 Å, respectively. These lead to atomic radii for carbon of 0.69 and 0.60 Å, both considerably smaller than 0.77 Å.

Despite this difficulty of definition, we can compare atomic radii of the elements if they are measured under circumstances that lead to essentially similar kinds of bonds between their atoms. In Figure 3.29 we illustrate the variation of atomic radius with atomic number. We see that as we proceed down within a group, the size of atoms generally increases, and that as we proceed from left to right across a period, a gradual decrease in size is observed.

In order to interpret these trends within the periodic table in terms of electronic structure, we must look at the factors that determine the size of the outer shells of atoms; that is, the average distance at which electrons in the outer shell occur. As we have discussed earlier (Section 3.16), one of these is the principal quantum number of the outer shell (you will recall that the electron occurs at increasingly larger distances from the nucleus with an increasing value of n).

Figure 3.29
A graph of atomic radius versus atomic number.

---

[7] We shall see how interatomic distances are obtained in Chapter 7.

The size of the outer shell also depends on the **effective nuclear charge** that an electron in that shell experiences. Electrons in inner shells tend to lie between the nucleus and those in the outer shell and thus shield the outer shell from the nuclear charge. In a sodium atom, for example, the 10 electrons of the neon core shield the outer $3s$ electron from the positive charge of the 11 protons in the nucleus so that the outer $3s$ electron feels an effective charge of only about $+1$. Electrons within the same shell also provide some shielding for one another; however, their ability to do so is not very great.

As we proceed from one atom to the next down within a group, each successive element has its outer electron in a shell with a larger value of n. The effective nuclear charge experienced by the outer electron(s) remains nearly the same so that the net effect is an increase in size with increase in atomic number within a group. For instance, among the alkali metals, Li to Cs, the principal quantum number increases from 2 for Li to 6 for Cs. The single outer electron in each of these elements, however, experiences a nearly constant effective nuclear charge of $+1$. Therefore the increase in size that occurs from Li to Cs is a result of the electron being in a shell with progressively higher n.

For the representative elements, as we move from left to right across a period, we add electrons to the same shell and simultaneously increase the nuclear charge. Since the outer shell electrons do not shield each other from the nucleus very well, the effective nuclear charge experienced by any one electron in the outer shell increases. This increase in effective nuclear charge leads to a greater attraction for the outer-shell electrons. As a result, they are pulled in closer to the nucleus and hence a decrease in the size of the atom occurs.

The variation in size as we pass through a row of transition or inner transition elements is much less than among the representative elements. This is so because electrons are being added to an inner shell as the nuclear charge gets larger. In the first row of the transition elements, for instance, the outer electrons occur in a $4s$ subshell, but each successive electron is added to the inner $3d$ subshell as we proceed across the table. The inner-shell electrons are nearly completely effective at shielding the outer shell from the nuclear charge so that the outer $4s$ electrons experience only a very gradual increase in effective nuclear charge across this region of the periodic table. Hence small changes in size occur.

The gradual decrease in size that occurs upon the filling of the $4f$ subshell in the lanthanides, which is termed the **lanthanide contraction,** has some very marked effects on the chemistry of the transition elements that follow the lanthanides in the sixth period. For example, because of the lanthanide contraction, the size of Hf is the same as the size of Zr. Since their outer-shell electron configurations are virtually identical, the chemistry of these two elements are very similar and it is extremely difficult to separate them from one another.

As we shall discuss in the next chapter, when many of the elements react to produce compounds, they do so by the formation of ions. We find that positive ions are smaller than the neutral atoms from which they are formed, while negative ions are larger than neutral atoms (Table 3.7). The decrease in size that accompanies the creation of a positive ion is often a result of the removal of all of the electrons from the outer shell of the atom so that a noble

Table 3.7
Atomic Ionic Radii (in angstroms)

Positive Ions

		Atomic Radius	Ionic Radius	Charge
Group IA	Li	1.35	0.60	(+1)
	Na	1.54	0.95	(+1)
	K	1.96	1.33	(+1)
	Rb	2.11	1.48	(+1)
	Cs	2.25	1.69	(+1)
Group IIA	Be	0.90	0.31	(+2)
	Mg	1.30	0.65	(+2)
	Ca	1.74	0.99	(+2)
	Sr	1.92	1.13	(+2)
	Ba	1.98	1.35	(+2)
Group IIIA	Al	1.43	0.50	(+3)
	Ga	1.22	0.62	(+3)
	In	1.62	0.81	(+3)

Negative Ions

		Atomic Radius	Ionic Radius	Charge
Group VIIA	F	0.64	1.36	(−1)
	Cl	0.99	1.81	(−1)
	Br	1.14	1.95	(−1)
	I	1.33	2.16	(−1)
Group VIA	O	0.66	1.40	(−2)
	S	1.04	1.84	(−2)
	Se	1.17	1.98	(−2)
	Te	1.37	2.21	(−2)
Group VA	N	0.70	1.71	(−3)
	P	1.10	2.12	(−3)

Elements that Form More Than One Ion

Fe	1.26	Fe^{2+}	0.76	Fe^{3+}	0.64
Co	1.25	Co^{2+}	0.78	Co^{3+}	0.63
Cu	1.28	Cu^{+}	0.96	Cu^{2+}	0.69

gas electron configuration is attained. For example, the sodium atom loses its single $3s$ electron to produce an Na^+ ion whose electronic structure consists of the neon core. The outer shell at this point has its principal quantum number equal to two and, thus, the outer shell electrons in the Na^+ ion are at a smaller average distance from the nucleus than the $3s$ electron in the Na atom.

When negative ions are produced from neutral atoms, electrons are added to the outer shell without any change in the nuclear charge. An additional electron will provide some degree of shielding for other electrons originally present; therefore the effective nuclear charge felt by any one electron in the outer shell will decrease. At the same time, the presence of an additional electron in the outer shell will increase interelectron repulsions (i.e., repulsions between electrons). Both of these factors tend to cause the outer shell to expand in size, causing the negative ion to be larger than the neutral atom.

IONIZATION ENERGY. The **ionization energy** is defined as the energy required to remove an electron from an isolated atom in its ground state. This is an endothermic process because the electron is attracted to the positive nucleus; therefore, energy must be supplied to remove it. Since all atoms other than hydrogen possess more than one electron, they also have more than one ionization energy. The amount of energy required to remove the first electron from the neutral atom is termed the first ionization energy and that required to remove the second is called the second ionization energy. As we might expect, successive ionization energies increase in magnitude because the species from which the electron is removed becomes progressively more positively charged. For example, the first ionization energy involves the removal of an electron from a neutral atom while the second ionization energy involves the removal of an electron from an ion whose charge is +1.

Table 3.8 contains successive ionization energies, measured in electron volts,[8] eV, for the first 20 elements in the periodic table. An examination of

Table 3.8
Ionization Energies of the First Twenty Elements (electron volts)

	First	Second	Third	Fourth	Fifth	Sixth	Seventh	Eighth
H	13.60							
He	24.6	54.4						
Li	5.4	75.6	121.8					
Be	9.3	18.1	153.1	216.6				
B	8.3	25.0	37.8	258.1	338.5			
C	11.3	24.3	47.6	64.2	390.1	490		
N	14.5	29.5	47.4	77.0	97.4	552	666.8	
O	13.6	34.9	54.9	77.0	113	137.5	739	871.1
F	17.3	34.8	62.4	86.7	113.7	156.4	184.3	953.6
Ne	21.6	40.9	63.2	97.2	126.4	158.0	–	–
Na	5.1	47.1	70.7	99.1	138.6	172.4	208.4	264.2
Mg	7.6	15.0	79.7	108.9	141.2	187	225	266
Al	6.0	18.7	28.3	119.4	153.4	190.4	241.9	285.1
Si	8.2	16.3	33.4	44.9	165.6	205.1	246.4	303.1
P	11.0	19.6	30.0	51.1	64.7	220.4	263.3	309.3
S	10.4	23.3	34.9	47.1	72.5	88.0	281.0	328.8
Cl	13.0	23.8	39.9	53.5	67.8	96.7	114.3	348.3
Ar	15.8	27.6	40.9	59.8	75.0	91.3	124.0	143.5
K	4.3	31.8	46.5	60.9	82.6	99.7	118	155
Ca	6.1	11.9	51.2	67	84.4	109	128	143

[8] The electron volt is a unit of energy corresponding to the kinetic energy that an electron acquires when it is accelerated from a negative to a positive electrode maintained at a potential difference of one volt.

the data in this table points out again the great stability associated with a noble gas electron configuration. We see, for example, that for a Group IA element the first ionization energy is relatively low and that the second ionization energy is very much greater. For the Group IIA elements a large increase in ionization energy occurs after two electrons have been removed while for Group IIIA elements the break occurs after the third electron has been lost. In fact, we see that in general a very large jump in ionization energy always occurs after an atom has lost a number of electrons that is numerically equal to its group number. Since a Group IA element contains one electron outside a noble gas electron configuration, a Group IIA element, two, etc., these large increases in ionization energy must reflect the extreme difficulty that is encountered in trying to break into the noble gas structure.

The variation of the first ionization energy across periods and down groups, illustrated in Figure 3.30, quite closely parallels the trends in atomic size. This really should not be too surprising since the energy required to remove an electron from an atom completely should depend in part upon how far away it is from the nucleus. In addition, the same factors that are responsible for causing an outer shell to contract in size as we proceed across a period will also lead to the electron being held more tightly. Thus, as we proceed down within a group (e.g., the alkali metals), the increase in size that occurs is accompanied by a decrease in ionization energy. As we move across a period, from left to right, the increased effective nuclear charge experienced by the outer-shell electrons causes the shell to shrink in size and also makes it more difficult to remove an electron.

If we examine more closely the trend in ionization energy across a period, we note some irregularities. In period 2, for example, we expect a uniform increase in ionization energy as we go from Li to Ne. We observe, however, that the ionization energy of beryllium is higher than that for boron; in addition, the energy required to remove an electron from nitrogen is

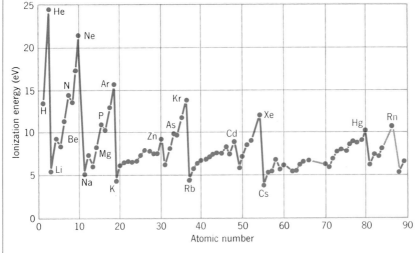

Figure 3.30
The variation of first ionization energy with atomic number.

Atomic Structure and the Periodic Table

greater than for oxygen. These reversals can also be explained by the electronic structures of the elements.

In the case of beryllium, the first electron that is removed (recall that we are comparing first ionization energies) lies in the filled $2s$ subshell while the electron that is removed first from boron lies in the singly occupied $2p$ subshell. The $2p$ subshell is higher in energy than the $2s$; thus the $2p$ electron of boron is more easily removed than a $2s$ electron of beryllium.

When we get to nitrogen, we find that we have a half-filled $2p$ subshell (electronic structure of nitrogen is $1s^2 2s^2 2p^3$), while in oxygen the $2p$ subshell is occupied by four electrons. As we have said earlier, apparently a half-filled subshell has some extra stability associated with it. This phenomenon seems to manifest itself here where it is more difficult to break up the half-filled subshell of nitrogen than to disturb the neither filled nor half-filled $2p$ subshell of oxygen. Note that the same inverted order of values for the ionization energy also occurs in periods 3 and 4 where the ionization energy of phosphorus is greater than for sulfur, and that for arsenic it is greater than for selenium.

ELECTRON AFFINITY. The electron affinity is the energy that is released or absorbed when an electron is added to a neutral gaseous atom. Such a process occurs, for example, when a chlorine atom picks up an electron to become a negative ion

$$Cl(g) + e^- \rightarrow Cl^-(g)$$

The electron affinity (like the ionization energy) applies to isolated atoms and usually represents an exothermic process. This is so because we are placing the electron into an environment where it experiences the attraction of the nucleus. We can see how the addition of an electron to an atom would release energy by considering the reverse process, pulling the electron away from the attractive force of the nucleus. If removing the electron requires work (i.e., is endothermic), the opposite process would release energy.

There are instances where more than one electron is added to the outer shell of the atom. For example, oxygen reacts to form the ion, O^{2-}, in which an oxygen atom picks up two electrons. The second electron that is added to give the ion a charge of -2 must be forced onto an already negative ion. This requires work; therefore we find that the second electron affinity of an atom is an endothermic quantity. Table 3.9 contains electron affinities for some of the representative elements. The table is not complete because electron affinities are difficult to measure and, for many elements, have not been determined.

As with the ionization energy, the variations in electron affinity generally parallel the variations in atomic size. This is because we are considering the placement of an electron into the outer shell of the atom. The closer the electron can get to the nucleus the greater will be the effect of the nuclear charge. Atoms that are very small and that have outer shells that experience a high effective nuclear charge (e.g., elements in the upper right of the periodic table) therefore have very large electron affinities. On the other hand, atoms that are large and whose outer shells feel the effect of a small effective nuclear charge (such as the elements in Groups IA and IIA) are expected to have small electron affinities.

If we examine the data in Table 3.9, we find that fluorine, which is

Table 3.9
Electron Affinities[a] for the Representative Elements (in electron volts)

IA	IIA	IIIA	IVA	VA	VIA	VIIA
H 0.75						
Li 0.61	Be ≈−0.6	B 0.3	C 1.25	N ≈−0.09	O 1.45	F 3.46
Na ≈0.5	Mg ≈−0.3	Al 0.52	Si 1.4	P 0.74	S 1.7	Cl 3.62
K ≈0.4	Ca —	Ga 0.18	Ge 1.2	As 0.60	Se 1.7	Br 3.37
Rb ≈0.2	Sr —	In —	Sn —	Sb —	Te —	I 3.07
Cs —	Ba —	Tl —	Pb —	Bi —	Po —	At —

[a] Negative values mean that the process, $M + e^- \longrightarrow M^-$, is endothermic.

smaller than chlorine, also has a smaller electron affinity. It appears that the greater interelectron repulsions experienced by an entering electron in the small outer shell of fluorine more than compensate for the fact that the added electron in fluorine lies closer to the nucleus than it does in chlorine.

Finally, carbon has a rather substantial electron affinity while that of nitrogen is actually negative (meaning that the process of adding an electron to nitrogen is endothermic). This once again illustrates the unusual stability of a half-filled shell, which carbon can obtain by acquiring of an electron and nitrogen can retain by *tending* to refuse additional electrons. (Under appropriate conditions nitrogen will acquire electrons.)

Review Questions and Problems

3.1 What are the properties of the electron, the proton, and the neutron?

3.2 Verify the value for the mass of the electron using the experimentally determined charge-to-mass ratio of -1.76×10^8 coulombs/gram and the measured charge, -1.60×10^{-19} coulombs.

3.3 In Millikan's experiment, the charge on the oil droplets was always found to be a multiple of -1.60×10^{-19} coulombs. Suppose this experiment were repeated and the following values were obtained:

3.20×10^{-19} coulombs
5.60×10^{-19} coulombs
6.40×10^{-19} coulombs
2.40×10^{-19} coulombs
7.20×10^{-19} coulombs

On the basis of these data, what is the charge on the electron? Give some examples of how this value affects our picture of the atom.

3.4 What are the numbers of protons, neutrons, and electrons in each of the following: $^{132}_{55}Cs$, $^{115}_{48}Cd^{2+}$, $^{194}_{81}Tl$, $^{105}_{47}Ag^{1+}$, $^{78}_{34}Se^{2-}$?

3.5 The element Eu occurs naturally as a mixture of 47.82% $^{151}_{63}$Eu, whose mass is 150.9 amu, and 52.18% $^{153}_{63}$Eu, whose mass is 152.9 amu. Calculate the average atomic mass of Eu.

3.6 The proton (the nucleus of a hydrogen atom) has a mass of 1.67×10^{-24} g and a diameter of 10^{-13} cm. Calculate the density of the nucleus assuming it to be spherical in shape.

3.7 The earth has a mass of 6.59×10^{21} tons and a diameter of approximately 8000 miles. What would be the diameter of the earth (in miles) if it had the same mass but was composed entirely of nuclear material? (Use the density of nuclear material calculated in question 3.6.)

3.8 The charge-to-mass ratio (e/m) of the proton (a hydrogen nucleus) is 9.65×10^4 coulombs/gram. The proton has a charge equal to 1.60×10^{-19} coulombs. Calculate the value of Avogadro's number.

3.9 Use the Rydberg equation (Equation 3.3) to calculate the wavelengths of the first two lines in the Pfund series of the hydrogen spectrum and compare your answers to the values in Table 2.4.

3.10 Use the Rydberg equation to calculate the wavelength of the spectral line in hydrogen that would result when an electron drops from the fourth Bohr orbit to the second.

3.11 How much energy must be supplied to raise an electron from the first Bohr orbit to the third?

3.12 Calculate the wavelength of light whose frequency is 8.0×10^{15} sec^{-1}. Calculate the frequency of light whose wavelength is 2000 Å.

3.13 Calculate the radii of the second and third Bohr orbits if the first has a radius of 0.529 Å.

3.14 Calculate the kinetic energy of an electron with a wavelength of 1.0 Å.

3.15 How long would it take a 2.0 gram bullet to travel the length of a 10-cm gun barrel if it had a wavelength of 1 Å?

3.16 What is a diffraction pattern? How is it produced?

3.17 Describe the contributions made by each of the following scientists toward the development of atomic theory: Faraday, Thomson, Rutherford, Millikan, Moseley.

3.18 Discuss the contributions made by each of the following scientists toward the classification of the elements: Döbereiner, Newlands, Mendeleev.

3.19 Find, in the periodic table, three "Döbereiner triads" (other than the Li, Na, K triad) that satisfy the atomic weight relationships described by Döbereiner. Keep in mind that the elements of any triad must have similar properties and therefore must be in the same group.

3.20 How many tons of water could be heated from 0 °C to 100 °C by converting 1.0 g of matter entirely into energy. Recall that it takes 1 calorie of energy to raise the temperature of 1.0 g of water by 1 °C. (See also Section 1.11.)

3.21 How many electrons can be accommodated in each of the following types of subshells: s, p, d, f, g, h? What is the lowest value of n for a shell that has an h subshell? What are the allowed values of m for an h subshell?

3.22 Give the values of n, l, m, and s for each electron in a filled L shell.

3.23 Draw orbital diagrams for each element in the first row of transition elements ($Z = 21$ to 30). Indicate which of these are paramagnetic and which are diamagnetic.

3.24 Use the periodic table as a guide in writing the complete electron configurations of these elements: P, Ni, As, Ba, Rh, Ho, Sn.

3.25 Use the periodic table to arrive at the electronic structure of the outer shells of the atoms of Si, Se, Sr, Cl, O, S, As, and Sb.

3.26 What is the major difference between a Bohr orbit and an orbital? In what way is the Heisenberg uncertainty principle involved in this comparison?

3.27 On a single set of Cartesian coordinate axes, sketch the "shapes" of the three p orbitals. Label them p_x, p_y, and p_z.

3.28 On the basis of interelectron repulsion and the spatial arrangement of the p orbitals, the observed orbital diagram for nitrogen seems quite reasonable. Why?

3.29 How does the shape of an s orbital differ from that of a p orbital?

3.30 Explain the variation in ionic size observed for the series, N^{3-}, O^{2-}, and F^- (Table 3.7) in terms of the effective nuclear charge and interelectron repulsions experienced by the outer-shell electrons.

3.31 In Table 3.7 we find some elements that form more than one positive ion. In each case the ion with the greater positive charge is smaller. Why is this so?

3.32 What is the lanthanide contraction? How might this be used to explain why the elements in the sixth period following the lanthanides have higher ionization energies than the elements directly above them in the fifth period (e.g., the ionization energy of $Pt = 9.0$ eV while that of Pd is 8.35 eV)?

3.33 Use the Rydberg equation (Equation 3.3) to calculate the ionization energy of hydrogen and compare your answer to that in Table 3.8 (1 eV $= 1.60 \times 10^{-12}$ erg).

3.34 How can we explain the variation of ionization energy across a period in the periodic table?

3.35 Draw a graph, on a set of axes like that below, of the ionization energy versus the number of electrons removed from the atom for each of the elements Li, C, O, S, and Ne.

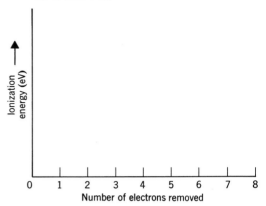

3.36 In the text a possible explanation was given as to why carbon has a relatively large electron affinity while that of nitrogen is very small. Use a similar line of reasoning to explain the very large electron affinities of the Group VIIA elements.

4
Chemical Bonding: General Concepts

In Chapter 3 we spent considerable time discussing electronic structures of atoms and their relationship to some properties of atoms. A property possessed by almost all atoms is their ability to combine with other atoms to produce more complex species. The forces of attraction that hold atoms together in their combined states are called chemical bonds.

The theories and language used to describe chemical bonds have evolved from very simple theories to more elaborate ones based on wave mechanics. In this chapter we examine some of the simpler ideas about chemical bonding that should serve to carry you through the discussions in the first part of this book. In Chapter 16 we delve deeper into the more modern, sophisticated theories about bonding.

4.1 Lewis Symbols

When atoms interact to form a bond only the outer portions of the atoms come in contact; consequently, only their outer electron configurations are usually important. To keep tabs on the outer shell (also called **valence shell**) electrons a special type of notation is used called **Lewis symbols,** named after the American chemist, G. N. Lewis (1875–1946). To construct the Lewis symbol for an element, we write down its atomic symbol surrounded by a number of dots (or x's or circles, etc.), each of which represents one electron in the atom's valence shell.

For example, the element hydrogen, which has one electron in its valence shell, is given the Lewis symbol, H·. Any atom, in fact, with one electron in its outer shell has a similar Lewis symbol. This includes any element in Group IA of the periodic table, so that each of the elements Li, Na, K, Rb, Cs, and Fr has a Lewis symbol that we might generalize as X· (where X = Li, Na, etc.). Generalized Lewis symbols for the representative elements are given in Table 4.1.

We shall see that Lewis symbols are useful in discussing bonds between atoms. The formulas we draw with them are called either Lewis structures or electron-dot formulas.

4.2 The Ionic Bond

Chemical bonds can be divided into two general categories: **ionic** (or **electrovalent**) bonds and **covalent** bonds.

Table 4.1

Lewis Symbols for A-Group Elements

Group	IA	IIA	IIIA	IVA	VA	VIA	VIIA
Symbol	$X\cdot$	$\cdot X\cdot$	$\cdot \overset{\cdot}{X}\cdot$	$\cdot \overset{\cdot}{\underset{\cdot}{X}}\cdot$	$\cdot \overset{\cdot}{\underset{\cdot}{X}}\colon$	$\cdot \overset{\cdot\cdot}{\underset{\cdot}{X}}\colon$	$\cdot \overset{\cdot\cdot}{\underset{\cdot\cdot}{X}}\colon$

An ionic bond occurs when one or more electrons are transferred from the valence shell of one atom to the valence shell of another. The atom that loses electrons becomes a positive ion (**cation**) while the atom which acquires electrons becomes negatively charged (an **anion**). The ionic bond results from the attraction between the oppositely charged ions.

An example of the formation of an ionic substance is the reaction between atoms of lithium and fluorine. The electronic structures of these are

$$Li \quad 1s^22s^1$$

and

$$F \quad 1s^22s^22p^5$$

When they react the lithium atom loses the electron from its $2s$ subshell to become Li^+, thereby assuming an electron configuration that is the same as the noble gas He.

$$Li(1s^22s^1) \longrightarrow Li^+(1s^2) + e^-$$

The electron lost by Li is picked up by the fluorine atom which thereby acquires an electron configuration identical to that of the noble gas Ne.

$$F(1s^22s^22p^5) + e^- \longrightarrow F^-(1s^22s^22p^6)$$

Once formed, the Li^+ and F^- ions attract one another because of their opposite charges. It is this attraction between the ions that constitutes the ionic bond.

The actual reaction that takes place between elemental lithium and fluorine is certainly not exactly as we have just pictured it, since lithium exists as a solid, not as isolated Li atoms, and fluorine occurs as a gas composed of F_2 molecules. Nevertheless, an important feature of the reaction is the transfer of electrons from the lithium to the fluorine. The Li^+ and F^- ions that are produced are attracted to each other and pack themselves together to form the ionic solid, LiF. It is important to remember than an ionic solid such as this does not contain discrete molecules, but instead contains ions packed so that the attractive forces between ions of opposite charge are maximized while repulsive forces between ions of the same charge are minimized. In LiF, for example, each cation (Li^+) is surrounded by, and attracted equally to six anions (F^-), as shown in Figure 4.1. In a similar fashion each anion is attracted equally to the six cations surrounding it.

The Lewis symbols that were introduced in the last section can be used to illustrate the transfer of electrons that occurs during the formation of an ionic compound. For instance, we can show the reaction of Li and F as

$$Li \overset{\frown}{+} \overset{..}{\underset{..}{F}}: \longrightarrow Li^+ \left[:\overset{..}{\underset{..}{F}}: \right]^-$$

The brackets appearing around the fluorine on the right are intended to show that all eight electrons are the exclusive property of the fluoride ion, F^-.

When Li and F react, electrons are lost or gained until a noble gas electron configuration is reached. Except for He, this corresponds to ns^2np^6 (a total of eight electrons in the outer shell). As you will recall from the last

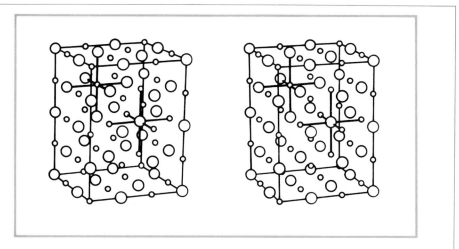

Figure 4.1
The structure of LiF. Small circles = Li$^+$; large circles = F$^-$.

chapter, the electronic structures of the noble gases possess a great deal of stability. The tendency for atoms to achieve this electronic arrangement forms the basis of the so-called **octet rule,** which simply states that *atoms tend to gain or lose electrons until there are eight electrons in their valence shell.*

The octet rule does not always hold, however. Other relatively stable outer-shell electronic configurations also exist and explain the stability of ions such as Zn^{2+} and Ag^+ ($ns^2np^6nd^{10}$ — often called the **pseudonoble gas con-figuration**) and ions such as Fe^{3+} and Mn^{2+} ($ns^2np^6nd^5$).

Some ions occur with none of the electronic structures just discussed. Since it is very difficult to form a highly charged ion (ions with a charge greater than +3 are rare), electron loss sometimes ceases before a noble gas electron configuration is achieved. Some examples are Ti^{2+} ([Ar]$3d^2$), V^{2+} ([Ar]$3d^3$), and Cr^{2+} ([Ar]$3d^4$). Ions of this type are common among the transition elements. A list of some common cations and anions, including the type of electron configuration they possess, is given in Table 4.2.

The ratio in which two elements react to form an ionic substance is usually determined by the numbers of electrons that must be lost or gained by the respective reactant atoms in order to attain a stable electron configuration. For instance, in the reaction between calcium (Group IIA) and chlorine (Group VIIA), each calcium atom must lose two electrons to achieve a noble gas structure whereas each chlorine atom needs to acquire only one electron to obtain an octet. As we can see,

$$\text{or}$$
$$\text{CaCl}_2$$

the result is that two chlorine atoms must react with each calcium atom to produce one Ca^{2+} and two Cl^- ions. The neutral compound has the formula

Table 4.2
Some Common Cations and Anions

(a) Ions with noble gas electron configuration

+1	+2	+3	−3	−2	−1
Li^+		Al^{3+}	N^{3-}	O^{2-}	F^-
Na^+	Mg^{2+}		P^{3-}	S^{2-}	Cl^-
K^+	Ca^{2+}			Se^{2-}	Br^-
Rb^+	Sr^{2+}			Te^{2-}	I^-
Cs^+	Ba^{2+}				

(b) Ions with pseudonoble gas electron
configurations

+1	+2	+3
Cu^+	Zn^{2+}	Ga^{3+}
Ag^+	Cd^{2+}	In^{3+}
Au^+	Hg^{2+}	Tl^{3+}

(c) Ions with other electron configurations

Cr^{2+}	[Ne]	$3s^2$	$3p^6$	$3d^4$
Cr^{3+}	[Ne]	$3s^2$	$3p^6$	$3d^3$
Mn^{2+}	[Ne]	$3s^2$	$3p^6$	$3d^5$
Fe^{2+}	[Ne]	$3s^2$	$3p^6$	$3d^6$
Fe^{3+}	[Ne]	$3s^2$	$3p^6$	$3d^5$
Co^{2+}	[Ne]	$3s^2$	$3p^6$	$3d^7$
Ni^{2+}	[Ne]	$3s^2$	$3p^6$	$3d^8$
Cu^{2+}	[Ne]	$3s^2$	$3p^6$	$3d^9$

$CaCl_2$. Similar reasoning leads us to expect a compound between Li and O to have the formula Li_2O.

$$\text{Li}\cdot\quad\ddot{\ddots}\quad\longrightarrow\ 2Li^+,\left[:\ddot{\ddot{O}}:\right]^{2-}$$

A slightly more complex situation occurs with Al and O. Aluminum, in Group IIIA, loses three electrons to achieve a noble gas structure and produces the ion Al^{3+}. Oxygen, on the other hand, forms the ion O^{2-}. To produce a neutral compound two Al^{3+} ions must be combined with three O^{2-} ions; hence aluminum oxide has the formula Al_2O_3.

At this point we might ask: What conditions most favor the formation of an ionic substance? To answer this question we have to consider the formation of an ionic compound as proceeding via a series of steps. These include, again choosing LiF as an example, the energy required to remove an electron from Li (ionization energy), the energy released when an electron is added to fluorine (electron affinity) and the energy released when the Li^+ ions and F^- ions come together to form the solid (called the lattice energy).

The first quantity, that is, the ionization energy of Li, is endothermic. The latter two processes are exothermic.[1] This means that the most stable ionic structures (i.e., those of lowest energy) generally result when atoms of low ionization energy combine with elements of high electron affinity, or when the lattice energy of the resulting compound is very large, or both. Under these conditions more energy is given off by the exothermic processes than is absorbed by the endothermic one, with the net result that the total energy contained within the reacting species decreases.

Since metals generally have rather low ionization energies and electron affinities, they tend to lose electrons to form cations while nonmetals, with their large ionization energies and electron affinities, usually acquire electrons to produce anions. For this reason, most compounds formed between metals and nonmetals are ionic, particularly the substances formed when an element from Group IA or IIA reacts with an element in the upper right corner of the periodic table (excluding Group O).

There are many substances that contain ions composed of more than one atom (i.e., **polyatomic ions**). The formulas of these compounds are determined by the relative numbers of cations and anions that must be present in order to achieve a neutral solid. In Table 4.3 some of the common polyatomic ions are listed. Some of these are highly colored and impart their characteristic colors to compounds (and aqueous solutions) containing them. Let us now look at some examples of how the formulas of ionic compounds containing this type of ion are obtained.

Example 4.1

What is the formula for the ionic substance containing the ions Na^+ and CO_3^{2-}?

Solution

In order for the compound to be neutral the number of positive charges must equal the number of negative charges. This requires two Na^+ ions per CO_3^{2-} ion. The compound (called sodium carbonate) therefore has a formula, Na_2CO_3.

Example 4.2

An ionic compound contains the ions Ca^{2+} and PO_4^{3-}. What is its formula?

Solution

The total number of positive or negative charges represented in the formula must be divisible by both 2 and 3. The smallest number that meets this requirement is $2 \times 3 = 6$. Thus, there must be six positive and six negative charges in the formula. This is achieved by taking three Ca^{2+} ions and two PO_4^{3-} ions. The formula of this compound (calcium phosphate), then, is $Ca_3(PO_4)_2$.

---

[1] Since energy is required to separate positive and negative ions, when they are brought together, energy must obviously be released. The reaction $Li^+ + F^- \rightarrow LiF$ is, therefore, exothermic.

Table 4.3

Some Common Polyatomic Ions

(a) Cations

NH_4^+	Ammonium
H_3O^+	Hydronium

(b) Anions (alternate names in parentheses)

CO_3^{2-}	Carbonate
HCO_3^-	Hydrogen carbonate (bicarbonate)
$C_2O_4^{2-}$	Oxalate
NO_3^-	Nitrate
NO_2^-	Nitrite
OH^-	Hydroxide
SO_4^{2-}	Sulfate
HSO_4^-	Hydrogen sulfate (bisulfate)
SO_3^{2-}	Sulfite
HSO_3^-	Hydrogen sulfite (bisulfite)
ClO_4^-	Perchlorate
ClO_3^-	Chlorate
ClO_2^-	Chlorite
ClO^-	Hypochlorite
PO_4^{3-}	Orthophosphate (phosphate)
HPO_4^{2-}	Hydrogen orthophosphate (hydrogen phosphate)
$H_2PO_4^-$	Dihydrogen orthophosphate (dihydrogen phosphate)
CrO_4^{2-}	Chromate
$Cr_2O_7^{2-}$	Dichromate
MnO_4^-	Permanganate
$C_2H_3O_2^-$	Acetate

4.3
The Covalent Bond

In many instances the formation of an ionic substance is not energetically favorable. For example, the creation of a cation may require too large an energy input (ionization energy) to be recovered by the energy released when the anion is formed and the ionic solid is produced (electron affinity and lattice energy). In these situations a **covalent bond** is formed.

A covalent bond results from the *sharing* of a *pair* of electrons between atoms. The binding force results from the attraction between these shared electrons and the positive nuclei of the atoms entering into the bond. In this sense, the electrons serve as a sort of glue cementing the atoms together.

The simplest covalent bond that we can examine exists between H atoms in the molecule H_2. Each H atom completes its valence shell by acquiring a share of an electron from another atom. We can indicate the formation of H_2, using Lewis symbols, as

$$H\cdot + H\cdot \longrightarrow H:H$$

in which the pair of electrons in the bond is shown as a pair of dots between the two H atoms. Often a dash is used instead of the pair of dots so that the H_2 molecule may be represented as H—H. The actual distribution of electronic charge in this molecule is illustrated in Figure 4.2. Notice that the electron density is concentrated between the positive nuclei.

As with the ionic bond, the number of covalent bonds that an atom will form can frequently be predicted by counting the number of electrons required to achieve a stable electron configuration (usually that of a noble gas). For example, the carbon atom has four electrons in its valence shell and, to attain a noble gas configuration, usually acquires, through sharing, four additional electrons. The carbon atom, therefore, is capable of forming four bonds with H atoms to form the molecule CH_4 (methane).

$$\cdot\overset{\displaystyle\cdot}{\underset{\displaystyle\cdot}{C}}\cdot + 4H\times \longrightarrow H\overset{\displaystyle\overset{H}{\overset{\cdot\times}{}}}{\underset{\displaystyle\underset{H}{\overset{\cdot\times}{}}}{\times C\cdot}}H$$

Nitrogen, which has five valence electrons, has to gain only three electrons through sharing to complete an octet; therefore, nitrogen forms three covalent bonds with hydrogen to form the ammonia molecule, NH_3. In a similar fashion it is easy to see why the formula for water is H_2O and hydrogen fluoride is HF.

$$H\overset{\displaystyle\overset{\cdot\cdot}{}}{\underset{\displaystyle\underset{H}{\overset{\cdot\times}{}}}{\times N \times}}H \qquad\qquad H\overset{\displaystyle\overset{\cdot\cdot}{}}{\underset{\displaystyle\underset{H}{\overset{\cdot\times}{}}}{\times O}}: \qquad\qquad H\overset{\displaystyle\overset{\cdot\cdot}{}}{\underset{\displaystyle\underset{}{\overset{\cdot\cdot}{}}}{\times F}}:$$

Unfortunately it is not always possible to predict the formula of a covalent molecule on the basis of these simple rules. There are many examples of covalent compounds that fail to obey the octet rule. For instance, the molecule $BeCl_2$ is formed by the pairing of the two Be valence electrons with

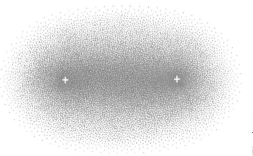

Figure 4.2

The electron distribution in the H_2 molecule.

electrons on two chlorine atoms[2]:

$$:\overset{..}{\underset{..}{Cl}}\cdot + {}_{\times}Be_{\times} + \cdot\overset{..}{\underset{..}{Cl}}: \longrightarrow :\overset{..}{\underset{..}{Cl}} \overset{}{\underset{\times}{:}} Be \overset{}{\underset{\times}{:}} \overset{..}{\underset{..}{Cl}}:$$

In this molecule the Be atom has only four electrons in its valence shell. Since this is less than the usual octet, $BeCl_2$ is said to be electron deficient. Another example of an electron deficient molecule is BCl_3:

$$:\overset{..}{\underset{..}{Cl}}:$$
$$:\overset{..}{\underset{..}{Cl}} \overset{\times}{\underset{\times}{:}} B \overset{}{\underset{\times}{:}} \overset{..}{\underset{..}{Cl}}:$$

Besides electron deficient compounds, there are many more examples of molecules in which the central atom has more than eight electrons in its valence shell. Two typical examples are PCl_5 and SF_6. To form covalent bonds between the central atom (generally the one written first in the formula) and each of the surrounding atoms, more than four pairs of electrons (an octet) are needed. In PCl_5, for example, there are five covalent bonds; in SF_6 there are six. The central atom in each of these molecules uses all of its valence electrons to form covalent bonds.

In these compounds, both phosphorus and sulfur have exceeded the number of electrons required for a noble gas electron configuration. This can occur with these elements because, in each case, the valence shell can accommodate more than eight electrons (both P and S are in the third period and the third shell can contain up to 18 electrons). Elements in the second period (e.g., Li to Ne) almost never form compounds with more than eight electrons in their valence shell because the second shell cannot accommodate more than an octet.

Writing Lewis structures for molecules containing only single bonds (i.e., one electron pair shared between two atoms) is usually relatively simple and straightforward. There are, however, a very large number of molecules in which atoms complete their octet by sharing more than one pair of electrons in a bond. For example, CO_2 has a Lewis structure that we can write as

$$\overset{..}{\underset{..}{O}}::C::\overset{..}{\underset{..}{O}} \quad \text{or} \quad \overset{..}{\underset{..}{O}}{=}C{=}\overset{..}{\underset{..}{O}}$$

Let us take a look at how to draw the electron-dot formulas of this type. Before we can proceed we must first know the relative positions of the atoms that are bound together. For example, in CO_2 we must know that there are two oxygen atoms bound to the carbon atom and that it does not have a

Chemical Bonding: General Concepts

[2] Even though $BeCl_2$ is formed from elements in Groups IIA and VIIA, it is covalent rather than ionic.

structure such as O—O—C. In many instances the arrangement of atoms can be inferred from the formula, since it is common practice to write the central atom of a molecule first in the formula, followed by the atoms that surround the central atom. This is so with CO_2, for example. It is also true for species such as NH_3, NO_2, NO_3^-, SO_3, CO_3^{2-}, and SO_4^{2-}. However it is not true for H_2O and H_2S (in which H atoms are bound to O and S, respectively). Nor is it true for molecules such as HClO (in which the Cl is the central atom) or ions like SCN^- (in which C is central). The structure of the molecule is therefore not always obvious. If you must guess, the most symmetrical arrangement of atoms has the greatest chance of being correct. Once we know the arrangement of atoms in the molecule, however, we can then go about distributing the valence electrons. This is done usually so that after the electrons are distributed each atom in the molecule is surrounded by an octet.

For CO_2 we proceed as follows. Oxygen is in Group VIA of the periodic table and each oxygen atom, therefore, supplies six valence electrons to the molecule; carbon, in Group IVA, provides four. The total number of valence electrons that we must account for, then, is 16. For the oxygen atoms to be bound to the carbon, at least one pair of electrons must be shared between the carbon and each oxygen. Placing these into the molecule gives us

$$O:C:O$$

This accounts for four electrons, leaving 12 more to play with. These 12 electrons must be distributed so that each atom is surrounded by eight electrons (an octet). One possible arrangement is

$$:\overset{..}{\underset{..}{O}}:C:\overset{..}{\underset{..}{O}}:$$

In this structure each oxygen atom is surrounded by an octet of electrons; however, the carbon atom has only four electrons around it. We can give carbon an octet by allowing it to share an additional pair of electrons from each oxygen atom. The oxygen atom still retains a share of this pair of electrons, but now the carbon atom shares it too. Here we have eight electrons about each atom and a satisfactory Lewis structure. Thus

$$:\overset{..}{O}:C:\overset{..}{O}: \qquad (:\overset{..}{O}::C::\overset{..}{O}:)$$

When two pairs of electrons are shared between two atoms, we call the bond a **double bond. Triple bonds** are also possible. In the N_2 molecule, for example, each N atom completes its valence shell by sharing three electrons with another nitrogen atom.

$$:N:::N: \qquad or \qquad :N\equiv N:$$

4.4
Resonance

It often happens that a single satisfactory electron-dot formula for a compound cannot be drawn. For example, two electron-dot structures that obey the octet rule can be drawn for sulfur dioxide (a substance that is a major air pollutant). These are shown below as structures **1** and **2**.

(1)   **(2)**

These structures have their nuclei in identical positions but differ in the arrangement of electrons. In both, one oxygen atom is bound through a single bond to the sulfur while the other oxygen atom is connected via a double bond.

All experimental evidence suggests that the two sulfur-oxygen bonds are identical; therefore neither of the electron-dot structures that we have drawn for SO_2 is satisfactory. In fact, it is impossible to draw a single electron-dot formula for SO_2 that obeys the octet rule and that is, at the same time, consistent with all of the experimental facts.

We circumvent this problem by the concept of **resonance.** We say that the actual electronic structure of SO_2 does not correspond to either **1** or **2** but, instead, to a structure somewhere in between that has properties of both. This true structure is known as a **resonance hybrid** of the contributing structures **1** and **2**.

It is really quite unfortunate that the term resonance was used to describe this phenomenon, because the impression is often received that the structure of SO_2 fluctuates between **1** and **2**. This is definitely not the case. The structure of SO_2 is never **1** or **2** but a structure in between that we cannot draw satisfactorily using Lewis symbols. The problem is somewhat like trying to describe the beast you would obtain if you were able to cross a cat and a dog. When you try to picture this hypothetical offspring, you visualize it having characteristics of both parents. However, you do not think of it as being a cat one instant and a dog the next.

Some species cannot be adequately explained with only two resonance structures. For example, even though SO_2 can be represented by two structures, the SO_3 molecule requires three.

The electronic structure of some ions must also be represented by resonance. For example, the nitrate ion, NO_3^-, and the carbonate ion, CO_3^{2-}, have the same number of valence electrons as SO_3 and therefore have similar resonance structures. Likewise, SO_2 and NO_2^- have the same number of valence electrons and thus also have similar resonance structures. Resonance is certainly not restricted to inorganic compounds. In proteins, for example, amino acids are linked together in long chains by "peptide bonds"

glycine   peptide bond

(an amino acid)   (a peptide)

There is evidence that the C—N bond in the peptide linkage actually lies somewhere between a single bond and a double bond. To explain this it is suggested that the peptide bond is a resonance hybrid of structures such as

(1)   and   **(2)**

Structure **2** is obtained by rearranging the electrons in structure **1** in this way:

4.5

Coordinate Covalent Bonds

When a nitrogen atom combines with three hydrogen atoms to form the molecule NH_3, the N atom has completed its octet. We might expect, therefore, that the maximum number of covalent bonds that we would observe an N atom to form is three. There are instances, however, where N may have more than three covalent bonds. In the ammonium ion, NH_4^+, which is formed in the reaction

$$H:\overset{\displaystyle H}{\underset{\displaystyle H}{N}}: + H^+ \longrightarrow \left[H:\overset{\displaystyle H}{\underset{\displaystyle H}{N}}:H \right]^+$$

the nitrogen is covalently bound to four hydrogen atoms. When the additional bond between the H^+ and the N atom is created, both of the electrons in the bond come from the nitrogen. *This type of bond, where a pair of electrons from one atom is shared by two atoms, is called either a* **coordinate covalent bond**, *or a* **dative bond**. It is important that you remember the coordinate covalent bond is really no different, once formed, than any other covalent bond and that our distinction is primarily aimed at keeping track of electrons, that is, it is "bookkeeping."

When Lewis structures are written using dashes to represent electron pairs, the coordinate covalent bond is sometimes indicated by means of an arrow pointing away from the atom supplying the electron pair. For example, the product of the reaction of boron trichloride, BCl_3, and ammonia, NH_3, is a substance known as an **addition compound** (since it is formed by the simple addition of two molecules).

To show that the electron pair shared between the B and N originates on the nitrogen, the Lewis structure of this addition compound can be written

Using this type of notation we are tempted to write the structure of the NH_4^+ ion as

This gives the impression that one of the N—H bonds is different from the other three. It has been shown experimentally, however, that all four N—H bonds are identical. Therefore, to avoid conveying false impressions, the NH_4^+ ion is simply written as

A word that often appears in discussions of chemical bonding is **valence,** which *refers to the combining capacity of an element.* Valence is a *positive integer,* originally defined as the number of hydrogen atoms that combine with one atom of an element. In modern usage it is the number of positive or negative charges on a monatomic ion, or the number of pairs of electrons shared by an atom in a covalently bound substance.

4.6
Valence

Thus, for example, the valences of Na^+ and Cl^- are 1 and the valences of Ca^{2+} and O^{2-} are 2. A divalent atom (valence = 2) will combine with two monovalent atoms (valence = 1). We would therefore predict formulas of $CaCl_2$ and Na_2O for calcium chloride and sodium oxide. When the valences of both elements are the same, they combine in a ratio of 1:1; therefore, CaO and NaCl are the formulas of calcium oxide and sodium chloride.

In the covalent substances,

(1) acetylene   (2) ethylene   (3) ethane

each carbon atom shares four electron pairs and we assign to carbon a valence of four. However, remnants of the original usage of the term valence remain. For example, compounds **1** and **2** can be reacted with hydrogen to produce compound **3**:

(2)   (3)

Compound **2** is referred to as **unsaturated,** in the sense that the carbon atom may bond to additional hydrogen atoms via chemical reaction. The compound C_2H_6 is said to be **saturated** since no further reaction with hydrogen can occur. For this reason, covalent molecules containing double bonds or triple bonds are often said to be unsaturated, while those possessing only single bonds are termed saturated.

4.7
Bond Order and Some Bond Properties

The term **bond order** refers to the number of covalent bonds that exist between a pair of atoms. For example, the carbon-carbon bond order in acetylene, C_2H_2, is 3; in ethylene, C_2H_4, it is 2; and in ethane, C_2H_6, it is 1 (see above). Fractional bond orders are also possible as in the case of SO_2. Each SO bond in the two resonance structures we draw for SO_2 is shown as a single bond in one structure and a double bond in the other. As we might expect, the bond order in sulfur dioxide is intermediate between 1 and 2.

The concept of bond order arises as a result of our description of the bonding in covalent molecules. Since we cannot view electrons directly, we have no firsthand way of knowing whether or not the theory is sound. As a result, we must examine properties that are directly related to bond order in an effort to test the validity of our theories. One of these properties is the **bond length,** *the distance between the nuclei of two bonded atoms.* As the bond order between a pair of atoms increases, additional electron density is placed between the two nuclei, causing them to be pulled together. Consequently, we expect bond length to decrease with increasing bond order, and we see that this is true in Table 4.4.

A second property related to bond order is the **bond energy,** *the energy*

Table 4.4
Variation of Bond Properties with Bond Order

Bond	Bond Order	Average Bond Length (angstroms)	Average Bond Energy (eV/bond)	Average Vibrational Frequency (sec^{-1})
C—C	1	1.54	3.8	3.0×10^{13}
C=C	2	1.37	7.2	4.9×10^{13}
C≡C	3	1.20	10	6.6×10^{13}
C—O	1	1.43	3.6	3.2×10^{13}
C=O	2	1.23	7.8	5.2×10^{13}
C—N	1	1.47	3.1	3.7×10^{13}
C≡N	3	1.16	7.6	6.8×10^{13}

required to break the bond to produce neutral fragments. For a diatomic molecule such as H_2 this represents the process

$$H:H(g) \longrightarrow H\cdot(g) + H\cdot(g)$$

while in a molecule such as C_2H_6 the carbon-carbon bond energy represents the energy needed to cause the reaction

In general the bond energy increases with increasing bond order. As the electron density between the nuclei is increased, the nuclei are held together more tightly; therefore, more energy (work) must be supplied to pull the nuclei apart. The variation of this bond property with bond order is also shown in Table 4.4.

The last property that we shall examine which is related to the bond order is the **vibrational frequency** of the atoms joined by the bond. The atoms within a molecule are not stationary; they are in constant motion. This motion can be resolved into two basic types; vibration in which a pair of atoms move toward and away from each other along a line joining their centers, much as two balls connected by a spring (Figure 4.3a), and bending in which the angle between the three atoms alternately increases and decreases (Figure 4.3b). For simplicity we shall restrict our discussion to vibrational motion.

There are two factors that affect the frequency of vibration (i.e., the number of vibrations per second). One of these is the masses of the atoms bonded together and the other is the bond order. *For a given pair of atoms, as the bond order increases the vibrational frequency does so too.* This is because increasing the bond order increases the attractive forces holding the nuclei together, in effect, stiffening the "spring" between the two atoms.

The measurement of the vibrational frequencies of bonds is really quite simple today. It happens that these vibrational frequencies are about the same as the frequency of infrared radiation and when IR light is shined on a substance, radiation having the same frequencies as the vibrational frequencies of the bonds is absorbed. By observing which frequencies are selectively removed from the IR spectrum, we can deduce these vibrational frequencies. The data recorded in the right column of Table 4.4 were obtained in this way.

In complex molecules there are many different vibrational modes available to the atoms and many different frequencies are absorbed from the infrared "rainbow." The infrared (IR) absorption spectra of any but the most simple molecules are therefore quite complicated. Nevertheless, an experienced chemist often finds such an absorption spectrum extremely valuable as an aid in deducing molecular structure. In addition, each molecule, because of its unique structure, gives rise to its own characteristic absorption spectrum, which can be used to identify the compound and thus serves as a sort of fingerprint. Examples of IR absorption spectra of some drugs are shown in Figure 4.4.

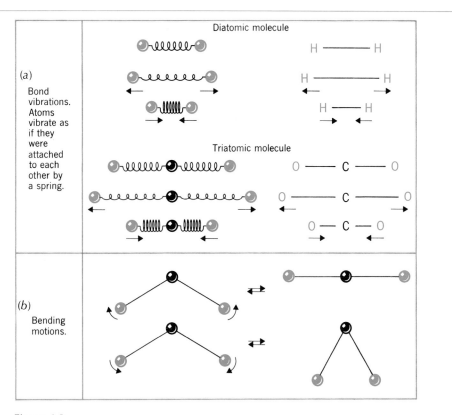

Figure 4.3
The motion of atoms within molecules.

4.8
Polar Molecules and Electronegativity

Generally, we expect atoms of different elements to have different abilities to attract electrons when they enter into a chemical bond. Consequently, we introduce the term, **electronegativity,** which is defined as *the attraction an atom has for electrons in a chemical bond.* It is important not to confuse this term with electron affinity, which is an energy term and refers to an isolated atom.

When two identical atoms combine, as for instance in H_2, both atoms have the same electronegativity. Since each atom is equally capable of attracting the electron pair in the bond, the pair will be shared equally and will spend, on the average, 50% of its time in the vicinity of each nucleus. Each H atom has around it, then, two electrons 50% of the time that when averaged out is the same as one electron all of the time. The "averaged" one electron will completely neutralize the positive charge on each nucleus, and each atom in H_2 carries a net charge of zero.

If the electronegativities of the two atoms in a bond are different, as is the case with HCl, the electron pair will spend more of its time around the more electronegative element (in this instance, Cl). For HCl, this means that the Cl atom acquires a slight negative charge and the H atom, a slight positive

Figure 4.4
Infrared absorption spectra of two drugs. (a) Heroin. (b) LSD.

charge. We indicate this as

$$\overset{\delta+}{H}\!\!-\!\!\overset{\delta-}{Cl}$$

where $\delta+$ and $\delta-$ are meant to indicate partial positive and negative charges, respectively.

Equal positive and negative charges separated by a distance constitute a **dipole.** The HCl molecule, with its centers of positive and negative charge, is, therefore, a dipole and is said to be **polar.** It follows that any diatomic molecule formed between two atoms of different electronegativity will be polar.

A dipole is defined quantitatively by its **dipole moment,** the product of the charge on either end of the dipole times the distance between the charges. For a given internuclear separation, a very polar molecule is one with a large dipole moment while a nonpolar molecule will have no dipole moment at all.

When three or more atoms are bonded together, it is possible to have a

Chemical Bonding: General Concepts

nonpolar molecule even though there are polar bonds present. CO_2 is an often-used example. The CO_2 molecule is linear and may be represented as

$$\overset{\delta-}{:\!\ddot{O}}\!=\!\overset{\delta+}{C}\!=\!\overset{\cdot\cdot}{\underset{\delta+\quad\;\delta-}{\ddot{O}:}}$$

The overall dipole moment of a molecule arises as a sum of the individual bond dipoles within the molecule, which add together like vectors. In CO_2 these bond dipoles are oriented in opposite directions, and they exactly cancel each other.

$$\underset{\longleftarrow\!+\quad+\!\longrightarrow}{O\!-\!C\!-\!O}$$

(An arrow with a plus sign on one end is used to represent the bond dipole.)

In the water molecule, which happens to have a bent or angular shape, the two bond dipoles do not cancel each other entirely, but rather are partially additive. As a result, the H_2O molecule does have a net dipole moment (the heavy arrow below) and is polar.

In general, a molecule AX_n (where n X's are bound to A) will have polar bonds if A and X differ in electronegativity. In order for such a molecule not to be polar, the individual bond dipoles must completely cancel. Some structures which satisfy this condition are listed in Table 4.5.

We would like to have some quantitative measure of electronegativity in order to be in a position to make predictions concerning the polarity of bonds. One approach toward this, taken by R. S. Mulliken in 1934, uses the average of the ionization energy and electron affinity. A very electronegative element has a very high ionization energy, so that it is difficult to remove its electrons, and a very high electron affinity, so that a very stable species results when electrons are added. On the other hand, an element of low electronegativity would have a low ionization energy and low electron affinity so that it loses electrons readily and has little tendency to pick them up. Unfortunately, it is very difficult to measure the electron affinity of an element. Therefore, this method of assigning electronegativities is not universally applicable.

Table 4.5
Symmetric Molecular Shapes

Formula	Example	Shape (see Appendix A)
AX_2	CO_2	Linear
AX_3	BCl_3	Planar triangle
AX_4	CH_4	Tetrahedral
AX_5	PCl_5	Trigonal bipyramidal
AX_6	SF_6	Octahedral

A system that has proven successful was developed by Linus Pauling and is based on the differences between calculated and experimental bond energies. Let us consider, as an example, the molecules H_2, Cl_2, and HCl. The bond energy of an H_2 molecule is equal to 4.52 eV/bond (Table 4.6). If we assume that each of the H atoms contributes equally to the stability of the H—H bond, then each H atom provides 2.26 eV toward the covalent bond energy. In the same manner, since the bond energy of the Cl_2 molecule is 2.51 eV/bond, each Cl atom contributes 1.26 eV toward the stability of a covalent bond.

If the H—Cl bond in hydrogen chloride were purely covalent, we might

Table 4.6
Bond Energies of Some Diatomic Molecules

Molecule	Bond Energy (eV/bond)
H—H	4.52
H—F	6.66
H—Cl	4.47
H—Br	3.79
H—I	3.10
H—Li	2.47
H—Na	2.1
H—K	1.9
H—Rb	1.7
H—Cs	1.8
F—F	1.64
Cl—Cl	2.51
Br—Br	2.00
I—I	1.56
Li—Li	1.14
Na—Na	0.790
K—K	0.553
Rb—Rb	0.507
Cs—Cs	0.490
Li—F	5.97
Li—Cl	4.85
Li—Br	4.35
Li—I	3.67

expect a bond energy that is the sum of contributions from the H and Cl atoms.

$\frac{1}{2}$ (H—H) bond energy = 2.26 eV
$\frac{1}{2}$ (Cl—Cl) bond energy = 1.26 eV
Calculated H—Cl bond energy = 3.52 eV
Experimental H—Cl bond energy = 4.47 eV
Difference (Δ) = 0.96 eV

When we perform this computation (see above), we find that the calculated bond energy is less than the actual bond energy, that is, the H—Cl bond is stronger by an amount Δ than we expect for a nonpolar covalent bond. This extra stability is attributed to the attraction between the oppositely charged ends of the polar HCl molecule. Since the value of Δ is assumed to be proportional to the polarity of the H—Cl bond and since the polarity of the bond depends on the difference in electronegativity between H and Cl, the value of Δ should be proportional to the electronegativity difference.

In Table 4.7 are tabulated the differences between the calculated and experimental bond energies for the molecules HF to HI. For comparison, the averages of the ionization energies (I.E.) and electron affinities (E.A.) of the elements F to I are also included. Since the values of (I.E. + E.A.)/2 decrease from F to I, we expect, as was pointed out above, that the electronegativities of the halogens decrease from F to I. This means that the HX bond becomes less polar from HF to HI, and we would also expect the extra stability associated with the ionic character of the bond to diminish. Since Δ does become smaller from HF to HI, as we predict, it seems that the value of Δ is, indeed, proportional to the difference in electronegativity ($X_A - X_B$) between two atoms. By *arbitrarily* assigning fluorine an electronegativity of 4.0, a complete table of electronegativities (Table 4.8) was prepared by Pauling based on the differences between experimental and calculated bond energies.

We have described electronegativity as an atom's ability to attract electrons in a bond. When the difference between the electronegativities of two combining atoms is very large, the electron pair will spend virtually 100% of its time about the more electronegative element. This is the same as saying that an electron is transferred from the atom of low electronegativity to that of high electronegativity. The result, of course, is an ionic bond. It is clear that

Table 4.7

Values of Δ for the Hydrogen Halides and Values of (EA + IE)/2 for Hydrogen and the Halogens

Molecule	Δ(eV/bond)	Atom	(IE + EA)/2 (eV)
HF	3.58	F	10.38
HCl	0.977	Cl	8.28
HBr	0.52	Br	7.6
HI	0.08	I	6.8
		H	7.13

Table 4.8
The Complete Electronegativity Scale

Li 1.0	Be 1.5	B 2.0					H 2.2							C 2.5	N 3.0	O 3.5	F 4.0
Na 0.9	Mg 1.2	Al 1.5												Si 1.8	P 2.1	S 2.5	Cl 3.0
K 0.8	Ca 1.0	Sc 1.3	Ti 1.5	V 1.6	Cr 1.6	Mn 1.5	Fe 1.8	Co 1.8	Ni 1.8	Cu 1.9	Zn 1.6	Ga 1.6	Ge 1.8	As 2.0	Se 2.4	Br 2.8
Rb 0.8	Sr 1.0	Y 1.2	Zr 1.4	Nb 1.6	Mo 1.8	Tc 1.9	Ru 2.2	Rh 2.2	Pd 2.2	Ag 1.9	Cd 1.7	In 1.7	Sn 1.8	Sb 1.9	Te 2.1	I 2.5
Cs 0.7	Ba 0.9	La–Lu 1.1–1.2	Hf 1.3	Ta 1.5	W 1.7	Re 1.9	Os 2.2	Ir 2.2	Pt 2.2	Au 2.4	Hg 1.9	Tl 1.8	Pb 1.8	Bi 1.9	Po 2.0	At 2.2
Fr 0.7	Ra 0.9	Ac 1.1	Th 1.3	Pa 1.5	U 1.7	Np–No 1.3										

Reprinted from Linus Pauling, *The Nature of the Chemical Bond.* Copyright 1939 and 1940 by Cornell University. Third edition © 1960 by Cornell University. Used by permission of Cornell University Press.

the degree of ionic character in a bond can vary between zero (e.g., H_2) to essentially 100% depending on the electronegativity difference between the bonded atoms. There is no sharp dividing line between ionic and covalent bonding.

In the formation of the ionic bond between Li and F we saw that an electron was transferred from Li to F to produce Li^+ and F^-. With HCl, we saw that a polar covalent bond was formed in which an electron was only partially transferred from the H to the Cl atom. Very many chemical reactions are of this type, that is, they involve some transfer of electronic charge from one atom to another. Because this is such a common and important process, we define terms that apply specifically to these changes. These are

4.9
Oxidation and Reduction

  Oxidation – a loss of electrons
  Reduction – a gain of electrons

    Thus in the formation of LiF, Li undergoes oxidation by losing an electron and F undergoes reduction by acquiring an electron. In a similar fashion, when the HCl molecule is formed by the reaction of hydrogen and chlorine, the H atom, by losing some electronic charge to the Cl atom, is oxidized while the Cl atom becomes reduced.
    Lithium, in its reaction with fluorine, is said to be a **reducing agent** since it supplies the electron that the fluorine requires in order to be reduced – that is, it is the agent which has allowed reduction to occur. The fluorine, on the other hand, by accepting the electron from Li, permits oxidation to take place and thus is said to be an **oxidizing agent.** In a similar manner, we would consider hydrogen the reducing agent and chlorine the oxidizing agent when these two elements react to produce HCl. In general, *oxidizing agents acquire electrons and become reduced while reducing agents lose electrons and become oxidized.*
    Oxidation and reduction are always discussed together because, in any

reaction, whenever one substance loses electrons, another substance picks them up. We know this is true because we never observe electrons as a product of a chemical reaction, nor are electrons ever consumed when a chemical change occurs. Thus, oxidation is always accompanied by reduction.

Chemists have devised a bookkeeping system using what are called **oxidation numbers** to keep track of electrons during chemical reactions. An oxidation number can be defined as *the charge that an atom would have if both of the electrons in each bond were assigned to the more electronegative element*. The term **oxidation state** is also used, interchangeably, with the term oxidation number.

In the substance LiF, since an electron has, in fact, been transferred to the F atom, the oxidation number assigned to Li^+ is $+1$. The oxidation number of fluorine in the F^- ion is -1.

In the HCl molecule, since Cl is more electronegative than hydrogen, we assign an oxidation number of $+1$ to H and -1 to Cl, as if the electron pair were in the sole possession of the Cl atom.

In a nonpolar molecule such as H_2, where both atoms are the same and therefore have the same electronegativity, it is senseless to assign the electron pair to either atom since no electron transfer has occurred. In this case each H atom is assigned an oxidation number of zero.

To aid us in assigning oxidation numbers to the various atoms in a compound the following set of rules has been developed.

1. The oxidation number of any element in its elemental form is zero, regardless of the complexity of the molecule in which it occurs. Thus the atoms in Ne, F_2, P_4, and S_8 all have oxidation numbers of zero.
2. The oxidation number of any simple ion (one atom) is equal to the charge on the ion. The ions Na^+, Al^{3+}, and S^{2-} have oxidation numbers of $+1$, $+3$, and -2, respectively.
3. The sum of all of the oxidation numbers of all of the atoms in a neutral compound is zero. For a complex ion (more than one atom) the algebraic sum of the oxidation numbers must be equal to the ion's charge.

In addition to these basic rules, the rules below also prove useful.

4. Fluorine, in compounds, always has an oxidation number of -1.
5. The elements in Group IA (except hydrogen) always have an oxidation number of $+1$ in compounds.
6. The elements in Group IIA always have an oxidation number of $+2$ in compounds.
7. A Group VIIA element has an oxidation number of -1 in binary compounds with metals (compounds that contain only two different elements). For example, Cl has an oxidation number of -1 in $FeCl_2$, $CrCl_3$, and NaCl.
8. Oxygen usually has an oxidation number of -2, with three exceptions:
   (a) In binary compounds with fluorine where it must have a positive oxidation number.
   (b) In peroxides (which contain an O—O bond, e.g., O_2^{-2} and H_2O_2) where it has an oxidation number of -1.
   (c) In the superoxide ion, O_2^-, in which the oxidation number is $-1/2$.

9. Hydrogen has an oxidation number of +1 except in binary compounds with metals where it has an oxidation number of −1.

Let us look at some examples of how these rules are applied.

Example 4.3

What are the oxidation numbers of all of the atoms in KNO_3 (potassium nitrate)?

Solution

We know that the sum of the oxidation numbers of all of the atoms must be equal to zero (the charge on KNO_3 rule 3).

$$
\begin{array}{lll}
K & 1 \times (+1) = +1 & \\
N & 1 \times (x) \ \ \ = x & \\
O & 3 \times (-2) = -6 & \text{(rule 8)} \\
\hline
\textit{Sum of oxidation numbers} = \ \ 0 & \text{(rule 3)}
\end{array}
$$

x must equal +5 in order for the sum to be zero.

Example 4.4

What is the oxidation number of sulfur in $Na_2S_4O_6$ (sodium tetrathionate)?

Solution

Again, the sum of the oxidation numbers must be zero.

$$
\begin{array}{lll}
Na & 2 \times (+1) = & +2 \\
S & 4 \times (x) \ \ = & 4x \\
O & 6 \times (-2) = & -12 \\
\hline
& \text{Sum} \ \ \ \ \ = & 0 \\
\end{array}
$$

$$
4x = +10 \quad \text{or} \quad x = \frac{+10}{4}
$$

Note that the oxidation number of an atom need not be an integer.

Example 4.5

What is the oxidation number of Cr in the $Cr_2O_7{}^{2-}$ ion?

Solution

This time the sum of the oxidation numbers must equal −2 (rule 3).

$$
\begin{array}{lll}
Cr & 2 \times (x) \ \ = & 2x \\
O & 7 \times (-2) = & -14 \\
\hline
& \text{Sum} \ \ \ \ \ = & -2 \\
\end{array}
$$

$$
\text{Therefore,} \ 2x = +12
$$
$$
x = \ +6
$$

Up to now we have been using chemical names simply as labels. When one chemist speaks to another, he finds it necessary to convey information about chemical substances by giving them names. Ideally, these should transmit a maximum amount of information about the composition and structure of his compounds. Toward this end a systematic procedure for naming chemical compounds has been developed and is presented, in a somewhat abbreviated

4.10

The Naming of Chemical Compounds

form, in Appendix B. Some substances are so familiar to chemists that their common (or **trivial**) names are always used, for example, water (H_2O) and ammonia (NH_3). Trivial names are often used for extremely complex compounds where the names derived on a systematic basis are very long, complex, and cumbersome.

4.11
Other Binding Forces

The ionic and covalent bond represent very strong interactions. In addition to these bonds, there are other, weaker attractive forces that exist between neutral atoms and molecules. We shall now take a brief look at the origin of these forces, and in later chapters we shall see how they may be used to explain a number of physical properties.

DIPOLE INTERACTIONS. When two polar molecules approach one another, they tend to line up so that the positive end of one dipole is directed toward the negative end of the other. When this occurs, there is an electrostatic attraction between the two dipoles. This is a much weaker attraction than between oppositely charged ions for several reasons: first, there are only partial charges on the ends of the dipoles; second, because atoms and molecules are in constant motion,[3] collisions prevent the dipoles from becoming perfectly aligned; and third, there is a repulsive force between the end of the dipoles that carry like charges (see Figure 4.5). Dipole interactions are only about 1% as strong as covalent and ionic bonds.

HYDROGEN BONDING. A particularly strong dipole-dipole attraction occurs when hydrogen is covalently bound to a very small electronegative element such as fluorine, oxygen, or nitrogen. In these instances very polar molecules result in which the extremely small hydrogen atom carries a substantial positive charge. Because the positive end of this dipole can approach close to the negative end of a neighboring dipole, the force of attraction between the two is quite large (Figure 4.6). This special kind of dipole interaction is called a hydrogen bond and is about 5 to 10% as strong as an ordinary covalent bond.

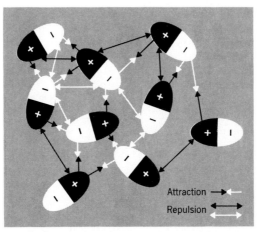

Attraction ▶—◀
Repulsion ◀——▶
◁——▷

Figure 4.5
Electrostatic interactions between dipoles.

[3] This is discussed more completely in Chapter 6.

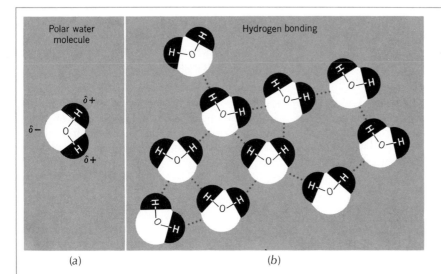

Polar water molecule	Hydrogen bonding
(a)	(b)

Figure 4.6

Hydrogen bonding in a substance such as H_2O. (a) Polar water molecule. (b) Hydrogen bonding.

Hydrogen bonds are one of the most important types of weak attractive forces. For instance, hydrogen bonds are responsible for water being a liquid at room temperature, rather than a gas. We shall also see (Chapter 21) that hydrogen bonds are responsible for the helical structures of protein molecules and the double helical structure of DNA, some of the primary building blocks of life as we know it.

VAN DER WAALS FORCES. Even uncombined atoms and nonpolar molecules experience weak attractions. These interactions, called van der Waals forces (after the Dutch physicist J. D. van der Waals) are considered to arise from the random motion of the electrons in an atom or molecule.

When electrons move about an atom, there is a chance that at some instant more electrons will be on one side of the nucleus than on the other and for this brief moment a dipole exists because of the imbalance in charge

Figure 4.7

Van der Waals forces. The instantaneous dipole in atom A induces a dipole in atom B.

(Figure 4.7). The positive end of the instantaneous dipole will attract electrons in a neighboring atom, thereby inducing another instantaneous dipole. These dipoles attract one another briefly before they disappear. While attractions between the instantaneous dipoles may be fairly strong, their duration is very short, and the average attraction over a period of time is generally very small.

The magnitude of van der Waals forces increases with increasing numbers of electrons and are present in all molecules, atoms, and ions. Their effect is generally felt, however, only when other forces are lacking or when there are many electrons present.

Review Questions and Problems

4.1 Write Lewis symbols for Se, Br, Al, K, Ba, Ge, and P.

4.2 Draw Lewis structures for the ionic compounds $AlCl_3$, BaO, and Na_2O.

4.3 Draw Lewis structures for the molecules PCl_3, SiH_4, BCl_3, H_2S, C_3H_8, and CO.

4.4 Which of the following contain coordinate covalent bonds? CCl_4, HNO_3 (see question 4.7), NO_3^-, BH_4^-, OF_2

4.5 Which of the following compounds do not obey the octet rule? ClF_3, OF_2, SF_4, SO_2, IF_7

4.6 Draw the resonance structures of SO_3; NO_3^- and CO_3^{2-}; SO_2 and NO_2^-.

4.7 Draw the resonance structures of HNO_3, SeO_2, SeO_3, and N_2O_4.

Structures: H—O—N , O—Se—O,

O—Se , and N—N

4.8 The C—C bond length in a series of compounds was found to be: compound 1, 1.54 Å; compound 2, 1.37 Å; compound 3, 1.46 Å; and compound 4, 1.40 Å. Arrange these in order of increasing C—C bond order. How would you expect the C—C bond energies to vary?

4.9 What would you expect the value of the bond order in SO_3 to be?

4.10 For the molecules SO_2 and SO_3, compare the SO bond energies, bond lengths, and S—O vibrational frequencies.

4.11 How would you expect the S—O bond distance in SO_2 to compare with that in SO_3?

4.12 Define polar, dipole, and dipole moment.

4.13 Arrange the following compounds in order of increasing ionic character of their bonds: SO_2, H_2S, SF_2, OF_2, ClF_3, and H_2Se.

4.14 Hydrogen is more electronegative than any of the Group IA elements. Based on this statement and the data in Table 4.6 show that electronegativities of the Group IA elements decrease from Li to Cs.

4.15 Use the data in Table 4.6 to show that the degree of ionic character decreases in the series LiF, LiCl, LiBr, and LiI. Does this agree with the conclusion you would draw based on the electronegativities in Table 4.8?

4.16 Define oxidation, reduction, oxidation state, oxidizing agent, and reducing agent.

4.17 Assign oxidation numbers to each atom in $KClO_2$, $BaMnO_4$, Fe_3O_4, O_2F_2, IF_5, $HOCl$, $CaSO_4$, $Cr_2(SO_4)_3$, $Na_2S_2O_8$ (contains one O—O bond), O_3, and Hg_2Cl_2.

4.18 Many biological processes involve oxidation and reduction. For example, ethyl alcohol (grain alcohol) is metabolized in a series of oxidation steps that involve the following carbon-containing compounds,

$$H-\overset{O}{\underset{H}{\overset{|}{\underset{|}{C}}}}-\overset{O}{\overset{||}{C}}-OH \longrightarrow CO_2$$

Follow the oxidation number of carbon as your body oxidizes the alcohol.

4.19 Name the following:

NaBr P_4O_6
CaO $AsCl_5$
$FeCl_3$ $Mn(HCO_3)_2$
$CuCO_3$ $NaMnO_4$
CBr_4 O_2F_2
(Refer to Appendix B)

4.20 Write chemical formulas for the following compounds.

Aluminum nitrate
Iron(II) sulfate
Ammonium dihydrogen phosphate
Iodine pentafluoride
Phosphorus(III) chloride
Dinitrogen tetroxide
Potassium permanganate
Magnesium hydroxide
Hydrogen selenide
Sodium hydride

4.21 Discuss the binding forces involved in (a) hydrogen bonding and (b) van der Waals' forces.

4.22 In Section 4.4 we noted that there was evidence that the C—N bond in the peptide linkage possessed some double-bond character. What kind of experimental evidence would be expected to confirm this?

5
Chemical Reactions in Aqueous Solution

As we might expect, in order for a chemical reaction to occur between two substances, the ions or molecules comprising the reactants must come into contact with one another. For this reason the speed at which a reaction takes place depends on how freely the reacting species are able to intermingle. For instance, if crystals of NaCl and AgNO$_3$ are mixed together, no noticeable chemical changes are observed. However, if the NaCl and AgNO$_3$ are first dissolved in water, and then their solutions mixed, a white solid, AgCl, is produced. Here, the formation of silver chloride requires that silver ions and chloride ions meet. When the two solids are mixed, this does not occur, except at the surfaces where the crystals touch one another. Because of the homogeneous nature of solutions, however, dissolved substances are intimately mixed at the molecular or ionic level, and chemical changes can occur rapidly. For this reason, chemists routinely use solutions to carry out chemical reactions.

Water is one of the most abundant chemicals in nature and serves as a good solvent for many substances, both ionic and molecular. The chemist's preoccupation with reactions in aqueous systems stems from the general availability of water as a solvent and, particularly in recent times, the importance of water as a medium in which biochemical reactions take place. In this chapter we shall discuss various types of chemical reactions that occur in aqueous solution and learn how the quantitative principles developed in Chapter 2 can be applied to such reactions.

5.1 Solution Terminology

There are certain terms that apply to all kinds of solutions and that should be understood before we proceed further. The words **solvent** and **solute** are two of these. The general practice is to refer to the substance present in greatest proportion in a solution as the solvent, with all of the other substances in the solution considered solutes. In solutions that contain water, however, the solvent is nearly always considered to be the water even when it is present in relatively small amounts. For example, a mixture of 96% H$_2$SO$_4$ and 4% H$_2$O by weight is called "concentrated sulfuric acid," which implies that a large quantity of sulfuric acid is *dissolved in* a small amount of water; that is, H$_2$O is taken to be the solvent and H$_2$SO$_4$ the solute.

It is often necessary to express the proportions of solute and solvent in a solution. This is done by specifying the concentration of the solute in the mixture. Concentration can be stated quantitatively in a variety of ways, as we shall see. The terms **concentrated** and **dilute** are used when we wish to speak, in qualitative terms, of the relative proportions of solvent and solute. In a concentrated solution there is a relatively large amount of solute present in the solvent; a dilute solution, on the other hand, possesses only a small quantity of solute. These two terms have meaning only in relationship to one another; they do not imply any specific quantities of solute in solvent. For example, concentrated sulfuric acid contains, as we said above, 96% H_2SO_4 and 4% H_2O. By comparison, a solution containing 20% H_2SO_4 would be dilute. This latter solution would be considered concentrated in comparison to a 5% H_2SO_4 solution.

In most cases there is a limit to the amount of solute that will dissolve in a fixed quantity of solvent at any particular temperature. For example, if we add sodium chloride to 100 ml of water at 0° C, only 35.7 g of the salt will dissolve, regardless of the total amount that we place into the water. A solution, such as this, which contains as much dissolved solute as it can hold, is said to be **saturated.** The **solubility** of the solute is taken to be the amount required to achieve a saturated solution. Thus the solubility of NaCl in water at 0° C is 35.7 g/100 ml; at 100° C its solubility is 39.1 g/100 ml. Solutions that contain less solute than required for saturation are said to be **unsaturated.**

The terms saturated and unsaturated are in no way directly related to the terms concentrated and dilute. A saturated solution of silver chloride, for example, contains only 0.000089 g AgCl per 100 ml of water and is certainly considered dilute. On the other hand, it would take 500 g of lithium chlorate, $LiClO_3$, to form a saturated solution in 100 ml of water. A solution containing 400 g of $LiClO_3$ in 100 ml of water is unsaturated but, nevertheless, is quite concentrated. Thus a saturated solution can be dilute and an unsaturated solution, concentrated.

Finally, there are some substances, such as sodium acetate, which frequently form **supersaturated** solutions — solutions that contain more solute than ordinarily required for saturation. Sodium acetate is soluble to the extent of 119 g/100 ml of water at 0° C, and becomes more soluble at higher temperatures. If an unsaturated hot solution containing more than 119 g of sodium acetate per 100 ml is cooled slowly to 0° C, the excess solute remains dissolved and the solution becomes supersaturated. Such solutions are un-

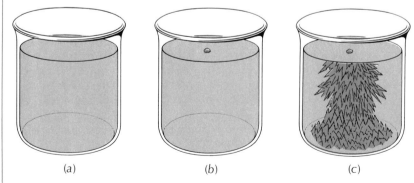

(a)   (b)   (c)

Figure 5.1

Supersaturation. (a) Supersaturated solution. (b) Introduction of a seed crystal. (c) Excess solute crystallizes upon the seed.

stable; if a small crystal of the solute is added, additional solute crystallizes upon this "seed" crystal until the concentration drops to the point of saturation (see Figure 5.1)

5.2
Electrolytes

As mentioned in the introduction to this chapter, water is generally a good solvent for ionic compounds. In the solid state these substances are composed of positive and negative ions held together in a rigid framework by electrostatic forces. When they dissolve in water they break apart, or **dissociate,** to produce ions that are more or less free to roam about in the solution. The presence of ions imparts to the water the ability to conduct electricity, which we can demonstrate using an apparatus such as that shown in Figure 5.2. When electrical contact is made across the two electrodes, an electric current can flow and the light bulb will light. If we immerse these electrodes in pure water, no conductivity is observed (i.e., the bulb will not light) because water is a very poor conductor of electricity. However, if we add NaCl, a typical ionic solid, to the water, the bulb will begin to burn brightly as soon as the NaCl begins to dissolve. Substances, like NaCl, which dissociate in solution to produce ions and, therefore, lead to solutions that conduct electricity, are called **electrolytes.**

The dissociation of NaCl that occurs when the solid is dissolved can be represented by the equation

$$NaCl(s) \longrightarrow Na^+(aq) + Cl^-(aq)$$

where the symbols s and aq in parentheses denote that the solid (s) produces ions in aqueous (aq) solution. As the solid dissolves, the ions become surrounded by water molecules and are said to be **hydrated.** We shall examine the solution process in more detail in Chapter 9.

Electrodes

Figure 5.2
Conductivity apparatus.

The production of ions in solution is not limited to ionic compounds. There are many covalent substances which, when they dissolve in water, react with the solvent to produce ions and therefore yield conducting solutions. Hydrogen chloride is a typical example. When HCl gas is dissolved in water the following reaction takes place:

$$HCl + H_2O \longrightarrow H_3O^+(aq) + Cl^-(aq)$$

This reaction occurs by the transfer of a proton from the HCl molecule to the water molecule to produce a **hydronium ion,** H_3O^+, and a chloride ion.

$$H\!:\!\overset{\cdot\cdot}{\underset{\cdot\cdot}{Cl}}\!: + :\!\overset{\cdot\cdot}{O}\!:\!H \longrightarrow \left[H\!:\!\overset{\cdot\cdot}{O}\!:\!H\right]^+ + \left[:\!\overset{\cdot\cdot}{\underset{\cdot\cdot}{Cl}}\!:\right]^-$$
$$\qquad\qquad\underset{H}{} \qquad\qquad \underset{H}{}$$

Thus, even though hydrogen chloride exists by itself as discrete molecules of HCl (liquid HCl does not conduct electricity), when it dissolves in water it produces ions and becomes an electrolyte.

The hydronium ion might be thought of as a proton (H^+) that has associated itself with a water molecule (we can think of the hydronium ion as a *hydrated proton*), and we will see shortly that when the hydronium ion reacts it loses a proton to regenerate the molecule of H_2O. In chemical reactions, therefore, the H_2O of the hydronium ion merely serves as a carrier for the H^+ ion and for that reason the H_3O^+ ion is very often written as H^+. Thus, the dissociation that occurs whe HCl is dissolved in water is often written as simply

$$HCl \longrightarrow H^+ + Cl^-$$

Even though we write H^+ always remember that there is at least one, and probably several more, H_2O molecules associated with the proton in aqueous solutions.[1]

The two examples of electrolytes that we have discussed above, NaCl and HCl, are essentially completely dissociated in aqueous solution, that is, one mole of NaCl gives one mole of Na^+ and one mole of Cl^-. Substances such as NaCl and HCl that, for all practical purposes are completely dissociated in aqueous solution, are said to be **strong electrolytes.** There are many compounds, such as acetic acid ($HC_2H_3O_2$, found in vinegar) for example, that dissociate to only a limited extent in water; that is, only a small fraction of all of the acetic acid molecules present in solution actually react with the solvent,

$$HC_2H_3O_2 + H_2O \longrightarrow H_3O^+ + C_2H_3O_2^-$$

Such substances are called **weak electrolytes** because their solutions contain relatively few ions. Additional examples are given in Table 5.1. In the table we see that water itself, by virtue of the reaction

$$H_2O + H_2O \dashrightarrow H_3O^+ + OH^-$$

is a *very* weak electrolyte. This slight dissociation of water plays a very important role in many chemical reactions in which water is the solvent. Special attention will be given to this topic in Chapter 13.

[1] There is, in fact, evidence that suggests that the H^+ ion may exist as $H_9O_4^+$ that is, $H_3O(H_2O)_3^+$, in aqueous solution.

Table 5.1
Some Weak Electrolytes

Substance	Dissociation Reaction	Percent Dissociation of Solution Containing 1.00 mole of Solute per Liter of Solution
Water	$H_2O + H_2O \longrightarrow H_3O^+ + OH^-$	1.8×10^{-7} (55.5 moles of H_2O per liter)
Acetic acid	$HC_2H_3O_2 + H_2O \longrightarrow H_3O^+ + C_2H_3O_2^-$	0.42
Ammonia	$NH_3 + H_2O \longrightarrow NH_4^+ + OH^-$	0.42
Hydrogen cyanide	$HCN + H_2O \longrightarrow H_3O^+ + CN^-$	2.0×10^{-3}
Mercury(II) chloride	$HgCl_2 \longrightarrow HgCl^+ + Cl^-$	1
Cadmium sulfate	$CdSO_4 \longrightarrow Cd^{2+} + SO_4^{2-}$	7

Finally, there are many molecular compounds that do not dissociate into ions at all when they are dissolved in water. Sugar and ethyl alcohol are two common examples. These are called **nonelectrolytes.** Since solutions of nonelectrolytes contain no ions, they do not conduct an electric current.

5.3
Chemical
Equilibrium

The reason for the limited degree of dissociation of weak electrolytes is worth discussing at this point because it illustrates one of the most important concepts in chemistry, one which we shall devote two chapters to at a later time (Chapters 12 and 14).

In a solution of acetic acid, molecules of $HC_2H_3O_2$ are constantly colliding with molecules of water and, in each encounter, there is a certain probability that a proton will be transferred from an $HC_2H_3O_2$ molecule to a water molecule to yield H_3O^+ and $C_2H_3O_2^-$ ions. There are also encounters, in this solution, between acetate ions and hydronium ions. When these ions meet, there is a high probability that an H_3O^+ ion will lose a proton to a $C_2H_3O_2^-$ ion to reform $HC_2H_3O_2$ and H_2O molecules. Thus, in this solution we have two reactions occurring simultaneously.

$$HC_2H_3O_2 + H_2O \longrightarrow H_3O^+ + C_2H_3O_2^-$$

and

$$H_3O^+ + C_2H_3O_2^- \longrightarrow HC_2H_3O_2 + H_2O$$

When the rate of production of the ions is equal to the rate at which they are disappearing to reform acetic acid and water molecules, the concentrations of all of the species in the solution will not change with time. Such a state of affairs is called **equilibrium.** It is said to be a dynamic equilibrium in that things are continually happening in the solution; two reactions are taking place, ions reacting to yield molecules and molecules reacting to produce ions.

To indicate chemical equilibrium in a reacting system, we use a double arrow in the chemical equation. Thus, the equilibrium that we have been discussing is expressed as

$$HC_2H_3O_2 + H_2O \rightleftharpoons H_3O^+ + C_2H_3O_2^-$$

The use of this notation implies that the forward reaction (as read from left to right) is occurring at the same rate as the reverse reaction (the reaction from right to left). In a solution of acetic acid this condition is met when only a small fraction of all of the acetic acid molecules has undergone dissociation and hence further dissociation does not occur. For strong electrolytes, the reverse reaction, that is, the production of the molecular form, has very little tendency to occur. The **position of equilibrium,** therefore, lies almost completely toward the ions, and the strong electrolyte is essentially fully dissociated in the solution.

Most of the chemical reactions encountered in the laboratory portion of an introductory chemistry course involve reactions between ions in solution. In fact, any chemist who uses water as a solvent eventually encounters such reactions. What we wish to consider now is what occurs when two solutions, each containing an electrolyte, are mixed.

**5.4
Ionic Reactions**

We shall begin by considering the reaction that takes place when a solution containing one mole of sodium chloride is added to a solution of one mole of silver nitrate. When these two are combined, one mole of the white solid, silver chloride, is formed (we refer to a solid that is formed in a solution as the result of chemical reaction as a **precipitate**) and the solution that remains contains one mole of sodium nitrate. If we wished, we could separate the AgCl from the solution by filtering the mixture. When the clear solution is evaporated, sodium nitrate remains behind.

The chemical equation that represents the change that has occurred during this reaction is

$$NaCl(aq) + AgNO_3(aq) \longrightarrow AgCl(s) + NaNO_3(aq)$$

This kind of reaction, in which cations and anions have changed partners, is known as **metathesis.** The equation that we have written to describe the reaction is called a **molecular equation,** because all of the reactants and products are written as if they were molecules. A more accurate representation of the reaction as it actually occurs in solution is given by the ionic equation. We know that a solution of NaCl does not contain molecules but, instead, consists of Na^+ and Cl^- ions dispersed throughout the solvent. Similarly, a silver nitrate solution contains Ag^+ and NO_3^- ions. When these two solutions are mixed, solid AgCl is formed by a combination of the Ag^+ and Cl^- ions. The solution of sodium nitrate that remains contains Na^+ and NO_3^- ions and we can write the **ionic equation** as

$$Na^+(aq) + Cl^-(aq) + Ag^+(aq) + NO_3^-(aq) \longrightarrow$$
$$AgCl(s) + Na^+(aq) + NO_3^-(aq)$$

In this equation we have shown all of the soluble ionic substances as being dissociated in solution.

If we examine the ionic equation that we have just written, we see that Na^+ and NO_3^- do not actually undergo any change during the course of the reaction. The same Na^+ and NO_3^- ions are present after the chemical reaction as before and they have, in a sense, just "gone along for the ride." For this reason, ions that do not change during reaction are frequently called **spectator ions.** Since they do not take part in the reaction, we eliminate them from the equation to arrive at the **net ionic equation,** that is, the equation for the net change that has taken place:

$$Ag^+(aq) + Cl^-(aq) \longrightarrow AgCl(s)$$

This net ionic equation is useful in more than one way. First, it focuses our attention on the species that participate in the important changes that are occurring in the solution. Second, it tells us that any substance which produces Ag^+ ions in solution will react with any other substance that gives Cl^- ions in solution to yield a precipitate of AgCl. Thus, after mixing together solutions of silver fluoride (AgF is soluble in water even though AgCl is not[2]) and potassium chloride, we would predict that a precipitate of AgCl would be formed. This is, in fact, precisely what takes place. The molecular, ionic, and net ionic equations for this reaction are, respectively,

$$AgF + KCl \longrightarrow AgCl(s) + KF$$
$$Ag^+ + F^- + K^+ + Cl^- \longrightarrow AgCl(s) + K^+ + F^-$$

and

$$Ag^+ + Cl^- \longrightarrow AgCl(s)$$

In the two metathesis reactions that we have considered, the driving force was provided by the formation of an insoluble precipitate of AgCl. If silver chloride were soluble in water, no chemical reaction of this type would occur. For instance, if solutions of KCl and $NaNO_3$ are combined, we might be tempted to write the equation,

$$KCl + NaNO_3 \longrightarrow KNO_3 + NaCl$$

However, both of these "would be" products are soluble and therefore totally dissociated into ions.[3] If we write this equation in ionic form we have

$$K^+ + Cl^- + Na^+ + NO_3^- \longrightarrow K^+ + NO_3^- + Na^+ + Cl^-$$

When we compare the left and right sides of this equation, we see that they are identical except for the sequence in which we have written the ions. If we cross out all spectator ions, there is nothing left; that is, there is no net chemical change when solutions of KCl and $NaNO_3$ are mixed.

We have seen now that when a precipitate is formed on mixing two solutions of electrolytes a net chemical change takes place. In principle, by knowing the solubilities of all of the compounds that can be formed between pairs of cations and anions, we could predict, based upon the formation of a precipitate, when chemical reactions would occur. This is made complicated because there is no sharp distinction between soluble and insoluble com-

---

[2] The solubility of AgCl is very low, 0.000089 g/100 ml, and AgCl can be considered for most purposes, to be insoluble; that is, the amount of AgCl in solution can generally be considered to be negligible.

[3] In almost all instances, ionic compounds (salts) undergo virtually complete dissociation when they dissolve in water.

pounds. Substances such as lithium chlorate or sodium chloride both of which were mentioned in Section 5.1, are certainly considered soluble. Silver chloride, on the other hand, would undoubtedly be termed insoluble. Some salts, such as $PbCl_2$ and $AgC_2H_3O_2$, are of intermediate solubility and are referred to as partially soluble, or slightly soluble compounds. Whether or not a precipitate of a particular salt will form upon mixing solutions of reactants depends on whether the concentration of the ions that comprise the salt exceeds that required to achieve a saturated solution of the salt. If dilute solutions of potential reactants are mixed, this might not occur and no precipitate would be formed in the reaction mixture. Generally, a compound is considered to be insoluble when the combination of even very dilute solutions of its constituent ions leads to the formation of a precipitate.

This entire discussion has been on a qualitative level. A quantitative treatment of the solubilities of ionic solids is reserved until Chapter 14. For now, we might use the following solubility rules as a rough guide in predicting the course of metathesis reactions.

1. All salts of the alkali metals are soluble.
2. All ammonium salts are soluble.
3. All salts containing the anions, NO_3^-, ClO_3^-, ClO_4^-, and $C_2H_3O_2^-$ are soluble ($AgC_2H_3O_2$ and $KClO_4$, however, are slightly soluble).
4. All chlorides, bromides, and iodides are soluble *except* those of Ag^+, Pb^{2+}, and Hg^{2+} (note that mercury in the $+1$ oxidation state exists as the ion Hg_2^{2+}). $PbCl_2$, you will recall, is slightly soluble.
5. All metal oxides *except* those of the alkali metals and Ca^{2+}, Sr^{2+}, and Ba^{2+} are *insoluble*. Metal oxides, when they dissolve, react with the solvent to form hydroxides; for example,

$$CaO + H_2O \longrightarrow Ca^{2+} + 2OH^-$$

6. All hydroxides are *insoluble* except those of the alkali metals, Ba^{2+} and Sr^{2+}. $Ca(OH)_2$ is slightly soluble.
7. All sulfates are *soluble* except those of Pb^{2+}, Sr^{2+}, and Ba^{2+}. The sulfates of Ca^{2+} and Ag^+ are slightly soluble.
8. All carbonates, phosphates, sulfides, and sulfites are *insoluble* except those of NH_4^+ and the alkali metals.

Example 5.1

Would a chemical reaction occur if solutions of $FeCl_3$ and KOH were mixed? If so, give the net ionic equation.

Solution

A metathesis reaction between these two salts would be represented by the equation

$$FeCl_3 + 3KOH \longrightarrow 3KCl + Fe(OH)_3$$

On the basis of our solubility rules, KCl is soluble but $Fe(OH)_3$ is not. Therefore, we would expect a precipitate of $Fe(OH)_3$ to form in the mixture and the KCl would remain dissociated in solution. The ionic equation for the reaction, then, is

$$Fe^{3+} + 3Cl^- + 3K^+ + 3OH^- \longrightarrow Fe(OH)_3 + 3K^+ + 3Cl^-$$

Cancelling spectator ions (K^+ and Cl^-) we have the net ionic equation

$$Fe^{3+} + 3OH^- \longrightarrow Fe(OH)_3$$

For ionic reactions that do not involve oxidation-reduction there is another factor, in addition to the formation of a precipitate, that can lead to a net chemical change. When HCl is added to a solution of $NaC_2H_3O_2$, the following reaction takes place:

$$H^+ + C_2H_3O_2^- \longrightarrow HC_2H_3O_2$$

Here the product of the reaction is the slightly dissociated weak electrolyte, acetic acid. $HC_2H_3O_2$ has very little tendency to dissociate; and when H^+ and $C_2H_3O_2^-$ ions are mixed, the molecular compound is formed. A net change has occurred because the particles present in the reaction mixture after the reaction is over are not the same as before it began. Thus the formation of a slightly dissociated product from ionic reactants leads to a net chemical change.

Frequently, the molecular species formed in a reaction can escape as a gas, either directly, or by decomposing to produce a gaseous product. For example, when HCl is added to Na_2S, the reaction that occurs is

$$2HCl + Na_2S \longrightarrow H_2S + 2NaCl$$

for which the net ionic equation is

$$2H^+ + S^{2-} \longrightarrow H_2S$$

Hydrogen sulfide gas has a limited solubility in water and when that solubility is exceeded, the H_2S bubbles out of the solution.

Another example is the reaction between HCl and Na_2CO_3,

$$2HCl + Na_2CO_3 \longrightarrow H_2CO_3 + 2NaCl$$

for which the net ionic equation is

$$2H^+ + CO_3^{2-} \longrightarrow H_2CO_3$$

Carbonic acid, H_2CO_3, is not stable and decomposes readily to produce CO_2 and H_2O,

$$H_2CO_3 \longrightarrow H_2O + CO_2(g)$$

Therefore, we can write the net ionic equation for the reaction between hydrogen ion and carbonate ion as

$$2H^+ + CO_3^{2-} \longrightarrow H_2O + CO_2(g)$$

A list of some other substances that are gases, having a limited solubility in water, or that decompose to produce gaseous products is given in Table 5.2.

5.5
Acids and Bases in Aqueous Solution

The concept of acids and bases is, without a doubt, one of the most important and useful in all of chemistry. In fact, nearly all chemical reactions can be broadly classified as either reactions between acids and bases, or as reactions involving oxidation and reduction. Because the acid-base concept is so im-

Table 5.2
Gases Partially Soluble in Water

Gas	Typical Reaction in Which It Is Produced
CO_2	$Na_2CO_3 + 2HCl \longrightarrow H_2CO_3 + 2NaCl$ $H_2CO_3 \longrightarrow H_2O + CO_2(g)$ Net Equation: $CO_3^{2-} + 2H^+ \longrightarrow H_2O + CO_2(g)$
SO_2	$Na_2SO_3 + 2HCl \longrightarrow H_2SO_3 + 2NaCl$ $H_2SO_3 \longrightarrow H_2O + SO_2(g)$ Net Equation: $SO_3^{2-} + 2H^+ \longrightarrow H_2O + SO_2(g)$
NH_3	$NH_4Cl + NaOH \longrightarrow NH_3(g) + H_2O + NaCl$ Net Equation: $NH_4^+ + OH^- \longrightarrow NH_3(g) + H_2O$
H_2S	$Na_2S + 2HCl \longrightarrow H_2S(g) + 2NaCl$ Net Equation: $S^{2-} + 2H^+ \longrightarrow H_2S(g)$
NO NO_2	$NaNO_2 + HCl \longrightarrow HNO_2 + NaCl$ $2HNO_2 \longrightarrow H_2O + NO_2(g) + NO(g)$ Net Equation: $2NO_2^- + 2H^+ \longrightarrow H_2O + NO_2(g) + NO(g)$

portant, an entire chapter (Chapter 13) is devoted to a detailed discussion of acid-base behavior. For now, we will limit ourselves to a simple definition of acids and bases, adequate for the treatment of reactions in aqueous solution.

We shall define an **acid** to be any substance that increases the concentration of hydronium ion in solution. Thus, HCl is an acid because when it is dissolved in water, it reacts with the solvent to produce H_3O^+.

$$HCl + H_2O \longrightarrow H_3O^+ + Cl^-$$

Aqueous solutions of hydrogen chloride are called hydrochloric acid.

Carbon dioxide is also an acid because its aqueous solutions contain more H_3O^+ than does pure water. In the last section we described the decomposition of H_2CO_3 that occurs when it is produced in large quantities as the result of a chemical reaction. The reverse reaction occurs to a limited extent when CO_2 is dissolved in water.

$$CO_2 + H_2O \longrightarrow H_2CO_3$$

Carbonic acid is able to dissociate slightly (actually by further reaction with water).[4]

$$H_2CO_3 + H_2O \longrightarrow H_3O^+ + HCO_3^-$$

We see that acids do not necessarily have to contain hydrogen; HCl does but CO_2 does not. The only requirement is that solutions of these substances contain more H_3O^+ than is found in pure H_2O. Table 5.3 contains a list of common acids and the reactions that they undergo to produce acidic solutions. You might note that nonmetal oxides, when dissolved in water, produce acids which dissociate by further reaction with the solvent. This pro-

[4] Carbonic acid, being a weak electrolyte, is also called a **weak acid**.

Table 5.3
Some Acids and Bases

Acids (Those followed by an asterisk are weak electrolytes and exist primarily as undissociated molecules in aqueous solution.)

Monoprotic Acids: $HX \longrightarrow H^+ + X^-$	HF HCl HBr HI HOCl $HClO_2$	Hydrofluoric acid (*) Hydrochloric acid Hydrobromic acid Hydriodic acid Hypochlorous acid (*) Chlorous acid (*)	$HClO_3$ $HClO_4$ HIO_4 HNO_3 HNO_2 $HC_2H_3O_2$	Chloric acid Perchloric acid Periodic acid Nitric acid Nitrous acid (*) Acetic acid (*)
Diprotic Acids: $H_2X \longrightarrow H^+ + HX^-$ $HX^- \longrightarrow H^+ + X^{2-}$	H_2SO_4 H_2SO_3 H_2CO_3 $H_2C_2O_4$	Sulfuric acid^a Sulfurous acid (*) Carbonic acid (*) Oxalic acid (*)	H_2S H_3PO_3	Hydrosulfuric acid (*) Phosphorous acid (only two hydrogens can be removed as protons) (*)
Triprotic Acids: $H_3X \longrightarrow H^+ + H_2X^-$ $H_2X^- \longrightarrow H^+ + HX^{2-}$ $HX^{2-} \longrightarrow H^+ + X^{3-}$	H_3PO_4 H_3BO_3	Orthophosphoric acid (*) Boric acid (*)		
Typical Acidic Oxides (Nonmetal Oxides)	SO_2 SO_3 N_2O_3 N_2O_5 P_4O_6 P_4O_{10}	$SO_2 + H_2O \longrightarrow H_2SO_3$ $SO_3 + H_2O \longrightarrow H_2SO_4$ $N_2O_3 + H_2O \longrightarrow 2HNO_2$ $N_2O_5 + H_2O \longrightarrow 2HNO_3$ $P_4O_6 + 6H_2O \longrightarrow 4H_3PO_3$ $P_4O_{10} + 6H_2O \longrightarrow 4H_3PO_4$		

Bases

Molecular Bases	NH_3 N_2H_4 NH_2OH	Ammonia^b Hydrazine Hydroxylamine	$(NH_3 + H_2O \rightleftharpoons NH_4^+ + OH^-)$ $(N_2H_4 + H_2O \rightleftharpoons N_2H_5^+ + OH^-)$ $(NH_2OH + H_2O \rightleftharpoons NH_3OH^+ + OH^-)$
Ionic Bases	Metal hydroxides $\qquad M(OH)_n \longrightarrow M^{n+} + nOH^-$ (see solubility rules—Section 5.4)		
Typical Basic Oxides (metal oxides)	Na_2O K_2O CaO SrO BaO	$M_2O + H_2O \longrightarrow 2MOH$ $MO + H_2O \longrightarrow M(OH)_2$	

^{a} The second dissociation of H_2SO_4 is that of a weak acid.
^{b} Aqueous solutions of ammonia are sometimes referred to as ammonium hydroxide. Commercially prepared solutions of NH_3, for example, are labeled in this way.

vides us with one chemical distinction between metals and nonmetals (we will shortly see that soluble metal oxides react with water to yield bases).

A **base** will be defined as any substance that increases the concentration of hydroxide ion in aqueous solutions. Sodium hydroxide, an example of an ionic compound containing metal ions and hydroxide ions, is a base because when it dissolves in water it dissociates to yield Na^+ and OH^- ions.

$$NaOH \longrightarrow Na^+ + OH^-$$

Ammonia, NH_3, is also a base because it reacts with water to produce hydroxide ions.

$$NH_3 + H_2O \longrightarrow NH_4^+ + OH^-$$

In this case, a proton is transferred from a water molecule to the ammonia molecule.

Metal oxides, when they are dissolved in water, produce basic solutions. We saw, for example, that when CaO is treated with water, it undergoes reaction to yield calcium hydroxide.

$$CaO + H_2O \longrightarrow Ca(OH)_2$$

When $Ca(OH)_2$ dissolves, it dissociates to Ca^{2+} and OH^-.

$$Ca(OH)_2 \longrightarrow Ca^{2+} + 2OH^-$$

With NH_3 and CaO, we see that a base does not necessarily have to contain hydroxide ions in order to produce a basic solution. Furthermore, the presence of the OH unit in a compound does not guarantee that the substance will behave as a base. For example, the compounds, sodium hydroxide, sulfuric acid, and methyl alcohol can be represented by the following structures and formulas:

Each of these contains an OH group; however, only the first compound is basic. Sulfuric acid dissociates by the breakage of O—H bonds and the transfer of protons to water molecules while methyl alcohol does not dissociate at all in water. Therefore, we must exercise care in drawing conclusions about the acid-base behavior of compounds based on their chemical formulas or structures.

Acids and bases, in general, exhibit certain properties that lead us to characterize them as acids or bases. One of these properties is that they react with one another in a process called **neutralization.** In aqueous solution the neutralization reaction takes the form of the net ionic equation,

$$H_3O^+ + OH^- \longrightarrow 2H_2O$$

As mentioned previously, the hydronium ion can be considered a hydrated proton; it is therefore common practice to leave out the water molecule that is attached to the proton and simply to write the equation

$$H^+ + OH^- \longrightarrow H_2O$$

A typical reaction between an acid and a base occurs between hydrochloric acid and sodium hydroxide.

$$HCl + NaOH \longrightarrow NaCl + H_2O$$

We see that the products of the neutralization reaction are a salt and water. The hydrochloric acid used in this reaction is said to be a **monoprotic acid** because one mole of HCl produces only one mole of H^+ ions in solution. There are also many acids that are capable of furnishing more than one mole of H^+ per mole of acid and, as a group, are called **polyprotic acids.** Examples are H_2SO_4 and H_3PO_4. These substances dissociate in a series of steps, each providing one proton. Sulfuric acid, for example, dissociates as follows:

$$H_2SO_4 \longrightarrow H^+ + HSO_4^-$$
$$HSO_4^- \longrightarrow H^+ + SO_4^{2-}$$

In the neutralization of a diprotic acid, such as H_2SO_4, it is possible, by controlling the quantity of base available for the reaction, to form either salts in which only one of the available protons has been neutralized or salts that are the product of complete neutralization. For example, when two moles of NaOH are reacted with one mole of H_2SO_4, two moles of H_2O and one mole of Na_2SO_4 are produced.

$$2NaOH + H_2SO_4 \longrightarrow 2H_2O + Na_2SO_4$$

However, if only one mole of NaOH is available to react with one mole of H_2SO_4, the products are one mole of water and one mole of $NaHSO_4$ (called sodium bisulfate or sodium hydrogen sulfate).

$$NaOH + H_2SO_4 \longrightarrow H_2O + NaHSO_4$$

Similar reactions can occur with a triprotic acid such as H_3PO_4, and salts such as the following can be obtained.

Na_3PO_4 Trisodium phosphate (sodium phosphate)
Na_2HPO_4 Disodium hydrogen phosphate
NaH_2PO_4 Sodium dihydrogen phosphate

Salts that are the product of the partial neutralization of a polyprotic acid are called acid salts because they can furnish protons for the neutralization of additional base. Thus $NaHSO_4$ can react with NaOH in a neutralization reaction.

$$NaHSO_4 + NaOH \longrightarrow Na_2SO_4 + H_2O$$

5.6
The Preparation of Inorganic Salts by Metathesis Reactions

In the past two sections we have described typical metathesis reactions that occur in aqueous solutions. These reactions can often be of use to us in preparing inorganic salts. For example, suppose we were faced with the problem of preparing relatively pure NaBr using other inorganic compounds as starting materials. How would we proceed? We cannot achieve success if we mix together solutions of NaCl and KBr because, although the mixture would contain the constituents of NaBr, all of the ions, Na^+, K^+, Cl^-, and Br^-, are in solution together, and it would be nearly impossible to isolate pure NaBr from the mixture. What we are looking for are reactions in which one product is easily separated from the other, a condition that is met if one of the products is either a precipitate, a gas, or the solvent, water.

PRECIPITATION REACTIONS. These reactions are particularly useful when the desired product is insoluble in water. For instance, if we wish to prepare silver chloride, we need only mix a solution containing a soluble silver salt, such as $AgNO_3$, with one containing a soluble chloride, such as KCl. The reaction

$$Ag^+ + Cl^- \longrightarrow AgCl(s)$$

gives us our product as a solid that can be separated from the reaction mixture by filtration, while the other product, KNO_3, remains in solution. In this preparation it is not necessary to worry too much about stoichiometry since, if we want to prepare one mole of AgCl, we can add one mole of $AgNO_3$ to a solution containing at least one mole of KCl. Any excess KCl beyond that required to react completely with the $AgNO_3$ remains dissolved and does not contaminate the AgCl precipitate.

When the desired product remains in solution, greater attention must be paid to the stoichiometry of the reaction. For example, to prepare NaBr (which is water soluble) by a precipitation reaction we might mix solutions of K_2SO_4 and $BaBr_2$, since $BaSO_4$ is insoluble. The net reaction is

$$Ba^{2+} + SO_4^{2-} \longrightarrow BaSO_4(s)$$

while the molecular equation is

$$Na_2SO_4 + BaBr_2 \longrightarrow BaSO_4 + 2NaBr$$

To prepare pure NaBr it is necessary to react equal numbers of moles of Na_2SO_4 and $BaBr_2$. The solid $BaSO_4$ could then be filtered from the mixture and the volume of the resulting solution could then be reduced by evaporation to the point where nearly all of the desired product crystallizes.

There are some instances where use can be made of relative degrees of insolubility in synthetic procedures. For instance, the silver halides decrease in solubility in this order: AgCl, AgBr, and AgI. If a solution containing KBr is stirred with solid AgCl, the solid is converted to the less soluble AgBr, and Cl^- replaces Br^- in solution. The net effect is that AgCl is converted to AgBr.

NEUTRALIZATION REACTIONS. Another useful method of preparing salts is by the neutralization of an acid with a base. For example, we could prepare our NaBr by the reaction of NaOH with HBr,

$$NaOH + HBr \longrightarrow H_2O + NaBr$$

If we are careful to react precisely one mole of NaOH with one mole of HBr, the only product in the aqueous solution is NaBr.

Metal oxides can be used in place of hydroxides in similar reactions. For example, the reaction

$$ZnO + 2HClO_4 \longrightarrow Zn(ClO_4)_2 + H_2O$$

could be used to prepare zinc perchlorate.

REACTIONS IN WHICH ONE PRODUCT IS A GAS. A very convenient way to prepare inorganic salts is provided by the reaction between an acid and a metal carbonate. Recall that the carbonate ion reacts with hydrogen ion to produce carbon dioxide and water,

Chemical Reactions in Aqueous Solution

$$CO_3^{2-} + 2H^+ \longrightarrow H_2O + CO_2(g)$$

This reaction proceeds readily regardless of whether the source of carbonate ion is a soluble or insoluble salt.

The fact that most metal carbonates are insoluble leads to a very simple method for the preparation of pure salts. The synthesis of copper(II) bromide, for example, can be accomplished by the addition of HBr to a suspension of the insoluble $CuCO_3$ until nearly all of the starting material reacts according to the equation

$$CuCO_3(s) + 2HBr(aq) \longrightarrow CuBr_2(aq) + H_2O + CO_2(g)$$

When the mixture is filtered to remove the undissolved $CuCO_3$, the solution that remains contains $CuBr_2$, which is virtually free of contaminants. The product can then be recovered by evaporation of the water.

The preparation of salts from soluble carbonates is also feasible; however, once again attention must be paid to the stoichiometry of the reaction to obtain pure products. For example, NaBr can be made from Na_2CO_3 and HBr,

$$Na_2CO_3(aq) + 2HBr(aq) \longrightarrow 2NaBr(aq) + H_2O + CO_2(g)$$

The quantity of *pure* NaBr that can be crystallized from the reaction mixture depends on how closely a 1:2 mole ratio is maintained between the reactants Na_2CO_3 and HBr.

5.7 Oxidation-Reduction Reactions

Oxidation-reduction processes (often referred to as **redox** reactions) form a very important class of chemical reactions. They take place between many inorganic and organic compounds and, as implied in the last chapter, they are extremely important in biochemical systems where they provide the mechanism for energy transfer in living organisms.

As a rule, the stoichiometry of redox reactions tends to be more complicated than for reactions that do not involve electron transfer. As a result, chemical equations for oxidation-reduction reactions are often complex and are difficult to balance by inspection. Fortunately, there are methods that can be applied to aid in balancing these equations.

OXIDATION-NUMBER-CHANGE METHOD. In Chapter 4 it was noted that during redox reactions electrons are never produced as a product, nor are they necessary as a reactant. When a chemical reaction involves oxidation-reduction, the total number of electrons lost in the oxidation process must equal the total number gained during reduction. We can use this fact to help us balance equations for this type of reaction. Here is one procedure that can be used, called the **oxidation-number-change** method.

1. Assign oxidation numbers to each atom in the equation (we shall write them directly below the symbol for the element to avoid confusing oxidation numbers and charges). For example,

$$\begin{array}{ccccccc} HCl + & K_2Cr_2O_7 & \longrightarrow & KCl + & Cl_2 + & CrCl_3 + & H_2O \\ {\scriptstyle +1\,-1} & {\scriptstyle +1\,+6\,-2} & & {\scriptstyle +1\,-1} & {\scriptstyle 0} & {\scriptstyle +3\,-1} & {\scriptstyle +1\,-2} \end{array}$$

2. Note which atoms change oxidation number and calculate the number of electrons transferred, per atom, during oxidation and reduction.

Each Cr atom gains three electrons during reduction

$$Cr \xrightarrow{+3e^-} Cr$$
$$+6 \qquad +3$$

Each Cl^- that changes oxidation number loses 1 electron during oxidation.

$$Cl^- \xrightarrow{-1e^-} Cl$$
$$-1 \qquad 0$$

3. When more than one atom of an element that changes oxidation number is present in a formula, we next calculate the number of electrons transferred per formula unit.

   $K_2Cr_2O_7$ contains two Cr atoms, each of which gains 3 electrons; a total of $6e^-$/formula unit.

$$+6e^-/K_2Cr_2O_7$$

The coefficient 2 must be placed in front of the $CrCl_3$ to balance the chromium, which is the species being reduced.

When one Cl_2 molecule is formed, two electrons are lost by two Cl^- ions. Therefore the number of electrons transferred per formula unit of Cl_2 produced $= 2e^-$/formula unit. Note that a 2 must be placed before the HCl to balance chlorine.

$$2HCl \longrightarrow Cl_2$$
$$2e^-/Cl_2 \text{ formed}$$

4. Our next step is to make the number of electrons gained equal to the number lost by realizing that *three* Cl_2 formula units must be formed for every *one* $K_2Cr_2O_7$ formula unit reduced. These numbers become coefficients in our equation. (Remember also that a 2 must be placed before the $CrCl_3$ to balance the chromium.) We then write

$$HCl + K_2Cr_2O_7 \longrightarrow KCl + 3Cl_2 + 2\,CrCl_3 + H_2O$$

(We have not placed the coefficient 6 in front of the HCl^- because some Cl has not undergone change and appears in several places on the right, thereby making the total number of Cl on the right greater than six.)

5. In step 4 we have established some of the coefficients. Once these have been obtained the remainder of the equation is balanced by inspection. In this example, the sequence in which coefficients are introduced could be (a) a 2 is placed in front of KCl to balance the potassium, (b) since there are 14 Cl atoms on the right, place 14 in front of the HCl, (c) 14HCl on the left requires 7 H_2O on the right to balance hydrogen (this also balances oxygen at the same time). The final equation is

$$14HCl + K_2Cr_2O_7 \longrightarrow 2KCl + 3Cl_2 + 2CrCl_3 + 7H_2O$$

The procedure outlined above will work on any oxidation-reduction equation, although simple equations (e.g., $H_2 + Cl_2 \longrightarrow 2HCl$) are more easily balanced by inspection.

ION ELECTRON METHOD. In addition to the oxidation-number-change method, outlined above, there is still another method that is particularly well suited for balancing net ionic equations for oxidation-reduction reactions in solution. The procedure is called the **ion-electron** method and involves breaking the overall reaction into two half-reactions, one for the oxidation step and one for reduction. Each half-reaction is balanced materially (i.e., in terms of atoms) and then electrically, by adding electrons to the side of the half-reaction deficient in negative charge. Finally, the balanced half-reactions are added together so that the electrons cancel from both sides of the final equation.

As a simple example, let us consider the reaction of Sn^{2+} with Hg^{2+} in the presence of chloride ion to produce Hg_2Cl_2 and Sn^{4+} as products. The unbalanced equation is,

$$Sn^{2+} + Hg^{2+} + Cl^- \longrightarrow Hg_2Cl_2 + Sn^{4+}$$

In this reaction, tin is undergoing a change in oxidation number. The unbalanced half-reaction corresponding to the change experienced by Sn is

$$Sn^{2+} \longrightarrow Sn^{4+}$$

To balance the half-reaction according to charge (it obviously is already balanced according to atoms), we must add two electrons to the right side of the equation so that the net charge on both sides is the same, $+2$.

$$Sn^{2+} \longrightarrow Sn^{4+} + 2e^- \qquad (5.1)$$

This is the balanced half-reaction for the oxidation of Sn^{2+} to Sn^{4+}.

The next step is to write a half-reaction for the reduction of mercury(II). We begin by writing the formulas of the reactants and products that contain mercury.

$$Hg^{2+} \longrightarrow Hg_2Cl_2$$

We balance the equation materially by placing two chloride ions on the left and by writing the coefficient 2 in front of the Hg^{2+},

$$2Hg^{2+} + 2Cl^- \longrightarrow Hg_2Cl_2$$

The next step is to balance the charge. On the left, the net charge is $+2$ $(2(+2) + 2(-1) = +2)$; on the right it is zero. We must therefore add two electrons, each with a charge of -1, to the left to give,

$$2Hg^{2+} + 2Cl^- + 2e^- \longrightarrow Hg_2Cl_2 \qquad (5.2)$$

This is the balanced half-reaction for the reduction of Hg^{2+}.

To obtain the final balanced equation we add Equations 5.1 and 5.2 together so that there are the same number of electrons gained as lost.

$$Sn^{2+} \longrightarrow Sn^{4+} + 2e^-$$
$$\underline{2Hg^{2+} + 2Cl^- + 2e^- \longrightarrow Hg_2Cl_2}$$
$$Sn^{2+} + 2Hg^{2+} + 2Cl^- + 2e^- \longrightarrow Sn^{4+} + Hg_2Cl_2 + 2e^-$$

Since two electrons appear on both sides of the equation, we can cancel them out so that the final balanced equation is

$$Sn^{2+} + 2Hg^{2+} + 2Cl^- \longrightarrow Sn^{4+} + Hg_2Cl_2$$

In many oxidation-reduction reactions that take place in aqueous solution, water plays an active role. Any aqueous solution contains the species H_2O, H^+, and OH^-. In acidic solution the predominant species are H_2O and H^+; in basic solution they are H_2O and OH^-. When balancing half-reactions that occur in solution, we can use these species to achieve material balance. For example, let us consider the reaction between dichromate ion, $Cr_2O_7^{2-}$, and hydrogen sulfide in acidic solution to produce chromium(III) ion and elemental sulfur.

$$Cr_2O_7^{2-} + H_2S \longrightarrow Cr^{3+} + S$$

To balance the equation we write individual half-reactions for the changes that occur for chromium and sulfur. For the sulfur we write

$$H_2S \longrightarrow S$$

The half-reaction is balanced materially by adding two H^+ ions to the right.

$$H_2S \longrightarrow S + 2H^+$$

Electrical balance is achieved by adding two electrons to the right so that the net charge on both sides is zero.

$$H_2S \longrightarrow S + 2H^+ + 2e^- \tag{5.3}$$

For chromium we begin by writing

$$Cr_2O_7^{2-} \longrightarrow Cr^{3+}$$

The chromium atoms are balanced by placing the coefficient 2 before the Cr^{3+} on the right. The oxygens that appear on the left are balanced by adding H_2O (seven of them) to the right and the hydrogen imbalance that results is removed by placing 14 H^+ on the left.

$$14H^+ + Cr_2O_7^{2-} \longrightarrow 2Cr^{3+} + 7H_2O$$

To balance the half-reaction electrically we must add six electrons to the left side to give

$$6e^- + 14H^+ + Cr_2O_7^{2-} \longrightarrow 2Cr^{3+} + 7H_2O \tag{5.4}$$

The next step is to add these half-reactions (Equations 5.3 and 5.4) so that all of the electrons in the final equation cancel. This is accomplished by multiplying Equation 5.3 through by 3 before adding.

$$3(H_2S \longrightarrow S + 2H^+ + 2e^-)$$
$$\underline{6e^- + 14H^+ + Cr_2O_7^{2-} \longrightarrow 2Cr^{3+} + 7H_2O}$$
$$3H_2S + 14H^+ + Cr_2O_7^{2-} + \cancel{6e^-} \longrightarrow 3S + \cancel{6}H^+ + 2Cr^{3+} + 7H_2O + \cancel{6e^-}$$
$$\quad\quad\quad\quad 8$$

The six electrons cancel as do six of the H^+, so that the final balanced equation reads

$$3H_2S + 8H^+ + Cr_2O_7^{2-} \longrightarrow 3S + 2Cr^{3+} + 7H_2O$$

By way of summary, when balancing half-reactions in acid solution:

(a) *To balance a hydrogen atom we add a hydrogen ion, H^+, to the other side of the equation.*

(b) *To balance an oxygen atom we add a water molecule to the side deficient in oxygen and then two H^+ ions to the opposite side to remove the hydrogen imbalance.*

Applying the ion-electron method to reactions that occur in basic solution is perhaps a bit more difficult. The reason for this is that the species which can be used to balance the half-reactions, H_2O and OH^-, both contain oxygen. We use the following rules.

(a) *To balance a hydrogen atom we add* **one** *H_2O molecule to the side of the half-reaction deficient in hydrogen, and to the other side we add* **one** *hydroxide ion.*

(b) *To balance one oxygen atom we add* **two** *hydroxide ions to the side deficient in oxygen and* **one** *H_2O molecule to the other side.* For example, consider the oxidation of Pb to PbO in basic media.

$$Pb \longrightarrow PbO$$

The left side of the equation needs an oxygen atom; so we add two OH^-.

$$2OH^- + Pb \longrightarrow PbO$$

Now we add one H_2O to the right.

As indicated above, $2OH^-$ ions become one H_2O molecule and one oxygen atom in some other formula.

As an example of how these rules are applied, let us balance the equation for the oxidation of methyl alcohol (wood alcohol), CH_4O, with permanganate ion, MnO_4^-, in basic solution, to yield CO_2 and MnO_2 as products.

$$CH_4O + MnO_4^- \longrightarrow MnO_2 + CO_2$$

As before, we break the equation into two half-reactions.

$$CH_4O \longrightarrow CO_2$$
$$MnO_4^- \longrightarrow MnO_2$$

The first equation is deficient in hydrogen on the right and deficient in oxygen on the left. To balance the hydrogen we place four H_2O on the right and four OH^- on the left.

$$4OH^- + CH_4O \longrightarrow CO_2 + 4H_2O$$

Now, to balance the oxygen we place two more OH^- on the left and another H_2O on the right.

$$6OH^- + CH_4O \longrightarrow CO_2 + 5H_2O$$

The half-reaction is now balanced materially. *Note that we balanced hydrogen and oxygen in separate operations.* Finally, it is necessary to balance the half-reaction electrically by placing six electrons on the right.

$$6OH^- + CH_4O \longrightarrow CO_2 + 5H_2O + 6e^- \qquad (5.5)$$

Next we turn our attention to the reduction of the MnO_4^- to MnO_2. This half-reaction is deficient in oxygen on the right, and balance is achieved by supplying four OH^- to the right and two H_2O to the left.

$$2H_2O + MnO_4^- \longrightarrow MnO_2 + 4OH^-$$

Electrical balance is provided by adding three electrons to the left.

$$3e^- + 2H_2O + MnO_4^- \longrightarrow MnO_2 + 4OH^- \qquad (5.6)$$

In order to obtain the final equation we must add the two balanced half-reactions together so that all of the electrons will cancel. This requires that Equation 5.6 be multiplied by two before they are added.

$$6OH^- + CH_4O \longrightarrow CO_2 + 5H_2O + 6e^-$$
$$\underline{2(3e^- + 2H_2O + MnO_4^- \longrightarrow MnO_2 + 4OH^-)}$$
$$6OH^- + CH_4O + 4H_2O + 2MnO_4^- \longrightarrow CO_2 + 5H_2O + 2MnO_2 + 8OH^-$$

We finally cancel six OH^- and four H_2O from both sides to obtain

$$CH_4O + 2MnO_4^- \longrightarrow CO_2 + 2MnO_2 + H_2O + 2OH^-$$

In Section 5.6 we saw that the preparation of pure compounds often depends on the mixing together of reactants in the proper ratio, as determined by the stoichiometry of the reaction. For example, if we wish to prepare $NaHSO_4$ from solutions containing NaOH and H_2SO_4, the two must be combined in such a way as to maintain a 1:1 mole ratio between the base and the acid. This means that we must have a means at our disposal of expressing the concentrations of reactants so that appropriate quantities of the various solutions may be mixed.

There are many ways of expressing the concentration of a solute in solution, and each has its advantages for specific applications. One way to express concentration, for instance, is percent composition by weight. Concentrated sulfuric acid, we have said, is composed of 96% H_2SO_4 and 4% H_2O. This tells us the composition of the solution in parts per hundred. A similar unit that is frequently used to express very small concentrations (e.g., the concentration of impurities or pollutants in air or water) is **parts per million, ppm.** For instance, a typical carbon monoxide level in heavy smog is approximately 40 ppm whereas a typical nitrogen oxide concentration is about 0.2 ppm.

Example 5.2

Atmospheric SO_2, produced by the combustion of high sulfur fuels, is an important air pollution problem. The amount of SO_2 in the air may be determined by bubbling the air through an acidic solution of $KMnO_4$ which oxidizes the SO_2 to SO_4^{2-}. In a typical analysis, 500 liters of air having a density of 1.20 g/liter are passed through the $KMnO_4$ solution and 1.50×10^{-5} moles of $KMnO_4$ are reduced to Mn^{2+}. What is the concentration of SO_2 in parts per million (by weight).

Solution

We first must have a balanced equation. The ion-electron method gives us

$$2MnO_4^- + 5SO_2 + 2H_2O \longrightarrow 2Mn^{2+} + 5SO_4^{2-} + 4H^+$$

If 1.50×10^{-5} mole of MnO_4^- is reduced, this requires (based on the stoichiometry of the reaction)

$$1.5 \times 10^{-5} \text{ mole } MnO_4^- \times \left(\frac{5 \text{ mole } SO_2}{2 \text{ mole } MnO_4^-}\right) = 3.75 \times 10^{-5} \text{ mole } SO_2$$

The weight of SO_2 in the air, then, is

$$3.75 \times 10^{-5} \text{ mole } SO_2 \times \left(\frac{64.1 \text{ g } SO_2}{1 \text{ mole } SO_2}\right) = 2.40 \times 10^{-3} \text{ g } SO_2$$

The 500 liter sample of air contained 2.40×10^{-3} g SO_2. The concentration in ppm is obtained as

$$\frac{\text{Weight of } SO_2}{\text{Total weight of air sample}} \times 10^6$$

The weight of the air sample can be obtained from its density

$$500 \text{ liter} \times \left(\frac{1.20 \text{ g}}{\text{liter}}\right) = 600 \text{ g air}$$

Therefore, the concentration of the SO_2 pollutant is

$$\frac{2.40 \times 10^{-3} \text{ g } SO_2}{600 \text{ g air}} \times 10^6 = 4.00 \text{ ppm}$$

Expressing concentration in percent (as in the H_2SO_4 solution described earlier) or in parts per million (as in the solution of gases in Example 5.2) conveys information about the relative proportions of solute and solvent. However, for many applications to stoichiometry these concentration units are inconvenient and therefore not very useful. This leads us, then, to define concentration units that more closely suit problems in solution stoichiometry.

MOLARITY. Molarity is defined as the number of moles of solute per liter of solution[5]; that is, it represents a *ratio* in which the quantity of solute, in moles, is divided by the total volume of the mixture, in liters. A solution prepared by dissolving 58.44 g of NaCl (one mole) in sufficient water to give a total volume of 1.00 liter, therefore, is a one molar solution of sodium chloride, written $1.00M$ NaCl. If we dissolve 0.500 mole of NaCl in enough water to make 0.500 liter of solution, the concentration is 0.500 mole/0.500 liter which, of course, is the same as 1.00 mole/liter, or $1.00M$.

---

[5] Formality, another concentration unit, is defined in essentially the same way as the number of gram formula weights of solute per liter of solution. Since one gram formula weight contains Avogadro's number of formula units and one mole of solute is taken to mean the same thing, molarity and formality are identical.

Example 5.3

How many grams of NaOH are required to prepare 350 ml of 1.25M NaOH solution?

Solution

We must first determine how many moles of NaOH are in 350 ml of 1.25M NaOH. This can be accomplished by multiplying molarity by volume (in liters).

$$\left(\frac{1.25 \text{ mole NaOH}}{\text{liter}}\right) \times 0.350 \text{ liter} = 0.437 \text{ mole NaOH}$$

The weight of NaOH can then be obtained from the formula weight.

$$0.437 \text{ mole NaOH} \times \left(\frac{40.0 \text{ g NaOH}}{1 \text{ mole NaOH}}\right) = 17.5 \text{ g NaOH}$$

Example 5.4

What is the molarity of a solution prepared by adding water to 10.0 g of KCl until 200 ml of solution is achieved?

Solution

The final solution contains 10.0 g/200 ml. Since our definition of molarity is the number of moles of solute per liter of solution, it is necessary to convert the 10.0 g of KCl into moles and the 200 ml of solution into liters.

We convert grams of KCl into moles of KCl by our usual method.

$$10.0 \text{ g KCl} \times \left(\frac{1 \text{ mole}}{74.6 \text{ g}}\right) = 0.134 \text{ mole of KCl}$$

The number of liters of solution is 0.200 liter, so that the molarity is

$$\frac{0.134 \text{ moles}}{0.200 \text{ liter}} = 0.670M$$

Example 5.5

What is the molar concentration of concentrated sulfuric acid that contains 96% by weight H_2SO_4 and has a density of 1.84 g/ml?

Solution

We must determine the number of moles of H_2SO_4 in one liter of the concentrated acid. We know, from the density, that

$$1.00 \text{ ml} \sim 1.84 \text{ g}$$

or

$$1000 \text{ ml} = 1.00 \text{ liter} \sim 1840 \text{ g}$$

Only 96.0% of this weight is contributed by H_2SO_4.

$$1840 \text{ g} \times 0.960 = 1770 \text{ g } H_2SO_4$$

The formula weight of $H_2SO_4 = 98.08$; therefore,

$$1770 \text{ g } H_2SO_4 \times \left(\frac{1 \text{ mole } H_2SO_4}{98.08 \text{ g } H_2SO_4}\right) = 18.0 \text{ mole of } H_2SO_4$$

The concentration, then, is

$$\frac{18.0 \text{ mole } H_2SO_4}{1.00 \text{ liter}} = 18.0M \ H_2SO_4$$

Molarity is a useful concentration unit because it allows us to meter out quantities of solute merely by dispensing volumes of solution. For instance, if we wished to use 0.300 mole of NaCl in some reaction and we had at our disposal a solution containing 1.00 mole of NaCl per liter (i.e., a 1.00M NaCl solution), we could measure out 0.300 liters of this solution and know that it contained the desired 0.300 mole of NaCl. It is this aspect of molarity that is useful in problems dealing with the stoichiometry of solutions.

Example 5.6
How many milliliters of 2.00M NaCl solution are required to react with exactly 5.37 g of $AgNO_3$ to form AgCl?

Solution
The first step in any problem involving the stoichiometry of a chemical reaction should be to write a chemical equation (balanced, of course).

$$NaCl + AgNO_3 \longrightarrow AgCl + NaNO_3$$

We see that one mole of NaCl reacts with one mole of $AgNO_3$. Therefore, 5.37 g $AgNO_3$ requires.

$$5.37 \text{ g } AgNO_3 \times \left(\frac{1 \text{ mole } AgNO_3}{170 \text{ g } AgNO_3}\right) \times \left(\frac{1 \text{ mole NaCl}}{1 \text{ mole } AgNO_3}\right) \sim 3.16 \times 10^{-2} \text{ mole NaCl}$$

Molarity has the units, mole/liter; therefore the concentration of the NaCl solution can be written as

$$2.00 \ M \ NaCl = \frac{2.00 \text{ mole NaCl}}{1.00 \text{ liter solution}}$$

This can be used as a conversion factor to find the volume of solution that contains 3.16×10^{-2} mole NaCl

$$3.16 \times 10^{-2} \text{ mole NaCl} \times \left(\frac{1.00 \text{ liter solution}}{2.00 \text{ mole NaCl}}\right) = 1.58 \times 10^{-2} \text{ liter solution}$$

Expressed in ml the volume is 15.8 ml.

Example 5.7
How many ml of 0.500M NaOH are required to precipitate all of the magnesium, as $Mg(OH)_2$, from 600 ml of a 0.450M $MgCl_2$ solution?

Solution
Again, we begin with a balanced chemical equation.

$$MgCl_2 + 2NaOH \longrightarrow Mg(OH)_2 + 2NaCl$$

The equation tells us that

$$1 \text{ mole } MgCl_2 \sim 2 \text{ mole NaOH}$$

In 600 ml of 0.450M MgCl$_2$ there are

$$\left(\frac{0.450 \text{ mole MgCl}_2}{\text{liter}}\right) \times 0.600 \text{ liter} = 0.270 \text{ mole MgCl}_2$$

From the stoichiometry of the chemical equation,

$$0.270 \text{ mole MgCl}_2 \times \left(\frac{2 \text{ mole NaOH}}{1 \text{ mole MgCl}_2}\right) \sim 0.540 \text{ mole NaOH}$$

The solution of NaOH contains 0.500 mole/1.00 liter. The volume of NaOH solution required in this problem is obtained as

$$0.540 \text{ mole NaOH} \times \left(\frac{1.00 \text{ liter}}{0.500 \text{ mole}}\right) = 1.08 \text{ liter NaOH solution}$$

Expressed in milliliters, the volume is 1080 ml of NaOH solution.

Many calculations concerning the stoichiometry of oxidation-reduction reactions may be performed without having to resort to balanced chemical equations. This is accomplished by using the fact that the number of electrons lost by the reducing agent must equal the number gained by the oxidizing agent. In oxidation-reduction reactions, we define the **equivalent weight** (or **gram-equivalent weight**) of a substance as the quantity that gains or loses one mole of electrons. Thus, one **gram-equivalent** (or simply, one equivalent) of a reducing agent loses one mole of electrons and, conversely, one equivalent of an oxidizing agent gains one mole of electrons. In this way, one equivalent of oxidizing agent reacts with one equivalent of reducing agent.

5.9
Equivalent Weights and Normality

Example 5.8
FeSO$_4$ reacts with KMnO$_4$ in H$_2$SO$_4$ to produce Fe$_2$(SO$_4$)$_3$ and MnSO$_4$. How many grams of FeSO$_4$ react with 3.71 g KMnO$_4$?

Solution
In this reaction,

1 mole FeSO$_4$ loses 1 mole of electrons;   Fe $\longrightarrow$ Fe
                                             +2   +3
1 mole KMnO$_4$ gains 5 moles of electrons;   Mn $\longrightarrow$ Mn
                                             +7   +2

The equivalent weight of FeSO$_4$ in this reaction would be the same as the gram-formula weight since each mole FeSO$_4$ loses 1 mole of electron.

$$1 \text{ g-eq FeSO}_4 = \frac{1 \text{ mole FeSO}_4}{1} = 152 \text{ g FeSO}_4$$

The equivalent weight of KMnO$_4$ in this reaction is $\frac{1}{5}$ of the weight of one mole since each mole of KMnO$_4$ gains five moles of electrons:

1 g-eq KMnO$_4$ = $\frac{1}{5}$ mole KMnO$_4$ = $\frac{1}{5}$ (158.0 g KMnO$_4$)
1 g-eq KMnO$_4$ = 31.6 KMnO$_4$

Since 1 g-eq FeSO$_4$ must react with precisely 1 g-eq KMnO$_4$,

$$152 \text{ g FeSO}_4 \sim 31.6 \text{ g KMnO}_4$$

The solution to the problem can then be obtained as

$$3.71 \text{ g KMnO}_4 \times \left(\frac{152 \text{ g FeSO}_4}{31.6 \text{ g KMnO}_4} \right) \sim 17.8 \text{ g FeSO}_4$$

The solution to this last example problem could have also been obtained simply by using moles. We know, for example, that if we were to balance an equation for this reaction, we would have to take five moles $FeSO_4$ for every one mole $KMnO_4$ so that the number of electrons lost would equal the number gained.

$$5 \text{ mole FeSO}_4 \sim 1 \text{ mole KMnO}_4$$

The solution to the problem could then have been obtained as

$$3.71 \text{ g KMnO}_4 \times \left(\frac{1 \text{ mole KMnO}_4}{158 \text{ g KMnO}_4} \right) \times \left(\frac{5 \text{ mole FeSO}_4}{1 \text{ mole KMnO}_4} \right)$$
$$\times \left(\frac{152 \text{ g FeSO}_4}{1 \text{ mole FeSO}_4} \right) \sim 17.8 \text{ g FeSO}_4$$

Both methods of solving the problem are actually equivalent.

Finally, note that the equivalent weight of a substance can vary from one reaction to another. Potassium permanganate, for example, has an equivalent weight equal to $\frac{1}{5}$ of a mole when it reacts, as above, to produce manganese in the +2 oxidation state. In some reactions, $KMnO_4$ produces MnO_2 as a product. In this case one mole of $KMnO_4$ gains only three moles of electrons, and its equivalent weight is $\frac{1}{3}$ of its gram-formula weight.

NORMALITY. Another concentration unit useful for the treatment of problems concerning stoichiometry is normality, defined as the number of gram-equivalents of solute per liter of solution. Again, we have a ratio—the quantity of solute, expressed in gram-equivalents (or simply equivalents), divided by the total volume of the solution. Above, the equivalent for oxidizing and reducing agents was defined as the quantity that either gains or loses one mole of electrons. We saw that when the permanganate ion in $KMnO_4$ is reduced to Mn^{2+}, five electrons are gained by each formula unit of $KMnO_4$. One mole of $KMnO_4$, therefore, consumes five moles of electrons and the quantity of $KMnO_4$ that acquires only one mole of electrons (i.e., an equivalent of $KMnO_4$) is one-fifth of a mole of the salt.

$$1 \text{ equivalent KMnO}_4 = \tfrac{1}{5} \text{ mole KMnO}_4$$

Thus a 1.00 normal (1.00N) solution of $KMnO_4$ contains 1.00 equivalent of $KMnO_4$ per liter, or 0.200 mole $KMnO_4$ per liter. A 1.00N $KMnO_4$ solution is also a 0.200M $KMnO_4$ solution. We see that there is a very simple relationship between normality and molarity—the normality is always an integral multiple of the molarity,

$$N = n \times M$$

where n is an integer. For substances undergoing oxidation or reduction the integer, n, corresponds to the number of electrons transferred per formula unit. When $KMnO_4$ is reduced to Mn^{2+}, $n = 5$. If the product of the reduction is MnO_2, only three electrons are acquired by each $KMnO_4$ formula unit, and

n would equal three. Similarly, when oxalic acid, $H_2C_2O_4$, is oxidized to produce CO_2, two electrons are lost by each $H_2C_2O_4$ formula unit and the normality of a solution of $H_2C_2O_4$ is twice its molarity.

For acids and bases the equivalent is defined, respectively, as the amount of the *acid* that furnishes one mole of hydrogen ion and as the amount of *base* that reacts with one mole of hydronium ion. One mole of HCl can supply but one mole of H^+; therefore one equivalent of HCl is the same as one mole of HCl. For H_2SO_4, or any other polyprotic acid, the number of equivalents per mole is equal to the number of hydrogen ions each molecule of the acid can furnish when it undergoes neutralization by a base. One mole of H_2SO_4, therefore, consists of two equivalents and a 1.00M H_2SO_4 solution could also be labeled 2.00N H_2SO_4. Similarly a 1.00M H_3PO_4 solution is also 3.00N H_3PO_4.

Each mole of a base such as NaOH or NH_3 (which reacts to give NH_4^+ and OH^-) is able to neutralize only one mole of H^+ and the normality of their solutions are the same as their molarity. $Ba(OH)_2$, on the other hand, is an example of a compound that supplies two moles of OH^- per mole of base. The normality of a $Ba(OH)_2$ solution is therefore twice its molarity.

Normality is useful because the equivalent is defined so that one equivalent of an oxidizing agent will react with *exactly* one equivalent of a reducing agent (one mole of electrons gained and one mole of electrons lost). Similarly, one equivalent of acid completely neutralizes precisely one equivalent of base, since one mole of H^+ reacts with one mole of OH^-.

$$H^+ + OH^- \longrightarrow H_2O$$

This means that mixing equal volumes of solutions having the same normality will lead to complete reaction between their solutes; one liter of one normal acid will completely neutralize one liter of one normal base because one equivalent of acid reacts with one equivalent of base. In the same manner, one liter of one normal oxidizing agent will completely oxidize the contents of one liter of one normal reducing agent.

Example 5.9
How many ml of 0.150N $K_2Cr_2O_7$ are required to react with 75.0 ml of 0.400N $H_2C_2O_4$?

Solution

In 75.0 ml of 0.400N $H_2C_2O_4$ there are

$$0.075 \text{ liter} \times \left(\frac{0.400 \text{ equiv } H_2C_2O_4}{1.00 \text{ liter}}\right) = 3.00 \times 10^{-2} \text{ equiv } H_2C_2O_4$$

This will require 3.00×10^{-2} equivalents of $K_2Cr_2O_7$ for complete oxidation. The volume of potassium dichromate that is contained is

$$3.00 \times 10^{-2} \text{ equiv} \times \left(\frac{1.00 \text{ liter}}{0.150 \text{ equiv}}\right) = 2.00 \times 10^{-1} \text{ liter}$$

or 200 ml

In Example 5.8 we used the idea that equal numbers of equivalents of oxidizing and reducing agent must react. Since the product of volume

$(V) \times$ normality (N) gives equivalents, when any two substances in solution undergo complete reaction (let's call them A and B),

$$V_A N_A = V_B N_B$$

which means, (number of equivalents A) = (number of equivalents B).

5.10 Chemical Analysis

A chemist must know the nature of the substances he has produced in a chemical reaction before he can theorize about their structure and bonding, for example. When a substance appears as the product of a reaction, it does not stand up and declare its composition — it must be analyzed. One very important facet of chemistry is the investigation of new and better ways to achieve chemical analyses, particularly of compounds present in mixtures where difficulties in separating the components present a major obstacle to success.

Chemical reactions in solution can frequently be used to advantage in performing chemical analyses. Example 5.9 illustrates how a precipitation reaction might be put to use.

Example 5.10

A mixture of NaCl and BaCl$_2$ weighing 0.200 g was dissolved in water, and sulfuric acid was added until no further precipitate was formed. The solid BaSO$_4$ was filtered, dried, and found to weigh 0.0643 g. What percent of the mixture was BaCl$_2$?

Solution

In this example the barium is precipitated as BaSO$_4$.

$$\text{BaCl}_2 + \text{H}_2\text{SO}_4 \longrightarrow 2\text{HCl} + \text{BaSO}_4$$

From the weight of BaSO$_4$ we can compute the weight of BaCl$_2$ in the mixture and hence the fraction that was BaCl$_2$.

$$0.0643 \text{ g BaSO}_4 \times \left(\frac{1 \text{ mole BaSO}_4}{233.4 \text{ g BaSO}_4}\right) = 2.75 \times 10^{-4} \text{ mole BaSO}_4$$

According to the chemical equation one mole of BaSO$_4$ is produced from one mole of BaCl$_2$; therefore the original mixture must have contained 2.75×10^{-4} mole BaCl$_2$. The weight of BaCl$_2$ in the mixture was

$$2.75 \times 10^{-4} \text{ mole BaCl}_2 \times \left(\frac{208.4 \text{ g BaCl}_2}{1 \text{ mole BaCl}_2}\right) = 5.73 \times 10^{-2} \text{ g BaCl}_2$$

Weight of BaCl$_2$ = 0.0573 g

$$\text{The percent BaCl}_2 = \frac{\text{weight BaCl}_2}{\text{total weight of mixture}} \times 100$$

$$\% \text{ BaCl}_2 = \frac{0.0573}{0.200} \times 100 = 28.7\%$$

Often, the quantitative relationships introduced in our discussion of concentration can be employed directly in what are termed **volumetric analyses.** For example, the quantity of base in some sample of an unknown can be determined by a procedure called a **titration,** in which a solution of an acid is

Figure 5.3

(a) A buret. (b) Titration of a base with an acid.

dispensed into the base from a buret (Figure 5.3) until reaction is complete, as signaled by some form of **indicator.** There are a series of organic dyes that have one color in acid solution and a different color in basic solution that can be used for this purpose. Litmus is a common example—red in acid solution and blue in the presence of base. In a typical titration, the base is dissolved and the indicator added. Acid is delivered gradually from the buret until the **end point** (or **equivalence point**) is marked by a color change signifying that all of the base has been neutralized. At the equivalence point an equal number of equivalents of acid and base have been combined. From the total volume of acid added and its concentration, the number of equivalents of acid and, hence, the number of equivalents of base can be deduced.

Example 5.11

A sample of 0.500 g of a pure compound, known to be either NaOH or KOH, requires 89.1 ml of 0.100N HCl in a titration. What is the identity of the compound?

Solution

From the volume of acid and its concentration we can determine the number of equivalents of acid used.

$$0.0891 \text{ liter} \times \left(\frac{0.100 \text{ equiv}}{\text{liter}}\right) = 8.91 \times 10^{-3} \text{ equiv acid}$$

The base also constituted 8.91×10^{-3} equiv.

$$1 \text{ equiv NaOH} = 40.0 \text{ g}$$
$$1 \text{ equiv KOH} = 56.1 \text{ g}$$

If the sample had been NaOH, its weight would have had to be

$$8.91 \times 10^{-3} \text{ equiv} \times \left(\frac{40.0 \text{ g NaOH}}{1 \text{ equiv}}\right) = 0.356 \text{ g}$$

while if it were KOH its weight would be

$$8.91 \times 10^{-3} \text{ equiv} \times \left(\frac{56.1 \text{ g KOH}}{1 \text{ equiv}}\right) = 0.500 \text{ g}$$

Since the sample weighed 0.500 g, the solid was KOH.

Similar titrations are frequently performed in which the reaction involves oxidation and reduction. $KMnO_4$ is a useful **titrant** (the solution delivered from the buret) to use because it has a deep purple color whereas the product of the reduction of the MnO_4^- ion in acid solution is usually the almost colorless Mn^{2+} ion. The end point of the titration is signaled by the appearance of a pale purple color in the reaction mixture due to excess MnO_4^- ion.

Example 5.12
A 1.000 gram sample of an iron ore containing Fe_2O_3 was dissolved in acid and all of the iron converted to Fe^{2+}. The solution was titrated with 90.4 ml of 0.100 N $KMnO_4$ to give Mn^{2+} and Fe^{3+} as products. What percent of the ore is Fe_2O_3?

Solution
The number of equivalents of $KMnO_4$ reduced was

$$9.04 \times 10^{-2} \text{ liter} \times \left(\frac{0.100 \text{ equiv } KMnO_4}{\text{liter}}\right) = 9.04 \times 10^{-3} \text{ equiv } KMnO_4$$

This amount of $KMnO_4$ has oxidized 9.04×10^{-3} equiv Fe^{2+} to Fe^{3+}. Since one mole of Fe^{2+} loses one mole of electrons,

$$1 \text{ mole } Fe^{2+} = 1 \text{ equiv } Fe^{2+}$$

Therefore, 9.04×10^{-3} equiv $Fe^{2+} = 9.04 \times 10^{-3}$ mole Fe^{2+}.
Each mole of Fe_2O_3 contains two moles of Fe.

$$1 \text{ mole } Fe_2O_3 \sim 2 \text{ mole Fe}$$

The number of moles of Fe_2O_3 in the sample was

$$9.04 \times 10^{-3} \text{ mole Fe} \times \left(\frac{1 \text{ mole } Fe_2O_3}{2 \text{ mole Fe}}\right) \sim 4.52 \times 10^{-3} \text{ mole } Fe_2O_3$$

The number of grams of Fe_2O_3 was

$$4.52 \times 10^{-3} \text{ mole } Fe_2O_3 \times \left(\frac{160 \text{ g } Fe_2O_3}{1 \text{ mole } Fe_2O_3} \right) = 0.723 \text{ g } Fe_2O_3$$

Since the sample weighed 1.000 g,

$$\% Fe_2O_3 = \frac{0.723}{1.000} \times 100 = 72.3\%$$

In the course of performing routine laboratory operations as well as volumetric analyses, it is not uncommon to have to dilute solutions. Many of the common laboratory reagents are bottled in concentrated form (Table 5.4) and usually must be diluted for use. The process of dilution actually simply involves spreading a given quantity of solute throughout a larger volume of solution; when a concentrated solution is diluted by the addition of solvent, the number of moles of solute in the mixture remains constant. This means that the solution's molar concentration multiplied by its volume must remain constant as solvent is added[6] since

$$M \cdot V = \frac{\text{moles}}{\text{liter}} \times \text{liter} = \text{moles}$$

In other words, the product of the initial molarity and volume (M_iV_i) must equal the product of the final molarity and volume (M_fV_f).

$$M_iV_i = M_fV_f$$

For dilution problems in which the concentration unit is normality the similar equation, $N_iV_i = N_fV_f$, applies.

Table 5.4
Concentrated Laboratory Reagents

Reagent	Density (g/ml)	Weight Percent	Molarity
Sulfuric acid	1.84	96	18
Hydrochloric acid	1.18	36.5	12.0
Phosphoric acid	1.7	85	15
Nitric acid	1.42	70.7	16.0
Acetic acid	1.05	100	17.5
Aqueous Ammonia	0.90	28	14.8

[6] *CAUTION! When diluting concentrated reagents, the proper procedure is to add the reagent to the solvent. Never add water to a concentrated acid.*

Example 5.13

How much water must be added to 25.0 ml of 0.500M KOH solution to produce a solution whose concentration is 0.350M?

Solution

Our equation is $M_iV_i = M_fV_f$

$$M_i = 0.500$$
$$V_i = 25.0 \text{ ml}$$
$$M_f = 0.350$$
$$V_f = ?$$

Solving for V_f and substituting, we get

$$V_f = \frac{(0.0500M)\ (25.0 \text{ ml})}{0.350M}$$

$$V_f = 35.7 \text{ ml}$$

Since the initial volume was 25.0 ml, we must *add* 10.7 ml. (We are assuming volumes are additive. When working with dilute solutions this assumption is generally quite valid.)

Example 5.14

How many ml of concentrated H_2SO_4 (18.0M) are required to prepare 750 ml of 3.0M H_2SO_4 solution?

Solution

Again

$$M_iV_i = M_fV_f$$
$$M_i = 18.0 \qquad M_f = 3.0M$$
$$V_i = ? \qquad\quad V_f = 750 \text{ ml}$$

Solving for V_i gives

$$V_i = \frac{M_fV_f}{M_i}$$

$$V_i = \frac{(3.0M)\ (750 \text{ ml})}{18.0M}$$

$$V_i = 125 \text{ ml}$$

To prepare the solution, 125 ml of the concentrated H_2SO_4 is diluted to a total final volume of 750 ml.

**Review
Questions
and Problems**

5.1 Why are solutions usually employed for carrying out chemical reactions?

5.2 How can one distinguish between a strong and a weak electrolyte? What are present in solutions of electrolytes that are absent in solutions of nonelectrolytes?

5.3 What are the meanings of polyprotic acid, monoprotic acid, acid salt, and neutralization?

5.4 A newly discovered element is found to react with oxygen to form an oxide that, when dissolved in water, causes litmus to turn blue. Would the element be classed as a metal or a nonmetal? Why?

5.5 Explain in your own words the concept of dynamic equilibrium. Write chemical equations for the equilibria that exist when the following weak electrolytes undergo dissociation: HCN, H_2S, NH_3, $HgCl_2$.

5.6 Write the equations for the two-step dissociation of the acid, H_2SeO_3.

5.7 Write net ionic equations for each of the following:

(a) $NaBr + AgNO_3 \rightarrow$
$AgBr + NaNO_3$

(b) $CoCO_3 + 2HNO_3 \rightarrow$
$Co(NO_3)_2 + CO_2 + H_2O$

(c) $NaC_2H_3O_2 + HNO_3 \rightarrow$
$NaNO_3 + HC_2H_3O_2$

(d) $Pb(NO_3)_2 + (NH_4)_2 SO_4 \rightarrow$
$PbSO_4 + 2NH_4NO_3$

(e) $H_2S + Cu(NO_3)_2 \rightarrow$
$2HNO_3 + CuS$

(f) $NaOH + NH_4Cl \rightarrow$
$NH_3 + H_2O + NaCl$

5.8 Write molecular, ionic, and net ionic equations for the reaction, if any, between (a) H_2SO_4 and $BaCl_2$, (b) KI and AgBr, (c) K_2SO_3 and HCl, (d) $(NH_4)_2SO_4$ and $Ba(OH)_2$, (e) $SrCl_2$ and KNO_3, (f) $Mg(OH)_2$ and $HC_2H_3O_2$, (g) $Na_2C_2O_4$ and HCl, (h) SO_3 and H_2O, (i) CO_2, H_2O and NaOH, (j) SO_2, H_2O, and $BaCl_2$.

5.9 Describe how you would prepare each of the following by a metathesis reaction in which one of the products was a precipitate: $CaCl_2$, NH_4NO_3, $Fe_3(PO_4)_2$, $CuCO_3$, Na_2SO_4, $Pb(NO_3)_2$, $Ba(ClO_4)_2$.

5.10 Which of the salts in Question 5.9 could most effectively be prepared from a carbonate and the appropriate acid?

5.11 Write chemical equations for the preparation of the salts in Question 5.9 by a neutralization reaction.

5.12 Balance the following equations by the oxidation-number-change method.

(a) $HNO_3 + Zn \rightarrow$
$Zn(NO_3)_2 + H_2O + NH_4NO_3$

(b) $K + KNO_3 \rightarrow N_2 + K_2O$

(c) $C_3H_7OH + Na_2Cr_2O_7 + H_2SO_4 \rightarrow$
$Cr_2(SO_4)_3 + Na_2SO_4 + H_2O + HC_3H_5O_2$

(d) $H_2S + HNO_3 \rightarrow S + NO + H_2O$

(e) $Fe(OH)_2 + O_2 + H_2O \rightarrow Fe(OH)_3$

5.13 Balance the following equations by the ion-electron method. All reactions are in acid solution.

(a) $Cu + NO_3^- \rightarrow Cu^{2+} + NO$

(b) $Zn + NO_3^- \rightarrow Zn^{2+} + NH_4^+$

(c) $P_4 + NO_3^- \rightarrow H_3PO_4 + NO + NO_2$

(d) $Cr_2O_7^{2-} + H_3AsO_3 \rightarrow$
$Cr^{3+} + H_3AsO_4$

(e) $I^- + SO_4^{2-} \rightarrow I_2 + H_2S$

(f) $Ag^+ + AsH_3 \rightarrow H_3AsO_4 + Ag$

(g) $Cu + Cl^- + As_4O_6 \rightarrow CuCl + As$

(h) $MnO_2 + Br^- \rightarrow Br_2 + Mn^{2+}$

(i) $S_2O_3^{2-} + I_2 \rightarrow I^- + S_4O_6^{2-}$

(j) $IO_3^- + SO_3^{2-} \rightarrow I^- + SO_4^{2-}$

(k) $Cr_2O_7^{2-} + CH_3CH_2OH \rightarrow$
$Cr^{3+} + CH_3CHO$

(l) $NO_2 \rightarrow NO_3^- + NO$

(m) $Mn^{2+} + BiO_3^- \rightarrow MnO_4^- + Bi^{3+}$

(n) $HSO_2NH_2 + NO_3^- \rightarrow N_2O + SO_4^{2-}$

(o) $PH_3 + I_2 \rightarrow H_3PO_2 + I^-$

5.14 Balance the following equations by the ion-electron method. All reactions are in basic solution.

(a) $Zn \rightarrow Zn(OH)_4^{2-} + H_2$

(b) $CrO_2^- + HO_2^- \rightarrow CrO_4^{2-}$

(c) $Br_2 \rightarrow Br^- + BrO_3^-$

(d) $ClO_2 \rightarrow ClO_2^- + ClO_3^-$

(e) $Zn + NO_3^- \rightarrow Zn(OH)_4^{2-} + NH_3$

(f) $Cu(NH_3)_4^{2+} + S_2O_4^{2-} \rightarrow$
$SO_3^{2-} + Cu + NH_3$

(g) $Fe(OH)_2 + O_2 \rightarrow Fe(OH)_3$

(h) $MnO_4^- + C_2O_4^{2-} \rightarrow MnO_2 + CO_2$

(i) $P_4 + OH^- \rightarrow PH_3 + H_2PO_2^-$

(j) $Al + OH^- \rightarrow AlO_2^- + H_2$

5.15 Describe in detail how you would prepare a $1.50M$ KCl solution.

5.16 Pure nitric acid has a density of 1.513 g/ml. What is its molar concentration?

5.17 A solution of $MgSO_4$ contains 22.0% $MgSO_4$ by weight and contains 273.8 g of the salt per liter. What is the density of the solution?

5.18 How many ml of $0.300M$ NaOH are required to react with 500 ml of $0.170M$?

H_3PO_4 to yield (a) Na_3PO_4, (b) Na_2HPO_4, (c) NaH_2PO_4?

5.19 If 380 ml of 0.273M $Ba(OH)_2$ is added to 500 ml of 0.520M HCl, will the mixture be acidic or basic? Calculate the concentration of H^+ (or OH^- if the solution is basic) in the final mixture. Assume volumes are additive.

5.20 Calculate the equivalent weight of the oxidizing agent in each oxidation-reduction reaction of Question 5.12.

5.21 What is the equivalent weight of $MnSO_4$ when it is oxidized to produce (a) Mn_2O_3, (b) MnO_2, (c) K_2MnO_4, (d) $KMnO_4$?

5.22 How many grams of $NaBiO_3$ are required to react with 0.500 g of $Mn(NO_3)_2$ to produce $NaMnO_4$ and $Bi(NO_3)_3$?

5.23 A 10.0 ml portion of a solution of HCl was diluted to exactly 50.0 ml. If 5.0 ml of this solution required 41.0 ml of 0.255N NaOH for complete neutralization, what was the concentration of the original HCl solution before dilution?

5.24 A volume of 129 ml of 0.850N $Ba(OH)_2$ was required to completely neutralize a 4.93 g sample of an acid. What is the equivalent weight of the acid?

5.25 What is the normality of each of the following solutions?
(a) 22.0 g of $Sr(OH)_2$ in 800 ml of solution.
(b) 500 ml of 0.25M H_2SO_4.
(c) 500 ml of 0.25M H_2SO_4 diluted to a final volume of 1200 ml.
(d) 41.7 g $K_2Cr_2O_7$ in 600 ml of solution when used in a reaction where one product is Cr^{3+}.
(e) 25.0 g Na_2O dissolved in sufficient water to give 1.50 liters of solution.

(f) 0.135 equivalents of H_2SO_4 in 400 ml of solution.

5.26 Mercury is an extremely toxic substance that deactivates enzyme molecules that promote biochemical reactions. A 25.0 g sample of tuna fish taken from a large shipment was analyzed for this substance and found to contain 2.1×10^{-5} moles of Hg. By law, foods having a mercury content above 0.50 ppm cannot be sold (they cannot even be given away!). Determine whether this shipment of tuna must be confiscated.

5.27 A 1.850 g sample of a mixture of $CuCl_2$ and $CuBr_2$ was dissolved in water and mixed thoroughly with a 1.800 g portion of AgCl. After the reaction the solid, which now consisted of a mixture of AgCl and AgBr, was filtered, washed, and dried. Its mass was found to be 2.052 g. What percent of the original mixture was $CuBr_2$?

5.28 In acid solution, 45.0 ml of $KMnO_4$ solution is required to react with 50.0 ml of 0.250N $H_2C_2O_4$ to give Mn^{2+} and CO_2 as products. How many ml of this same $KMnO_4$ solution is required to oxidize 25.0 ml of 0.250N $K_2C_2O_4$ to yield, in basic solution, MnO_2 and CO_2 as products?

5.29 A 0.249 g sample of a compound containing titanium and chlorine was dissolved in water and treated with silver nitrate solution. The silver chloride that formed was found to weigh 0.694 g after being filtered, washed, and dried. What is the empirical formula of the original compound?

6
Gases

Matter is capable of existing in three different physical forms or **states:** solid, liquid, and gas. In the next three chapters we will examine the physical and chemical characteristics of these states and the transformations that occur among them. This chapter deals with the gaseous state in which the intermolecular forces of attraction (the attraction one molecule experiences toward another) are sufficiently small to allow rapid, independent movement of the molecules.

The physical behavior of a gas is, to a first approximation, independent of its chemical composition and is determined, instead, by the variables: volume, pressure, temperature, and the number of moles of the substance. Since these variables are of paramount importance, we shall begin our discussion by taking a close look at them as they apply to the gaseous state.

When a gas is introduced into a container the molecules move freely within the walls, occupying its entire volume. The volume of a gas, then, is given simply by specifying the volume of the vessel in which it is held. Since gases mix freely with one another, if several gases were present each would occupy the same volume, that is, the entire volume of the container.

Pressure is defined as force per unit area; thus it is an intensive quantity formed as a ratio of two extensive quantities, force and area. For example, if a 100 lb force is exerted on a piston (Figure 6.1) whose total area is 100 in.², the pressure acting on each square inch is only 100 lb/100 in.² or 1 psi. If the same force is applied to a piston whose total area is 1.0 in.², the pressure exerted by the piston is 100 lb/1 in.², or 100 psi. In these examples we have considered the unit area to be one square inch. The dependence of pressure on both force and the area over which it is spread has been experienced firsthand by anyone who has had someone step on his or her toe with a sharp heel. A 110 lb woman wearing heels with a diameter of one-half inch can, with the weight of her body alone, exert a pressure of *560 psi!*

If the pressure generated by a piston is applied to a fluid (a gas or liquid), as illustrated in Figure 6-1, it is transmitted uniformly in all directions so that all of the walls of the container experience the same pressure. If the piston is supported by the fluid, then the fluid also exerts an equal pressure on the piston as well as the other walls of the container.

The ability of trapped gases to exert a pressure is demonstrated every time an automobile tire is inflated. In most cases the four tires that support the car are inflated to a pressure of approximately 28 lb/in.² (psi). The reason a tire becomes flat when it "springs a leak" is because gases flow from a region

6.1
Volume and Pressure

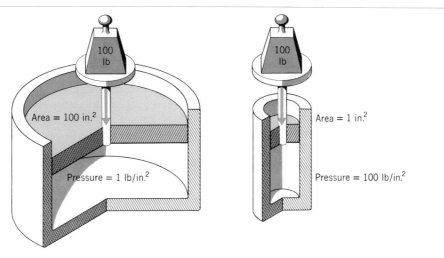

Figure 6.1
Pressure.

of high pressure to a region of lower pressure, in our example, this flow is from inside the tire to the atmosphere.

The atmosphere of the earth is a mixture of gases that exert a pressure which, logically enough, we call the atmospheric pressure. We measure this pressure using a device called a **barometer.** A barometer, Figure 6.2, can be constructed by filling a glass tube with mercury and inverting it (without spilling any) into a mercury reservoir so that the open end is submerged. In Figure 6.2 we see that the mercury in the tube does not completely pour out when it is inverted; rather it maintains a particular height (h) above the reser-

Figure 6.2
A barometer. The atmospheric pressure supports a column of mercury of height h.

voir. The height of this column is found to be independent of the diameter and length of the glass tube, as long as a space appears over the mercury in the column. This space, for all practical purposes, is a vacuum ($P \approx 0$).

To determine atmospheric pressure we compare the various pressures acting along some reference level, which we choose as the surface of the reservoir. At this level outside the inverted tube, the pressure is caused by the downward force of the gases in the atmosphere (P_{atm}). Inside the tube the pressure at the reference level is caused by the downward pull of gravity on the mercury in the column (P_{Hg}). When these two opposing pressures are exactly equal ($P_{Hg} = P_{atm}$), the mercury in the column remains stationary. Atmospheric pressure, then, is directly related to the length (h) of the column of mercury in a barometer and can, therefore, be expressed in units of centimeters (cm Hg) or, more commonly, millimeters of mercury (mm Hg). A standard unit of pressure, the **standard atmosphere** (**atm**), is defined as the pressure that will support a column of mercury 760 mm in length measured at 0° C and at sea level; that is, 1 atm = 760 mm Hg.[1]

Another unit of pressure is the **torr,** named after Evangelista Torricelli, the inventor of the barometer. One torr is equal to the pressure exerted by a column of mercury 1 mm high; that is, 1 torr = 1 mm Hg.

A barometer similar to that described above could also be constructed using water as a liquid. The length of the column of water would be considerably greater than that of a column of mercury because the atmospheric pressure would be supporting a less dense liquid ($d_{water} \approx 1.00$ g/ml, $d_{Hg} = 13.6$ g/ml).

Example 6.1
If water were the liquid in a barometer, what would be the length (h) of the water column at one atmosphere of pressure?

Solution
The density of Hg is 13.6 times that of H_2O. This means that in order to have the same mass of mercury and water, the volume of water must be 13.6 times as great as the volume of mercury. Since we are comparing columns of the same diameter, the water column would have to be 13.6 times as long as the Hg column to contain 13.6 times as much volume. Thus,

$$1 \text{ mm Hg} = 13.6 \text{ mm } H_2O$$

Then

$$1 \text{ atm} = 760 \text{ mm Hg} \times \left(\frac{13.6 \text{ mm } H_2O}{1 \text{ mm Hg}} \right) = 1.03 \times 10^4 \text{ mm } H_2O$$

In general, as the density of the liquid being supported in a column by some external pressure increases, the length of the column of liquid decreases.

---

[1] The length of the column of mercury, which is supported by atmospheric pressure, varies with both the density of the mercury and the pull of gravity on the mercury in the column. Since density varies with temperature and the pull of gravity varies with altitude, then, in the definition of the standard atmosphere, it is necessary to specify a reference temperature (0° C) as well as a reference altitude (sea level).

Often it is desirable to know the pressure of a gas present in a closed system (e.g., the pressure of gases produced during a chemical reaction). The instrument used for these pressure measurements is called a manometer. An open-end manometer (Figure 6.3) is simply a U-shaped tube containing some liquid, such as mercury. One arm of the tube is connected to a system whose pressure is to be measured while the other arm remains open to the atmosphere. When the pressure of the gas inside the system (P_{gas}) is equal to P_{atm}, the level of the liquids in both arms will be the same as shown in Figure 6.3a. If the pressure of the gas is greater than P_{atm}, the mercury in the left arm will be forced downward causing the mercury in the right arm to rise (Figure 6.3b). We obtain the pressure of the gas in this system by comparing the pressures exerted in both arms at some reference level, h_0 (in a manometer this reference level is chosen to be the height of the shortest column). The pressure exerted on the left column when $P_{gas} > P_{atm}$ is simply P_{gas}, while at the same level in the right arm the pressure is P_{atm} plus the pressure exerted by the column of mercury that rises above the reference level, P_{Hg}. When the levels are stationary,

$$P_{gas} = P_{atm} + P_{Hg}$$

The atmospheric pressure (P_{atm}) is found with the aid of a barometer, and P_{Hg} is simply the difference in the heights of the two mercury columns (Δh). Similarly when $P_{gas} < P_{atm}$, shown in Figure 6.3c, the pressure in the left arm at the reference level is $P_{gas} + P_{Hg}$ while in the right column the pressure is P_{atm}. In this case, when the columns are stationary,

$$P_{gas} + P_{Hg} = P_{atm}$$

so that

$$P_{gas} = P_{atm} - P_{Hg}$$

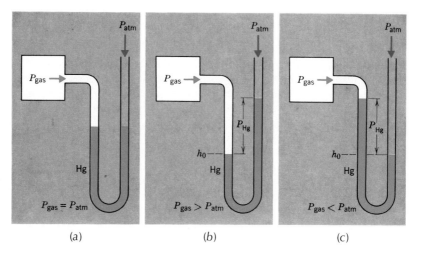

Figure 6.3
Open end manometer.

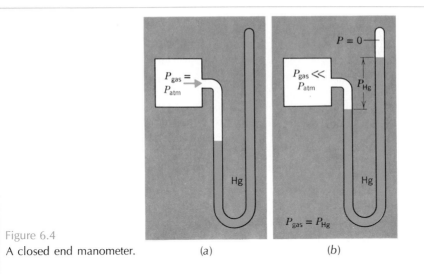

Figure 6.4

A closed end manometer.

(a)   (b)

Therefore, when $P_{gas} < P_{atm}$ the pressure of the gas in the system is found by subtracting the difference in the heights of the columns from atmospheric pressure.

A closed-end manometer is generally used for the measurement of low pressures (usually much smaller than atmospheric pressure). This manometer consists of a U-shaped tube with one arm closed and the other connected to the system as shown in Figure 6.4. When the pressure of the gas in the system is equal to P_{atm}, the right arm is completely filled while the left arm is only partially filled. If the pressure of the gas in the system is reduced, the level in the left arm will increase, which causes the level in the right arm to decrease, as shown in Figure 6.4b. At the reference level the pressure exerted on the left arm is P_{gas}, while on the right arm the pressure is P_{Hg} (the space above the mercury is a vacuum). When the columns are stationary, $P_{gas} = P_{Hg}$, and the pressure exerted by the gas in the system is simply found as the difference in the heights of liquid in the two arms of the manometer.

By using an apparatus similar to that in Figure 6.5, Robert Boyle found that at constant temperature the volume of a quantity of trapped gas decreases as the pressure on the gas is increased. Anyone who has ever used a bicycle pump is aware of this inverse relationship between the pressure and volume of a gas. As the piston of the pump is forced downward, the gas is compressed into a smaller volume while its pressure is raised (Figure 6.6). If we allow the compressed gas to escape, we can, for instance, use it to inflate a tire.

Boyle repeated his experiments many times with several different gases and found his initial observation to be a universal property of all gases. The results of his experiments can be formulated into **Boyle's law** which states that *at a constant temperature, the volume occupied by a fixed quantity of gas is inversely proportional to the applied pressure.* This can be expressed mathematically as

6.2
Boyle's Law

$$V \propto \frac{1}{P} \tag{6.1}$$

Trapped
gas

Figure 6.5

Boyle's law apparatus. Pressure on the trapped gas is increased by adding mercury to U-tube. The volume of trapped gas decreases as the pressure is raised.

The proportionality can be made into an equality by the introduction of a proportionality constant. Thus

$$V = \text{constant} \cdot 1/P$$

or

$$PV = \text{constant} \tag{6.2}$$

Figure 6.6

A bicycle pump, an example of Boyle's law.

Equation 6.2 means that for a given quantity of any gas at a constant temperature the product of its volume and pressure is equal to a constant. Table 6.1 shows the P-V relationship for 1 mole of H_2, He, and Ar at several different temperatures. We can see from Table 6.1 that the PV product is not perfectly constant at any given temperature and, in fact, undergoes greater variations at lower temperatures. It also happens that at constant temperature the changes in the PV product become pronounced at very high pressures. This behavior is observed for all real gases (N_2, O_2, etc.) and Boyle's law is actually obeyed well only at relatively high temperatures and low pressures. When a gas does obey Boyle's law, we call its behavior **ideal,** and a hypothetical gas that would obey this law at all temperatures and pressures is called an **ideal gas.** Deviations from Boyle's law that occur with real gases therefore represent nonideal behavior.

Under conditions usually encountered in the laboratory, and to the degree of precision of most of our calculations, gases generally behave ideally; that is, they obey Boyle's law. Equation 6.1, therefore, is useful in calculating the effect of a pressure change on the volume of a gas at constant temperature.

Example 6.2
If 100 ml of a gas, originally at 760 torr, is compressed to a pressure of 800 torr, at a constant temperature, what would be its final volume?

Solution 1
We have two conditions for the gas, initial and final.

Initially	$V_i = 100$ ml	and	$P_i = 760$ torr
Finally	$V_f = ?$	and	$P_f = 800$ torr

According to Boyle's law the PV products in both initial and final states are equal to a constant; thus, they must be equal to each other. This means that

$$P_i V_i = P_f V_f$$

or

$$V_f = V_i \left(\frac{P_i}{P_f} \right)$$

If we substitute our values into this equation, we get

$$V_f = 100 \text{ ml} \times \left(\frac{760 \text{ torr}}{800 \text{ torr}} \right)$$

$$V_f = 95.0 \text{ ml}$$

Solution 2
The same result can be obtained following the reasoning of Boyle's law. If the pressure is increased on a quantity of gas at constant temperature, its volume must decrease. We have just seen that the V_f is related to V_i by a ratio of pressures,

$$V_f = V_i \times (\text{ratio of pressures})$$

We know that in this problem the final volume must be smaller than the ini-

Table 6.1
PV Data for One Mole of Gas at Three Different Temperatures

Element	Temperature (°C)	P (atm)	V (liter)	P × V
Ar	−100° C	5	2.779	13.899
		25	0.510	12.750
		50	0.231	11.571
		75	0.142	10.677
		100	0.101	10.069
	0° C	5	4.465	22.326
		25	0.878	21.960
		50	0.431	21.564
		75	0.283	21.234
		100	0.209	20.969
	100° C	5	6.122	30.161
		25	1.222	30.546
		50	0.610	30.506
		75	0.407	30.501
		100	0.306	30.562
H_2	−100° C	5	2.848	14.240
		25	0.577	14.434
		50	0.294	14.701
		75	0.199	14.992
		100	0.153	15.309
	0° C	5	4.491	22.457
		25	0.910	22.739
		50	0.462	23.097
		75	0.313	23.461
		100	0.238	23.830
	100° C	5	6.132	30.662
		25	1.239	30.973
		50	0.627	31.363
		75	0.423	31.752
		100	0.321	32.141
He	−100° C	5	2.851	14.253
		25	0.579	14.493
		50	0.296	14.797
		75	0.201	15.106
		100	0.154	15.420
	0° C	5	4.489	22.447
		25	0.907	22.684
		50	0.459	22.980
		75	0.310	23.276
		100	0.236	23.573
	100° C	5	6.124	30.621
		25	1.234	30.848
		50	0.623	31.133
		75	0.419	31.417
		100	0.317	31.702

tial volume because the pressure on the gas in increasing as we go from the initial to the final state. This requires that the initial volume be multiplied by a factor smaller than one; that is,

$$V_f = V_i \times \left(\frac{760 \text{ torr}}{800 \text{ torr}} \right)$$

since

$$V_i = 100 \text{ ml}$$

then

$$V_f = 95.0 \text{ ml}$$

Jacques Charles investigated the effect of temperature changes on the volume of a given quantity of gas held at constant pressure. He found that, if a gas is heated so that the pressure remains constant, the gas will expand. If we plot data gathered in such experiments, we obtain a graph similar to that shown in Figure 6.7, where the volume is plotted against temperature in degrees Celsius. Each line represents the results of a series of measurements performed on a different quantity of gas. The straight lines obtained show that there is a direct proportionality between the volume of a gas and its temperature. All gases ultimately condense if they are cooled to a sufficiently low temperature, and the solid portions of the lines in Figure 6.7 represent the temperature region above the condensation (liquefaction) point, where the substance is in the gaseous state. When these straight lines are extended back (extrapolated) until they intersect with the temperature axis, it is found that they all meet at the same point, $t = -273.15°$ C. It is also found that, regardless of the gas, similar extrapolations yield the same temperature. This point represents a temperature at which all gases, if they did not condense, would have a volume of zero, and below which they would have a negative volume. Since negative volumes are impossible, this temperature, therefore, must represent the lowest temperature possible and is called **absolute zero.**

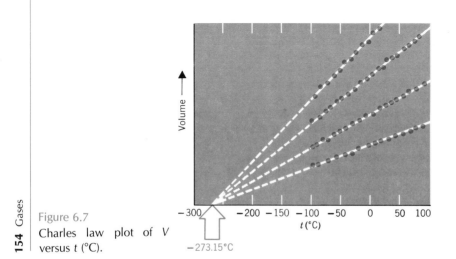

Figure 6.7
Charles law plot of V versus t (°C).

Another temperature scale, called the **Kelvin scale** (named after an English physicist, Lord Kelvin) or the **absolute scale,** has been developed where zero degrees Kelvin corresponds to $-273.15°$ C. For general purposes we shall use only three significant figures; that is, we will take $0°$ K $= -273°$ C. The size of the Kelvin degree is equal to the size of the Celsius degree and any temperature in degrees Celsius (t) can be expressed in degrees Kelvin (T) by merely adding 273 to the Celsius temperature; that is, $T(°K) = t(°C) + 273$. The absolute temperature scale is *always* used when the temperature enters numerically into a computation involving pressures and volumes of gases.

The relationship that exists between the volume of a gas and its absolute temperature, represented in Figure 6.7, is summarized by **Charles' law,** which states that, *at constant pressure, the volume of a given quantity of a gas varies directly with its absolute temperature*. Writing Charles' law mathematically we have

$$V \propto T \tag{6.3}$$

Making the proportionality an equality and rearranging, we obtain

$$\frac{V}{T} = \text{constant} \tag{6.4}$$

If Charles' law were strictly obeyed, gases would not condense when they are cooled. Condensation is, therefore, considered to be nonideal behavior, and all real gases behave more and more nonideally as their condensation temperatures are approached. This means that gases behave in an ideal fashion only at relatively high temperatures and low pressures, the same conclusion drawn in our discussion of Boyle's law. An example of the application of Charles' law is shown below.

Example 6.3
What would be the new volume of a gas, if 250 ml, originally at $27°$ C, is heated at a constant pressure to $35°$ C?

Solution
We know, from Charles' law, that the final volume must be larger than the initial volume because the temperature is increasing at constant pressure. From Charles' law we can see that the initial and final volumes are related as

$$V_f = V_i \times \text{(ratio of temperatures)}$$

Following the same reasoning that was applied to the previous example, the ratio of temperatures must be larger than one in order to have the final volume larger than the initial volume. Thus

$$V_f = 250 \text{ ml} \times \left(\frac{308° \text{ K}}{300° \text{ K}}\right)$$

$$V_f = 257 \text{ ml}$$

When working with gases it is useful to define a reference set of conditions of temperature and pressure. These conditions, known as **standard tem-**

perature and pressure, or simply **STP,** are 0° C (273° K) and one standard atmosphere (760 torr).

Example 6.4

What would be the volume of a gas at STP, if it was found to occupy a volume of 255 ml at 25° C and 650 torr?

Solution

In this problem the pressure is increasing ($P_i = 650$ torr and $P_f = 760$ torr) the final volume, therefore, we have to combine Boyle's law with Charles' law. In this case the final volume equals the initial volume times a ratio of pressures and a ratio of temperatures.

$$V_f = V_i \times \text{(ratio of pressures)} \times \text{(ratio of temperatures)}$$

In this problem the pressure is increasing ($P_i = 650$ torr and $P_f = 760$ torr) and, according to Boyle's law, the volume must decrease. The ratio of pressures, therefore, must be less than one.

$$\text{Ratio of pressures} = \frac{P_i}{P_f} = \frac{650 \text{ torr}}{760 \text{ torr}}$$

The temperature in this problem is decreasing ($T_i = 298°$ K and $T_f = 273°$ K), and according to Charles' law the volume must also decrease. Therefore the ratio of temperatures must also be less than one.

$$\text{Ratio of temperatures} = \frac{T_f}{T_i} = \frac{273° \text{ K}}{298° \text{ K}}$$

The final volume, therefore, is

$$V_f = V_i \left(\frac{P_i}{P_f}\right) \left(\frac{T_f}{T_i}\right)$$

$$V_f = 255 \text{ ml} \times \left(\frac{650 \text{ torr}}{760 \text{ torr}}\right) \times \left(\frac{273° \text{ K}}{298° \text{ K}}\right)$$

$$V_f = 200 \text{ ml}$$

When two or more gases, which do not react chemically, are placed in the same container, the pressure exerted by each gas in the mixture is the same as it would be if it were the only gas in the container. The pressure exerted by each gas in a mixture is called its **partial pressure** and, as observed by John Dalton, the total pressure exerted by a mixture of gases is equal to the sum of the partial pressures of each gas in the mixture. This statement, known as **Dalton's law of partial pressures,** can be expressed as

$$P_T = p_a + p_b + p_c + \ldots$$

where P_T is the total pressure of the mixture (which could be measured with a manometer) and p_a, p_b, and p_c are the partial pressures of gases a, b, and c, respectively. For example, if nitrogen, oxygen, and carbon dioxide were placed in the same vessel, the total pressure of the mixture would be

$$P_T = p_{N_2} + p_{O_2} + p_{CO_2}$$

In this mixture, if the partial pressure of nitrogen were 200 torr, that of oxygen 250 torr, and that of carbon dioxide 300 torr, then the total pressure of the mixture is

$$P_T = 200 \text{ torr} + 250 \text{ torr} + 300 \text{ torr}$$
$$P_T = 750 \text{ torr}$$

Dalton's law can be useful in determining the pressure resulting from the mixing of two gases that were originally in separate containers, as shown by the following example.

Example 6.5
If 200 ml of N_2 at 25° C and a pressure of 250 torr is mixed with 350 ml of O_2 at 25° C and a pressure of 300 torr, so that the resulting volume is 300 ml, what would be the final pressure of the mixture at 25° C?

Solution
We know, from Dalton's law, that we can treat each gas in a mixture as if it were the only gas present. We can, therefore, independently calculate the new pressures of N_2 and O_2 when they are placed into the 300 ml container at a constant temperature.

Following the method outlined in Examples 2 to 4, we see that the final pressure of N_2 is equal to the initial pressure times some ratio of the volumes; that is,

$$p_f = p_i \times \text{(ratio of volumes)}$$

Since the volume of N_2 is increasing, then p_f must be smaller than p_i, and the ratio of volumes must be smaller than one. This means that

$$p_f = p_i \times \left(\frac{V_i}{V_f}\right)$$

or

$$p_{N_2} = 250 \text{ torr} \times \left(\frac{200 \text{ ml}}{300 \text{ ml}}\right)$$

Therefore

$$p_{N_2} = 167 \text{ torr}$$

For O_2 we see that the volume is decreasing and, therefore, the final pressure must be greater than the initial pressure, which means that

$$p_{O_2} = p_i \times \left(\frac{V_i}{V_f}\right)$$

or

$$p_{O_2} = 300 \text{ torr} \times \left(\frac{350 \text{ ml}}{300 \text{ ml}}\right)$$

Therefore

$$p_{O_2} = 350 \text{ torr}$$

The total pressure of the mixture is

$$P_T = p_{N_2} + p_{O_2} = 167 \text{ torr} + 350 \text{ torr}$$

or

$$P_T = 517 \text{ torr}$$

Gases prepared in the laboratory are quite often collected by the displacement of water, as shown in Figure 6.8. A gas collected in this manner always becomes "contaminated" with gaseous water molecules that have escaped from the liquid. The pressure caused by the presence of these gaseous water molecules is called the **vapor pressure** of water and changes with temperature. (Vapor pressure will be discussed in more detail in Chapter 8.) Since these gaseous water molecules also exert a partial pressure, then the total pressure exerted by a gas collected in this fashion would be

$$P_T = p_{gas} + p_{H_2O}$$

If the level of the water is the same inside the collection flask as outside, then the pressure inside must also be the same as outside, namely, atmospheric pressure. The atmospheric pressure can be determined with the aid of a barometer, and the vapor pressure of water at various temperatures is given in Table 6.2. The partial pressure of the pure gas, therefore, is

$$p_{gas} = P_T - p_{H_2O}$$

Example 6.6
A student generates oxygen gas in the laboratory and collects it in a manner similar to that shown in Figure 6.8. He collects the gas at 25° C until the levels of the water inside and outside the flask are equal. If the volume of the gas is 245 ml and the atmospheric pressure is 758 torr:
  (a) What is the partial pressure of O_2 gas in the "wet" gas mixture at 25° C?
  (b) What would be the volume of dry oxygen at STP?

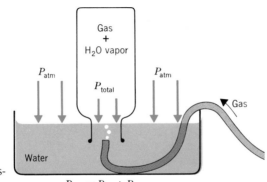

Figure 6.8
Collection of a gas by displacement of water.

$$P_{total} = P_{gas} + P_{H_2O \text{ vapor}}$$

Table 6.2
Vapor Pressure of Water in Torr as a Function of Temperature

Temperature (°C)	Vapor Pressure (torr)
−10	2.2
−5	3.2
0	4.6
5	6.5
10	9.2
15	12.8
20	17.5
25	23.8
30	31.8
35	42.2
40	55.3
45	71.9
50	92.5
55	118.5
60	149.4
65	187.5
70	233.7
75	289.1
80	355.1
85	433.6
90	525.8
95	633.9
100	760.0
105	906.1

Solution

(a) For gases collected in this manner we know that

$$P_T = p_{gas} + p_{H_2O}$$

Substituting p_{O_2} for p_{gas} and rearranging, we have

$$p_{O_2} = P_T - p_{H_2O}$$

The partial pressure of water at 25° C is 23.8 torr, according to Table 6.2, and the atmospheric pressure was given as $P_{atm} = 758$ torr. Therefore the partial pressure of O_2 is

$$p_{O_2} = 758.0 \text{ torr} - 23.8 \text{ torr}$$
$$p_{O_2} = 734.2 \text{ torr}$$

(b) From Boyle's law we know that the final volume must be smaller than the initial volume since the pressure is increasing, and from Charles' law, that the final volume is less than the initial volume if the temperature is decreasing. We should be able to write

$$V_f = V_i \times (\text{ratio of pressures}) \times (\text{ratio of temperatures})$$

where, in our problem, the pressure ratio and the temperature ratio must both be less than one. Therefore,

$$V_f = V_i \times \left(\frac{P_i}{P_f}\right) \times \left(\frac{T_f}{T_i}\right)$$

$$V_f = 245 \text{ ml} \times \left(\frac{734.2 \text{ torr}}{760 \text{ torr}}\right) \times \left(\frac{273° \text{ K}}{298° \text{ K}}\right)$$

$$V_f = 218 \text{ ml at STP}$$

In experiments similar to those performed by Charles, Joseph Gay-Lussac observed that if the volume of a given quantity of gas is held constant (e.g., in any closed vessel), the pressure of the gas varies directly with the absolute temperature; that is,

6.5
Laws of Gay-Lussac

$$P \propto T$$

or

$$\frac{P}{T} = \text{constant} \tag{6.5}$$

This is a mathematical statement of the **law of Gay-Lussac.**

Example 6.7
What would be the pressure of a gas, originally at 760 torr if the temperature is lowered from 35° C to 25° C at a constant volume?

Solution
We should now be able to write that

$$P_f = P_i \times (\text{ratio of temperatures})$$

From Gay-Lussac's law this means that $P_f < P_i$ when the temperature decreases, which means

$$P_f = 760 \text{ torr} \times \left(\frac{298° \text{ K}}{308° \text{ K}}\right)$$

$$P_f = 735 \text{ torr}$$

Gay-Lussac also studied the volume changes that occurred when two or more gases were allowed to react with one another to form gaseous products at a constant temperature and pressure. He observed that the volumes of the gases that reacted and of those that were produced were related to one another in a simple fashion. For example, when hydrogen and oxygen are placed in a vessel and allowed to react with each other to form gaseous water, two volumes of hydrogen always react with one volume of oxygen to form two volumes of gaseous water. This can be expressed in the form of an equation as

2 volumes hydrogen + 1 volume oxygen $\longrightarrow$ 2 volumes gaseous water

Similarly when one volume of hydrogen reacts with one volume of chlorine, two volumes of hydrogen chloride gas are produced; that is,

1 volume hydrogen + 1 volume chlorine $\longrightarrow$ 2 volumes hydrogen chloride

It could further be shown that the reaction between hydrogen and nitrogen to form ammonia is

1 volume nitrogen + 3 volumes hydrogen $\longrightarrow$ 2 volumes ammonia

Such observations form the basis of **Gay-Lussac's law of combining volumes,** which states that *when reactions take place in the gaseous state, under conditions of constant temperature and pressure, the volumes of reactants and products can be expressed as ratios of small whole numbers.*

The significance of Gay-Lussac's observation was later recognized by Amadeo Avogadro. He proposed what is now known as **Avogadro's principle:** *under conditions of constant temperature and pressure, equal volumes of gas contain equal numbers of molecules.* Since equal numbers of molecules mean equal numbers of moles (recall from Chapter 2 that there are 6.02×10^{23} molecules in one mole of *any* substance), the number of moles of any gas is directly related to its volume:

$$V \propto n \qquad\qquad (6.6)$$

where n is the number of moles of gas.

6.6
The Ideal Gas Law

We have thus far discussed three volume relationships that an ideal gas obeys. These are

Boyle's law	$V \propto \dfrac{1}{P}$
Charles' law	$V \propto T$
Avogadro's law	$V \propto n$

We can combine these to obtain

$$V \propto n(1/P)(T)$$

or

$$V \propto \frac{nT}{P} \qquad\qquad (6.7)$$

Equation 6.7 reduces to an expression of Boyle's law during a process where the volume changes as a result of a pressure change only. Since n and T remain constant, volume is only proportional to the pressure; that is,

$$V \propto 1/P \qquad \text{at constant } n \text{ and } T$$

Similarly we can see that Equation 6.7 becomes Charles' law for a volume change at constant n and P.

$$V \propto T \qquad \text{at constant } n \text{ and } P$$

Avogadro's principle is seen as

$$V \propto n \qquad \text{at constant } T \text{ and } P$$

The proportionality in Equation 6.7 can be made into an equality by the introduction of a proportionality constant, R, called the **universal gas constant.** Equation 6.7 then becomes

$$V = \frac{nRT}{P}$$

or

$$PV = nRT \qquad (6.8)$$

Equation 6.8 is obeyed only by the hypothetical ideal gas (when a real gas comes very close to obeying this equation its behavior is said to be ideal) and is, therefore, a mathematical statement of the **ideal gas law** or **perfect gas law.** It is also called the **equation of state for an ideal gas** because it relates those variables (P, V, n, T) that specify the physical properties of the gas. If any three of these are given, the fourth variable can have only one value as determined by Equation 6.8.

For most real gases under conditions of temperature and pressure encountered in the laboratory the ideal gas law can be used quite accurately to describe their behavior. If extremely accurate computations are desired, however, Equation 6.7 cannot be used.

In order to use the ideal gas law in calculations, it is first necessary to evaluate the gas constant R. To do this it is necessary to insert into the equation, values for P, V, n, and T. One mole of gas at STP has been found to occupy, on the average, 22.4 liters. We shall assume this to be the volume occupied by one mole of an ideal gas at STP and refer to it as the **molar volume.** For real gases, the molar volume actually fluctuates about this average as shown in Table 6.3. Rearranging Equation 6.8 to solve for R and substituting into the equation the molar volume at STP, we have, for one mole of an ideal gas

$$R = \frac{PV}{nT}$$

$$R = \frac{(1 \text{ atm})(22.4 \text{ liters})}{(1 \text{ mole})(273° \text{ K})}$$

$$R = 0.0821 \, \frac{\text{liter-atm}}{\text{mole°K}}$$

The constant R may have other numerical values depending on the units of pressure and volume. For example, if the standard pressure were expressed in units of torr and the molar volume in milliliters, then R would be

$$R = 6.23 \times 10^4 \, \frac{\text{ml torr}}{\text{mole° K}}$$

Table 6.3
Molar Volumes of Several Real Gases at STP

Substance	Molar Volume (liters)
Oxygen	22.414
Nitrogen	22.413
Hydrogen	22.410
Helium	22.398
Argon	22.401
Carbon dioxide	22.414
Ammonia	22.423

The most useful values of R with their corresponding units of pressure and volume are listed in Table 6.4.

The ideal gas law has many laboratory applications as shown by the following examples.

Example 6.8
What volume would 25.0 g of O_2 occupy at STP?

Solution
We can find the number of moles of oxygen gas by dividing the mass of oxygen present by the molecular weight of O_2. The number of moles of oxygen is

$$\text{Number of moles of } O_2 \; (n_{O_2}) = \frac{\text{grams of } O_2}{\text{molecular weight of } O_2}$$

$$n_{O_2} = \frac{25.0 \text{ grams}}{32.0 \text{ grams/mole}}$$

$$n_{O_2} = 0.781 \text{ moles}$$

Table 6.4
The Gas Constant R in Various Units

Units of PV	Value of R ($R = PV/nT$)
Atmosphere-liter	$0.0821 \; \dfrac{\text{atmosphere liter}}{\text{mole}°K}$
Torr-liter	$62.36 \; \dfrac{\text{torr-liter}}{\text{mole}°K}$
Calorie	$1.987 \; \dfrac{\text{calorie}}{\text{mole}°K}$

Rearranging the ideal gas equation to solve for volume, we have

$$V = \frac{nRT}{P}$$

The volume at STP can be obtained by substituting $n = 0.781$ mole, $R = 0.0821$ l-atm/mole $°$ K, $T = 273°$ K, and $P = 1$ atm.

$$V = \frac{(0.781 \text{ mole}) \times \left(\frac{0.0821 \text{ l-atm}}{\text{mole } ° \text{ K}}\right) \times (273° \text{ K})}{1 \text{ atm}}$$

$$V = 17.5 \text{ liter}$$

Example 6.9
A student collected natural gas from a laboratory gas jet at 25° C in a 250 ml flask until the pressure of the gas was 550 torr. He then determined that the gas sample weighed 0.118 grams at a temperature of 25°C. From these data, calculate the molecular weight of the gas.

Solution
    We can determine the number of moles of gas present using the ideal gas law.

$$n = \frac{PV}{RT}$$

$$n = \frac{\left(\frac{550 \text{ torr}}{760 \text{ torr/atm}}\right) \times (0.250 \text{ liter})}{\left(0.0821 \frac{\text{l-atm}}{° \text{ K mole}}\right) \times (298° \text{ K})}$$

$$n = 0.00739 \text{ moles}$$

The molecular weight is numerically equal to the weight, in grams, of one mole. We know that 0.00739 moles weigh 0.118 grams; therefore one mole would weigh

$$1 \text{ mole} \times \left(\frac{0.118 \text{ grams}}{0.00739 \text{ mole}}\right) = 16.0 \text{ grams}$$

Hence the molecular weight is 16.0 amu.

Example 6.10
A student determined the density of a gas to be 1.340 g/liter at 25° C and 760 torr. In a separate experiment he determined that the gas was composed of 79.8% carbon and 20.2% hydrogen.
(a) What is the empirical formula of the compound?
(b) What is its molecular weight?
(c) What is the molecular formula of the compound?

Solution
    (a) Following the procedure outlined in Section 2.5 we find the empirical formula of the carbon-hydrogen compound to be

$$79.8 \text{ g C} \times \left(\frac{1 \text{ mole C}}{12.0 \text{ g C}}\right) = 6.65 \text{ mole C}$$

$$20.2 \text{ g H} \times \left(\frac{1 \text{ mole H}}{1.01 \text{ g H}}\right) = 20.0 \text{ mole H}$$

The empirical formula is $C_{\frac{6.65}{6.65}}H_{\frac{20.0}{6.65}}$ or CH_3 which would give an empirical formula weight of $CH_3 = 15$.

(b) We know that one mole of gas occupies 22.4 liters at STP. The conditions at which the density was determined were standard pressure but not standard temperature. We must, therefore, calculate the volume that one liter would occupy at 0° C. We know, from Charles' law, that a temperature decrease brings about a volume decrease. Since the final volume must be smaller than the initial volume, then

$$V_f = V_i \times \left(\frac{T_f}{T_i}\right)$$

$$V_f = 1 \text{ liter} \times \left(\frac{273° \text{ K}}{298° \text{ K}}\right)$$

$$V_f = 0.916 \text{ liter}$$

Therefore, at STP the density of the gas would be

$$d = \frac{1.340 \text{ g}}{0.916 \text{ liter}}$$

The weight of gas in 22.4 liters would be

$$\left(\frac{1.340 \text{ g}}{0.916 \text{ liter}}\right) \times 22.4 \text{ liters} = 32.8 \text{ g}$$

Since this is the weight of one mole, the molecular weight of the gas must be 32.8 amu

(c) We see that the molecular weight is approximately twice the empirical formula weight, which means that the molecular formula must be

$$(CH_3)_2 \quad \text{or} \quad C_2H_6$$

This is a substance called ethane.

**6.7
The
Determination of
Atomic Weights:
Cannizzaro's
Method**

In Section 1.9 we saw that in the early part of the nineteenth century accurate atomic weights could not be determined because no one had yet found a sure method of determining chemical formulas. The first hint of success came from Avogadro who interpreted the results of Gay-Lussac's investigations of the combining volumes of gases in chemical reactions. According to Gay-Lussac, in the reaction between hydrogen and chlorine, one volume of hydrogen totally reacts with one volume of chlorine to produce two volumes of hydrogen chloride. We know, from Avogadro's principle, that equal volumes of gas at the same temperature and pressure contain equal numbers of molecules and, therefore, two volumes of hydrogen chloride contain twice as many molecules as one volume of hydrogen. From the law of conservation of matter the total number of hydrogen atoms must be the same in both the two volumes of hydrogen chloride and the one volume of hydrogen. Since hydrogen chloride contains at least one hydrogen atom per molecule, the only way one hydrogen molecule could react to produce two hydrogen chloride molecules would be if each hydrogen molecule contained at least two hydrogen atoms. Thus, we see that from Avogadro's principle we can deduce that the hydrogen molecule must be at least diatomic. Similar reasoning leads to the conclusion that chlorine is also diatomic. We can now confidently write the equation for the reaction between hydrogen and chlorine to produce hydrogen chloride as

$$H_2(g) \ + \ Cl_2(g) \longrightarrow 2HCl(g)$$

1 volume   1 volume   2 volumes

The reasoning that we have just applied does not rule out the possibility of hydrogen molecules having the formula H_4, chlorine, Cl_4 and the product of their reaction, the formula H_2Cl_2. It was Stanislao Cannizzaro, a countryman of Avogadro, who demonstrated that the principle of Avogadro could be used to determine atomic weights, and hence chemical formulas. If the density of a number of gases are measured at the same temperature and pressure, the weights of one liter of the various gases can be obtained (Table 6.5). Since, according to Avogadro, at the same temperature and pressure equal volumes contain equal numbers of molecules, these densities correspond to the weights of equal numbers of gas molecules in each case.

It is also possible to determine, by chemical analysis, the percent composition of each gas. If we multiply the weight percent of each element by the weight of one liter of the gas, we obtain the weight contributed by each element toward the total mass of the collection of molecules in one liter. Let us take a look at some of the numbers obtained for hydrogen-containing compounds. We see in Table 6.5 that the weight of hydrogen in one liter of hydrogen gas is twice that in one liter of hydrogen chloride. Similarly, the weights of hydrogen in one liter of chloromethane, ethylene, ethane, and propane are equal to three, four, six, and eight times the weight of hydrogen in one liter of hydrogen chloride. If we assign hydrogen an atomic weight of 1.00 amu (as Cannizzaro did), then the molecules of hydrogen chloride, hydrogen, chloromethane, ethylene, ethane, and propane contain 1, 2, 3, 4, 6, and 8 hydrogen atoms respectively.

Have we, in fact, ruled out the possibility that the formula of the hydrogen molecule is H_4 or that of hydrogen chloride is H_2Cl_2? The answer is no. However, if we examine a very large number of hydrogen containing compounds, we always find the weight of hydrogen in one liter of a gas to be an integral multiple of that in one liter of hydrogen chloride. If hydrogen chloride is H_2Cl_2, then the formulas of all of these other gas molecules must contain an integral multiple of two hydrogen atoms (i.e., an even number of H atoms). Chloromethane, ethylene, ethane, and propane, for example, would then contain 4, 6, 8, 12, and 16 hydrogen atoms, respectively. Since we would expect, sooner or later, to encounter some compound with an odd number of hydrogen atoms, it seems reasonable to assume that the multiples that are observed are multiples of the weight of a single hydrogen atom.

Table 6.5
Data for the Determination of Atomic Weights by Cannizzaro's Method

Gas	Density at STP (g/liter)	Percent Composition	Weight in Grams of Each Element in One Liter	Weight of Hydrogen in One Liter of Gas / Weight of Hydrogen in One Liter of Hydrogen Chloride
Hydrogen	0.0899	100	0.0899	$1.99 \approx 2$
Chlorine	3.17	100	3.17	—
Hydrogen chloride	1.63	H; 2.76 Cl; 97.3	0.0450 1.59	$1.00 = 1$
Carbon dioxide	1.96	C; 27.3 O; 72.7	0.535 1.42	—
Methane	0.714	C; 74.9 H; 25.1	0.535 0.179	$3.97 \approx 4$
Ethane	1.34	C; 80.0 H; 20.0	1.07 0.268	$5.96 \approx 6$
Ethylene	1.25	C; 85.6 H; 14.4	1.07 0.180	$4.00 = 4$
Propane	1.96	C; 81.8 H; 18.2	1.61 0.358	$7.96 \approx 8$
Carbon tetrachloride	6.88	C; 7.8 Cl; 92.2	0.54 6.34	—
Methyl chloride	2.25	C; 23.8 Cl; 70.2 H; 5.99	0.537 1.58 0.135	$3.00 = 3$

Hydrogen chloride therefore has the formula, HCl, hydrogen gas is H_2, and chlorine gas is Cl_2.

A similar examination of the weights of carbon in many gaseous carbon compounds leads to the conclusion that the atomic weight of carbon is 12.0 amu (based on hydrogen having an atomic weight of 1.00 amu).

Cannizzaro's method of determining the atomic weights of the elements represented one of the major breakthroughs in the development of chemistry because it permitted, for the first time, the unambiguous assignment of chemical formulas. With this stumbling block removed, rapid advances in the science of chemistry were made possible.

6.8
Graham's Law of Effusion

The ability of a gas to mix spontaneously with and spread throughout another gas, a process known as diffusion, is demonstrated every time we drive near a skunk that didn't quite make it across the road. It isn't long before the occupants of the car are in need of "a little fresh air." Effusion, on the other hand, is the process by which a gas, under pressure, escapes from one

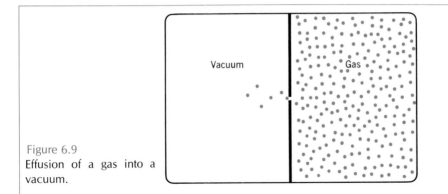

Figure 6.9
Effusion of a gas into a vacuum.

chamber of a vessel to another by passing through a very small opening, or orifice. This process is demonstrated in Figure 6.9.

Thomas Graham demonstrated that when the rates of effusion of several gases are compared, the less dense gases (lighter gases) always effuse faster than the more dense ones. When the rates are compared under identical conditions of temperature and pressure the best agreement between the rate of effusion and density is obtained when the rate is expressed as being inversely proportional to the square root of the density of the gas. This statement, known as **Graham's law,** can be expressed mathematically as

$$\text{Rate of effusion} \propto \sqrt{\frac{1}{d}}$$

The rate of effusion of two gases (labeled simply A and B) can be compared by dividing the rate of one by the other; that is,

$$\frac{\text{Rate of effusion } (A)}{\text{Rate of effusion } (B)} = \sqrt{\frac{d_B}{d_A}}$$

With the aid of the ideal gas equation we can show that the density of a gas is directly proportional to its molecular weight. We know from Example 6.8 that the number of moles of any substance present is obtained by dividing the weight, in grams, of that substance by its molecular weight; that is,

$$\text{Number of moles } (n) = \frac{\text{weight in grams } (g)}{\text{molecular weight } (M)}$$

or simply

$$n = \frac{g}{M} \tag{6.9}$$

Making this substitution into the ideal gas equation, we obtain

$$PV = \left(\frac{g}{M}\right) RT$$

which can be rearranged to solve for M.

$$M = \left(\frac{g}{V}\right) \frac{RT}{P}$$

Density (d) is equal to g/V (mass/volume) and, therefore, the above equation can be written as

$$M = d\,\frac{RT}{P}$$

This equation[2] demonstrates that at any particular temperature and pressure the molecular weight of a gas is indeed directly proportional to its density. Graham's law can now be written to include this relationship, as

$$\frac{\text{Rate of effusion } (A)}{\text{Rate of effusion } (B)} = \sqrt{\frac{d_B}{d_A}} = \sqrt{\frac{M_B}{M_A}}$$

where M_A and M_B are the molecular weights of gases A and B, respectively.

Example 6.11
Which gas will effuse faster, ammonia or carbon dioxide? What are their relative rates of effusion?

Solution
    The molecular weight of CO_2 is 44 and that of NH_3 is 17. Therefore NH_3 will effuse faster. We can calculate how much faster with the aid of the last equation.

$$\frac{\text{Rate of effusion } (NH_3)}{\text{Rate of effusion } (CO_2)} = \sqrt{\frac{M_{CO_2}}{M_{NH_3}}} = \sqrt{\frac{44}{17}} = 1.6$$

Therefore the rate of effusion of NH_3 is 1.6 times faster than the rate of CO_2.

6.9
The Kinetic Molecular Theory

Thus far in our discussion of the gaseous state we have seen that several laws governing gaseous behavior were formulated to account for experimental observations. We now need a theory to explain how gases function in obeying these laws. The kinetic molecular theory, which treats gases as being composed of randomly moving molecules, adequately explains these laws.
    The observations of Robert Brown in 1827 laid the groundwork for the idea that the molecules comprising a gas are in continuous random motion. Brown observed that when macromolecular particles (particles large enough to be seen only with the aid of a microscope) were suspended in water, the particles underwent continuous erratic motion. He first observed this motion by following the path of grains of pollen that were suspended in water. Similar random motion of any macromolecular particle when suspended in a liquid or gas is called **Brownian motion.** This motion, which becomes even more violent with increasing temperature, is believed to be caused by the suspended particle being constantly battered about by the molecules of the medium (liquid or gas). Brownian motion, therefore, is strong support for the idea that gases are composed of molecules that are constantly and randomly moving.

POSTULATES OF THE KINETIC MOLECULAR THEORY. The following are the five basic postulates of the kinetic molecular theory. Since the real test of

[2] $M = d\,(RT/P)$ and $M = g\,RT/PV$ are also useful in solving problems of the type discussed in Examples 6.9 and 6.10.

any theory manifests itself in how well it agrees with experimental observations, we shall look at the foundation upon which each postulate is based and then see how the theory accounts for the observed gas laws.

Postulate 1. A gas is composed of tiny molecules that have a negligibly small volume and that are separated from each other by relatively large distances. Gases therefore are mostly empty space.

Basis. One of the fundamental characteristics of a gas is its very high compressibility (particularly when compared to liquids and solids). According to this postulate, it is this empty space that can be occupied when gases are compressed.

Postulate 2. There are no intermolecular forces of attraction (the attraction of one molecule toward another) between the molecules of a gas. The molecules in a gas, therefore, move about independently of each other.

Basis. This postulate finds its support in the fact that even in those situations where the intermolecular forces of attraction are considered to be a maximum, for example, a highly compressed gas, the container holding them is completely filled. Since all gases spontaneously fill a container, there must be no appreciable intermolecular forces of attraction between the molecules.

Postulate 3. The molecules of a gas undergo constant motion, and their normally straight line paths are interrupted only by collision with each other or with the walls of the container. All collisions occur with no net loss in kinetic energy of the molecules, and are said to be **elastic collisions.**

Basis. Since it has been postulated that there are no intermolecular forces present to alter the paths of molecules, they should travel in straight lines. Because there are so many molecules present that are traveling extremely fast (the average speed of an O_2 molecule can be calculated to be nearly 1000 miles per hour at $0°$ C), many collisions between the molecules occur each second. A single gas molecule can undergo several million collisions each second, and each collision could cause the molecule to change its direction. If randomly moving molecules were constantly bombarding a macromolecular particle, they could cause it to move randomly. Brownian motion, then, is explained by considering that the suspended particle is being hit by several gas molecules at the same time, as shown in Figure 6.10. If the particle is struck equally on all sides (Figure 6.10a), it will not move. However, there is a great probability that, because of its small size, the particle will receive an unequal number of impacts on opposite sides (Figure 6.10b) thereby causing the particle to move. The next instant its direction could similarly be changed, and the process continues with the particle moving in a random fashion.

The fact that molecular motion is continuous means that the intermolecular collisions are *elastic.* When an elastic collision occurs, there is no net loss in the kinetic energies of the two molecules. This means that the total kinetic energy of the two colliding molecules before the collision must equal their total kinetic energy after the collision. If, as a result of a collision, one molecule loses energy, the other molecule must gain it in order to keep the total kinetic energy constant. In one mole of gas at STP approximately 10^{30} collisions occur each second, and

(a) Equal number of collisions on all sides — no displacement of particle.

(b) Unequal number of collisions on opposite sides — particle moves as shown by large arrow.

Figure 6.10
Brownian motion.

each could involve an exchange of kinetic energy. However, *overall* there is no net change in the total kinetic energy of the gas. If this were not the case and each collision resulted in a decrease in kinetic energy of the two molecules (inelastic collision), then each molecule would slow down upon each collision, since K.E. $= \frac{1}{2}mv^2$. For example, an inelastic collision occurs when a golf ball is thrown against a sidewalk. Initially the ball bounces very high in the air. But with each successive collision with the cement it loses energy and before long comes to rest. In a gas, inelastic collisions mean that the molecules would slow down and settle to the bottom of the container. Since this never occurs, the intermolecular collisions must be elastic.

Postulate 4. The pressure of a gas is the result of collisions between the gas molecules and the walls of the container.

Basis. We have defined pressure as a force per unit area. When the molecules in a gas collide with the walls of a container, they each exert a force on that wall. The total force acting on a unit area is directly related to both the force of each collision and the frequency of the collisions with the wall. According to this postulate, an increase in pressure can be achieved by increasing either the force or frequency of molecule-wall collisions.

An increase in the number of collisions per second can be brought about by packing the molecules closer together so that there are more molecules in the vicinity of any given area of a wall. A decrease in the volume of a container does just this by forcing the molecules closer together; therefore, a decrease in volume is accompanied by an increase in pressure — Boyle's law.

Postulate 5. In a gas individual molecules possess different kinetic energies; however, the average kinetic energy of all the molecules collectively is directly proportional to the absolute temperature of the gas; that is, K.E. $\propto T$.

Basis. This postulate leads directly to Graham's law. If two gases, which have different molecular weights, are at the same temperature, then, according to Postulate 5, their average kinetic energies are the same. This means that

$$\overline{K.E.}_A = \overline{K.E.}_B$$

or

$$\tfrac{1}{2}m_A\overline{v_A^2} = \tfrac{1}{2}m_B\overline{v_B^2} \tag{6.10}$$

where $\overline{v^2}$ is called the **mean square velocity** of the molecules, and is the average of the velocities-squared of all the molecules; that is,

$$\overline{v^2} = \frac{v_1^2 + v_2^2 + v_3^2 + \cdots}{n_T}$$

where v_1, v_2, v_3, etc., represent the velocities of molecules 1,2,3, etc. and n_T is the total number of molecules present. Equation 6.10 can be rearranged to give

$$\frac{\overline{v_A^2}}{\overline{v_B^2}} = \frac{m_B}{m_A}$$

Taking the squase root of both sides, we have that

$$\frac{\overline{v_A}}{\overline{v_B}} = \sqrt{\frac{m_B}{m_A}} \tag{6.11}$$

where $\overline{v}$ is called the root-mean-square velocity. We have seen earlier that for a given number of moles of a gas that the weight of the gas present is directly related to the molecular weight by Equation 6.9, which means that

$$M \propto m$$

Therefore, we can substitute M_A and M_B for m_A and m_B into Equation 6.11 and arrive at

$$\frac{\overline{v_A}}{\overline{v_B}} = \sqrt{\frac{M_B}{M_A}}$$

The rate at which gases effuse should be directly proportional to the velocity of their molecules, with faster molecules effusing at a higher rate. Thus we are led to conclude that

$$\frac{\text{Rate of effusion (A)}}{\text{Rate of effusion (B)}} = \frac{\overline{v_A}}{\overline{v_B}} = \sqrt{\frac{M_B}{M_A}}$$

or simply

$$\frac{\text{Rate of effusion (A)}}{\text{Rate of effusion (B)}} = \sqrt{\frac{M_B}{M_A}}$$

which is Graham's law.

We have seen that the kinetic molecular theory accounts for Boyle's law and Graham's law. The other gas laws can be interpreted also in terms of this theory.

LAW OF GAY-LUSSAC ($P \propto T$ AT CONSTANT V). According to Postulate 5 the average kinetic energy of a gas increases with increasing temperature. This means that the molecules of the gas are, on the average, moving faster at a higher temperature and hence collide with the walls of a container with greater force and with greater frequency. If the volume of the gas is held constant, therefore, the pressure must increase when the temperature of the gas is raised.

CHARLES' LAW (V $\propto$ T AT CONSTANT P). In order for the pressure to remain constant when the temperature of a gas is increased, the volume must increase as well in order to reduce the number of molecules colliding with any given area of the wall. In other words, if we allow the gas to expand as the temperature is raised, there are fewer collisions per square centimeter, and even though each occurs with greater force, the pressure, that is, the total force per unit area, remains constant.

DALTON'S LAW OF PARTIAL PRESSURES ($P_T = p_A + p_B + \ldots + p_N$). This gas law follows as a direct consequence of Postulate 2. If there are no intermolecular forces of attraction, each gas molecule behaves independently of all of the others in the container. The pressure exerted by a particular gas in a mixture, therefore, depends only on the number of molecules of that gas present in the container. Thus, each gas exerts a partial pressure that is independent of the other gases and the total pressure of the mixture is the sum of the pressures exerted by each gas.

It also follows that the partial pressure of a particular gas in a mixture depends upon its **mole fraction** (usually given the symbol, X); that is, the fraction of the total number of moles of gas that are contributed by the gas in question. For example, in a mixture of 1 mole of O_2 and 3 mole of N_2 at a total pressure of 2 atm, the mole fraction of O_2 is $\frac{1}{4} = 0.25$ and the partial pressure of O_2 is $(0.25)(2 \text{ atm}) = 0.50$ atm. Similarly, for N_2 the mole fraction is $\frac{3}{4} = 0.75$ and its partial pressure is $(0.75)(2 \text{ atm}) = 1.50$ atm. In general, we can write

$$p_A = X_A P_T$$

where p_A is the partial pressure of A, P_T is the total pressure, and X_A is the mole fraction of gas A, given as

$$X_A = \frac{\text{number of moles of } A}{\text{total number of moles of gas in the mixture}}$$

AVOGADRO's PRINCIPLE (V $\propto$ n AT CONSTANT P AND T). If a gas is contained in a given volume at a particular temperature and pressure, there are a certain number of collisions per second with each square centimeter of the walls. If more gas were introduced into the container, the number of collisions per second per square centimeter must increase. The only way for the pressure to stay the same is if there is a volume increase accompanying the addition of gas.

6.10
Distribution of Molecular Velocities

One of the most important and useful concepts to evolve from the kinetic theory is the idea that in a gas (or, in fact, in a liquid or a solid) there is a distribution of kinetic energies, and hence molecular velocities, that is dependent on temperature. Figure 6.11 is a graphical display of this distribution (called a Maxwell-Boltzmann distribution) for a gas at three different temperatures, in which the fraction of the total number of molecules possessing a particular kinetic energy is plotted against kinetic energy.

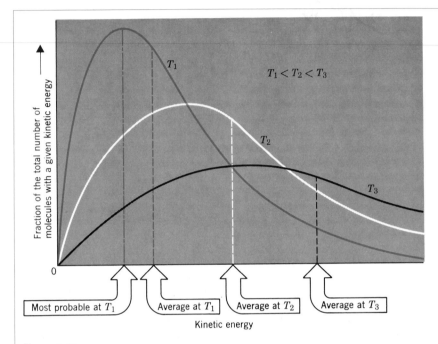

Figure 6.11
Distribution of kinetic energies.

At zero K.E., which corresponds to molecules standing still, this fraction is essentially zero since very few, if any, molecules are standing still at any instant. The fraction increases as we move to higher energies (higher velocities) and eventually becomes a maximum. At still higher kinetic energies the fraction decreases and gradually approaches zero again at kinetic energies corresponding to very fast-moving molecules. The curve does not go all the way to zero, however, because there is virtually no upper limit, other than the speed of light, to molecular speeds.

The maximum on this curve represents the kinetic energy possessed by the largest fraction of molecules in the gas and is found with the greatest probability if we examine molecules at random—hence it is called the *most probable kinetic energy*. The average kinetic energy occurs at a higher value than the most probable kinetic energy because the curve is not symmetrical. Just as a few "curve breakers" in a chemistry class tend to raise the class average on exams, the high velocity molecules shift the average K.E. above the most probable K.E.

When the temperature of a gas is raised, the curve changes so that the average K.E. increases, as shown in Figure 6.11. More molecules have high velocities and fewer have low ones and, on the average, the molecules of the gas move faster. Thus when heat is added to a substance to raise its temperature, the energy goes into increasing the average kinetic energy and increases the motion of the particles.

The relationship between kinetic energy and temperature also leads quite naturally to the concept of an absolute zero. As kinetic energy is removed from a substance, its molecules move more and more slowly. If all of the molecules were to cease moving, their average kinetic energy would be zero and, since negative kinetic energies are impossible (a molecule cannot be going slower than when it is standing still), the temperature of the substance would also be at its lowest point. It is this temperature that we refer to as absolute zero—the temperature when all molecular motion has ceased.[3] It should be understood, however, that electronic motion would still continue at absolute zero. Even though the molecules would be motionless, the electrons would still be "whizzing" about their respective nuclei.

<table>
<tr><td>6.11
Real Gases</td><td>We have seen, earlier in this chapter, that real gases deviate from ideal behavior at high pressure and low temperature. This is especially true near their condensation point. In order to explain nonideal behavior of gases we must examine the postulates of the kinetic molecular theory more closely.</td></tr>
</table>

**6.11
Real Gases**

We have seen, earlier in this chapter, that real gases deviate from ideal behavior at high pressure and low temperature. This is especially true near their condensation point. In order to explain nonideal behavior of gases we must examine the postulates of the kinetic molecular theory more closely.

According to Postulate 1, the volume of the gas molecules is negligibly small compared to the volume of the container in which they are held. This is not completely valid. Let us imagine, for simplicity, that gas molecules could be stopped in a container and allowed to settle to the bottom. We would see, contrary to Postulate 1, that part of the volume of the container is occupied by the gas molecules. The remaining free space is somewhat less than the volume of the container. If, in this hypothetical case, another gas molecule were introduced, it could move in the free space, but not the entire volume of the container. The volume that the molecules cannot move in is called the **excluded volume.** This same situation exists when the molecules are moving and, therefore, the real volume is actually slightly higher than the ideal volume (the volume the gas would occupy if the molecules themselves occupied no space). According to J. D. van der Waals the real volume is equal to

$$V_{real} = V_{ideal} + nb$$

where b is the correction due to the excluded volume per mole and n, as usual, is the number of moles of gas. Solving for the ideal gas volume we have

$$V_{ideal} = V_{real} - nb \qquad (6.12)$$

Van der Waals also included a correction for pressure, necessitated by the attractive forces between gas molecules that exist contrary to Postulate 2 of the kinetic molecular theory. These intermolecular forces of attraction are of the van der Waals type discussed in Chapter 4. We know that such attractions exist because gases condense. Lowering the temperature of a gas causes the molecules to slow down to a point where the attractive forces between the molecules are strong enough to cause "clumps" of molecules to form and we observe condensation.

A molecule in a gas that is just about to make a collision with the wall

---

[3] In fact, even at absolute zero there is still a residual molecular motion that is required by the Heisenberg uncertainty principle (Chapter 3). This principle states that we cannot simultaneously know the position and momentum of a particle. We shall see in the next chapter that the average positions of particles in the solid state can be determined, and if the particles are motionless we would also know that their momentum was zero, thereby violating the uncertainty principle.

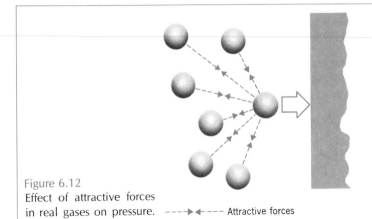

Figure 6.12
Effect of attractive forces
in real gases on pressure. ---▶◀--- Attractive forces

feels the attraction of all the molecules surrounding it. Since there are no molecules in front of it (or relatively very few), the greatest concentration of these forces is in a direction away from the wall (Figure 6.12). When the collision does take place, it is less energetic than it would be if there were no attractive forces present. The overall effect of these forces is a lowering of the pressure. The extent to which the pressure will be reduced is directly proportional to (1) the number of impacts with the wall, which is directly proportional to the concentration of the molecules (n/V), and (2) the decrease in the force of each impact, which is also proportional to the concentration of the molecules. The decrease in the pressure, therefore, is directly proportional to the square of the concentration, or n^2/V^2. The ideal pressure, that is, the pressure the gas could exert in the absence of intermolecular attractive forces, is higher than the actual pressure by an amount that is directly proportional to n^2/V^2 or

$$P_{ideal} = P_{real} + \frac{n^2 a}{V^2} \qquad (6.13)$$

where a is a proportionality constant that depends on the strength of the intermolecular attractions.

Substituting these corrected pressures and volumes (from Equations 6.13 and 6.12, respectively) into the ideal gas equation gives us

$$\left(P + \frac{n^2 a}{V^2}\right)\left(V - nb\right) = nRT \qquad (6.14)$$

which is the van der Waals equation of state for a real gas.

The values of the constants a and b depend on the nature of the gas because the molecular volumes and the molecular attractions vary from gas to gas. Some typical values of a and b are found in Table 6.6. This equation is much more complex than the ideal equation, but does work well for many gases over fairly wide ranges of temperature and pressure.

Another manifestation of the attractive forces between gas molecules is the cooling that occurs when most gases[4] at ordinary temperatures are al-

---

[4] Exceptions are H_2 and He where the repulsive forces are so great that their temperatures actually increase upon their expansion into a vacuum.

Table 6.6
Van der Waals Constants for Real Gases

	a (liter²atm/mole²)	b (liter/mole)
He	0.034	0.0237
O_2	1.36	0.0318
CH_4	2.25	0.0428
NH_3	4.17	0.0371
H_2O	5.46	0.0305
C_2H_5OH	12.0	0.0841

lowed to expand freely into a vacuum. As the gas expands, the average distance of separation between the molecules increases. Since there are forces of attraction between them, moving the molecules apart requires work (energy). The source of this energy is the kinetic energy of the gas—in the process of expansion kinetic energy is converted to potential energy. This removal of kinetic energy leads, of course, to a decrease in the average kinetic energy of the gas and, since the average kinetic energy is directly related to temperature, the gas becomes cooler. An example of this phenomenon is observed when a CO_2 fire extinguisher is discharged, producing a fine snow of dry ice. The rapid expansion of the CO_2 cools the gas to the point where it condenses to a solid.

Review Questions and Problems

6.1 If a gas, originally in a 50 ml container at a pressure of 645 torr, is transferred to another container whose volume is 65 ml, what would be its new pressure if
(a) There were no temperature change?
(b) The temperature of the first container was 25° C and that of the second was 35° C?

6.2 What would be the final volume of a 2.00 liter sample of a gas that is heated from 26 to 100° C at constant pressure?

6.3 The density of CO_2 is 1.96 g/l at 0° C and 1 atm. Determine its density at 650 torr and 25° C.

6.4 A 2.00 liter sample of a gas originally at 25° C, and a pressure of 700 torr is allowed to expand to a volume of 5.00 liters. If the final pressure of the gas is 585 torr, what is its final temperature?

6.5 What would be the total pressure of a mixture prepared by adding 20.0 ml of N_2 at 0° C and 740 torr plus 30 ml of O_2 at 0° C and 640 torr to a 50 ml container at 0° C?

6.6 A mixture of N_2 and O_2 has a volume of 100 ml at a temperature of 50° C and a pressure of 800 torr. It was prepared by adding 50 ml of O_2 at 60° C and 400 torr with X ml of N_2 at 40°C and 400 torr. What is the value of X?

6.7 A 50.0 ml sample of gas exerts a pressure of 450 torr at 35° C. What is its volume at STP?

6.8 A man digs a well and finds water 35 ft below the ground. If the average atmospheric pressure at the well is 1 atm, will the man be able to draw water from the well using a pump, mounted at ground level, that works by suction? Explain your answer.

6.9 A gas is collected by the displacement of water until the total pressure inside a 100 ml flask is 700 torr at 25° C. Calculate the volume of dry gas at STP.

6.10 A mixture of N_2 and O_2 in a 200 ml vessel exerts a pressure of 720 torr at 35° C. If there are 0.0020 moles of N_2 present,
(a) What is the mole fraction of N_2?
(b) What is the partial pressure of N_2?
(c) What is the partial pressure of O_2?
(d) How many moles of O_2 are present?

6.11 Compare the rates of effusion of He and Ne. Which gas effuses faster, and how much faster?

6.12 If, at a particular temperature, the average speed of CH_4 molecules is 1000 mph, what would be the average speed of CO_2 molecules at the same temperature?

6.13 The rate of effusion of an unknown gas was determined to be 2.92 times faster than that of NH_3. What is the approximate molecular weight of the unknown gas?

6.14 Three gases were added to the same 10 liter container to give a total pressure of 800 torr at 30°C. If the mixture contained 8.0 g of CO_2, 6.0 g of O_2, and an unknown amount of N_2, calculate the following.
(a) The total number of moles of gas in the container.
(b) The mole fraction of each gas.
(c) The partial pressure of each gas.
(d) The weight of N_2 in the container.

6.15 If a gas, occupying a volume of 500 ml over water at 35° C, is compressed to a volume of 250 ml, also over water at 35° C, what will be the final partial pressure of the dry gas?

6.16 In the laboratory a student filled a 250 ml container with an unknown gas until a pressure of 760 torr was obtained. He then found that the sample of gas weighed 0.164 g. Calculate the molecular weight of the gas if the temperature in the laboratory was 25° C.

6.17 Calculate the pressure, in torr and in atmospheres, which would be exerted by 25 kg of steam (H_2O) in a 1000 liter boiler at 200° C assuming ideal gas behavior.

6.18 The density of a gas was found to be 1.81 g/liter of 30° C and 760 torr. What is its molecular weight?

6.19 A chemist observed a gas being evolved in a chemical reaction and collected some of it for analysis. It was found to contain 80.0% carbon and 20.0% hydrogen. It was also observed that 500 ml of the gas at 760 torr and 0° C weighed 0.6695 g.
(a) What is the empirical formula of the gaseous compound?
(b) What is its molecular weight?
(c) What is its molecular formula?

6.20 Oxygen gas, generated in the reaction, $KClO_3 \rightarrow KCl + O_2$ (unbalanced), was collected over water at 30° C in a 150 ml vessel until the total pressure was 600 torr.
(a) How many grams of dry O_2 were produced?
(b) How many grams of $KClO_3$ were consumed in the reaction?

6.21 Calculate the maximum volume of CO_2, at 750 torr and 28° C, that could be produced by reacting 500 ml of CO, at 760 torr and 15° C, with 500 ml of O_2 at 770 torr and 0° C.

6.22 Ozone, O_3, is an important species in the chain of reactions that lead to the production of smog. In an ozone analysis, 2.0×10^4 liters of air at STP were drawn through a solution of NaI where the O_3 undergoes the reaction

$$O_3 + 2I^- + H_2O \longrightarrow O_2 + I_2 + 2OH^-$$

The I_2 formed was titrated with 0.0100 M $Na_2S_2O_3$ with which it reacts

$$I_2 + 2S_2O_3{}^{2-} \longrightarrow 2I^- + S_4O_6{}^{2-}$$

In the analysis, 0.042 ml of the $Na_2S_2O_3$ solution were required to completely react with all of the I_2.
(a) Calculate the number of moles of I_2 that were reacted with the $S_2O_3{}^{2-}$ solution.

(b) How many moles of I_2 were produced in the first reaction?

(c) How many moles of O_3 were contained in the 20,000 liters of air?

(d) What volume would the O_3 occupy at STP?

(e) What is the concentration of O_3, in ppm by volume, in the air sample?

6.23 Use the data in Table 6.5, and the reasoning of Cannizzaro, to deduce the atomic weight of carbon if the atomic weight of hydrogen is taken to be 1.00 amu. Can we prove, on the basis of these data alone, that the atomic weight of carbon is *not* 6 amu?

6.24 In Table 6.5, the smallest weight of carbon in any of the compounds listed is 0.535 grams. What would be the atomic weights of hydrogen and chlorine if carbon is assigned an atomic weight of precisely 12.000 amu?

6.25 Without referring to the text, state the five basic postulates of the kinetic molecular theory.

6.26 If a warm object is placed in contact with a cool object, heat transfer occurs until they both come to the same temperature. How can the kinetic molecular theory account for this heat transfer and these temperature changes that occur?

6.27 What is meant by nonideal behavior of a gas? Under what conditions is this behavior most evident? What errors are present in the postulates of the kinetic molecular theory that are exposed by the nonideal character of real gases?

6.28 What physical significance do the constants *a* and *b* have in the van der Waals equation of state for real gases?

6.29 Explain, on the basis of the kinetic molecular theory, why most gases cool upon expansion into a vacuum.

7
Solids

As you know, solids differ from gases in many ways: (1) they retain their shape and volume when transferred from one container to another, (2) they are virtually incompressible and (3) they exhibit extremely slow rates of diffusion. What features do solids possess that cause them to be so different from gases?

We saw in the last chapter that the particles comprising a gas move about freely and are essentially independent of one another. Just the opposite is true in a solid where the attractive forces between the atoms, molecules, or ions are relatively strong. Here the particles are held in a rigid structural array, wherein they exhibit only vibrational motion.

There are two types of solids, **amorphous** and **crystalline.** If the temperature at which the solid is formed is approached slowly so as to allow the array of particles to become well ordered, a crystalline solid results. If, on the other hand, the temperature is lowered very rapidly, there is a chance that the particles will be "frozen" in a chaotic state. In this case the particles are arranged in a random fashion and the resulting solid is said to be amorphous (i.e., without form).

7.1 Crystalline Solids

The surface of a well-formed crystalline solid reveals, upon examination, flat planes that intersect at angles characteristic of the particular substance under investigation (Figure 7.1). These flat planes are called **faces,** and the characteristic angles are called **interfacial angles.** These faces and characteristic interfacial angles are present no matter how the crystal was formed and are completely independent of the size of the crystal. Furthermore, if a crystal is cleaved, or even crushed into a powder, each resulting particle will possess identical interfacial angles.

This may be contrasted with what happens when glass, a typical amorphous solid, is broken. The surfaces of the broken pieces are generally not flat and intersect at random angles.

7.2 X-Ray Diffraction

It is clear, even from a visual observation, that there is much external order to a crystal. How can the internal order of a crystal, which is implied by this external regularity, be confirmed?

A German physicist, Max von Laue, pointed out in 1912 that a crystal could serve as a three-dimensional diffraction grating if the wavelength of the incident radiation were of the same order of magnitude as the distance between particles in the solid. This condition is fulfilled by X-rays, whose wavelength is approximately 1 Å.

Figure 7.1
Photograph of some typical crystals. On the left, potassium alum, $KAl(SO_4)_2 \cdot 12H_2O$. On the right, calcite, $CaCO_3$.

When a crystal is bathed in X-rays, each atom of the crystal within the path of an X-ray absorbs some of its energy and then reemits it in all directions. Thus each atom is a source of secondary wavelets, and the X-rays are said to be scattered by the atoms. These secondary wavelets from the different sources interfere with each other, either by reinforcing or by cancelling each other. In certain directions the waves emanating from nearly all of the atoms in any orderly array are in phase (i.e., the peaks and troughs of the waves coincide as shown in Figure 7.2a) and intense beams of X-rays are observed in these directions. In all other directions the waves from various atoms are out of phase (Figure 7.2b) and cancel each other; thus no intensity is detected.

Two English scientists, William Bragg and his son Lawrence, treated the diffraction of X-rays as if the process were reflection. In Bragg's treatment the X-rays that penetrate a crystal are reflected by successive layers of particles within the substance (Figure 7.3). We can see from this diagram that beams reflected from deeper layers must travel further to reach the detector. For there to be any intensity at the detector these waves have to be in phase (see above), which must mean that the extra distance traveled by the more penetrating beam has to be some integral multiple of the wavelength of the X-rays. This requires that, in Figure 7.3

$$AY + YB = \lambda n \tag{7.1}$$

where $n = 1,2,3 \ldots$ and λ is the wavelength. From geometry, θ, the angle of incidence (which equals the angle of reflection) is related to the interplanar distance d by

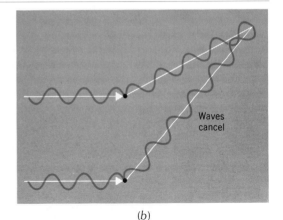

(a)   (b)

Figure 7.2

Diffraction. Scattered waves reinforce each other only at certain angles. (a) In phase. (b) Out of phase.

$$\sin \theta = \frac{AY}{d}$$

or

$$AY = d \sin \theta \qquad (7.2)$$

Since $AY = YB$, it follows from Equation 7.1 that

$$2AY = \lambda n \qquad (7.3)$$

By combining Equations 7.2 and 7.3, we arrive at

$$2d \sin \theta = \lambda n \qquad (7.4)$$

Equation 7.4 is called the Bragg equation and serves as the basis for the study of crystalline structure by X-ray diffraction.

In practice, X-rays of known wavelength are directed at the crystal; and the angles at which they are reflected are recorded, for example, on a piece of photographic film, Figure 7.4. From these data the spacings between layers of particles may be computed.

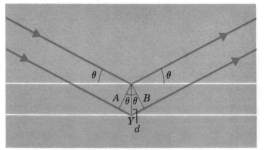

Figure 7.3

Reflection of X-rays from layers of crystal planes.

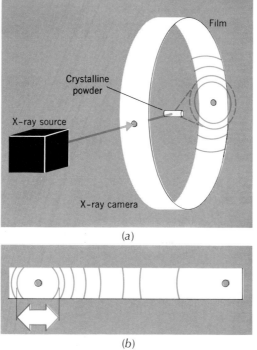

Figure 7.4
(a) Recording an X-ray diffraction pattern produced by a crystalline powder. (b) Film after development. Spacings between lines on the film are related to the angles of reflection and therefore can be used to calculate interplanar spacings in the crystalline material.

(a)

(b)

7.3
Lattices

We have seen that X-ray diffraction yields information about the distances between planes of atoms in a crystal. This knowledge permits a crystallographer to deduce, through a rather complex procedure, the actual positions of atoms within the solid. In this way the molecular structures of many substances have been found. In recent years, X-ray diffraction has become a powerful tool in biochemistry where the structures of even very complex molecules have been investigated. Watson and Crick, for example, shared the Nobel Prize with Wilkins in 1962 for their use of X-ray data in the deduction of the double helix structure of DNA.

There are only a very limited number of ways in which atoms, molecules, or ions may be packed together in a crystalline substance. We can characterize these packing arrangements by what are called **space lattices.**

In the strictest sense, a lattice is merely a regular or repetitive pattern of points that may be either one-, two-, or three-dimensional. A two-dimensional lattice is illustrated in Figure 7.5a. What we see there is called a square lattice since it consists of a set of points arranged so that each lattice point lies at the corner of a square.

If we were wallpaper designers we could create, with this one simple square lattice, an unlimited number of wallpaper patterns. For example, each lattice point could be assigned some geometrical design such as a diamond or a flower (Figure 7.5b). By varying the size of the squares defined by the lattice points and by changing the motif about each point, an infinite number of designs could be formed.

We can think of a lattice as extending to infinity in all directions.

(a) Square lattice.	(b) Design based on square lattice.

Figure 7.5
Two-dimensional lattice. (a) Square lattice. (b) Design based on square lattice.

However, in order to completely describe the lattice, or any pattern derived from it, it is not necessary to specify explicitly the positions of each and every point. Instead, we only have to describe a portion of the lattice, called the **unit cell,** which can be used to generate the entire lattice. In the lattice of Figure 7.5a the unit cell corresponds to a square drawn by connecting four points. By moving this unit cell repeatedly to the left and right and up and down by a distance equal to the length of the unit cell, we can create the entire square lattice.

The extension of the lattice concept to three dimensions is quite straightforward. In Figure 7.6 we see an example of a simple cubic space lattice in which the unit cell is drawn with heavy lines. By associating a particular chemical environment with each lattice point on the three-dimensional lattice we can arrive at a chemical structure, and by varying the chemical environment about each point we can create an infinite number of chemical structures all based on the same lattice. It is easy to see, then, how a very small number of lattices could be sufficient to describe the crystal structures of all known chemical substances.

In addition to simple, or primitive lattices, which have lattice points located only at the corners of the unit cell, it is also possible to have body-

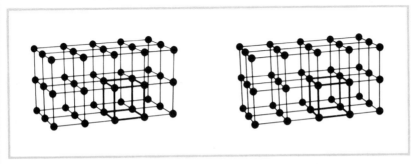

Figure 7.6
A simple cubic space lattice.

centered, face-centered, and end-centered lattices (Figure 7.7). In a body-centered lattice there are points at each corner of the unit cell plus an additional point in the very center. A face-centered lattice has an additional lattice point in the center of each of the six faces of the unit cell while an end-centered lattice has additional lattice points in only two opposite faces. It was shown by A. Bravais in 1848 that the total number of possible space lattices is 14. They belong to seven basic crystal systems: cubic, tetragonal, orthorhombic, monoclinic, triclinic, rhombohedral, and hexagonal.

The unit cell for each lattice may be described by specifying the quantities a, b, and c, which correspond to the lengths of the edges of the cell, and the angles α, β, and γ, which are the angles at which these edges intersect one another (Figure 7.8). The properties of the 14 Bravais lattices are given in Table 7.1.

In the various unit cells that we have just seen, there are three kinds of lattice points: points located at the corners, points in the face-centers, and points that lie entirely within the unit cell. In a crystal, atoms located at the corner and face-center of a unit cell are shared by other cells, and only a portion of such an atom actually lies within a given unit cell. In certain instances we also find a fourth type of lattice point where an atom or ion lies along an edge of a unit cell. We might wish to know how many chemical units (atoms, molecules, or groups of ions) are found in one unit cell of a crystalline substance. This is related to the density of the substance and is controlled by the attractive and repulsive forces that the particles experience in the solid.

A point that lies at the corner of a unit cell is shared between eight unit cells and, therefore, only $\frac{1}{8}$ of each such point lies within one given unit cell (see Figure 7.6).

A point along an edge is shared by four unit cells and only $\frac{1}{4}$ lies within any one cell.

A face-centered point (which we have in both face-centered and end-centered lattices) lies partly in two cells and only $\frac{1}{2}$ is present in a given unit cell.

Table 7.1
Properties of the Seven Crystal Systems

System	Edge Lengths	Angles
Cubic	$a = b = c$	$\alpha = \beta = \gamma = 90°$
Tetragonal	$a = b \neq c$	$\alpha = \beta = \gamma = 90°$
Orthorhombic	$a \neq b \neq c$	$\alpha = \beta = \gamma = 90°$
Monoclinic	$a \neq b \neq c$	$\alpha = \beta = 90° \neq \gamma$
Triclinic	$a \neq b \neq c$	$\alpha \neq \beta \neq \gamma$
Rhombohedral	$a = b = c$	$\alpha = \beta = \gamma \neq 90°$
Hexagonal	$a = b \neq c$	$\alpha = \beta = 90°; \gamma = 120°$

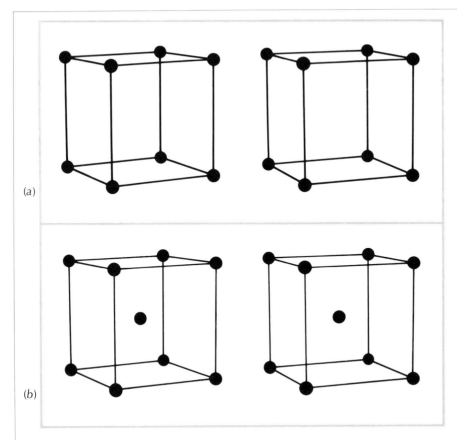

(a)

(b)

Figure 7.7
The four types of unit cells. (a) Simple cubic, (b) Body-Centered cubic, (c) Face-centered cubic, (d) End-centered orthorhombic.

A body-centered point lies entirely within the unit cell, and contributes one complete point to the cell.

Type of Lattice Point	Contribution to One Unit Cell
Corner	$\frac{1}{8}$
Edge	$\frac{1}{4}$
Face-center	$\frac{1}{2}$
Body-center	1

A simple (primitive) lattice has points only at the eight corners of the unit cell (Figure 7.7a). The number of lattice points that lie within the cell is

$$8 \; \text{corners} \times \frac{1 \; \text{point}}{8 \; \text{corner}} = 1 \; \text{point}$$

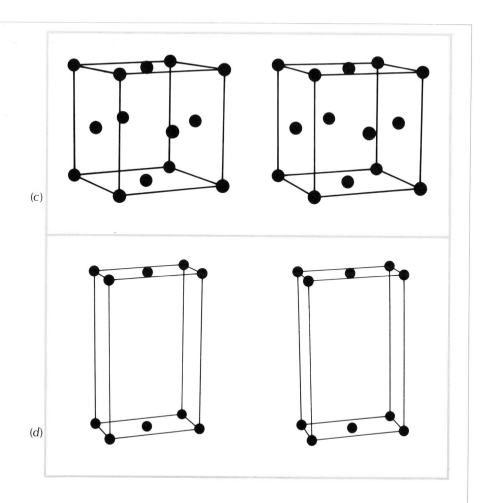

(c)

(d)

A face-centered unit cell has points at the eight corners and, in addition, points centered in the six faces (Figure 7.7c). It, therefore, contains a total of four lattice points.

$$8 \text{ corners} \times \frac{1}{8} \frac{\text{point}}{\text{corner}} = 1 \text{ point}$$

$$6 \text{ faces} \times \frac{1}{2} \frac{\text{point}}{\text{face}} = \underline{3 \text{ points}}$$

$$\text{Total} = 4 \text{ points}$$

You can quite easily verify for yourself that the body-centered and end-centered unit cells each contain two lattice points.

7.4
Avogadro's
Number

Knowledge of the crystalline structure of a substance allows us to compute the value of Avogadro's number. For example, metallic copper is found to crystallize with a face-centered cubic lattice in which a copper atom is

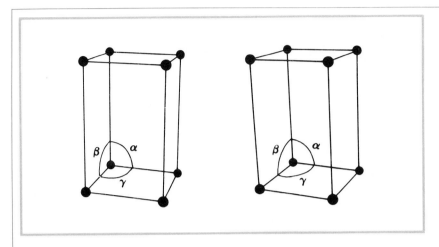

Figure 7.8
The unit cell. Edges intersect at characteristic angles α, β, and γ.

located at each lattice point. The length of the edge of the unit cell is found to be 3.615 Å (from X-ray diffraction).

Copper has an atomic mass of 63.54 and a density of 8.936 g/cm³. The volume occupied by one mole of copper is

$$1 \text{ mole Cu} \times \left(\frac{63.54 \text{ g Cu}}{1 \text{ mole Cu}}\right) \times \left(\frac{1 \text{ cm}^3}{8.936 \text{ g Cu}}\right) = 7.111 \text{ cm}^3$$

The volume occupied by one unit cell is found by multiplying the height times the width times the depth of the cube.

$$\text{Volume of unit cell} = (3.615 \times 10^{-8} \text{ cm})^3$$
$$1 \text{ unit cell} = 4.724 \times 10^{-23} \text{ cm}^3$$

The number of unit cells in one mole of copper is

$$1 \text{ mole Cu} \times \left(\frac{7.111 \text{ cm}^3}{1 \text{ mole Cu}}\right) \times \left(\frac{1 \text{ unit cell}}{4.724 \times 10^{-23} \text{ cm}^3}\right) = 1.505 \times 10^{23} \text{ unit cells}$$

Since there are four atoms of copper in the face-centered unit cell and since there are 1.505×10^{23} unit cells per mole, the number of atoms of Cu per one mole of Cu is

$$\left(\frac{4 \text{ atoms}}{\text{unit cell}}\right) \times \left(\frac{1.505 \times 10^{23} \text{ unit cells}}{1 \text{ mole}}\right) = \frac{6.02 \times 10^{23} \text{ atoms}}{1 \text{ mole}}$$

The value of the Avogadro number, therefore, is 6.02×10^{23}.

If we know the crystal structure of a substance, we are also able to determine the radii of the particles in the solid. For example, we can calculate the atomic radius of the Cu atom using the information in the last section.

In the face-centered cell of Cu, Figure 7.9, the length $AC = 3.615$ Å. Copper atoms are in contact along the line joining points A and B (face

7.5

Atomic and Ionic Radii

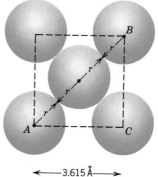

Figure 7.9
Copper atoms in the face of a unit cell.

←——3.615 Å——→

diagonal). This distance corresponds to four times the radius of a copper atom. From geometry we know,

$$\overline{AB} = \sqrt{2} \cdot \overline{AC}$$
$$\overline{AB} = 5.12 \text{ Å}$$

Therefore,

$$4r = 5.12 \text{ Å}$$
$$r = 1.28 \text{ Å}$$

The radius of a copper atom in metallic copper is therefore 1.28 Å. In a similar fashion we can also use the results of X-ray diffraction to determine the radii of ions in ionic crystals.

7.6
The Face-Centered Cubic Lattice

As mentioned, a single lattice can characterize a variety of different chemical structures. The face-centered cubic lattice serves as a convenient example of this. We know that copper crystallizes with a face-centered structure, and, in fact, so do many other metals, including such familiar ones as aluminum, lead, silver, and gold.

Sodium chloride (ordinary table salt) is characteristic of many ionic solids in which the ratio of cations to anions is 1:1. The structure of NaCl, termed the **rock salt structure,** is shown in Figure 7.10. The solid is constructed of positive and negative ions arranged so that each positive ion is surrounded by six negative ions at the vertices of an octahedron; similarly, each negative ion is surrounded octahedrally by positive ions. If the solid spheres in Figure 7.10 represent Na^+ ions located at the lattice points in a face-centered lattice, then there are 4 Na^+ ions per unit cell. Chloride ions are located along each of the 12 edges and in the center of the unit cell. Only $\frac{1}{4}$ of each edge ion is within the cube and therefore these contribute a total of three chloride ions to the unit cell. The chloride ion in the center adds one more, so that there are a total of four Cl^- ions and, therefore, the ratio of Na^+ to Cl^- is 4:4 which, of course, is the same as 1:1. Other substances possessing the rock salt structure are:

(a) All alkali halides except CsCl (below 450° C), CsBr and CsI.
(b) AgF, AgCl, AgBr.
(c) Oxides and sulfides of the alkaline earth metals.
(d) NiO.
(e) Some alloys.

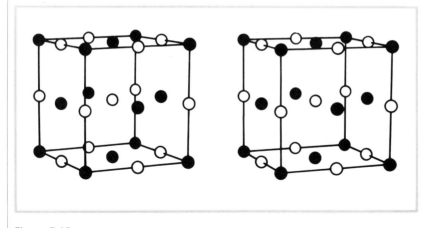

Figure 7.10
Sodium chloride structure.

In the **zinc-blende structure** (Figure 7.11), which is observed for ZnS, ZnO, CuCl, CuBr, and BeO, we also have a 1:1 ratio of cations to anions. In this case, we have one type of ion, for example an anion, located at the lattice points in a face-centered cubic unit cell. Each of the cations, shown as open circles within the unit cell, are surrounded tetrahedrally by four anions. Each anion is also surrounded tetrahedrally by four cations.

The **fluorite structure** (Figure 7.12), as exemplified by CaF_2, is similar to the zinc-blende structure. Here we have cations at lattice points in the face-centered cube and eight anions, surrounded tetrahedrally by cations, located entirely within the cube. The cations at the lattice points contribute a total of four positive ions to the cube. Since all eight anions are within the unit cell, they contribute eight negative ions to the cube. The ratio of positive to negative ions in the fluorite structure is therefore 1:2. Other solids exhibiting the

Figure 7.11
Zinc-blende structure.

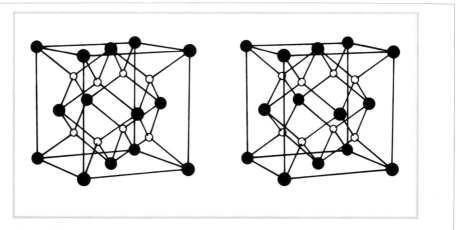

Figure 7.12
Fluorite structure.

fluorite structure are BaF_2, $SrCl_2$, and ThO_2. Another structure, called the **antifluorite structure** is precisely the same as the fluorite structure except that the positions of positive and negative ions are reversed to give a cation to anion ratio of 2 : 1. Solids displaying this type of structure include the oxides, sulfides, selenides, and tellurides of sodium and potassium (e.g., M_2S, M_2Se and M_2Te, where $M = Na^+$ or K^+).

7.7 Closest-Packed Structures

When all atoms in a substance are the same, its crystal structure is often determined by the most efficient packing of spheres, that is, the arrangement of atoms that gives rise to the smallest amount of unoccupied space. There are two ways in which spheres can be packed most efficiently; the resulting arrangements are called **closest-packed structures.** One of these is called **cubic closest-packed** (*ccp*) and the other, **hexagonal closest-packed** (*hcp*). We can consider them to be constructed in the following way.

A first layer of spheres is packed together as tightly as possible (Figure 7.13*a*). A second layer of spheres is placed upon the first with each sphere in the second layer resting in a depression between spheres in the first layer (Figure 7.13*b*). There are now two ways of arranging the spheres in the third layer. One is to place them directly above those in the first layer (Figure 7.13*c*). If we arrange them in this manner, we arrive at the *hcp* structure. We can continue this process with the fourth layer directly above the second, the fifth above the third, etc., and we can think of this as an alternating *ABAB* . . . pattern.

The second way of arranging the third layer of spheres is shown in Figure 7.13*d* and leads to the *ccp* structure. In this case the third layer lies above holes in the first layer; if we continue building the structure by laying the fourth layer directly above the first, the fifth above the second, etc., we arrive at an *ABCABC* . . . pattern. The lattice that corresponds to this arrangement is face-centered cubic, although it is not particularly evident from Figure 7.13*d*. Actually, in this case we are looking down at a corner of the face-centered cube along a body diagonal (the line joining one corner of the cube, the center, and an opposite corner).

In both the *ccp* and the *hcp* structures each atom is in contact with 12

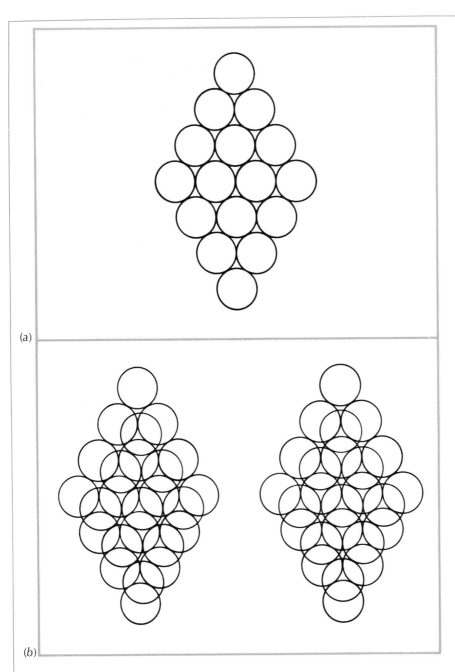

(a)

(b)

Figure 7.13

Closest-packing of spheres. (a) First layer of tangent spheres. (b) Second layer of spheres resting in depressions in first layer. (c) Third layer of spheres resting in depressions in second layer over spheres in first layer—hcp structure. (d) Third layer of spheres resting in depressions in second layer over unused depressions in first layer—ccp structure.

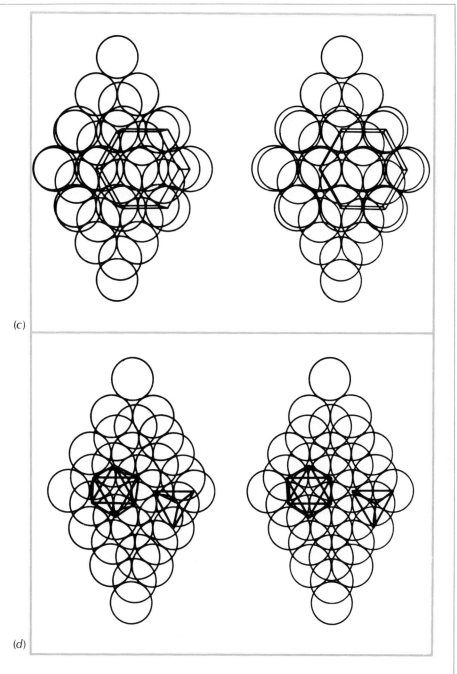

(c)

(d)

other "nearest neighbors." We can compare this with a body-centered cubic structure where each atom would be in contact with its eight nearest neighbors and with a simple cubic structure where each atom would have but six nearest neighbors.

There are a relatively large number of substances that crystallize with these closest-packed structures (e.g., most metals). It is also possible to interpret a variety of other structures as being derived from the closest-packed arrangements. For example, the *ccp* structure, we have said, has a face-centered cubic lattice. We have also discussed several ionic crystals in terms of this same lattice. These substances can be considered to be derived from the *ccp* structure as follows.

In the *ccp* structure there are two types of vacant sites: tetrahedral and octahedral. A tetrahedral site is surrounded by four spheres and an octahedral site by six spheres. These are shown in Figure 7.13*d*. There are twice as many tetrahedral sites as there are octahedral sites. These vacant holes may be used to accommodate other atoms and, when filled in this way, give rise to a variety of different structures.

Sodium chloride, for example, may be considered as a *ccp* arrangement of Cl^- ions with Na^+ ions located in octahedral sites (recall that in NaCl each cation is surrounded octahedrally by anions).

The zinc-blende structure may be looked upon as a *ccp* arrangement of anions with cations in tetrahedral sites (Figure 7.11). In this structure only $\frac{1}{2}$ of the available tetrahedral sites are filled. In the fluorite structure, all of the tetrahedral sites are filled, and a $1:2$ ratio of cation to anions is achieved.

We have just seen that there are only a limited number of ways of arranging particles in a crystalline solid. The particular arrangements, as well as the physical properties of the solid, are determined by the types of particles present at the lattice points and the nature of the attractive forces between them. There are four types of crystals: molecular, ionic, covalent, and metallic.

7.8
Types of Crystals

MOLECULAR CRYSTALS. In molecular crystals either molecules or atoms occupy lattice points, and the attractive forces between them are generally much weaker than the covalent bonds that exist within an individual molecule. Van der Waals forces are present in crystals of nonpolar substances such as Ar, O_2, napthalene (moth crystals) and CO_2 (dry ice). In substances like solid SO_2 the dominant forces are a result of dipole-dipole interactions and in solids such as ice (H_2O), NH_3 or HF the molecules are held in place by hydrogen bonds (see Chapter 4). Since there are relatively weak forces in molecular solids, these substances are usually soft and low melting. Where the binding forces are very weak, the molecules (or atoms) often arrange themselves in one of the closest-packed structures. However, in substances like water the directional nature of the hydrogen bonds causes H_2O to crystallize with a very open structure. We also find molecular crystals to be poor conductors of electricity because electrons may not move freely through the solid.

IONIC CRYSTALS. In an ionic crystal such as NaCl there are ions located at lattice points and the bonding between them is mainly electrostatic (which is essentially nondirectional). Because the electrostatic forces are strong, these crystals are characterized by high melting points and are quite hard. Ionic solids are also poor conductors of electricity because the ions are held rigidly in place. However, when an ionic substance is melted, it becomes a good conductor because the ions are then free to move about.

COVALENT CRYSTALS. There is a network of covalent bonds existing between the atoms in a covalent crystal. The highly directional nature of these covalent bonds usually prevents these substances from assuming one of the closest-packed structures, and we usually observe somewhat open structures like those found in diamond (Figure 7.14) and zinc blende (Figure 7.11). Other common examples of covalent crystals are carborundum (SiC) and quartz (SiO$_2$ shown in Figure 18.16). Because of this interlocking framework of strong covalent bonds, these substances are generally very hard and have extremely high melting points. Covalent crystals also tend to be poor conductors of electricity.

METALLIC CRYSTALS. The simplest picture of a metallic crystal has positive ions (nuclei plus core electrons) situated at lattice points, with the valence electrons belonging to the crystal as a whole instead of to any single atom. The solid is held together by the electrostatic attraction between the lattice of positive ions and this sort of "sea of electrons." These electrons may move freely; hence, we find metals to be good conductors of electricity. Since the melting points and hardness of metals vary over wide ranges (see Table 7.2) there must, at least in some cases, also be some degree of covalent bonding between atoms in the solid.

7.9
Band Theory of Solids

We have remarked that certain types of crystals (molecular, ionic, and covalent) are poor conductors of electricity while metallic crystals are very good conductors. There are some substances, such as germanium and silicon, that have conductivities between these two extremes. To explain these properties the **band theory** of solids was developed.

In a solid an energy band is composed of a very large number of closely spaced energy levels, formed by combining atomic orbitals (of similar energy) from each of the atoms within the substance. For example, in sodium the 1s

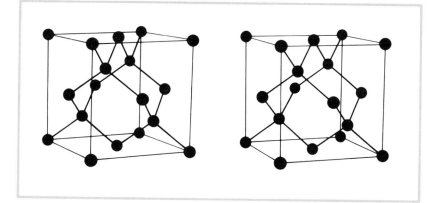

Figure 7.14
Diamond

Table 7.2
Properties of Some Typical Metals

Metal	Melting Point (°C)	Brinell Hardness Number (Room Temperature)
Mercury	−38.4	(liquid)
Sodium	97.8	0.07
Lead	327	4.2
Magnesium	650	30
Aluminum	660	16
Nickel	1453	90
Iron	1536	77
Platinum	1769	64
Molybdenum	2610	150
Tungsten	3410	350

atomic orbitals, one from each atom, combine to form a single 1s band that extends in three dimensions throughout the entire solid. The same thing occurs with the 2s, 2p, etc., orbitals, so that we also have 2s, 2p, etc., bands within the lattice.

Sodium atoms have filled 1s, 2s, and 2p orbitals; therefore the corresponding bands in the solid are also filled. The 3s orbital of sodium, however, is only half-filled, which leads to a half-filled 3s band. Following the same logic, it is clear that the 3p and higher energy bands are completely empty.

When a voltage is applied across sodium, electrons in filled bands may not move through the solid because orbitals in the same band on neighboring atoms are already filled and thus cannot acquire an additional electron. In the 3s band, which is half-filled, an electron may hop from atom to atom with ease. In Figure 7.15 the localized nature of electrons in filled bands is illustrated. We can also see that the 3s band extends continuously through the solid, which accounts for the high conductivity of Na.

We refer to the band containing the outer-shell (valence shell) electrons as the **valence band.** Any band that is either vacant or partially filled and is uninterrupted throughout the lattice is called a **conduction band.**

In metallic sodium the valence band and conduction band are the same. In magnesium the 3s valence band is filled and therefore cannot be used to transport electrons. However, the vacant 3p conduction band actually overlaps the valence band and can easily be populated by electrons when a voltage is applied (Figure 7.16a). This causes Mg to be a conductor.

In an **insulator** (e.g., glass, diamond, or rubber) the energy separation between the filled valence band and the conduction band is very large (Figure 7.16b). This large separation prevents electrons from populating the conduction band under an applied voltage.

A **semiconductor,** such as Si and Ge, has a relatively small gap between the valence and conduction bands (Figure 7.16c). Thermal energy promotes

Empty

Filled

Figure 7.15
Bands in metallic sodium.

some electrons to the conduction band where they are then able to move through the solid. This is why the conductivity of these substances increases with temperature.

We have described a crystal as an ordered array of particles. In all real crystals this order is not perfect, and any deviation from perfection is termed a **lattice defect.**

If only a few atoms are causing this disorder, a **point defect** results. There exists in crystals two main types of point defects. The first, called a **Frenkel defect,** is caused by cations that are not present in their normal lattice sites

Figure 7.16
Energies of the valence band and conduction band in three kinds of solids. (a) Magnesium — a conductor. (b) Insulator. (c) Semiconductor.

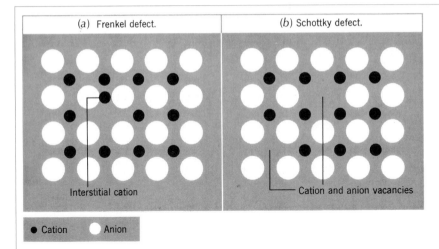

| (a) Frenkel defect. | (b) Schottky defect. |

Interstitial cation

Cation and anion vacancies

● Cation   ○ Anion

Figure 7.17

Defects in crystals. (a) Frenkel defect. (b) Schottky defect.

but are found in between layers in some **interstitial** positions, as shown in Figure 7.17. This type of defect can only occur in solids where the difference in size between the cation and anion provides interstitial sites large enough to accommodate the cation. Frenkel defects are found in such substances as AgCl and AgBr.

The second major type of point defect is the **Schottky defect.** This results when an equal number of anionic and cationic sites are left vacant in a crystal (an equal number so as to maintain electrical neutrality). This type of defect is common in the alkali halides (e.g., NaCl and KBr).

Another variety of defect occurs in nonstoichiometric compounds. The formula for iron(II) sulfide is best represented as $Fe_{(1-x)}S$, where x is the fraction of unoccupied cationic sites. In order to maintain an electrically neutral solid, some of the iron is present as Fe^{3+}, which compensates for the loss of Fe^{2+}. This type of defect is often found in solids where the cation may have more than one oxidation state (other examples are FeO, Cu_2O, and CuO).

Other nonstoichiometric compounds can be prepared from substances that are ordinarily stoichiometric. For example, NaCl can be reacted with Na vapor to form $Na_{(1+x)}Cl$. In this case the extra sodium is present as Na^+ ions in cationic sites, and the extra anionic sites are occupied by the electrons liberated from the sodium. These trapped electrons in anionic sites are called **F-centers.**

Perhaps the most important type of defect results when an impurity, often amounting to merely several parts per million, is introduced into a crystal. Here we find that a certain number of atoms in a solid are replaced by atoms of a different kind. For example, Ga or As atoms may be introduced into very pure germanium to yield substances with enhanced semiconductor properties.

Germanium has the diamond structure (Figure 7.14) with four covalent bonds to each Ge atom. When a Ga atom $(4s^24p^1)$ replaces a Ge atom $(4s^24p^2)$, one of the Ga—Ge bonds is deficient in electrons. Under an applied

Si doped with B
(p-type)

Sunlight

Si doped with As
(n-type)

Figure 7.18
A solar battery.

voltage an electron from a neighboring atom may move to fill this deficiency and thus leave a positive "hole" behind. When this "hole" is filled, another is created elsewhere and electrical conduction results from the migration of this positive "hole" through the crystal. Because of the positive nature of the charge carrier, the substance is said to be a **p-type semiconductor.**

Arsenic ($4s^2 4p^3$) has one more electron in its valence shell than germanium. When arsenic is added to germanium as an impurity (we say the Ge is "doped" with As), the extra electrons supplied are able to move through the solid when a voltage is applied. Because the conduction is due to negative electrons, the crystal is called an **n-type semiconductor.**

Since their discovery, n- and p-type semiconductors have served as the nucleus for the explosive growth of solid-state electronics. The transistor, used in so many electronic gadgets that we now take for granted, is made of n- and p-type semiconductors. Perhaps of even more current interest, these materials may also be used to harness the sun's energy in solar batteries.

A solar battery is composed of a silicon wafer doped with arsenic (giving an n-type semiconductor) over which there is placed a thin layer of silicon doped with boron (a p-type semiconductor), as illustrated in Figure 7.18. When exposed to sunlight, energy is absorbed that upsets the equilibrium between electrons and "holes" that exists at the p-n junction between the layers. This produces a small voltage which, coupled with the voltage from other solar cells, can be used to accomplish useful work.

Review Questions and Problems

7.1 Show that there are two atoms per unit cell in a substance that crystallizes in (a) a body-centered lattice and (b) an end-centered lattice.

7.2 How many formula units are there in the unit cell of CaF_2 (fluorite structure)?

7.3 LiBr has the rock salt structure in which Br^- ions, centered at lattice points, are in contact. Calculate the ionic radii of Br^- and Li^+ if the unit cell edge is 5.50 Å. Why is the accepted value for the ionic radius of Li^+ (0.60 Å) smaller than the value that you just computed?

7.4 CsCl crystallizes with a cubic unit cell of edge length 4.123 Å. The density of CsCl is 3.99 g/cm³. Is the unit cell primi-

Review Questions and Problems 199

tive, body-centered, or face-centered? (Hint. Calculate the number of formula units/unit cell.)

7.5 From the list of angles below, determine the angles at which X-rays of wavelength 1.4 Å, diffracted from planes of atoms 2.0 A apart, are in phase. $\theta = 17.3°$, 20.5°, 44.4°, and 55.3°.

7.6 Gold has an atomic radius of 1.46 Å. What would be the density of Au if it were to crystallize in the following structures? (a) A ccp structure. (b) A body-centered cubic lattice. (c) A simple cubic lattice. What is the correct structure for gold?

7.7 Calculate the amount of vacant (unoccupied) space in a primitive cubic, a body-centered cubic, and a face-centered cubic packing of identical spheres of diameter 1.0 Å.

7.8 Indicate which type of crystal (ionic, covalent, etc.) each of the following would form upon solidification: (a) O_2 (b) H_2S (c) Pt (d) KCl (e) Ge.

7.9 (a) Give two elements that make Si a p-type semiconductor.
(b) Name two that would make Si an n-type semiconductor.

7.10 Construct a diagram to illustrate the band structure of potassium.

7.11 Prove that there are twice as many tetrahedral sites as octahedral sites in the face-centered cubic lattice.

8
Liquids
and Changes
of State

When heat is added to a solid, the energy supplied causes an increase in the kinetic energy of the molecules or ions occupying the lattice sites. If this increase in kinetic energy is large enough to allow the molecules or ions to break away from their rigid lattice positions and move about, the solid melts, forming a liquid. Furthermore, a gas can be made to condense into a liquid by either sufficiently lowering its temperature or, under appropriate conditions, sufficiently increasing its pressure. Lowering the temperature of a gas brings about a decrease in the kinetic energy of the molecules. This causes the molecules to slow down and, at the condensation temperature, the intermolecular attractive forces are able to allow groups of molecules to cling together. Increasing the pressure on a gas causes the molecules to move closer together; if the attractive forces become strong enough, condensation occurs.

8.1 General Properties of Liquids

A liquid is composed of molecules that are constantly and randomly moving about, each undergoing many billions of collisions per second. These molecules, however, because of the relatively strong intermolecular forces of attraction between them (these forces are discussed in Section 4.11; i.e., dipole interactions, hydrogen bonding and/or van der Waals forces), are not as far apart or moving as freely as they are in a gas. On the other hand, the molecules of a liquid are not as close together or as structured as they are in a solid.[1] For these reasons liquids exhibit characteristics that place them somewhere between the completely chaotic gaseous state and the well-ordered solid state.

VOLUME AND SHAPE. In a liquid the attractive forces are strong enough to restrict the molecules to move about within a definite volume, but they are not strong enough to cause the molecules to maintain a definite position within the liquid. In fact the molecules, within the limits of the liquid's volume, are free to move over and around one another, thus allowing liquids to flow. Liquids, therefore, maintain a definite volume but, because of their ability to flow, their shape depends on the contour of the container holding them.

COMPRESSION AND EXPANSION. In a liquid the attractive forces hold the molecules close together, and increasing the pressure has little effect on the volume because there is little free space into which the molecules may be crowded. Liquids are therefore virtually incompressible. Similarly, changes in

---

[1] One exception to this is ice, which is less dense than water (see Section 8.8).

temperature cause only small volume changes (compared to a gas). The increasing molecular motion that accompanies rising temperature tends to increase the intermolecular distances, but this is opposed by the strong attractive forces.

DIFFUSION. When two liquids mix, the molecules of one liquid diffuse throughout the molecules of the other liquid at a rate much slower than is observed when two gases are mixed. We can observe the diffusion of two liquids by dropping a small quantity of ink into some water. As demonstrated in Figures 8.1a and b, when the ink drop strikes the water we see it as a concentrated "dot," which slowly spreads throughout the water. Diffusion takes place because the molecules in both liquids are able to move throughout the container. However, because the molecules in both liquids are so close together, each molecule undergoes billions of collisions before traveling very far. The average distance between collisions, called the *mean free path*, is much shorter in liquids than in gases, where the molecules are relatively far apart. Because of the constant interruptions in their molecular paths, liquids diffuse much more slowly than gases.

SURFACE TENSION. Each molecule in a liquid moves about, always under the influence of its neighboring molecules. A molecule near the middle of a quantity of liquid feels its attracting neighbors nearly equally in all directions (Figure 8.2a). A molecule at the surface of the liquid, however, is not completely surrounded and, as a result, feels attractions only by those molecules below and beside it (Figure 8.2b). The molecules along the surface thus feel an attraction in a direction toward the interior of the liquid, which causes the surface molecules to be drawn in. The most stable situation arises when the number of molecules experiencing these unequal attractive forces at the surface is a minimum, a condition that is fulfilled when the surface area of the liquid is as small as possible. The tendency of liquids to minimize their surface area explains why water, for example, beads when splashed on a clean polished solid surface and why the shape of raindrops is spherical. The amount of work needed to expand the surface of a liquid

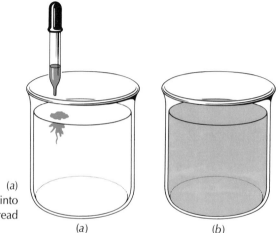

Figure 8.1
Diffusion in liquids. (a) Ink drop placed into water. (b) Ink has spread throughout the liquid.

(a)   (b)

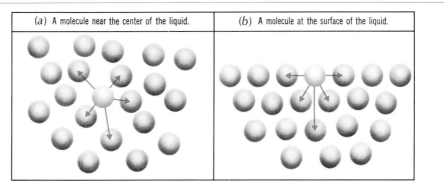

| (a) A molecule near the center of the liquid. | (b) A molecule at the surface of the liquid. |

Figure 8.2
Intermolecular attractive forces in liquids. (a) A molecule near the center of the liquid. (b) A molecule at the surface of the liquid.

is dependent on the strength of the inward forces and is called the liquid's **surface tension.** Surface tension is also a function of the temperature of the liquid. Since increasing the temperature (which increases the kinetic energy of the individual molecules) decreases the effectiveness of the inter-molecular attractive forces, surface tension decreases as the temperature is raised.

EVAPORATION. In a liquid the molecules are constantly undergoing rela-tively elastic collisions, giving rise to a distribution in individual molecular velocities and, of course, kinetic energies. Following reasoning similar to that described in Section 6.10, it follows that in a liquid, even at room tempera-ture, a small percentage of the molecules are moving with relatively high kinetic energies. If some of these faster moving molecules possess enough kinetic energy to overcome the attractive forces operating within the liquid and can escape through the surface into the gaseous state, the liquid evap-orates. Figure 8.3 represents a typical distribution of the kinetic energies of the molecules in a liquid; the shaded area corresponds to the fraction of the total number of molecules that possess sufficient kinetic energy to evaporate. The loss of the higher energy fraction because of evaporation leads to a lowering of the average kinetic energy of the remaining molecules. Since the

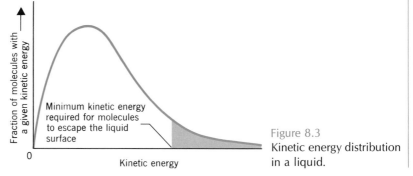

Figure 8.3
Kinetic energy distribution in a liquid.

temperature is directly proportional to the average kinetic energy, this results in a decrease in the temperature of a liquid as it evaporates. For example, we have all felt cool after a bath, because the evaporation of water from the body has drawn heat from us. In fact, evaporation of perspiration provides the body with a mechanism for controlling body temperature.

If a liquid, such as water, is to continue to evaporate from a container, heat must constantly be absorbed from the surroundings in order to replenish the energy taken away by the molecules leaving the liquid. If the surroundings are at a high temperature, heat can be supplied faster than when they are cool. Thus water evaporates faster on dry hot days than on dry cold days.

8.2
Heat of Vaporization

The **molar heat of vaporization,** which we shall refer to as $\Delta H_{\text{vaporization}}$ (or simply ΔH_{vap}), *represents the amount of energy that must be supplied to one mole of liquid to convert it into one mole of vapor at the same temperature.* The Greek letter, Δ, is usually used to symbolize a change, in this case, a change in the heat content (i.e., the total amount of heat energy) of a substance as it undergoes a change from liquid to vapor. This change in heat content is equal to the energy contained in the substance in its final state (vapor) minus the energy that the substance possessed in its initial state (liquid). Thus,

$$\Delta H_{\text{vaporization}} = H_{\text{vapor}} - H_{\text{liquid}}$$

In actual practice, neither H_{vapor} nor H_{liquid} can be measured; however, their difference (ΔH_{vap}) can be.

The magnitude of ΔH_{vap} provides a good measure of the strengths of the attractive forces operative in a liquid. In Table 8.1 we find values of ΔH_{vap} for several substances. If we look at the series of hydrocarbons, CH_4 through $C_{10}H_{22}$, we observe a steady increase in ΔH_{vap} with an increase in molecular weight. These compounds are nonpolar; therefore the only attractive forces that exist between their molecules are van der Waals forces. In Chapter 4 it was mentioned that the strengths of the van der Waals attractions increase with increasing numbers of electrons. As we proceed to higher molecular weights protons and neutrons, in the form of atomic nuclei, and electrons are added; hence an increase in molecular weight should be accompanied by an increase in the magnitude of the van der Waals forces of attraction. The trend in ΔH_{vap} is, therefore, consistent with the increase in attractive forces that is expected to parallel an increase in molecular weight.

When we look at the hydrogen halides, HF through HI, however, we find that the expected variation of ΔH_{vap} with molecular weight is reversed between HF and HCl. In fact, HF has a considerably higher heat of vaporization than any of the other HX compounds. This anomalous behavior is attributed to the presence of hydrogen bonding. As discussed in Chapter 4, this bonding is a particularly strong dipole-dipole interaction that can occur when hydrogen is bound covalently to a small, very electronegative element. We see the same inverted order of ΔH_{vap} for H_2O and H_2S and for NH_3 and PH_3, where again, hydrogen bonding is significant for H_2O and NH_3 but not for H_2S and PH_3. Oxygen, fluorine, and nitrogen are all very small and are the most electronegative elements in the periodic table, while the elements below them are much larger and much less electronegative. Thus we predict that hydrogen bonding is only important for H_2O, HF, and NH_3. "Normal" behav-

Table 8.1

Heats of Vaporization and Boiling Points

Compound	ΔH_{vap} (kcal/mole)	Boiling Point (°C)
CH_4	2.20	−161
C_2H_6	∼3.3	−89
C_3H_8	4.32	−30
C_4H_{10}	5.32	0
C_6H_{14}	6.83	68
C_8H_{18}	8.10	125
$C_{10}H_{22}$	8.56	160
HF	7.21	17
HCl	3.60	−84
HBr	3.90	−70
HI	4.34	−37
H_2O	9.71	100
H_2S	4.49	−61
NH_3	5.63	−33
PH_3	3.49	−88
SiH_4	2.95	−112

ior is reached in Group IVA hydrids where ΔH_{vap} for CH_4 is less than ΔH_{vap} for SiH_4. Here neither CH_4 nor SiH_4 have any tendency to hydrogen bond.

8.3

Vapor Pressure

If a liquid evaporates in an open container, eventually all the liquid will disappear because the molecules that have escaped from the liquid into the vapor phase diffuse readily into the atmosphere. If the same quantity of liquid at the same temperature is placed in a closed container, what will happen? In this case the volume of the liquid would initially decrease and then eventually become constant. If we monitored the pressure of the gas above the liquid, we would find that it initially increases and then it too levels off at a constant value. These observations can be explained in the following way. The molecules with the higher kinetic energies begin to leave the liquid, evaporating into the vapor phase, where they become trapped. The loss of molecules from the liquid must, of course, be accompanied by a volume decrease. In time the space above the liquid becomes occupied with more and more gaseous molecules, and the pressure of the vapor increases. With the increasing number of chaotically moving gas molecules the number of collisions with the walls in this restricted volume also increases. One of these walls is the surface of the liquid itself, which will trap any bombarding molecules having low kinetic energies. Thus evaporation (vaporization) as well as condensation take place at the surface of the liquid. Eventually the number of molecules in the vapor becomes large enough so that the rate at which the gas condenses will exactly equal the rate at which the liquid evaporates, and no further change in either the volume of the liquid or the pressure exerted by its vapor is observed. To emphasize this point again, vaporization and condensation are still taking place but with no change in the liquid volume or the vapor pressure. At this point the liquid is said to be in **dynamic equilibrium** with its vapor. The pressure exerted by the quantity of vapor above the liquid, when equilibrium is established, is called the **equilibrium vapor pressure** of the liquid. The vapor pressure of a liquid, quite naturally, depends on the ease with which its molecules can leave the liquid and enter the vapor state. In liquids where the intermolecular attractive forces are strong, the vapor pressure will be low, and in liquids where the attractive forces are weak, the vapor pressure will be high. Since increasing the temperature of a liquid increases the number of molecules possessing sufficient energy to overcome the attractive forces, then vapor pressure must increase with increasing temperature. Thus, whenever the vapor pressure is given, the temperature at which it was measured must also be specified.

One way that we can determine the vapor pressure of a liquid is by the use of a barometer, as shown in Figure 8.4. The height of the mercury in the barometer, before any liquid is added, is accurately measured. A liquid whose vapor pressure is to be determined is carefully added to the barometer by means of an eye dropper, and allowed to rise to the top of the mercury in the column, as shown in Figures 8.4b, c, and d (most liquids are less dense than mercury and will therefore float on the mercury surface). The space above the mercury column in Figure 8.4a is, for all practical purposes a vac-

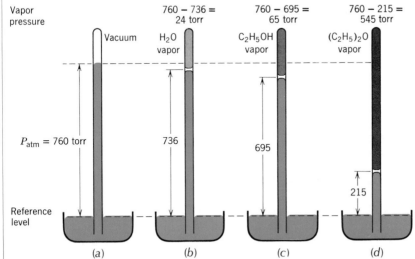

Figure 8.4

Measurement of vapor pressure. When a small amount of liquid is introduced above the mercury in a barometer, the vapor pressure of the liquid forces the mercury down. (a) no liquid above the mercury (b) H_2O (c) ethyl alcohol, C_2H_5OH, (d) diethyl ether, $(C_2H_5)_2O$.

uum² and exerts no downward force. The space above the mercury in Figures 8.4b, c, and d is filled with a small amount of liquid and its vapor. As the liquid begins to evaporate, the pressure of the trapped vapor causes the level of the mercury in the column to decrease; when the liquid and vapor are in equilibrium, the height of the mercury column becomes stationary. The total pressure exerted at the reference level (Figure 8.4) outside each barometer will be the atmospheric pressure, P_{atm}. The total pressure exerted within the barometer is P_{Hg}, the pressure due to the pull of gravity on the mercury in the column, plus P_{vapor}, the pressure exerted by the vapor in equilibrium with its liquid. The additional pressure exerted by the weight of the small amount of liquid on top of the column is negligibly small. Therefore, at equilibrium in each barometer

$$P_{atm} = P_{Hg} + P_{vapor}$$

In Figure 8.5a, $P_{vapor} = 0$; therefore, $P_{atm} = P_{Hg} = 760$ torr. In Figure 8.4b, c, and d, $P_{Hg} = 736$ mm, 695 mm, and 215 mm, respectively. Therefore, at 25° C the vapor pressure of water is 24 torr, that of ethyl alcohol is 65 torr, and that of diethyl ether is 545 torr. Water has the lowest vapor pressure of the three liquids in our example; therefore, it must have the strongest intermolecular attractive forces. Diethyl ether, on the other hand, has the highest vapor pressure of the three liquids, which means relatively weak attractive forces exist in it.

We have seen that the vapor pressure of a liquid is dependent on the nature of the liquid and its temperature. What happens to the vapor pressure when the volume or pressure of the vapor is changed? In Figure 8.5 we see an illustration of an apparatus that can be used to demonstrate the effect of volume and pressure changes on a liquid-vapor equilibrium. At a constant temperature equilibrium is established, as shown by Figure 8.5a. If, at this same constant temperature we allow the vapor to expand, by rapidly raising the piston (see Figure 8.5b), then the system is no longer in equilibrium. As the piston is withdrawn, creating the larger volume, the number of molecule-wall collisions decreases, causing a decrease in the pressure exerted by the vapor. The rate of condensation at the surface, which depends on the number of collisions between the vapor molecules and the surface of the liquid, must also decrease. The rate of evaporation, however, remains essentially the same. This means that, at a constant temperature, an increase in the volume of a vapor in equilibrium with its liquid causes more molecules to leave the liquid state than return to it. As the process continues, more and more molecules enter the vapor phase, causing an increase in the pressure exerted by the vapor and a corresponding increase in the rate of condensation. After a while enough molecules will be present in the vapor phase, so that the rate of condensation will again exactly equal the rate of evaporation, and equilibrium will be reestablished. At this newly established equilibrium the larger volume of gas is now occupied by more molecules. The vapor pressure will be the same as before the volume change occurred, but the volume of the liquid is slightly smaller.

---

² Mercury itself does have a finite vapor pressure (about 10^{-3} torr at room temperature) and, therefore, should never be left in an open container because of its high toxicity.

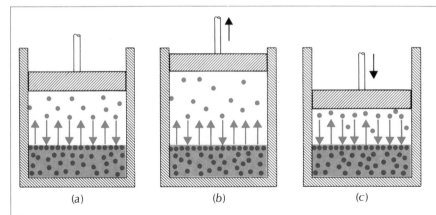

Figure 8.5

Effect of volume changes on vapor pressure. (a) Equilibrium between liquid and vapor. (b) No equilibrium. Rate of evaporation is greater than rate of condensation. (c) No equilibrium. Rate of condensation is greater than rate of evaporation.

Decreasing the volume of the vapor by lowering the piston (Figure 8.5c) will also disturb the equilibrium. Increasing the pressure of the vapor will cause an increase in the number of molecule-wall collisions. This, in turn, will lead to an increase in the rate of condensation but will have very little effect on the rate of evaporation. The rate at which the molecules leave the vapor phase, then, will be greater than the rate at which molecules leave the liquid phase. This imbalance in rates causes the pressure exerted by the vapor to decrease and the volume of the liquid to increase. Eventually the rate of condensation will decrease to a point where it exactly equals the rate of evaporation, reestablishing equilibrium. At this new equilibrium the smaller vapor volume, caused by the movement of the piston, will be occupied by fewer gaseous molecules. The vapor pressure will have returned to its initial value and the volume of the liquid will have increased slightly.

LE CHATELIER'S PRINCIPLE. The dynamic equilibrium between a liquid and its vapor can be represented by the equation,

$$\text{Liquid} \rightleftharpoons \text{Vapor} \tag{8.1}$$

Here the double half-arrows mean that the rates of evaporation and condensation are equal. If we in any way disturb this system so that it is no longer at equilibrium (we say a "stress" is applied to the system), a change occurs that will, if possible, bring the system back to equilibrium. In our example above, an increase in the volume of the vapor caused the system to no longer be at equilibrium. We saw that more liquid evaporated until equilibrium was reestablished. In Equation 8.1 above, this corresponds to the process as read from left to right, that is, Liquid $\rightarrow$ Vapor, and results in a new "position of equilibrium" in which there is less liquid and more vapor. In this sense the position of equilibrium has shifted to the right when we applied the stress. The action taken by any system at equilibrium when a stress is applied can be described by the **principle of Le Chatelier**[3] which states that, *when a system in a state of dynamic equilibrium is acted upon by some outside stress, the system will shift to a new position of equilibrium in order to minimize the effect of the stress.*

For example, let us apply Le Chatelier's principle to describe what effect pressure changes have upon a liquid-vapor equilibrium. When the stress applied is a pressure decrease (volume increase), the system attempts to undergo a change that will return the pressure up to its initial value. In this example, the pressure can be increased if more molecules enter the vapor phase, that is, if some additional liquid evaporates. After equilibrium has been reestablished, there will be less liquid and more vapor present in the container; and we say that the position of equilibrium represented by Equation 8.1 has shifted to the right. However, if the volume is increased sufficiently; for example, if the piston in Figure 8.5 is removed entirely so that the liquid is open to the atmosphere, all of the liquid will evaporate and equilibrium will not be reestablished.

In a similar fashion, we predict that raising the pressure leads to a decrease in the quantity of vapor and, of course, a corresponding increase in the amount of liquid. Thus, we might conclude in general that an increase in pressure upon a system at equilibrium favors the production of the more dense phase, while a decrease in pressure favors the formation of the less dense phase.

The effect of temperature changes on an equilibrium can also be described using Le Chatelier's principle. Increasing the temperature on a system at equilibrium favors the absorption of energy (an endothermic change). In a liquid-vapor equilibrium system this means that a temperature increase causes more liquid molecules to evaporate because this process absorbs heat. Decreasing the temperature, on the other hand, favors the release of energy (exothermic). As the temperature is decreased in a liquid-vapor equilibrium, more molecules condense into the liquid phase, releasing heat and thereby minimizing the effect of the applied stress.

In summary, Le Chatelier's principle predicts that a temperature increase will shift the position of equilibrium to favor the endothermic process. Similarly, a decrease in temperature will favor the exothermic change.

VAPOR PRESSURE CURVES FOR LIQUIDS. We can determine the vapor pressure of liquids as a function of temperature by using the same apparatus described in Figure 8.4 and varying the surrounding temperature. Data accumulated in such experiments performed on water, ethyl alcohol, and diethyl ether are illustrated graphically in Figure 8.6. We see from the shapes of the curves in the figure that at lower temperatures vapor pressure changes relatively slowly with increasing temperature while at higher temperatures the changes are more rapid. Points along the curve in Figure 8.6 represent combinations of pressures and temperatures that must be satisfied in order for the liquid to be in equilibrium with its vapor. These curves terminate at a temperature, called the **critical temperature** (T_c), above which molecular motion is

[3] Henry Le Chatelier, a professor in Paris, proposed his important "law of reaction" in 1888. He died in 1936.

Figure 8.6
Vapor pressure curves.

so violent that the substance can only exist as a gas. In other words, *the critical temperature is that temperature above which a substance can no longer exist as a liquid, regardless of the applied pressure.* The pressure that must be applied to a substance at its critical temperature in order to achieve a liquid-vapor equilibrium is called its **critical pressure.** In Table 8.2 a few substances are listed with their corresponding critical temperatures and critical pressures.

Table 8.2
Some Critical Temperatures and Pressures

Compound	T_c (°C)	P_c (atm)
Methane (CH_4)	−82.1	45.8
Ethane (C_2H_6)	32.2	48.2
Benzene (C_6H_6)	288.9	48.6
Ammonia	132.5	112.5
Carbon dioxide	31	72.9
Water	374.1	217.7
Helium	−267.8	2.3

That temperature at which the vapor pressure of a liquid is equal to the atmospheric pressure is known as the **boiling point** of the liquid. At this temperature the vapor pressure is high enough to cause vaporization to occur at various points throughout the interior of the liquid. Thus boiling is accompanied by the formation of bubbles, which simultaneously form at many spots in the liquid.[4]

When a bubble is formed within the liquid, the liquid that originally occupied this space is pushed aside and the level of the liquid in the container is forced to rise against the downward pressure exerted by the atmosphere. In other words, it is the pressure exerted by the vapor inside the bubble that pushes the surface of the liquid up against the atmospheric pressure. This can only occur when the vapor pressure of the liquid becomes equal to the prevailing atmospheric pressure. If it were less, the atmospheric pressure would cause the bubble to collapse.

As long as bubbles are forming within the liquid, that is, as long as the liquid is boiling, the vapor pressure of the liquid is equal to the atmospheric pressure. Since the vapor pressure remains constant, the temperature of the boiling liquid also stays the same. An increase in the rate at which heat is supplied to the boiling liquid simply causes bubbles to form more rapidly but does not increase the temperature.

It is obvious, from the discussion above, that the boiling point of a liquid depends on the prevailing atmospheric pressure. The boiling point of a liquid at one standard atmosphere, 760 torr, is referred to as its **standard** or **normal boiling point.** For water, the normal boiling point is 100° C. At higher pressures its boiling point is greater; at lower pressures (e.g., on a mountain top) its boiling point is less. Boiling points given in reference tables are always normal boiling points, unless otherwise stated.

The constant temperature maintained by a boiling liquid is utilized when we employ water for cooking foods. Once water boils its temperature remains at 100° C, which is ideal for cooking foods evenly and at a rapid rate. The pressure cooker also takes advantage of the fact that the boiling point changes with pressure. These cookers are time-savers because they enable foods to be prepared at a much faster rate than they could be in an open pot. The lid on a pressure cooker forms a tight seal on the pot and is equipped with a pressure relief valve to prevent the pot from exploding. The heat supplied by the stove causes more and more liquid water to evaporate; as a result, the pressure inside the kettle increases until steam begins to exit from the relief valve. Since the pressure inside the cooker at this point is higher than 760 torr, the water boils at a higher temperature and foods cook faster.

The temperature at which liquids boil is another example of a property that gives a good estimation of the strength of the attractive forces operating within a liquid. Liquids whose attractive forces are relatively high have correspondingly high boiling points, while liquids with weak attractive forces boil at a relatively low temperature. This is demonstrated by Figure 8.7 in which the boiling points of some hydrogen compounds of the elements in

---

[4] When heat is first applied to a liquid many small bubbles begin to form. These small bubbles are due to the expulsion of dissolved gases and do not mean that the liquid has begun to boil. We refer here to the larger bubbles formed during boiling (see Section 9.5).

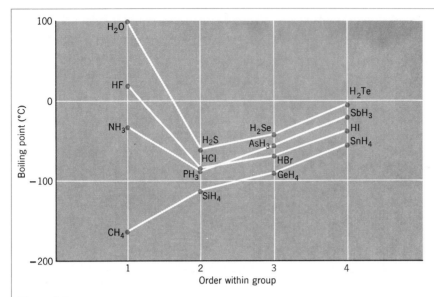

Figure 8.7

Boiling points of hydrogen compounds of Group IVA, VA, VIA, and VIIA Elements.

Groups IVA, VA, VIA, and VIIA are compared. Let us look at the compounds of Group IVA first because they form a nearly ideal pattern. We see from the figure that as the atomic weights of the elements in Group IVA increase, so do the boiling points of their hydrogen compounds. Using reasoning similar to that developed in Section 8.2 in the discussion of the molar heat of vaporization, we know that as the number of electrons in the compounds increase, that is, from CH_4 to SnH_4, so also do the van der Waals forces of attraction. We expect, therefore, that the boiling points of this series of compounds increase, as they actually do, in the direction of increasing molecular weights.

Except for the first members of the hydrogen compounds of Groups VA, VIA, and VIIA, the same trend is also observed—that is, increasing boiling point with increasing atomic weight of the element in the group. The first member of each of these groups, however, has a relatively high boiling point, much higher than expected from molecular weights alone. They, therefore, must possess attractive forces in addition to the those of the van der Waals type. The position of each of the first members in relation to the rest of the group can be attributed to the presence of hydrogen bonding. We saw in Section 8.2 that the strongest hydrogen bonds form when hydrogen is bonded to a very small electronegative element. Therefore, the contribution of hydrogen bonding is expected to be the strongest for the first member of Groups VA, VIA, and VIIA, and becomes relatively unimportant for the remaining members. Methane, CH_4, which is nonpolar and cannot hydrogen bond (the electronegativity of carbon is too low and there are no lone electron pairs on the carbon to which hydrogen bonds can be formed) follows the normal, nearly straight-line pattern, within its group.

The fact that water has a higher boiling point than HF, even though fluorine is more electronegative than oxygen, seems to be because each water

Liquids and Changes of State

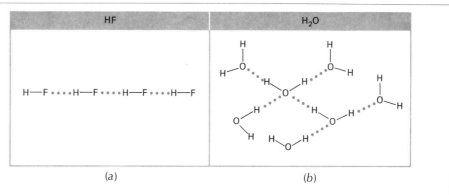

| HF | H₂O |

(a)   (b)

Figure 8.8

Hydrogen bonding in HF and H₂O. (a) Each HF molecule has only one hydrogen atom that can hydrogen bond to something else. (b) Each H₂O molecule has two hydrogen atoms that can hydrogen bond to other H₂O molecules.

molecule is capable of forming four hydrogen bonds with other H_2O molecules while an HF molecule forms only two hydrogen bonds with two other HF molecules (Figure 8.8). The strength of the four hydrogen bonds in water exceeds that of the two hydrogen bonds in HF even though an HF hydrogen bond is stronger than a single hydrogen bond between H_2O molecules. The hydrogen bonding in NH_3 is much weaker than in H_2O or HF because of the considerably lower electronegativity of nitrogen. Thus, even though NH_3 could conceivably form three or four hydrogen bonds, their total strength is so small that NH_3 has a lower boiling point than either HF or H_2O.

8.5 Freezing Point

Liquids become solids by the removal of heat. As the temperature is lowered, more molecules in the liquid slow down. If the temperature is sufficiently lowered so that the attractive forces cause the slower moving molecules to become rigidly held in position, the liquid begins to freeze. As pointed out in Chapter 7, if the molecules are "frozen" in a well-ordered lattice a crystalline solid is formed, while if the molecules are frozen in a random fashion an amorphous solid is produced. As a solid forms, the average kinetic energy of the molecules remaining in the liquid increases. This is because the molecules with low kinetic energies have been lost to the solid. Therefore, heat must continually be removed if freezing is to continue.

In a solid, as well as in a liquid, we have a distribution of kinetic energies. When a solid is in contact with its liquid, high energy molecules at its surface can break away from the solid and enter the liquid state. At some particular temperature, called the **freezing point** (or **melting point**), the rate at which molecules leave the solid to enter the liquid is the same as the rate at which molecules are leaving the liquid state to become part of the solid. Thus at the melting point of a solid or the freezing point of a liquid equilibrium exists between the liquid and solid.

The total amount of heat that must be removed in order to freeze one mole of a liquid is called its **molar heat of crystallization.** *The* **molar heat of fusion,** ΔH_{fus}, *on the other hand, is equal in magnitude but opposite in sign to the molar heat of crystallization and is defined as the amount of heat that* *must be supplied to melt one mole of a solid.* Since heat must be added to melt a solid, the attractive forces must be slightly higher in a solid than in a liquid. This is not surprising since it is expected that greater attractive forces are necessary to hold the particles rigidly in place in a solid than are required to keep them within the liquid where they are free to roam about. The magnitude of the molar heat of fusion provides us with a measure of the differences between the intermolecular attractive forces in the solid and the liquid, that is,

$$\Delta H_{\text{fus}} = H_{\text{liquid}} - H_{\text{solid}}$$

As before, we cannot actually measure H_{liquid} or H_{solid} but, instead, only their difference, ΔH_{fus}. ΔH_{fus} is always much smaller than the molar heat of vaporization, as shown in Table 8.3, because the relative changes in attractive forces experienced by the molecules on going from a solid to a liquid are very small compared to the differences in attractive forces between a liquid and a gas. This means that the amount of energy (ΔH_{fus}) required to cause the molecules of a solid to overcome their attractive forces and form a liquid is small compared to the energy (ΔH_{vap}) required for liquid molecules to move apart, forming a gas.

The data obtained by following the temperature of a solid as the heat supplied to it causes it to melt, forming a liquid, and continuing until the liquid boils into a gas, gives rise to a heating curve. A graph produced from such an experiment is represented in Figure 8.9 where we have plotted temperature versus time for one mole of a substance as it is heated. If the rate of heating is constant during the entire experiment, then any length of time (distance along the abscissa) also represents an amount of heat that has been added. As heat is added to the solid, the average kinetic energy of the molecules in the solid increases; that is, they vibrate more vigorously about their lattice positions, and an increase in the temperature of the solid is observed. This is represented as line A in Figure 8.9. When enough heat has been added to cause the molecules to break away from their lattice positions, the solid begins to melt and we have arrived at point B on the curve. The temperature corresponding to point B is the melting point, or freezing point, T_f.

As the solid melts along line BC in Figure 8.9, the temperature of both the solid and liquid remains constant. This means that as the solid melts the average kinetic energy of the system does not change; therefore, the energy

Table 8.3

Heats of Fusion and Vaporization

Substance	ΔH_{fus} (kcal/mole)	ΔH_{vap} (kcal/mole)
Water	1.43	9.71
Benzene	2.37	7.35
Chloroform	2.97	7.62
Diethyl ether	1.64	6.21
Ethanol	1.82	9.22

Figure 8.9

A typical heating curve for one mole of a substance.

added between points B and C must raise the potential energy of the substance. Where does this potential energy appear?

Recall that potential energy depends on position and, in this case, the position of the molecules of the substance relative to one another. In general, as a substance melts, there is an increase in the average distance of separation between the particles as they pass from the solid to the liquid state.[5] Energy must be absorbed for the particles to move apart against the attractive forces that exist between them. In other words, the energy that is absorbed during the melting process increases the potential energy of the particles with respect to one another as they become separated. The total amount of energy that must be absorbed to melt the one mole of material, its molar heat of fusion, is represented as the interval from points B to C.

Once the entire sample has melted, that is, point C has been reached, the supply of heat is used to increase the kinetic energy of the liquid molecules. Since the average kinetic energy is increasing, we see a corresponding increase in the temperature from points C to D. At point D the liquid begins to boil. As the liquid boils the temperature of both the liquid and its gas remain constant (line DE), which means the heat supplied during the boiling process does not further increase the kinetic energy of the molecules. Therefore, when point D is reached, as when point B was reached, the energy supplied goes to increasing the potential energy of the molecules as they pass from the liquid to the gas. The total amount of energy absorbed to boil the one mole of material, its molar heat of vaporization, is represented as the interval between points D and E, and the boiling point (T_b) is the temperature corresponding to line DE on the graph. When the supply of energy is sufficient to cause the entire one mole of sample to be in the gaseous state, any additional heat causes an increase in the average kinetic energy of the gaseous molecules and an increase in the temperature of the gas.

A cooling curve, shown in Figure 8.10, can be obtained by following the temperature of a gas as it is cooled. The decrease in the temperature of the gas (line A) is caused by a decrease in the average kinetic energy of the gaseous molecules as heat is removed. At point B the kinetic energy has been lowered to the extent that the attractive forces can cause condensation to occur. During condensation (line BC) the temperature remains constant, which means once again that the average kinetic energy of the molecules in

[5] The exception to this, of course, is ice in which the solid is less dense (more expanded) than the liquid. Here energy supplied to the solid disrupts some of the hydrogen bonding that exists in the solid ice.

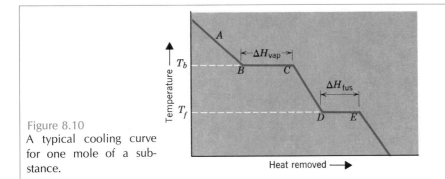

Figure 8.10
A typical cooling curve for one mole of a substance.

both the gas and the liquid must also stay the same. The removal of heat between points B and C, therefore, goes to decreasing the potential energy of the molecules as the gas condenses into a liquid. The total amount of heat that must be removed for the one mole of gas to condense completely, as pointed out in Section 8.2, is equal in magnitude (but opposite in sign) to its molar heat of vaporization and is given by the length of the line BC. The temperature at which condensation occurs is the same as the boiling point (T_b) of the liquid.

When point C is reached the material is entirely in the liquid state. Any further cooling will bring about a decrease in the average kinetic energy of the molecules and a corresponding decrease in its temperature. At point D enough heat has been removed to cause the liquid to begin to crystallize. While crystallization is taking place (line DE) there is no further decrease in the temperature, which means that the average kinetic energy of the molecules, both liquid and solid, must remain the same. The removal of heat during the crystallization process must, therefore, go to decreasing the potential energy of the molecules as the solid forms from the liquid. The total amount of heat that must be removed in order to crystallize the one mole of material is numerically equal to (but opposite in sign to) the molar heat of fusion. The temperature corresponding to line DE is the freezing point of the liquid and, of course, is the same as the melting point of the solid. Once point E is reached any further removal of heat causes a decrease in the average kinetic energy of the molecules in the solid, and the temperature decreases.

Some liquids do not follow a smooth transition into the solid state, but instead give rise to a cooling curve such as that shown in Figure 8.11. As the

Figure 8.11
Supercooling. As the liquid is cooled, its temperature drops below the freezing point. After a short time freezing begins and the temperature rises to the freezing point.

temperature of the liquid drops, it eventually reaches point A, the expected freezing point of the substance. The molecules, however, may not be oriented properly to fit into the crystalline lattice and random motion continues as heat is further withdrawn from the liquid. Consequently the temperature of the liquid drops below its expected freezing point and the liquid is said to be **supercooled.** Once a small number of molecules have achieved the correct pattern, a tiny crystal is formed that serves as a seed upon which additional molecules may rapidly accumulate. Potential energy is suddenly released as this crystal quickly grows and the energy that is evolved increases the average kinetic energy of the molecules in the liquid and solid. As a result, the temperature of the system rises again until it returns to the freezing point, after which the substance behaves normally. Further removal of heat eventually leads to complete conversion of the liquid to a solid.

Some substances, such as glass, rubber, and many plastics, never do achieve a crystalline state when their liquids solidify upon cooling. These compounds consist of long chainlike molecules that intertwine in the liquid. As they are cooled their molecules move so slowly that they never do find the proper orientation to form a crystalline solid and an amorphous solid results instead. Thus, these amorphous solids are actually supercooled liquids and in fact continue to flow, although very slowly to be sure, even at room temperature. For example, very old glass shows greater crystallinity, when examined by X-ray diffraction, than does freshly formed glass, showing that molecules are slowly finding their way into a crystalline lattice. Furthermore, a glass rod suspended horizontally between supports at opposite ends will, in time, acquire a permanent bend because of this gradual liquid flow.

8.7
Vapor Pressure of Solids

Like liquids, solids too undergo evaporation and therefore exhibit a vapor pressure. The molecules in a crystalline solid are vibrating about their lattice positions and are continually undergoing collisions with their nearest neighbors, giving rise to a distribution of kinetic energies. A small fraction of the molecules at the surface of a solid possess large enough kinetic energies for them to overcome the attractive forces within the solid and break away from the surface, entering the gaseous phase above. The process whereby molecules go directly from the solid into the gaseous state is known as **sublimation.** Anyone who has seen ''dry ice'' (solid carbon dioxide) disappear knows that no liquid puddles are left behind because the CO_2 evaporates directly from the solid to the gas.

Freeze dried coffee is manufactured by first freezing a batch of brewed coffee and then removing the ice component by vacuum. The vacuum creates an atmosphere of diminished pressure where ice readily sublimes, thereby removing the water vapor rapidly. Once the water is removed, the dried solid (''brewed'') coffee that remains is ready to be placed into a cup, along with some hot water and perhaps a little cream and sugar (or anisette or brandy) for your consumption.

When sublimation takes place in a closed container, more and more molecules enter the gaseous state and the pressure exerted by the vapor increases. The slower moving gaseous molecules that are colliding with the surface of the solid become trapped and return to the solid state. In time the leaving rate will exactly equal the returning rate, and a dynamic equilibrium will be established. The pressure exerted by a vapor in equilibrium with its solid is known as the **equilibrium vapor pressure of the solid.** Just as in liquids, the vapor pressure of a solid is dependent on the ease with which its molecules enter the gaseous state. For example, the attractive forces are stronger in ionic solids than in molecular solids and, as expected, we find that the vapor pressures of ionic solids are generally much lower than those of molecular solids.

The vapor pressure of a solid, like that of a liquid is a function of its temperature. Increasing the temperature on a solid-vapor equilibrium, according to Le Chatelier's principle, leads to a shift in the position of the equilibrium that will occur with the absorption of heat. The production of vapor from the solid is an endothermic process; therefore as the temperature rises more of the solid will evaporate and more of the vapor will be produced until equilibrium is once again attained. Hence, the equilibrium vapor pressure of a solid increases with increasing temperature until eventually a temperature is reached at which the solid melts. Further increases in temperature, beyond this point, will then give rise to a liquid-vapor equilibrium curve that terminates at the critical temperature of the substance. If, using water as an example, we plotted the vapor pressure versus temperature for the solid-vapor equilibrium and for the liquid-vapor equilibrium on the same graph, we would produce Figure 8.12. Each point along the "solid" curve represents the specific combinations of temperatures and pressures that must be achieved in order for the solid to be in equilibrium with its vapor. Likewise, every point along the "liquid" curve represents combinations of temperatures and pressures required for the liquid to be in equilibrium with its vapor. The point of intersection of these two curves, the **triple point,** corresponds to a unique temperature and pressure where all *three* states of matter (solid, liquid, and gas) coexist in equilibrium with each other. The triple point occurs at a temperature and pressure that depends on the nature of the substance in question. For example, the triple point of water occurs at a temperature of 0.01° C and a pressure of 4.58 torr, while the triple point temperature of carbon dioxide is −57° C and the triple point pressure is 5.2 atm.

There is still another equilibrium that can be represented on the same graph. This line corresponds to the combinations of temperatures and pressures that must be maintained in order to achieve a solid-liquid equilibrium.

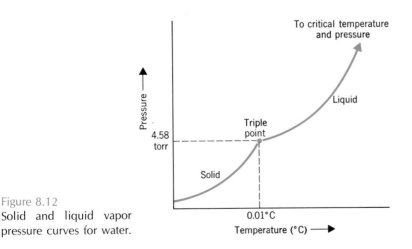

Figure 8.12
Solid and liquid vapor pressure curves for water.

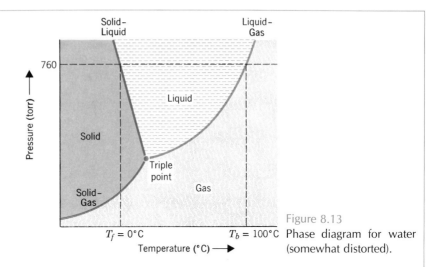

Figure 8.13

Phase diagram for water (somewhat distorted).

At a pressure of one atmosphere, the melting point of water is 0° C; therefore, the solid-liquid equilibrium line passes through both the triple point and the normal melting point as shown in Figure 8.13. The resulting drawing is called a **phase diagram** because it allows us to pinpoint temperatures and pressures at which the various phases exist, as well as those conditions under which equilibrium can occur. For instance, at a pressure of one atmosphere water exists as a solid at all temperatures below 0° C, and in fact the region bounded by the solid-liquid[6] and solid-vapor equilibrium lines corresponds to all of the temperatures and pressures at which water exists as a solid. Similarly, in the region bounded by the solid-liquid and liquid-vapor equilibrium lines, the substance can only exist as a liquid, while to the right of both the solid-vapor and liquid-vapor lines the substance must be a gas.

In Table 8.4 are some randomly chosen temperatures and pressures and the physical states of water that we can predict from its phase diagram. You might verify these predictions to illustrate how to use a phase diagram.

Table 8.4
Physical State of Water at Random Temperatures and Pressures

Temperature (°C)	Pressure (atm)	State
25	1.0	Liquid
0	2.0	Liquid
0	0.5	Solid
100	0.5	Gas

[6] The slope of this line, as drawn in Figure 8.14, is exaggerated for our discussion. The actual slope is much less to the left (33 atm is required to lower the melting point of ice by only one degree).

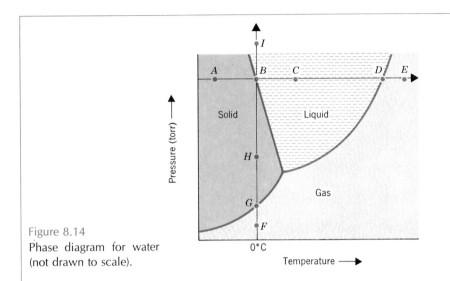

Figure 8.14
Phase diagram for water
(not drawn to scale).

To gain a further insight into the meaning of a phase diagram, let us follow the changes that take place as we move along a line of constant pressure, say 1 atm, by varying the temperature. In Figure 8.14, point A lies in the region of the diagram where a sample of the substance would exist entirely as a solid, as shown in Figure 8.15a. When the temperature rises to point B in Figure 8.14 the solid begins to melt and an equilibrium between the solid and

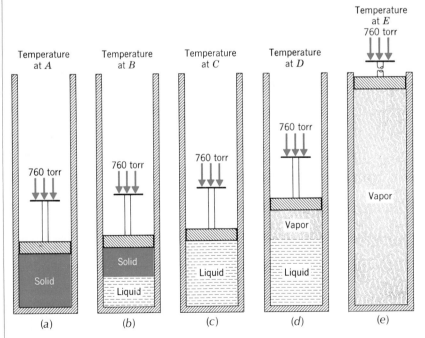

Figure 8.15
Raising the temperature at a constant pressure of 760 torr. Temperatures correspond to points A to E in Figure 8.14.

liquid can occur (Figure 8.15b). At still a higher temperature, point C, all of the solid will have been converted to a liquid (Figure 8.15c); and, when the liquid-vapor line is encountered at point D in Figure 8.14, vapor may at last begin to form and an equilibrium can exist (Figure 8.15d). Finally at a sufficiently high temperature, such as point E, all of the water will exist in the vapor state (Figure 8.15e).

We could also proceed with a similar analysis in which the temperature is held constant and the pressure is permitted to change. For example, at point F in Figure 8.14 the water would exist entirely as a gas (Figure 8.16a). At a higher pressure, point G in Figure 8.15, a solid-vapor equilibrium would exist (Figure 8.16b) and above that pressure, at point H, all of the water would be converted to a solid (Figure 8.16c). As the pressure is increased further we encounter the solid-liquid line at point B in Figure 8.14 where we again have an equilibrium as represented by Figure 8.16d. At still higher pressures the water will melt so that at point I all of the water is present in the liquid state (Figure 8.16e).

In the phase diagram for water we see that the solid-liquid equilibrium line slants to the left. This is a direct consequence of the fact that liquid water at 0° C has a higher density than does the solid. Le Chatelier's principle requires that an increase in pressure on a system at equilibrium will lead to the production of the more dense phase; that is, a rise in pressure favors the packing together of molecules — quite a reasonable expectation. This means that if we have solid and liquid water at equilibrium and we increase the pressure while holding the temperature at 0° C, we should produce the

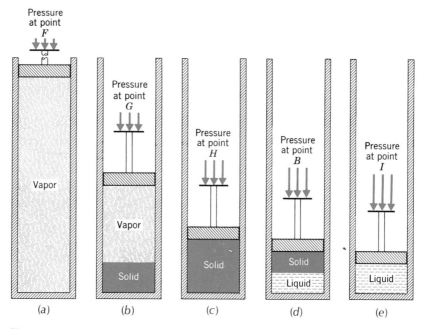

Figure 8.16
Raising the pressure at a constant temperature of 0°C. Pressures correspond to points on Figure 8.14.

Figure 8.17
Phase diagram for carbon
dioxide.

higher density, liquid phase.[7] On the phase diagram a rise in pressure at constant temperature amounts to moving upward along a vertical line. We can
move from the solid-liquid equilibrium line upward into a region of all liquid
only if the solid-liquid line leans to the left.

Water is quite an unusual substance. For nearly all other compounds the
solid phase is more dense than the liquid and for these substances the solid-
liquid line slants to the right, as is shown in the phase diagram for CO_2 that
appears in Figure 8.17. An interesting feature of this phase diagram is that the
entire liquid range lies above a pressure of one atmosphere; therefore, it is
impossible to form liquid CO_2 at atmospheric pressure. Instead, as the gas is
cooled, the solid-vapor equilibrium line is encountered at $-78°C$ and the
vapor is converted directly to the solid. This also explains why "dry ice"
sublimes rather than melts at ordinary temperatures and pressures.

**Review
Questions
and Problems**

8.1   State how the following physical properties differ for the three states of
matter—solid, liquid, and gas: (a) density, (b) rate of diffusion, (c) compressibility, (d) ability to flow.

8.2   If you were asked to compare the
strengths of the attractive forces operative in liquid A with those in liquid B,
what types of data would you collect?

8.3   Trouton's rule states that the ratio of the
heat of vaporization to the boiling point
(in degrees Kelvin) is approximately a
constant. Verify this for the hydrocarbons CH_4 through $C_{10}H_{22}$ in Table
8.1. What conclusions can you draw
concerning the relationship between
ΔH_{vap} and boiling point?

8.4   Suppose two substances, X and Y, have

[7] In fact, it had long been thought that this melting of water that occurs at high pressure is responsible for our ability to skate on ice. It was believed that the high pressure produced by the skat-
er's weight concentrated on the sharp edge of a blade caused the ice just beneath the blade to
melt, producing a thin film of liquid water that serves as a lubricant and allows the skate to slide
smoothly on the ice. Current feeling, however, is that this film of water is most probably the
result of melting due to friction between the moving skate blade and the ice.

heats of vaporization equal to 9.0 and 6.5 kcal/mole, respectively. Which compound would you expect to have the highest boiling point? Which compound would be least likely to exhibit hydrogen bonding?

8.5 Explain why, for any given substance, ΔH_{fus} is smaller than ΔH_{vap}.

8.6 Using Le Chatelier's principle, predict the effect of a change in temperature and pressure on the equilibria,

(a) solid + heat $\rightleftharpoons$ liquid
(b) liquid + heat $\rightleftharpoons$ vapor

8.7 At a pressure of 760 torr a new compound was found to melt at 25° C and boil at 95° C. The triple point of the substance was determined to occur at a pressure of 150 torr and at a temperature of 20° C. Sketch the phase diagram for this substance. Label, on your drawing, the solid, liquid, and vapor regions as well as the solid-liquid, liquid-vapor, and solid-vapor equilibrium lines.

8.8 On the basis of the phase diagram in Question 8.7, describe the changes that you would observe if, at a constant temperature of 22° C, the pressure on a sample of the compound is gradually increased from 10 to 1000 torr. What would be observed if the same process were to occur at a constant temperature of 10° C?

8.9 Sketch the heating curve that you would expect to find when one mole of the compound described in Question 8.7 is heated at a constant rate under a constant pressure of 1.00 atm. On your drawing, indicate the melting point and

boiling point of the substance. Also, label the intervals that correspond to ΔH_{fus} and ΔH_{vap}.

8.10 What can we conclude about the relative densities of the liquid and solid phases of the compound in Question 8.7?

8.11 Use Le Chatelier's principle to predict how variations in pressure will affect the melting point of: (a) water, (b) carbon dioxide.

8.12 Explain, on a molecular level, why the temperature remains constant as heat is added to vaporize a liquid at its boiling point.

8.13 Why will a decrease in the volume of a vapor in equilibrium with its liquid *not* cause a permanent change in the pressure?

8.14 With the aid of the phase diagram in Figure 8.17, predict the physical state of carbon dioxide under the following conditions of temperature and pressure:

Temperature (°C)	Pressure (atm)
−80	1.0
−60	1.0
−56	10.0
−56	2.0
−65	5.0
−40	10.0

8.15 Define surface tension, melting point, boiling point, vapor pressure, sublimation, and supercooling.

9
Properties
of Solutions

In the previous three chapters we discussed the properties associated with the three states of matter. These discussions, however, applied, for the most part, to pure substances and it is only rarely that a chemist works with pure materials. Usually his chemicals occur in mixtures and very often these are solutions. The presence of a solute in a solution has very marked effects on the properties of the substance in which it is dissolved, and many times these effects can provide us with useful information about the way substances interact with one another.

In this chapter we shall take a close look at the solution process to explore the changes that occur when one substance dissolves in another. We shall also focus our attention on the way in which the solute affects the physical properties of the solution. Many of these phenomena have very practical laboratory applications, such as separating mixtures into their components and the determination of molecular weights.

In the most common type of solution that we come across in the laboratory, a solute is dissolved in a liquid, and for this reason most of our attention will be directed toward this. Liquid solutions can be prepared by dissolving a solid in a liquid (e.g., NaCl in water), a liquid in a liquid (e.g., ethyl alcohol in water — a highball) or a gas in a liquid (e.g., any carbonated beverage contains dissolved carbon dioxide).

In addition to liquid solutions it is possible to have solutions of gases, such as the atmosphere that surrounds the earth, and solid solutions, formed when a substance is dissolved in a solid. The properties of gaseous solutions were discussed in Section 6.4 under the heading Dalton's Law of Partial Pressures, and nothing more need be said about them here. Solid solutions, of which many alloys are examples, are of two types. **Substitutional solid solutions** exist in which atoms, molecules, or ions of one substance take the place of particles of another substance in a crystalline lattice, as shown in Figure 9.1a. Zinc sulfide and cadmium sulfide form such mixtures in which cadmium ions randomly replace zinc ions in the ZnS lattice. Another example is provided by brass, which is a substitutional solid solution of copper and zinc.

Interstitial solid solutions constitute the other type and are formed by placing atoms of one kind into voids, or **interstices,** that exist between atoms in the host lattice, as illustrated by Figure 9.1b. These voids occur, for example, as the tetrahedral or octahedral holes in the closest-packed structures described in Chapter 7. Tungsten carbide, WC, an extremely hard substance that has found many uses in cutting tools designed for machining steels, is an example of an interstitial solid solution in which the tungsten atoms are ar-

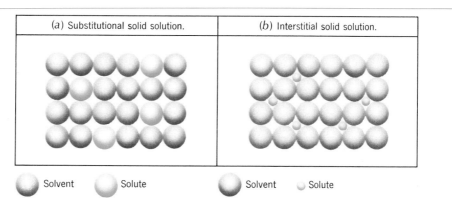

| (a) Substitutional solid solution. | (b) Interstitial solid solution. |

Solvent   Solute   Solvent   Solute

Figure 9.1
Solid solutions. (a) Substitutional solid solution. (b) Interstitial solid solution.

ranged in a cubic closest-packed pattern with carbon atoms in octahedral holes.

The physical properties of solutions are determined by the relative proportions of the various components of which they are composed. We have already seen that there are a variety of ways of expressing the concentration of one substance in another. For example, in Chapter 5 we discussed the concentration units, molarity and normality, and we saw that they are useful for the purpose of dealing with the stoichiometry between reactants in solution. Molarity and normality were defined to satisfy a need and, in a similar fashion, it has been found that the physical properties of solutions can be most readily interpreted with the aid of the following concentration units.

MOLE FRACTION. This unit of concentration appeared in our discussion of Dalton's law of partial pressures in Section 6.9. It is defined as the number of moles of a particular component of the solution divided by the total number of moles of all of the substances present in the mixture,

$$X_A = \frac{n_A}{n_A + n_B + n_c + \dots}$$

For example, a solution composed of two moles of water and three moles of ethanol (C_2H_5OH) has a mole fraction of water given by

$$X_{H_2O} = \frac{2 \text{ moles } H_2O}{2 \text{ mole } H_2O + 3 \text{ mole } C_2H_5OH} = \frac{2 \text{ moles}}{5 \text{ moles}}$$

$$X_{H_2O} = 0.40$$

Similarly, the mole fraction of ethanol in the mixture is

$$X_{C_2H_5OH} = \frac{3 \text{ moles}}{5 \text{ moles}} = 0.60$$

We see that the sum of all of the mole fractions is equal to one, as of course it must be.

Another frequently used term is **mole percent,** which is simply equal to

100 × mole fraction. Thus, the mixture above is composed of 40 mole percent water and 60 mole percent ethanol.

WEIGHT FRACTION. This quantity specifies the fraction of the total weight of a mixture that is contributed by a particular component. A mixture composed of 25.0 g of water and 75.0 g of ethanol has a weight fraction of water, w_{H_2O}, given by

$$w_{H_2O} = \frac{25.0 \text{ g } H_2O}{25.0 \text{ g } H_2O + 75.0 \text{ g } C_2H_5OH} = \frac{25.0 \text{ g}}{100.0 \text{ g}}$$

$$w_{H_2O} = 0.250$$

In a similar fashion we find that the weight fraction of ethanol in the mixture is 0.750.

Weight percent, which is equal to the weight fraction multiplied by 100, is more frequently used than weight fraction. The solution above, then, is described as composed of 25.0% water and 75.0% ethanol, by weight.

MOLALITY. Molality is defined as the number of moles of solute per 1000 g (1.00 kg) of the solvent; that is, it is a ratio of *moles of solute* to *mass of solvent.* A 1.00 molal (written 1.00 *m*) solution would therefore contain 1.00 mole of solute for every 1.00 kg of the solvent. It is very important not to confuse molality with molarity—they are quite different. To see this difference let us consider how to prepare typical 1.00M and 1.00 *m* solutions using, for example, sucrose ($C_{12}H_{22}O_{11}$) as the solute and water as the solvent.

To prepare the 1.00M solution we place exactly 1.00 mole of sucrose (342 g) into a flask that is calibrated to contain precisely 1.00 liter when filled to a line around its neck (see Figure 9.2). Water would be added, while the

Flask contains 1.00 liter when filled to mark

1.00 liter

Figure 9.2
A volumetric flask.

mixture is stirred to dissolve the solute, until the flask was filled to the mark. At this point we have exactly 1.00 mole of solute in a total volume of 1.00 liter of solution and the concentration is 1.00 mole/liter, or 1.00 molar (1.00M).

To prepare the 1.00 molal solution, we place 1.00 mole of sucrose into a flask or beaker and *add to it* 1000 g of water. Since the density of water is nearly 1 g/ml (at room temperature its density is 0.9982 g/ml) we would be adding very nearly one liter of water to the 342 g (1 mole) of solute; however, the total final volume of this 1.00 molal solution is somewhat larger than one liter (it is actually 1110 ml) because part of the volume of the final solution is taken up by molecules of the sucrose. The molarity of this one molal solution would be 1.00 mole/1.110 liter = 0.901M. Because the mole of solute is contained in a larger volume in this one molal solution, a 1 ml portion would contain a smaller amount of solute than a similar 1 ml portion of the 1M solution of sucrose.

The difference between molarity and molality becomes even more apparent if we choose a solvent whose density is far from 1 g/ml. For instance, as illustrated in Figure 9.3, a 1M solution of a solute in carbon tetrachloride[1] contains 1 mole of solute in a total volume of one liter; however, since CCl_4 has a density of 1.59 g/ml (considerably larger than water), a 1 m solution would contain the one mole of solute in a volume of only about 630 ml. In other words it takes much less than one liter of CCl_4 to weigh 1000 g.

The three concentration units, mole fraction, weight fraction, and molality are closely related and can easily be converted from one to another as shown in Example 9.1.

Example 9.1
An aqueous solution is composed of 20.0% by weight magnesium sulfate. What is: (a) the molality of the solution; (b) the mole fraction of $MgSO_4$ and H_2O?

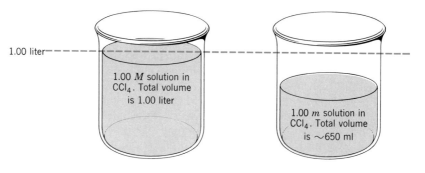

1.00 liter

1.00 M solution in CCl_4. Total volume is 1.00 liter

1.00 m solution in CCl_4. Total volume is ~650 ml

Figure 9.3
Molar and molal solutions in CCl_4.

Solution

(a) We know that 100 g of the solution contains 20.0 g $MgSO_4$ and 80.0 g of water. Since we wish a ratio of moles of solute to mass of solvent, let us first convert 20.0 g $MgSO_4$ to moles.

$$20.0 \text{ g } MgSO_4 \times \left(\frac{1 \text{ mole } MgSO_4}{120.4 \text{ g } MgSO_4}\right) = 0.166 \text{ mole } MgSO_4$$

Thus, there is 0.166 mole $MgSO_4$ in 80 g of water. All we must do now is calculate how many moles of the solute are in 1.00 kg of water.

$$\left(\frac{0.166 \text{ mole } MgSO_4}{80.0 \text{ g } H_2O}\right) \times \left(\frac{1000 \text{ g } H_2O}{1.00 \text{ Kg } H_2O}\right) = \frac{2.08 \text{ mole } MgSO_4}{1.00 \text{ kg } H_2O}$$

Hence the concentration of this solution is 2.08 m.

(b) To calculate mole fraction we must know the number of moles of each component in the solution. We have already found that 100 g of the solution contains 0.166 mole $MgSO_4$. The number of moles of water in the solution is

$$80.0 \text{ g } H_2O \times \left(\frac{1 \text{ mole } H_2O}{18.0 \text{ g } H_2O}\right) = 4.44 \text{ moles } H_2O$$

The mole fraction of $MgSO_4$ is found by dividing the number of moles of $MgSO_4$ by the total number of moles of both the solute and solvent.

$$X_{MgSO_4} = \frac{0.166 \text{ mole}}{4.44 \text{ moles} + 0.166 \text{ moles}} = \frac{0.166 \text{ moles}}{4.61 \text{ moles}}$$

$$X_{MgSO_4} = 0.0360$$

It follows that the mole fraction of water must be

$$X_{H_2O} = 1.00 - 0.0360 = 0.9640$$

Experience has taught us that substances differ widely in their solubilities in various solvents. For instance, we all know that oil and water "don't mix," and that to remove an oil stain from clothing a solvent such as carbon tetrachloride must be used. It is also generally known that sodium chloride (table salt) will dissolve in water but not in gasoline. What accounts for these differences in behavior? The answer lies in a close examination of the solution process.

When one substance dissolves in another, particles of the solute must be distributed throughout the solvent and, in a sense, the solute particles in the solution are occupying positions that are normally taken by solvent molecules. In a liquid (we restrict this discussion to liquid solutions), molecules are packed together very closely and interact strongly with their neighbors. The ease with which a solute particle may replace a solvent molecule depends on the relative forces of attraction of solvent molecules for each other, solute particles for each other, and the strength of the solute-solvent interactions. For example, in a solution formed between benzene (C_6H_6) and carbon tetrachloride (CCl_4), both species are nonpolar and, therefore, experience only relatively weak van der Waals attractions. As it happens, the strengths of the attractive forces between pairs of benzene molecules and between pairs of carbon tetrachloride molecules are of nearly the same magnitude as between molecules of benzene and carbon tetrachloride. For this reason, molecules of benzene can replace CCl_4 molecules in solution with ease; as a consequence, these two substances are completely miscible (soluble in all proportions).

What happens when we attempt to dissolve water in CCl_4? Water is a very polar substance that interacts with other water molecules through the formation of hydrogen bonds. By comparison, the strength of the attractive forces between water and the nonpolar CCl_4 molecules is much weaker. If we attempt to disperse water molecules throughout CCl_4, we find that when the water molecules encounter one another, they tend to stick together simply because they attract each other much more strongly than they do molecules of the solvent. This "clumping together" continues until the two substances have formed two distinct phases, one consisting of water with a very small amount of CCl_4 in it and the other, CCl_4 containing a small quantity of H_2O.

When two polar substances, such as ethanol and water, are mixed we again have a situation in which the solute-solute forces of interaction are of comparable strength to the solvent-solvent attractive forces, and where the solute and solvent molecules interact strongly with each other. Once again a condition exists where the solute particles can readily replace those of the solvent and, hence, water and ethanol are miscible.

Between the two extremes of complete miscibility (benzene–carbon tetrachloride; water–ethanol) and virtually total immiscibility (CCl_4–H_2O) we have many substances that are only partially soluble in one another. For instance, in Table 9.1 the solubilities of a series of different alcohols are listed in moles of solute (alcohol) per 100 g of water. As we proceed to higher molecular weights the polar OH group represents an ever smaller portion of the molecules; as a result, alcohol molecules become less polar as they become larger. Paralleling the decrease in polarity we observe a corresponding decrease in their solubilities in water.

When a solid dissolves in a liquid, somewhat different factors must be considered. In a solid, the molecules or ions are arranged in a very regular pattern and the attractive forces are at a maximum. In order for the solute particles to enter into a solution the solute-solvent forces of attraction must be

Table 9.1
Solubilities of Some Alcohols in Water

Substance	Formula	Solubility (moles solute/ 100 g H_2O)
Ethanol	C_2H_5OH	∞
Propanol	C_3H_7OH	∞
Butanol	C_4H_9OH	0.12
Pentanol	$C_5H_{11}OH$	0.031
Hexanol	$C_6H_{13}OH$	0.0059
Heptanol	$C_7H_{15}OH$	0.0015

sufficient to overcome the attractive forces that hold the solid together. In molecular crystals these attractive forces are relatively weak, being of dipole-dipole or van der Waals type, and are rather easily overcome. Substances whose crystals are held together by van der Waals forces will, therefore, dissolve to appreciable extents in nonpolar solvents. They are not, however, soluble to any great degree in polar solvents for the same reasons that nonpolar liquids are not soluble in polar solvents—that is, the polar solvent molecules attract each other too strongly to be replaced by the molecules to which they are only weakly attracted. For example, solid iodine, which is composed of nonpolar I_2 molecules, is appreciably soluble in CCl_4 (giving rise to a beautiful violet solution) but only very slightly soluble in water (where it produces a pale yellow-brown solution).

By similar reasoning we expect that very polar solutes and ionic solids are not soluble in nonpolar solvents. The weak solute-solvent interactions, compared to the attractions between solute particles, are not sufficient to tear apart the lattice. Ionic solids, in particular, are held together by the very strong electrostatic forces between ions; therefore a very polar solvent, such as water, is required to rip apart an ionic lattice. Hence NaCl is soluble in water but not in gasoline, which is a mixture of nonpolar hydrocarbons (compounds containing only C and H).

When an ionic substance dissolves in water, the ions that are adjacent to one another in the solid become separated and are surrounded by water molecules, as shown in Figure 9.4. In the immediate vicinity of a positive ion, the surrounding water molecules are oriented so that the negative ends of their dipoles point in the direction of the positive charge, while surrounding a negative ion the water molecules have their positive ends directed at the ion. An ion enclosed within this "cage" of water molecules is said to be **hydrated** and, in general, when a solute particle becomes surrounded by molecules of a solvent we say that it is **solvated;** hydration is a special case of the more general phenomenon of solvation.

The layer of oriented water molecules that surrounds an ion (actually this orientation may extend through several layers) helps to neutralize its charge and serves to keep ions of opposite charge from attracting each other strongly over large distances within the solution. Nonpolar solvents do not dissolve

Figure 9.4

Hydration of ions in solution.

ionic compounds because neither can they tear an ionic lattice apart nor do they offer any shielding for the ions. In a nonpolar solvent ions quickly congregate and separate from the solution as the solid.

In summary, substances that exhibit *similar* intermolecular attractive forces tend to be soluble in one another. This observation is often stated very simply as "like dissolves like." Nonpolar substances are soluble in nonpolar solvents while polar or ionic compounds dissolve in polar solvents.

<table>
<tr><td>9.4
Heats of Solution</td><td>

The solution process nearly always occurs with either an absorption or release of energy. For example, if potassium iodide is dissolved in water the mixture becomes cool, indicating that for potassium iodide the solution process is endothermic. On the other hand, when lithium chloride is added to water the mixture becomes warm, signifying that, in this case, the solution process evolves heat and is therefore exothermic. The quantity of energy that is absorbed or released when a substance enters solution is called the **heat of solution** and is given the symbol, ΔH_{soln}. As in our definitions of the heat of vaporization and heat of fusion, ΔH_{soln} represents a difference between the energy possessed by the solution after it has been formed and the energy that the components of the solution possessed before they were mixed, that is,
</td></tr>
</table>

$$\Delta H_{soln} = H_{soln} - H_{components}$$

Neither H_{soln} nor $H_{components}$ can actually be measured; however, their difference, ΔH_{soln}, can be. When energy is evolved during the solution process, the resulting solution possesses less energy than did the components from which it was prepared and the difference represented by ΔH_{soln} is a negative number. Conversely, an endothermic solution process would have a positive ΔH_{soln}. Heats of solution for some typical ionic solids in water are shown in Table 9.2.

Table 9.2
Heats of Solution

Substance	Heat of Solution[a] (kcal/mole of solute)
KCl	4.25
KBr	4.75
KI	4.86
LiCl	−8.85
LiI	−14.1
$LiNO_3$	−0.3
$AlCl_3$	−76.8
$Al_2(SO_4)_3 \cdot 6H_2O$	−56
NH_4Cl	−3.9
NH_4NO_3	6.2

[a]At "infinite" dilution. The heat of solution depends, to an extent, on the concentration of the solution produced. A negative sign signifies an exothermic process.

The magnitude of the heat of solution provides us with information about the relative forces of attraction between the various particles that make up a solution. To analyze the various factors that contribute to this absorption or evolution of energy, let us imagine that we could create the solution in a stepwise fashion.

SOLUTIONS OF LIQUIDS IN LIQUIDS. When one liquid dissolves in another, we can imagine that the particles of the solvent are caused to move apart so as to allow room for the solute particles. Similarly, for the solute to enter solution, its molecules must also become separated so that they can take their places in the mixture. Since there are attractive forces between molecules in both the solvent and solute, the process of separating their molecules requires an input of energy; that is, work must be done on both the solute and solvent to separate their molecules from one another. Finally, as the solute and solvent, in their "expanded" states, are brought together, energy is released because of the attractions that exist between the solute and solvent molecules.[2] This sequence of steps we have just described is illustrated in Figure 9.5.

In some substances, such as benzene and carbon tetrachloride, the intermolecular attractive forces are of very nearly the same magnitude; therefore, these compounds form solutions with virtually no evolution or absorption of heat. Solutions in which the solute-solute, solute-solvent, and

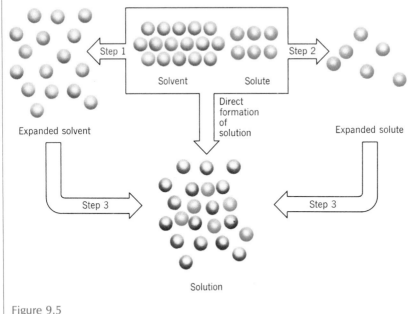

Figure 9.5
Formation of a liquid-liquid solution.

[2] As already noted, when particles that attract one another are pulled apart, potential energy (work) must be supplied. When these particles are brought together again, the same amount of energy is released.

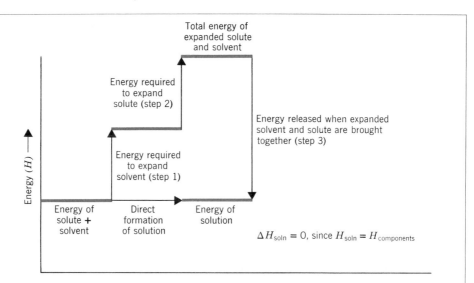

$\Delta H_{\text{soln}} = 0$, since $H_{\text{soln}} = H_{\text{components}}$

Figure 9.6
Energy changes to produce an ideal solution.

solvent-solvent interactions are all the same are called **ideal solutions.**
The energy changes that occur along the series of steps that we have devised
to arrive at the solution are shown graphically in Figure 9.6. We see that
for an ideal solution the energy released in the final step is the same as that
absorbed in the first two; thus the net change is zero.

When the solute and solvent molecules are strongly attracted to each
other, more energy can be released in the final step, when the solution is
formed, than is initially required to separate the molecules of the solute and
the molecules of the solvent (Figure 9.7). Under these circumstances the
overall solution process results in the evolution of heat and is therefore exothermic. On the other hand, when the solute-solvent attractive forces are

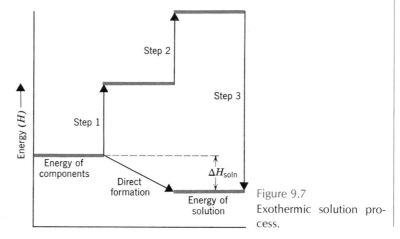

Figure 9.7
Exothermic solution pro-
cess.

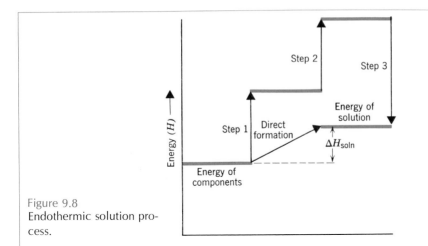

Figure 9.8
Endothermic solution process.

weak, the formation of a solution requires a net input of energy because more energy is absorbed in separating the molecules in the first two steps than is recovered when the solute and solvent are brought together in the third (Figure 9.8). In this case the solution becomes cool as it is formed, signifying that an endothermic change has occurred.

SOLUTIONS OF SOLIDS IN LIQUIDS. We can approach the energetics of the solution of a solid in a liquid in much the same way as we have for a liquid in a liquid. We recognize that the act of dissolving a compound such as KI in water involves removing the K^+ and I^- ions from the solid and placing them in an environment where they are surrounded by water molecules. Suppose, now, that we could separate these two steps. The first step involves separating the ions so that they are infinitely far apart, and the second amounts to taking these now isolated ions and placing them into water where they become surrounded by molecules of the solvent. To accomplish the first step, it is necessary to break apart the lattice by pulling the ions away from one another, so that the process,

$$KI(s) \longrightarrow K^+(g) + I^-(g)$$

requires an input of energy; that is, it is endothermic. The amount of energy required to tear apart the solid to obtain isolated particles is called the **lattice energy.** Ionic substances, because of the very strong electrostatic attractions between oppositely charged ions, have quite large lattice energies. Molecular solids, however, have small lattice energies due to their relatively weak intermolecular attractive forces.

In the second step of our solution process we imagine that we place the K^+ and I^- ions into water where they become hydrated. As described in the last section, an ion in water is surrounded by a "cage" of oriented water molecules; therefore a hydrated ion experiences a net attraction for the solvent dipoles. Because of this attraction, a quantity of energy, known as the **hydration energy,** is released when an ion is placed into water.[3] We indicate

---

[3] In general, for a solvent other than water, this energy is called the **solvation energy.**

the change that occurs in this hydration step as,

$$K^+(g) + I^-(g) + xH_2O \longrightarrow K^+(aq) + I^-(aq)$$

which tells us that some number, x, of water molecules become "bound" to the potassium and iodide ions as they become hydrated in aqueous solution.

The overall change that takes place when KI dissolves in water can be represented as the sum of the two steps that we have just considered,

$$KI(s) \longrightarrow K^+(g) + I^-(g)$$
$$KI(s) \longrightarrow K^+(g) + I^-(g)$$
$$\underline{K^+(g) + I^-(g) + xH_2O \longrightarrow K^+(aq) + I^-(aq)}$$
$$KI(s) + K^+(g) + I^-(g) + xH_2O \longrightarrow K^+(g) + I^-(g) + K^+(aq) + I^-(aq)$$

or, after cancelling species that appear on both sides,

$$KI(s) + xH_2O \longrightarrow K^+(aq) + I^-(aq)$$

The heat of solution, ΔH_{soln}, corresponds to the net energy change that occurs and is equal to the difference between the amount of energy supplied during the first step and the amount of energy evolved during the second (Figure 9.9). If the lattice energy is greater than the hydration energy, a net input of energy is required when the substance dissolves and the process is endothermic. This occurs with KI. Conversely, when the hydration energy exceeds the lattice energy, a situation that occurs with LiCl, more energy is released when the ions become hydrated than is required to break up the ionic lattice, and an exothermic change is observed when the solid dissolves.

This relationship between the lattice energy and hydration energy can be seen in Table 9.3. Note that the agreement between the calculated and experimentally determined heats of solution is far from perfect. This clearly shows that errors[4] creep into all experimentally measured quantities, as you

Figure 9.9
Energy changes that occur when KI dissolves in H_2O.

---

[4] In this case, and indeed in many others, errors also arise from the approximations and inadequacies of the models used in performing the theoretical calculations, not only in the experimentally measured ΔH_{soln}.

Table 9.3
Lattice Energy, Hydration Energy, and Heats of Solution for Some Alkali Halides

Compound	Lattice Energy (kcal/mole)	Hydration Energy (kcal/mole)	Calculated ΔH_{soln} (kcal/mole)	Measured ΔH_{soln} (kcal/mole)
LiCl	+199	−211	−12	−8.85
LiBr	+188	−204	−16	−11.7
NaCl	+183	−184	−1	+0.93
NaBr	+174	−177	−3	−0.144
KCl	+165	−164	+1	+4.12
KBr	+159	−157	+2	+4.75
KI	+151	−148	+3	+4.86

may have already discovered for yourself in the laboratory portion of this course.

The preceding explanations allow us to understand the energy changes that take place during the solution process. Unfortunately, however, it is very difficult to predict ahead of time, in any particular case, whether the formation of a solution will be exothermic or endothermic, since the same factors leading to a high hydration energy also tend to produce a high lattice energy. The extent to which an ion is attracted to a solvent dipole increases as the ion becomes smaller, because a smaller ion can get closer to a solvent molecule than can a larger one. The interaction of the solvent with an ion also increases as the charge on the ion becomes greater. However, the degree to which ions attract each other in the solid also grows with decreasing size and increasing charge, so that as the hydration energy becomes larger so does the lattice energy. Thus we have two factors that are affected in the same way by changes in size and charge, and it is virtually impossible to predict in advance which effect will predominate.

In Chapter 5 we defined the solubility of a substance as the amount of solute required to produce a saturated solution in some particular quantity of solvent. At a given temperature a saturated solution in contact with undissolved solute represents another example of a state of dynamic equilibrium. As illustrated in Figure 9.10, particles of the solute are constantly passing into the solution and, at the same time, the solute particles already in the solution are continually colliding with and sticking to the undissolved solute. Although we show this equilibrium here for a solid dissolved in a liquid, the same concept applies to any type of solution (except gases—all gases are completely miscible).

9.5 Solubility and Temperature

Since a saturated solution in contact with excess solute constitutes a state of dynamic equilibrium, when the system is disturbed the effect of the disturbance can be predicted on the basis of Le Chatelier's principle. A change in temperature corresponds to such a disturbance; in Chapter 8 we saw that a rise in temperature favors a shift in the position of equilibrium in a direction that will absorb heat. Therefore, if the dissolving of additional solute into an

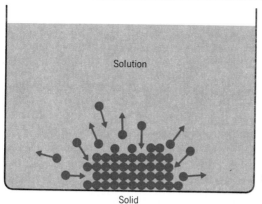

Figure 9.10
Solubility as an equilibrium state. Solute dissolves at the same rate as it crystallizes.

Solution

Solid

already saturated solution will absorb energy, the solubility of that substance increases as the temperature is raised. Conversely, if placing additional solute into the saturated solution is an exothermic process, the solute will become less soluble as the temperature is increased.

In general, the solubility of most solid and liquid substances in a liquid solvent increases with increasing temperature. For gases in liquids, the opposite behavior has been observed. The solution process for a gas in a liquid is nearly always exothermic, since the solute particles are already separated from each other and the dominant heat effect arises from the solvation that occurs when the gas dissolves. Le Chatelier's principle predicts that a rise in temperature will favor an endothermic change which, for a gas, takes place when it leaves solution. Therefore, we expect gases to become less soluble as the temperature of the liquid in which they are dissolved becomes higher. For example, in bringing water to a boil, tiny bubbles appear on the surface of the pot before boiling begins. These bubbles contain air that is driven out of solution as the water becomes hot. We also use this general solubility behavior of gases when we store opened bottles of carbonated beverages in the refrigerator. These liquids retain their dissolved CO_2 longer when they are kept cold because CO_2 is more soluble in them at low temperatures. Analysis of the quantity of dissolved gases in streams, lakes, and rivers reveals still another example of this phenomenon. The concentration of dissolved oxygen, which is imperative to marine life, decreases in the summer months, compared to when similar analyses are performed during the winter months—all other conditions being equal of course.

**9.6
Fractional
Crystallization**

Figure 9.11 graphically illustrates the way in which solubility changes with temperature for a variety of typical solids in water. From these solubility curves it is evident that the variation of solubility with temperature is quite different for different substances. For some, such as KNO_3, the solubility changes very rapidly, while for others the change is much less. These differences in solubility behavior provide the basis for a useful laboratory technique, called **fractional crystallization,** which is frequently used for the separation of impurities from the products of a chemical reaction.

In this technique the impure product is first dissolved in a small amount of hot solvent (generally one in which the desired product is less soluble than

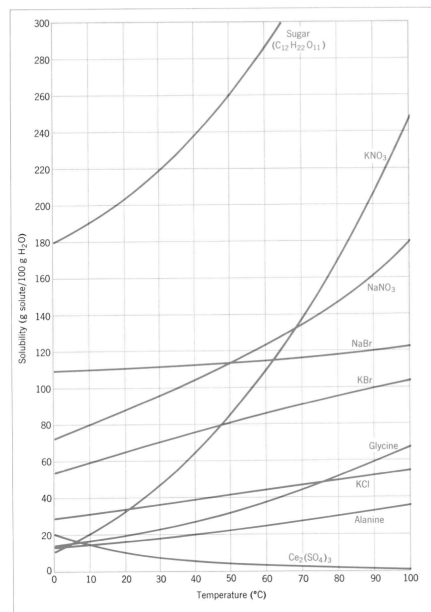

Figure 9.11
Solubility curves for typical solids in H_2O.

the impurities). As the hot solution is allowed to cool, the pure product separates from the mixture, leaving the impurities behind. Finally, the crystals of the product are filtered from the cool solution and dried. The quantity of pure product that can be recovered in this fashion depends on the concentration of the impurities and their solubility relative to that of the desired material.

To illustrate how this technique might be applied, let us suppose that we had a mixture of two amino acids[5] and that the composition of this mixture was 80% glycine (Gly) and 20% alanine (Ala). How much pure Gly could we expect to be able to recover from 100 g of this mixture?

The first step in the procedure is to dissolve the solids in hot water. After the solution has been cooled we want all of the Ala to remain in solution and as much solid Gly as possible to have been formed. According to Figure 9.11 the solubilities of alanine and glycine at 0° C and 100° C are as follows.

Amino Acid	Solubility in g/100 g H_2O at:	
	0° C	100° C
Alanine	13	37
Glycine	14	67

Our mixture contains 20 g of Ala so that the minimum amount of water required to keep the Ala dissolved at 0° C is

$$20 \text{ g Ala} \times \left(\frac{100 \text{ g } H_2O}{13 \text{ g Ala}} \right) = 154 \text{ g } H_2O$$

We must next check to see if this is enough water to also dissolve, when hot, the 80 g of Gly that is present in the mixture. At 100° C the solubility of Gly is 67 g/100 g of H_2O; therefore, 154 g of water at 100° C is capable of dissolving

$$\left(\frac{67 \text{ g Gly}}{100 \text{ g } H_2O} \right) \times 154 \text{ g } H_2O = 103 \text{ g Gly}$$

Thus we see that 154 g of hot water is *more* than sufficient to dissolve not only the alanine but all of the glycine as well.

The amount of pure Gly that will separate from the solution after it has been cooled to 0° C is the difference between the amount of Gly in the original mixture, 80 g, and the amount that will remain in the solution at 0° C. The solubility of Gly at 0° C is 14 g/100 g H_2O and therefore in 154 g of water at 0° C we have

$$\left(\frac{14 \text{ g Gly}}{100 \text{ g } H_2O} \right) \times 154 \text{ g } H_2O = 22 \text{ g Gly}$$

The quantity of solid Gly that is precipitated, then, is 80 g − 22 g = 58 g. This represents the maximum amount of pure glycine that we can recover from our mixture. If we use more than 154 g H_2O, we will obtain less than 58 g of Gly when the solution is cooled. If we use *less than* 154 g of water, more solid will be obtained but it will be contaminated by some alanine, which will also precipitate out.

9.7
The Effect of Pressure on Solubility

In general, pressure has very little effect on the solubility of liquids or solids in liquid solvents. The solubility of gases, however, always increases with in-

[5] Amino acids are constituents of the biologically important compounds, proteins, and are discussed in detail in Chapter 21.

creasing pressure. Carbonated beverages, for example, are bottled under pressure to insure a high concentration of CO_2; once the bottle has been opened, the beverage quickly loses its carbonation unless it is recapped.

The solubility behavior of gases with respect to pressure can easily be explained in terms of Le Chatelier's principle. Let us imagine that a liquid is saturated with a gaseous solute and that this solution is in contact with the gas at some particular pressure. Once again we have a dynamic equilibrium where molecules of the solute are leaving the solution and entering the vapor phase at the same rate at which molecules from the gas are entering the solution, as shown in Figure 9.12a. As we might expect, the rate at which molecules go into solution depends on the number of collisions per second that the gas experiences with the surface of the liquid and, similarly, the rate at which the solute molecules leave the solution depends on their concentration.[6] If we suddenly increase the pressure on the gas, we pack the molecules closer together and the number of collisions per second that the gas molecules make with the surface of the liquid gets larger. When this occurs, the rate at which molecules of the solute (gas) enter the solution also gets larger without a corresponding increase in the rate at which they leave (Figure 9.12b). As a result, the concentration of solute molecules in solution rises until the rate at which they are leaving the solution once again equals the rate at which they enter; at this point we have reestablished equilibrium (Figure 9.12c).

We might represent the equilibrium by the following equation:

$$\text{Solute } (g) + \text{solvent } (l) \rightleftharpoons \text{solution } (l)$$

According to Le Chatelier's principle, an increase in pressure upon this system at equilibrium favors a shift in the position of equilibrium that leads to

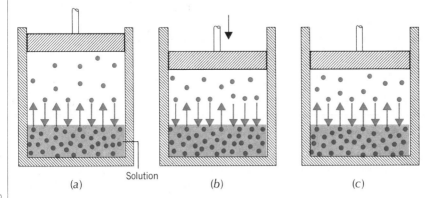

(a)   Solution   (b)   (c)

Figure 9.12

Effect of pressure on the solubility of a gas. (a) Solution equilibrium. (b) System not at equilibrium. (c) Equilibrium again.

[6] At a given temperature and concentration, the rate at which gaseous solute molecules leave a solution also depends on the nature of the gas. Some gases are very soluble, and at a given concentration their rate of escape from the liquid is much less than that for a gas of low solubility.

a decrease in pressure. If the reaction were to proceed to the right, so that more of the gaseous solute dissolves, the quantity of solute in the gas phase would decrease. As you recall from Chapter 6, a decrease in the number of moles of gas leads to a drop in pressure. Thus, an externally imposed increase in pressure favors an increase in the solubility of a gas because it is this process that tends to bring the pressure down toward its initial value.

Quantitatively, the influence of pressure on the solubility of a gas is given by **Henry's law,** which states that the concentration of the gaseous solute in the solution C_g, is directly proportional to the partial pressure of the gas above the solution, that is,

$$C_g = k_g p_g$$

where the proportionality constant, k_g, is called the **Henry's law constant.** This relationship allows us to compute the solubility of a gas at some particular pressure, provided we know its solubility at some other pressure, as demonstrated in Example 9.2. Actually, Henry's law is only accurate for relatively low concentrations and pressures.

Example 9.2

At 25° C, oxygen gas collected over water at a *total* pressure of 1 atm is soluble to the extent of 0.0393 g/liter. What would its solubility be if its partial pressure over water were 800 torr?

Solution

The data given us permits the calculation of the Henry's law constant if we know the partial pressure of oxygen above the solution. The total pressure is the sum of the partial pressures of the H_2O vapor and the oxygen,

$$P_{total} = p_{H_2O} + p_{O_2}$$

From Table 6.2 we find the vapor pressure of water to be 23.8 torr at 25° C; therefore, the partial pressure of oxygen is

$$p_{O_2} = P_{total} - p_{H_2O}$$
$$p_{O_2} = 760 \text{ torr} - 24 \text{ torr} = 736 \text{ torr}$$

The Henry's law constant is obtained as the ratio

$$k_{O_2} = \frac{C_{O_2}}{p_{O_2}}$$

$$k_{O_2} = \frac{0.0393 \text{ g/liter}}{736 \text{ torr}} = 5.34 \times 10^{-5} \frac{g}{\text{liter torr}}$$

Now we can use Henry's law to determine that at a partial pressure of 800 torr the solubility of oxygen is

$$C_{O_2} = \left(5.34 \times 10^{-5} \frac{g}{\text{liter torr}}\right) (800 \text{ torr})$$

$$C_{O_2} = 0.0427 \frac{g}{\text{liter}}$$

We have discussed, thus far, some of the factors that affect the solubilities of solutes in various types of solvents. When a solution is formed, its physical properties are no longer the same as the solvent or solute but, instead, depend on the concentrations of the components that make up the mixture. One property that we might examine, which is rather easily measured, is the vapor pressure of the solution.

For a solution in which a nondissociating, nonvolatile solute is dissolved in a solvent (i.e., the solute itself has very little tendency to dissociate or to escape from the solution and enter the gas phase), the vapor pressure is due only to the vapor of the *solvent* above the solution. This vapor pressure is given by **Raoult's law,** which states that the vapor pressure of the solution at a particular temperature is equal to the mole fraction of the solvent in the liquid phase multiplied by the vapor pressure of the pure solvent at the same temperature, that is,

$$P_{\text{solution}} = X_{\text{solvent}} P^{\circ}_{\text{solvent}}$$

Thus, for example, a solution that contains 95 mole % water and 5 mole % of a nonvolatile solute such as sugar will have a vapor pressure only 95% as great as would the pure solvent. Stated qualitatively, the vapor pressure of the solution is lowered by the addition of a nonvolatile solute.

Let us see why Raoult's law holds. Figure 9.13a illustrates the condition in which the pure solvent is in equilibrium with its vapor. This vapor, as described in Section 8.3, exerts a pressure (which we call the vapor pressure) that is ultimately determined by the fraction of the total number of molecules at the surface that have enough kinetic energy to escape the liquid and enter the gas phase. If we now look at a solution containing a nonvolatile solute (Figure 9.13b), we find that a portion of the solvent molecules at the surface have been replaced by molecules of the solute. Since the entire system, solvent plus solute, is at a single temperature, all of the molecules in the solution belong to a single distribution of kinetic energies. In both the solution and the pure solvent, the same fraction of surface molecules has more than that minimum kinetic energy that solvent molecules need in order to break away from the liquid, but in the solution only a *portion* of that fraction is actually

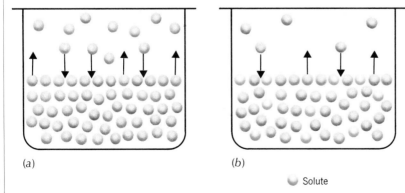

(a)   (b)

   Solute

Figure 9.13

Molecular view of Raoult's law. (a) Pure solvent (b) Solution.

composed of molecules of the solvent. The others are solute molecules. The result is that there are fewer molecules at the surface of the solution capable of leaving than at the surface of the pure solvent. Consequently, the rate of evaporation of solvent molecules from a solution is less when a solute is present.

The magnitude of the equilibrium vapor pressure is determined by the rate of evaporation from the surface of the liquid. If the rate is high, a large concentration of molecules must be present in the vapor at equilibrium so that the rate of return to the liquid can also be high. Conversely, if the rate of evaporation is low, so must be the concentration in the vapor phase. Since the rate of evaporation from the solution is less than from the pure solvent, the equilibrium concentration of molecules in the vapor is smaller over the solution than over the solvent alone; therefore, the equilibrium vapor pressure is lower for the solution than for the pure solvent.

Because only the solvent can evaporate, the fraction of molecules at the surface of the solution that can escape from the liquid depends on that fraction of all of the molecules at the surface that are solvent molecules—in other words, the ratio of the number of moles of solvent particles to the total number of moles of particles that comprise the surface. This, of course, is the mole fraction of the solvent. If the solution were composed of 95 mole % solvent (we expect to find only 95% of the molecules at the surface to belong to the solvent), then the rate of evaporation from the solution is expected to be only 95% of that for the solvent alone. The equilibrium vapor pressure should therefore be reduced to 95% of that for the pure solvent, which is the same result that we obtain by the application of Raoult's law. The linear relationship, predicted by Raoult's law, between vapor pressure and mole fraction is shown in Figure 9.14.

In many solutions, such as benzene and carbon tetrachloride, for example, both solute and solvent have appreciable tendencies to undergo evaporation. In this case, the vapor will contain both solute and solvent molecules, and the vapor pressure of the solution will be the sum of the partial pressures exerted by each component. If we follow the same line of reasoning as above, we conclude that the partial pressure of any component above such a mixture is also given by Raoult's law, that is, the partial pressure of component A, p_A, is given by

$$p_A = X_A P_A^\circ$$

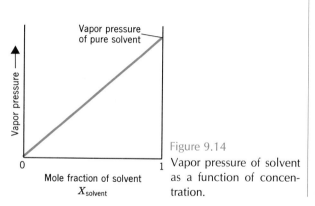

Vapor pressure of pure solvent

Vapor pressure →

0   1

Mole fraction of solvent
$X_{solvent}$

Figure 9.14
Vapor pressure of solvent as a function of concentration.

where P_A° is the vapor pressure of pure A and X_A is its mole fraction in the solution. In the same manner, the partial pressure of a second component, p_B, is given as

$$p_B = X_B P_B^\circ$$

Finally, the total vapor pressure of a binary mixture of A and B, according to Dalton's law, is

$$P_T = p_A + p_B$$

Figure 9.15 is a plot of the partial pressures of A and B, and the total vapor pressure as a function of solution composition for such a two-component mixture.

In actual fact, very few mixtures really obey Raoult's law very closely over wide ranges of composition. Benzene and carbon tetrachloride, a pair of substances that do form such mixtures, are said to yield ideal solutions. Mixtures that deviate from Raoult's law are called *nonideal*. When the vapor pressure of a mixture is greater than that predicted, it is said to exhibit a *positive deviation* from Raoult's law; conversely, when a solution gives a lower vapor than we would expect from Raoult's law, it is said to show a *negative deviation*.

The origin of nonideal behavior lies in the relative strengths of the interactions between molecules of the solute and solvent. When the attractive forces between the solute and solvent molecules are weaker than those between solute molecules or between solvent molecules, neither the solute nor solvent particles are held as tightly in the solution as they are in the pure substances. The escaping tendency of each is therefore greater in the solution than in the solute or solvent alone; as a result, the partial pressures of both over the solution are greater than predicted by Raoult's law. Consequently, the total effect is that the solution exhibits a larger vapor pressure than expected.

Just the opposite effect is produced when the solute-solvent interactions

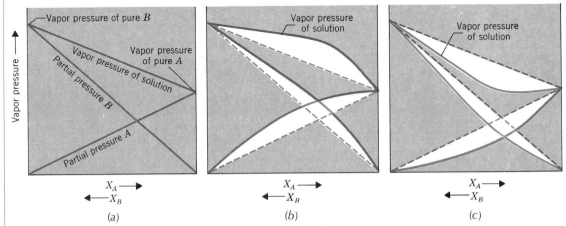

(a) (b) (c)

Figure 9.15

Vapor pressure of a two-component system (a) Ideal solution. (b) Positive deviations from Raoult's law. (c) Negative deviations from Raoult's law.

are stronger than the solute-solute or solvent-solvent interactions. Each substance, in the presence of the other, is held more tightly than in the pure materials, and their partial pressures over a solution are therefore less than Raoult's law would predict. The result is that such a solution exhibits a negative deviation from ideality.

Since, in a solution that shows positive deviations from ideal behavior, the forces of attraction between solute and solvent are weaker than those between both solute molecules and solvent molecules, the formation of these solutions occurs with the absorption of energy (Section 9.4). Conversely, of course, mixtures that exhibit negative deviations from Raoult's law are formed with the evolution of heat.

9.9 Fractional Distillation

In a simple distillation process, one that could be used to separate sodium chloride and water, for example, a volatile solvent is vaporized from a solution and subsequently condensed to provide a pure liquid (see Figure 1.4). If the process is continued, eventually all of the solvent will be removed and only the solid solute will remain.

The separation of mixtures of volatile liquids into their components presents more of a problem. A technique that can frequently be used successfully to accomplish this task is called **fractional distillation.**

Let us suppose that we had a mixture of two volatile liquids, A and B. This mixture will boil when the sum of the partial pressures of A and B equals the prevailing atmospheric pressure; that is, when

$$P_{atm} = p_A + p_B$$

The boiling points of various mixtures of A and B will increase gradually from that of the more volatile component (let us say, A) to that of the less volatile one, B, as shown in Figure 9.16.

Suppose, now, that when one mole of A is mixed with two moles of B, the resulting mixture boils (at 1 atm) at a temperature at which the vapor pressure of pure A is 1140 torr and that of pure B is 570 torr. Under these conditions the partial pressure of A is

$$p_A = X_A P_A^\circ$$

$$p_A = \left(\frac{1 \text{ mole } A}{1 \text{ mole } A + 2 \text{ mole } B} \right) 1140 \text{ torr}$$

$$p_A = \left(\frac{1 \text{ mole}}{3 \text{ mole}} \right) 1140 \text{ torr}$$

$$p_A = (0.333)(1140 \text{ torr}) = 380 \text{ torr}$$

Figure 9.16 Boiling point curve for a mixture of A and B.

Similarly, the partial pressure of B would be

$$p_B = \left(\frac{2 \text{ mole}}{3 \text{ mole}}\right) 570 \text{ torr}$$

$$p_B = 380 \text{ torr}$$

The sum of P_A and P_B is 760 torr as, of course, it must be if the solution is to boil.

What can we say about the composition of the vapor? In Section 6.9, under our discussion of Dalton's law of partial pressures, it was stated that the partial pressure of a gas in a mixture is equal to its mole fraction multiplied by the total pressure *exerted by the gas.*

$$p_A = X_A P_T$$

In the vapor over our solution the partial pressure of each gas is 380 torr and the total pressure is 760 torr. This means that the mole fraction of both A and B *in the vapor* must be 0.500. Now the mole fraction of A in the liquid was 0.333; thus we see that the vapor contains a greater amount of the more volatile component (A) than does the solution. In fact, anytime we boil a mixture of these two substances the vapor will be richer than the solution in the more volatile compound. On our boiling point diagram we can indicate the composition of the vapor by the upper curve drawn in Figure 9.17. Here, points corresponding to the composition of liquid and vapor in equilibrium can be obtained by drawing a horizontal line, called a **tie line,** between the curve for the liquid and that of the vapor. When the composition of the mixture is X_1, it boils at a temperature T_1 to provide a vapor that has a composition X_2. If this vapor is condensed and then reheated, it will boil at a temperature T_2 and give a vapor whose composition is X_3. Repetition of this process will produce fractions ever richer in A. This procedure is called fractional distillation. This

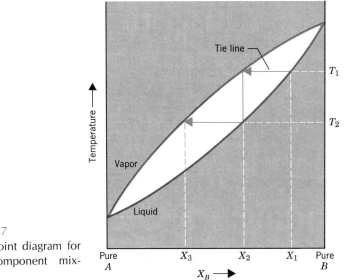

Figure 9.17
Boiling point diagram for a two-component mixture.

technique is useful not only in the laboratory, where it is employed for purifying the products of chemical reactions, but also industrially. For instance, the petroleum industry uses fractional distillation to separate crude oil into its various components, which include gasoline, kerosene, oils, and paraffin.

There are some solutions that exhibit very large deviations from ideality; as a result they cannot be totally separated into their components even by fractional distillation. Ethyl alcohol (grain alcohol) and water form such a mixture. Solutions of these two substances have such large positive deviations from Raoult's law that there is a maximum in the vapor pressure curve and hence a minimum in the boiling point diagram as shown in Figure 9.18. A solution with such a minimum boiling point is called an **azeotrope.** Fractional distillation of solutions lying on either side of this azeotropic composition is capable of separating them into, at best, one pure component plus a solution having the minimum boiling point. As any "moonshiner" will agree, ethyl alcohol—water mixtures obtained by fermentation of sugars, for example—are rich in water. Fractional distillation is able to concentrate the alcohol to, at best, the azeotropic composition of approximately 95% ethyl alcohol.[7] Once this composition has been achieved, the liquid and vapor have the same composition, and no additional fractionation takes place.

There are also solutions that show large negative deviations from ideality and therefore have a minimum in their vapor pressure curves. This leads to a maximum on the boiling point diagram and hence to a **maximum boiling azeotrope** (water-ethyl alcohol forms a minimum boiling azeotrope). Hydrochloric acid, for instance, forms a maximum boiling azeotrope having the approximate composition, 20% HCl and 80% H_2O, with a boiling point of 109° C.

9.10 Colligative Properties of Solutions

Properties that depend on the number of particles of solute in a solution, in-instead of on their specific chemical nature, are called **colligative properties.**[8] Vapor pressure is one of these. We have seen that, according to Raoult's law,

Figure 9.18

Boiling point diagram for water-ethanol mixtures (not drawn to scale).

[7] This is too strong to consume without dilution. A 95% solution of ethyl alcohol is 190 proof. Good aged whiskey that is 86 proof is only 43% alcohol, by volume.

[8] Derived from the Latin, *colligare,* to collect. These properties are determined by the entire "collection" of particles and not by what they are composed of.

the addition of a nonvolatile solute to a substance causes its vapor pressure to be lowered. In our explanation of how this happened, nothing was said about the specific nature of the solute other than that it was incapable of escaping from the solution and that it was undissociated.

What effect does this vapor pressure lowering have upon the phase diagram of a solvent such as water? In Figure 9.19, we see that the vapor pressure of the solution lies below that of the pure solvent at every temperature, as indicated by the dotted line. Because of this the solution must attain a higher temperature in order to have the same vapor pressure as water alone; therefore the normal boiling point of the solution is higher, by an amount ΔT_b, than that of water by itself.

We also find, in Figure 9.19, that the vapor pressure curve of the solution intercepts the solid-vapor line of water at a temperature below the triple point of the solvent (there is almost never a solid-vapor curve for the solution because when it freezes, the solid that is nearly always formed is composed of the pure solvent—solute particles can only rarely be accommodated in the solid solvent lattice). The solid-liquid equilibrium line of the solution must pass through this new triple point and consequently lies to the left of the solid-liquid line for pure water. As a result, the freezing point of the solution, at 1 atm, lies below that of the solvent by ΔT_f.

In summary, then, a nonvolatile solute increases the liquid range of a solution and results in a boiling point elevation and freezing point depression. In effect the solute reduces the escaping tendency of the solvent molecules both in the direction of the vapor and the solid. A common example of this is the effect that automotive antifreeze has on the liquid range of antifreeze solution in the radiator of an automobile. The solute (usually ethylene glycol) lowers the freezing point so that the radiator does not freeze in cold weather and it also raises the boiling point so that the radiator does not "boil over" as readily in very hot weather.

The extent to which the boiling point is raised, and the freezing point lowered, is found to depend on the *molality* of the solute in the solution,

$$\Delta T_b = K_b m \tag{9.1}$$

Figure 9.19
Effect of a nonvolatile solute on the phase diagram of H_2O.

Table 9.4

K_b and K_f for Some Solvents

Solvent	Boiling Point(°C)	K_b(°C/m)	Melting Point(°C)	K_f(°C/m)
Water	100.0	0.51	0.0	1.86
Benzene	80.1	2.53	5.5	5.12
Camphor	—	—	179	39.7
Acetic acid	118.2	2.93	17	3.90

and

$$\Delta T_f = K_f m \qquad (9.2)$$

where K_b and K_f are referred to as the **molal boiling point elevation constant** and the **molal freezing point depression constant,** respectively. The magnitudes of K_b and K_f are characteristic of each solvent. Table 9.4 contains a list of some typical solvents and their values of K_b and K_f.

If we know the concentration of the solution, in moles of solute per kilogram of solvent, the relationships expressed in Equations 9.1 and 9.2 permit us to calculate the extent to which the boiling point and freezing point are changed. For example, a solution containing one mole of sugar in 1000 g of water will have its freezing point lowered by 1.86° C and its boiling point raised by 0.51° C. The solution will therefore freeze at $-1.86°$ C and boil at 100.51° C when the atmospheric pressure is one atm. Example 9.3 provides another sample calculation.

Example 9.3

What would be the freezing point and boiling point of a solution containing 6.50 g of ethylene glycol ($C_2H_6O_2$), commonly used as an automotive antifreeze, in 200 g of water?

Solution

To determine ΔT_f and ΔT_b we must know the molality of the solution — the number of moles of $C_2H_6O_2$ per one kilogram of water. Since the molecular weight of $C_2H_6O_2$ is 62.0 amu, the concentration of our solution is

$$\left(\frac{6.50 \text{ g } C_2H_6O_2}{200 \text{ g } H_2O}\right) \times \left(\frac{1 \text{ mole } C_2H_6O_2}{62.0 \text{ g } C_2H_6O_2}\right)$$

$$\times \left(\frac{1000 \text{ g } H_2O}{1.00 \text{ kg } H_2O}\right) = \frac{0.525 \text{ mole } C_2H_6O_2}{1.00 \text{ kg } H_2O}$$

or, 0.525 m.

For H_2O, $K_f = 1.86°$ C/m and $K_b = 0.51°$ C/m. Hence the changes in the freezing point and boiling point are

$$\Delta T_f = (1.86 \frac{°C}{m}) \times (0.525m) = 0.98° \text{ C}$$

$$\Delta T_b = (0.51 \frac{°C}{m}) \times (0.525 \text{ m}) = 0.27° \text{ C}$$

The freezing and boiling points of the solution are then -0.98 and $100.27°$ C, respectively. Thus, we see that solutions considerably more concentrated than this (approximately 3%) are necessary to protect an automobile's cooling system in frigid weather.

If a knowledge of the molal concentration permits us to determine the extent to which the boiling point and freezing point differ from those of the pure solvent, then it should also be possible to calculate the molal concentration of a solution from ΔT_b and ΔT_f. This aspect of these colligative properties proves particularly useful because with it we can measure molecular weights experimentally. This, combined with analytical data on percent composition, enables us to find molecular formulas for the substances prepared in the laboratory. Example 9.4 illustrates how to apply this concept.

Example 9.4

A 5.50 g sample of a compound, whose empirical formula is C_3H_3O, dissolved in 250 g of benzene gives a solution whose freezing point is $1.02°$ C below that of pure benzene. Determine (a) the molecular weight and (b) the molecular formula of this compound.

Solution
(a) From Table 9.4, K_f for benzene is $5.12°$ C/m. If we solve Equation 9.2 for the concentration, we obtain

$$m = \frac{\Delta T_f}{K_f}$$

Upon substituting the values for the freezing point depression and K_f, we have

$$m = \frac{1.02° \text{ C}}{5.12° \text{ C/m}} = 0.200 \ m$$

or, 0.200 moles solute/1.00 kg benzene.
    From the data given we have 5.50 g of solute per 250 g of benzene, from which we can obtain the number of grams of solute per kilogram of solvent.

$$\left(\frac{5.50 \text{ g solute}}{250 \text{ g benzene}}\right)\left(\frac{1000 \text{ g benzene}}{1.00 \text{ kg benzene}}\right) = \frac{22.0 \text{ g solute}}{1.00 \text{ kg benzene}}$$

Since the concentration of our solution is 0.200 mole of solute per kilogram of benzene and *also* 22.0 g of solute per kilogram of benzene, we say that

$$0.200 \text{ mole solute} = 22.0 \text{ g solute}$$

Therefore the weight of one mole (which is numerically equal to the molecular weight) is

$$1 \text{ mole solute} \times \left(\frac{22.0 \text{ g solute}}{0.200 \text{ mole solute}}\right) = 110 \text{ g solute}$$

(b) Now that we know the molecular weight we can determine the molecular formula as in Chapter 2. The molecular formula must contain the empirical formula repeated an integral number of times. The molecular weight is therefore an integral multiple of the empirical formula weight, which for C_3H_3O is 55.0 amu. Since the molecular weight that we have found is twice this value, the molecular formula must be $(C_3H_3O)_2$ or $C_6H_6O_2$.

In practice, molecular weights cannot be determined by this method as accurately as we have implied. However, an error even as large as 10% (i.e., measured molecular weights ranging, in this case, from about 100 to 120 amu) certainly still permits us to choose between the possibilities, C_3H_3O, $C_6H_6O_2$, and $C_9H_9O_3$, with their corresponding molecular weights, 55.0, 110, and 165 amu. For this reason, this technique has proven to be an invaluable tool for the chemist.

SOLUTIONS OF ELECTROLYTES. For simplicity, we have limited our discussion thus far to solutions that do not contain an electrolyte. The reason for this is that the freezing point depression and boiling point elevation depend on the number of particles present in the solution. One mole of a nonelectrolyte, such as sugar, when placed in water yields one mole of particles, and a solution labeled "1 m sucrose" would, therefore, have a freezing point 1.86° C lower than pure water. However, a solution containing one mole of an electrolyte such as NaCl contains two moles of particles—one mole of Na$^+$ ions and one mole of Cl$^-$ ions. As a result, a solution labeled "1 m NaCl" actually contains two moles of particles per 1000 g of water and theoretically should have a freezing point depression of $2 \times 1.86° \text{ C} = 3.72°$ C. In a similar fashion, a 1 m solution of $CaCl_2$, which contains three moles of ions per 1000 g of water, would have a freezing point depression three times as great as a 1 m solution of sucrose. In fact, neither the prediction for NaCl or for $CaCl_2$ is entirely accurate (see the next section). These numbers then refer to the extremes in freezing point depression for these solutions.

For a weak electrolyte, such as acetic acid, we expect the freezing point depression and boiling point elevation to be intermediate between that of a nonelectrolyte and that of a strong electrolyte. Acetic acid, which we have said in Chapter 5 undergoes reaction with water to establish the equilibrium

$$HC_2H_3O_2 + H_2O \rightleftharpoons H_3O^+ + C_2H_3O_2^-$$

is only partially dissociated. Thus at equilibrium one mole of this solute exists as more than one mole of particles but less than two. We can use the boiling point elevation and freezing point depression to determine the extent to which such a weak electrolyte is dissociated.

Example 9.5

Very careful measurement reveals that a 1.00 m solution of HF has a freezing point of $-1.91°$ C. What percent of the HF is dissociated into H$^+$ and F$^-$ ions in this solution?

Solution

The dissociation of hydrogen fluoride can be represented by

$$HF \rightleftharpoons H^+ + F^-$$

If HF were a nonelectrolyte, a 1.00 m solution would have a freezing point of $-1.86°$ C; if it were a strong electrolyte, the solution should freeze at

$-3.72°$ C. Because the measured freezing point lies between these two extremes, only a part of the one mole of HF has dissociated. We want to know how much. Since this is presently unknown, let us give it a name—that is, let us call the amount of HF that has undergone dissociation in 1.00 kg of water by the name, x.

Number of moles HF dissociated $= x$

The number of moles of HF remaining, then, must be the difference between the amount of HF that we put into the solution and the amount dissociated; that is, $1.00 - x$

Number of moles HF remaining at equilibrium $= 1.00 - x$

We see from the chemical equation above that for every mole of HF that dissociates we produce 1 mole of H^+ and 1 mole of F^-. When x moles of HF dissociate, we must therefore form x moles of H^+ and x moles of F^-, so that at equilibrium we also have

Number of moles $H^+ = x$
Number of moles $F^- = x$

The *total* number of moles of particles in one kilogram of solvent is the sum of the moles contributed by the HF, H^+, and F^-. This total is

$$\text{Total number of moles of particles} = (1.00 - x) + x + x$$
$$= 1.00 + x$$

Now, from the freezing point depression of this solution, $1.91°$ C, we can calculate the molal concentration of *particles*.

$$m = \frac{\Delta T_f}{K_f} = \frac{1.91° \text{ C}}{1.86° \text{ C/m}} = 1.03 \ m$$

This number *also* represents the total number of moles of particles per 1.00 kg of water. Therefore,

1.03 moles particles $= (1.00 + x)$ mole particles

and thus

$$x = 0.03 \text{ moles}$$

The fraction of HF dissociated is equal to the number of moles that have broken apart (0.03 mole) divided by the total number of moles of HF placed in the solution (1.00 mole).

$$\text{Fraction dissociated} = \frac{0.03}{1.00} = 0.03$$

The percentage of the HF dissociated, then, is 3%.

Solutions of electrolytes have larger values of ΔT_f and ΔT_b than we might have initially expected because of dissociation. There are also instances where the freezing point depression and boiling point elevation are smaller than we would at first predict. This occurs when **association** (the opposite of dissociation) takes place between solute particles in the solution. For instance, a solution of one mole (122 g) of benzoic acid in 1.00 kg of benzene produces a freezing point depression only slightly more than half the expected depression, implying that there are only about half as many particles in the solution as we anticipated. Since this approximately half mole of particles weighs 122 g, the apparent molecular weight is about 240. Therefore, when association takes place, the measured molecular weights are actually higher than we would predict.

In this particular example, the anomalous behavior of benzoic acid in benzene is attributed to hydrogen bonding between benzoic acid molecules to form a **dimer** (a particle created from two identical simpler units).

benzoic acid   benzoic acid dimer — hydrogen bonds shown as
                                    dotted lines

9.11
Osmotic Pressure

Osmosis is a process whereby a solvent passes from a region of low concentration of solute to a region of high concentration through a thin film that permits only the passage of the solvent. Such films are called **semipermeable membranes,** some typical examples of which include certain types of parchment paper, some gelatinlike inorganic substances, and the cell walls of living organisms.

In the process of osmosis there is a drive toward equalization of concentrations between the two solutions in contact with one another across the membrane. The rate of passage of solvent molecules through the membrane into the more concentrated solution is greater than their rate of passage in the opposite direction, presumably because, at the surface of the membrane, the solvent concentration is greatest in the more dilute solution (Figure 9.20). We observe a similar effect if two solutions, with unequal concentrations of a nonvolatile solute, are placed in a sealed enclosure, as shown in Figure 9.21. The rate of evaporation from the dilute solution is greater than that from the concentrated solution, but the rate of return to each is the same (both solutions are in contact with the same gas phase). As a result, neither solution is in equilibrium with the vapor. In the dilute solution molecules are evaporating faster than they are condensing, while in the concentrated solution the reverse occurs. Consequently there is a gradual net transfer of solvent from the dilute solution into the more concentrated one until they both achieve the same concentration.

If we perform an osmosis experiment using the apparatus in Figure 9.22, in which we have a solution in compartment A and pure water in B, the passage of solvent from B to A will slowly increase the volume of A and decrease

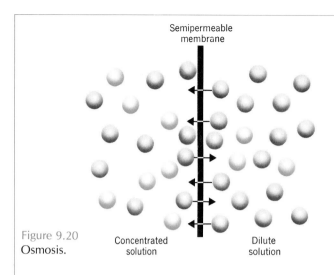

Figure 9.20
Osmosis.

Concentrated
solution

Dilute
solution

the volume of *B*. As this occurs the height of the liquid in the capillary of compartment *A* will rise while the height of the liquid in the other capillary will drop, and there will be a pressure difference between the two solutions that depends on the difference in these heights, Δh. Now, the ease with which a solvent molecule can be transferred from *B* into *A*, and from *A* to *B*, also depends on this pressure difference. As the pressure on side *A* increases, it becomes increasingly more difficult to squeeze another solvent molecule into this solution. It also becomes easier to "pop out" a solvent molecule and push it into the pure water in *B*. Hence as Δh increases, the rate of transfer of solvent from *B* to *A* decreases while that from *A* to *B* gets larger, until finally a value of Δh is achieved at which both rates are the same and equilibrium is established. The pressure difference between the two compartments at equilibrium is called the **osmotic pressure** of the *solution* and is symbolized by the Greek letter, π.

The magnitude of the osmotic pressure developed in a solution is directly proportional to the *molar* concentration of the solute.

$$\pi \propto M$$

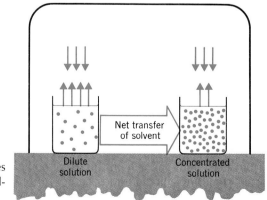

Figure 9.21
Unequal vapor pressures lead to a net transfer of solvent.

Net transfer
of solvent

Dilute
solution

Concentrated
solution

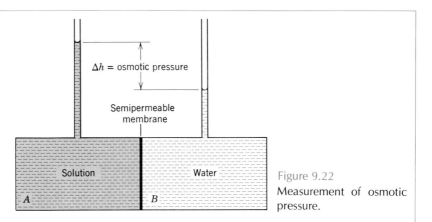

Figure 9.22
Measurement of osmotic pressure.

or,

$$\pi = kM$$

Since molarity has the units, moles/liter, we can write $M = n/V$ where n is the number of moles of solute particles and V is the volume in liters. Therefore

$$\pi = k\left(\frac{n}{V}\right)$$

or

$$\pi V = nk$$

The proportionality constant, k, is a function of temperature and, for very dilute solutions, can be quite closely approximated by RT, where R is the universal gas constant and T is the absolute temperature. Thus we obtain the van't Hoff equation,

$$\pi V = nRT \qquad (9.3)$$

Note the very close similarity between this equation and the ideal gas law:

$$PV = nRT$$

The magnitude of the osmotic pressure, even in very dilute solutions, is quite large. For instance, with a concentration of 0.010 moles of solute particles per liter (0.010M) at room temperature (298° K) the osmotic pressure would be

$$\pi = \left(\frac{0.010 \text{ mole}}{1 \text{ liter}}\right) \times \left(\frac{0.082 \text{ liter atm}}{\text{mole °K}}\right) 298° \text{ K}$$

$$\pi = 0.24 \text{ atm}$$

This pressure is sufficient to support a column of water 8.1 feet high!

Because the osmotic pressure that can be developed between solutions of only slightly different concentrations is so great, it is very important that fluids added to the body intravenously not alter significantly the osmotic pressure of the blood. If the blood fluids become too dilute, the osmotic pressure that develops within the blood cells can cause them to rupture. On the other hand, if the fluids are too concentrated, water will diffuse out of the cells and they will no longer function properly. For this reason care is taken to use solutions with the same osmotic pressure as the solution within the cells. Solutions that have the same osmotic pressure are called **isotonic** solutions.

The large differences in pressure developed between solutions of very similar concentrations provide us with a method of measuring the very large molecular weights of polymers (both of synthetic and biological origin). Freezing point lowering and boiling point elevation just won't work in these cases. For instance, a solution containing even as much as 150 g of a solute, whose molecular weight is 30,000, in 1000 g of water produces a solution with a concentration of only 0.005 m. The freezing point depression of this solution is approximately 0.009 degrees. Moreover, in most cases it is not even possible to dissolve this much solute, so that in practice the freezing point changes are virtually undetectable.

A solution containing but 15 g of this solute per 1000 g of water, (5.0×10^{-4} m), however, would nevertheless have a measurable osmotic pressure.

$$\pi = \left(\frac{5.0 \times 10^{-4} \text{ mole}}{\text{liter}}\right) \times \left(\frac{0.082 \text{ liter atm}}{\text{mole } °K}\right) \times 298° \text{ K}$$

$$\pi = 1.2 \times 10^{-2} \text{ atm}$$

This is 9.1 mm Hg or 9.1 torr. If the solution is assumed to have a density of 1.0 g/ml, this pressure will support a column of the liquid 12.3 cm high (about 4.8 in.).

The osmosis process can be stopped and even reversed by the application of pressure equal to or greater than the osmotic pressure of the solution. This reversal of osmosis is used to desalinate sea water. An apparatus for this process is shown in Figure 9.23. If pressure were not applied to this system, osmosis would occur from left to right, that is, molecules would be transferred from fresh water to salt water, Figure 9.23a. However, with the application of pressure exceeding π, we drive the osmosis in the reverse direction, Figure 9.23b, forcing water molecules out of the saline solution, thereby leaving the impurities behind. One type of membrane, which is strong enough to withstand these pressures, is a film of cellulose acetate placed over a suitable support. Cellulose acetate is permeable to water but impermeable to the ions and impurities in the saline solution. Desalination plants have been constructed thus far that can produce as much as 45,000 gallons of fresh water daily.

In Section 9.10 it was stated that a solution of an electrolyte such as NaCl will produce a freezing point depression twice that caused by a nonelectrolyte at the same concentration. This is not entirely correct. Very careful measurements made by the Dutch scientist, van't Hoff, on the colligative properties of solutions of electrolytes revealed that as these solutions become more and more concentrated the ions in a solution become less independent. As a result, their effectiveness at altering the properties of the solution (boiling point, freezing point, osmotic pressure) diminishes as the concentration of the solute gets larger. Ionic compounds, then, behave as if they are less fully dissociated in concentrated solutions than when they are dilute.

9.12 Interionic Attractions

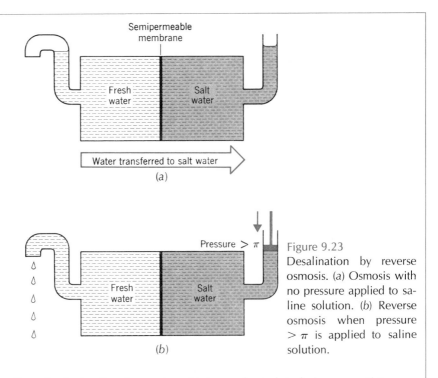

Figure 9.23
Desalination by reverse osmosis. (a) Osmosis with no pressure applied to saline solution. (b) Reverse osmosis when pressure $> \pi$ is applied to saline solution.

Quantitatively, the degree to which an electrolyte behaves as if it were dissociated can be expressed by the van't Hoff factor, i. This quantity may be defined as the ratio of the observed freezing point depression produced by a solution to the freezing point that the solution would exhibit if the solute were a nonelectrolyte.

$$i = \frac{(\Delta T_f) \text{ measured}}{(\Delta T_f) \text{ calculated as nonelectrolyte}}$$

For NaCl, the i factor approaches a value of 2.00 at very high dilutions (essentially infinite dilution). As Table 9.5 reveals, however, the i factor diminishes at higher concentrations. For nonelectrolytes, of course, the i factor has a value of 1.00.

Table 9.5
Values of the van't Hoff Factor at Various Concentrations

Salt	Concentration (mole/kg H$_2$O)			i Factor If Completely Dissociated
	0.1	0.01	0.001	
NaCl	1.87	1.94	1.97	2.00
KCl	1.85	1.94	1.98	2.00
K$_2$SO$_4$	2.32	2.70	2.84	3.00
MgSO$_4$	1.21	1.53	1.82	2.00

9.1 A mixture is prepared from 45.0 g of benzene (C_6H_6) and 80.0 g of toluene (C_7H_8). Calculate (a) the weight percent of each component, (b) the mole fraction of each component, (c) the molality of the solution if toluene is taken to be the solvent.

9.2 A solution containing 121.8 g of $Zn(NO_3)_2$ per liter has a density of 1.107 g/ml. Calculate (a) the weight percent of $Zn(NO_3)_2$ in the solution, (b) the molality of the solution, (c) the mole fraction of $Zn(NO_3)_2$.

9.3 What is the mole fraction, molality, and weight percent of a solution prepared by dissolving 0.30 mole of $CuCl_2$ in 40.0 moles of H_2O?

9.4 As applied to solubility, what is the significance, on a molecular level, of the phrase "like dissolves like"?

9.5 Discuss the relationship between lattice energy and hydration energy in determining the magnitude and sign of the heat of solution.

9.6 On the basis of size and charge, choose the ion in each of the following pairs with the largest hydration energy:
(a) Na^+ or K^+ (d) Fe^{2+} or Fe^{3+}
(b) F^- or Cl^- (e) S^{2-} or Cl^-
(c) K^+ or Ca^{2+}

9.7 On the basis of the information provided in Figure 9.11, predict which solid will separate first from a solution containing equal weights of KNO_3 and KBr when the solution is gradually evaporated at a temperature of 70° C. What will occur if the solution is gradually evaporated at 20° C?

9.8 Describe, qualitatively, the procedure called fractional crystallization.

9.9 What is the maximum amount of pure KNO_3 that can be obtained by fractional crystallization of a mixture containing 65 g of KNO_3 and 25 g of KBr?

9.10 The partial pressure of ethane over a saturated solution containing 6.56×10^{-2} g of ethane is 751 torr. What is its par-

tial pressure when the saturated solution contains 5.00×10^{-2} g of ethane?

9.11 The vapor pressure of benzene (C_6H_6) at 25° C is 93.4 torr. What will be the vapor pressure, at 25° C, of a solution prepared by dissolving 56.4 g of the nonvolatile solute, $C_{20}H_{42}$, in 1000 g of benzene?

9.12 At 25° C the vapor pressures of benzene (C_6H_6) and toluene (C_7H_8) are 93.4 and 26.9 torr, respectively. At what applied pressure will a solution prepared from 60 g of benzene and 40 g of toluene boil at 25° C?

9.13 What will be the freezing point and boiling point of an aqueous solution containing 55.0 g of glycerol, $C_3H_5(OH)_3$, dissolved in 250 g of water? Glycerol is a nonvolatile, undissociated solute.

9.14 What is the molecular weight and molecular formula of a nondissociating compound whose empirical formula is C_4H_2N if 3.84 g of the compound in 500 g of benzene gives a freezing point depression of 0.307° C?

9.15 A solution containing 15.9 g of a nondissociating substance in 250 g of water has a freezing point of −0.744° C. The substance is composed of 57.2% C, 4.77% H, and 38.1% O. What is the molecular formula of the compound?

9.16 A solution containing 8.3 g of a nonvolatile nondissociating substance dissolved in one mole of chloroform, $CHCl_3$, has a vapor pressure of 511 torr. The vapor pressure of pure $CHCl_3$ at the same temperature is 526 torr. Calculate (a) the mole fraction of the solute, (b) the number of moles of solute, (c) the molecular weight of the solute.

9.17 The vapor pressure of a mixture containing 400 g of carbon tetrachloride and 43.3 g of an unknown substance is 137 torr at 30° C. The vapor pressure of pure carbon tetrachloride at 30° C is 143 torr while that of the pure unknown is

85 torr. What is the approximate molecular weight of the unknown?

9.18 We found that the addition of a nonvolatile solute to a solvent reduces the escaping tendency of the solvent from the solution, and in Section 9.10 we saw that this leads to a boiling point elevation. On a molecular level, account for the fact that the presence of a solute also reduces the tendency of the solvent to escape from the liquid into the solid. Explain why a lower temperature must be achieved to establish equilibrium between the solid solvent and the solution than between the pure solid and liquid solvent.

9.19 Calculate the freezing point of a 0.100 m aqueous solution of a weak electrolyte that is 7.5% dissociated.

9.20 What is the percent dissociation of a weak electrolyte if a 0.250 m solution of it has a freezing point of −0.500° C?

9.21 Calculate the osmotic pressure, in torr, of an aqueous solution containing 5.0 g of sucrose, $C_{12}H_{22}O_{11}$, per liter at 25° C.

9.22 What would be the osmotic pressure of a 0.010M aqueous solution of the electrolyte, NaCl at 25° C?

9.23 A solution of 0.40 g of a polypeptide in 1.00 liter of an aqueous solution has an osmotic pressure at 27° C of 3.74 torr. What is the approximate molecular weight of this polymer?

9.24 Describe, qualitatively, the procedure called fractional distillation.

9.25 Referring to Figure 9.17, approximately how many times must boiling, followed by condensation of the resulting vapor, be repeated in order to obtain a portion of liquid having a mole fraction of A of

at least 0.80 if the original mole fraction of A was 0.20?

9.26 Based on the data in Table 9.5, which 1 : 1 electrolyte (one positive ion to one negative ion) appears to be *least* fully dissociated in concentrated solutions? How does this agree (or disagree) with what might be predicted based on the charges on the ions involved?

9.27 Calculate the i factor for the weak electrolyte, HF, in Example 9.5. What conclusions would you draw about the i factors for weak electrolytes?

9.28 On the basis of what you have learned in this chapter, how would you interpret an i factor having a value less than 1.00?

9.29 Below is a list of the most abundant ions in sea water.

Ion	Molality
Chloride	0.566
Sodium	0.486
Magnesium	0.055
Sulfate	0.029
Calcium	0.011
Potassium	0.011
Bicarbonate	0.002

Calculate the weight, in grams, of each component contained in a 3.78 liters (1.00 gallon) of sea water having a density of 1.024 g/ml. What is the total weight of ions in this sample?

10
Chemical
Thermodynamics

Eventually, in the course of his investigations, a chemist begins to wonder why certain chemical reactions take place and why others do not. Certainly, he would be happy if he could predict what will occur when a group of chemicals are mixed. This chapter and the next discuss the two factors that ultimately determine whether one is able to observe a particular chemical reaction, either in the laboratory or elsewhere. A study of thermodynamics reveals whether or not a given process can occur spontaneously and what will be the position of equilibrium after reaction has ceased. Chemical kinetics, the subject of Chapter 11, is concerned with the speeds at which chemical changes take place. Both of these factors, spontaneity and speed, must be in our favor if we hope to observe the formation of products of a chemical change. For example, from thermodynamics we know that the reaction

$$2H_2 + O_2 \longrightarrow 2H_2O$$

is spontaneous. However, a mixture of hydrogen and oxygen gases is stable virtually indefinitely (provided nobody strikes a match). This is so because, at room temperature, hydrogen and oxygen react at such an extremely slow rate that even though their reaction to produce water is spontaneous it takes nearly forever for the reaction to proceed to completion.

The subject, thermodynamics, is basically concerned with the energy changes that accompany chemical and physical processes. Historically, it evolved without a detailed knowledge of the structure of matter; in fact, this is one of its strongest points. In this chapter we shall take a rather informal approach to the subject in an effort to avoid mathematical formalism, and we shall develop many of the concepts of thermodynamics by considering changes that take place on a molecular level.

Before we proceed, let us establish the meaning of some frequently used terms. A word that has been used rather loosely in previous sections is **system.** By system we mean *that particular portion of the universe upon which we wish to focus our attention.* Everything else we call the **surroundings.** For example, if we wished to consider the changes taking place in a solution of sodium chloride and silver nitrate, our system is the solution, while the beaker and everything else around the solution is considered the surroundings.

If a change occurs so that heat cannot be transferred across the interface, or boundary, between the system and its surroundings, we speak of it as an **adiabatic** process. An example is a reaction carried out in an insulated con-

**10.1
Some Commonly
Used Terms**

tainer, such as a "Thermos" bottle. Explosive reactions are also examples of adiabatic processes. Such reactions occur so rapidly that the heat energy produced cannot be readily dissipated. The heat build-up that occurs therefore raises the products to very high temperatures, and these products fly apart rapidly, pushing walls, ceilings, etc. (the surroundings) before them.

When thermal contact is maintained between system and surroundings, it is frequently possible to keep the system at a constant temperature while a change takes place. In this case the process is called **isothermal.**

To discuss the changes that occur in a system it is necessary to define its properties very precisely before and after the change occurs. We do this by specifying the **state** of the system, that is, some particular set of conditions of pressure, temperature, number of moles of each component, and their physical form (e.g., gas, liquid, solid, or crystalline form). When these variables are specified all of the properties of the system are fixed. Thus a knowledge of these quantities permits us to define unambiguously the properties of our system. For instance, if we have two samples of pure liquid water, each consisting of one mole and each at the same temperature and pressure, we know that all of the properties of each sample will be identical (volume, density, surface tension, vapor pressure, etc.).

The quantities P, T (and V) are called **state functions.** This is because (1) they serve to determine the state of any given system, and (2) in a particular state their values do not depend on the prior history of the sample. Furthermore, upon going from one state to another the changes in these quantities do not depend on how the sample is treated. For example, the volume of one mole of water at 25° C and 1 atm does not depend on what its temperature or pressure might have been at some time in the past. For the same reason, if the temperature of this sample is changed to 35° C, it does not matter if the sample were first cooled to 0° C and then warmed to 35° C, or whether the temperature were increased directly from 25 to 35° C. In the final state the temperature is the same regardless of the path taken between the initial and final conditions, and the change in temperature, ΔT, is therefore dependent *only* on the temperatures of those initial and final states.

There are some instances where the interrelationships between the state functions can be expressed in equation form, to give an **equation of state.** The equation of state for an ideal gas, $PV = nRT$, is an example. We have also seen the van der Waals equation of state that can be applied with reasonable success to real gases.

Another quantity that we shall use is called the **heat capacity:** this is the amount of heat energy required to raise the temperature of a given quantity of a substance one degree Celsius. The **specific heat** represents the heat capacity per gram, that is, it is the amount of heat necessary to raise the temperature of one gram of a substance by 1.0°C. The specific heat of water is 1.00 cal/g°C. We also speak of the **molar heat capacity:** the heat necessary to raise the temperature of *one mole* of a substance one degree.

10.2
The First Law of Thermodynamics

In thermodynamics we study the energy changes that occur when systems pass from one state to another. Repeated observations by many scientists over many years have led to the conclusion that in any process, energy is neither created nor destroyed. Another way of saying this is that energy is conserved. The **first law of thermodynamics** merely puts this meaning into the form of a simple equation,

$$\Delta E = q - w \tag{10.1}$$

Here, E represents the **internal energy** of the system; the total of all of the energies possessed by the system as a consequence of the kinetic energy of its atoms, ions, or molecules, plus the potential energy that arises from the binding forces between the particles that make up the system. ΔE is the difference between the energy contained in a system in some final state and the energy it possessed in an initial state. It corresponds to the change in the internal energy of a system that occurs when the system goes from an initial to a final state.

$$\Delta E = E_{final} - E_{initial}$$

Note that we have used the same convention here as in our previous discussions of energy changes (e.g., heats of vaporization, heats of solution). Here too, we cannot actually determine E, but instead only ΔE.

The quantity q in Equation 10.1 represents the amount of heat that is *added* to the system as it passes from the initial to the final state, and w denotes the work done *by* the system upon its surroundings. Thus Equation 10.1 simply states that the change in the internal energy is equal to the difference between the energy *supplied to* the system as heat and the energy *removed from* the system as work performed upon the surroundings.[1]

Since the first law deals with the transfer of quantities of energy, it is necessary to establish sign conventions to avoid confusion in our bookkeeping. Heat *added to* a system and work *done by* a system are considered positive quantities. Thus, if a certain change is accompanied by the absorption of 50 cal of heat and the expenditure of 30 cal of work, $q = +50$ cal and $w = +30$ cal. The change in internal energy is

$$\Delta E = (+50 \text{ cal}) - (+30 \text{ cal})$$

or

$$\Delta E = +20 \text{ cal}$$

Thus the system has undergone a net increase in energy amounting to $+20$ cal. How about the surroundings?

When the system gains 50 cal, the surroundings lose 50 cal; therefore, $q = -50$ cal for the surroundings. When the system performs work it does so upon the surroundings. We say that the surroundings have done negative work, and $w = -30$ cal for the surroundings. The change in the internal energy of the surroundings is thus

$$\Delta E = (-50 \text{ cal}) - (-30 \text{ cal})$$
$$\Delta E = -20 \text{ cal}$$

The change in internal energy of the system is thus equal, but opposite in sign, to ΔE for the surroundings, so that the net change for the universe (system plus surroundings) is zero. This is what we mean by the law of conservation of energy.

[1] Energy, you recall, is the capacity to do work. When the system performs work, its capacity to do additional work diminishes, which means that its energy has diminished. Energy, equal to the work performed, has been lost by the system and, in the process, gained by the surroundings.

In summary,

q positive ($q > 0$); heat is added to the system
q negative ($q < 0$); heat is evolved by (removed from) the system
w positive ($w > 0$); the system performs work — energy is removed
w negative ($w < 0$); work is done on the system — energy is added

The internal energy happens to be a state function and the magnitude of ΔE therefore depends only on the initial and final states of the system and not on the path taken between them. This is very much the same as the change in your bank balance that occurs between the beginning and the end of a month. During any given month the change in the balance is brought about as the combined results of some number of deposits and withdrawals. If the total number of dollars provided by the deposits exceeds those removed by the withdrawals, your balance increases. However, the net change in your balance at the end of the month depends only on the initial and final amounts of money in the bank, not on the individual transactions during the month. There is an infinite number of combinations of deposits and withdrawals that could lead to the same change in your balance. The same sort of relationship exists between ΔE, q, and w. The sign and magnitude of ΔE is controlled only by the values of E in the initial and final state. For any given change, ΔE, there are many different paths that can be followed with their own characteristic values of q and w. However, for the same initial and final states the difference between q and w is always the same.

To illustrate the preceding point, let us consider some concrete examples. We shall first look at the change that takes place when 1.0 liter of a gas at an initial pressure of 10 atm is permitted to push back the piston (assumed to be frictionless) in the cylinder shown in Figure 10.1, isothermally, against a uniform opposing pressure of 1.00 atm. Clearly, this is a spontaneous process, and the expansion continues until the internal pressure of the gas

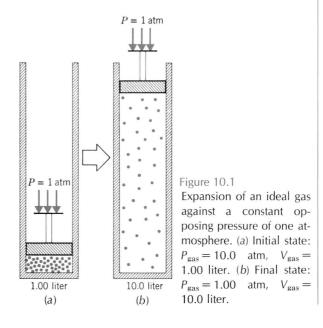

$P = 1$ atm

$P = 1$ atm

1.00 liter

(a)

10.0 liter

(b)

Figure 10.1
Expansion of an ideal gas against a constant opposing pressure of one atmosphere. (a) Initial state: $P_{gas} = 10.0$ atm, $V_{gas} = 1.00$ liter. (b) Final state: $P_{gas} = 1.00$ atm, $V_{gas} = 10.0$ liter.

is the same as the external pressure exerted on the piston, 1.00 atm. From Boyle's law we find that the final volume is 10 liters.

According to our definition of an ideal gas, there are no attractive forces present between the gas particles. Consequently, as these particles move apart during the expansion, there is no change in their potential energy. Since the temperature is held constant during the expansion, the average kinetic energy of the gas also stays the same. Thus, during the expansion there is no change in either kinetic or potential energy; therefore, for an isothermal expansion (or compression) of an *ideal gas*, $\Delta E = 0$. This means that

$$q - w = 0$$

or

$$q = w$$

In other words, when the expansion takes place, any heat absorbed by the system is returned to the surroundings by way of the system performing work on the surroundings.

Before we can proceed further we must ask ourselves, how does the system do work? We know that work is accomplished by moving an opposing force through some distance.

$$\text{Work} = \text{force} \times \text{distance}$$

Pressure is defined as force per area. In Figure 10.2 the external pressure on the piston corresponds to a certain total force, F, spread over the area of the piston, A.

$$P = F/A$$

The volume of the gas in the cylinder is equal to its cross-sectional area, A, multiplied by the height of the column of gas, h.

$$V = Ah$$

Chemical Thermodynamics

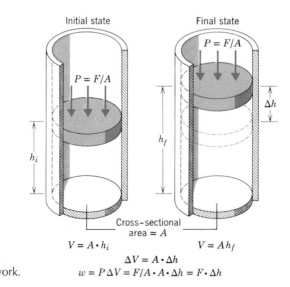

Figure 10.2
Pressure-volume work.

Initial state

Final state

$P = F/A$

$P = F/A$

Δh

h_i

h_f

Cross-sectional area = A

$V = A \cdot h_i$

$V = Ah_f$

$\Delta V = A \cdot \Delta h$

$w = P \Delta V = F/A \cdot A \cdot \Delta h = F \cdot \Delta h$

When the gas expands, A remains the same but h changes. The volume change is therefore

$$\Delta V = V_f - V_i$$
$$\Delta V = Ah_f - Ah_i$$
$$\Delta V = A(h_f - h_i) = A(\Delta h)$$

The product of pressure times volume change, then, is

$$P \,\Delta V = \frac{F}{A} \, A(\Delta h) = F \,\Delta h$$

We see that $P \,\Delta V$ is equivalent to force (F) times distance (Δh) and, therefore, equals work.

This expansion work[2] is performed by a system (any system — it doesn't have to be an ideal gas) when it expands against an external pressure imposed by the surroundings. Conversely, when the system contracts under the influence of an external pressure, work is performed *on* the system. If the volume change is measured in liters and the pressure in atmospheres, $P \,\Delta V$ has the units, liter atmosphere (liter-atm). We could, if we wish, convert this to the more familiar unit of energy, the calorie, by using the relationship: 1 liter-atm = 24.2 cal.

In our example, the ideal gas expands from its initial volume of 1.0 liter to a final volume of 10 liters, hence $\Delta V = 9$ liters. The external pressure is constant at 1.00 atm; therefore, the gas does work on the surroundings amounting to

$$w = P \,\Delta V$$
$$w = (1.00 \text{ atm}) \times (9 \text{ liters}) = 9 \text{ liter-atm}$$

Since $\Delta E = 0$, the system must simultaneously absorb heat from the surroundings in an amount precisely equal to the energy it expends by doing work. Consequently, $q = +9$ liter-atm.

Let us now consider a second path to take us between the same initial and final states. Suppose the apparatus in Figure 10.1 is modified so that a perfect vacuum ($P = 0$) exists above the piston in the cylinder. Once again the gas will expand by pushing back the piston (Figure 10.3). However, this time there is no resistance to the expansion because there is a zero opposing pressure (assuming, again, the piston is frictionless). Since no pressure opposes the expansion, $P = 0$ and the $P \,\Delta V$ product is equal to zero. No work is performed by the system. Since $\Delta E = 0$ (constant T)

$$q = w = 0$$

Thus no heat is exchanged with the surroundings either.

We have now considered two processes whereby an ideal gas is taken from an initial state of $P = 10$ atm, $V = 1$ liter to a final state of $P = 1$ atm and $V = 10$ liters. In both cases ΔE was the same, namely zero. However, q and w were not identical along the two paths; following the first, $q = +9$ liter-atm

---

[2] $P \,\Delta V$ work is only one kind of work that a physical or chemical change can produce. It is also possible (depending on conditions) to obtain other kinds of work from a changing system; for example, electrical work from the discharge of a dry cell.

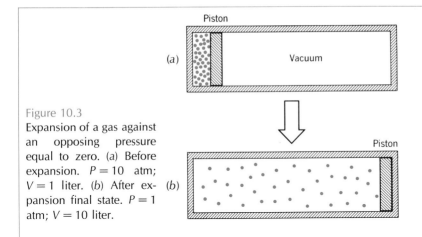

Figure 10.3
Expansion of a gas against an opposing pressure equal to zero. (a) Before expansion. $P = 10$ atm; $V = 1$ liter. (b) After expansion final state. $P = 1$ atm; $V = 10$ liter.

and $w = +9$ liter-atm, while along the second path, $q = 0$ and $w = 0$. Thus, while E is a state function, q and w are not.

In the examples just presented we saw that the work obtained from the expansion of an ideal gas depends on the external pressure resisting the expansion. When $P = 0$, the work done was also zero, while, when $P = 1$ atm, work equal to 9 liter-atm was performed as the gas expanded from 1 to 10 liters. What would be the *maximum* amount of work that we could obtain when a gas expands from an initial state of $P = 10$ atm and $V = 1$ liter to a final state in which $P = 1$ atm and $V = 10$ liters?

10.3
Reversible and Irreversible Processes

Let's suppose that we carried out the process in two steps; first with the opposing pressure equal to 5 atm and then, in the second step, with an opposing pressure equal to 1 atm (Figure 10.4). During the first step the gas expands until its pressure drops from 10 atm to that of the opposing pressure, 5 atm. At this point the gas would occupy a volume of 2 liters (why?). The work performed, let's call it w_1, is

$$w_1 = P\,\Delta V = 5 \text{ atm (1 liter)}$$
$$w_1 = 5 \text{ liter-atm}$$

In the second step the gas expands from a volume of 2 liters to the final volume, 10 liters, ($\Delta V = 8$ liters) against an opposing pressure of 1 atm. The work performed here is

$$w_2 = P\,\Delta V = 1 \text{ atm (8 liters)}$$
$$w_2 = 8 \text{ liter-atm}$$

The total work performed by the gas is the sum of that performed in each step along the way.

$$w_{\text{total}} = w_1 + w_2$$
$$w_{\text{total}} = (5 + 8) \text{ liter-atm} = 13 \text{ liter-atm}$$

Note that we have obtained more work in this two-step expansion than when the pressure is kept constant at 1 atm throughout the entire change. We might conclude (correctly) that even more work could be obtained by carrying out the expansion in still more steps, in which the opposing pressure is kept as

P = 5 atm

P = 5 atm

Figure 10.4
Two-step expansion. (a) Initial state. External pressure equals 5 atm. $V = 1.00$ liter; $P = 10.0$ atm (b) After step 1. $V = 2.00$ liter; $P = 5.0$ atm (c) After step 2. $V = 10.0$ liter; $P_{gas} = 1.00$ atm.

(a)   (b)   (c)

high as possible. We might also conclude that the greatest amount of work could be extracted if an infinite number of steps were employed in which the external pressure is always just barely below that exerted by the gas. This process could be approximated if we used a piston-cylinder apparatus such as that shown in Figure 10.5. As the water slowly evaporates, the external pressure gradually decreases and the gas gradually expands. In this case we have to use calculus to derive the expression for the total work. The result for one mole of a gas is

$$w_{maximum} = 2.3 \ RT \log \frac{V_f}{V_i}$$

H_2O

Gas

Figure 10.5
Reversible expansion. As H_2O gradually evaporates, pressure decreases and the gas expands. Process will be reversed if water molecules begin to condense into the liquid instead of evaporate.

where R is the universal gas constant (0.082 liter-atm/mole°K). In this example the maximum work is (at 0° C)

$$W_{max} = 2.3 \ (0.082 \text{ liter-atm/mole°K}) \ (273°K) \ \log \left(\frac{10 \text{ liters}}{1 \text{ liters}} \right)$$

$$W_{max} = 51.5 \text{ liter-atm/mole}$$

The number of moles of gas is (10 liters/22.4 liter-mole^{-1}) = 0.446 mole Therefore

$$W_{max} = \left(51.5 \ \frac{\text{liter-atm}}{\text{mole}} \right) (0.446 \text{ mole}) = 23.0 \text{ liter-atm}$$

The expansion of a gas in the manner just described, where the opposing pressure is virtually equal to the pressure exerted by the gas is one example of a **reversible process.** It is reversible because any slight increase in the external pressure will reverse the process and cause compression to occur. Any change that is resisted by an opposing "force" essentially equal to the driving force of the process constitutes a reversible change. Because it is reversible, we can obtain the maximum amount of work from it. However, because an infinite number of steps are required, a reversible process takes forever to occur. All real, spontaneous changes are therefore not really reversible, and the work that we derive from an irreversible change is not equal to the theoretical maximum.

When chemical reactions occur, they do so with either the absorption or evolution of energy. These changes reflect the differences between the potential energies associated with the bonds in the reactants and the products. For instance, when two hydrogen atoms come together to form an H_2 molecule, energy is evolved because the total potential energy of the nuclei and electrons in the H_2 molecule is lower than the total potential energy of these particles in two isolated H atoms. This same energy, when added to an H_2 molecule, can break it apart; it thus represents the bond energy discussed in Chapter 4. We see, therefore, that the measurement of the amount of energy evolved or absorbed when a chemical reaction occurs has the potential of providing us with very fundamental information concerning the stability of molecules and the strengths of chemical bonds.

If we carry out a chemical reaction in a closed container of fixed volume, the system undergoing reaction cannot perform pressure-volume work on the surroundings because $\Delta V = 0$; hence $P \Delta V = 0$. Any heat absorbed or evolved under these circumstances (let's call it q_v) is precisely equal to the change in the internal energy of the system,

$$\Delta E = q_v$$

Stated another way, ΔE is equal to the heat absorbed or evolved by the system under conditions of constant volume; therefore ΔE is called the *heat of reaction at constant volume*. When the reaction is endothermic, both q and ΔE are positive. For an exothermic process ΔE is negative.

Experimentally ΔE can be measured by using a device called a bomb calorimeter, (Figure 10.6). The apparatus consists of a strong steel "bomb" into which the reactants (e.g., H_2 and O_2) are placed. The bomb is then immersed into an insulated bath containing a precisely known quantity of water.

Thermometer

Stirrer

Electrical
connections
to ignition
wire

Insulated
container

Reactants

Steel
"bomb"

Water

Figure 10.6
Bomb calorimeter.

There the reaction is set off by a small heater wire within the bomb, and heat is evolved. The entire system is permitted to come to thermal equilibrium, at which point the calorimeter (bomb and water) will be at a higher temperature than before reaction. By carefully measuring the temperature of the water before and after reaction, and by knowing the heat capacity of the calorimeter (including the bomb and the water), the quantity of heat evolved by the chemical reaction can be computed.

Example 10.1
Hydrogen (0.100 g) and oxygen (0.800 g) are compressed into a 1.00 liter bomb, which is then placed into the water in the calorimeter. Before the reaction is set off the temperature of the water was 25.000° C; after reaction the temperature rose to 25.155° C. The heat capacity of the calorimeter was 21,700 cal/°C. What is ΔE for this reaction?

Solution
The change in temperature that occurs is 0.155° C. From the heat capacity we can find the number of calories evolved.

$$\text{Heat } evolved,\ q = \left(21{,}700\ \frac{\text{cal}}{°C}\right)(0.155°\ C)$$

Therefore

$$q_v = -3360\ \text{cal}$$

Hence,

$$\Delta E = -3.36\ \text{kcal}$$

In Example 10.1, the magnitude of ΔE depends on the quantity of H_2 and O_2 reacted; that is, ΔE is an extensive quantity. We can convert this to an intensive property, one which is characteristic of the reaction between any amounts of H_2 and O_2, by calculating the heat evolved *per mole* of product formed. In Example 10.1 we produced 0.0500 mole of H_2O. Therefore, we say that $\Delta E = -3.36$ kcal/0.0500 mole H_2O or $\Delta E = -67.2$ kcal/mole.

The heat evolved at constant volume permits us to compute ΔE. However, most changes that are of practical interest to the chemist take place in open containers at essentially constant atmospheric pressure. Under these conditions rather sizable volume changes can occur. For example, when two moles of gaseous H_2 react with one mole of gaseous O_2 to produce two moles of liquid water, at a constant pressure of 1 atm, the volume changes from about 67 liters to 0.036 liter. Imagine that this change takes place in a cylinder with a piston exerting a constant pressure of 1 atm; as the reaction proceeds to completion, the surroundings perform work on the system, the magnitude of which is the $P \Delta V$ product of very nearly 67 liter-atm.

In order to avoid the necessity of considering PV work when heats of reaction are measured at constant pressure, we define a new thermodynamic function called the **heat content,** or **enthalpy** (from the German, *enthalen*, to contain). This is denoted by the symbol H, as

$$H = E + PV$$

For a change at constant pressure,

$$\Delta H = \Delta E + P \Delta V \qquad (10.2)$$

If only PV work is involved in the change, we know that

$$\Delta E = q - P \Delta V$$

Substituting this into Equation 10.2 we have

$$\Delta H = (q - P \Delta V) + P \Delta V$$
$$\Delta H = q_p$$

Thus we see that ΔH is the heat, q_p, absorbed or evolved at constant pressure.

The enthalpy, like the internal energy, is a state function and thus the magnitude of ΔH depends only on the heat contents of the initial and final states. Thus we can write

$$\Delta H = H_{final} - H_{initial}$$

Here we use the same symbolism as in our discussion of ΔH_{vap}, ΔH_{fus}, etc. Those quantities, in fact, correspond to enthalpy changes associated with vaporization, fusion, etc.

In many instances the differences between ΔH and ΔE are small, particularly for chemical reactions. When a reaction occurs in which the reactants and products are liquids or solids, only very small changes in volume take place. As a result, $P \Delta V$ is very small and ΔH has very nearly the same magnitude as ΔE. When chemical reactions occur in which gases are either consumed or produced, much larger volume changes occur, and the $P \Delta V$ product is also much greater. Even in these cases, however, ΔE is usually so large compared to the $P \Delta V$ term that ΔE and ΔH are still nearly the same.

Example 10.2
When two moles of H_2 and one mole of O_2, at 100° C and 1 atm, react to produce two moles of gaseous water at 100° C and 1 atm, a total of 115.8 kcal are evolved. What are (a) ΔH, and (b) ΔE for the production of a single mole of $H_2O(g)$?

Solution

(a) Since the reaction

$$2H_2(g) + O_2(g) \longrightarrow 2H_2O(g)$$

is occurring at constant pressure,

$$q = \Delta H = \frac{-115.8 \text{ kcal}}{2 \text{ moles } H_2O}$$

The minus sign, remember, signifies that the reaction is exothermic. For the production of one mole of water,

$$\Delta H = -57.9 \text{ kcal/mole}$$

(b) If we assume ideal behavior of the gaseous reactants and products, we have for the reactants (the initial state) at a given P and T

$$PV_i = n_i RT$$

where n_i corresponds to the number of moles of gaseous reactants. In a similar fashion, for the final state,

$$PV_f = n_f RT$$

The pressure-volume work in the process is given by

$$PV_f - PV_i = P(V_f - V_i) = P\,\Delta V$$

This is equal to

$$P\,\Delta V = n_f RT - n_i RT$$
$$P\,\Delta V = (n_f - n_i)RT = \Delta nRT$$

The quantity, $\Delta n = $ (number of moles of *gaseous* products) $-$ (number of moles of *gaseous* reactants).
In this example

$$\Delta n = 2 \text{ moles} - 3 \text{ moles} = -1 \text{ mole}$$

Therefore, using $R = 1.99$ cal/mole°K

$$P\,\Delta V = (-1.0 \text{ mole})(1.99 \text{ cal/mole°K})(373°K)$$
$$P\,\Delta V = -742 \text{ cal} = -0.742 \text{ kcal}$$

For each *one* mole of H_2O produced, $P\,\Delta V = -0.371$ kcal
Hence, solving Equation 10.2 for the change in internal energy,

$$\Delta E = \Delta H - P\,\Delta V$$
$$\Delta E = -57.9 \text{ kcal/mole} - (-0.371 \text{ kcal/mole})$$
$$\Delta E = -57.5 \text{ kcal/mole}$$

From this last example we see that the $P\,\Delta V$ product can be calculated in a very simple fashion for a reaction in which a gas is either consumed or produced:

$$P\,\Delta V = \Delta nRT$$

This relationship, however, cannot be applied when all of the reactants and products are either liquids or solids.

Since enthalpy is a state function, the magnitude of ΔH for a chemical reaction does not depend on the path taken by the reactants as they proceed to form the products. Consider, for example, the conversion of one mole of liquid water at 100° C and 1 atm to one mole of vapor at 100° C and 1 atm. This process absorbs 9.7 kcal of heat for each mole of H_2O vaporized and, hence, $\Delta H = +9.7$ kcal. We can represent this "reaction" as

$$H_2O(l) \longrightarrow H_2O(g) \qquad \Delta H = +9.7 \text{ kcal}$$

An equation written in this manner, in which the energy change is also shown, is called a **thermochemical equation** and is nearly always interpreted on a mole basis. Here, for instance, we see that one mole of $H_2O(l)$ is converted to one mole of $H_2O(g)$ by the absorption of 9.7 kcal.

The value of ΔH for this process will always be +9.7 kcal, provided we refer to the same pair of initial and final states. We could even go so far as to first decompose the one mole of liquid into gaseous hydrogen and oxygen and then recombine the elements to produce $H_2O(g)$ at 100° C and 1 atm. The net change in enthalpy would still be the same, +9.7 kcal. Consequently, it is possible to look at some overall change as the net result of a sequence of chemical reactions. The net value of ΔH for the overall process is merely the sum of all of the enthalpy changes that take place along the way. These last statements constitute **Hess' law of heat summation.**

Thermochemical equations serve as a useful tool for applying Hess' law. For example, the thermochemical equations that correspond to the indirect path just described for the vaporization of water are,[3]

$$H_2O(l) \longrightarrow H_2(g) + \tfrac{1}{2}O_2(g) \qquad \Delta H = +67.6 \text{ kcal}$$
$$H_2(g) + \tfrac{1}{2}O_2(g) \longrightarrow H_2O(g) \qquad \Delta H = -57.9 \text{ kcal}$$

These equations tell us that 67.6 kcal are required to decompose one mole of $H_2O(l)$ into its elements and that 57.9 kcal are evolved when they recombine to produce one mole of $H_2O(g)$. The sum of the two equations, after cancelling quantities that appear on both sides of the arrow, gives us the equation for the vaporization of one mole of water,

$$H_2O(l) + \cancel{H_2(g)} + \cancel{\tfrac{1}{2}O_2(g)} \longrightarrow H_2O(g) + \cancel{H_2(g)} + \cancel{\tfrac{1}{2}O_2(g)}$$

or

$$H_2O(l) \longrightarrow H_2O(g)$$

We also find that the heat of the overall reaction is equal to the algebraic sum of the heats of reaction for the two steps.

$$\Delta H = +67.6 \text{ kcal} + (-57.9 \text{ kcal})$$
$$\Delta H = +9.7 \text{ kcal}$$

Thus, *when we add thermochemical equations to obtain some net change, we also add their corresponding heats of reaction.*

---

[3] Note that fractional coefficients are permitted in thermochemical equations. This is because a coefficient such as 1/2 is taken to mean 1/2 mole. In ordinary equations fractional coefficients are avoided because they are meaningless on a molecular level. One cannot have half an atom or molecule and still retain the chemical identity of the species.

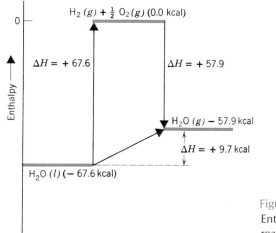

To illustrate the nature of these thermochemical changes, we can also demonstrate them graphically (Figure 10.7). This type of figure is frequently called an **enthalpy diagram.** Notice that we have chosen as our zero on the energy scale the enthalpy of the free elements. This choice is entirely arbitrary since we are only interested in determining differences in H. In fact, we have no way at all of knowing absolute enthalpies; we can only measure ΔH.

HEATS OF FORMATION. A particularly useful type of thermochemical equation corresponds to the formation of a substance from its elements. The energies associated with these reactions are called **heats of formation** and are denoted as ΔH_f. For example, thermochemical equations for the formation of liquid and gaseous water at 100° C and 1 atm are, respectively,

$$H_2(g) + \tfrac{1}{2}O_2(g) \longrightarrow H_2O(l) \qquad \Delta H_f = -67.6 \text{ kcal/mole}$$
$$H_2(g) + \tfrac{1}{2}O_2(g) \longrightarrow H_2O(g) \qquad \Delta H_f = -57.9 \text{ kcal/mole}$$

How can we use these equations to obtain the heat of vaporization of water? Clearly, we must reverse the first equation and then add it to the second. When we reverse this equation, we must also change the sign of ΔH, since if the formation of $H_2O(l)$ is exothermic, as indicated by a negative ΔH_f, the reverse process must be endothermic.

(Exothermic) $\qquad H_2(g) + \tfrac{1}{2}O_2(g) \longrightarrow H_2O(l) \qquad \Delta H = \Delta H_f = -67.6 \text{ kcal}$
(Endothermic) $\qquad H_2O(l) \longrightarrow H_2(g) + \tfrac{1}{2}O_2(g) \qquad \Delta H = -\Delta H_f = +67.6 \text{ kcal}$

When this last equation is added to that for the formation of $H_2O(g)$, we obtain

$$H_2O(l) \longrightarrow H_2O(g)$$

and the heat of reaction is

$$\Delta H = \Delta H_f[H_2O(g)] - \Delta H_f[H_2O(l)]$$
$$\Delta H = -57.9 \text{ kcal} - (-67.6 \text{ kcal}) = +9.7 \text{ kcal}$$

Notice that the heat of reaction for the overall change is equal to the heat of formation of the product *minus* the heat of formation of the reactant. In general, we can write that for any overall reaction

$$\Delta H_{reaction} = \text{(sum of } \Delta H_f \text{ of products)} - \text{(sum of } \Delta H_f \text{ of reactants)} \qquad (10.3)$$

The magnitude of ΔH_f depends on the conditions of temperature, pressure, and the physical form (gas, liquid, solid, crystalline form) of the reactants and products. For instance, at 100° C and 1 atm the heat of formation of liquid water is −67.6 kcal/mole, while at 25° C and 1 atm ΔH_f for $H_2O(l)$ is −68.3 kcal/mole. To avoid the necessity of always having to specify the conditions for which ΔH_f is recorded, and to permit comparisons between ΔH_f for various compounds, a standard set of conditions is chosen, namely 25° C[4] and one atmosphere of pressure. Heats of formation of substances in their standard states are indicated as ΔH_f°. For example, the standard heat of formation of liquid water, $\Delta H_f^\circ[H_2O(l)] = -68.3$ kcal/mole, and represents the heat liberated when H_2 and O_2, each in their natural form at 25° C and 1 atm, react to produce $H_2O(l)$ at 25° C and 1 atm.

Table 10.1 contains standard heats of formation for a variety of different substances. Such a table is very useful because it permits us to calculate, using Equation 10.3, the standard heats of reaction, ΔH°, for a very large number of different chemical changes. In performing these calculations, we arbitrarily take the ΔH_f° for an element in its natural, most stable form at 25° C and 1 atm to be equal to zero. Thus, in our computations, the zero point on the energy scale is again chosen to be that of the free elements. As mentioned before, since we only speak of changes in energy, the actual location of this zero point is unimportant. The following examples illustrate how the principles developed in the preceding two sections can be applied.

Example 10.3
Determine ΔH° for the reaction,

$$2Na_2O_2(s) + 2H_2O(l) \longrightarrow 4NaOH(s) + O_2(g)$$

Solution
Equation 10.3 implies that

$$\Delta H^\circ = \text{(sum } \Delta H_f^\circ \text{ products)} - \text{(sum } \Delta H_f^\circ \text{ reactants)}$$

This means that we must add up all of the heat evolved during the formation of the products from their elements and then subtract the heat evolved by the formation of the reactants from their elements.
For the products the total enthalpy of formation is

$$4 \text{ mole NaOH} \times \left(\frac{-102.0 \text{ kcal}}{1 \text{ mole NaOH}} \right) = -408.0 \text{ kcal}$$

$$\underline{1 \text{ mole } O_2 \times \left(\frac{0.0 \text{ kcal}}{1 \text{ mole } O_2} \right) = 0.0 \text{ kcal}}$$

$$\text{Total of } \Delta H_f^\circ \text{ products} = -408.0 \text{ kcal}$$

[4] Note that this temperature differs from the standard temperature of 0° C used for calculations involving gases in Chapter 6.

Table 10.1

Standard Heats of Formation of Some Substances at 25° C and 1 Atmosphere

Substance	ΔH_f° (kcal/mole)	Substance	ΔH_f° (kcal/mole)
$Al_2O_3(s)$	-400.5	$HF(g)$	-64.8
$Br_2(l)$	0.00	$HCl(g)$	-22.1
$Br_2(g)$	$+7.39$	$HBr(g)$	-8.7
$C(s, \text{diamond})$	$+0.45$	$HI(g)$	$+6.3$
$CO(g)$	-26.4	$KCl(s)$	-104.2
$CO_2(g)$	-94.1	$LiCl(s)$	-97.7
$CH_4(g)$	-17.9	$MgCl_2(s)$	-153.4
$C_2H_6(g)$	-20.2	$MgCl_2 \cdot 2H_2O(s)$	-306.0
$C_2H_4(g)$	$+12.4$	$Mg(OH)_2(s)$	-221.0
$C_2H_2(g)$	$+54.2$	$NH_3(g)$	-11.0
$C_3H_8(g)$	-24.8	$N_2O(g)$	$+24.8$
$C_6H_6(l)$	$+11.7$	$NO(g)$	-21.6
$CH_3OH(l)$	-57.0	$NO_2(g)$	$+9.1$
$HCOOH(g)$	-86.7	$NaF(s)$	-136.5
$CS_2(l)$	$+21.4$	$NaCl(s)$	-98.6
$CS_2(g)$	$+28.0$	$NaBr(s)$	-86.0
$CCl_4(l)$	-32.1	$NaI(s)$	-68.8
$C_2H_5OH(l)$	-66.4	$Na_2O_2(s)$	-120.6
$CH_3CHO(g)$	-39.8	$NaOH(s)$	-102.0
$CH_3COOH(l)$	-116.4	$PbO_2(s)$	-66.3
$CaO(s)$	-151.9	$PbSO_4(s)$	-219.9
$Ca(OH)_2(s)$	-235.8	$SO_2(g)$	-70.9
$CaSO_4(s)$	-342.4	$SO_3(g)$	-94.6
$CuO(s)$	-37.1	$H_2SO_4(l)$	-194.5
$Fe_2O_3(s)$	-196.5	$SiO_2(s)$	-217.7
$H_2O(l)$	-68.3	$SiH_4(g)$	$+8.2$
$H_2O(g)$	-57.8	$ZnO(s)$	-83.2
		$Zn(OH)(s)$	-153.5

For the reactants we have

$$2 \text{ mole Na}_2\text{O}_2 \times \left(\frac{-120.6 \text{ kcal}}{1 \text{ mole Na}_2\text{O}_2}\right) = -241.2 \text{ kcal}$$

$$2 \text{ mole H}_2\text{O(l)} \times \left(\frac{-68.3 \text{ kcal}}{1 \text{ mole H}_2\text{O(l)}}\right) = -136.6 \text{ kcal}$$

Total of ΔH_f° reactants $= -377.8$ kcal

We have said that

$$\Delta H^\circ = (\text{sum } \Delta H_f^\circ \text{products}) - (\text{sum } \Delta H_f^\circ \text{reactants})$$

Therefore

$$\Delta H^\circ = -408.0 \text{ kcal} - (-377.8 \text{ kcal})$$

or

$$\Delta H^\circ = -30.2 \text{ kcal}$$

Example 10.4
On the basis of the previous example, how many kilocalories of heat are evolved when 25.0 g of Na_2O_2 are treated with water to produce NaOH and O_2?

Solution
    In the preceding example we found that 30.2 kcal are evolved when two moles of Na_2O_2 are reacted with water. Thus we write for this reaction,

$$2 \text{ mole Na}_2\text{O}_2 \sim 30.2 \text{ kcal}$$

or $$1 \text{ mole Na}_2\text{O}_2 \sim 15.1 \text{ kcal}$$

Since the formula weight of Na_2O_2 is 78.0,

$$25.0 \text{ g Na}_2\text{O}_2 \times \left(\frac{1 \text{ mole Na}_2\text{O}_2}{78.0 \text{ g Na}_2\text{O}_2}\right) \times \left(\frac{15.1 \text{ kcal}}{1 \text{ mole Na}_2\text{O}_2}\right) \sim 4.84 \text{ kcal}$$

Hence the consumption of 25.0 g Na_2O_2 releases 4.84 kcal.

It is frequently impossible to measure directly the heat of formation of a compound. For example, we cannot get hydrogen, oxygen, and graphite (the most stable crystalline form of carbon) to react directly together to produce ethyl alcohol, C_2H_5OH. This is also true with many, if not most, other compounds. In order to determine ΔH_f° for these substances, then, an indirect method must be applied. One technique, which can be applied to most organic materials, is to burn the substance to produce products whose heats of formation are known, as shown in Example 10.5.

Example 10.5
The combustion of one mole of benzene, $C_6H_6(l)$, to produce $CO_2(g)$ and $H_2O(l)$ liberates 781.2 kcal when the products are returned to 25°C and 1 atm. What is the standard heat of formation of $C_6H_6(l)$?

Solution

The equation for the combustion of one mole of C_6H_6 is

$$C_6H_6(l) + 7\tfrac{1}{2}O_2(g) \longrightarrow 6CO_2(g) + 3H_2O(l)$$

The standard heat of reaction, $\Delta H° = -781.2$ kcal. From Equation 10.3 we know

$$\Delta H° = 6\ \Delta H_f[CO_2(g)] + 3\ \Delta H_f[H_2O(l)] - \Delta H_f[C_6H_6(l)]$$

Solving for the heat of formation of benzene

$$\Delta H_f°[C_6H_6(l)] = 6\ \Delta H_f°[CO_2(g)] + 3\ \Delta H_f°[H_2O(l)] - \Delta H°$$

From Table 10.1 we can obtain the heats of formation of CO_2 and H_2O. Therefore,

$$\Delta H_f°[C_6H_6(l)] = 6(-94.1)\ \text{kcal} + 3(-68.3)\ \text{kcal} - 781.2\ \text{kcal}$$
$$\Delta H_f°[C_6H_6(l)] = +11.7\ \text{kcal}$$

10.7
Bond Energies

We stated earlier that it should be possible to relate heats of reaction to changes in the energies associated with chemical bonds. Strictly speaking, we should use ΔE for this purpose; however, since $P\ \Delta V$ contributions to ΔH are relatively small for chemical reactions, we can use ΔH in place of ΔE and still expect to obtain quite reasonable results.

In Chapter 4 the bond energy was defined as the energy required to break a bond to produce neutral fragments. For a complex molecule, the energy needed to reduce the molecule entirely to neutral atoms, called the **atomization energy,** is the sum of all of the bond energies in the molecule. Simple diatomic molecules, such as H_2, O_2, Cl_2, or HCl possess only one bond; therefore the atomization energy is the same as the bond energy. For these simple cases the atomization energy can be obtained by studying the spectra produced when these molecules absorb or emit light. For more complex molecules, however, we employ an indirect method that makes use of measured heats of formation.

As an example, let's consider the molecule CH_4. If we use the same approach that was followed in Example 10.5, the standard heat of formation of $CH_4(g)$ can be experimentally determined to be -17.9 kcal/mole. This corresponds to the enthalpy change, $\Delta H_f°$, for the reaction

$$C(s, \text{graphite}) + 2H_2(g) \longrightarrow CH_4(g)$$

We can envision an alternate path to take us from the free elements to the compound, methane, that follows the succession of reactions

(1) $C(s, \text{graphite}) \longrightarrow C(g)$ ΔH_1
(2) $2H_2(g) \longrightarrow 4H(g)$ ΔH_2
(3) $C(g) + 4H(g) \longrightarrow CH_4(g)$ ΔH_3

The sum of these three will give us our desired overall reaction.

Steps 1 and 2 each involve the heat of formation of gaseous atoms from an element in its standard state. For hydrogen the heat of formation of each mole of gaseous hydrogen atoms is half the atomization energy of $H_2(g)$. For carbon it amounts to the sublimation energy of graphite. Table 10.2 contains some heats of formation of gaseous atoms from typical elements in their standard states.

Applying Hess' law, we know that ΔH for the overall reaction, ΔH_f° for $CH_4(g)$, can be obtained by adding up the enthalpy changes for each step.

$$\Delta H_f^\circ = \Delta H_1 + \Delta H_2 + \Delta H_3 \qquad (10.4)$$

From the data in Table 10.2 we find that $\Delta H_1 = +170.9$ kcal, the heat of formation of gaseous carbon atoms. Similarly, $\Delta H_2 = 4(+52.1$ kcal), that is, four times the heat of formation of one mole of H(g). The quantity ΔH_3 is the negative of the atomization energy of $CH_4(g)$

$$\Delta H_3 = -\Delta H_{atom}[CH_4(g)]$$

Solving for the atomization energy of methane in Equation 10.4 gives

$$\Delta H_{atom}[CH_4(g)] = \Delta H_1 + \Delta H_2 - \Delta H_f^\circ$$

Substituting numerical values we have

$$\Delta H_{atom}[CH_4(g)] = (+170.9 + 208.4 + 17.9) \text{ kcal}$$
$$\Delta H_{atom}[CH_4(g)] = +397.2 \text{ kcal}$$

This quantity is the total amount of energy that must be absorbed to break all four moles of C—H bonds in one mole of CH_4. Division by four, then, provides us with an *average* bond energy of 99.3 kcal/mole of C—H

Table 10.2
Heats of Formation of Gaseous Atoms from the Elements in Their Standard States

Atom	ΔH_f(kcal/mole of atoms)
H	52.1
Li	38.4
Be	78.2
B	132.6
C	170.9
N	113.0
O	59.6
F	18.9
Na	25.8
Si	108.4
Cl	28.9
Br	26.7
I	25.5

Table 10.3
Average Bond Energies

Bond	Bond Energy (kcal/mole)
H—C	99.3
H—O	110.6
H—N	93.4
H—F	134.6
H—Cl	103.2
H—Br	87.5
H—I	71.4
C—O	85.0
C=O	173
C—N	69.7
C=N	148
C≡N	210
C—C	83.1
C=C	145
C≡C	199

bonds. This value, along with some other bond energies appears in Table 10.3.

A very significant fact is that the average bond energies found in Table 10.3 can be used, in many cases, to compute heats of formation with a fair degree of accuracy, as illustrated below in Example 10.6. Apparently a C—H bond, for instance, has very nearly the same strength in one molecule as it does in another, implying that nearly all C—H bonds are pretty much alike. This is also true of many other bonds. This phenomenon has greatly simplified the development of the modern theories about chemical bonding that we shall discuss in Chapter 16.

Example 10.6
Use the data in Tables 10.2 and 10.3 to compute the heat of formation of liquid ethyl alcohol. This compound has a heat of vaporization, $\Delta H^\circ_{vap} = 10.0$ kcal/mole and the structural formula,

Solution

We wish to determine $\Delta H°$ for the reaction

$$2C(s, \text{graphite}) + 3H_2(g) + \tfrac{1}{2}O_2(g) \longrightarrow C_2H_5OH(l)$$

From the data in Tables 10.2 and 10.3 we can find $\Delta H°$ for the reaction to produce gaseous C_2H_5OH, that is,

$$2C(s, \text{graphite}) + 3H_2(g) + \tfrac{1}{2}O_2(g) \longrightarrow C_2H_5OH(g)$$

Applying Hess' law, we write the following equations whose sum gives us our desired overall change.

$2C(s, \text{graphite}) \longrightarrow 2C(g)$	$\Delta H_1 = 2\,\Delta H_f[C(g)]$
	$\Delta H_1 = 2(+\,170.9 \text{ kcal}) = +\,341.8 \text{ kcal}$
$3H_2(g) \longrightarrow 6H(g)$	$\Delta H_2 = 6\,\Delta H_f[H(g)]$
	$\Delta H_2 = 6(+52.1 \text{ kcal}) = +312.6 \text{ kcal}$
$\tfrac{1}{2}O_2(g) \longrightarrow O(g)$	$\Delta H_3 = \Delta H_f[O(g)] = +59.6 \text{ kcal}$
$2C(g) + 6H(g) + O(g) \longrightarrow C_2H_5OH(g)$	$\Delta H_4 = -\,\Delta H_{\text{atom}}[C_2H_5OH(g)]$

The heat of atomization of C_2H_5OH is equal to the sum of the energy of five C—H bonds, one C—C bond, one C—O bond, and one O—H bond. From Table 10.3 we have

$$\Delta H_{\text{atom}}[C_2H_5OH(g)] = 5(99.3 \text{ kcal}) + (83.1 \text{ kcal}) + (85.0 \text{ kcal}) + (110.6 \text{ kcal})$$

$$\Delta H_{\text{atom}}[C_2H_5OH(g)] = 775.2 \text{ kcal}$$

and therefore

$$\Delta H_4 = -775.2 \text{ kcal}$$

The heat of formation of gaseous ethyl alcohol is equal to the sum of ΔH_1 to ΔH_4. Performing the arithmetic we get

$$\Delta H_f°[C_2H_5OH(g)] = -61.2 \text{ kcal/mole}$$

Since we want $\Delta H_f°$ for the liquid, we must use the heat of vaporization, which is the enthalpy change for the reaction

$$C_2H_5OH(l) \longrightarrow C_2H_5OH(g) \qquad \Delta H_{\text{vap}} = +10.0 \text{ kcal/mole}$$

From Hess' law we know

$$\Delta H_{\text{vap}} = \Delta H_f°(\text{vapor}) - \Delta H_f°(\text{liquid})$$

and thus

$$\Delta H_f°[C_2H_5OH(l)] = -61.2 \text{ kcal/mole} - 10.0 \text{ kcal/mole}$$

or

$$\Delta H_f° = -71.2 \text{ kcal/mole}$$

Comparing this to the value reported in Table 10.1

$$\Delta H_f° = -66.4 \text{ kcal/mole}$$

we see that the agreement is not really too bad, considering that we have assumed that any particular bond has the same energy in *all* compounds.

10.8
Spontaneity of Chemical Reactions

We stated in the introduction to this chapter that any chemist would like to answer this question: Will a particular chemical reaction occur for him spontaneously? If it will not, he can sit around forever waiting for his chemicals to react and yet never observe the products that he wishes to be formed. As it happens, thermodynamics is capable of providing an answer to this question.

To arrive at this answer, we must begin by finding out what factors are involved in determining whether a particular process, either physical or chemical, will occur spontaneously. One spontaneous process that we have all observed is a ball rolling down a hill. When it finally comes to rest at the bottom, its potential energy has decreased and it is in a more stable, lower energy state than before. We might conclude therefore that a process leading to a decrease in the energy of a system (here, a ball) should tend to be spontaneous. Indeed, many processes that are spontaneous do occur with the evolution of energy. For example, a mixture of hydrogen and oxygen, when ignited, reacts very rapidly to produce water. This chemical change is accompanied by the release of a large quantity of heat, so much, in fact, that the hot water vapor produced expands explosively.

Evolution of energy, however, is not the only criterion to be considered. There are many examples of processes that occur with the absorption of energy and yet are spontaneous. In the last chapter, for example, we discussed heats of solution and saw that in many instances when a salt dissolves in water (e.g., KI) energy is absorbed. The formation of the solution, although endothermic, is nevertheless spontaneous. What, then, is the driving force for this process that is capable of outweighing the endothermic energy effect that occurs?

When a solid such as KI dissolves in water, the particles of the solute leave the well-ordered crystalline state and gradually diffuse throughout the liquid to produce a solution. In this final state the particles of the solute are in a more disordered, random condition than they were before they dissolved, as shown in Figure 10.8. Similarly, the solvent is in a more disordered state in the solution because the solvent molecules are, in a sense, dispersed throughout those of the solute as well.

In any process there is a natural tendency or drive toward disorder because a disordered state represents a condition of higher statistical probability than an ordered one. To see this, suppose that we had one mole of a gas

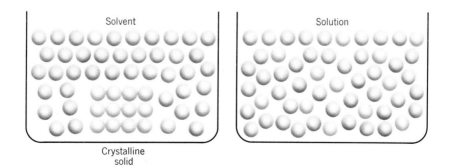

Figure 10.8

Increase in disorder occurs when solution is formed.

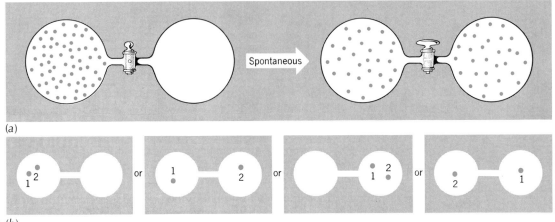

(a)

(b)

Figure 10.9

(a) A gas spontaneously fills a vacuum. (b) Two molecules with four possibilities.

in the left compartment of the apparatus in Figure 10.9, and a vacuum in the other compartment. If the stopcock between them is opened, we know intuitively that the gas will rush into the right compartment, and that after a time each side will contain equal numbers of molecules, provided the volume of each is the same. Why does this change occur spontaneously?

To understand this, imagine that there are only two molecules in the compartment on the left. When the stopcock is opened 'both molecules are free to wander into the other bulb and, at this point, there are four different particle distributions possible for the system (Figure 10.9b). Since both particles reside in the left compartment in only one of these four distributions, the probability of finding both on the left is one out of four, or 1/4. Similarly, since there are two ways of having an even distribution of particles between the two compartments, the probability of an even distribution is two out of four, or 1/2.

Even in the two-particle case we feel intuitively that when the stopcock is opened the gas will expand in such a way as to occupy the entire volume of both compartments. In light of the discussion above, this simply means that the system proceeds from a state of low statistical probability (both particles on one side) to a state of higher probability.

When there are more particles in our system, the difference between these two types of molecular distributions is even more dramatic. With six particles, for instance, the probability of finding all of them in one compartment is only 1/64; whereas, an even 3 to 3 distribution has a probability of 20/64.

In general, the probability of finding all of the molecules in a gas in one compartment of this apparatus is $1/2^n$, where n is the number of particles. For one mole of gas the probability that all of the gas will remain in one bulb (i.e., the probability that the gas will *not* spontaneously expand) is

$$p = \frac{1}{2^{(6.02 \times 10^{23})}}$$

This can be shown to be the same as

$$p = \frac{1}{10^{(1.81 \times 10^{23})}} = 10^{(-1.81 \times 10^{23})}$$

It is nearly impossible to appreciate how small this probability is. If we were to write this as a decimal fraction it would be,

$$p = \underbrace{0.0000 \ldots\ldots\ldots\ldots 00001}_{1.81 \times 10^{23} \text{ zeros}}$$

and if each zero were 0.10 inches in diameter, we would need a piece of paper 2,800,000,000,000,000,000 *miles* long just to write the number! Therefore, the probability that the mole of gas will not expand is extremely small, and the spontaneous expansion that does occur is simply a manifestation of the system proceeding from a state of low probability to a state of higher probability. The final state is one of greater disorder, or randomness, because the particles of the gas are more widely distributed and have a greater degree of freedom of movement.

The degree of disorder of a system is represented by a thermodynamic quantity called the **entropy,** denoted by the symbol S; the greater the disorder, the greater the entropy. A change in entropy, given by ΔS, can be brought about by the addition of heat to a system. For example, consider a perfect crystal of carbon monoxide at $0°$ K in which all of the C—O dipoles are aligned in the same direction (Figure 10.10). Because of the perfect alignment of the dipoles there is essentially perfect order in the crystal, and the entropy of the system is at a minimum. When heat is added to this crystal, the temperature rises above $0°$ K and thermal motion (vibrations) within the lattice cause some of the dipoles to become oriented in the opposite direction (Figure 10.10b). As a result, there is less order (more disorder) and the entropy of the crystal has obviously increased. Logically the more heat added to the system, the greater will be the extent of disorder afterwards; therefore we ex-

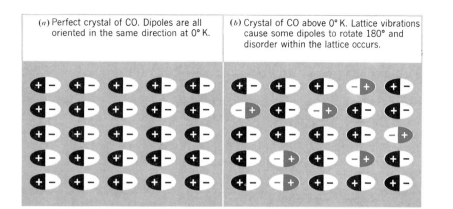

Figure 10.10
(a) Perfect crystal of CO at 0°K. (b) Crystal of CO above 0°K.

pect the entropy change, ΔS, to be directly proportional to the quantity of heat, q_{rev}, added to the system,[5]

$$\Delta S \propto q_{rev}$$

The magnitude of ΔS is also inversely proportional to the temperature at which the heat is added. At low temperatures a given quantity of heat makes a large change in the relative degree of disorder. Near absolute zero the system goes from essentially perfect order to some extent of disorder—a very substantial change. If the same amount of heat is added to a system at high temperature, the system goes from an already highly disordered state to one just slightly more disordered. This constitutes only a very slight change in the *relative* degree of disorder and, hence, to only a small entropy change. Thus, we find that a change in entropy is finally given as

$$\Delta S = \frac{q_{rev}}{T}$$

where T is the absolute temperature at which q is transferred to the system. The entropy, like E and H, is also a state function; ΔS only depends on the initial and final entropies of the system.

The second law of thermodynamics provides us with a way of comparing the effects of the two driving forces involved in a spontaneous process—changes in energy and changes in entropy. One statement of the second law is that *in any spontaneous process there is always an increase in the entropy of the universe* ($\Delta S_{total} > 0$). This increase takes into account entropy changes in both the system and its surroundings,

$$\Delta S_{total} = \Delta S_{system} + \Delta S_{surroundings}$$

The entropy change that occurs in the surroundings is brought about by the heat added to the surroundings divided by the temperature at which it is transferred. For a process at constant P and T, the heat added to the system is ΔH_{system} and is equal to the *negative* of the heat added to the surroundings. In other words,

$$-q_{surroundings} = \Delta H_{system}$$

or

$$q_{surroundings} = -\Delta H_{system}$$

The entropy change for the surroundings is therefore

$$\Delta S_{surroundings} = \frac{-\Delta H_{system}}{T}$$

The total entropy change for the universe is thus

$$\Delta S_{total} = \Delta S_{system} - \frac{\Delta H_{system}}{T}$$

or

[5] For reasons that we shall not discuss, the q used in computations of ΔS must be the q for a reversible change (q_{rev}).

$$\Delta S_{\text{total}} = \frac{T \, \Delta S_{\text{system}} - \Delta H_{\text{system}}}{T}$$

This can be rearranged to give

$$T \, \Delta S_{\text{total}} = -(\Delta H_{\text{system}} - T \, \Delta S_{\text{system}})$$

Since ΔS_{total} must be a positive number for a spontaneous change, the product $T \, \Delta S_{\text{total}}$ must also be positive. This means that the quantity in parentheses on the right, $(\Delta H_{\text{system}} - T \, \Delta S_{\text{system}})$, must be negative so that $-(\Delta H_{\text{system}} - T \, \Delta S_{\text{system}})$ may be positive. Thus, in order for a spontaneous change to take place, the expression, $\Delta H_{\text{system}} - T \, \Delta S_{\text{system}}$, must be negative.

At this point it is convenient to introduce another thermodynamic state function, G, called the **Gibbs free energy,** which we define as

$$G = H - TS$$

For a change at constant T and P, we write

$$\Delta G = \Delta H - T \, \Delta S$$

From the argument presented in the preceding paragraph, we see that ΔG must be less than zero for a spontaneous process; that is, ΔG must have a negative value at constant T and P.

The Gibbs free energy change, ΔG, represents a composite of the two factors contributing to spontaneity, ΔH and ΔS. For systems in which ΔH is negative (exothermic) and ΔS is positive (increased disorder accompanying the change) both factors favor spontaneity and the process will occur spontaneously at all temperatures. Conversely, if ΔH is positive (endothermic) and ΔS is negative (increase in order), ΔG will always be positive and the change cannot occur spontaneously at any temperature.

In situations where ΔH and ΔS are both positive, or both negative, temperature plays the determining role in controlling whether or not a reaction will take place. In the first case (ΔH, $\Delta S > 0$), ΔG will only be negative at high temperatures where $T \, \Delta S$ is greater in magnitude then ΔH; as a consequence, the reaction will only be spontaneous at elevated temperatures. On the other hand, when ΔH and ΔS are both negative, ΔG will only be negative at low temperatures. An example of this is the freezing of water. We know that heat must be removed from the liquid to produce ice; hence the process is exothermic with a negative ΔH. Freezing is also accompanied by an ordering of the water molecules as they leave the disordered liquid state and become part of the crystal. As a result, ΔS is also negative. The sign of ΔG is determined both by ΔH, which in this case is negative, and by $T \, \Delta S$, which is also negative. To compute ΔG we must subtract a negative $T \, \Delta S$ from a negative ΔH. The result will only be negative at low temperature. Consequently, at one atm we only observe H_2O to freeze spontaneously below 0° C. Above 0° C the magnitude of $T \, \Delta S$ is greater than ΔH, and ΔG becomes positive. As a result, freezing is no longer spontaneous. Instead, the reverse process (melting) occurs.

10.10
Free Energy and Useful Work

One of the most important applications of chemical reactions is in the production of energy in the form of useful work. This can, for example, take the form of combustion, in which the heat generated is used to create steam for the production of mechanical work, or perhaps electrical work drawn from a dry cell or storage battery. The quantity G is called the *free energy* because ΔG represents the *maximum* amount of energy released in a process that is free to perform useful work. We have already associated ΔG with the factors that lead to a drive for spontaneity. What we see now is that this driving force in a chemical change can be harnessed to perform work for us.

The actual amount of work obtained from any real spontaneous process is always somewhat less than the maximum predicted by ΔG. This is because real processes are always irreversible, and we saw earlier that the maximum work can only be extracted from a truly reversible change. Nevertheless, ΔG gives us a goal to aim at. The closer a given process is to reversibility, the greater will be the amount of available work that can be used.

10.11
Free Energy and Equilibrium

In the last paragraph we said that ΔG determines the maximum amount of energy that is available to perform useful work as a system passes from one state to another. As a reaction proceeds, its capacity to perform work, as measured by G, diminishes until finally, at equilibrium, the system is no longer able to supply additional work. This means that both reactants and products possess the same free energy and therefore $\Delta G = 0$. We see, then, that the value of ΔG for a particular change determines the approach toward equilibrium. When ΔG is negative, the reaction is spontaneous. When ΔG is zero, the system is in a state of dynamic equilibrium, and when ΔG is positive, the reaction is really spontaneous in the reverse direction.

At this point, it should be reemphasized that although ΔG may predict that a particular process is spontaneous, nothing is implied about how rapid the change will be.

10.12
Standard Entropies and Free Energies

The **third law of thermodynamics** states that the entropy of any pure crystalline substance at absolute zero is equal to zero. This makes sense because in a perfect crystal at absolute zero there is perfect order. Because of this it is possible, by summing q_{rev}/T increments from $0°$ K to $298°$ K($25°$ C), to determine the absolute entropy of a substance in its standard state. Table 10.4 contains a number of such **standard entropies.**

From standard heats of formation and standard entropies we can also calculate standard free energies of formation, $\Delta G_f°$. For example, consider the formation of CO_2 from the elements, all reactants and products in their standard states,

$$C(s,\ graphite) + O_2(g) \longrightarrow CO_2(g)$$

Table 10.1 gives us the standard enthalpy of formation, $\Delta H_f°$, as -94.1 kcal/mole. From the data in Table 10.4 we can calculate $\Delta S_f°$.

$$\Delta S_f° = S_{CO_2}° - (S_C° + S_{O_2}°)$$
$$\Delta S_f° = 51.1 - (1.4 + 49.0)\ \text{cal/mole°K}$$
$$\Delta S_f° = +0.7\,\text{cal/mole°K}$$

We can then obtain $\Delta G_f°$ as

$$\Delta G_f° = \Delta H_f° - T\,\Delta S_f°$$

At $25°$ C($298°$ K), then,

$$\Delta G_f° = -94.1\ \text{kcal/mole} - (298°\ \text{K})(0.7\ \text{cal/mole°\ K})$$
$$\Delta G_f° = (-94.1 - 0.2)\ \text{kcal/mole}$$
$$\Delta G_f° = -94.3\ \text{kcal/mole}$$

Table 10.4

Absolute Entropies at 25° C and 1 Atmosphere

Substance	S°(cal/degree mole)		Substance	S°(cal/degree mole)
$Al(s)$	6.77		$N_2(g)$	45.77
$Al_2O_3(s)$	12.19		$NH_3(g)$	46.01
$Br_2(l)$	36.38		$N_2O(g)$	52.58
$Br_2(g)$	58.65		$NO(g)$	50.34
$C(s, graphite)$	1.36		$NO_2(g)$	57.47
$C(s, diamond)$	0.58		$Na(s)$	12.2
$CO(g)$	47.30		$NaF(s)$	12.3
$CO_2(g)$	51.06		$NaCl(s)$	17.4
$CH_4(g)$	44.50		$NaBr(s)$	20.0
$C_3H_8(g)$	64.51		$NaI(s)$	21.8
$CCl_4(l)$	51.25		$O_2(g)$	49.00
$Cl_2(g)$	53.29		$Pb(s)$	15.5
$F_2(g)$	48.44		$PbO_2(s)$	16.4
$H_2(g)$	31.21		$PbSO_4(s)$	35.5
$H_2O(l)$	16.72		$S(s, rhombic)$	7.60
$H_2O(g)$	45.11		$SO_2(g)$	59.3
$HF(g)$	41.47		$SO_3(g)$	61.3
$HCl(g)$	44.62		$H_2SO_4(l)$	37.5
$HBr(g)$	47.44		$Si(s)$	4.5
$HI(g)$	49.3		$SiO_2(s)$	10.0
$I_2(s)$	27.76		$Zn(s)$	10.0
$Mg(s)$	7.77		$ZnO(s)$	10.4
$Mg(OH)_2(s)$	15.09			

This and other standard free energies of formation are given in Table 10.5.

Earlier in this chapter we saw that $\Delta H°$ for a reaction can be computed from standard heats of formation. The same rules also apply for the calculation of $\Delta G°$ using standard free energies of formation, that is,

$$\Delta G° = \text{(sum of } \Delta G_f° \text{ of products)} - \text{(sum of } \Delta G_f° \text{ of reactants)}$$

For example, consider the reaction

$$SiH_4(g) + 2\ O_2\ (g) \longrightarrow SiO_2(s) + 2\ H_2O(g)$$

for which we would write

$$\Delta G° = (\Delta G_f°[SiO_2(s)] + 2\ \Delta G_f°[H_2O(g)]) - (\Delta G_f°[SiH_4(g)] + 2\ \Delta G_f°[O_2(g)])$$

Table 10.5
Standard Free Energies of Formation at 25° C and 1 Atmosphere

Substance	ΔG_f° (kcal/mole)	Substance	ΔG_f° (kcal/mole)
$Al_2O_3(s)$	-376.8	$HBr(g)$	-12.7
$AgNO_3(s)$	-7.7	$HI(g)$	$+0.31$
$C(s, \text{diamond})$	$+0.69$	$H_2O(l)$	-56.7
$CO(g)$	-32.8	$H_2O(g)$	-54.6
$CO_2(g)$	-94.3	$MgCl_2(s)$	-141.6
$CH_4(g)$	-12.1	$Mg(OH)_2(s)$	-199.3
$C_2H_6(g)$	-7.9	$NH_3(g)$	-4.0
$C_2H_4(g)$	$+16.3$	$N_2O(g)$	$+24.8$
$C_2H_2(g)$	$+50.0$	$NO(g)$	$+20.7$
$C_3H_8(g)$	-5.6	$NO_2(g)$	$+12.4$
$CCl_4(l)$	-15.6	$HNO_3(l)$	-19.1
$C_2H_5OH(l)$	-41.8	$PbO_2(s)$	-52.3
$CH_3COOH(l)$	-93.8	$PbSO_4(s)$	-193.9
$CaO(s)$	-144.4	$SO_2(g)$	-71.8
$Ca(OH)_2(s)$	-214.3	$SO_3(g)$	-88.5
$CaSO_4(s)$	-315.6	$H_2SO_4(l)$	-164.9
$CuO(s)$	-30.4	$SiO_2(s)$	-192.4
$HF(g)$	-64.7	$SiH_4(g)$	-9.4
$HCl(g)$	-22.8	$ZnO(s)$	-76.1

As with enthalpy calculations, we take ΔG_f° for any free element to be equal to zero. Therefore, using the data in Table 10.5 we have

$$\Delta G^\circ = 1 \text{ mole} \times \left(-192.4 \ \frac{\text{kcal}}{\text{mole}}\right) + 2 \text{ moles} \times \left(-54.6 \ \frac{\text{kcal}}{\text{mole}}\right)$$
$$- 1 \text{ mole} \times \left(-9.4 \ \frac{\text{kcal}}{\text{mole}}\right) = -292.2 \text{ kcal}$$

The reason that this type of calculation is important is because the value of ΔG° for a reaction determines the position of equilibrium, that is, the relative numbers of moles of reactants and products that will be present when the chemical system achieves equilibrium. We shall deal with this quantitatively in Chapter 12, but for now let us look qualitatively at the relationship between the computed value of ΔG° and the free energy of a chemical system as it passes from reactants to products, as illustrated in Figure 10.11. This graph allows us to make a very important distinction between ΔG°, the free energy difference between the reactants and products in their standard states,

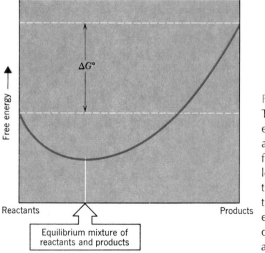

Figure 10.11
The variation of free energy in a chemical system as the reaction proceeds from pure reactants on the left to pure products on the right. The minimum in the curve represents the extent of reaction required for the system to achieve equilibrium.

and the free energy change that occurs at various intervals along the reaction path.

In Figure 10.11 we see that there is a minimum in the free energy curve that lies below both $G°_{reactants}$ and $G°_{products}$. Since a system always seeks the minimum free energy, some reaction always occurs, whether we begin with pure reactants and proceed in the direction of the products or vice versa. In each case the chemical change is accompanied by a free energy decrease as G heads toward the minimum; hence ΔG for *reaction toward the minimum* is negative. A spontaneous chemical change, therefore, always takes place regardless of whether we begin with pure reactants or products.

Another important aspect of Figure 10.11 is the minimum itself. Once this minimum free energy has been achieved by the system the composition of the system can no longer change, since such a change involves going "uphill" on the free energy curve (i.e., ΔG positive). We already know that this is not spontaneous. As a result, the minimum on the curve represents the system in a state of equilibrium.

A third important point to note is that the direction in which the reaction will proceed for a given composition (i.e., a given number of moles of reactants and products) is not controlled by the sign of $\Delta G°$ but, instead, depends on the slope of the free energy curve at that point along the reaction path. For the reaction depicted by Figure 10.11, for instance, if we were to begin with pure reactants and proceed in the direction of the products, ΔG would be negative but $\Delta G° = G°_{prod} - G°_{react}$ would be positive. It is the sign of ΔG, not $\Delta G°$, that determines spontaneity.

Whereas the slope of the free energy curve determines the direction in which the reaction will proceed at a given point along the reaction path, it is $\Delta G°$ that controls the position of the minimum in the curve. This is shown in Figure 10.12. Notice that the position of the minimum in the free energy curve depends on the relative values of $G°_{react}$ and $G°_{prod}$, always lying closer to the one that is lowest in energy. Since these minima correspond to different system compositions, we are therefore observing different positions of equilib-

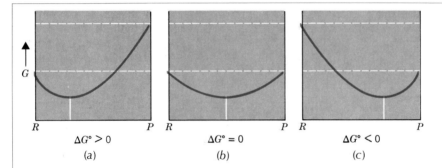

R P R P R P

$\Delta G° > 0$ $\Delta G° = 0$ $\Delta G° < 0$

(a)   (b)   (c)

Figure 10.12
The position of equilibrium changes as the value of $\Delta G°$ changes. (a) Position of equilibrium in favor of reactants. (b) Position of equilibrium intermediate between reactants and products. (c) Position of equilibrium in favor of products.

rium that depend on the sign and magnitude of $\Delta G°$. Whereas the computed value of $\Delta G°$ for a reaction is not an indicator of spontaneity, it does, however, serve as a guide to the *feasibility* of a reaction. Consider, for example, the compounds listed in Table 10.5. Some of these compounds possess positive standard free energies of formation. In these cases the reaction

$$\text{elements} \rightleftharpoons \text{compound}$$

will not proceed far to the right (i.e., not much product will be formed) because the position of equilibrium favors the reactants. Consequently, substances having positive free energies of formation are not usually prepared directly from the free elements because the equilibrium reaction mixture contains only a very small quantity of the desired product. From a practical standpoint, then, such a reaction is not very feasible. Since we can use values of $\Delta G_f°$ to compute $\Delta G°$ for any reaction, we therefore have a guide to feasibility.

Example 10.7
Under standard conditions, would the reaction,

$$2HI(g) + Cl_2(g) \longrightarrow 2HCl(g) + I_2(s)$$

be thermodynamically feasible?

Solution
To answer this question, we must compute $\Delta G°$ for the reaction.

$$\Delta G° = 2 \ \Delta G_f°[HCl(g)] - 2 \ \Delta G_f°[HI(g)]$$

Using the data in Table 10.5,

$$\Delta G° = 2 \text{ mole} \left(-22.8 \ \frac{\text{kcal}}{\text{mole}}\right) - 2 \text{ mole} \left(+0.31 \ \frac{\text{kcal}}{\text{mole}}\right)$$

$$\Delta G° = -46.22 \text{ kcal}$$

From the sign of $\Delta G°$ we conclude that the position of equilibrium in this system lies far to the right and that when HI and Cl_2 gases are mixed, we ought to observe a significant amount of I_2 being formed. This, in fact, is precisely what is observed when these two gases are mixed.

<table>
<tr><td>

10.13
Applications of the Principles of Thermodynamics

</td><td>

The thermodynamic concepts of enthalpy, entropy, and spontaneity that have been developed in this chapter have wide applicability to all areas of chemistry. We shall have occasion to apply them to chemical equilibrium, electrochemistry (where we look at the relationship between electrical energy and chemical change), and to many of the chemical characteristics of the elements and their compounds.

The impact of thermodynamics is also manifest in the practical world around us. A classic example of this is in the production of synthetic diamonds. People had been fascinated by this problem ever since 1797 when it was found that diamond was simply a form of carbon. Over the years many experiments were devised in an attempt to convert graphite, the common form of carbon, into its much more valuable counterpart. However, as of 1938 no one had yet been able to accomplish this feat. At that time a careful thermodynamic analysis of the problem was performed, the results of which are summarized in Figure 10.13.

In this figure, $-\Delta G/T$ is plotted along the vertical axis and temperature, in degrees Kelvin, is plotted along the horizontal axis. Since ΔG for the reaction,

$$C(s, graphite) \longrightarrow C(s, diamond)$$

must be negative in order for the process to be spontaneous, diamond can only be produced at temperatures and pressures that lie *above* the zero on the $-\Delta G/T$ scale. For example, at 470° K the conversion of graphite into dia-

</td></tr>
</table>

Figure 10.13
Thermodynamics of graphite to diamond conversion. From Chemical and Engineering News, April 5, 1971, p. 51. Used by permission.

mond can only take place at pressures greater than or equal to 20,000 atmospheres. We can also conclude that at a constant pressure of 20,000 atm the reaction is not spontaneous above 470° K.

This analysis, then, served to define the limits of temperature and pressure that would permit the conversion to take place. That was not the end of the problem, however, since suitable materials had to be found that would enable the reaction to proceed at a measurable rate. In fact, it was not until 1955 that success was finally achieved.

The conversion of graphite to diamond represents but one example of how thermodynamics can serve as a guide to answering practical questions. The principles of thermodynamics have been applied over the years to such diverse problems as the design of steam engines to the development of fuel cells that are now used as a source of electric power in spacecraft. Recently, Dr. Frederick Rossini, a noted thermodynamicist, has pointed out that the balance between the simultaneous opposing drives toward security (low energy) and freedom (high entropy) that control chemical equilibrium also seem to determine equilibrium in a stable society, illustrating, perhaps, the truly wide scope of thermodynamics.

Another illustration of the impact of thermodynamics on our lives is the problem of thermal pollution that has become so important in our efforts to save what is left of our environment. It is a consequence of the second law of thermodynamics that any heat engine that is designed to convert heat into useful work must absorb heat from a high temperature source. Some of this heat can be converted to work; however, some of it must also be deposited in a low temperature reservoir. Therefore, any device that tries to convert heat into useful work also discharges some heat into the environment. An automobile engine burns gasoline (when it's available), converts part of the resultant heat energy into work that propels the car, while the remainder is discharged to the radiator. Large power stations burn fuels (either fossil fuels like coal or oil, or nuclear fuels such as uranium) to generate electrical power. In the process they too deposit relatively large quantities of heat into the environment.

Cooling water from such power plants is heated to such an extent that it has been known to raise the temperature of surrounding rivers or lakes by as much as 25° F or more. These higher temperatures decrease the solubility of oxygen and kill some species of fish. It has also been found, however, that clams and oysters thrive in these warmer waters.

Recently, a nuclear power station in New Jersey was forced to shut down for repairs during the winter. It is interesting that in this case, the halting of thermal pollution caused a large fish kill because many fish that normally live in warm water were suddenly "put out in the cold."

Review
Questions
and Problems

10.1 Define state, system, state function, and heat capacity.

10.2 Describe what is meant by adiabatic process, isothermal process, and reversible process.

10.3 A gas, initially under a pressure of 15.0 atm and having a volume of 10.0 liter, is permitted to expand isothermally in two steps. In the first the external pressure is held constant at 7.50 atm, and in the second step the external pressure is maintained at 1.00 atm. What are q and w for each step? What is the net change in the internal energy for both the system and its surroundings?

10.4 It was pointed out in this chapter that for an isothermal expansion of an ideal gas into an evacuated container ΔE, q, and w are all equal to zero. In Section 6.12 we saw that real gases cool upon expansion into a vacuum. What conclusions can you draw about ΔE, q, and w for the isothermal expansion of a real gas into a vacuum?

10.5 At 25° C and a constant pressure of 1.00 atm the reaction of one half mole of OF_2 with water vapor, according to the equation,

$$OF_2(g) + H_2O(g) \longrightarrow O_2(g) + 2HF(g)$$

liberates 38.6 kcal. Calculate ΔH and ΔE per mole of OF_2.

10.6 At 25° C and 1.00 atm the reaction of 1.00 mole of CaO with water (shown below) evolves 15.6 kcal.

$$CaO(s) + H_2O(l) \longrightarrow Ca(OH)_2(s)$$

What are ΔH and ΔE, per mole of CaO, for this process given that the densities of CaO(s), $H_2O(l)$, and $Ca(OH)_2(s)$ at 25° C are 3.25 g/ml, 0.997 g/ml, and 2.24 g/ml, respectively?

10.7 The combustion of 0.20 mole of H_2 with 0.1 mole O_2 to produce $H_2O(l)$ in a bomb calorimeter raises the temperature of the apparatus 0.880° C. When 0.0100 mole of toluene, C_7H_8, are burned in this calorimeter, the temperature is raised by 0.615° C. The equation for the combustion reaction is

$$C_7H_8(l) + 9\ O_2(g) \longrightarrow$$
$$7\ CO_2(g) + 4\ H_2O(l)$$

Calculate ΔE for this reaction.

10.8 Based on the results of Question 10.7, compute the standard heat of formation, ΔH_f^o, of toluene (assume the reaction was carried out at 25° C).

10.9 Use Hess' law to calculate $\Delta H°$ for each of the following reactions:
(a) $2Al(s) + Fe_2O_3(s) \rightarrow$
$Al_2O_3(s) + 2Fe(s)$
(b) $SiH_4(g) + 2O_2(g) \rightarrow$
$SiO_2(s) + 2H_2O(g)$
(c) $CaO(s) + SO_3(g) \rightarrow CaSO_4(s)$
(d) $CuO(s) + H_2(g) \rightarrow Cu(s) + H_2O(g)$
(e) $C_2H_4(g) + H_2(g) \rightarrow C_2H_6(g)$

10.10 Given the following thermochemical equations,

$Fe_2O_3(s) + 3CO(g) \longrightarrow 2Fe(s) + 3CO_2(g)$
$3Fe_2O_3(s) + CO(g) \longrightarrow 2Fe_3O_4(s) + CO_2(g)$
$Fe_3O_4(s) + CO(g) \longrightarrow 3FeO(s) + CO_2(g)$
$\Delta H = -6.6$ kcal
$\Delta H = -14.0$ kcal
$\Delta H = +9.1$ kcal

Calculate ΔH for the reaction

$FeO(s) + CO(g) \longrightarrow Fe(s) + CO_2(g)$

without referring to the data in Table 10.1

10.11 Use the results of Question 10.10 and the data in Table 10.1 to compute the standard heat of formation of FeO.

10.12 Explain why $\Delta E = 0$ for the isothermal expansion of an ideal gas.

10.13 The heat of vaporization ΔH_{vap} of H_2O at 25° C is 10.5 kcal/mole. Calculate q, w, and ΔE for the process.

10.14 Use the data in Table 10.1 to determine how many calories are evolved in the combustion of 45.0 g of $C_2H_6(g)$ to produce $CO_2(g)$ and $H_2O(g)$ under a constant pressure of 1.00 atm.

10.15 Explain why $\Delta G = 0$ for a system that is in a state of equilibrium.

10.16 ΔH and ΔS are nearly independent of temperature. Why is this not true for ΔG?

10.17 Use the average bond energies in Table 10.3 to compute the standard heat of formation of C_3H_8. Its structure

is H—C—C—C—H How well does

your computed value compare with that reported in Table 10.1?

10.18 Explain why the entropy of a pure substance is zero at 0° K.

10.19 What two criteria must be met for a reaction to be spontaneous at all temperatures?

10.20 The heat of fusion of water at 0° C is 1.44 kcal/mole; its heat of vaporization is 9.72 kcal/mole at 100° C. What are ΔS for the melting and boiling of one mole of water? Can you explain why ΔS_{vap} is greater than $\Delta S_{melting}$?

10.21 Which of the following reactions could *potentially* serve as a practical method for the preparation of NO_2? (*Note.* The equations are not balanced.)
(a) $N_2(g) + O_2(g) \rightarrow NO_2(g)$
(b) $HNO_3(l) + Ag(s) \rightarrow$
     $AgNO_3(s) + NO_2(g) + H_2O(l)$
(c) $NH_3(g) + O_2(g) \rightarrow$
     $NO_2(g) + NO(g) + H_2O(g)$
(d) $CuO(s) + NO(g) \rightarrow NO_2(g) + Cu(s)$
(e) $NO(g) + O_2(g) \rightarrow NO_2(g)$
(f) $H_2O(g) + N_2O(g) \rightarrow$
     $NH_3(g) + NO_2(g)$

10.22 What is the sign of the entropy change for each of the following processes?
(a) A solute crystallizes from a solution.
(b) Water evaporates.
(c) A deck of playing cards is shuffled.
(d) A card player is dealt 13 spades.
(e) Pure carbon and oxygen react to produce CO.
(f) $^{235}_{92}U$ is extracted from a mixture of $^{235}_{92}U$ and $^{238}_{92}U$.

10.23 Which of the following reactions is accompanied by the greatest entropy change?
(a) $SO_2(g) + \frac{1}{2}O_2(g) \rightarrow SO_3(g)$
(b) $CO(g) + \frac{1}{2}O_2(g) \rightarrow CO_2(g)$

10.24 What is the maximum amount of useful work that could be obtained by the oxidation of propane, C_3H_8, according to the equations:

$C_3H_8(g) + 5O_2(g) \longrightarrow$
$$3CO_2(g) + 4H_2O(g)$$

Why is it that we always get less than this maximum amount of work in any real process that uses propane as a fuel?

10.25 Referring to Figure 10.13., what is the minimum pressure necessary for the conversion of graphite to diamond at a temperature of 200° K? Is it theoretically possible to change graphite to diamond at one atmosphere?

10.26 The ionization energy of Na is 118.1 kcal/mole and the electron affinity of Cl is 83.3 kcal/mole. Use this information, along with the heats of formation of gaseous Na and Cl atoms as well as the heat of formation of NaCl, to calculate ΔH for the reaction

$$NaCl(s) \longrightarrow Na^+(g) + Cl^-(g)$$

The answer corresponds to the lattice energy of sodium chloride. (*Hint.* It will help if you write thermochemical equations for each process described in the problem.)

10.27 From the data in Tables 10.1 and 10.4, calculate the boiling point of liquid bromine (i.e., the temperature at which $Br_2(l)$ and $Br_2(g)$ can coexist in equilibrium with each other).

10.28 Describe the relationship between $\Delta G°$ and the position of equilibrium in a chemical reaction.

11
Chemical Kinetics

It does not take long to find a reaction that thermodynamics predicts should proceed nearly to completion but yet is not observed to occur. We know from the last chapter that hydrogen and oxygen can be kept in contact with one another almost forever without forming water, even though their reaction to produce water is accompanied by a free energy decrease. This is an example of a chemical change where the speed of the reaction governs whether the formation of the products will or will not be observed.

Chemical kinetics, also referred to as chemical dynamics, concerns itself with the speed, or rates, of chemical reactions. In this area of chemistry we study the factors that control how fast chemical change occurs. These include the following.

1. The nature of the reactants and products.
2. The concentration of reacting species (related to this is the pressure of gaseous reactants).
3. The effect of temperature.
4. The influence of outside agents (catalysts).

By studying the factors that influence rates of reaction we can begin to obtain an understanding of the detailed sequence of steps (called the **mechanism** of the reaction) that are followed along the path from reactants to products. Such information, clearly, is of very fundamental interest to the chemist.

11.1 Reaction Rates and Their Measurement

Before we examine the factors that influence rates of reaction let's be sure we know what is meant by "rate." To determine the rate of a given chemical reaction, we must measure how fast the concentration of a reactant or product changes during the course of the investigation. In practice, the species whose concentration is easiest to follow is determined at various time intervals. The simplest example is a reaction where only one reactant undergoes a change to form a single product. An example of this type of reaction is the conversion of cyclopropane H_2C—CH_2 (with CH_2) into the molecule propylene:

In general the balanced equation for this type of reaction would be

$$A \longrightarrow B \qquad (11.1)$$

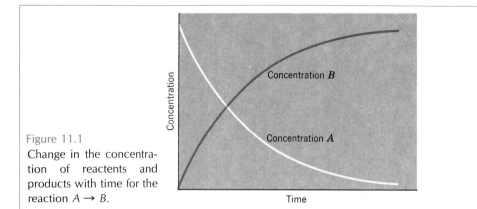

Figure 11.1
Change in the concentration of reactents and products with time for the reaction $A \rightarrow B$.

When the reaction is carried out, no product (B) is present initially and, as time goes on, the concentration of B increases with a corresponding decrease in the concentration of A (Figure 11.1).

In general, the rate (or speed) of any chemical reaction can be expressed as the ratio of the change in the concentration of a reactant (or product) to a change in time. This is exactly analogous to giving the speed of an automobile as the change in position (i.e., the distance traveled) divided by its time of travel. Here the speed might be given in the units, miles per hour. With chemical reactions the rate is usually expressed in the units, moles/liter per second:

$$\text{Speed of auto} = \frac{\text{distance}}{\text{time}} = \frac{\text{miles}}{\text{hour}}$$

$$\text{Rate of chemical reaction} = \frac{\text{change in concentration}}{\text{time}} = \frac{\text{mole/liter}}{\text{sec}}$$

An inspection of Figure 11.1 reveals that the rate of this chemical reaction changes with time. For instance, near the start of the reaction the concentration of A is decreasing rapidly and the concentration of B is rising rapidly. Much later during the reaction, however, only small changes in concentration occur with time, and the rate is therefore much less. In general, this type of behavior is observed with nearly every chemical reaction; as the reactants are consumed the rate of reaction gradually decreases.

An accurate, quantitative estimate of the rate of reaction at any given moment during the reaction can be obtained from the slope of the tangent to the concentration-time curve at that particular instant. This is shown in Figure 11.2. Square brackets, [ ], are used here to denote molar concentration. From the tangent to the curve we can write that

$$\text{Rate} = \frac{\Delta[B]}{\Delta t} \qquad (11.2)$$

We can also express the rate of the above reaction in terms of the concentration of the reactant A, since its concentration is also changing with time. The rate measured in terms of the concentration of A would be

$$\text{Rate} = \frac{-\Delta[A]}{\Delta t} \qquad (11.3)$$

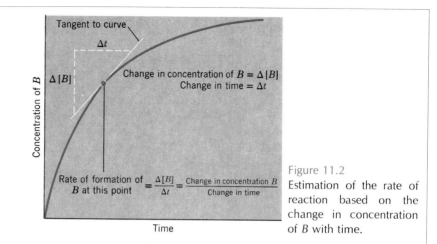

Figure 11.2
Estimation of the rate of reaction based on the change in concentration of B with time.

The minus sign indicates that the concentration of A is decreasing with time. A minus sign is always used whenever reactants are employed to express the rate.

When measuring the rate of any chemical reaction the concentration that is monitored and the technique that is used to measure the change depends on the nature of the reaction. Whatever method of analysis is employed, it must be fast, accurate, and in no way interfere with the normal course of the reaction being studied.

11.2
Rate Laws

As we have seen, if we follow a chemical reaction over a period of time we find that its rate gradually decreases as the reactants are consumed: we might conclude that the rate is related, in some way, to the concentration of the reacting species. In fact, the rate is nearly always directly proportional to the concentration of the reactants raised to some power. This means that for the general reaction

$$A \longrightarrow B$$

the rate can be written as

$$\text{Rate} \propto [A]^x \qquad (11.4)$$

where the exponent, x, called the **order of the reaction,** can be either a whole number or a fraction. **The value of x can only be determined experimentally.**

The rate of a somewhat more complicated reaction, such as

$$A + B \longrightarrow \text{products}$$

is usually dependent on both the concentration of A and the concentration of B. Decreasing the concentration of either reactant will therefore usually decrease the rate of the reaction. The rate, then, is proportional to both the concentration of A and B, each raised to some power. Consequently we write, for this reaction,

$$\text{Rate} \propto [A]^x [B]^y \qquad (11.5)$$

From this equation we say that the order of the reaction with respect to A is x

and that the order with respect to B is y, and that the overall order (i.e., the sum of the individual orders) is $x + y$. Again, the values of x and y can only be determined experimentally.

The proportionality represented by Equation 11.5 can be converted to an equality by introducing a proportionality constant, which we call the **rate constant.** The resulting equation, termed the **rate law** for the reaction, is

$$Rate = k[A]^x [B]^y$$

For example, the rate law for the reaction between ICl and H_2

$$2 \ ICl(g) + H_2(g) \longrightarrow I_2(g) + 2 \ HCl(g)$$

at 230° C has been found experimentally to be

$$Rate = \left(0.163 \ \frac{liter}{mole \ sec}\right) [ICl] \ [H_2]$$

This reaction, therefore, is *first order* with respect to both ICl and H_2 (hence *second order,* overall) and has as its rate constant, $k = 0.163$ liter/mole sec.

How can a rate law such as this be determined? One way is to perform a series of experiments in which the initial concentration of each reactant is systematically varied. Once again we can use as our example the simple reaction

$$A \longrightarrow B$$

The rate law for this reaction would take the form

$$Rate = k[A]^x$$

If the reaction were first order, the value of x would be one, and the rate expression would then be

$$Rate = k[A]$$

This means that the rate of the reaction varies directly with the concentration of A raised to the first power. As a result, if we were to double the concentration of A from one experiment to another, we would also find that the rate would increase by a factor of two. We conclude, then, that when the reaction rate is doubled by doubling the concentration of a reactant, the order with respect to that reactant is one.

Suppose, now, that the rate law were, instead,

$$Rate = k[A]^2$$

In this instance, a twofold increase in the concentration would cause a *four-fold* increase in rate. To see this, let's imagine that the initial rate were measured with the concentration of A equal to, say m moles/liter. This rate would be given by

$$Rate = k(m)^2$$

Now, if the reaction were repeated with $[A] = 2m$, the rate would be

$$Rate = k(2m)^2$$

or

$$Rate = 4km^2$$

which is four times the previous rate. Thus, if the rate is increased by a factor of four when the concentration of a reactant is doubled, the reaction is second order with respect to that component. In a similar fashion we predict that the rate of a third order reaction would undergo an eightfold increase in rate when the concentration is doubled ($2^3 = 8$).

The following example illustrates how we can use these ideas to obtain the rate law for a reaction by varying the concentration of reactants.

Example 11.1
Below are some data collected in a series of experiments on the reaction of nitric oxide with bromine:

$$2NO(g) + Br_2(g) \longrightarrow 2NOBr(g)$$

at 273° C.

Experiment	Initial Concentration (mole/liter)		Initial Rate (mole/liter sec)
	NO	Br$_2$	
1	0.1	0.1	12
2	0.1	0.2	24
3	0.1	0.3	36
4	0.2	0.1	48
5	0.3	0.1	108

Determine the rate law for the reaction and compute the value of the rate constant.

Solution
In experiments 1 to 3 the concentration of NO is constant and the concentration of Br$_2$ is varied. When the concentration of Br$_2$ is doubled (experiments 1 and 2), the rate is increased by a factor of two; when it is tripled (experiments 1 and 3) the rate is increased by a factor of 3. We conclude, therefore, that the concentration of Br$_2$ appears to the first power in the rate law.

Comparing experiments 1 and 4 we see that by holding the Br$_2$ concentration constant, the rate increases by a factor of 4 when the NO concentration is multiplied by 2. Similarly, raising the concentration of NO by a factor of three causes a ninefold increase in rate (experiments 1 and 5). Thus the exponent of the NO concentration in the rate law is two. Therefore

$$Rate = k[NO]^2[Br_2]$$

The rate constant can be evaluated using the data from any of these experiments. Working with experiment 1 we have

$$12 \frac{mole}{liter \ sec} = k(0.10 \ mole/liter)^2(0.10 \ mole/liter)$$

$$12 \frac{mole}{liter \ sec} = k(0.0010 \ mole^3/liter^3)$$

Solving for k, we get

$$k = \frac{12 \text{ mole/liter sec}}{1.0 \times 10^{-3} \text{ mole}^3/\text{liter}^3} = 1.2 \times 10^4 \text{ liter}^2/\text{mole}^2 \text{ sec}$$

You might wish to verify for yourself that the same rate constant is obtained from the other data.

In this example, unlike the reaction between $ICl(g)$ and $H_2(g)$, the exponents in the rate law happen to be the same as the coefficients in the chemical equation. In general, however, there need be no direct correlation between these two. For instance the reaction between nitric oxide and hydrogen, whose balanced equation is

$$2NO(g) + 2H_2(g) \longrightarrow N_2(g) + 2H_2O(g)$$

has been experimentally determined to have the rate law

$$\text{Rate} = k[NO]^2[H_2]$$

which means that the reaction is third order overall. Again, please remember that the exponents of the rate equation can *only* be found experimentally.

Thus far we have discussed reactions obeying rate laws that are first or second order with respect to the individual reactants. This does not mean, however, that reactions that are zero or fractional order with respect to a single reactant do not exist. For instance, a reaction whose rate does not depend on the initial concentration of a reactant is zero order with respect to that species. In general, if the reaction

$$A \longrightarrow B$$

is zero order, the rate law would be written

$$\text{Rate} = k[A]^0$$

or

$$\text{Rate} = k$$

An example of this type of reaction is the decomposition of ammonia on a platinum or tungsten metal surface. The rate at which the ammonia decomposes is always the same regardless of its concentration. Another example of a zero order process (which perhaps strikes closer to home) is the elimination of ethyl alcohol by the body. Regardless of how much alcohol is present in the bloodstream its rate of expulsion from the body is constant. Thus the rate is independent of concentration.

Examples of reactions whose rate laws contain reactants with fractional order are briefly discussed later in this chapter under the heading of chain reactions.

In order for a chemical reaction to occur, the reacting molecules must collide with each other. This idea forms the basis of the **collision theory** of chemical kinetics. Basically, this theory states that the rate of a reaction is proportional to the number of collisions occurring each second between the reacting molecules.

11.3 Collision Theory

$$\text{Rate} \propto \frac{\text{number of collisions}}{\text{sec}} \qquad (11.6)$$

As we shall see shortly, this permits us to explain the dependence of reaction rate on the concentration of the reactants. In Section 11.5 we shall also see that not all of these collisions are effective in producing chemical change and that the number of collisions that are effective is dependent on the nature of the reactants and the temperature.

At this point, let's see how collision theory accounts for the concentration dependence observed in the last section. Suppose that we have a reaction that occurs by the collision of two molecules, such as

$$A + B \longrightarrow \text{products}$$

In this case we are assuming that we know precisely what occurs between A and B, that is, we assume for the sake of this discussion that the products are formed in collisions between A and B.

According to our theory, the rate of the reaction is proportional to the number of collisions each second between molecules of A and B. If the concentration of A is doubled, then the number of A-B collisions would also be doubled because there would be twice as many A molecules that can collide with B. Hence the rate is increased by a factor of two. Similarly, if the concentration of B were doubled, there would be a twofold increase in the number of A-B collisions and the rate would increase by a factor of two. From our previous discussion we conclude that the order with respect to each reactant is one and the rate law for this **bimolecular** (two molecule) collision process is

$$\text{Rate} = k[A][B]$$

What would happen if we had a reaction of the type,

$$2A \longrightarrow \text{products}$$

where reaction occurs by the collision of two A molecules? In this instance, if we double the concentration of A, we double the number of collisions that each *single* A molecule makes with its neighbors, because we have doubled the number of neighbors. We have also, however, doubled the number of A molecules that are colliding. The number of A-A collisions has therefore "doubly doubled," that is, increased by a factor of two squared. Consequently, the rate law for this bimolecular reaction between identical molecules is

$$\text{Rate} = k[A]^2$$

What we find, then, is that *if* we know what collision process is involved in the production of products, we can predict, on the basis of collision theory, what the rate law for that process will be. The exponents in the rate law are equal to the coefficients in the balanced equations for that simple reaction. We might ask, then, why is it necessary to determine the rate law for a reaction experimentally? Why can we not simply use the coefficients of the balanced overall equation to deduce the rate law?

11.4
Reaction Mechanism

The overall balanced equation for a reaction represents the net chemical change that occurs as the reaction proceeds to completion. This does not mean, however, that all of the reactants must come together simultaneously to undergo a change that produces the products. In fact the net change can actually represent the sum of a series of simple reactions. These simple reactions are referred to as **elementary processes.** The sequence of elementary processes that ultimately leads to the formation of the products is called the **reaction mechanism.**

For example, it appears that the reaction

$$2NO + 2H_2 \longrightarrow 2H_2O + N_2$$

proceeds by the three-step mechanism

$$2NO \longrightarrow N_2O_2$$
$$N_2O_2 + H_2 \longrightarrow N_2O + H_2O$$
$$N_2O + H_2 \longrightarrow N_2 + H_2O$$

The sum of these steps in the sequence does give us the overall balanced reaction. Such a mechanism is arrived at by bringing together both theory and experiment.

Suppose now that we wished to study the general reaction

$$2A + B \longrightarrow C + D$$

We first determine the rate law, perhaps by varying the concentration of A and B; let us say it turned out to be

$$\text{Rate} = k[A]^2[B]$$

Next, we attempt to propose a mechanism that, by the application of the principles of collision theory, gives us a predicted rate law that is the same as the one found by experiment.

Since we are beginners at proposing mechanisms, we might be tempted to propose a one-step mechanism in which two molecules of A and one of B come together simultaneously, that is, a three-body or **termolecular** collision. This process

$$2A + B \longrightarrow C + D$$

indeed leads to the rate law

$$\text{Rate} = k[A]^2[B]$$

which is the same as that found from experiment. We must now ask ourselves, is this a realistic mechanism? A simultaneous three-body (termolecular) collision is a very unlikely event, and it has been generally found that reactions that must proceed by such a path are very slow. As a result, a third-order reaction such as this, if it is fairly rapid, is usually interpreted as taking place by way of a series of simple bimolecular processes. (Back to the drawing board!)

One possible sequence of reactions is

$$2A \longrightarrow A_2$$
$$A_2 + B \longrightarrow C + D$$

Here we have two steps in which we propose that some relatively unstable intermediate, A_2, is first formed by the collision of two molecules of A. In a second step a reaction between A_2 and B produces the products C and D. Again the sum of these elementary processes gives us our net overall change.

Both of these reactions are unlikely to occur at the same rate. Let us suppose that the first reaction was slow and that once the intermediate, A_2, is formed it rapidly reacts with B in the second step to produce the products. If this were true, the rate at which the final products appear is actually determined by how fast A_2 is produced. This first step, then, serves as a "bottleneck" in the reaction path. We refer to this slowest step as the **rate determining step** in the reaction since it governs how rapidly the overall reaction takes place. Because the rate determining step is an elementary process (in this instance a bimolecular collision between two A molecules), we can predict, with the aid of collision theory, that the rate law should be

$$\text{Rate} = k[A]^2$$

If this is the rate law for the rate determining step, it will also be the rate law for the overall reaction. However, this rate law cannot be the correct one since it is not the same as the one determined from experiment. This does not necessarily mean that our mechanism is wrong. Let us see what we would expect to observe if the second step, instead of the first, were the slow step. In this instance the rate law is

$$\text{Rate} = k[A_2]\,[B]$$

However, this rate law contains the concentration of the proposed intermediate (A_2) and the experimental rate law contains only the concentration of reactants A and B. How can we express the concentration of A_2 in terms of A and B?

Once A_2 has been formed, it can react in either of two ways. Since we propose that A_2 is unstable (if it were stable we could isolate it and there would be no question at all about the path of the overall reaction) it can undergo decomposition to reform two molecules of A. The other possibility is that it undergoes a collision with B that leads to the formation of the products, C and D. Our mechanism therefore should include a reaction that allows A_2 to decompose, that is,

$$A_2 \longrightarrow 2A$$

Our total mechanism is now

$$2A \longrightarrow A_2$$
$$A_2 \longrightarrow 2A$$
$$A_2 + B \longrightarrow C + D$$

If the rate at which the intermediate is formed from reactant A is equal to the rate at which A is formed from intermediate A_2, then these two reactions represent a state of dynamic equilibrium. We could therefore write our first two equations as an equilibrium, which would take the form

$$2A \rightleftharpoons A_2$$

We are now back to a two-step mechanism in which the first step is an equilibrium. Our mechanism is now

$$2A \rightleftharpoons A_2 \qquad \text{fast}$$
$$A_2 + B \longrightarrow C + D \qquad \text{slow}$$

Since, in an equilibrium situation, the rate of the forward reaction ($Rate_f$) is equal to the rate of the reverse reaction ($Rate_r$),

$$Rate_f = k_f[A]^2 = Rate_r = k_r[A_2]$$

or simply

$$k_f[A]^2 = k_r[A_2]$$

Solving this equation for A_2 we have

$$[A_2] = \frac{k_f[A]^2}{k_r}$$

We can now substitute this expression for $[A_2]$ into our rate equation by assuming that reaction two is the rate-determining step; we find that the rate is

$$Rate = \frac{kk_f}{k_r}[A]^2[B]$$

Combination of these constants yields still another constant, let us say k'; therefore our rate expression is

$$Rate = k'[A]^2[B]$$

which does agree with the rate law found from experiment. Our proposed mechanism, therefore, appears to be a good one. However, a mechanism is, in essence, a theory. It is a sequence of steps that we dream up to explain the chemistry and to provide a rate law that agrees with experiment. It frequently happens, though, that more than one mechanism can be written to satisfy both criteria, so that we can never be certain we have truly discovered the actual path of the reaction. We can only hope to gather further information that either supports (or proves wrong) our guess.

If all collisions that take place in a reaction vessel were effective in producing chemical change, all chemical reactions, including biochemical ones, would be over almost instantaneously. Since living creatures have finite life spans, it is clear that some factor (or factors) must intervene to decrease reaction rates to a reasonable level. Consider, for example, the decomposition of hydrogen iodide.

$$2HI(g) \longrightarrow H_2(g) + I_2(g)$$

At a concentration of only 10^{-3} mole/liter of HI there are approximately 3.5×10^{28} collisions per liter per second at 500° C. This is equivalent to 5.8×10^4 moles of collisions per liter per second; if each of these collisions were effective, we would expect a rate of reaction of 5.8×10^4 mole/liter sec. Actually the rate under these conditions is only about 1.2×10^{-8} mole/liter sec; smaller by a factor of approximately 5×10^{12} than we would observe if all collisions led to reaction! Clearly, not all encounters between HI molecules result in the production of H_2 and I_2. In fact, only a very small fraction of the total number of collisions are effective. If we let Z be the total number of collisions that occur per second and f be the fraction of the total number of collisions that are effective, the rate of a reaction, according to collision theory, is

$$Rate = fZ \qquad (11.7)$$

The fraction, f, is determined by the energies of the molecules that collide and, as we shall see shortly, a certain minimum energy is required in order to cause reaction to occur. In addition to this, in many instances the molecules must also collide with the proper orientation. The decomposition of the hypothetical AB molecule, whose collisions result in the formation of A_2 and B_2, can serve as an example.

$$2AB \longrightarrow A_2 + B_2$$

In order to have the products A_2 and B_2 produced, the two atoms of A and two atoms of B must approach each other very closely so that A—A and B—B bonds can be formed. Suppose, now, that two A—B molecules come together in a collision oriented as shown below.

$$
\begin{array}{cc}
A & B \\
| & | \\
B & A
\end{array}
$$

We certainly do not expect this collision to be effective in forming the products. However, a collision in which the AB molecules are aligned as

$$
\begin{array}{cc}
A & A \\
| & | \\
B & B
\end{array}
$$

can lead to the creation of A—A and B—B bonds and, hence, to a net chemical change. Thus the number of effective collisions and, therefore, the rate of the reaction, is further decreased by a factor, p, that is a measure of the importance of the molecular orientations during collision.

$$\text{Rate} = pfZ \tag{11.8}$$

We have already seen that Z, the collision frequency, is proportional to the concentrations of the reacting molecules; therefore, in general,

$$Z = Z_0[A]^n[B]^m \text{ etc.}$$

where Z_0 is the collision frequency when all of the reactants are at unit concentration. Substituting this into Equation 11.8 gives us

$$\text{Rate} = pfZ_0[A]^n[B]^m \text{ etc.}$$

or

$$\text{Rate} = k[A]^n[B]^m \text{ etc.}$$

where $k = pfZ_0$. This then is the rate law derived from the principles of collision theory.

**11.6
Energy of
Activation**

To have a chemical change occur during a collision between two molecules, some bonds must be broken and new ones formed. This requires that the electron clouds of the reacting molecules interpenetrate one another very substantially so that the needed electron reshuffling can take place. A collision between two molecules is quite unlike a collision between two billiard balls. The electron cloud of a molecule has no sharp boundary and, as molecules approach each other, they experience a gradual increase in their mutual repulsion that causes them to slow dow, stop, and then fly apart again. Let us follow the kinetic and potential energies of a pair of molecules in a gas as they undergo such a collision.

As the molecules near one another, they begin to slow down and their kinetic energy decreases, with a corresponding rise in their potential energy. If the molecules were not moving very fast when they entered into this collision, they would stop and reverse direction before any appreciable interpenetration of their electron clouds has taken place. As a result, slow-moving molecules simply bounce off one another without reacting. Very rapidly moving molecules, on the other hand, can penetrate each other so that the products are able to be formed. These high-speed molecules have large kinetic energies that yield large increases in potential energy during collision. When the products fly apart, this potential energy decreases as the product molecules gain velocity (and hence kinetic energy). Thus only fast-moving molecules are able to react. In fact, there must be some minimum kinetic energy possessed jointly by the two molecules that is available to be transformed into potential energy. The minimum energy that must be available in a collision is called the **activation energy, E_a.**

The change in potential energy that takes place during the course of a reaction is shown in Figure 11.3. The horizontal axis is called the **reaction coordinate** and positions along this axis represent the extent to which the reaction has progressed toward completion. On the left of this potential energy diagram we find two molecules of AB. As they approach each other, their potential energy increases to a maximum. As we continue toward the right along the reaction coordinate, the potential energy of the system decreases as the products, A_2 and B_2, move apart. When the A_2 and B_2 molecules are finally separated from one another, the total potential energy drops to essentially a constant value.

The activation energy for the decomposition of AB corresponds to the difference between the energy of the reactants and the maximum on the potential energy curve. Slow-moving molecules of AB do not possess sufficient energy to overcome this potential energy barrier, while fast-moving ones do.

In Figure 11.3 we have drawn the potential energy of the products to be lower than that of the reactants. The difference between them corresponds to the heat of reaction. In this case, since the products are at a lower energy than the reactants, the reaction is exothermic. The energy released appears as an

Figure 11.3

Potential energy diagram for an exothermic reaction.

Figure 11.4

Potential energy diagram for an endothermic reaction.

increase in the kinetic energy of the products; therefore, the temperature of the system rises as the reaction progresses.

In an endothermic reaction, depicted in Figure 11.4, the products are at a higher potential energy than the reactants. The net absorption of energy that takes place as the products are formed occurs at the expense of kinetic energy. Consequently there is a net overall decrease in the average kinetic energy as the reaction proceeds and the reaction mixture becomes cool.

11.7
Effect of Temperature on Reaction Rate

In nearly every instance an increase in temperature causes an increase in the rate of reaction and, as a very general rule of thumb, the rate is about doubled, for many reactions, by a 10-degree rise in temperature. How can this behavior be explained?

According to kinetic theory, in any system there is a distribution of kinetic energies. In the last section we interpreted the activation energy to be the minimum kinetic energy required for a collision to be effective. All molecules having kinetic energies higher than this minimum are, therefore, capable of reacting. This can be illustrated for the kinetic energy distribution in a system, as shown in Figure 11.5. The total fraction of all of the molecules having energies equal to or greater than E_a corresponds to the shaded portion of the area under the curve. If we compare this area for two different temperatures, we see that the total fraction of molecules with sufficient kinetic energy to undergo effective collisions is greatest at the higher temperature. As a result, the number of molecules that are capable of undergoing reaction

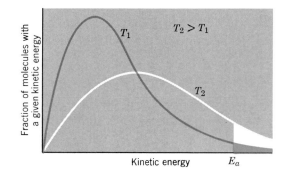

Figure 11.5

The effect of temperature on the number of molecule having kinetic energies greater than E_a.

increases with increasing temperature and, consequently, so does the reaction rate.

The magnitude of the rate constant, which is the rate of the reaction when all of the concentrations have a value of one, depends on the size of E_a and also on the absolute temperature. For instance, k is small when the activation energy is very large or when the temperature of the reaction mixture is low. Quantitatively, k is related to E_a and T by the equation

$$k = Ae^{-E_a/RT} \qquad (11.9)$$

where A is a proportionality constant, R is the gas constant, and e is the base of the natural logarithms (Appendix C). This relationship is known as the Arrhenius equation after its discoverer, the Swedish chemist, Svanté Arrhenius.[1]

To measure the activation energy the rate constant must be measured at at least two different temperatures. Taking the natural logarithm of Equation 11.9 gives

$$\ln k = \ln A - \frac{E_a}{RT} \qquad (11.10)$$

We can compare this equation to the equation for a straight line.

$$\ln k = \ln A - \frac{E_a}{R}\left(\frac{1}{T}\right)$$
$$\begin{array}{ccccc} \updownarrow & & \updownarrow & \updownarrow & \updownarrow \\ y & = & b & +m & x \end{array}$$

Thus, a plot of $\ln k$ versus $1/T$ gives a straight line whose slope is equal to $-E_a/R$ and whose intercept with the ordinate (i.e., the vertical axis) is $\ln A$. (Figure 11.6).

We can also obtain E_a from k at two temperatures by direct computation. For any temperature, T_1, Equation 11.9 becomes

$$k_1 = Ae^{-E_a/RT_1}$$

and for any other temperature, T_2, we can write

$$k_2 = Ae^{-E_a/RT_2}$$

Dividing k_1 by k_2, we have

$$\frac{k_1}{k_2} = \frac{Ae^{-E_a/RT_1}}{Ae^{-E_a/RT_2}}$$

or

$$\frac{k_1}{k_2} = e^{-E_a/R}\left(\frac{1}{T_1} - \frac{1}{T_2}\right)$$

which can be rewritten as

$$\frac{k_1}{k_2} = e^{-E_a/R}\left(\frac{T_2 - T_1}{T_2 T_1}\right)$$

[1] Arrhenius received the third Nobel prize ever awarded in chemistry in 1903.

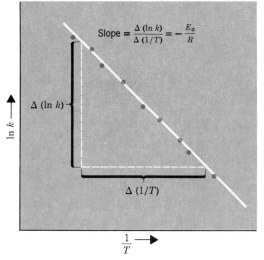

$$\text{Slope} = \frac{\Delta(\ln k)}{\Delta(1/T)} = -\frac{E_a}{R}$$

Figure 11.6
Graphical determination of the activation energy, E_a. Points on the line represent the natural logarithms of experimentally measured rate constants at various temperatures. We determine the slope of the straight line that best fits the experimental data.

Taking the natural logarithm of both sides, we see that

$$\ln \frac{k_1}{k_2} = -\frac{E_a}{R}\left(\frac{T_2 - T_1}{T_2 T_1}\right)$$

and converting to common logarithms (base 10 logarithms) we have

$$\log \frac{k_1}{k_2} = \frac{-E_a}{2.303\,R}\left(\frac{T_2 - T_1}{T_2 T_1}\right)$$

Multiplying both sides by -1 and inverting the fraction on the left side, we obtain

$$\log \frac{k_2}{k_1} = \frac{E_a}{2.303\,R}\left(\frac{T_2 - T_1}{T_1 T_2}\right) \tag{11.11}$$

Chemists are interested in measuring E_a because it provides them with information about the energy changes that take place during an effective collision. In addition, Equation 11.11 can be used to calculate the rate constant at some specific temperature if E_a and k at some other temperature are known.

Example 11.2
At 300° C the rate constant for the reaction

is 2.41×10^{-10} sec^{-1}. At 400° C k equals 1.16×10^{-6} sec^{-1}. What are the values of E_a and A for this reaction?

Solution

Solving Equation 11.12 for E_a we have

$$E_a = 2.303 \, R \left(\frac{T_1 T_2}{T_2 - T_1} \right) \log \frac{k_2}{k_1}$$

If we take

$$T_1 = 300 + 273 = 573° \, K \qquad k_1 = 2.41 \times 10^{-10}$$

and

$$T_2 = 400 + 273 = 673° \, K \qquad k_2 = 1.16 \times 10^{-6}$$

Substituting into the equation for E_a (using $R = 1.99$ cal/mole °K)

$$E_a = 2.303(1.99 \text{ cal/mole °K}) \left(\frac{573 \times 673}{100} \right) \log \left(\frac{1.16 \times 10^{-6}}{2.41 \times 10^{-10}} \right)$$

$$E_a = 65.4 \text{ kcal/mole}$$

We can compute A from the equation

$$k = Ae^{-E_a/RT}$$

Taking the natural logarithm of both sides of the equation

$$\ln k = \ln A - \frac{E_a}{RT}$$

or, in terms of common logarithms,

$$2.303 \log k = 2.303 \log A - E_a/RT$$

Solving for $\log A$,

$$\log A = \log k + \frac{E_a}{2.303 \, RT}$$

Substituting the values for 300° C,

$$\log A = \log (2.41 \times 10^{-10}) + \frac{65,400 \text{ cal/mole}}{2.303(1.99 \text{ cal/mole° K})(573° \text{ K})}$$

$$\log A = (0.382 - 10.000) + (24.90)$$
$$\log A = 15.28$$

If we take the antilog,

$$A = 1.9 \times 10^{15} \text{ sec}^{-1}$$

Note that A must have the same units as k.

The species that exists at the top of the potential energy barrier (Figure 11.3) during an effective collision corresponds to neither the reactants nor the products but, instead, to some highly unstable combination of atoms that we speak of as the **activated complex.** This activated complex is said to exist in a **transition state** along the reaction coordinate (hence the name **transition state theory**).

**11.8
Transition State
Theory**

Transition state theory views chemical kinetics in terms of the energy and geometry of the activated complex which, once it has formed, can come apart to yield the reactants again or go on to produce the products. For example, let us examine again the decomposition of hypothetical AB molecules to produce A_2 and B_2. The change that takes place along the reaction coordinate can be represented as

$$\begin{matrix} A \\ | \\ B \end{matrix} + \begin{matrix} A \\ | \\ B \end{matrix} \rightleftharpoons \begin{bmatrix} A\cdots A \\ \vdots \quad \vdots \\ B\cdots B \end{bmatrix} \longrightarrow \begin{matrix} A-A \\ + \\ B-B \end{matrix}$$

where we have used solid dashes to denote ordinary covalent bonds and dotted lines to symbolize the partially broken and partially formed bonds in the transition state (which is enclosed within brackets). Figure 11.7 illustrates this change as it occurs on our potential energy diagram for the reaction.

If the potential energy of the transition state is very high, then a great deal of energy must be available in a collision to form the activated complex. This results in a high activation energy and consequently to a slow reaction. If it were possible somehow to produce an activated complex whose energy is closer to that of the reactants, the decreased activation energy would lead to a faster reaction rate.

11.9
Catalysts

A **catalyst** is a substance that, by its participation in a chemical reaction, provides a low energy alternate pathway (mechanism) for the production of the products, and that is recovered afterward from the reaction mixture chemically unchanged. For any given reaction the activation energy with the catalyst present is lower than when the catalyst is absent, as shown in Figure 11.8. Note that the energy curve of the catalyzed reaction is drawn along a different reaction coordinate to emphasize that a different mechanism is involved. In addition, the energy barrier for the catalyzed path is lower than for the uncatalyzed reaction. This smaller activation energy means that in the reaction mixture there is a greater total fraction of molecules possessing sufficient kinetic energy to react (Figure 11.9). Therefore in the presence of the catalyst there are an increased number of effective collisions. Of course, an increased number of effective collisions means a greater reaction rate.

Since a catalyst emerges chemically unchanged from a chemical reac-

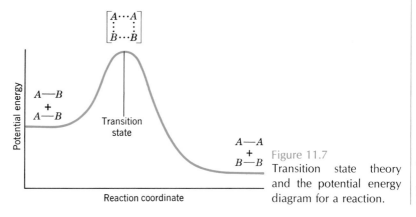

Figure 11.7
Transition state theory and the potential energy diagram for a reaction.

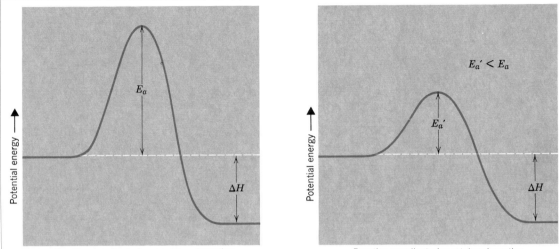

Figure 11.8

Effect of a catalyst on the potential energy diagram. The catalyst changes the reaction mechanism by providing a different, low energy mechanism for the formation of the products. ΔH is same for each path (it must be so because ΔH is a state function).

tion, it does not appear either as a reactant or a product in the overall balanced chemical equation. Instead, its presence is indicated by writing its name or formula over the arrow. For example, in the preparation of oxygen by the thermal decomposition of $KClO_3$, manganese dioxide (MnO_2) is added to speed up the reaction and to allow the decomposition to proceed rapidly at a relatively low temperature. By contrast, in the absence of MnO_2 the reaction is slow and the $KClO_3$ must be heated to high temperatures to cause it to decompose. Analysis of the reaction mixture after the evolution of oxygen has ceased reveals that all of the MnO_2 added initially is still present, showing that MnO_2 has served as a catalyst. The equation for the catalyzed reaction is given as

$$2KClO_3 \xrightarrow{\ \ MnO_2\ \ } 2KCl + 3O_2$$

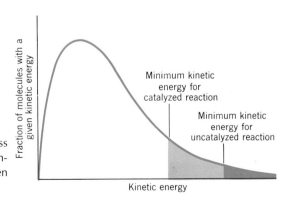

Figure 11.9

More molecules possess the minimum kinetic energy for a reaction when the catalyst is present.

Even though a catalyst does not change the overall stoichiometry of a reaction, it does participate chemically by being consumed at one stage in the mechanism and being produced again at a later stage. This regeneration of the catalyst permits the same catalyst to be used over and over again; therefore even a small amount of catalyst can have very profound effects on reaction rate. This phenomenon is particularly significant in biological systems where practically every reaction is catalyzed by very small quantities of highly specific biochemical catalysts called enzymes.

Catalysts may be broadly classified into two categories: homogeneous and heterogeneous catalysts. A **homogeneous catalyst** is present in the same phase as the reactants and can serve to speed up the reaction by forming a reactive intermediate with one of the reactants. For example, the decomposition of t-butyl alcohol, $(CH_3)_3COH$, to produce water and isobutene, $(CH_3)_2C = CH_2$,

$$(CH_3)_3COH \longrightarrow (CH_3)_2C = CH_2 + H_2O$$

is catalyzed by the presence of small amounts of HBr. In the absence of HBr the activation energy for the reaction is 65.5 kcal/mole and below 450° C the reaction takes place at a barely perceptible rate. In the presence of HBr an activation energy of only 30.4 kcal/mole is found, and it is possible that the catalyzed reaction proceeds by an attack of HBr on the alcohol

$$(CH_3)_3COH + HBr \longrightarrow (CH_3)_3CBr + H_2O$$

followed by the rapid decomposition of the t-butyl bromide

$$(CH_3)_3CBr \longrightarrow (CH_3)_2C = CH_2 + HBr$$

Thus HBr provides an alternative low energy path for the reaction and the mechanism for the reaction when HBr is present is different than when it is absent.

A **heterogeneous catalyst** is not in the same phase as the reactants, but provides a favorable surface upon which the reaction can take place. An example of a reaction whose rate is increased by the presence of a heterogeneous catalyst is the reaction between hydrogen and oxygen to produce water. In the introduction to this chapter we pointed out that this reaction proceeds at a very slow rate when the two gases are mixed at room temperature. However, it has been found that the reaction proceeds at an appreciable rate when metals such as nickel, copper, or silver are present.

Heterogeneous catalysts appear to function through a process whereby reactant molecules are adsorbed upon a surface where the reaction then takes place. The high reactivity of hydrogen, in the presence of certain metals, for example, is thought to occur by the adsorption of H_2 molecules on the catalytic surface. On the surface of the metal the bonds between hydrogen atoms are apparently stretched or broken, as shown in Figure 11.10, so that the metal surface actually behaves as if it contains highly reactive hydrogen atoms.

Unless the reactant molecules can be adsorbed on the catalyst, no increase in reaction rate can occur. A substance whose presence during a reaction interferes with the adsorption process will therefore reduce the effectiveness of the catalysts and is thus called an **inhibitor.** These substances, by

Figure 11.10

Production of H atoms on a metal surface. H_2 molecule collides with the surface where it is adsorbed and dissociates to produce H atoms.

being strongly adsorbed on the catalytic surface, decrease the available space upon which the reaction can occur. In some cases the catalyst eventually becomes useless and is said to be poisoned. The destruction of catalytic activity by poisoning is very important in biological systems, as we shall see in Chapter 21.

There have been many commercial and industrial applications of heterogeneous catalysis. For example, small portable flameless heaters can be purchased (for heating camping tents and the like) in which the fuel and oxygen combine on a catalytic surface. The flameless combustion evolves the same amount of heat that would be generated if the fuel were burned directly. Since no flame is involved, however, the heaters are much safer to operate.

Another very important application is in the control of auto exhaust emissions. Catalytic mufflers are in use that employ a mixed metal oxide bed over which the exhaust gases pass after they are mixed with additional air (Figure 11.11). The catalyst quite effectively promotes the oxidation of CO and hydrocarbons to harmless CO_2 and H_2O. Unfortunately it does not reduce the emission of nitrogen oxides, another important pollutant. Catalytic mufflers suffer from the disadvantage that they are poisoned by lead. As a result, leadfree fuels must be employed in autos fitted with this type of antipollution device.

The reactions that we have discussed up to now have been rather simple and straightforward with uncomplicated rate laws. There are some reactions,

Figure 11.11

The catalytic muffler.

usually with very complex kinetics, that take place by way of an extremely reactive intermediate, such as a free atom or a **free radical** (a neutral group of atoms or an ion containing one or more unpaired electrons). These species can be produced either thermally (at high temperature) or by the absorption of light of an appropriate wavelength; once created they can sometimes react with other molecules to produce the products *plus* yet another free atom or radical. This process once initiated, can be repeated over and over, making the reaction self-propagating. The entire series of reactions following the production of the very reactive intermediate is called a **chain reaction.**

A chain reaction mechanism has been suggested to explain the rate equation observed for the reaction between hydrogen and bromine. The overall reaction is

$$H_2 + Br_2 \longrightarrow 2HBr$$

If this reaction proceeded simply by a bimolecular collision between H_2 and Br_2 the expected rate law would be

$$Rate = k \, [H_2] \, [Br_2]$$

However, the actual rate law turns out to be

$$Rate = k \, \frac{[H_2] \, [Br_2]^{1/2}}{1 + [HBr]/k'[Br_2]}$$

which is very complex indeed. A mechanism that has been proposed to account for this rate law is the chain reaction

1.	$Br_2 \longrightarrow 2Br$	initiation
2.	$Br + H_2 \longrightarrow HBr + H$ ⎱	propagation
3.	$H + Br_2 \longrightarrow HBr + Br$ ⎰	
4.	$H + HBr \longrightarrow H_2 + Br$	inhibition
5.	$2Br \longrightarrow Br_2$	termination

Reaction 1 is the thermal decomposition of diatomic bromine molecules to produce bromine atoms (the reactive intermediate). The overall reaction proceeds very slowly when the two reactants are mixed at room temperature. However, at high temperature reaction 1 takes place to an appreciable extent and rapidly sets off the remaining reactions. Step 1 is, therefore, the **initiation step.** In steps 2 and 3 the product HBr is formed as well as additional free atoms that serve to keep the reaction going. These steps then are **propagation steps** in the chain. Step 5, which leads only to the formation of a stable species, serves to end the chain and is known as the **termination step.** Step 4 is called an **inhibition step** since its occurrence removes product and thus decreases the overall rate of production of HBr. It is included in the mechanism because the presence of HBr decreases the reaction rate (note the appearance of [HBr] in the denominator of the rate law).

In general, chain reactions are very rapid; in fact, many explosive reactions appear to occur by chain mechanisms. The production of a single reactive intermediate produces many product molecules before the chain is terminated. Consequently the rate of production of the products is many times greater than the rate of the initiation step alone.

11.1 What are meant by reaction rate, order of reaction, rate law, and rate constant?

11.2 For each of the following reactions, how would we express the reaction rate in terms of the disappearance of the reactants and the appearance of the products? Predict the role of the coefficients in the balanced overall equation in determining the relative rates of disappearance of reactants and formation of products.
(a) $2H_2 + O_2 \rightarrow 2H_2O$
(b) $2NOCl \rightarrow 2NO + Cl_2$
(c) $NO + O_3 \rightarrow NO_2 + O_2$
(d) $H_2O_2 + H_2 \rightarrow 2H_2O$

11.3 The rate constant for the reaction

$$2\ ICl + H_2 \longrightarrow I_2 + 2\ HCl$$

is 1.63×10^{-1} liter/mole sec. The rate law is given by

$$Rate = k\ [ICl]\ [H_2]$$

What is the rate of the reaction for each of the sets of concentrations given below?

ICl Concentration (mole/liter)	H$_2$ Concentration (mole/liter)
0.25	0.25
0.25	0.50
0.50	0.50

11.4 What is the order with respect to each reactant and the overall order of the reactions described by the following rate laws:
(a) $Rate = k_1\ [A]\ [B]$
(b) $Rate = k_2\ [E]^2$
(c) $Rate = k_3\ [G]^2\ [H]^2$

11.5 For the decomposition of dinitrogen pentoxide,

$$2N_2O_5 \longrightarrow 4NO_2 + O_2$$

the following data were collected.

N$_2$O$_5$ Concentration (mole/liter)	Time (sec)
5.00	0
3.52	500
2.48	1000
1.75	1500
1.23	2000
0.87	2500
0.61	3000

(a) Make a graph of the concentration of N$_2$O$_5$ versus time. Draw tangents to the curve at $t = 500$, 1000, and 1500 seconds. Determine the rate at these different reaction times.
(b) Determine the value of the rate constant at 500, 1000, and 1500 seconds, given the rate law, Rate = $k[N_2O_5]$.

11.6 The reaction of ICl with H$_2$ follows the equation,

$$2\ ICl + H_2 \longrightarrow I_2 + 2\ HCl$$

The following data were collected at 230° C.

Initial ICl Concentration (mole/liter)	Initial H$_2$ Concentration (mole/liter)	Initial Rate (mole/liter sec)
0.10	0.10	1.6×10^{-3}
0.10	0.20	3.2×10^{-3}
0.20	0.10	6.4×10^{-3}
0.30	0.10	1.4×10^{-2}

(a) What is the rate law for the reaction?
(b) What is the value of the rate constant (be sure to give the proper units)?

11.7 At 27° C, the reaction, $2NOCl \rightarrow 2NO + Cl_2$, is observed to exhibit the following dependence of rate on concentration.

Initial NOCl Concentration (mole/liter)	Initial Rate (mole/liter sec)
0.30	3.60×10^{-9}
0.60	1.44×10^{-8}
0.90	3.24×10^{-8}

(a) What is the rate law for the reaction?
(b) What is the rate constant?
(c) By what factor would the rate increase if the initial concentration of NOCl were increased from 0.30 to 0.45 *M*?

11.8 What is meant by reaction mechanism?

11.9 A mechanism for the reaction, $2NO + Br_2 \rightarrow 2NOBr$, has been suggested to be

Step 1 $NO + Br_2 \longrightarrow NOBr_2$
Step 2 $NOBr_2 + NO \longrightarrow 2NOBr$

(a) What would be the rate law for the reaction if the first step in this mechanism were slow and the second fast?
(b) What would be the rate law if the second step were slow, with the first reaction being a rapidly established dynamic equilibrium?
(c) Experimentally, the rate law has been found to be Rate = $k[NO]^2[Br]$. What can we conclude about the relative rates of steps 1 and 2?
(d) Why do we not prefer a simple, one-step mechanism,
$NO + NO + Br_2 \longrightarrow 2NOBr$
(e) Can we, on the basis of the experimental rate law, definitely exclude the mechanism in part (d)?

11.10 How does collision theory account for the dependence of rate on the concentration of the reactants?

11.11 How do we know that not all collisions between reactant molecules lead to chemical change? What determines whether a particular collision will be effective?

11.12 Explain qualitatively, in terms of the kinetic theory, why an increase in temperature leads to an increase in reaction rate.

11.13 Draw a potential energy diagram for an endothermic reaction. Indicate on the drawing (a) the potential energy of the reactants, (b) the potential energy of the products, (c) the energies of activation for the forward and reverse reaction, (d) the heat of reaction.

11.14 What are meant by the terms, transition state and activated complex? Where on the potential energy diagram for a reaction will we find the transition state?

11.15 The rate constants for the reaction between ICl and H_2 (question 11.6) at 230 and 240° C have been found to be 0.163 and 0.348 liter²/mole² sec, respectively. What are the values of E_a and A for this reaction?

11.16 A chemist was able to determine that the rate of a particular reaction at 100° C was four times faster than at 30° C. Calculate the approximate energy of activation for the reaction.

11.17 The activation energy for the reaction, $HI + CH_3I \rightarrow CH_4 + I_2$ is 33.4 kcal/mole. At 200° C the rate constant has a value of 7.9×10^{-4} liter/mole sec. What is the rate constant at 300° C?

11.18 What effect does a catalyst have on
(a) The heat of reaction.
(b) The potential energy of the reactants.
(c) The transition state.
(d) The activation energy.

11.19 What is a heterogeneous catalyst? How does it function? What is an inhibitor?

11.20 The decomposition of acetaldehyde, CH_3CHO, follows the overall reaction

$$CH_3CHO \longrightarrow CH_4 + CO$$

with small amounts of H_2 and C_2H_6 also being produced. The reaction is thought to proceed by a chain reaction involving free radicals (a free radical is indicated using a dot to represent its unpaired electron). A proposed mechanism is

(1) $CH_3CHO \rightarrow CH_3 \cdot + CHO\cdot$
(2) $2CH_3 \cdot \rightarrow C_2H_6$
(3) $CHO \cdot \rightarrow H\cdot + CO$
(4) $H \cdot + CH_3CHO \rightarrow H_2 + CH_3CO\cdot$
(5) $CH_3 \cdot + \quad CH_3CHO \rightarrow CH_4 + CH_3CO\cdot$
(6) $CH_3CO \cdot \rightarrow CH_3 \cdot + CO$

Identify (a) the initiation step, (b) the propagation step(s), and (c) the termination step(s).

11.21 Insects, which are cold-blooded animals whose changes in body temperature tend to follow changes in the temperature of their environment, become quite sluggish in cool weather. On the basis of chemical kinetics, explain this phenomenon.

12
Chemical
Equilibrium

When chemical reactions occur spontaneously, they continue to proceed until a state of dynamic equilibrium is achieved. At equilibrium both the forward and reverse reactions are taking place at the same rate, and the concentrations of the reactants and products no longer change with time, as illustrated in Figure 12.1.

All chemical systems tend toward equilibrium. In this chapter we shall explore the quantitative relationships that can be used to describe the equilibrium state, and we shall see how the principles of kinetics and thermodynamics can be applied to a description of equilibrium.

12.1
The Law of Mass
Action

It has been found experimentally that a very simple relationship governs the relative proportions of reactants and products present in an equilibrium system. For the general reaction

$$aA + bB \rightleftharpoons eE + fF$$

it is observed that, at constant temperature, the condition that is fulfilled at equilibrium is

$$\frac{[E]^e[F]^f}{[A]^a[B]^b} = K \tag{12.1}$$

where, the quantities written within square brackets denote equilibrium molar concentration. The quantity, K, is a constant, called the **equilibrium constant** and the entire relationship, discovered in 1866 by the Norwegian chemists Guldberg and Waage, is known as the **law of mass action.**

The fraction appearing to the left of the equal sign in Equation 12.1 is called the **mass action expression,** and it is constructed using the coefficients in the balanced chemical equation as exponents on the appropriate concentrations. For instance, if we consider the nitrogen fixation reaction used industrially in the production of ammonia and nitrogen fertilizers,

$$N_2(g) + 3H_2(g) \rightleftharpoons 2NH_3(g)$$

the mass action expression would be written as

$$\frac{[NH_3]^2}{[N_2][H_2]^3}$$

This fraction will, of course, always have some numerical value for a system containing these three gases. For example, if N_2 and H_2 are introduced into a container and are permitted to react, we would initially have no NH_3 and the

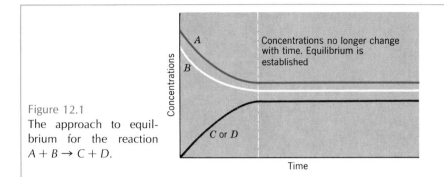

Figure 12.1
The approach to equilibrium for the reaction $A + B \rightarrow C + D$.

value of the mass action expression would be zero. As NH_3 is produced the fraction grows larger until, when equilibrium is reached, the fraction becomes equal to a value that we call the equilibrium constant, K.

The important point about the law of mass action, however, is that, at a given temperature, any system containing N_2, H_2, and NH_3 in equilibrium will have its mass action expression equal to the *same* number. *There are no restrictions on the individual concentrations of any reactant or product.* The only requirement for equilibrium is that when these concentrations are substituted into the mass action expression, the fraction is equal to K. The data in Table 12.1 illustrate this point.

We have written the mass action expression with the concentration of the products in the numerator and those of the reactants in the denominator. Since this fraction is equal to a constant at equilibrium, certainly its reciprocal also is. Thus

$$\frac{[NH_3]^2}{[N_2][H_2]^3} = K$$

and

$$\frac{[N_2][H_2]^3}{[NH_3]^2} = \frac{1}{K} = K'$$

Table 12.1
Equilibrium Concentrations (in moles per liter) at 500° C and the Mass Action Expression for the Reaction: $3H_2(g) + N_2(g) \rightleftharpoons 2NH_3(g)$

$[H_2]$	$[N_2]$	$[NH_3]$	$[NH_3]^2/[H_2]^3[N_2]$
0.15	0.75	1.23×10^{-2}	5.98×10^{-2}
0.50	1.00	8.66×10^{-2}	6.00×10^{-2}
1.35	1.15	4.12×10^{-1}	6.00×10^{-2}
2.43	1.85	1.27	6.08×10^{-2}
1.47	0.75	3.76×10^{-1}	5.93×10^{-2}

Either equation is a valid description of the equilibrium state. However, chemists have chosen, somewhat arbitrarily, always to write the equilibrium expression with the concentration of the products appearing in the numerator. This allows us then to tabulate equilibrium constants without the necessity of always having to state explicitly the form of the mass action expression.

12.2 The Equilibrium Constant

The equilibrium constant is a quantity that must be calculated from experimental data. One method, involving the use of standard free energies of formation to determine a *thermodynamic equilibrium constant,* is outlined in Section 12.4. Another method involves the direct measurement of equilibrium concentrations that can then be inserted into the equilibrium expression to obtain a numerical value for K. We shall look at a sample calculation of this type in Section 12.7.

Simply examining the magnitude of K provides us with information about the extent to which a reaction proceeds toward completion. For example, the reaction

$$H_2(g) + Cl_2(g) \rightleftharpoons 2HCl$$

has an equilibrium constant, at 27° C, of 3.2×10^{16}. The equilibrium expression for the reaction is therefore

$$\frac{[HCl]^2}{[H_2][Cl_2]} = K = 3.2 \times 10^{16}$$

Because K is so very large, the numerator in the mass action expression must be much larger than the denominator. This can only be true if the concentration of HCl is much, much greater than the concentrations of H_2 and Cl_2. We conclude, therefore, that when this reaction occurs it proceeds far toward completion.

Similarly, we conclude that the decomposition of water at 50° C

$$2H_2O(g) \rightleftharpoons 2H_2(g) + O_2(g)$$

which has an equilibrium constant of about 6×10^{-28}, takes place to only a very small degree; because, in order to have such a very small value of K the concentration of the products must also be very small. It is a general rule, then, that when K is large the position of equilibrium lies far to the right. Conversely, when K is small only relatively small amounts of the products are present in the system at equilibrium.

12.3 Kinetics and Equilibrium

Guldberg and Waage, in 1866, realized that the rate of a chemical reaction depends on the concentration of the reactants. In fact, they believed that the rate of a chemical reaction was proportional to the concentrations of the reactants raised to powers *equal* to their coefficients in the balanced equation. That is, they believed that for the reaction

$$aA + bB \rightleftharpoons eE + fF$$

the rate of the forward reaction is

$$\text{Rate}_f = k_f[A]^a[B]^b$$

while the rate of the reverse reaction is

$$\text{Rate}_r = k_r[E]^e[F]^f$$

Once a dynamic equilibrium is established in this system, the rate of the forward and reverse reactions must, of course, be equal so that we can write

$$k_f[A]^a[B]^b = k_r[E]^e[F]^f$$

This can be easily rearranged to give

$$\frac{k_f}{k_r} = \frac{[E]^e[F]^f}{[A]^a[B]^b}$$

Since the ratio of two constants, k_f/k_r, is also a constant, let's call it K, we obtain the equilibrium condition for the reaction

$$K = \frac{[E]^e[F]^f}{[A]^a[B]^b}$$

which is the law of mass action.

We know now, of course, that one cannot predict the rate law for a chemical reaction based on the balanced overall equation. Does this affect our kinetic interpretation of the law of mass action?

The answer to this question is no. Let us suppose we had the reaction

$$2A + B \longrightarrow C + D$$

The equilibrium condition, we know, would be written as

$$\frac{[C][D]}{[A]^2[B]} = K$$

We also know that, *if* the reaction proceeded by a one-step mechanism, the rate of the forward reaction is

$$\text{Rate}_f = k_f[A]^2[B]$$

while the rate of the reverse reaction would be given as

$$\text{Rate}_r = k_r[C][D]$$

Following the same argument as before, we find that

$$\frac{[C][D]}{[A]^2[B]} = \frac{k_f}{k_r} = K$$

Suppose, now, that this reaction proceeded in two steps, each of which is reversible.

Step 1 $\qquad\qquad 2A \underset{k_{1r}}{\overset{k_{1f}}{\rightleftharpoons}} A_2$

Step 2 $\qquad\quad A_2 + B \underset{k_{2r}}{\overset{k_{2f}}{\rightleftharpoons}} C + D$

At equilibrium the rates of the forward and reverse reactions in *each* step must be the same; therefore, for step 1 we have

$$K_1 = \frac{k_{1f}}{k_{1r}} = \frac{[A_2]}{[A]^2}$$

and similarly for step 2

$$K_2 = \frac{k_{2f}}{k_{2r}} = \frac{[C][D]}{[A_2][B]}$$

If we now multiply K_1 and K_2 together, we can eliminate the concentration of the intermediate A_2 and obtain

$$K_1 K_2 = K = \frac{[\cancel{A_2}]}{[A]^2} \times \frac{[C][D]}{[\cancel{A_2}][B]}$$

or

$$K = \frac{[C][D]}{[A]^2[B]}$$

Thus we see that regardless of the mechanism of the reaction, we always arrive at the same requirement for equilibrium.

12.4
Thermodynamics and Chemical Equilibrium

In Chapter 10 we found that the free energy change for any process becomes equal to zero at equilibrium. As applied to chemical equilibrium this means that the total free energy of the products is precisely equal to the total free energy of the reactants. Let us see how we can use this principle to arrive at a quantitative expression of equilibrium.

Again let us consider the general reaction,

$$aA + bB \longrightarrow eE + fF$$

As this reaction proceeds, A and B are consumed and E and F are formed until finally equilibrium is reached, at which point the reactants and products will most likely *not* be in their standard states.

The free energy of a substance varies in a more or less logarithmic fashion with its concentration; the relationship being exact if we employ *effective* concentrations, or **activities,** instead of actual stoichiometric concentrations. The free energy per mole of a substance in a state other than its standard state is

$$G_X = G_X^\circ + 2.303\,RT \log a_X$$

or

$$G_X = G_X^\circ + RT \ln a_X$$

where G_X° is the standard state free energy of X, R, and T have their usual meaning, and $\ln a_X$ is the natural logarithm (see Appendix C) of the activity of X. For n moles of X under these conditions, then, the total free energy would be

$$nG_X = nG_X^\circ + nRT \ln a_X$$

which can also be written as

$$nG_X = nG_X^\circ + RT \ln a_X^n$$

With this as background let us turn to our problem of chemical equilibrium. In our reaction we have a moles of A and b moles of B reacting to produce e moles of E and f moles of F. The total free energy of each of the reactants being consumed and products being formed is therefore the number of moles of each multiplied by their free energy per mole.

$$aG_A = aG_A^\circ + RT \ln a_A{}^a \tag{12.2}$$

$$bG_B = bG_B^\circ + RT \ln a_B{}^b \tag{12.3}$$

$$eG_E = eG_E^\circ + RT \ln a_E{}^e \tag{12.4}$$

$$fG_f = fG_F^\circ + RT \ln a_F{}^f \tag{12.5}$$

To obtain the total free energy change, we must add up the total free energy of the products and then subtract the total free energy of the reactants. This is equivalent to adding together Equations 12.4 and 12.5, and then subtracting both Equations 12.2 and 12.3. On the left of the equal sign we get

$$\Delta G = eG_E + fG_F - aG_A - bG_B$$

On the right of the equal sign there are two terms, one obtained by combining the standard free energies and one that results from the combination of the logarithm terms. The first of these is

$$\Delta G^\circ = eG_E^\circ + fG_F^\circ - aG_A^\circ - bG_B^\circ$$

while the second is

$$RT \ln a_E{}^e + RT \ln a_F{}^f - RT \ln a_A{}^a - RT \ln a_B{}^b$$

which can be rearranged to give

$$RT \ln \left(\frac{a_E{}^e a_F{}^f}{a_A{}^a a_B{}^b} \right)$$

Putting all of this together, we can write the total free energy change as

$$\Delta G = \Delta G^\circ + RT \ln \left(\frac{a_E{}^e a_F{}^f}{a_A{}^a a_B{}^b} \right)$$

This final equation tells us how the free energy difference between reactants and products varies with temperature and with the effective concentration of each component.

At equilibrium, $\Delta G = 0$; therefore we find that

$$0 = \Delta G^\circ + RT \ln \left(\frac{a_E{}^e a_F{}^f}{a_A{}^a a_B{}^b} \right)$$

or

$$\Delta G^\circ = -RT \ln \left(\frac{a_E{}^e a_F{}^f}{a_A{}^a a_B{}^b} \right) \tag{12.6}$$

The standard free energy change, ΔG°, is a constant since, for any given reaction, it has a certain fixed numerical value that is determined solely by the difference between the standard free energies of the reactants and products at temperature, T. This means that the quantity appearing on the right of the equal sign must also be a constant. As a result, the fraction appearing in the logarithm term can only have one value when the system is at equilibrium. In other words, this fraction is equal to a constant (an equilibrium constant) at constant temperature.

$$\frac{a_E{}^e a_F{}^f}{a_A{}^a a_B{}^b} = K$$

This relationship is the law of mass action in which activities appear, instead of stoichiometric concentrations. We can now rewrite Equation 12.6 in the form

$$\Delta G° = -RT \ln K \qquad (12.7)$$

In many cases activities cannot be easily determined and in those instances we often must make do with stoichiometric concentrations or partial pressures (in the case of gases) instead. Fortunately, in dilute solutions, or mixtures of gases at relatively low pressures, the activity is nearly the same as the concentration and the mass action expression that we first saw, written with concentrations, fairly closely approximates the thermodynamic equilibrium expression.

Example 12.1
What is the thermodynamic equilibrium constant for the reaction

$$2SO_2(g) + O_2(g) \rightleftharpoons 2SO_3(g)$$

at 25° C?

Solution
From the data in Table 10.5 we can obtain the standard free energies of formation of SO_3 and SO_2.

$$\Delta G_f°(SO_3) = -88.5 \ \frac{kcal}{mole}$$

$$\Delta G_f°(SO_2) = -71.8 \ \frac{kcal}{mole}$$

By definition, $\Delta G_f°(O_2) = 0.0$ kcal/mole
Using these data we can compute $\Delta G°$ for the reaction:

$$\Delta G° = 2\left(-88.5 \ \frac{kcal}{mole}\right) - 2\left(-71.8 \ \frac{kcal}{mole}\right)$$

$$\Delta G° = -33.4 \ \frac{kcal}{mole}$$

Equation 12.7,

$$\Delta G° = -RT \ln K$$

can be rewritten in terms of logarithms to the base 10 as

$$\Delta G° = -2.303 \ RT \log K$$

Solving for log K,

$$\log K = \frac{\Delta G°}{-2.303 \ RT}$$

At 25° C (298° K), using $R = 1.99$ cal/mole° K

$$\log K = \frac{-33,400 \text{ cal/mole}}{-2.303 \ (1.99 \text{ cal/mole° K}) \ (298° \text{ K})}$$

$$\log K = 24.4$$

Taking the antilog gives us K

$$K = 2.5 \times 10^{24}$$

The magnitude of K for this reaction tells us that the position of equilibrium in the system should lie far in the direction of SO_3 and that at room temperature SO_2 should react almost completely with oxygen to form SO_3. Without a catalyst, however, this reaction is extremely slow at room temperature. With a catalyst, this reaction is important in the industrial preparation of H_2SO_4.

The conversion of SO_2 to SO_3 in catalytic mufflers, installed on automobiles to reduce air pollution, presents a potential obstacle to their widespread use. The H_2SO_4 mist produced in the auto exhaust presents a health problem and may cause the abandonment of the use of catalytic mufflers.

Now go back and calculate the thermodynamic equilibrium constants for the HCl and H_2O equilibria in Section 12.2.

12.5 Ways of Expressing Equilibrium Constants

Frequently chemists are faced with the problem of dealing with equilibria between gaseous reactants and products. The activity of a gaseous substance is approximately equal to its partial pressure in atmospheres, so that if our hypothetical reaction between A and B to produce E and F occurs entirely in the gas phase, we can write that

$$\frac{a_E^e a_F^f}{a_A^a a_B^b} = \frac{p_E^e p_F^f}{p_A^a p_B^b} = K_P$$

where we have used K_P to indicate that the equilibrium constant is expressed by the partial pressures of the reactants and products.

Earlier we found that we could also write the equilibrium condition for *any* reaction in terms of the concentration of the reactants and products, so that for this same reaction we have

$$\frac{[E]^e[F]^f}{[A]^a[B]^b} = K_c$$

where K_c is the equilibrium constant expressed in concentration units. How are these two ways of writing the equilibrium condition related?

Concentration, you recall, has the units moles per liter, or n/V. Assuming ideal gas behavior, we can use the ideal gas law,

$$PV = nRT$$

to obtain the concentration of a gas, X, in a mixture as

$$[X] = \frac{n_X}{V} = \frac{p_X}{RT}$$

where p_X is its partial pressure. From this it follows that

$$p_X = [X]RT$$

Substituting this relationship into the expression for K_P, we have

$$K_P = \frac{p_E^e p_F^f}{p_A^a p_B^b} = \frac{[E]^e (RT)^e [F]^f (RT)^f}{[A]^a (RT)^a [B]^b (RT)^b}$$

This can be rearranged to give

$$K_P = \frac{[E]^e [F]^f}{[A]^a [B]^b} (RT)^{(e+f) - (a+b)}$$

or

$$K_P = K_c (RT)^{\Delta n_g} \qquad (12.8)$$

where Δn_g is the change in the number of moles of gas upon going from reactants to products. Thus, K_P and K_c are related in a very simple fashion for reactions between ideal gases; a relationship that also holds reasonably well for real gases.

Example 12.2
In Example 12.1 we determined the value of the equilibrium constant (actually K_P) for the reaction of SO_2 with O_2 to produce SO_3. What is K_c for this equilibrium at 25° C?

Solution
Solving Equation 12.7 for K_c we obtain

$$K_c = \frac{K_P}{(RT)^{\Delta n_g}} = K_P (RT)^{-\Delta n_g}$$

For the reaction in Example 12.1 we find $\Delta n_g = -1$, since there are three moles of gaseous reactants and only two moles of gaseous products. Using the result of Example 12-1, we have

$$K_P = \frac{p_{SO_3}^2}{p_{SO_2}^2 p_{O_2}} = 2.5 \times 10^{24}$$

From the equilibrium constant expression we see that K_P in this case has the units 1/atm, or atm^{-1}. As a result we must use $R = 0.0821$ liter atm mole^{-1} °K^{-1} to obtain the proper units for K_c (why?). Thus

$$K_c = (2.5 \times 10^{24} \text{ atm}^{-1}) \, [(0.0821 \text{ liter atm mole}^{-1} \text{ °K}^{-1}) \, (298° \text{ K})]^{-(-1)}$$

Therefore

$$K_c = 6.2 \times 10^{25} \text{ liter mole}^{-1}$$

**12.6
Heterogeneous
Equilibria**

Up to now our discussion has been focused on homogeneous reactions in which all of the reactants and products are in the same phase. Heterogeneous reactions, of which there are many examples, also eventually arrive at a state of equilibrium. A typical reaction that we might consider is the decomposition of solid $Ca(HCO_3)_2$ to produce solid $CaCO_3$,[1] gaseous CO_2, and gaseous H_2O.

$$Ca(HCO_3)_2(s) \rightleftharpoons CaCO_3(s) + CO_2(g) + H_2O(g)$$

Applying the law of mass action we would write the equilibrium expression as

$$\frac{[CaCO_3(s)][CO_2(g)][H_2O(g)]}{[Ca(HCO_3)_2(s)]} = K_c'$$

For reasons that will be apparent shortly, we have temporarily indicated the equilibrium constant as K_c'.

In this reaction we have an equilibrium between the gases, CO_2 and H_2O, and the two pure solid phases, $Ca(HCO_3)_2$ and $CaCO_3$. We know that a pure solid substance such as $Ca(HCO_3)_2$ is characterized by a density that is the same for all samples of $Ca(HCO_3)_2$, regardless of their size. In addition, this density is unaffected by the nature of the chemical reaction. This means that even during a chemical reaction the amount of $Ca(HCO_3)_2$ in a given volume of the pure solid is always the same; as a result, the concentration of $Ca(HCO_3)_2$ in pure solid $Ca(HCO_3)_2$ is a constant. We cannot alter the number of mole/liter of $Ca(HCO_3)_2$ in the pure solid, nor can we change the concentration of $CaCO_3$ in pure $CaCO_3$. Consequently, the concentrations of these two substances in the equilibrium expression above take on constant numerical values and can be incorporated into the equilibrium constant.

$$[CO_2(g)][H_2O(g)] = K_c' \frac{[Ca(HCO_3)_2(s)]}{[CaCO_3(s)]}$$

or

$$[CO_2(g)][H_2O(g)] = K_c$$

Thus we find that for heterogeneous reactions *the equilibrium constant expression does not include the concentrations of pure solids.* Similarly, in reactions in which a reactant or product occurs as a pure liquid phase, the concentration of that substance in the pure liquid is also constant. As a result, *the concentrations of pure liquid phases also do not appear in an equilibrium constant expression.* These simplifications *only* apply when we are dealing with *pure* condensed phases. When substances occur in liquid or solid solutions, their concentrations are variable and their concentration terms in the mass action expression therefore cannot be incorporated into K.

If we wish to work with K_P rather than K_c, we again need only take into account the substances present in the gas phase. For the decomposition of $Ca(HCO_3)_2$, therefore, we have

$$K_P = p_{CO_2(g)} p_{H_2O(g)}$$

As noted in the last section, if we know K_c we can evaluate K_P as

$$K_P = K_c(RT)^{\Delta n_g}$$

where, for this reaction, $\Delta n_g = +2$.

---

[1] $CaCO_3$ is the chief component of limestone and can be further decomposed by heating to form $CO_2(g)$ and quick lime $CaO(s)$.

Example 12.3

What are the values of K_P and K_c for the "reaction"

$$H_2O(l) \rightleftharpoons H_2O(g)$$

at 25° C given that the vapor pressure of water at 25° C equals 23.8 torr?

Solution

Since liquid water is a pure liquid phase, we can write

$$K_P = p_{H_2O(g)}$$

and

$$K_c = [H_2O(g)]$$

(a) If we express the vapor pressure of water in atm,

$$p_{H_2O} = 23.8 \text{ torr} \left(\frac{1 \text{ atm}}{760 \text{ torr}}\right) = 0.0313 \text{ atm}$$

and thus

$$K_P = p_{H_2O} = 3.13 \times 10^{-2} \text{ atm}$$

Note that this equilibrium expression states that the partial pressure of water must be a constant when the liquid and vapor are in equilibrium.

(b) We can evaluate K_c as

$$K_c = K_P(RT)^{-\Delta n_g}$$

For this "reaction", $\Delta n_g = 1$; therefore,

$$K_c = K_P(RT)^{-1} = \frac{K_P}{RT}$$

$$K_c = \frac{3.13 \times 10^{-2} \text{ atm}}{(0.0821 \text{ liter atm/mole° K})(298° K)}$$

or

$$K_c = 1.28 \times 10^{-3} \frac{\text{mole}}{\text{liter}}$$

**12.7
Equilibrium
Calculations**

This section is intended to illustrate the type of computations that one might perform either to evaluate an equilibrium constant from measured concentrations or to use the equilibrium constant to calculate the concentrations of the reactants and products in a particular equilibrium mixture. First let us see how we might evaluate K in a typical experiment.

Example 12.4

The brown gas NO_2 and the colorless gas N_2O_4 exist in equilibrium as indicated by the equation

$$2NO_2 \rightleftharpoons N_2O_4$$

In an experiment, 0.625 moles of N_2O_4 were introduced into a 5.0 liter vessel and permitted to come to equilibrium with NO_2. At equilibrium the concentration of N_2O_4 was 0.0750 M. What is K_c for this reaction?

Solution

The equilibrium constant expression for this reaction is

$$\frac{[N_2O_4]}{[NO_2]^2} = K_c$$

In order to calculate K_c we must know the equilibrium concentrations of N_2O_4 and NO_2. Initially the concentration of N_2O_4 was 0.625 mole/5.0 liter $= 0.125M$; the initial concentration of NO_2 was zero. From the equilibrium concentration of N_2O_4 given to us in the problem we can conclude that in arriving at equilibrium 0.050 mole/liter of N_2O_4 was decomposed. Based on the stoichiometry of the equation, this means that 0.10 mole/liter of NO_2 (2 × 0.050 mole/liter of N_2O_4) was formed.

	Initial Concentrations	Equilibrium Concentrations
N_2O_4	0.125M	0.075M
NO_2	0.0M	0.10M

Substituting the equilibrium concentrations into the mass action expression, we have

$$\frac{(0.075M)}{(0.100M)^2} = K_c$$

and, finally,

$$K_c = 7.5 \text{ liter/mole}$$

Knowledge of the equilibrium constant for a reaction allows us to calculate the concentrations of the substances present in a reaction mixture at equilibrium. The ease with which these computations can be carried out depends on the complexity of the mass action expression, the concentrations of the various species in the reaction mixture, and the magnitude of the equilibrium constant. We shall look at only some of the more simple examples of problems of this type. The following sample problems, however, illustrate the type of reasoning that is employed in these computations, as well as some of the concepts that have been presented up to this point.

Example 12.5

At a temperature of 500° C the equilibrium constant, K_c, for the nitrogen fixation reaction for the production of ammonia

$$3H_2(g) + N_2(g) \rightleftharpoons 2NH_3(g)$$

has a value of 6.0×10^{-2} liter^2/mole^2. If, in a particular reaction vessel at this temperature, there are 0.250 mole/liter H_2 and 0.0500 mole/liter NH_3 present at equilibrium, what is the concentration of N_2?

Solution

The first step in the solution of any problem dealing with equilibrium is to write down the equilibrium constant expression. For this reaction we have

$$K = \frac{[NH_3]^2}{[H_2]^3[N_2]} = 6.0 \times 10^{-2} \frac{liter^2}{mole^2}$$

We wish to calculate the concentration of N_2. This can be accomplished if we know the values of the equilibrium concentrations of both NH_3 and H_2 and, in this problem, these are given to us. If we now let x be equal to the molar concentration of N_2 at equilibrium we have

$$[NH_3] = 0.0500M$$
$$[H_2] = 0.250M$$
$$[N_2] = x$$

Substituting these into the mass action expression gives us

$$\frac{(0.0500)^2}{(0.250)^3(x)} = 6.0 \times 10^{-2}$$

If we solve for x,

$$x = \frac{(0.0500)^2}{(0.250)^3(6.0 \times 10^{-2})}$$
$$x = 2.7$$

The equilibrium concentration of N_2 is thus 2.7 mole/liter.

Example 12.6

At 440° C the equilibrium constant for the reaction,

$$H_2(g) + I_2(g) \rightleftharpoons 2HI(g)$$

is 49.5. If 0.20 moles of H_2 and 0.20 moles of I_2 are placed into a 10 liter vessel and permitted to react at this temperature, what will be the concentration of each substance at equilibrium?

Solution

Our equilibrium expression is

$$\frac{[HI]^2}{[H_2][I_2]} = 49.5$$

In this example we are given the *initial* concentrations of the reactants and products. These are

$$[H_2] = \frac{0.20 \text{ mole}}{10 \text{ liters}} = 0.020M$$

$$[I_2] = \frac{0.20 \text{ mole}}{10 \text{ liters}} = 0.020M$$

$$[HI] = 0.0M$$

Since no HI is present initially, we know that it will be formed from the violet-colored mixture of the H_2 and I_2 (the color being due to the I_2). Let us approach the problem, then, by allowing x to be equal to the number of moles per liter of H_2 that react. From the stoichiometry of the reaction we realize that this x mole/liter of H_2 will react with x mole/liter of I_2 to produce $2x$ mole/liter of HI. At equilibrium, the violet color will have faded as we have $(0.020 - x)$ mole/liter of H_2 remaining unreacted as well as $(0.020 - x)$ mole/liter of I_2. The *equilibrium concentrations* are thus

$$[H_2] = 0.020 - x$$
$$[I_2] = 0.020 - x$$
$$[HI] = 2x$$

Substituting these into the mass action expression gives

$$\frac{(2x)^2}{(0.020 - x)(0.020 - x)} = 49.5$$

or

$$\frac{(2x)^2}{(0.020 - x)^2} = 49.5$$

In this case, we can take the square root of both sides of the equation to obtain

$$\frac{2x}{0.020 - x} = 7.03$$

Solving for x,

$$2x = 7.03(0.020 - x) = 0.140 - 7.03x$$
$$2x + 7.03x = 0.140$$
$$9.03x = 0.140$$
$$x = 0.016$$

Finally, the equilibrium concentrations are

$$[H_2] = 0.020 - 0.016 = 0.004M$$
$$[I_2] = 0.020 - 0.016 = 0.004M$$
$$[HI] = 2(0.016) = 0.032M$$

In this last problem we employed some relatively simple algebra to help us arrive at the solution. Let us look at another example of this type.

Example 12.7
A 10 liter vessel is filled with 0.40 moles of HI at 440° C. What will be the concentration of H_2, I_2, and HI at equilibrium?

Solution
In this example we are concerned with the same equilibrium as in the previous problem. Initially we have no H_2 or I_2 and the reaction mixture is colorless, and as HI decomposes the reaction mixture begins to acquire a pale violet color. Since we know HI must break down, let x be equal to the number of moles per liter of HI that decompose. Since one mole each of H_2 and I_2 are produced from every two moles of HI that breaks down, we have at

equilibrium

$$[H_2] = 0.50x$$
$$[I_2] = 0.50x$$

The initial concentration of HI is 0.40 mole/10 liters $= 0.040M$. At equilibrium its concentration will be $0.040M$ minus the number of mole/liter that have decomposed. Hence

$$[HI] = 0.040 - x$$

Substituting these quantities into the mass action expression gives us

$$\frac{(0.040 - x)^2}{(0.50x)(0.50x)} = 49.5$$

or

$$\frac{(0.040 - x)^2}{(0.50x)^2} = 49.5$$

Taking the square root of both sides of the equation, we have

$$\frac{0.040 - x}{0.50x} = 7.03$$

Solving for x we get

$$0.040 - x = 0.50x(7.03)$$
$$x = 0.00887$$

We now calculate the equilibrium concentrations to be

$$[H_2] = 0.50x = 0.0044M$$
$$[I_2] = 0.50x = 0.0044M$$
$$[HI] = 0.040 - x = 0.031M$$

Observe that we have obtained essentially the same answers in both Examples 12.7 and 12.8. If all of the H_2 and I_2 in Example 12.7 had completely reacted, it would have produced 0.40 mole of HI, the same amount of HI that we began with in Example 12.8. We find, therefore, that the same position of equilibrium can be approached from either direction.

In cases where the equilibrium constant is either extremely large, or extremely small, it is frequently possible to make some approximations that greatly simplify the kind of calculations we have just seen.

Example 12.8
The equilibrium constant for the decomposition of water at 500° C is 6.0×10^{-28}. If 2.0 moles of H_2O are placed into a 5.0 liter container, what will be the equilibrium concentrations of H_2, O_2, and $H_2O(g)$ at 500° C?

Solution
The equation for the reaction is

$$2H_2O \rightleftharpoons 2H_2 + O_2$$

Therefore we can write

$$\frac{[H_2]^2[O_2]}{[H_2O]^2} = 6.0 \times 10^{-28}$$

The initial H_2O concentration is 2.0 mole/5.0 liter = 0.40M. If we let x equal the number of mole/liter of H_2O that decomposes, we will get x moles of H_2 and $x/2$ moles O_2. At equilibrium, therefore,

$$[H_2] = x$$
$$[O_2] = 0.50x$$
$$[H_2O] = 0.40 - x$$

and substituting these into the mass action expression gives

$$\frac{(x)^2(0.50x)}{(0.40 - x)^2} = 6.0 \times 10^{-28}$$

Unless we can somehow simplify this equation, we have a real mess on our hands. Fortunately, in this case the problem can be made easy to solve.

Because of the size of K we know that very little H_2O will actually have decomposed when the system reaches equilibrium. Therefore x will be very small. If x is *much* smaller than 0.40, then when we subtract it from 0.40 and round the answer to two significant figures, we shall *still* have 0.40. Let us, then, assume that x is very small. Our equilibrium concentrations become

$$[H_2] = x$$
$$[O_2] = 0.50x$$
$$[H_2O] = 0.40 - x \approx 0.40$$

Substituting these into the mass action expression gives

$$\frac{(x)^2(0.50x)}{(0.40)^2} = 6.0 \times 10^{-28}$$

from which we get

$$\frac{0.50x^3}{0.16} = 6.0 \times 10^{-28}$$

or

$$x^3 = \frac{0.16}{0.50}(6.0 \times 10^{-28}) = 1.9 \times 10^{-28}$$

At this point, x can be obtained by extracting the cube root. Note that we must first make the exponent divisible by three.

$$x^3 = 190 \times 10^{-30}$$
$$x = 5.7 \times 10^{-10}$$

We see that x is, in fact, very much smaller than 0.40, thus justifying our initial assumption. The final equilibrium concentrations are

$$[H_2] = x = 5.7 \times 10^{-10}M$$
$$[O_2] = 0.50x = 2.8 \times 10^{-10}M$$
$$[H_2O] = 0.40 - (5.7 \times 10^{-10}) = 0.40M$$

As useful as it may be to be capable of performing the type of equilibrium calculations illustrated in the preceding section, frequently it is desirable simply to be able to predict how some disturbance imposed on a system from outside will influence the position of equilibrium. For instance, we may wish to predict, in a qualitative way, the conditions that favor the greatest production of products. Should we run our reaction at high or low temperature? Should the pressure on the system be high or low? These are questions that we would like to answer quickly without having to perform tedious computations. We have already seen how Le Chatelier's principle can be applied to dynamic equilibria involving such phenomena as the vapor pressure of a liquid and solubility. Changes in the position of equilibrium in chemical systems can also be understood by applying the same concepts.

CHANGES IN THE CONCENTRATION OF A REACTANT OR PRODUCT. In a system such as

$$H_2 + I_2 \rightleftharpoons 2HI$$

any change in the concentration of a reactant or product will cause the system to no longer be at equilibrium. As a result, a chemical reaction will occur that will return the system to equilibrium. From Le Chatelier's principle we know that if a system at equilibrium is disturbed it will attempt to undergo some change to diminish the effect of the disturbance. For example, the addition of H_2 to an equilibrium mixture of H_2, I_2, and HI upsets the equilibrium, and the system responds by using up part of the additional H_2 by reaction with I_2 to produce more HI. When equilibrium has finally been reestablished, there will be a greater concentration of HI than before, and we say that for this reaction the position of equilibrium has been shifted to the right.

We can arrive at this same conclusion by considering the effect of added H_2 on the value of the mass action expression. For this reaction at equilibrium we have

$$\frac{[HI]^2}{[H_2][I_2]} = K$$

If H_2 were suddenly added to the system, the value of the denominator of the mass action expression would become larger, and the entire fraction would therefore become smaller than the equilibrium constant. The reaction that occurs to return the system to equilibrium must increase the value of the mass action expression until it once again is equal to K. In order for this to take place, the numerator must become larger and the denominator smaller. In other words, more HI will be formed at the expense of H_2 and I_2, and again we conclude that the addition of H_2 shifts the position of equilibrium in this reaction to the right.

By applying Le Chatelier's principle we can also predict the effect that removing a reactant or product will have on a system at equilibrium. For instance, if H_2 is somehow removed from the reaction vessel, the system will adjust by having some HI decompose in an effort to replenish the lost reactant. Consequently we conclude that the position of equilibrium is shifted to the left when H_2 is removed.

Therefore we may conclude that to drive a reaction far toward completion we can either add a large excess of one of the reactants or remove the products as they are formed. Recall that it is the latter that serves as the driving force in ionic reactions (Chapter 5) where the product is either a precipitate, a gas, or a weak electrolyte. The creation of these products removes ions from solution and therefore forces the reaction to proceed toward completion.

THE EFFECT OF TEMPERATURE ON EQUILIBRIUM. Up to now we have been careful to imply that the equilibrium constant for a reaction has a fixed numerical value only as long as the temperature remains constant. This is because temperature, as well as the concentrations of reactants and products, affects the position of equilibrium. However, the temperature, unlike the concentrations of reactants and products, affects the value of the equilibrium constant itself.

The reaction between H_2 and N_2 to form NH_3 is exothermic, and the equation for the formation of ammonia is written as

$$3H_2(g) + N_2(g) \rightleftharpoons 2NH_3(g) + 22.0 \text{ kcal}$$

where the heat of reaction is indicated as a product. If we have a system of these gases in equilibrium and wish to raise its temperature, we do so by adding heat to it from the surroundings. Le Chatelier's principle tells us that when we add this heat, the system will attempt to undergo a change that tends to use some of it up and, as before, we conclude that a rise in temperature should cause an endothermic change to take place. Since the production of NH_3 is exothermic, its decomposition is endothermic. Consequently, raising the temperature of this system will cause the position of equilibrium to shift to the left. An increase in temperature, then, drives the reaction to a new position of equilibrium in which there is more N_2 and H_2 and less NH_3. *In general, an increase in temperature causes the position of equilibrium of an exothermic reaction to be shifted to the left while that of an endothermic reaction is shifted to the right.*

In Section 12.1, it was stated that at equilibrium the mass action expression is always equal to the same K provided that the temperature remains the same. A change in temperature, however, alters K. We have just seen that an increase in temperature leads to a decrease in the concentration of NH_3 and an increase in the concentration of both N_2 and H_2. This means that at equilibrium at the higher temperature the value of the mass action expression,

$$\frac{[NH_3]^2}{[H_2]^3[N_2]}$$

will have decreased. Thus we find that for this exothermic reaction K decreases with rising temperature. By the same token, for a reaction that is endothermic in the forward direction, K increases with increasing temperature.

EFFECT OF PRESSURE AND VOLUME CHANGES ON EQUILIBRIUM. At constant temperature a change in the volume of a system also causes a change in pressure. We quite logically expect, therefore, that an increase in the external pressure on a system should favor any change that leads to a smaller volume (recall Boyle's law). We would not expect pressure changes to have any marked effect on the position of equilibrium in reactions where all of the reactants and products are either solids or liquids, since these phases are virtually incompressible. However, pressure changes can have a very dramatic effect on equilibria that involve reactions in which gases are consumed or produced.

Let us again choose as an example the reaction for the formation of NH_3. If we have this system at equilibrium and suddenly decrease the volume of the container, we know that the pressure will go up. By applying Le Chatelier's principle we expect that a change in the system should occur that will reduce the pressure. How can this be brought about?

Let us suppose that we had initially placed into our reaction vessel one mole of nitrogen and three moles of hydrogen. Thus four moles of gas are initially in the container. If the reaction were to proceed entirely to completion, two moles of NH_3 would be produced; therefore as the reaction

$$N_2(g) + 3H_2(g) \rightleftharpoons 2NH_3(g)$$

proceeds from left to right, the number of moles of gas in the container decreases. Since fewer moles of gas means lower pressure, it is clear that the production of additional NH_3 in a vessel of fixed volume leads to a reduction in the pressure. Consequently, when the volume of an equilibrium mixture of H_2, N_2, and NH_3 is reduced, and then held constant, more NH_3 will be formed and the position of equilibrium will shift to the right.[2] Note, however, that the value of K remains unchanged.

In short, increasing the pressure upon a system at equilibrium will cause a shift in the position of equilibrium in the direction of the fewest number of moles of *gaseous* reactants or products.

Finally, note when there are the same number of moles of gaseous reactants and products, as in the reaction between H_2 and I_2,

$$H_2(g) + I_2(g) \rightleftharpoons 2HI(g)$$

pressure changes will not influence the quantities of the various substances present in the reaction mixture at equilibrium.

ADDITION OF AN INERT GAS. If an inert (i.e., nonreacting) gas is introduced into a reaction vessel containing other gases at equilibrium, it will cause an increase in the total pressure within the container. This kind of pressure increase, however, will not affect the position of equilibrium since it will not alter the partial pressures or the concentrations of any of the substances already present.

EFFECT OF A CATALYST ON THE POSITION OF EQUILIBRIUM. In Chapter 11 we saw that a catalyst affects a chemical reaction by lowering the activation energy barrier that must be overcome in order for the reaction to proceed. A catalyst affects the rate of a chemical change. It does not, however, affect the heat of reaction, and it is the heat of reaction, $\Delta H°$, along with the

[2] From application of Le Chatelier's principle we know that the production of ammonia from H_2 and N_2 is favored by high pressures and low temperatures. At low temperatures, however, the reaction is very low; therefore in the industrial preparation of NH_3 pressures of 10^2 to 10^3 atm and temperatures from 400 to 550° C are employed. Even though there is less NH_3 produced at equilibrium at these high temperatures, the speed of reaction is boosted to the point where the production of NH_3 is economically worthwhile.

entropy change, $\Delta S°$, that determine $\Delta G°$, which in turn fixes the position of equilibrium at any given temperature. A catalyst merely speeds the approach to the position of equilibrium that is determined by $\Delta G°$.

12.1 What is meant by a dynamic equilibrium?

12.2 Write the mass action expression for each of the following reactions:
(a) $N_2(g) + O_2(g) \rightleftharpoons 2NO(g)$
(b) $2NO(g) + O_2(g) \rightleftharpoons 2NO_2(g)$
(c) $2H_2(g) + S_2(g) \rightleftharpoons 2H_2S(g)$
(d) $2N_2O_5(g) \rightleftharpoons 4NO_2(g) + O_2(g)$
(e) $P_4O_{10}(g) + 6PCl_5(g) \rightleftharpoons 10POCl_3(g)$
(f) $5CO(g) + I_2O_5(s) \rightleftharpoons I_2(g) + 5CO_2(g)$

12.3 Arrange the following reactions in order of their increasing tendency to proceed toward completion:
(a) $4NH_3(g) + 3O_2(g) \rightleftharpoons$
$2N_2(g) + 6H_2O(g)$ $K = 1 \times 10^{228}$
(b) $N_2(g) + O_2(g) \rightleftharpoons$
$2NO(g)$ $K = 5 \times 10^{-31}$
(c) $2HF(g) \rightleftharpoons$
$H_2(g) + F_2(g)$ $K = 1 \times 10^{-13}$
(d) $2NOCl(g) \rightleftharpoons$
$2NO(g) + Cl_2(g)$ $K = 4.7 \times 10^{-4}$

12.4 Show that the following data, obtained for the reaction

$$PCl_5(g) \rightleftharpoons PCl_3(g) + Cl_2(g)$$

demonstrate the law of mass action. What is K_c for this reaction?

Experiment	$[PCl_5]$	$[PCl_3]$	$[Cl_2]$
1	0.0023	0.23	0.055
2	0.010	0.15	0.37
3	0.085	0.99	0.47
4	1.00	3.66	1.50

12.5 At a certain temperature, the equilibrium constant (K_c) for the reaction

$$2SO_2(g) + O_2(g) \rightleftharpoons 2SO_3(g)$$

is 35.5 liter/mole. Give the value of K_c for the reaction

$$2SO_3(g) \rightleftharpoons 2SO_2(g) + O_2(g)$$

12.6 The reaction $2NO(g) + 2H_2(g) \rightarrow N_2(g) + 2H_2O(g)$ is believed to occur by the mechanism
$2NO(g) \rightleftharpoons N_2O_2(g)$ fast
$N_2O_2(g) + H_2(g) \longrightarrow$
$N_2O(g) + H_2O(g)$ slow
$N_2O(g) + H_2(g) \longrightarrow$
$N_2(g) + H_2O(g)$ fast
On the basis of kinetics, derive the mass action law for the overall reaction.

12.7 At 700° K the reaction $CO(g) + 2H_2(g) \rightleftharpoons CH_3OH(g)$ has a standard free energy change, $\Delta G°_{700°K} = -3.22$ kcal. What is the value of K_P for this reaction at 700° K?

12.8 The equilibrium constant, K_p, for the reaction, $COCl_2(g) \rightleftharpoons CO(g) + Cl_2(g)$ has a value of 4.56×10^{-2} atm at 395° C. What is the value of $\Delta G°_{668°K}$ for this reaction?

12.9 Use the data in Table 10.5 to calculate K at 25° C for the reaction

$$2HCl(g) + F_2(g) \rightleftharpoons 2HF(g) + Cl_2(g)$$

12.10 Use the data in Tables 10.1 and 10.4 to compute $\Delta G°$ and K at 500° C for the reaction $2HCl(g) \rightleftharpoons H_2(g) + Cl_2(g)$.

12.11 At 25° C, in a mixture of N_2O_4 and NO_2 in equilibrium at a total pressure of 0.844 atm, the partial pressure of N_2O_4 is 0.563 atm. Calculate for the reaction $N_2O_4(g) \rightleftharpoons 2NO_2(g)$:
(a) K_P, (b) K_c, (c) $\Delta G°_{298°K}$

12.12 The following thermodynamic data apply at 25° C:

Substance	ΔG_f° (kcal/mole)
$NiSO_4 \cdot 6H_2O(s)$	-531.0
$NiSO_4(s)$	-184.9
$H_2O(g)$	-54.6

(a) What is ΔG° for the reaction
$NiSO_4 \cdot 6H_2O(s) \rightleftharpoons$
$\qquad NiSO_4(s) + 6H_2O(g)$?
(b) What is K_P for this reaction?
(c) What is the equilibrium vapor pressure of H_2O over solid $NiSO_4 \cdot 6H_2O$?

12.13 Why is it *not* necessary to include the concentration of pure liquid or solid phases in the equilibrium constant expression?

12.14 For the reaction $PCl_5(g) \rightleftharpoons PCl_3(g) + Cl_2(g)$, $K = 33.3$ at $760°$ C. In a container at equilibrium there are 1.29×10^{-3} mole/liter PCl_5 and 1.87×10^{-1} mole/liter Cl_2. Calculate the equilibrium concentration of PCl_3 in the vessel.

12.15 At $100°$ the equilibrium constant for the reaction $CO(g) + Cl_2(g) \rightleftharpoons COCl_2(g)$ has a value of 4.6×10^9 liter/mole. If 0.20 mole of $COCl_2$ are placed into a 10.0 liter flask at $100°$ C, what will be the concentration of all species at equilibrium?

12.16 For the reaction $H_2(g) + CO_2(g) \rightleftharpoons CO(g) + H_2O(g)$, $K_c = 0.771$ at $750°$ C. If 1.00 mole H_2 and 1.00 mole CO_2 are placed into a 5.00 liter container and permitted to react, what will be the equilibrium concentration of all species?

12.17 In a 10.0 liter mixture of H_2, I_2, and HI at equilibrium at $425°$ C there are 0.100 mole H_2, 0.100 mole I_2, and 0.740 mole HI. If 0.50 mole of HI are now added to this system, what will be the concentration of H_2, I_2, and HI once equilibrium has been reestablished?

12.18 Indicate how each of the following changes affects the concentration of H_2 in the system,

9.9 kcal $+ H_2(g) + CO_2(g)$
$\qquad \rightleftharpoons H_2O(g) + CO(g)$

(a) addition of CO_2, (b) addition of H_2O, (c) addition of a catalyst, (d) increase in temperature, (e) decrease in the volume of the container.

12.19 How will the equilibrium constant for the reaction in question 12.18 be affected by each of the changes?

12.20 Sketch a graph to show how the concentrations of H_2, N_2, and NH_3 would change with time after N_2 had been added to a mixture of these gases initially at equilibrium.

13
Acids
and Bases

In Chapter 5 the concept of acids and bases was introduced, although at that time we restricted our discussion to reactions in aqueous solution. It was pointed out, however, that the acid-base concept is extremely useful since it permits the correlation of large amounts of what at first glance may appear to be widely different types of chemical reactions. This idea is so useful, in fact, that chemical reactions can be broadly classified into two categories, acid-base reactions and oxidation-reduction reactions. Often there is an overlap between them, so that for some reactions it is sometimes convenient to view them as acid-base reactions, while at other times it seems best to describe them in terms of oxidation-reduction.

In this chapter we shall see that whether a particular substance behaves as an acid or a base depends on the way in which acids and bases are defined. There are a variety of ways of approaching this problem, with some definitions being more restrictive than others. Nevertheless, several properties are characteristic of acids and bases in general. These include

1. *Neutralization.* Acids and bases react with one another so as to cancel, or neutralize, their acidic and basic characters.
2. *Reaction with indicators.* Certain organic dyes, called indicators, give different colors depending upon whether they are in an acidic or basic medium.
3. *Catalysis.* Many chemical reactions are catalyzed by the presence of acids or bases.

In its modern version, the Arrhenius concept of acids and bases defines an acid as *any substance that can produce hydronium ion, H_3O^+, in aqueous solution.* A *base*, on the other hand, *is a substance that increases the hydroxide ion concentration in water.* Thus, it is this concept of acids and bases that was first presented in Section 5.5. Let us briefly review some of the ideas that were developed there. You will recall, for example, that HCl is an acid because it reacts with water according to the equation

**13.1
The Arrhenius
Definition of
Acids and Bases**

$$HCl + H_2O \longrightarrow H_3O^+ + Cl^-$$

Similarly, CO_2 is an acid because it reacts with water to form carbonic acid, H_2CO_3,

$$CO_2 + H_2O \rightleftharpoons H_2CO_3$$

which then undergoes further reaction to produce H_3O^+ and HCO_3^-.

$$H_2CO_3 + H_2O \rightleftharpoons H_3O^+ + HCO_3^-$$

In general, nonmetal oxides react with water to yield acidic solutions and are said to be **acid anhydrides** (Greek, *anydros*, waterless).

An example of an Arrhenius base is NaOH, an ionic compound containing Na^+ and OH^- ions. In water it undergoes dissociation.

$$NaOH(s) \xrightarrow{H_2O} Na^+(aq) + OH^-(aq)$$

Other examples of bases include substances such as NH_3 and N_2H_4 that react with water to produce OH^-.

$$NH_3 + H_2O \rightleftharpoons NH_4^+ + OH^-$$

$$N_2H_4 + H_2O \rightleftharpoons N_2H_5^+ + OH^-$$
(hydrazine)   **(hydrazinium ion)**

You will also recall that metal oxides (**basic anhydrides**) undergo reaction with water to give the corresponding hydroxides,

$$Na_2O + H_2O \longrightarrow 2NaOH$$
$$BaO + H_2O \longrightarrow Ba(OH)_2$$

Finally, in aqueous solution the neutralization of an acid by a base takes the form of the ionic reaction

$$H_3O^+ + OH^- \longrightarrow 2H_2O$$

13.2 Brønsted-Lowry Definition of Acids and Bases

The definition of acids and bases in terms of the hydronium ion and hydroxide ion in water is very restricted since it limits us to discussing acid-base phenomena in aqueous solutions only. A somewhat more general approach was that proposed independently in 1923 by the Danish chemist, J. N. Brønsted, and the British chemist, T. M. Lowry. They defined an *acid* as *a substance that is able to donate a proton* (i.e., a hydrogen ion, H^+) *to some other substance*. A *base*, then, is defined as *a substance that is able to accept a proton from an acid*. Stated more simply, an acid is a proton donor and a base is a proton acceptor.

A typical example of a Brønsted-Lowry acid-base reaction occurs when HCl is added to water.

$$HCl + H_2O \longrightarrow H_3O^+ + Cl^-$$

In this reaction HCl is functioning as an acid because it is donating a proton to the water molecule. Water, on the other hand, is behaving as a base by accepting a proton from the acid.

If we have a solution of concentrated HCl and heat it, we drive off HCl gas. In other words, we can reverse the above reaction so that H_3O^+ and Cl^- react with each other to produce HCl and H_2O. This reverse reaction is also a Brønsted-Lowry reaction, with hydronium ion serving as an acid by giving up its proton, and with the chloride ion functioning as a base by acquiring it. Thus, we might view our reaction as an equilibrium where we have two acids and two bases, one of each on either side of the arrow.

$$HCl + H_2O \rightleftharpoons H_3O^+ + Cl^-$$
acid   **base**   **acid**   **base**

When the acid, HCl, reacts it yields the base, Cl⁻. These two substances are related to one another by the loss or gain of a single proton and constitute a **conjugate acid-base pair.** We say that Cl⁻ is the **conjugate base** of the acid, HCl, and similarly that HCl is the **conjugate acid** of the base, Cl⁻. In this reaction we also find that H_2O and H_3O^+ form a conjugate pair. Water is the conjugate base of H_3O^+, and H_3O^+ is the conjugate acid of H_2O.

Another example of a Brønsted-Lowry acid-base reaction occurs in aqueous solutions of ammonia.

$$NH_3 + H_2O \rightleftharpoons NH_4^+ + OH^-$$

In this case water serves as an acid by giving up a proton to a molecule of NH_3, which thereby acts as a base. In the reverse reaction, on the other hand, NH_4^+ is the acid and OH^- is the base. Again we have two acid-base conjugate pairs: NH_3 and NH_4^+ plus H_2O and OH^-.

In general, we can represent any Brønsted-Lowry acid-base reaction as,

$$Acid(X) + Base(Y) \rightleftharpoons Base(X) + Acid(Y)$$

where Acid(X) and Base(X) represent one conjugate pair and Acid(Y) and Base(Y) the other.

In the two examples that we examined above, water, in one instance, functioned as a base, and in the other it behaved as an acid. Such a substance, which can serve in either capacity depending on conditions, is said to be **amphiprotic** or **amphoteric.** Water is not the only substance to behave in this fashion. For example, the autoionization of water, pure acetic acid, and liquid ammonia can be represented by the equations:

$$H_2O \;+\; H_2O \;\rightleftharpoons\; H_3O^+ \;+\; OH^-$$

$$HC_2H_3O_2 + HC_2H_3O_2 \rightleftharpoons H_2C_2H_3O_2^+ + C_2H_3O_2^-$$

$$NH_3(l) \;+\; NH_3(l) \;\rightleftharpoons\; NH_4^+ \;+\; NH_2^-$$

$$\textbf{(acid)} \;+\; \textbf{(base)} \;\rightleftharpoons\; \textbf{(acid)} \;+\; \textbf{(base)}$$

These autoionization reactions can also be diagramed using Lewis structures as follows.

(base)   (acid)

In each case the solvent (either water, acetic acid or ammonia) is playing the role of both acid and base.

The Brønsted-Lowry concept is more useful than the Arrhenius concept because it does not limit us to water solutions. In fact we can find acid-base reactions that occur in the absence of any solvent whatsoever. For instance, when HCl and NH_3 gases are mixed they react immediately to form the white ionic solid, NH_4Cl.

$$H:\overset{H}{\underset{H}{\overset{..}{N}}}: + H:\overset{..}{\underset{..}{Cl}}: \longrightarrow \left[H:\overset{H}{\underset{H}{\overset{..}{N}}}:H\right]^+ + \left[:\overset{..}{\underset{..}{Cl}}:\right]^-$$

Since a proton is transferred from HCl to NH_3, this is clearly an acid-base reaction in the Brønsted-Lowry sense. However, since neither hydronium ion nor hydroxide ion ever enter the picture, the Arrhenius view of acids and bases ignores this reaction completely.

13.3
Strengths of Acids and Bases

Any Brønsted-Lowry acid-base reaction can be viewed as two opposing or competing reactions between acids and bases and, in one sense, the two bases can be considered to be competing for a proton. When HCl reacts with water, for example, we find from conductivity measurements and freezing point depression that essentially all of the HCl has reacted with water. This means that the position of equilibrium in the reaction

$$HCl + H_2O \rightleftharpoons H_3O^+ + Cl^-$$

lies very far to the right. In turn, this tells us that H_2O has a much stronger affinity for a proton than does a chloride ion because the water is able to capture essentially all of the available H^+. We express this relative ability to pick up a proton by saying that water is a stronger base than chloride ion.

We can also speak of the relative strengths of the two acids in this reaction, HCl and H_3O^+. Here we see that HCl is better able to donate its proton than is H_3O^+ since all of the HCl has lost its protons to give H_3O^+.

The position of equilibrium in an acid-base reaction tells us of the relative strengths of the acids and bases involved. Hydrogen chloride is a strong acid in water because the position of equilibrium in the ionization reaction lies far to the right. Hydrogen fluoride, on the other hand, is said to be a weak acid because in water it is only very slightly dissociated (actually about 3% for a $1\,M$ solution at room temperature).

$$HF + H_2O \rightleftharpoons H_3O^+ + F^-$$

We can easily compare the relative strengths of HCl and HF because we are reacting them with the same reference base, water. We conclude that HCl is a stronger acid than HF because HCl is better able to protonate the base than is HF. With a given base, however, it is not possible to compare the strengths of *all* acids. For example, HCl, HNO$_3$, and HClO$_4$ all appear to be 100% ionized in water. With the reference base water, then, each of these acids appear to be of equal strength. Their differences are removed, or leveled out, and we speak of this phenomenon as the **leveling effect.**

If acetic acid (the substance that gives vinegar its sour taste) is used as a solvent instead of water, we find that there is an appreciable difference between the extents to which the reactions,

$$HCl + HC_2H_3O_2 \rightleftharpoons H_2C_2H_3O_2^+ + Cl^-$$
$$HNO_3 + HC_2H_3O_2 \rightleftharpoons H_2C_2H_3O_2^+ + NO_3^-$$
$$HClO_4 + HC_2H_3O_2 \rightleftharpoons H_2C_2H_3O_2^+ + ClO_4^-$$

proceed toward completion. In this instance, acetic acid is a much weaker base than water and is not as easily protonated. Consequently, by using acetic acid as the solvent it is possible to distinguish among the strengths of these three acids and, in fact, we find that their acidity increases in the order, HNO$_3 <$ HCl $<$ HClO$_4$. For these substances water is a **leveling solvent** while acetic acid serves as a **differentiating solvent.**

The leveling effect is not restricted to acids alone. Strong bases such as the oxide ion, O^{2-}, amide ion, NH$_2^-$, and hydride ion, H$^-$, react completely with water to give hydroxide ion.

$$O^{2-} + H_2O \longrightarrow OH^- + OH^-$$
$$NH_2^- + H_2O \longrightarrow NH_3 + OH^-$$
$$H^- + H_2O \longrightarrow H_2 + OH^-$$

In these examples, water is such a strong acid that it is able completely to protonate all three bases. Therefore, with water as a solvent, it is impossible to differentiate among their base strengths.

In general, any basic solvent tends to exert a leveling effect on acids while an acidic solvent tends to level the strengths of bases. For example, in water HF and HCl are clearly of different strengths as measured by their ability to protonate water molecules. However, in liquid ammonia (a basic solvent) the reaction

$$HX + NH_3 \longrightarrow NH_4^+ + X^-$$

proceeds essentially to completion for $X =$ Cl and $X =$ F. In ammonia, then, HF and HCl appear to be of equal strength.

A similar phenomenon is observed with ammonia in the solvents H$_2$O and HC$_2$H$_3$O$_2$. In water ammonia is only slightly protonated while it becomes completely protonated when it is added to pure acetic acid. In the latter solvent, NH$_3$ behaves as a strong base while in water it is weak.

By this time you have perhaps begun to wonder why some acids are strong and why others are weak. How do we explain the trends in acid strength that we can observe experimentally?

Let us first consider the oxoacids such as H$_2$SO$_4$ and H$_2$SO$_3$. The structures of these two particular compounds can be represented as

13.4
Factors Influencing the Strengths of Acids

Each undergoes ionization in water by transferring its protons, in two successive steps, to water molecules. Thus, for H_2SO_4 we have in the first step

With H_2SO_4 this step proceeds essentially to completion while for H_2SO_3, on the other hand, the first step

takes place to only a very limited degree (about 11% for a $1M$ solution). In each of these substances the sulfur is bonded to two O—H groups. However, in H_2SO_4 the sulfur is also attached to two other lone oxygen atoms, while the sulfur in H_2SO_3 is only bonded to one such O atom. We know that oxygen is a very electronegative element, and we expect these S—O bonds to be polar, with the negative charge concentrated about the oxygen end of the dipole. In other words, the oxygen atoms, attached to the sulfur, draw electron density away so that the sulfur acquires a partial positive charge, the magnitude of which will be greater in H_2SO_4 than in H_2SO_3.

$$O^{\delta-} \qquad\qquad O^{\delta-}$$
$$\uparrow \qquad\qquad\qquad \uparrow$$
$$H—O—\underset{\downarrow}{\overset{}{S}}{\scriptstyle 2\delta+}—O—H \qquad H—O—\underset{\delta+}{S}—O—H$$
$$O^{\delta-}$$

This positive charge on the sulfur tends to draw electron density from the S—OH bonds. Since the electronegative oxygen does not wish to lose electrons, the overall net effect is that some electronic charge will be withdrawn from the O—H bonds. This electron withdrawing effect will be greater in H_2SO_4 than in H_2SO_3 because the sulfur in H_2SO_4 bears a greater positive charge and is therefore better able to draw electrons to itself. Consequently, in H_2SO_4 the hydrogen atoms carry a higher positive charge than those in H_2SO_3 and are thus more easily removed as H^+. As a result, H_2SO_4 is a stronger acid than H_2SO_3.

This phenomenon is illustrated even more graphically if we consider the oxoacids of chlorine.

hypochlorous acid HOCl H—Ö—C̈l:

chlorous acid HClO₂ H—Ö—C̈l—Ö:

chloric acid HClO₃ H—Ö—C̈l—Ö:
                              |
                             :O:

perchloric acid HClO₄
                     :O:
                      |
                 H—Ö—C̈l—Ö:
                      |
                     :O:

As the number of oxygen atoms surrounding the chlorine increases, so does the acidity. Thus HOCl is a relatively weak acid, $HClO_2$ is stronger, $HClO_3$ is fully dissociated in water, and $HClO_4$ is just about the strongest protonic acid there is.

This reasoning can also be extended to other compounds. For example, the substance ethanol (ethyl alcohol) is, for all practical purposes, undissociated in water. Its structure is

and there is virtually no tendency for the O—H bond to split off a proton. However, if the two hydrogen atoms that are attached to the carbon, which is also bound to the OH group, are replaced by oxygen to give

the acidity of the molecule is greatly enhanced (the compound is acetic acid). This new oxygen atom draws electron density from the carbon atom which, in turn, draws electronic charge from the C—OH bond. As before, the oxygen atom, not wishing to lose electron density, removes negative charge from the O—H bond and the hydrogen becomes more positively charged. As a result it can be more readily removed as H⁺. This acidity can, in fact, be further increased by substituting a very electronegative element, such as Cl or F, for one or more of the hydrogen atoms of the CH_3 group. Thus chloroacetic acid is a stronger acid than ordinary acetic acid

chloroacetic acid   acetic acid

Following this same reasoning we can also understand why certain solutions of metal salts show acid properties. The acidity of these solutions is explained by the ability of the metal ion to polarize the water molecules that surround it in solution. For instance, in solutions of aluminum salts there is evidence that the Al^{3+} ion is surrounded by six water molecules that are quite tightly bound to the metal ion, as shown in Figure 13.1. The high positive charge on the metal tends to draw electron density from the oxygen atoms of the water molecules and, as we have already seen, this leads to an increase in the positive charge on the hydrogen atoms. Consequently, when this hydrated aluminum ion collides with a water molecule, a proton can be rather easily transferred to H_2O to give the hydronium ion

$$Al(H_2O)_6^{3+} + H_2O \rightleftharpoons Al(H_2O)_5OH^{2+} + H_3O^+$$

This reaction is sometimes viewed as simply a reaction of the metal ion with a water molecule,

$$Al^{3+} + H_2O \rightleftharpoons AlOH^{2+} + H^+$$

In this last equation we have left out the five water molecules that have not changed, as well as the one that picked up the proton. Such a reaction is called **hydrolysis** (reaction with water). In general, *when a substance undergoes reaction with the solvent, the process is referred to as **solvolysis.***

In summary, the acidity of an OH bond in M—O—H depends on the ability to the substance M to draw electrons to itself. It is not surprising to find, then, that as the oxidation number of an element covalently bound to an OH group increases, so does the acidity of the compound. In H_2SO_3 the oxidation number of S is +4; in H_2SO_4 it is +6. As we have said, H_2SO_4 is a stronger acid than H_2SO_3. Similarly, in the sequence HOCl, $HClO_2$, $HClO_3$, $HClO_4$ the acid strength parallels the increase in oxidation number of the chlorine. With metal ions, too, the acidity of their solutions depends on the charge on the metal ion. Aqueous solutions of $FeCl_2$ are only slightly acidic while $FeCl_3$ solutions show extensive hydrolysis.

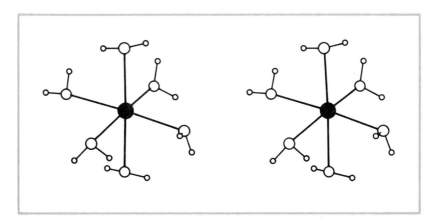

Figure 13.1
The hydrated aluminum ion, $Al(H_2O)_6^{3+}$. The aluminum ion (large solid sphere) is surrounded octahedrally by six water molecules.

Thus far we have followed the trends in acidity for several series of acids in which the central atom remains the same as we vary those surrounding it. We can also examine the change in acid strength as the atom to which the OH groups are bound is varied. For instance, in the series H_2SO_4, H_2SeO_4, H_2TeO_4 the acidity decreases as we go from sulfur to selenium to tellurium. In this case, as we proceed down within Group VIA, the atoms become larger and the positive charge induced by the oxygen atoms is spread out over a larger volume. As a result, the central atom becomes less effective in polarizing the O—H bonds and the acidity of the compounds decreases. It is thus a general observation that as we go down within a group in the periodic table, the strengths of the oxoacids decrease.

In addition to the oxoacids we also must try to account for the relative strengths of the binary, or hydro acids in which one or more hydrogen atoms are bound directly to a nonmetal; for example, HF, HCl, and H_2S. When we consider binary acids derived from elements belonging to the same group, an increase in acidity with increasing atomic number is observed experimentally. Thus the hydrohalides increase in acidity in the order $HF \ll HCl < HBr < HI$.

At first glance this order of acidity seems opposite to what we might predict. We know, for instance, that fluorine is more electronegative than chlorine; therefore the HF bond is more polar than the HCl bond. Consequently, the H in HF is more positively charged than the H in HCl. Thus we are tempted to predict that HF should lose a proton more readily than HCl. This, however, is precisely the reverse of their acid strengths in water where HF is weak and HCl is essentially 100% ionized.

The resolution of this dilemma can be understood by realizing that there are actually two opposing factors that contribute to the acidity of these compounds. One is, indeed, the ionic character of the H—X bond; the other is the H—X bond strength. As we proceed down within a group, the nonmetal becomes progressively larger and there is an accompanying rapid decrease in the strength of the H—X bond. This weakening of the H—X bond turns out to be more than sufficient to compensate for the decrease in the polarity of the bonds and a net increase in acid strength is observed.

When we look at the acidity of the binary hydrogen compounds of elements in the same period, for example NH_3, H_2O, and HF, the dominant factor becomes the polarity of the H—X bond. As we go from left to right within the period there is little change in the size of the nonmetal and relatively little change in the H—X bond energy. There is, however, a very dramatic increase in the ionic character of the H—X bond, which is reflected in a rapid increase in acidity from ammonia to hydrogen fluoride.

The Brønsted-Lowry definition of acids and bases is superior to the Arrhenius definition because it removes the restriction of having to deal with reactions in aqueous solution. However, even the Brønsted-Lowry concept is restricted in scope, since it limits discussion of acid-base phenomena to proton transfer reactions. There are many reactions that have all the earmarks of acid-base reactions but that do not fit the Brønsted-Lowry mold. The approach taken by G. N. Lewis further extends the acid-base concept to cover these cases.

In the Lewis definition of acids and bases, primary attention is focused on the base. A base is defined as a substance that can donate a *pair* of electrons

13.5
Lewis Acids and Bases

to the formation of a covalent bond. An acid, then, is a substance that can accept a pair of electrons to form the bond.

A simple example of an acid-base neutralization is the reaction of a proton with hydroxide ion.

The hydroxide ion is the Lewis base because it furnishes the pair of electrons that become shared with the hydrogen. The hydrogen ion, on the other hand, is the Lewis acid because it accepts a share of the pair of electrons when the O—H bond is created.

Another example of neutralization is the reaction between BF_3 and ammonia.

In this case the NH_3 functions as the base and BF_3 serves as the acid. Compounds containing elements with incomplete valence shells, such as BF_3 or $AlCl_3$, tend to be Lewis acids while compounds or ions that have lone pairs of electrons can behave as Lewis bases.

Still other examples of Lewis acid-base reactions are provided by the reaction of metal oxides with nonmetal oxides. Recall that metal oxides, in water, produce hydroxides. For instance,

$$Na_2O + H_2O \longrightarrow 2NaOH$$

Nonmetal oxides react to form acids as illustrated by the reaction

$$SO_3 + H_2O \longrightarrow H_2SO_4$$

When these two solutions are mixed, neutralization occurs with the production of the solvent plus a salt,

$$2NaOH + H_2SO_4 \longrightarrow 2H_2O + Na_2SO_4$$

The production of Na_2SO_4 from Na_2O and SO_3 can take place directly without the introduction of any water whatsoever as shown by the equation

$$Na_2O + SO_3 \longrightarrow Na_2SO_4$$

According to the Lewis definition this too is a neutralization reaction between a Lewis base (oxide ion) and a Lewis acid (sulfur trioxide).

base   acid   sulfate ion

In this case we find that some electronic rearrangement must take place as the oxygen becomes attached to the sulfur. Nevertheless, the overall change can be viewed as a neutralization reaction. A reaction quite analogous to that just described has been used on spacecraft to remove carbon dioxide from the air breathed by the astronauts. In this case carbon dioxide is reacted with LiOH

$$CO_2(g) + LiOH(s) \longrightarrow LiHCO_3(s)$$
(lithium bicarbonate)

(Lithium hydroxide is used because of the very low atomic weight of Li. This results in many moles of LiOH per pound.) This reaction too can be viewed as a Lewis acid-base reaction.

bicarbonate ion

The reaction between Na_2O and SO_3 illustrates the limitations of the Brønsted-Lowry concept. Since no protons are involved in the reaction, it would never have been classified as an acid-base reaction under the Brønsted-Lowry definition.

Thus far we have looked only at simple acid-base neutralizations. The acid-base reactions discussed in Sections 13.2 and 13.3 can also be treated from the Lewis point of view. In the Brønsted-Lowry theory these reactions were looked upon as competitions in which the strongest acid prevails by losing its proton. Under the Lewis definition these reactions are looked upon as constituting the displacement of one base (the weaker one) by another. Referring to our reaction of HCl with H_2O, for example,

$$HCl + H_2O \longrightarrow H_3O^+ + Cl^-$$

the Lewis theory interprets the change as the result of replacing the Cl^- ion in HCl by the stronger base, H_2O.

In other words, the stronger base, H_2O, pushes out the weaker one, Cl^-. Here we interpret the acid to be the H^+ ion rather than the entire HCl molecule.

By its definition a Lewis base is a substance that in its reactions seeks a nucleus with which it can share a pair of electrons; it is thus said to be a **nucleophile** (nucleus loving). A base displacement reaction therefore is called a **nucleophilic displacement,** since one Lewis base (nucleophile) pushes out another. From this point of view a very large number of chemical reactions can be looked on simply as acid-base reactions in which a stronger base displaces a weaker one.

There are also acid-base reactions in which one Lewis acid displaces another. For example, consider the reaction

$$AlCl_3 + COCl_2 \longrightarrow COCl^+ + AlCl_4^-$$

Using electron-dot formulas we can analyze this reaction as shown below.

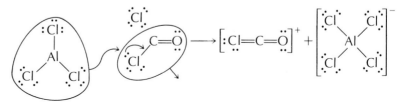

Notice that we view this reaction to take place by the displacement of the acid, $COCl^+$, by the stronger acid $AlCl_3$. In other words, we imagine the molecule, $COCl_2$, to be the product of the Lewis acid $COCl^+$ and the Lewis base Cl^-, and it is the base that changes partners when the displacement reaction occurs.

$$AlCl_3 + COCl_2 \longrightarrow [Cl_3Al \cdots\cdots Cl^- \cdots\cdots COCl^+] \longrightarrow AlCl_4^- + COCl^+$$

Since Lewis acids seek substances having electron pairs to which they can become bound, Lewis acids are said to be **electrophilic** (electron loving) in character and an acid displacement reaction is said to be an **electrophilic displacement.**

With a little practice it becomes relatively simple to distinguish between nucleophilic and electrophilic displacement processes. Often, in a nucleophilic displacement one of the products is derived from a reactant species by the loss of one or more atoms and an *increase in negative charge*. For instance, in the reaction of H_2O with HCl the product Cl^- is derived from the reactant, HCl, by the loss of a proton. The product species, Cl^-, has a charge of minus one while the reactant from which it is formed is neutral. In this example, the chloride ion is the base that has been displaced.

By comparison, in an electrophilic displacement reaction it frequently happens that one of the products is derived from a reactant by the loss of one or more atoms and an *increase in positive charge*. Thus $COCl^+$ is produced from $COCl_2$ by the loss of Cl^-. The species $COCl^+$ carries a charge of plus one while the $COCl_2$ molecule from which it is formed is neutral. In this case our analysis has identified $COCl^+$ as the acid that has been displaced. We can also conclude that the Cl^- ion is the base that is changing partners.

13.6
The Solvent System Approach to Acids and Bases

Because water is so plentiful and is such a good solvent for so many substances, much of the solution chemistry that we meet in the laboratory uses water as the solvent. Chapter 5 dealt with the various kinds of reactions that are encountered in aqueous solutions and the definitions of acids and bases that were presented there were given in terms of the ions, H_3O^+ and OH^-. As mentioned before, water undergoes a limited degree of autoionization that can be described by the equation

$$H_2O + H_2O \rightleftharpoons H_3O^+ + OH^-$$

Thus we see that an acid in water is a substance that yields, in some way, the cation characteristic of the solvent. Similarly, a base is a substance that gives the anion derived from the solvent. These last two observations form the basis of the **solvent system** approach to dealing with acid-base reactions. This approach recognizes that water is really not unique in its solvent properties and that many reactions in other solvents can be viewed as analogs of reactions that take place in aqueous media.

LIQUID AMMONIA AS A SOLVENT. Probably the most thoroughly studied solvent besides water is liquid ammonia. At atmospheric pressure ammonia exists as a gas at room temperature, it is therefore necessary to cool the substance to its boiling point, $-33°$ C, or below, in order to work with the liquid, which leads to some minor experimental difficulties.

Liquid ammonia, as a solvent, has many properties that are similar to water. Like water, it undergoes a slight degree of autoionization. When it does so, it produces ammonium ion and amide ion, NH_2^-.

$$NH_3 + NH_3 \rightleftharpoons NH_4^+ + NH_2^-$$

By analogy with water we would predict that an acid in liquid NH_3 is any substance capable of forming ammonium ion. Thus any ammonium salt, such as NH_4Cl for example, should show acid properties. In fact, in liquid ammonia NH_4Cl is exactly analogous to the product of the reaction of HCl with H_2O in aqueous solution, that is, $(H_3O)Cl$.

Following the solvent system definition, a base in liquid ammonia is any substance capable of providing amide ion—for example, KNH_2, potassium amide.

In water the neutralization of an acid and a base occurs by the reaction

$$H_3O^+ + OH^- \longrightarrow 2H_2O$$

Thus the cation and anion of the solvent combine in neutralization to produce the solvent. In liquid ammonia this corresponds to the reaction

$$NH_4^+ + NH_2^- \longrightarrow 2NH_3$$

and when solutions of NH_4Cl and KNH_2 are mixed, the neutralization process

$$NH_4Cl + KNH_2 \longrightarrow KCl + 2NH_3$$

takes place. This is analogous to the reaction, in water, between HCl and KOH

$$HCl + KOH \longrightarrow KCl + H_2O$$

or

$$(H_3O)Cl + KOH \longrightarrow KCl + 2H_2O$$

A further similarity between the acid-base phenomena in water and liquid ammonia is revealed by the behavior of indicators. For example, in water phenolphthalein is pink in basic solutions and colorless in acid solutions. This fact is employed in acid-base titrations where there is a sharp color change when neutralization is achieved. This same indicator can also be used for performing titrations with liquid ammonia as a solvent. In basic solutions containing an excess of NH_2^- ion the phenolphthalein is pink while in acid solutions it is colorless.

A further similarity between acid-base behavior in these two solvents is provided by the amphoteric behavior of certain metals. For example, if an aqueous solution of ZnI_2 is treated with KOH, a precipitate of $Zn(OH)_2$ is formed that redissolves upon further addition of base to produce $K_2[Zn(OH)_4]$, as indicated below.

$$ZnI_2 + 2KOH \longrightarrow Zn(OH)_2 \ (s) + 2KI.$$
$$Zn(OH)_2 + 2KOH \longrightarrow K_2[Zn(OH)_4]$$

In liquid ammonia we find precisely the same behavior.

$$ZnI_2 + 2KNH_2 \longrightarrow Zn(NH_2)_2 \ (s) + 2KI$$
$$Zn(NH_2)_2 + 2KNH_2 \longrightarrow K_2[Zn(NH_2)_4]$$

In each of these cases the addition of acid (H_3O^+ in water or NH_4^+ in ammonia) causes the reprecipitation of the zinc hydroxide, or amide; further addition of acid will cause these precipitates to dissolve, thereby regenerating the solvated zinc ion.

The similarities between the neutralization reactions in water and ammonia have also suggested other chemical parallels. By analogy we have established that H_3O^+ and NH_4^+ occupy corresponding positions in the water (aquo) and ammonia (ammono) systems. Similarly, OH^- and NH_2^- stand opposite one another. This leads us to postulate further that the oxide ion, O^{2-}, in the aquo system should be equivalent to the *imide ion*, NH^{2-}, or the *nitride ion*, N^{3-}, in the ammono system. In fact, extension of this reasoning leads us to a whole series of compounds that are analogous to each other in the two solvent systems. Some examples are given in Table 13.1.

In addition to acid-base neutralization there are many other parallels in the reactions that take place in water and liquid ammonia. For instance, we know that in water metal oxides react with the solvent to produce hydroxides

$$Li_2O + H_2O \longrightarrow 2LiOH$$

In liquid ammonia we find the similar reaction

$$Li_3N + 2NH_3 \longrightarrow 3LiNH_2$$

Another example is the reaction of a metal hydride with the solvent.

$$NaH + H_2O \longrightarrow H_2 + Na^+ + OH^- \qquad \text{(aquo system)}$$
$$NaH + NH_3 \longrightarrow H_2 + Na^+ + NH_2^- \qquad \text{(ammono system)}$$

Still another similarity is the reaction of an active metal with an acid to yield hydrogen plus a salt.

$$Ca + 2(H_3O)Cl \longrightarrow CaCl_2 + H_2 + 2H_2O \qquad \text{(aquo system)}$$
$$Ca + 2NH_4Cl \longrightarrow CaCl_2 + H_2 + 2NH_3 \qquad \text{(ammono system)}$$

In Chapter 5 we saw that metathesis reactions provide a convenient route for the synthesis of certain compounds. In nonaqueous solvents too we can have these kinds of chemical changes. Because of different solubility relationships in the aqueous and nonaqueous media, however, it is sometimes possible to prepare compounds by metathesis in a solvent such as ammonia

Table 13.1
Analogous Compounds in the Aquo and Ammono Systems

Aquo Compound	Ammono Compound
H_2O	NH_3
$(H_3O)Cl$ or HCl	NH_4Cl
KOH	KNH_2
Li_2O	Li_3N
P_2O_5 (actually P_4O_{10})	P_3N_5
$PO(OH)_3$ or H_3PO_4	$[PN(NH_2)_2]_3$ $P(NH)(NH_2)_3$
CH_3OH	CH_3NH_2
C_2H_5OH	$C_2H_5NH_2$
$(CH_3)_2O$	$(CH_3)_3N$, $(CH_3)_2NH$
H_2O_2	N_2H_4
$HONO_2$ or HNO_3	HNN_2
$HOCl$	H_2NCl
$CO(OH)_2$ or H_2CO_3	$C(NH)(NH_2)_2$
$CH_3COOC_2H_5$	$CH_3C(NH)(NHC_2H_5)$
CH_3COOH	$CH_3C(NH)(NH_2)$
$Cu(H_2O)_4{}^{2+}$	$Cu(NH_3)_4{}^{2+}$
$Si(OH)_4$	$Si(NH_2)_4$
$B(OH)_3$	$B(NH_2)_3$

that cannot be made in the same way in water. For example, in liquid ammonia $Ba(NO_3)_2$ and $AgCl$ are both soluble. When their solutions are mixed, a white precipitate of $BaCl_2$ is produced.

$$Ba(NO_3)_2 + 2AgCl \xrightarrow{\text{NH}_{3(l)}} BaCl_2(s) + 2AgNO_3$$

This is, of course, exactly the reverse of the reaction that takes place in aqueous solution where $BaCl_2$ and $AgNO_3$ are soluble and form a precipitate of $AgCl$ when their aqueous solutions are mixed.

The examples in the preceding section demonstrate that the solvent system approach to acids and bases has some useful and appealing features. One of these expands our thinking to reactions in solvents other than water. However, this view of acids and bases suffers from some of the same faults as the Arrhenius theory. For example, it does not allow us to consider, as acid-base interactions, reactions that take place in the absence of a solvent.

13.7 Summary

The Brønsted-Lowry approach frees us from the restriction of having to have a solvent present, although it limits us to dealing with systems in which there is proton transfer. Clearly, the most general approach that we have considered is the Lewis theory. All of the acid-base reactions that we have examined under the various headings in this chapter can be interpreted from the Lewis point of view.

The choice of which definition of acids and bases one wishes to use in a particular instance depends largely on the sort of chemistry that is studied. For example, in a practical sense the Arrhenius definition is perfectly satisfactory for dealing with the reactions in aqueous solution that you will encounter in the laboratory.

Review Questions and Problems

13.1 Give three properties that are characteristic of acids and bases in general.

13.2 What is the Arrhenius definition of an acid and a base?

13.3 Identify each of the following as Arrhenius acids or bases. For each, write a chemical equation to show its reaction with water. If necessary, refer to Table 5.3.
(a) P_4O_{10} (d) HBr
(b) CaO   (e) H_2O
(c) NH_3OH^+ (f) $Al(H_2O)_6^{3+}$

13.4 What is the Brønsted-Lowry definition of an acid and a base?

13.5 Identify the two acid-base conjugate pairs in each of the following reactions;
(a) $C_2H_3O_2^- + H_2O \rightleftharpoons$ $OH^- + HC_2H_3O_2$
(b) $HF + NH_3 \rightleftharpoons NH_4^+ + F^-$
(c) $Zn(OH)_2 + 2OH^- \rightleftharpoons$ $ZnO_2^- + 2H_2O$
(d) $Al(H_2O)_6^{3+} + OH^- \rightarrow$ $Al(H_2O)_5OH^{2+} + H_2O$
(e) $N_2H_4 + H_2O \rightleftharpoons N_2H_5^+ + OH^-$
(f) $NH_2OH + HCl \rightleftharpoons NH_3OH^+ + Cl^-$
(g) $O^{2-} + H_2O \rightleftharpoons 2OH^-$
(h) $H^- + H_2O \rightleftharpoons H_2 + OH^-$
(i) $NH_2^- + N_2H_4 \rightleftharpoons NH_3 + N_2H_3^-$
(j) $HNO_3 + H_2SO_4 \rightleftharpoons$ $H_3SO_4^+ + NO_3^-$

13.6 Hydrogen sulfide is a stronger acid than phosphine, PH_3. What may we conclude about the strengths of their conjugate bases, HS^- and PH_2^-?

13.7 What is the leveling effect? How would it apply to the strong acids HCl and HBr? Suggest a substance that might serve as a differentiating solvent for these two acids.

13.8 Predict the order of acid strength for each of the following protonic acids.
(a) H_2S and H_2Se
(b) H_2Se and HBr
(c) PH_3 and NH_3
(d)

(g) $HClO_3$ and $HBrO_3$

(h) $HOBr$ and $HBrO_3$

13.9 Why is HNO_3 a stronger acid than HNO_2?

13.10 Why is HCl a stronger acid than HF?

13.11 What is the Lewis definition of an acid and a base?

13.12 Boron trichloride, BCl_3, reacts with diethyl ether, $(C_2H_5)_2O$ to form an *addition compound*, which we can write as $Cl_3B \leftarrow O(C_2H_5)_2$. Use electron-dot formulas to explain why this reaction is interpreted as a Lewis acid-base neutralization.

13.13 Interpret each of the acid-base reactions in Question 13.5 above as displacement reactions according to the Lewis concept.

13.14 Explain why the reaction of CO_2 with H_2O to produce H_2CO_3, which we can also write as $CO(OH)_2$, can be viewed as a Lewis acid-base neutralization.

13.15 Analyze the following as either electrophilic or nucleophilic displacement reactions:

(a) $N_2O_4 + BF_3 \rightarrow NO_2^+ + NO_2BF_3^-$

(b) $SiCl_4 + 4H_2O \rightarrow Si(OH)_4 + 4HCl$

(c) $(CH_3)_3CCl + OH^- \rightarrow$
$(CH_3)_3COH + Cl^-$

(d) $SnCl_4 + 2SeOCl_2 \rightarrow$
$SnCl_6^{2-} + SeOCl^+$

(e) $Co(NH_3)_6^{3+} + Cl^- \rightarrow$
$Co(NH_3)_5Cl^{2+} + NH_3$

(f) $Br_2 + FeBr_3 \rightarrow Br^+ + FeBr_4^-$

13.16 What is the basic concept behind the solvent system approach to acids and bases?

13.17 How do the Arrhenius, Brønsted-Lowry, Lewis and Solvent System concepts of acids and bases differ in terms of their general applicability?

13.18 Concentrated sulfuric acid undergoes the autoionization reaction

$$2H_2SO_4 \rightleftharpoons H_3SO_4^+ + HSO_4^-$$

In this solvent, acetic acid behaves as a

base and perchloric acid behaves as an *acid*. Write chemical equations to show the following.

(a) The reaction of $HC_2H_3O_2$ with the solvent, H_2SO_4.

(b) The reaction of $HClO_4$ with the solvent.

(c) The neutralization reaction that occurs when solutions of $HC_2H_3O_2$ and $HClO_4$ in H_2SO_4 are mixed.

13.19 From the data in Table 13.1, write equations for reactions in liquid ammonia that are analogous to the following reactions that take place in aqueous solution.

(a) $P_4O_{10} + 6H_2O \rightarrow 4H_3PO_4$

(b) $Cl_2 + H_2O \rightarrow HCl + HOCl$

(c) $CH_3COOC_2H_5 + H_2O \xrightarrow{H_3O^+}$
$CH_3COOH + C_2H_5OH$

(d) $Zn + 2HCl \rightarrow ZnCl_2 + H_2$

(e) $H_2CO_3 + 2OH^- \rightarrow CO_3^{2-} + 2H_2O$

(f) $SiCl_4 + 4H_2O \rightarrow Si(OH)_4 + 4HCl$

(g) $Zn + 2OH^- + 2H_2O \rightarrow$
$Zn(OH)_4^{2-} + H_2$

(h) $Cu^{2+} + 4H_2O \rightarrow Cu(H_2O)_4^{2+}$

(i) $BCl_3 + 3H_2O \rightarrow 3HCl + B(OH)_3$

13.20 Liquid hydrogen cyanide can be thought to undergo the autoinization

$$2HCN \rightleftharpoons H_2CN^+ + CN^-$$

(a) Would KCN be considered an acid or a base in this solvent?

(b) H_2SO_4 is an acid in $HCN(l)$. Write the equation for the ionization of H_2SO_4 in this solvent.

(c) $(CH_3)_3N$ is a base in $HCN(l)$. Write an equation for the reaction of $(CH_3)_3N$ with the solvent. Using electron-dot formulas, show how this reaction takes place.

(d) What is the ionic equation for the neutralization of H_2SO_4 by $(CH_3)_3N$ in liquid HCN?

(e) What is the net ionic equation for the neutralization reaction in this solvent?

14
Ionic
Equilibria

In Chapter 5 we saw that many of the reactions that are of concern to the chemist take place in aqueous solution. We now realize that chemical changes do not, in general, proceed entirely to completion but instead approach a state of dynamic equilibrium. In this chapter we are concerned with the equilibria involving ionic species in solution, including such things as the dissociation (or ionization) of water and other weak electrolytes and also the solubility of substances that we have considered, until now, to be essentially insoluble.

14.1

Ionization of Water, pH

In the last chapter we saw that some solvents can be considered to undergo autoionization. Recall, for example, that the autoionization of pure water is written as

$$H_2O + H_2O \rightleftharpoons H_3O^+ + OH^-$$

Since this is an equilibrium, we can write an equilibrium expression for this reaction. Following the concepts developed in Chapter 12, this can be represented as

$$K = \frac{[H_3O^+][OH^-]}{[H_2O][H_2O]}$$

The molar concentration of water, which appears in the denominator of this expression, is very nearly constant ($\approx 55.6M$) in both pure water and also in dilute aqueous solutions. Therefore, $[H_2O]^2$ can be included with the equilibrium constant, K, on the left side of the above equation. We would then write

$$K \cdot [H_2O]^2 = [H_3O^+][OH^-]$$

The left side of this expression is the product of two constants which, of course, is also equal to a constant that we shall call K_w.

$$K_w = K[H_2O]^2$$

Our equilibrium condition therefore becomes

$$K_w = [H_3O^+][OH^-]$$

Since $[H_3O^+][OH^-]$ is the product of ionic concentrations, K_w is called the **ion product constant** for water, or frequently simply the **ionization constant** or **dissociation constant** of water.

The equation for the autoionization of water can be simplified somewhat by omitting the solvent from the expression. The dissociation reaction for water is then

$$H_2O \rightleftharpoons H^+ + OH^-$$

and the simplified expression for the dissociation constant of water is written as

$$K_w = [H^+][OH^-] \qquad (14.1)$$

The equilibrium dissociation of H_2O is present in any aqueous solution and Equation 14.1 must always be fulfilled regardless of what other equilibria may also be taking place in solution.

The ionization constant of water at 25° C has been found to have a value of 1.0×10^{-14} and can be used to calculate the molar concentrations of both the H^+ and OH^- ions in pure water. From the stoichiometry of the dissociation we see that, whenever one mole of H^+ ions are formed, one mole of OH^- ions are also produced. This means that, at equilibrium, $[H^+] = [OH^-]$. If we let x equal the hydrogen ion concentration, then

$$x = [H^+] = [OH^-]$$

Substituting into Equation 14.1 gives

$$K_w = x \cdot x = x^2$$

or, since $K_w = 1.0 \times 10^{-14}$,

$$x^2 = 1.0 \times 10^{-14}$$

Taking the square root yields

$$x = 1.0 \times 10^{-7}$$

which means that the concentrations of hydrogen ions and hydroxide ions in pure water are

$$[H^+] = [OH^-] = 10^{-7}M$$

Whenever the hydrogen ion concentration equals the hydroxide ion concentration, as it does in pure water, the solution is said to be *neutral*. The addition to this neutral solution of any substance that raises the H^+ concentration above the concentration of OH^- is an acid; conversely, any substance that raises the OH^- concentration above the H^+ concentration is a base.

In an aqueous solution of an acid it is safe to assume that the hydrogen ion concentration is derived almost entirely from the solute. For example, in a 0.001 M solution of HCl we have 1×10^{-3} mole per liter of H^+, from the virtually complete dissociation of the strong acid, plus an additional small amount of H^+ from the dissociation of water. By applying Le Chatelier's principle (Section 12.8) to the reaction,

$$H_2O \rightleftharpoons H^+ + OH^- \qquad (14.2)$$

we know that an external source of either H^+ or OH^- is going to suppress the dissociation of water. Therefore, in the 10^{-3} M HCl solution, the quantity of H^+ produced by the dissociation of water is much less than that present in pure H_2O. In other words, we make the very valid assumption that in solutions of acids the amount of H^+ in the mixture that comes from water is negligible. Similarly, in solutions of bases we assume that the quantity of OH^- that is derived from the dissociation of water is negligible compared to the quantity produced in solution by the base. Therefore we will take the H^+ concentration in a $0.001M$ HCl solution, for example, to be $0.001M$. Likewise, a $0.001M$ solution of the strong base NaOH would have a hydroxide ion concentration equal to $0.001M$.

Example 14.1
(a) What is the OH^- concentration in a $0.001M$ HCl solution?
(b) What is the H^+ concentration derived from the dissociation of the solvent?

Solution
   (a) At equilibrium we must have,

$$[H^+][OH^-] = 1.0 \times 10^{-14}$$

We have said that in a $0.001M$ HCl solution, the total H^+ concentration can be represented as $0.001M$. If we solve for the hydroxide ion concentration,

$$[OH^-] = \frac{1.0 \times 10^{-14}}{[H^+]}$$

and therefore,

$$[OH^-] = \frac{1.0 \times 10^{-14}}{1 \times 10^{-3}} = 1 \times 10^{-11}M$$

   (b) The hydroxide ion in part (a) comes entirely from the dissociation of water. The amount of H^+ derived from H_2O must therefore *also* be $1 \times 10^{-11}M$, as can be seen from the stoichiometry of Equation 14.2. Note that this value ($1 \times 10^{-11}M$) is indeed negligible compared to the H^+ concentration produced by the HCl ($1 \times 10^{-3}M$)

Hydrogen ion and hydroxide ion enter into many equilibria in addition to the dissociation of water; therefore it is frequently necessary to specify their concentrations in aqueous solutions. These concentrations may range from relatively high values to very small ones (e.g., $1M$ to $10^{-14}M$) and a logarithmic notation has been devised to simplify the expression of these quantities. In general, for some quantity X,

$$pX = \log \frac{1}{X} = -\log X \tag{14.3}$$

For example, if we wish to denote the hydrogen ion concentration in a solution, we speak of pH, defined as

$$pH = \log \frac{1}{[H^+]} = -\log [H^+]$$

In a solution where the hydrogen ion concentration is $10^{-3}M$ we therefore have[1]

---

[1] A discussion on the use of logarithms can be found in Appendix C.

$$pH = -\log (10^{-3}) = -(-3.0)$$

or

$$pH = 3.0$$

Similarly, if the hydrogen ion concentration is $10^{-8}M$, the pH of the solution would be 8.0.

Following the same approach for the hydroxide ion concentration we can define the pOH of a solution as

$$pOH = -\log [OH^-]$$

From the equilibrium expression for the dissociation of water

$$\log K_w = \log [H^+] + \log [OH^-]$$

Multiplying through by -1 gives

$$(-\log K_w) = (-\log [H^+]) + (-\log [OH^-])$$

or simply

$$pK_w = pH + pOH$$

Since $K_w = 1.0 \times 10^{-14}$, $pK_w = 14.0$, therefore we find that

$$pH + pOH = 14.0$$

In a neutral solution, $[H^+] = [OH^-] = 10^{-7}M$, and $pH = pOH = 7.0$, so that in a neutral solution we say the $pH = 7.0$. In an acidic solution the hydrogen ion concentration is greater than $10^{-7}M$ (e.g., $10^{-3}M$) and the pH is less than 7.0. By the same token, in basic solutions the $[H^+]$ is less than $10^{-7}M$ (e.g., $10^{-10}M$) and the pH is greater than 7.0.

Let us now look at a few sample problems dealing with typical calculations involving pH.

Example 14.2
What is the pH of a 0.0020M HCl solution?

Solution
Since HCl is a strong acid, we can write

$$[H^+] = 0.0020M = 2.0 \times 10^{-3}M$$

We know that

$$pH = -\log [H^+]$$

and for this problem,

$$pH = -\log (2.0 \times 10^{-3})$$
$$pH = -(\log 2.0 + \log 10^{-3})$$
$$pH = -(0.30 + (-3))$$
$$pH = -(-2.7)$$
$$pH = 2.7$$

Example 14.3
What is the pH of a $5.0 \times 10^{-4}M$ NaOH solution?

Solution

In this example the OH^- concentration is given and we are asked to compute the pH. We can solve this problem in either of two ways; (1) knowing K_w for water and $[OH^-]$ for this solution, we can calculate $[H^+]$ using Equation 14.1 and then proceed as we did in Example 14.1, or (2) we can calculate pOH from the OH^- concentration and subtract it from pK_w to obtain pH.

Method 1. We know that

$$K_w = [H^+][OH^-]$$

and therefore

$$[H^+] = \frac{1.0 \times 10^{-14}}{5.0 \times 10^{-4}}$$

$$[H^+] = 0.2 \times 10^{-10} \quad \text{or} \quad 2.0 = 10^{-11}M$$

Now proceeding as we did in Example 14.2,

$$pH = -\log (2.0 + 10^{-11})$$
$$pH = -(0.3 + (-11))$$
$$pH = 10.7$$

Method 2. By definition we have

$$pOH = -\log [OH^-]$$

For this problem

$$pOH = -\log (5 \times 10^{-4})$$
$$pOH = -(0.7 + (-4))$$
$$pOH = 3.3$$

The pH would be

$$pH = pK_w - pOH$$
$$pH = 14 - 3.3$$
$$pH = 10.7$$

Example 14.4

What are the hydrogen ion and hydroxide ion concentrations in a solution whose pH is 4.40?

Solution

In order to compute $[H^+]$ from pH we must reverse the procedure that we followed in Example 14.2. We know that

$$pH = -\log [H^+] = 4.40$$

or

$$\log [H^+] = -4.40$$

Let us write this as the sum of a decimal fraction plus a negative integer. Thus, -4.40 is the same as 0.60 plus -5.00

$$\log [H^+] = 0.60 + (-5.00)$$

Since 0.60 is the logarithm of 4, and -5 is the logarithm of 10^{-5}, we can write

$$\log [H^+] = \log 4.0 + \log 10^{-5}$$

or

$$\log [H^+] = \log (4.0 \times 10^{-5})$$

Hence

$$[H^+] = 4.0 \times 10^{-5} M$$

To calculate $[OH^-]$ we can divide K_w by $[H^+]$

$$[OH^-] = \frac{1.0 \times 10^{-14}}{4.0 \times 10^{-5}}$$
$$[OH^-] = 2.5 \times 10^{-10} M$$

Alternatively, we could first have computed the pOH of the solution

$$pOH = 14.00 - 4.40$$
$$pOH = 9.60$$

From pOH we can obtain $[OH^-]$

$$pOH = -\log [OH^-] = 9.60$$
$$\log [OH^-] = -9.60 = 0.40 - 10.0$$
$$\log [OH^-] = \log (2.5) + \log (10^{-10})$$
$$[OH^-] = 2.5 \times 10^{-10} M$$

As a class, weak electrolytes include weak acids and bases as well as certain salts, such as $HgCl_2$ and $CdSO_4$, that are not fully dissociated in aqueous solution. In solutions of these materials, there is an equilibrium between the undissociated species and its corresponding ions. For example, acetic acid ionizes according to the equation

$$HC_2H_3O_2 + H_2O \rightleftharpoons H_3O^+ + C_2H_3O_2^-$$

The equilibrium expression for this reaction would be

$$K = \frac{[H_3O^+][C_2H_3O_2^-]}{[HC_2H_3O_2][H_2O]}$$

As before, the concentration of H_2O is considered a constant that can be included with K on the left side of the equal sign. That is,

$$K \times [H_2O] = K_a = \frac{[H_3O^+][C_2H_3O_2^-]}{[HC_2H_3O_2]}$$

where we have used K_a to represent an **acid dissociation constant.** This same equilibrium expression can be obtained if we simplify the dissociation by omitting the solvent. Thus for the dissociation of acetic acid we would write

$$HC_2H_3O_2 \rightleftharpoons H^+ + C_2H_3O_2^-$$

and the equilibrium expression would then be

$$K_a = \frac{[H^+][C_2H_3O_2^-]}{[HC_2H_3O_2]}$$

In general, then, for any weak acid, HA, the simplified dissociation reaction would be written as

$$HA \rightleftharpoons H^+ + A^-$$

and its acid dissociation constant would be given by

$$K_a = \frac{[H^+][A^-]}{[HA]}$$

This same approach can also be applied to weak bases. According to the Arrhenius theory, when a weak base ionizes in aqueous solution, it does so by picking up an H^+ ion from the solvent. For example, with NH_3 we have

$$NH_3 + H_2O \rightleftharpoons NH_4^+ + OH^-$$

If we omit the solvent, the base ionization constant, K_b, for this reaction is

$$K_b = \frac{[NH_4^+][OH^-]}{[NH_3]}$$

The extent to which a weak acid or base undergoes ionization, as well as the value of the ionization constant, must be determined experimentally. One way of doing this is to measure the pH of a solution prepared by dissolving a known quantity of the weak acid or base in a given volume of solution, as illustrated in Example 14.5. Dissociation constants of a number of weak acids and bases are listed in Table 14.1.

Example 14.5
A student prepared a $0.10M$ acetic acid solution and experimentally measured the pH of this solution to be 2.88. Calculate the K_a for acetic acid and its percent dissociation.

Solution
To evaluate K_a we must have equilibrium concentrations to substitute into the expression

$$K_a = \frac{[H^+][C_2H_3O_2^-]}{[HC_2H_3O_2]}$$

From pH we can obtain the H^+ concentration

$$pH = -\log [H^+] = 2.88$$
$$\log [H^+] = 0.12 - 3$$
$$\log [H^+] = \log 1.3 + \log 10^{-3}$$
$$[H^+] = 1.3 \times 10^{-3}M$$

This $[H^+]$ comes from the dissociation of $HC_2H_3O_2$

$$HC_2H_3O_2 \rightleftharpoons H^+ + C_2H_3O_2^-$$

and, from the stoichiometry of the equation we see that

$$[H^+] = [C_2H_3O_2^-] = 1.3 \times 10^{-3}M$$

Table 14.1
Ionization Constants for Some Weak Acids and Bases at 25° C

Weak Acids	Ionization	K_a	pK_a
Chlorous acid	$HClO_2 \rightleftharpoons H^+ + ClO_2^-$	1.1×10^{-2}	1.96
Chloroacetic acid	$HC_2H_2ClO_2 \rightleftharpoons H^+ + C_2H_2ClO_2^-$	1.4×10^{-3}	2.85
Formic acid	$HCHO_2 \rightleftharpoons H^+ + CHO_2^-$	1.8×10^{-4}	3.74
Hydrofluoric acid	$HF \rightleftharpoons H^+ + F^-$	6.5×10^{-4}	3.19
Acetic acid	$HC_2H_3O_2 \rightleftharpoons H^+ + C_2H_3O_2^-$	1.8×10^{-5}	4.74
Hypochlorous acid	$HOCl \rightleftharpoons H^+ + OCl^-$	3.1×10^{-8}	7.51
Hydrocyanic acid	$HCN \rightleftharpoons H^+ + CN^-$	4.9×10^{-10}	9.31

Weak Bases	Ionization	K_b	pK_b
Methyl amine	$CH_3NH_2 + H_2O \rightleftharpoons CH_3NH_3^+ + OH^-$	3.7×10^{-4}	3.43
Ammonia	$NH_3 + H_2O \rightleftharpoons NH_4^+ + OH^-$	1.8×10^{-5}	4.74
Hydrazine	$N_2H_4 + H_2O \rightleftharpoons N_2H_5^+ + OH^-$	9.5×10^{-7}	6.02
Pyridine	$C_5H_5N + H_2O \rightleftharpoons C_5H_5NH^+ + OH^-$	1.8×10^{-9}	8.74
Aniline	$C_6H_5NH_2 + H_2O \rightleftharpoons C_6H_5NH_3^+ + OH^-$	4.2×10^{-10}	9.38

The concentration of undissociated $HC_2H_3O_2$ at equilibrium is equal to the original concentration, $0.10M$, *minus* the number of moles per liter of acetic acid that have dissociated. At equilibrium, then,

$$[H^+] = 1.3 \times 10^{-3}$$
$$[C_2H_3O_2^-] = 1.3 \times 10^{-3}$$
$$[HC_2H_3O_2] = 1.0 \times 10^{-1} - 0.013 \times 10^{-1} = 1.0 \times 10^{-1}$$

Note that when we compute the acetic acid concentration to the *proper number of significant figures*, the quantity that has dissociated is negligible compared to the amount initially present. Thus

$$0.10 - 0.0013 = (0.0987) = 0.10$$

Substituting the equilibrium concentrations into the expression for K_a, we have

$$K_a = \frac{(1.3 \times 10^{-3})(1.3 \times 10^{-3})}{(1.0 \times 10^{-1})}$$
$$K_a = 1.7 \times 10^{-5}$$

The percent dissociation of acetic acid in this solution is found by dividing the number of moles per liter of $HC_2H_3O_2$ that have dissociated by the quantity of acetic acid that was available initially, all multiplied by 100.

$$\text{Percent dissociation} = \frac{\text{(mole/liter HC}_2\text{H}_3\text{O}_2 \text{ dissociated)}}{\text{(mole/liter HC}_2\text{H}_3\text{O}_2 \text{ available)}} \times 100$$

$$\text{Percent dissociation} = \frac{1.3 \times 10^{-3}M}{1.0 \times 10^{-1}M} \times 100 = 1.3\%$$

Example 14.6
A student prepared a 0.010M NH$_3$ solution and, by a freezing point lowering experiment, determined that the NH$_3$ had undergone 4.2% ionization. Calculate the K_b for NH$_3$.

Solution
Ammonia ionizes in water according to the reaction

$$NH_3 + H_2O \rightleftharpoons NH_4^+ + OH^-$$

for which we write

$$K_b = \frac{[NH_4^+][OH^-]}{[NH_3]}$$

From the stoichiometry of the ionization we see that at equilibrium

$$[NH_4^+] = [OH^-]$$

Since the 0.010M solution undergoes 4.2% ionization, the number of moles per liter of these ions present at equilibrium is

$$[NH_4^+] = [OH^-] = 0.042 \times 0.010M = 4.2 \times 10^{-4}M$$

The number of moles per liter of NH$_3$ at equilibrium would be

$$[NH_3] = 1.0 \times 10^{-2} - 0.04 \times 10^{-2} = 0.96 \times 10^{-2} = 1.0 \times 10^{-2}M$$

When these concentrations are substituted into the equation for K_b, we have

$$K_b = \frac{(4.2 \times 10^{-4})(4.2 \times 10^{-4})}{(1 \times 10^{-2})}$$

or

$$K_b = 1.8 \times 10^{-5}$$

From Examples 14.5 and 14.6 we know that K can be computed from a knowledge of equilibrium concentrations. We can also use our knowledge of K to calculate the concentrations in an equilibrium mixture. For example, let's calculate the concentrations of all of the species present in a 0.50M HC$_2$H$_3$O$_2$ solution.

For acetic acid, K has a value of 1.8×10^{-5}; therefore

$$\frac{[H^+][C_2H_3O_2^-]}{[HC_2H_3O_2]} = 1.8 \times 10^{-5}$$

We know that at equilibrium some of the acetic acid in the solution will have dissociated. If we therefore let x equal the number of moles/liter of HC$_2$H$_3$O$_2$ that dissociate, at equilibrium we shall have produced x mole/liter H$^+$, x

mole/liter $C_2H_3O_2^-$, and we shall have $(0.50 - x)$ mole/liter of $HC_2H_3O_2$ remaining undissociated.

Species	Equilibrium Concentration
H^+	x
$C_2H_3O_2^-$	x
$HC_2H_3O_2$	$0.50 - x$

Substituting these values into the equilibrium expression gives us

$$\frac{(x)(x)}{(0.50 - x)} = 1.8 \times 10^{-5}$$

Without simplification this expression leads to a quadratic equation that can be solved using the quadratic formula. However, in Example 12.8 we saw that it is sometimes possible to make simplifying assumptions that greatly reduce the effort required to obtain solutions to problems of this type. Because K is small, very little $HC_2H_3O_2$ will have actually undergone dissociation; hence x will be small. Let us assume that x will be negligible compared to 0.50, that is,

$$0.50 - x \approx 0.50$$

Our equation then becomes

$$\frac{x^2}{0.50} = 1.8 \times 10^{-5}$$

or

$$x = 3.0 \times 10^{-3}$$

If we look back on our assumption we see that x is in fact small compared to 0.50 and that *when rounded to the proper number of significant figures,*

$$0.50 - 0.003 = 0.50$$

Therefore the equilibrium concentrations of the species involved in the dissociation of the acid are

Species	Equilibrium Concentration (M)
H^+	3.0×10^{-3}
$C_2H_3O_2^-$	3.0×10^{-3}
$HC_2H_3O_2$	0.50

There is also a small amount of hydroxide ion present which can be obtained from K_w.

$$[OH^-] = \frac{K_w}{[H^+]}$$

$$[OH^-] = \frac{1.0 \times 10^{-14}}{3.0 \times 10^{-3}}$$

$$[OH^-] = 3.3 \times 10^{-12}$$

We see from Table 14.1 that the values of K_a and K_b for weak acids and bases are quite small, ranging from 10^{-2} to 10^{-10}. Recalling that numbers this small can be simplified by applying the logarithmic notation of Equation 14.3, we can write for any K_a

$$pK_a = -\log K_a$$

and for any K_b

$$pK_b = -\log K_b$$

For example the pK_a of acetic acid is

$$pK_a = -\log K_a = -\log 1.8 \times 10^{-5}$$
$$pK_a = 4.74$$

and for pyridine

$$pK_b = -\log 1.5 \times 10^{-9}$$
$$pK_b = 8.88$$

We know that the smaller the value of K_a and K_b the weaker will be the acid or base. Relative strengths of acids and bases can also be indicated by their pK_a's and pK_b's. In this case the smaller the value of pK_a or pK_b the *stronger* is the acid or base. Let us compare, for example, the pK_a's for acetic, chloroacetic, and dichloroacetic acids. Their pK_a's are

$$HC_2H_3O_2 \qquad pK_a = 4.74$$
$$HC_2H_2ClO_2 \qquad pKa = 2.85$$
$$HC_2HCl_2O_2 \qquad pK_a = 1.30$$

The order of increasing acidity is, therefore,

$$acetic < chloroacetic < dichloroacetic\ acid$$

A discussion as to why the acidity of these substances increases in this fashion was included in Section 13.4.

Acids containing more than one atom of hydrogen that can be lost upon dissociation are known as polyprotic acids. Some examples of polyprotic acids are H_2SO_4 and H_2S, both of which contain two ionizable hydrogens, and H_3PO_4, which contains three. We consider these acids to lose their hydrogens, one at a time, in a stepwise fashion. Thus we write two steps for the dissociation of sulfuric acid, each with a corresponding equation for K_a:

$$H_2SO_4 \rightleftharpoons H^+ + HSO_4^- \qquad K_{a_1} = \frac{[H^+][HSO_4^-]}{[H_2SO_4]}$$

$$HSO_4^- \rightleftharpoons H^+ + SO_4^{2-} \qquad K_{a_2} = \frac{[H^+][SO_4^{2-}]}{[HSO_4^-]}$$

and for H_2S

$$H_2S \rightleftharpoons H^+ + HS^- \qquad K_{a_1} = \frac{[H^+][HS^-]}{[H_2S]}$$

$$HS^- \rightleftharpoons H^+ + S^{2-} \qquad K_{a_2} = \frac{[H^+][S^{2-}]}{[HS^-]}$$

The three steps in the dissociation of H_3PO_4 are written

$$H_3PO_4 \rightleftharpoons H^+ + H_2PO_4^- \qquad K_{a_1} = \frac{[H^+][H_2PO_4^-]}{[H_3PO_4]}$$

$$H_2PO_4^- \rightleftharpoons H^+ + HPO_4^{2-} \qquad K_{a_2} = \frac{[H^+][HPO_4^{2-}]}{[H_2PO_4^-]}$$

$$HPO_2^{2-} \rightleftharpoons H^+ + PO_4^{3-} \qquad K_{a_3} = \frac{[H^+][PO_4^{3-}]}{[HPO_4^{2-}]}$$

Table 14.2 contains a list of some polyprotic acids and their stepwise dissociation constants. We see from this table that for sulfuric acid the first dissociation goes very nearly to completion, while the second dissociation occurs only to a relatively limited degree. Because of the near completion of its first dissociation, sulfuric acid is considered to be a strong acid. We also see that, in general, the first dissociation step of these acids occurs with the largest value of K_a and that each successive step occurs with an ever-decreasing value of K_a. This decreasing trend in K_a is reasonable considering that it should be easiest to remove a H^+ ion from an uncharged species and that it becomes progressively more difficult to do so as the negative charge on the ion increases.

As a result of the more complex equilibria that occur during the dissociation of a polyprotic acid, the calculation of the concentration of each species present at equilibrium is somewhat more complicated. To show this let's compute the concentration of the species present in a $0.01 M$ aqueous solution of the weak polyprotic acid H_2S. From the above discussion the spe-

Table 14.2
Stepwise Dissociation of Some Polyprotic Acids at 25° C

Acid	Stepwise Dissociation	Dissociation Constant for Each Step	pK_a
Phosphoric	$H_3PO_4 \rightleftharpoons H^+ + H_2PO_4^-$ $H_2PO_4^- \rightleftharpoons H^+ + HPO_4^{2-}$ $HPO_4^{2-} \rightleftharpoons H^+ + PO_4^{3-}$	$K_{a_1} = 7.5 \times 10^{-3}$ $K_{a_2} = 6.2 \times 10^{-8}$ $K_{a_3} = 2.2 \times 10^{-12}$	2.13 7.21 11.66
Sulfuric	$H_2SO_4 \rightleftharpoons H^+ + HSO_4^-$ $HSO_4^- \rightleftharpoons H^+ + SO_4^{2-}$	K_{a_1} — very large $K_{a_2} = 1.2 \times 10^{-2}$	<0 1.92
Sulfurous	$SO_2 + H_2O \rightleftharpoons H^+ + HSO_3^-$ $HSO_3^- \rightleftharpoons H^+ + SO_3^{2-}$	$K_{a_1} = 1.5 \times 10^{-2}$ $K_{a_2} = 1.0 \times 10^{-7}$	1.82 7.00
Hydrosulfuric	$H_2S \rightleftharpoons H^+ + HS^-$ $HS^- \rightleftharpoons H^+ + S^{2-}$	$K_{a_1} = 1.1 \times 10^{-7}$ $K_{a_2} = 1.0 \times 10^{-14}$	6.96 14.00
Carbonic	$CO_2 + H_2O \rightleftharpoons H^+ + HCO_3^-$ $HCO_3^- \rightleftharpoons H^+ + CO_3^{2-}$	$K_{a_1} = 4.3 \times 10^{-7}$ $K_{a_2} = 5.6 \times 10^{-11}$	6.37 10.26

cies whose equilibrium concentrations must be calculated are H_2S, HS^-, H^+, and S^{2-}.

Because K_{a_1} is *so much greater* than K_{a_2} we can safely assume that nearly all of the hydrogen ion in the solution is derived from the first step in the dissociation. In addition, only very little of the HS^- formed will undergo further dissociation. On the basis of this assumption we can calculate the H^+ and HS^- concentrations using the expression for K_{a_1} alone.

$$K_{a_1} = \frac{[H^+][HS^-]}{[H_2S]}$$

If we let x equal the number of mole/liter of H_2S that dissociate, we obtain, from the stoichiometry of the first step, x mole/liter of H^+ and x mole/liter of HS^-. At equilibrium we will have $(0.01 - x)$ mole/liter of H_2S remaining.

$$[H^+] = x$$
$$[HS^-] = x$$
$$[H_2S] = 0.01 - x$$

As before, since K_{a_1} is very small, we may assume that x will be negligible compared to 0.01 and write

$$[H_2S] = 0.01 - x \approx 0.01\,M$$

Substituting these equilibrium quantities into the expression for K_{a_1}, we have

$$\frac{(x)(x)}{0.01} = 1.1 \times 10^{-7}$$

$$x^2 = 1.1 \times 10^{-9}$$
$$x = 3.3 \times 10^{-5}$$

The equilibrium concentrations from this first dissociation are then

$[H^+]$	$3.3 \times 10^{-5}\,M$
$[HS^-]$	$3.3 \times 10^{-5}\,M$
$[H_2S]$	$1.0 \times 10^{-2} - 3.3 \times 10^{-5} = 1.0 \times 10^{-2}\,M$

We see that our approximation in the H_2S concentration is valid because 3.3×10^{-5} is in fact negligible compared to 1×10^{-2}.

By employing K_{a_2} we can now calculate the equilibrium concentration of S^{2-},

$$HS^- \rightleftharpoons H^+ + S^{2-}$$

and

$$K_{a_2} = \frac{[H^+][S^{2-}]}{[HS^-]}$$

If we let y equal the number of moles per liter of HS^- that dissociate, then, from the stoichiometry of this second dissociation step, the number of moles per liter of H^+ and S^{2-} produced would also be y. Thus the total hydrogen ion concentration from both the first and second dissociations will be $[H^+] = 3.3 \times 10^{-5} + y$ and the concentration of HS^- that remains at equilibrium will be $3.3 \times 10^{-5} - y$. For this second dissociation then

$$[H^+] \quad 3.3 \times 10^{-5} + y$$
$$[S^{2-}] \quad y$$
$$[HS^-] \quad 3.3 \times 10^{-5} - y$$

These quantities may be simplified if we recall that we have assumed that the hydrogen ion concentration and the HS^- concentration at equilibrium are determined by the first step in the dissociation. Our concentrations therefore become

$$[H^+] \quad 3.3 \times 10^{-5} M$$
$$[S^{2-}] \quad y$$
$$[HS^-] \quad 3.3 \times 10^{-5} M$$

and we obtain

$$K_{a_2} = \frac{(3.3 \times 10^{-5})(y)}{(3.3 \times 10^{-5})} = 1.0 \times 10^{-14}$$

or

$$[S^{2-}] = 1.0 \times 10^{-14}$$

In summary then, the concentration of all species present at equilibrium when a $0.01 M$ H_2S solution dissociates is

$$[H^+] \quad 3.3 \times 10^{-5} M$$
$$[HS^-] \quad 3.3 \times 10^{-5} M$$
$$[S^{2-}] \quad 1.0 \times 10^{-14} M$$
$$[H_2S] \quad 1.0 \times 10^{-2} M$$

Note that in any solution containing only H_2S the concentration of the sulfide ion is equal to K_{a_2}. In fact for any polyprotic acid where $K_{a_2} \ll K_{a_1}$, the concentration of the anion formed in the second dissociation will always equal K_{a_2}, provided, of course, that the acid is the only solute. For example, K_{a_2} for H_3PO_4 has a value of 6.2×10^{-8}, and in a solution containing only H_3PO_4 and H_2O, the concentration of HPO_4^{2-} is $6.2 \times 10^{-8} M$.

Saturated solutions of H_2S are sometimes used in chemical analysis to detect the presence of certain cations by the formation of an insoluble sulfide precipitate. In these analyses the concentration of the S^{2-} is critical and, therefore, must be controlled. Following the above sample calculation, the S^{2-}, H^+, and HS^- concentrations in a saturated $(0.1 M)$ H_2S solution are 1.0×10^{-14}, 1.0×10^{-4}, and $1.0 \times 10^{-4} M$, respectively. The pH of this saturated solution would then be 4. By applying Le Chatelier's principle to the dissociation of H_2S, we see that any increase in the H^+ concentration (perhaps by the addition of a strong acid) will cause a shift in the equilibrium to the left, favoring the formation of more H_2S, and decreasing the concentrations of both the S^{2-} and HS^- species. Thus, we can control the concentration of the S^{2-} in a saturated H_2S solution by varying the H^+ concentration. A useful equation expressing the relationship that exists between H^+ and S^{2-} in an H_2S solution can be derived by multiplying K_{a_1} by K_{a_2} for H_2S. Thus

$$K_a = K_{a_1} \times K_{a_2} = \frac{[H^+][HS^-]}{[H_2S]} \times \frac{[H^+][S^{2-}]}{[HS^-]}$$

or

$$K_a = \frac{[H^+]^2[S^{2-}]}{[H_2S]} = 1.1 \times 10^{-21} \qquad (14.4)$$

A word of caution about the use of Equation 14.4 is in order. This equation is useful *only* in situations where two of the three concentrations are given and we wish to calculate the third. It cannot be used to determine, for example, both the H^+ and S^{2-} concentrations in solutions of known H_2S concentrations. You can verify this for yourself by using it to calculate the concentration of H^+ and S^{2-} that are present in a $0.01M$ H_2S solution and then comparing your answers with those that we calculated using the two dissociation constants.

A practical example of the use of this equation is illustrated in Example 14.7.

Example 14.7

Calculate the S^{2-} concentration in a saturated solution of H_2S whose pH was adjusted to 2 by the addition of HCl.

Solution

Since this is a saturated H_2S solution with a known H^+ concentration we can use Equation 14.4.

$$K_a = \frac{[H^+]^2[S^{2-}]}{H_2S}$$

Rearranging this equation to solve for $[S^{2-}]$ we have

$$[S^{2-}] = \frac{K_a[H_2S]}{[H^+]^2}$$

From the pH of this acidic solution we calculate that $[H^+] = 1 \times 10^{-2}M$; since the solution is saturated with H_2S, we know that $[H_2S] = 0.1M$. Substituting these values and the value of K_a into our equation, we have

$$[S^{2-}] = \frac{(1.1 \times 10^{-21})(1.0 \times 10^{-1})}{(1.0 \times 10^{-2})^2}$$

or

$$[S^{2-}] = 1.1 \times 10^{-18}$$

We see very clearly that the concentration of S^{2-} is dependent upon the pH of the saturated solution.

**14.4
Titration,
Indicators,
and pH**

In Chapters 5 and 13 it was pointed out that a titration is a useful and accurate way of determining the concentrations of acids and bases provided the end point, or equivalence point, can be detected. During a typical titration a base is slowly added to a known amount of an acid. Initially, before any base is added, the pH of the solution is from the acid alone. As more and more base is added, the OH^- ions of the base react with and neutralize the H^+ ions of the acid, and the pH of the solution rises. At the end point sufficient OH^- ions have been added to neutralize completely all of the H^+ ions of the acid. Further addition of the base, beyond the end point, brings about an excess of OH^- ions in the solution and the pH indicates this change.

If we were to draw a graph of pH versus milliliters of base added for the titration of a strong acid with a strong base we would obtain a plot similar to that shown in Figure 14.1. Although the curve represents a strong acid-strong base titration, the same general shape is observed in weak acid-strong base and weak acid-weak base titrations. In these latter two cases, however, the pH at the equivalence point is not the same as that indicated in Figure 14.1. This is discussed further in Section 14.10.

One method that can be employed to detect the end point in a titration makes use of an indicator. Indicators, as we pointed out in Chapters 5 and 13, are weak organic acids or bases that change color on going from an acid medium to a basic medium. Not all indicators, however, change color at the same pH. Some common indicators, their color changes, and the pH ranges in which the color changes occur are listed in Table 14.3. At the equivalence point when a strong acid is titrated with a strong base a very large and rapid change in pH occurs with the addition of a very small quantity of base. As a result, any indicator that changes color in the pH range of about 4 to 10 (e.g., phenolphthalein) will be suitable for detecting the end point in the titration.

If we denote an indicator by the general formula HIn, we have the dissociation reaction

$$HIn \rightleftharpoons H^+ + In^-$$

Applying Le Chatelier's principle to this equilibrium we see that in an acid solution (excess H^+) the species that is present, in excess, is HIn. On the other hand, in basic solutions the equilibrium is shifted to the right and the predominant species is In^-. Therefore, HIn is said to be the "acid form" and In^- the "basic form" of the indicator. The ability of HIn to function as an indicator is based on the fact that the acid and basic forms differ in color.

The dissociation constant, K_I, for an indicator is

$$K_I = \frac{[H^+][In^-]}{[HIn]} \qquad (14.5)$$

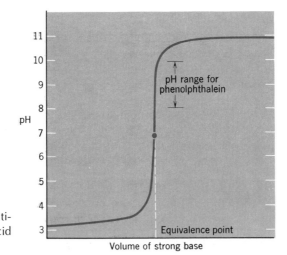

Figure 14.1

Titration curve for the titration of a strong acid with a strong base.

Table 14.3
Some Common Indicators

Indicator	Color-Change	pH Range in Which Color Change Occurs
Thymol blue	Red to yellow	1.2–2.8
Bromophenol blue	Yellow to blue	3.0–4.6
Congo red	Blue to red	3.0–5.0
Methyl orange	Red to yellow	3.2–4.4
Bromocresol green	Yellow to blue	3.8–5.4
Methyl red	Red to yellow	4.8–6.0
Bromocresol purple	Yellow to purple	5.2–6.8
Bromothymol blue	Yellow to blue	6.0–7.6
Cresol red	Yellow to red	7.0–8.8
Thymol blue	Yellow to blue	8.0–9.6
Phenolphthalein	Colorless to pink	8.2–10.0
Alizarin yellow	Yellow to red	10.1–12.0

The reciprocal of this equation is

$$\frac{1}{K_I} = \frac{[HIn]}{[H^+][In^-]}$$

Taking the logarithm of both sides we have

$$\log \frac{1}{K_I} = \log \frac{1}{[H^+]} + \log \frac{[HIn]}{[In^-]}$$

and following the notation introduced by Equation 14.3 we obtain

$$pK_I = pH + \log \frac{[HIn]}{[In^-]} \qquad (14.6)$$

From this equation we can see that, since pK_I is a constant, any large change in the pH of a solution must be accompanied by a large change in the ratio $[HIn]/[In^-]$, but in the opposite direction. We can better appreciate the magnitude of this change by rewriting Equation 14.5 in the form

$$\frac{K_I}{[H^+]} = \frac{[In^-]}{[HIn]}$$

If we had an indicator for which $K_I = 10^{-6}$, we could use this expression to estimate the change in the ratio $[In^-]/[HIn]$ accompanying the increase from pH = 5 to pH = 9 in Figure 14.1. At a pH of 5 we have

$$\frac{10^{-6}}{10^{-5}} = \frac{1}{10} = \frac{[In^-]}{[HIn]}$$

and at a pH of 9 we have

$$\frac{10^{-6}}{10^{-9}} = \frac{1000}{1} = \frac{[In^-]}{[HIn]}$$

In other words, at a pH of 5 the HIn concentration is 10 times that of In^-, and at a pH of 9 the concentration of In^- is 1000 times that of HIn. Thus, at the end point of a titration, when only a very small amount of titrant causes a drastic change in the relative concentrations of acidic and basic forms of the indicator, a marked and sudden change in the color of the solution is seen.

Any solution that contains both a weak acid and a weak base has the property that the addition of small quantities of a strong acid are neutralized by the weak base while small quantities of a strong base are neutralized by the weak acid. Such solutions are said to be **buffers,** since they have the ability to absorb these small additions of concentrated acids and bases without giving rise to a significant change in the pH of the solution.

A buffer whose pH is less than seven can be prepared by adding together a weak acid with the salt of that weak acid, for example, acetic acid and sodium acetate, while a buffer whose pH is greater than seven can be prepared by adding together a weak base with the salt of that weak base, for example, ammonia and ammonium chloride. When H^+ or OH^- are added to our acid buffer (pH < 7), the following neutralization reactions take place:

$$H^+ + C_2H_3O_2^- \longrightarrow HC_2H_3O_2$$
$$OH^- + HC_2H_3O_2 \longrightarrow H_2O + C_2H_3O_2^-$$

Similarly, for the NH_3, NH_4Cl, alkaline buffer we have

$$H^+ + NH_3 \longrightarrow NH_4^+$$
$$\text{and} \quad OH^- + NH_4^+ \longrightarrow H_2O + NH_3$$

In order to understand quantitatively how a buffer works, we must be able to calculate its pH before and after the addition of an acid or a base. To do this we consider the dissociation equilibria of the weak acid in the acid buffer, or of the weak base in the alkaline buffer. In an acetic acid-acetate buffer, for example, we have the dissociation of acetic acid

$$HC_2H_3O_2 \rightleftharpoons H^+ + C_2H_3O_2^-$$

which has the dissociation constant

$$K_a = \frac{[H^+][C_2H_3O_2^-]}{[HC_2H_3O_2]}$$

This expression can be rearranged to solve for the H^+ concentration

$$[H^+] = K_a \frac{[HC_2H_3O_2]}{[C_2H_3O_2^-]} \tag{14.7}$$

Following reasoning similar to that used in the development of Equation 14.6, we see that

$$pH = pK_a + \log \frac{[C_2H_3O_2^-]}{[HC_2H_3O_2]} \tag{14.8}$$

Therefore, in order to calculate the H^+ concentration (Equation 14.7) or the pH directly (Equation 14.8), we must know the K_a of the weak acid as well as the ratio of the concentrations of the weak acid and the anion. Although we can find the dissociation constant of the weak acid from Table 14.1, we must calculate the concentration of $HC_2H_3O_2$ and $C_2H_3O_2^-$.

Since salts completely dissociate in aqueous solution, the number of moles of the anion from this source is determined by the formula of the salt and the number of moles of salt dissolved. Thus a $1.0M$ $NaC_2H_3O_2$ solution contains 1.0 mole of $C_2H_3O_2^-$ while a $1.0M$ $Ca(C_2H_3O_2)_2$ solution contains 2.0 moles of $C_2H_3O_2^-$. In our buffer there will also be an additional amount of acetate ion coming from the dissociation of $HC_2H_3O_2$. The amount of H^+ and $C_2H_3O_2^-$ stemming from this source is very small even in solutions containing only acetic acid, and this small amount is reduced even further in the buffer because of the presence of the large concentration of $C_2H_3O_2^-$ from the salt.[2] The total anion concentration in the buffer, then, is essentially determined by the salt concentration alone, since the contribution from the dissociation of the weak acid is negligible. As in our previous calculations on weak acids, the concentration of $HC_2H_3O_2$ will not be reduced appreciably by its dissociation and, in a mixture of $1.0M$ $NaC_2H_3O_2$ and $1.0M$ $HC_2H_3O_2$, for example, the concentrations of both the molecular acid and the anion are $1.0M$. The H^+ concentration in such a buffer could be found from Equation 14.7 using $K_a = 1.8 \times 10^{-5}$ (Table 14.1).

$$[H^+] = (1.8 \times 10^{-5}) \left(\frac{1.0}{1.0} \right)$$

$$[H^+] = 1.8 \times 10^{-5}M$$

The pH of this solution is 4.74.

Notice that when the concentrations of the acid and anion are the same in a buffer, the H^+ concentration in that solution is equal to the K_a of the weak acid. Thus, if a buffer was prepared by mixing 0.1 mole of formic acid ($K_a = 1.8 \times 10^{-4}$ from Table 14.1) and 0.1 mole of sodium formate into a liter of solution, the resulting H^+ concentration would be

$$[H^+] = 1.8 \times 10^{-4}M$$

and

$$pH = 3.74$$

We can also use Equation 14.7 to calculate the concentrations of acid and salt that would be needed to achieve a certain pH buffer. This is illustrated in Example 14.8.

Example 14.8

What ratio of acetic acid to sodium acetate concentration is needed to achieve a buffer whose pH is 5.70?

---

[2] This can easily be seen by applying Le Chatelier's principle to the dissociation of $HC_2H_3O_2$. The presence of acetate ion from a salt causes the dissociation equilibrium of the acid to be shifted to the left and actually suppresses the dissociation.

Solution

To solve this problem we need to rearrange Equation 14.7 to solve for the ratio of concentrations. Thus

$$\frac{[HC_2H_3O_2]}{[C_2H_3O_2^-]} = \frac{[H^+]}{K_a}$$

The H^+ concentration when the pH is 5.70 is

$$[H^+] = 2.0 \times 10^{-6}M$$

Therefore

$$\frac{[HC_2H_3O_2]}{[C_2H_3O_2^-]} = \frac{2.0 \times 10^{-6}}{1.8 \times 10^{-5}} = \frac{2.0 \times 10^{-6}}{18 \times 10^{-6}}$$

or

$$\frac{[HC_2H_3O_2]}{[C_2H_3O_2^-]} = \frac{1}{9}$$

As long as this ratio is maintained, the pH of an acetic acid-sodium acetate buffer is 5.70. For example, if 0.2 mole of $HC_2H_3O_2$ and 1.8 moles of $NaC_2H_3O_2$ are placed in one liter of solution, the pH is 5.70.

We have seen that by adjusting the ratio of concentration of the weak acid to that of the salt, a buffer of almost any desired pH can be achieved. For example, a buffer composed of 0.01 mole of $HC_2H_3O_2$ and 1.00 mole of $NaC_2H_3O_2$ would have a pH of 6.74. However, when as little as 0.01 moles of base is added to this buffer, all of the acetic acid is neutralized and a large change in the pH of the buffer results. Therefore, we say that the most effective pH range for any buffer is at or near the pH where the acid and salt concentrations are equal (i.e., pK_a).

Let us now see what happens when a small quantity of acid or base is added to an acetic acid-sodium acetate buffer whose initial pH is 4.74. In this buffer the concentrations of $HC_2H_3O_2$ and $C_2H_3O_2^-$ were the same and equal to $1.0M$ and the concentration of H^+ was calculated to be 1.8×10^{-5}. Suppose that 0.2 mole of H^+ is added to one liter of this buffer. The following neutralization reaction occurs:

$$H^+ + C_2H_3O_2^- \longrightarrow HC_2H_3O_2$$

Below are the number of moles per liter of all species before and after the addition.

Initial	Final
$[H^+] = 1.8 \times 10^{-5}M$	$[H^+] = xM$
$[C_2H_3O_2^-] = 1.0M$	$[C_2H_3O_2^-] = 1.0 - 0.2M$
$[HC_2H_3O_2] = 1.0M$	$[HC_2H_3O_2] = 1.0 + 0.2M$

Substituting these final concentrations into Equation 14.7 we have

$$[H^+] = 1.8 \times 10^{-5} \left(\frac{1.2}{0.8}\right)$$

$$[H^+] = 2.7 \times 10^{-5} M$$

The pH of the buffer after the 0.2 mole of H^+ was added is 4.57. The pH of the buffer changed by only 0.17 pH units.

Suppose now that we were to add 0.2 mole of H^+ to one liter of a solution of HCl whose pH = 4.74 (i.e., a $1.8 \times 10^{-5} M$ HCl solution). Since Cl^- has virtually no tendency to react with H^+, the final H^+ concentration will be $0.20 M$ ($0.20 + 1.8 \times 10^{-5} = 0.20$), and the pH of the solution will be 0.70. The change in pH in this case is 4.04 pH units, as opposed to a change of only 0.17 pH units when the same quantity of H^+ is added to the buffer.

Additions of strong base are also absorbed by the buffer. When 0.2 mole of OH^- is added to the one liter of our original buffer, it is neutralized according to the reaction

$$OH^- + HC_2H_3O_2 \longrightarrow H_2O + C_2H_3O_2^-$$

The number of moles per liter before the addition are the same as above, but the number of moles per liter after 0.2 mole of OH^- is added would be:

Concentrations after Addition of 0.2 Mole OH^-
$[H^+] = x = xM$
$[C_2H_3O_2^-] = 1.0 + 0.2 = 1.2M$
$[HC_2H_3O_2] = 1.0 - 0.2 = 0.8M$

Substituting these values into Equation 14.7, we find that the H^+ concentration after the addition is $1.2 \times 10^{-5} M$ and the resulting pH would be 4.92, or once again a change of 0.17 pH units.

Example 14.9

A buffer was prepared by adding together 200 ml of a $0.6 M$ NH_3 solution to 300 ml of a $0.3 M$ NH_4Cl solution. (a) What is the pH of this buffer, assuming a final volume of 500 ml? (b) What would be the pH after 0.02 mole of H^+ is added?

Solution

The number of moles of NH_3 added to this solution is

$$0.6 \frac{mole}{liter} \times 0.2 \text{ liter} = 0.12 \text{ mole}$$

and the number of moles of NH_4^+ added would be

$$0.3 \frac{mole}{liter} \times 0.3 \text{ liter} = 0.99 \text{ mole}$$

Therefore the concentrations of these ions in the 500 ml would be

$$[NH_3] = \frac{0.12 \text{ mole}}{0.50 \text{ liter}} = 0.24 M$$

$$[NH_4^+] = \frac{0.09 \text{ mole}}{0.50 \text{ liter}} = 0.18 M$$

(a) The OH^- concentration for this alkaline buffer is found by using the K_b for NH_3:

$$NH_3 + H_2O \rightleftharpoons NH_4^+ + OH^-$$

$$K_b = \frac{[NH_4^+][OH^-]}{[NH_3]}$$

Rearranging and solving for $[OH^-]$ we have

$$[OH^-] = K_b \frac{[NH_3]}{[NH_4^+]}$$

Substituting K_b for NH_3 from Table 14.1 and the concentrations of NH_3 and NH_4^+ for this buffer into this equation gives

$$[OH^-] = 1.8 \times 10^{-5} \left(\frac{0.24}{0.18}\right)$$

$$[OH^-] = 2.4 \times 10^{-5}$$
$$pOH = 4.62$$
$$pH = 9.38$$

(b) The neutralization reaction for H^+ in this buffer is

$$H^+ + NH_3 \longrightarrow NH_4^+$$

We are adding 0.02 mole H^+ to 500 ml or 0.04 mole H^+ per liter. Therefore, the concentrations before and after the addition of the acid are as below

Initial	Final
$[OH^-] = 2.4 \times 10^{-5} M$	$[H^+] = xM$
$[NH_3] = 0.24 M$	$[NH_3] = 0.24 - 0.04 M = 0.20 M$
$[NH_4^+] = 0.18 M$	$[NH_4^+] = 0.18 + 0.04 M = 0.22 M$

and the OH^- concentration is

$$[OH^-] = 1.8 \times 10^{-5} \left(\frac{0.20}{0.22}\right)$$

$$[OH^-] = 1.6 \times 10^{-5} M$$
$$pOH = 4.80$$

The pH would therefore be 9.20, a change of 0.18 pH units.

Buffers are important in analytical chemistry and also to life itself. Many of our bodily fluids are buffered to maintain a certain pH. For example, blood is composed of, among other things, a H_2CO_3/HCO_3^- buffer system that maintains the pH at 7.4.

In the laboratory many inorganic and organic chemical reactions are performed in buffered solutions to minimize any adverse effects caused by acids or bases that might be consumed or produced during reaction.

14.6 Solubility Product Constant

We now turn our attention to aqueous solutions of salts. It was pointed out in Chapter 5 that, according to our list of solubility rules, some salts are considered soluble in water while others are said to be quite insoluble. However, even the most insoluble salts are soluble in water to at least some degree, and nearly all salts completely dissociate when they dissolve in aqueous solutions. Two examples of salts that do not dissociate completely upon dissolving are $HgCl_2$ and $CdSO_4$. Our discussion will not include this type of solid.

When a saturated solution of a salt is prepared, a dynamic equilibrium is established between the dissociated ions and the insoluble solid at the bottom of the container. For example, in a saturated solution of silver chloride we have the equilibrium

$$AgCl(s) \rightleftharpoons Ag^+(aq) + Cl^-(aq)$$

for which we can write

$$K = \frac{[Ag^+][Cl^-]}{[AgCl(s)]}$$

As we saw in Section 12.6 the concentration of a solid is constant and, therefore, can be included with the constant K, so that

$$K[AgCl(s)] = K_{sp} = [Ag^+][Cl^-]$$

The equilibrium constant K multiplied by the concentration of solid $AgCl$ is still another constant called the **solubility product constant,** given the label K_{sp}. For example, we can obtain the expression for the K_{sp} of silver acetate from its solubility equilibrium,

$$AgC_2H_3O_2(s) \rightleftharpoons Ag^+ + C_2H_3O_2^-$$

The equilibrium condition is therefore,

$$K_{sp} = [Ag^+][C_2H_3O_2^-]$$

In the case of an insoluble solid such as $Mg(OH)_2$ the coefficients in the dissociation equilibrium are not all equal to one:

$$Mg(OH)_2(s) \rightleftharpoons Mg^{2+}(aq) + 2OH^-(aq)$$

The K_{sp} for $Mg(OH)_2$ is then given by

$$K_{sp} = [Mg^{2+}][OH^-]^2$$

Thus the solubility product constant is equal to the product of the concentration of the ions produced in a saturated solution, each raised to a power equal to its coefficient in the balanced equation. A list of some ionic solids and their K_{sp}'s at temperatures ranging between 18 to 25° C is given in Table 14.4.

Table 14.4
Solubility Product Constants

Compound	K_{sp}	Compound	K_{sp}
$Al(OH)_3$	2×10^{-33}	PbS	7×10^{-27}
$BaCO_3$	8.1×10^{-9}	$Mg(OH)_2$	1.2×10^{-11}
$BaCrO_4$	2.4×10^{-10}	MgC_2O_4	8.6×10^{-5}
BaF_2	1.7×10^{-6}	$Mn(OH)_2$	4.5×10^{-14}
$BaSO_4$	1.5×10^{-9}	MnS	7×10^{-16}
CdS	3.6×10^{-29}	Hg_2Cl_2	2×10^{-18}
$CaCO_3$	9×10^{-9}	HgS	1.6×10^{-54}
CaF_2	1.7×10^{-10}	NiS	2×10^{-21}
$CaSO_4$	2×10^{-4}	$AgC_2H_3O_2$	2.3×10^{-3}
CoS	3×10^{-26}	Ag_2CO_3	8.2×10^{-12}
CuS	8.5×10^{-36}	AgCl	1.7×10^{-10}
Cu_2S	2×10^{-47}	AgBr	5×10^{-13}
$Fe(OH)_2$	2×10^{-15}	AgI	8.5×10^{-17}
$Fe(OH)_3$	1.1×10^{-36}	Ag_2CrO_4	1.9×10^{-12}
FeC_2O_4	2.1×10^{-7}	AgCN	1.6×10^{-14}
FeS	3.7×10^{-19}	Ag_2S	2×10^{-49}
$PbCl_2$	1.6×10^{-5}	$Sn(OH)_2$	5×10^{-26}
$PbCrO_4$	1.8×10^{-14}	SnS	1×10^{-26}
PbC_2O_4	2.7×10^{-11}	$Zn(OH)_2$	4.5×10^{-17}
$PbSO_4$	2×10^{-8}	ZnS	1.2×10^{-23}

There are various kinds of calculations that can be performed having to do with solubility equilibria. Knowledge of the solubility of a salt allows us to compute K_{sp}, as shown in Example 14.10. Conversely, K_{sp} can be used to calculate solubility, as illustrated in Examples 14.11 and 14.12. Still another type of calculation allows us to determine whether or not a precipitate will form when a solution is prepared containing ions at some specified concentrations. Computations of this nature are demonstrated by Example 14.13. Knowledge of K_{sp} also permits us to selectively precipitate one salt from a solution containing more than one kind of ion, as seen in Example 14.14.

Example 14.10
It was experimentally determined that at 25° C the solubility of $BaSO_4$ is 0.0091 g/liter. What is the value of K_{sp} for barium sulfate?

Solution

From the solubility it is clear that the number of moles of $BaSO_4$ that are dissolved in one liter of solution is

$$0.0091 \; \frac{g}{liter} \left(\frac{1 \; mole}{233 \; g} \right) = 3.9 \times 10^{-5} \; \frac{mole}{liter}$$

The solubility equilibrium for $BaSO_4$ is

$$BaSO_4(s) \rightleftharpoons Ba^{2+}(aq) + SO_4^{2-}(aq)$$

so that for every mole of $BaSO_4$ that dissolves, 1 mole of Ba^{2+} and 1 mole of SO_4^{2-} are produced. Therefore the molar concentrations of Ba^{2+} and SO_4^{2-} in this saturated solution at 25° C are

$$[Ba^{2+}] = 3.9 \times 10^{-5}$$
$$[SO_4^{2-}] = 3.9 \times 10^{-5}$$

and the K_{sp} would be

$$K_{sp} = [Ba^{2+}][SO_4^{2-}]$$
$$K_{sp} = (3.9 \times 10^{-5})(3.9 \times 10^{-5})$$
$$K_{sp} = 1.5 \times 10^{-9}$$

Example 14.11

What is the molar solubility of AgCl in water at 25° C?

Solution

We are concerned with the equilibrium,

$$AgCl(s) \rightleftharpoons Ag^+ + Cl^-$$

for which

$$K_{sp} = [Ag^+][Cl^-] = 1.7 \times 10^{-10}$$

If we let x equal the number of moles per liter of AgCl dissolved, then, since AgCl is completely dissociated in solution, the concentrations of Ag^+ and Cl^- would be

$$[Ag^+] = x$$
$$[Cl^-] = x$$

Substituting these quantities into the equilibrium expression,

$$K_{sp} = (x)(x) = 1.7 \times 10^{-10}$$
$$x^2 = 1.7 \times 10^{-10}$$
$$x = 1.3 \times 10^{-5}$$

Therefore, the molar solubility of AgCl in water is $1.3 \times 10^{-5} M$.

Example 14.12

What are the concentrations of Ag^+ and CrO_4^{2-} in a saturated solution of Ag_2CrO_4 at 25° C?

Solution

Ag_2CrO_4 dissolves in water according to the equilibrium,

$$Ag_2CrO_4(s) \rightleftharpoons 2Ag^+(aq) + CrO_4^{2-}(aq)$$

Thus there are 2 moles of Ag^+ produced for every 1 mole of CrO_4^{2-}. If x is the number of moles per liter of Ag_2CrO_4 that dissolve, then

$$[CrO_4^{2-}] = x$$
$$[Ag^+] = 2x$$

and the K_{sp} expression would be

$$K_{sp} = [Ag^+]^2[CrO_4^{2-}] = (2x)^2(x)$$

Substituting the value of K_{sp} (From Table 14.4) into this equation and solving for x, we have

$$K_{sp} = (x)(2x)^2 = 1.9 \times 10^{-12}$$
$$x(4x^2) = 4x^3 = 1.9 \times 10^{-12}$$
$$x^3 = 0.48 \times 10^{-12}$$

and

$$x = 7.8 \times 10^{-5}$$

Therefore

$$[Ag^+] = 2(7.8 \times 10^{-5}) = 1.6 \times 10^{-4}M$$
$$[CrO_4^{2-}] = 7.8 \times 10^{-5}M$$

From our earlier discussion it should be clear that a saturated solution exists *only* when the *ion product*, that is, the product of the concentrations of the dissolved ions each raised to its proper power, is exactly equal to K_{sp}. When the ion product is less than K_{sp}, the solution is unsaturated, since more salt would have to dissolve in order to raise the ion concentrations to the point where the ion product equals K_{sp}. On the other hand, when the ion product exceeds K_{sp}, a supersaturated solution exists because some of the salt would have to precipitate in order to lower the ion concentrations until the ion product is equal to K_{sp} once again.

In a solution a precipitate will only be formed when the mixture is supersaturated. Consequently, we may use the value of the ion product in a solution to tell us whether or not precipitation will occur. In summary, we find that

$$\left. \begin{array}{l} \text{Ion product} < K_{sp} \\ \text{Ion product} = K_{sp} \end{array} \right\} \text{no precipitate will form}$$

$$\text{Ion product} > K_{sp} \quad : \quad \text{precipitation will occur}$$

Example 14.13 demonstrates how this may be applied.

Example 14.13
Will a precipitate of $PbSO_4$ form when 100 ml of a $0.003\,M$ $Pb(NO_3)_2$ solution is mixed with 400 ml of $0.04\,M$ Na_2SO_4?

Solution
For a saturated solution of $PbSO_4$ we would have

$$K_{sp} = [Pb^{2+}][SO_4^{2-}]$$

To determine whether a precipitate will form, we must calculate the concentration of Pb^{2+} and SO_4^{2-} in our total volume of 500 ml. The 100 ml of the $0.003 M$ $Pb(NO_3)_2$ solution contains

$$0.1 \text{ liter} \left(0.003 \frac{\text{mole}}{\text{liter}}\right) = 0.0003 \text{ mole of } Pb^{2+}$$

(This solution would also contain 0.0006 mole of NO_3^-, but this species is unimportant in this calculation.)

The 400 ml of Na_2SO_4 contains

$$0.4 \text{ liter} \left(0.04 \frac{\text{mole}}{\text{liter}}\right) = 0.016 \text{ mole of } SO_4^{2-}$$

(This solution also contains 0.032 mole of Na^+, but it too is unimportant in this problem.)

The concentration of the Pb^{2+} in the 500 ml is then

$$\frac{0.0003 \text{ mole}}{0.5 \text{ liter}} = 0.0006 \text{ or } 6.0 \times 10^{-4} M$$

and the concentration of SO_4^{2-} in the 500 ml is

$$\frac{0.016 \text{ mole}}{0.5 \text{ liter}} = 0.032 = 3.2 \times 10^{-2} M$$

The ion product in the final solution is therefore

$$[Pb^2][SO_4^{2-}] = (6.0 \times 10^{-4})(3.2 \times 10^{-2}) = 19.2 \times 10^{-6}$$

When we compare the ion product to the K_{sp} for $PbSO_4$ (2×10^{-8}), we find that the ion product is greater than K_{sp} and therefore a precipitate will form.

From the solubility rules presented in Chapter 5 we know that it is possible to separate certain ions from each other when they are present together in solution. For instance, the addition of chloride ion to a solution containing both Na^+ and Ag^+ yields a precipitate of $AgCl$, thereby removing most of the Ag^+ from the mixture. In this case one possible product, $NaCl$, is soluble while the other, $AgCl$, is quite insoluble.

Even when both products are "insoluble," it is still frequently possible to achieve some degree of separation. Consider, for example, the salts $CaSO_4$ and $BaSO_4$. Although both have very low solubilities, as evidenced by their respective K_{sp}'s, we compute $CaSO_4$ to be about 1000 times more soluble, on a mole basis, than $BaSO_4$. As a result, if we had a solution containing equal concentrations of Ca^{2+} and Ba^{2+}, we would find that as the SO_4^{2-} concentration was increased in the solution, $BaSO_4$ would precipitate first. Conceivably, then, one could separate Ca^{2+} and Ba^{2+} by appropriately adjusting the SO_4^{2-} concentration so that the Ca^{2+} would remain in solution while nearly all of the Ba^{2+} would be removed as $BaSO_4$. This general concept is used often in the separation of ions in qualitative analysis.

When the anion employed in a separation is derived from a weak acid, it is possible to control its concentration by appropriately adjusting the hydrogen ion concentration. This is illustrated for the selective precipitation of metal sulfides in Example 14.14.

Example 14.14

A solution containing $0.1 M$ Sn^{2+} and $0.1 M$ Zn^{2+} is saturated with H_2S. What values of the hydrogen ion concentration will allow only one of these ions to be precipitated as its sulfide?

Solution

From Table 14.4 we have

$$SnS \qquad K_{sp} = 1 \times 10^{-26}$$
$$ZnS \qquad K_{sp} = 1 \times 10^{-23}$$

In this problem the sulfide ion concentration must be controlled so that one ion will precipitate while the other remains in solution. Let us, therefore, calculate for each salt the value of $[S^{2-}]$ that will make the ion product equal to K_{sp}. For tin we have

$$K_{sp} = [Sn^{2+}][S^{2-}] = 1 \times 10^{-26}$$

Substituting the Sn^{2+} concentration into the expression gives

$$(0.1)[S^{2-}] = 1 \times 10^{-26}$$
$$[S^{2-}] = 1 \times 10^{-25}M$$

In a similar fashion for zinc, we obtain

$$[S^{2-}] = 1 \times 10^{-22}M$$

These numbers tell us that if the sulfide ion concentration is *greater* than 1×10^{-25} but *less than or equal to* 1×10^{-22}, only SnS will precipitate.

We saw in Equation 14.4 that the sulfide ion concentration is directly related to the hydrogen ion concentration, that is,

$$\frac{[H^+]^2[S^{2-}]}{[H_2S]} = K_{a_1}K_{a_2} = 1.1 \times 10^{-21}$$

We can use this expression to calculate the $[H^+]$ that gives us our desired $[S^{2-}]$.

For the lower limit, $[S^{2-}] = 1 \times 10^{-25}$. Since the solution is saturated with H_2S, we have $[H_2S] = 0.1 M$. Substituting gives

$$\frac{[H^+]^2(1 \times 10^{-25})}{(0.1)} = 1.1 \times 10^{-21}$$

$$[H^+]^2 = \frac{(1.1 \times 10^{-21})(0.1)}{1 \times 10^{-25}}$$

$$[H^+]^2 = 1.1 \times 10^3$$
$$[H^+] = 3.3 \times 10^1 = 33 M$$

This calculation implies that in order to prevent SnS from precipitating, the H^+ concentration must be $33 M$. This concentration is impossible to achieve; therefore, SnS *must* precipitate.

To prevent ZnS from forming, the S^{2-} concentration cannot be larger than $1 \times 10^{-22}M$. Using this value for $[S^{2-}]$ in Equation 14.4, we have

$$\frac{[H^+]^2(1 \times 10^{-22})}{0.1} = 1.1 \times 10^{-21}$$

$$[H^+]^2 = 1.1$$
$$[H^+] = 1.05\,M$$

Thus when $[H^+] = 1.05\,M$ the S^{2-} concentration will be $1 \times 10^{-22}\,M$, the highest value it can have without causing the Zn^{2+} to precipitate. A hydrogen ion concentration greater than $1.05\,M$ will produce a sulfide ion concentration less than $1 \times 10^{-22}\,M$. In summary, then, in order to achieve a separation,

$$[H^+] < 33\,M$$

and
$$[H^+] \geqslant 1.05\,M$$

14.7
Common Ion Effect and Solubility

In the last section we discussed the solubility of salts when only one salt was present in the solution. We now turn our attention to the effect of other dissolved ions on the solubility of an insoluble salt. If one of these foreign ions is the same as one of the ions produced by the salt, it is called a **common ion.** For example, when NaCl is added to a solution of AgCl, the common ion is Cl^-. The equilibrium that exists between solid AgCl and its ions can be represented as

$$AgCl(s) \rightleftharpoons Ag^+(aq) + Cl^-(aq)$$

and the K_{sp}, from Table 14.4, is 1.7×10^{-10}. When additional Cl^- ions are added to this system, in the form of NaCl, for example, the concentrations of Ag^+ and Cl^- must, according to Le Chatelier's principle, shift so as to minimize the effect of this added ion.[3] In this case some of the excess Cl^- ions are picked up by the Ag^+, and more solid AgCl is produced. Thus, in order to preserve the value of the K_{sp}, the Ag^+ concentration must decrease whenever the Cl^- concentration increases and vice versa. The net result is that the salt becomes less soluble with the addition of a common ion. This phenomenon, called the **common ion effect,** can be examined quantitatively as shown by the next two example problems.

Example 14.15
What is the solubility of AgCl in a $0.010\,M$ solution of NaCl?

Solution
 For this salt we have

$$K_{sp} = [Ag^+][Cl^-] = 1.7 \times 10^{-10}$$

If we again let x equal the number of moles per liter of AgCl dissolved, we find that

$$[Ag^+] = x\,M$$
$$[Cl^-] = x + 0.010\,M$$

Note that there are two sources of chloride ion. Since the solubility of AgCl is

---

[3] A similar effect was seen in the section on buffers when a common ion was added to the equilibrium dissociation of a weak acid.

so very small, we may assume that x will be negligible compared to 0.010 and write

$$[Cl^-] \approx 0.010M$$

Substituting the concentrations into the expression for K_{sp}, we have

$$(x)(0.010) = 1.7 \times 10^{-10}$$

or

$$x = 1.7 \times 10^{-8}M$$

Thus, because of the way we defined x, the molar solubility of AgCl is $1.7 \times 10^{-8}M$. We might compare this to the molar solubility of AgCl in pure water, which we found in Example 14.11 to be $1.3 \times 10^{-5}M$. The solubility of AgCl is indeed much less in a solution containing a common ion.

Example 14.16
What is the solubility of $Mg(OH)_2$ in $0.10M$ NaOH?

Solution
The K_{sp} for $Mg(OH)_2$ is

$$K_{sp} = [Mg^{2+}][OH^-]^2$$

As usual, we let x equal the number of moles per liter of $Mg(OH)_2$ that dissolve. Therefore at equilibrium the ion concentrations would be

$$[Mg^{2+}] = xM$$
$$[OH^-] = 2x + 0.1 \approx 0.1M$$

Substituting these concentrations and the K_{sp} for $Mg(OH)_2$ into the solubility product expression, we have

$$(x)(0.10)^2 = 1.2 \times 10^{-11}$$

or

$$x = \frac{1.2 \times 10^{-11}}{(0.10)^2}$$

$$x = 1.2 \times 10^{-9}M$$

Thus 1.2×10^{-9} mole per liter of $Mg(OH)_2$ dissolves in a $0.10M$ solution of NaOH.

A situation quite opposite to that observed in the common ion effect can take place whenever a substance that forms a complex ion with the metal cation is added to the solution of an insoluble salt. These substances, which are called ligands in complex ion chemistry, are usually Lewis bases and can be either (a) neutral molecules such as H_2O and NH_3, (b) polyatomic anions such as CN^- and $C_2O_4^{2-}$, or (c) monatomic anions such as Cl^- and Br^-. When one or more of these ligands bond to a single metal ion, the resulting species is called a **complex ion.** One example of a complex ion is the species

**14.8
Complex Ions**

$Al(H_2O)_6^{3+}$ (see Chapter 13). Another example containing a fewer number of ligands is formed when Ag^+ and NH_3 react to yield $Ag(NH_3)_2^+$.

Before we see how these complex ions are involved in the solubility of a salt, we should say a few words about their stability. We can express the stability of these ions in the usual fashion, that is, through their dissociation equilibria. For example, the overall reaction for the equilibrium dissociation of $Ag(NH_3)_2^+$ can be written as

$$Ag(NH_3)_2^+ \rightleftharpoons Ag+ + 2NH_3$$

The dissociation constant, which is called the **instability constant,** K_{inst}, in complex ion chemistry, would be

$$K_{inst} = \frac{[Ag^+][NH_3]^2}{[Ag(NH_3)_2^+]}$$

The value of the instability constant for this complex is observed to be 6.0×10^{-8}. We can see by the size of this constant that this particular complex is quite stable and will readily form whenever Ag^+ and NH_3 are added to the same solution. Other examples of complex ions and their instability constants can be seen in Table 14.5.

When a complex ion is formed in a solution of an insoluble salt, the resulting stable complex reduces the concentration of free metal ion. As a result more solid must dissolve in order to replenish the amount of metal ion lost, until that concentration required by the K_{sp} of the salt is achieved. Thus

Table 14.5
Instability Constants at 25° C

Complex Ion	K_{inst}
AlF_6^{3-}	1.5×10^{-20}
$Cd(CN)_4^{2-}$	1.3×10^{-17}
$Co(NH_3)_6^{2+}$	1.3×10^{-5}
$Co(NH_3)_6^{3+}$	2.0×10^{-34}
$Cu(NH_3)_4^{2+}$	2.1×10^{-13}
$Cu(CN)_2^{-}$	1.0×10^{-16}
$Fe(CN)_6^{4-}$	1.0×10^{-35}
$Fe(CN)_6^{3-}$	1.1×10^{-42}
$Ni(NH_3)_4^{2+}$	1.1×10^{-8}
$Ni(NH_3)_6^{2+}$	2.0×10^{-9}
$Ag(NH_3)_2^+$	6.0×10^{-8}
$Ag(CN)_2^{-}$	1.9×10^{-19}
$Zn(OH)_4^{2-}$	3.6×10^{-16}

the solubility of an insoluble salt generally increases when complex ions are formed. To see this more clearly, let us see what effect adding NH_3 has on a saturated solution of AgCl. Before any NH_3 is added, we have the equilibrium

$$AgCl(s) \rightleftharpoons Ag^+(aq) + Cl^-(aq)$$

Because NH_3 forms such a stable complex with the free silver ion, when NH_3 is added to this system a second equilibrium is established, namely

$$Ag^+(aq) + 2NH_3(aq) \rightleftharpoons Ag(NH_3)_2^+ (aq)$$

The creation of this new equilibrium upsets the first by removing some of the Ag^+, thereby causing the first equilibrium to shift to the right. As a result some of the solid AgCl dissolves.

In the preceding discussion you might notice that the equilibrium reaction between Ag^+, NH_3, and its complex ion was not written as a dissociation, as shown earlier, but instead as a reaction for the formation of the complex ion. The resulting equilibrium constant, which is called the **formation constant,** K_{form}, of the complex, is expressed in the usual fashion, that is,

$$K_{form} = \frac{[Ag(NH_3)_2^+]}{[Ag^+][NH_3]^2}$$

and is simply equal to the reciprocal of the instability constant of the complex ion. This can be shown as

$$K_{form} = \frac{1}{K_{inst}} = \frac{1}{[Ag^+][NH_3]^2/[Ag(NH_3)_2^+]} = 1.7 \times 10^7$$

We can express the two equilibrium reactions of our example as one overall reaction that is obtained by simply adding together the two equilibrium reactions.

$$\begin{array}{r} AgCl(s) \rightleftharpoons Ag^+(aq) + Cl^-(aq) \\ Ag^+(aq) + 2NH_3 \rightleftharpoons Ag(NH_3)_2^+ \\ \hline AgCl(s) + 2NH_3 \rightleftharpoons Ag(NH_3)_2^+ + Cl^- \end{array}$$

The equilibrium constant for this overall reaction is

$$K_{eq} = \frac{[Ag(NH_3)_2^+][Cl^-]}{[NH_3]^2}$$

We can obtain this same expression by multiplying the K_{sp} of AgCl with the K_{form} of the complex ion. Thus

$$K_{sp} \times K_{form} = [\cancel{Ag^+}][Cl^-] \times \frac{[Ag(NH_3)_2^+]}{[\cancel{Ag^+}][NH_3]^2} = K_{eq}$$

Therefore, with a knowledge of K_{sp} of the salt, K_{form} of the complex ion (or K_{inst}), and the concentration of NH_3, it is possible to calculate the concentrations of Ag^+ and Cl^- present at equilibrium and thus determine the solubility of AgCl in NH_3, as shown by the next example.

Example 14.17
What is the molar solubility of AgCl in one liter of $1M$ NH_3 at $25°$ C?

Solution

As we have seen, the overall equilibrium reaction for this problem is

$$AgCl(s) + 2NH_3 \rightleftharpoons Ag(NH_3)_2^+ + Cl^-$$

for which we write

$$K_{eq} = \frac{[Ag(NH_3)_2^+][Cl^-]}{[NH_3]^2}$$

where

$$K_{eq} = K_{sp} \times K_{form} = (1.7 \times 10^{-10}) \times (1.7 \times 10^7) = 2.9 \times 10^{-3}$$

If we let x equal the number of moles/liter of AgCl that dissolves, then we have the following initial and equilibrium concentrations.

Initial	At Equilibrium
$[NH_3] = 1M$	$[NH_3] = 1 - 2xM$
$[Ag(NH_3)_2^+] = 0$	$[Ag(NH_3)_2^+] = xM$
$[Cl^-] = 0$	$[Cl^-] = xM$

Substituting the concentrations at equilibrium into the K_{eq} equation, we have

$$K_{eq} = \frac{(x)(x)}{(1 - 2x)^2} = \frac{x^2}{(1 - 2x)^2} = 2.9 \times 10^{-3}$$

Taking the square root of both sides we have

$$\frac{x}{1 - 2x} = 5.4 \times 10^{-2}$$

from which we obtain

$$x = 0.049$$

Therefore, we find that 0.049 mole of AgCl will dissolve in 1 liter of $1M$ NH_3.

In Example 14.17 we assumed that when the AgCl dissolves in the ammonia solution, essentially all of the Ag^+ becomes complexed by NH_3. In other words, we said that the chloride ion concentration was equal to the concentration of $Ag(NH_3)_2^+$. Note that this assumption is valid only if K_{form} is very large, indicating that the complex is very stable.

Example 14.18

How many moles of solid NaOH must be added to 1.0 liter of H_2O in order to dissolve 0.10 mole of $Zn(OH)_2$ according to the reaction,

$$Zn(OH)_2 + 2OH^- \rightleftharpoons Zn(OH)_4^{2-}$$
$$Zn(OH)_2 \qquad K_{sp} = 4.5 \times 10^{-17}$$
$$Zn(OH)_4^{2-} \qquad K_{inst} = 3.6 \times 10^{-16}$$

Solution

The two equilibria involved in this system are

$$Zn(OH)_2(s) \rightleftharpoons Zn^{2+} + 2OH^- \qquad K_{sp} = 4.5 \times 10^{-17}$$

$$Zn^{2+} + 4OH^- \rightleftharpoons Zn(OH)_4{}^{2-} \qquad K_{form} = \frac{1}{3.6 \times 10^{-16}} = 2.8 \times 10^{15}$$

As before we can write for the reaction

$$Zn(OH)_2(s) + 2OH^- \rightleftharpoons Zn(OH)_4{}^{2-}$$
$$K_{eq} = K_{sp} \times K_{form} = (2.8 \times 10^{15})(4.5 \times 10^{-17}) = 1.3 \times 10^{-1}$$

Therefore

$$1.3 \times 10^{-1} = \frac{[Zn(OH)_4{}^{2-}]}{[OH^-]^2}$$

In this problem we know that 0.10 mole of Zn goes into solution where it is present as either free Zn^{2+} or $Zn(OH)_4{}^{2-}$. Because K_{inst} is so very small, essentially all of the Zn will be present as the complex ion; therefore we can write

$$[Zn(OH)_4{}^{2-}] = 0.10M$$

Substituting this into the equilibrium expression gives

$$1.3 \times 10^{-1} = \frac{0.10}{[OH^-]^2}$$

and

$$[OH^-]^2 = \frac{0.10}{1.3 \times 10^{-1}} = 7.7 \times 10^{-1}$$

or

$$[OH^-] = 0.88M$$

This corresponds to the equilibrium concentration of free OH^-. In this solution, however, we also have 0.10 mole of $Zn(OH)_4{}^{2-}$, which contains an additional 0.40 mole of OH^-, 0.20 mole of which was contained in the original 0.10 mole of $Zn(OH)_2$ that dissolved. Consequently, the total number of moles of NaOH which must be added to the water is $0.88 + 0.20 = 1.08$ moles.

It was pointed out in Section 5.6 that salts are produced during an acid-base neutralization reaction. For example, NaCl is considered to be the salt of the strong acid HCl and the strong base NaOH. In a similar fashion we say that $NaC_2H_3O_2$ is the salt of a weak acid ($HC_2H_3O_2$) and a strong base (NaOH), and NH_4Cl is the salt of a strong acid (HCl) and a weak base (NH_3). A salt such as $NH_4C_2H_3O_2$ is the salt of a weak acid-weak base. When these salts are added to water, the pH of the resulting solution is found experimentally to be dependent on the type of salt dissolved. For example, the pH of an aqueous solution of a salt of a strong acid-strong base is always very close to

14.9 Hydrolysis

7 while the pH for a salt of a weak acid-strong base is greater than 7. The pH that results when each type of salt is dissolved in water is summarized below.

Type of Salt	pH of Aqueous Solution
Strong acid-strong base	7
Weak acid-strong base	>7
Strong acid-weak base	<7
Weak acid-weak base	Depends on salt

How might we account for these differences in pH?

When a salt dissolves in water, it dissociates fully to produce cations and anions that may subsequently react chemically with the solvent in a process called **hydrolysis.** For example, the cation of a salt undergoes the reaction

$$M^+ + H_2O \rightleftharpoons MOH + H^+$$

while an anion reacts according to the equation

$$X^- + H_2O \rightleftharpoons HX + OH^-$$

Since the H^+ and OH^- ions produced in these reactions influence the pH of the salt solution, the extent to which the hydrolysis reactions take place determines whether the pH will be greater than, less than, or equal to 7.

Consider, for example, NaCl, the salt of a strong acid-strong base. The NaOH and HCl produced by hydrolysis are both strong electrolytes and, of course, immediately dissociate. The result is that there will be the same amount of H^+ and OH^- in the solution, a condition that is fulfilled only when their concentrations are each $10^{-7}M$. Consequently the pH of the NaCl solution is 7. Since this is the same as pure water, the net effect is that no hydrolysis actually takes place at all. In general, then, we may conclude that *anions and cations of strong acids and bases, respectively, do not undergo hydrolysis, and the salts derived from strong acids and bases yield neutral solutions.* This however, is not the case for other types of salts.

SALTS OF WEAK ACIDS-STRONG BASES: ANION HYDROLYSIS. For this type of salt we are only concerned with anion hydrolysis because, in light of the previous discussion, cations of strong bases do not undergo hydrolysis. With $NaC_2H_3O_2$ as our example of these salts, we would write for the hydrolysis reaction of the anion

$$C_2H_3O_2^- + H_2O \rightleftharpoons HC_2H_3O_2 + OH^-$$

The acetate anion pulls some protons away from the water molecules, so that when equilibrium is established excess OH^- is present in solution. As a result the solution is basic. We can write an equilibrium constant (K_h') for this equilibrium in the usual fashion, which is

$$K_h' = \frac{[HC_2H_3O_2][OH^-]}{[C_2H_3O_2^-][H_2O]}$$

As usual the concentration of H_2O may be included with the equilibrium constant K'_h, and we obtain

$$K'_h \times [H_2O] = K_h = \frac{[HC_2H_3O_2][OH^-]}{[C_2H_3O_2^-]}$$

where, K_h is called the **hydrolysis constant.**

This same equation can be derived simply by dividing the equation for K_w by the K_a of the weak acid (acetic acid in this case). Thus

$$\frac{K_w}{K_a} = \frac{[H^+][OH^-]}{[H^+][C_2H_3O_2^-]/[HC_2H_3O_2]}$$

or

$$\frac{K_w}{K_a} = \frac{[HC_2H_3O_2][OH^-]}{[C_2H_3O_2^-]}$$

We see then, for the salt of a weak acid-strong base that the hydrolysis constant is equal to

$$K_h = \frac{K_w}{K_a}$$

Thus we can calculate the hydrolysis constant of the anion in this type of salt from a knowledge of K_w and K_a of the weak acid of that anion.

From K_h and the concentration of the anion in the salt solution, we may then calculate the concentration of OH^- and eventually determine the pH of the salt solution, as shown in Example 14.19.

Example 14.19

Calculate the pH of a $0.1\,M$ $NaC_2H_3O_2$ solution.

Solution

Since this is the salt of a weak acid-strong base, only the anion undergoes hydrolysis and that reaction is

$$C_2H_3O_2^- + H_2O \rightleftharpoons HC_2H_3O_2 + OH^-$$

for which we can write the equilibrium expression

$$K_h = \frac{[HC_2H_3O_2][OH^-]}{[C_2H_3O_2^-]}$$

Before this equation becomes useful, we must first calculate K_h for this salt. We can do this with the aid of the equation

$$K_h = \frac{K_w}{K_a}$$

Substituting the values of K_w and K_a (from Table 14.1) into this equation, we obtain

$$K_h = \frac{1 \times 10^{-14}}{1.8 \times 10^{-5}} = 5.6 \times 10^{-10}$$

From the magnitude of this K_h we see that the position of equilibrium in this hydrolysis lies far to the left in favor of $C_2H_3O_2^-$. Therefore, if we let x equal the number of moles of $C_2H_2O_2^-$ that undergo hydrolysis, then the equilibrium concentrations would be

$$[HC_2H_3O_2] = xM$$
$$[OH^-] = xM$$
$$[C_2H_3O_2^-] = 0.1 - x \approx 0.1M$$

Substituting these values along with K_h into the hydrolysis equation, we have

$$K_h = \frac{(x)(x)}{0.1} = 5.6 \times 10^{-10}$$
$$x^2 = 5.6 \times 10^{-11}$$
$$x = 7.5 \times 10^{-6}M$$

Thus

$$[OH^-] = 7.5 \times 10^{-6}M$$
$$pOH = 5.1$$

and

$$pH = 8.9$$

Indeed the pH of this solution indicates that it is basic.

SALTS OF A STRONG ACID-WEAK BASE: CATION HYDROLYSIS. From our previous discussion we know that only the cation in this type of salt undergoes hydrolysis. For example the hydrolysis reaction that takes place in an aqueous solution of NH_4Cl is

$$NH_4^+ + H_2O \rightleftharpoons H_3O^+ + NH_3$$

As a result of the hydrolysis of the cation some of the H_2O molecules are converted into H_3O^+ which, of course, make the solution acidic. The equation for this hydrolysis equilibrium is written as

$$K_h = \frac{[H_3O^+][NH_3]}{[NH_4^+]}$$

which can also be derived by dividing K_w by K_b of the weak base, NH_3. Therefore,

$$K_h = \frac{K_w}{K_b} = \frac{[H_3O^+][\cancel{OH^-}]}{[NH_4^+][\cancel{OH^-}]/[NH_3]} = \frac{[H_3O^+][NH_3]}{[NH_4^+]}$$

For this type of salt then, in order to calculate the pH, we must know K_w, K_b, and the concentration of the salt, as illustrated in Example 14.20.

Example 14.20
What is the pH of a $0.01M$ NH_4Cl solution?

Solution
We first must recognize that this is the salt of a strong acid (HCl) and a weak base (NH_3); therefore, only the cation undergoes hydrolysis. The hydrolysis reaction is

$$NH_4^+ + H_2O \rightleftharpoons H_3O^+ + NH_3$$

for which we can write for K_h

$$K_h = \frac{[H_3O^+][NH_3]}{[NH_4^+]}$$

For this salt we know that

$$K_h = \frac{K_w}{K_b}$$

so that, using K_b from Table 14.1,

$$K_h = \frac{1 \times 10^{-14}}{1.8 \times 10^{-5}} = 5.6 \times 10^{-10}$$

Once again, because of the size of the K_h, the hydrolysis equilibrium lies mainly to the left. If x equals the number of moles of NH_4^+ that undergo hydrolysis, then at equilibrium

$$[H_3O^+] = xM$$
$$[NH_3] = xM$$
$$[NH_4^+] = 0.01 - x \approx 0.01M$$

Substituting these concentrations and the derived value of K_h into the above equation, we have

$$K_h = \frac{(x)(x)}{0.01} = 5.6 \times 10^{-10}$$

$$x^2 = 5.6 \times 10^{-12}$$
$$x = 2.4 \times 10^{-6}M$$

Thus

$$[H_3O^+] = 2.4 \times 10^{-6}M$$

and

$$pH = 5.62$$

We see that the aqueous solution of this type of salt is indeed acidic.

SALT OF A WEAK ACID-WEAK BASE: CATION AND ANION HYDROLYSIS. Solutions of this type of salt can either be acidic, neutral, or basic since both the cation and the anion of the salt undergo hydrolysis. The pH of such a salt solution is determined by the relative extent of the hydrolysis reactions of each ion. By applying what we have learned from the last two types of salts to these salts we should be able to predict, at least qualitatively, the pH of their aqueous solutions. If the K_a of the weak acid and the K_b of the weak base of the salt are identical, then the extent of cation and anion hydrolysis is exactly the same (the K_h for the cation is exactly equal to the K_h for the anion) and the solution will be neutral. For example, in the case of $NH_4C_2H_3O_2$, where the K_b of NH_3 is 1.8×10^{-5} and the K_a of $HC_2H_3O_2$ is 1.8×10^{-5}, the value of K_h for both ions is 5.6×10^{-10}, and an aqueous solution of this salt, regardless of the concentration, is neutral. On the other hand, we would predict that an aqueous solution of NH_4CN would be basic because of the relative values of the K_h's of the cation and anion. For NH_4^+ we have

$$K_h = \frac{K_w}{K_b} = \frac{1.0 \times 10^{-14}}{1.8 \times 10^{-5}} = 5.6 \times 10^{-10}$$

and for CN^- we have

$$K_h = \frac{K_w}{K_a} = \frac{1.0 \times 10^{-14}}{4.9 \times 10^{-10}} = 2.0 \times 10^{-5}$$

Since the CN^- undergoes more extensive hydrolysis, this means that the equilibrium

$$CN^- + H_2O \rightleftharpoons HCN + OH^-$$

goes further to completion than does the reaction

$$NH_4^+ + H_2O \rightleftharpoons H_3O^+ + NH_3$$

and the excess OH^- ions make the solution basic.

In a similar fashion we predict that an aqueous solution of ammonium formate, NH_4CHO_2 is acidic because of the relative values of the K_h's for the cation and anion. The K_h for NH_4^+ from above is 5.6×10^{-10}, while the K_h for the formate ion, CHO_2^-, is

$$K_h = \frac{K_w}{K_a} = \frac{1.0 \times 10^{-14}}{1.8 \times 10^{-4}}$$

$$K_h = 5.6 \times 10^{-11}$$

Thus the hydrolysis reaction

$$NH_4^+ + H_2O \rightleftharpoons H_3O^+ + NH_3$$

occurs to a slightly greater extent than does the reaction

$$CHO_2^- + H_2O \rightleftharpoons HCHO_2 + OH^-$$

Therefore, there is a small excess of H_3O^+ ions and solutions of this salt are acidic.

HYDROLYSIS OF SALTS OF POLYPROTIC ACIDS. An example of this type of salt is Na_2S, the salt of a weak acid, H_2S, and a strong base, $NaOH$. Since only the anion undergoes hydrolysis, the equilibrium reaction for this salt would be

$$S^{2-} + H_2O \rightleftharpoons HS^- + OH^-$$

We see that in this reaction another anion, HS^-, that can also undergo hydrolysis is produced. Its equilibrium reaction is

$$HS^- + H_2O \rightleftharpoons H_2S + OH^-$$

The hydrolysis constant, K_{h_1}, for the first reaction is

$$K_{h_1} = \frac{K_w}{K_{a_2}}$$

where K_{a_2} is the acid dissociation constant for the weak acid, HS^-. The equilibrium constant for the second step in the hydrolysis is

$$K_{h_2} = \frac{K_w}{K_{a_1}}$$

where K_{a_1} in this case is the dissociation constant for the weak acid, H_2S. Substituting the values of the K_a's from Table 14.2 into these equations, we obtain

$$K_{h_1} = \frac{1.0 \times 10^{-14}}{1.0 \times 10^{-14}} = 1$$

and

$$K_{h_2} = \frac{1.0 \times 10^{-14}}{1.1 \times 10^{-7}} = 9.1 \times 10^{-8}$$

Surely the magnitude of these two equilibrium constants indicate to us that the second hydrolysis reaction occurs to a negligible extent compared to the first; therefore, only K_{h_1} need be used to determine the pH of an aqueous solution of this salt, as shown in our next example.

Example 14.21
What is the pH of a $0.2M$ solution of Na_2S?

Solution
     From our previous discussion we know that only the first hydrolysis reaction

$$S^{2-} + H_2O \rightleftharpoons HS^- + OH^-$$

is important in determining the pH of this solution.
     The equation for the hydrolysis constant is

$$K_h = \frac{[HS^-][OH^-]}{[S^{2-}]} = 1.0$$

Because of the magnitude of the K_h we see that the hydrolysis reaction occurs to a large extent. Therefore, if we let x equal the number of moles/liter of S^{2-} that undergo hydrolysis, then the equilibrium concentrations would be

$$[HS^-] = xM$$
$$[OH^-] = xM$$
$$[S^{2-}] = 0.2 - xM$$

Substituting these equilibrium concentrations into the expression for K_h gives

$$K_h = \frac{(x)(x)}{(0.2 - x)} = 1.0$$

     In this instance, because K_h is so large, we *cannot* make the assumption that x is negligible compared to 0.2, and we are forced to obtain x by applying the quadratic formula (Appendix C). Multiplying both sides by $(0.2 - x)$ gives

$$x^2 = (0.2 - x)1.0$$

which can be rearranged as

$$x^2 + x - 0.2 = 0$$

The quadratic formula then gives us

$$x = \frac{-1 \pm \sqrt{(1)^2 - 4(1)(-0.2)}}{(2)(1)}$$

$$x = \frac{-1 \pm \sqrt{1.8}}{2} = \frac{-1 \pm 1.34}{2}$$

Notice that two values of x are obtained.

$$x = \frac{-2.34}{2} = -1.17M$$

$$x = \frac{0.34}{2} = 0.17M$$

The first value of x is absurd. It has no physical meaning since it tells us that the concentration of HS^- and OH^- are negative. We cannot have less than nothing! The second value of x is meaningful, and we conclude that

$$x = 0.17M$$

and consequently

$$[OH^-] = 0.17M$$

from which we obtain

$$pOH = 0.47$$

The pH of the solution, therefore, is 13.5.

14.10
Acid-Base Titrations: The Equivalence Point

In Section 14.4 we mentioned that the pH at which the equivalence point or end point of an acid-base titration occurs is 7 only in the case of a strong acid-strong base titration. By applying our knowledge of hydrolysis discussed in the last section, we are now able to calculate the pH at the equivalence point for strong acid-strong base, weak acid-strong base, and for strong acid-weak base titrations.

STRONG ACID-STRONG BASE. A typical example of such a titration occurs when 25 ml of a 0.1M HCl solution is titrated with 0.1M NaOH. We can mathematically determine the pH throughout the titration by calculating the H^+ concentration present in the flask each time a quantity of NaOH is added to the HCl. For example, the number of moles of H^+ present in the 25 ml of a 0.1M HCl solution is

$$\frac{0.1 \text{ mole}}{1000 \text{ ml}} \times 25 \text{ ml} = 2.5 \times 10^{-3} \text{ mole of } H^+$$

When 10 ml of the 0.1 M NaOH are added, we in fact have added

$$\frac{0.1 \text{ mole}}{1000 \text{ ml}} \times 10 \text{ ml} = 1.0 \times 10^{-3} \text{ mole of } OH^-$$

The neutralization reaction

$$H^+ + OH^- \longrightarrow H_2O$$

occurs, and the amount of H^+ left after neutralization is

$$2.5 \times 10^{-3} - 1.0 \times 10^{-3} = 1.5 \times 10^{-3} \text{ mole of } H^+$$

The molar concentration of H^+ is now

$$[H^+] = \frac{1.5 \times 10^{-3} \text{ mole}}{0.035 \text{ liter}} = 4.3 \times 10^{-2} M$$

and the pH is calculated to be 1.4. The concentration of H^+ after further additions of NaOH have occurred are summarized in Table 14.6.

Our calculations show that the pH starts increasing slowly at first, then rises very rapidly near the equivalence point, and then finally levels off after the equivalence point is reached.

Drawing a graph of pH versus the volume of base added, we obtain for this titration, the plot shown in Figure 14.1. The pH at the equivalence point of this titration is the pH at which the H^+ and OH^- concentrations are identical, which was defined as a pH of 7. We did not have to employ our knowledge of hydrolysis in this type of titration because the ions produced during neutralization do not undergo hydrolysis. This is not true, however, for the remaining two types of titrations that we shall consider.

WEAK ACID-STRONG BASE. An example quite typical of this type of titration is the titration of 25 ml of a $0.1 M$ $HC_2H_3O_2$ with a $0.1 M$ NaOH solution. Initially, when the acetic acid is the only species present, the $[H^+]$ and pH are calculated in the usual fashion from the K_a, and we find that the pH is 2.9. When we begin to add the strong base, NaOH, to this solution, in effect we are making a buffer, and Equation 14.7 can be employed to calculate the H^+ concentration and eventually the pH. For example, when 10 ml of $0.1 M$ NaOH (1×10^{-3} mole) are added to the 25 ml of the $0.1 M$ $HC_2H_3O_2$ (2.5×10^{-3} mole), all of the NaOH is neutralized as well as 1.0×10^{-3} mole of $HC_2H_3O_2$. In the total volume of 35 ml we have $2.5 \times 10^{-3} - 1.0 \times 10^{-3} = 1.5 \times 10^{-3}$ mole of unreacted $HC_2H_3O_2$ as well as the $1.0 \times$

Table 14.6
Titration of 25 ml of $0.1 M$ HCl with a $0.1 M$ NaOH Solution

Volume HCl	Volume NaOH	Volume Total	Moles H^+	Moles OH^-	Molarity of Ion in Excess	pH
25.00	0.00	25.00	2.5×10^{-3}	0	0.1 (H^+)	1.0
25.00	10.00	35.00	2.5×10^{-3}	1.0×10^{-3}	$4.3 \times 10^{-2} (H^+)$	1.4
25.00	24.99	49.99	2.5×10^{-3}	2.499×10^{-3}	$2.0 \times 10^{-5} (H^+)$	4.7
25.00	25.00	50.00	2.5×10^{-3}	2.50×10^{-3}	0	7.0
25.00	25.01	50.01	2.5×10^{-3}	2.501×10^{-3}	$2.0 \times 10^{-5} (OH^-)$	9.3
25.00	26.00	51.00	2.5×10^{-3}	2.60×10^{-3}	$2.0 \times 10^{-3} (OH^-)$	11.3

10^{-3} mole of acetate ion produced in the neutralization. The concentrations of $HC_2H_3O_2$ and $C_2H_3O_2^-$ are, therefore,

$$[HC_2H_3O_2] = \frac{1.5 \times 10^{-3} \text{ moles}}{0.035 \text{ liter}} = 4.2 \times 10^{-2}M$$

$$[C_2H_3O_2^-] = \frac{1.0 \times 10^{-3} \text{ moles}}{0.035 \text{ liter}} = 2.9 \times 10^{-2}M$$

Substituting these concentrations, as well as the K_a for acetic acid, into Equation 14.7, we have

$$[H^+] = 1.8 \times 10^{-5} \left(\frac{4.2 \times 10^{-2}}{2.9 \times 10^{-2}}\right)$$

$$[H^+] = 2.6 \times 10^{-5}M$$

therefore,

$$pH = 4.58$$

From the time of the first addition of base, until the equivalence point is reached, the solution contains both acetic acid and acetate ion, and the pH may be computed in this fashion.

When 25 ml of NaOH have been added, all of the acetic acid is neutralized, and we have produced 2.5×10^{-3} mole of $NaC_2H_3O_2$, which will undergo hydrolysis. Only the anion of this salt undergoes hydrolysis, the equilibrium being

$$C_2H_3O_2^- + H_2O \rightleftharpoons HC_2H_3O_2 + OH^-$$

From the last section we know that

$$K_h = \frac{[HC_2H_3O_2][OH^-]}{[C_2H_3O_2^-]} = \frac{K_w}{K_a} = 5.6 \times 10^{-10}$$

We can calculate the OH^- concentration in this solution as we did previously by letting x equal the number of moles of $C_2H_3O_2^-$ that undergoes hydrolysis so that at equilibrium we have

$$[HC_2H_3O_2] = [OH^-] = x$$

$$[C_2H_3O_2^-] = \frac{(2.5 \times 10^{-3} - x) \text{ mole}}{0.050 \text{ liter}} = 0.05\,M$$

Substituting these quantities into the above equation, we have

$$K_h = \frac{(x)(x)}{5 \times 10^{-2}} = 5.6 \times 10^{-10}$$

$$x^2 = 28.0 \times 10^{-12}$$

$$x = 5.3 \times 10^{-6}M$$

$$[OH^-] = 5.3 \times 10^{-6}M$$

$$pOH = 5.28$$

and finally,

$$pH = 8.72$$

Thus the pH at which the equivalence point occurs is greater than 7. We also find that this is true for any weak acid-strong base titration.

Table 14.7

Titration of 25 ml of 0.1 M HC$_2$H$_3$O$_2$
with 0.1 M NaOH

Ml of Base Added	Concentration of Species in Excess	pH
0.0	1.3×10^{-3}	2.9
10.0	2.5×10^{-5}(H$^+$)	4.6
24.99	7.2×10^{-9}(H$^+$)	8.1
25.0	5.3×10^{-6}(OH$^-$)	8.7
25.01	2.0×10^{-5}(OH$^-$)	9.3
26.0	1.9×10^{-3}(OH$^-$)	11.3

Thus far we have only discussed the first half of the titration (see Table 14.7). What takes place beyond the equivalence point? As soon as all of the weak acid has been neutralized, any further addition of NaOH suppresses the hydrolysis of the anion and the pH is then solely dependent on the concentration of OH$^-$ coming from the added NaOH. Thus we generate the last half of Table 14.7 in the same manner as we did Table 14.6 in the HCl-NaOH titration.

A graph of these data is shown in Figure 14.2, where we have plotted pH versus volume of base added. From both Table 14.7 and Figure 14.2 we can see that the change in pH near the equivalence point is not as drastic as in the case of the HCl-NaOH titration. This leveling off near the equivalence becomes even more pronounced for weaker acids such as HCN.

WEAK BASE-STRONG ACID. When a weak base is titrated with a strong acid, the titration curve that is generated is very similar in shape to that obtained by reaction of a weak acid with a strong base. During the initial addition of acid

Figure 14.2
Titration of 25 ml of 0.1M acetic acid with 0.1M sodium hydroxide.

Figure 14.3

Titration of 25 ml of 0.1M
NH$_3$ with 0.1M HCl.

the solution contains unreacted weak base and its salt; it therefore constitutes a buffer. At the equivalence point the solution contains the salt of the weak base, and the pH of the mixture is determined by the hydrolysis of the cation. Finally, beyond the equivalence point the pH of the solution is controlled by the excess hydrogen ion from the strong acid. The shape of the titration curve for such a titration is shown in Figure 14.3 for the titration of 25 ml of 0.10M NH$_3$ with 0.10M HCl. We can show that the pH at the equivalence point is less than 7 by considering the hydrolysis of the NH$_4$Cl produced during the reaction.

From the last section we recall that the K_h for NH$_4{}^+$ is written as

$$K_h = \frac{[H_3O^+][NH_3]}{[NH_4{}^+]} = \frac{K_w}{K_b} = 5.6 \times 10^{-10}$$

All of the NH$_3$ is neutralized in this titration when exactly 25 ml of HCl (2.5×10^{-3} mole) have been added. At this point, the concentration of NH$_4{}^+$ would be

$$\frac{2.5 \times 10^{-3} \text{ mole}}{0.050 \text{ liter}} = 5.0 \times 10^{-2} M$$

If we let x equal the number of moles of NH$_4{}^+$ that undergo hydrolysis, then at equilibrium we have

$$[H_3O^+] = xM$$
$$[NH_3] = xM$$
$$[NH_4{}^+] = 5.0 \times 10^{-2} - x = 5.0 \times 10^{-2}$$

Substituting these concentrations into the above equation for K_h gives

$$K_h = \frac{(x)(x)}{5 \times 10^{-2}} = 5.6 \times 10^{-10}$$
$$x^2 = 28.0 \times 10^{-12}$$
$$x = 5.3 \times 10^{-6} M$$
$$[H_3O^+] = 5.3 \times 10^{-6} M$$

and

$$pH = 5.28$$

The pH at the equivalence point of this titration is less than 7, which is typical for all weak base-strong acid titrations.

14.1 Calculate the H^+ and OH^- concentrations and the pH of the following solutions of strong acids and bases:
(a) $0.0010M$ HCl
(b) $0.125M$ HNO_3
(c) $0.0031M$ NaOH
(d) $0.012M$ $Ba(OH)_2$
(e) $2.1 \times 10^{-4}M$ $HClO_4$
(f) $1.3 \times 10^{-5}M$ HCl
(g) $8.4 \times 10^{-3}M$ NaOH
(h) $4.8 \times 10^{-2}M$ KOH

14.2 Calculate the H^+ and OH^- concentrations in a solution having a pH equal to:
(a) 1.30   (d) 7.80
(b) 5.73   (e) 10.94
(c) 4.00   (f) 12.61

14.3 Calculate the concentrations of all species at equilibrium in the following solutions of weak electrolytes:
(a) $0.010M$ $HC_2H_3O_2$
(b) $0.010M$ NH_3
(c) $2.5M$ $HCHO_2$
(d) $1.2M$ HF
(e) $0.20M$ N_2H_4
(f) $3.0 \times 10^{-2}M$ $C_6H_5NH_2$
(g) $0.045M$ HOCl
(h) $0.020M$ HCN

14.4 What is the pH in each of the solutions in question 14.3?

14.5 What is the percent ionization of each of the electrolytes in question 14.3?

14.6 Calculate the concentrations of all species and the values of K_a or K_b for the following solutions of weak electrolytes, given the percent ionization of each:
(a) $0.10M$ H_3AsO_4 is 5.0% dissociated
(b) $0.05M$ H_3BO_3 is $1.34 \times 10^{-2}\%$ dissociated
(c) $0.20M$ HNO_2 is 4.5% dissociated

14.7 What is the molarity of a solution of acetic acid whose pH is 2.5?

14.8 What molar concentration of hydrazine, N_2H_4, yields a solution whose $pH = 10.64$?

14.9 A $0.010M$ solution of a weak acid, HA, is found to have a pH of 4.55. What is the value of K_a for this acid?

14.10 Calculate the concentrations of all species present in a $0.010M$ H_3PO_4 solution.

14.11 What must the H^+ concentration of a saturated H_2S solution be in order to give a sulfide ion concentration of $8.4 \times 10^{-15}M$?

14.12 What is the sulfide ion concentration in a saturated H_2S solution whose pH is 4.60?

14.13 The pH of a $0.012M$ solution of a weak base, BOH, was experimentally determined to be 11.40. Calculate K_b for the base.

14.14 An indicator, HIn, has an ionization constant, K_a, equal to 1×10^{-5}. If the molecular form of the indicator is yellow and the In^- ion is green, what is the color of a solution containing this indicator when its pH is 7.0?

14.15 Calculate the pH of each of the following buffers prepared by placing, in 1.0 liter of solution,
(a) 0.10 mole NH_3 and 0.10 mole NH_4Cl
(b) 0.20 mole $HC_2H_3O_2$ and 0.40 mole $NaC_2H_3O_2$
(c) 0.15 mole N_2H_4 and 0.10 mole N_2H_5Cl
(d) 0.20 mole HCl and 0.30 mole NaCl

14.16 How many grams of $NaC_2H_3O_2$ must be added to 1.00 mole of $HC_2H_3O_2$ in order to prepare 1.00 liter of a buffer whose pH equals 5.15?

14.17 What must the ratio of NH_3 to NH_4^+ be to have a buffer with a pH of 10.0?

14.18 How many moles of HCl must be added to 1.0 liter of a mixture containing $0.010M$ $HC_2H_3O_2$ and $0.010M$ $NaC_2H_3O_2$ in order to give a solution whose pH = 3.0?

14.19 How much would the pH change if 0.10 mole of HCl were added to 1.0 liter of a formic acid-sodium formate buffer containing 0.45 mole of $HCHO_2$ and 0.55 mole of $NaCHO_2$?

14.20 How much would the pH change if 0.20 mole of NaOH were added to the buffer in question 14.19?

14.21 Write the equilibrium solubility product expression for each of the following substances:
(a) Ag_2S (d) MgC_2O_4
(b) CaF_2 (e) Bi_2S_3
(c) $Fe(OH)_3$ (f) $BaCO_3$

14.22 The molar solubility of $CaCrO_4$ is 1.0×10^{-2} mole per liter. What is K_{sp} for $CaCrO_4$?

14.23 The solubility of lead iodide is 1.4×10^{-3} mole per liter. Calculate its K_{sp}.

14.24 The solubility of Bi_2S_3 is 2.5×10^{-12} g/liter. What is K_{sp} for Bi_2S_3?

14.25 Using the data in Table 14.4, calculate the molar solubility of each of the following:
(a) PbS
(b) $Fe(OH)_2$
(c) $BaSO_4$
(d) Hg_2Cl_2 (which yields Hg_2^{2+} and $2Cl^-$)
(e) $Al(OH)_3$
(f) MgC_2O_4

14.26 The pH of a saturated solution of $Ni(OH)_2$ is 8.83. Calculate K_{sp} for $Ni(OH)_2$.

14.27 What is the molar solubility of CaF_2 in $0.010M$ NaF?

14.28 What is the molar solubility of $Mg(OH)_2$ in $0.10M$ NH_3 solution? Remember that NH_3 is a weak base.

14.29 What volume of saturated HgS solution contains a single Hg^{2+} ion?

14.30 The solubility of MgF_2 is 7.6×10^{-2} g/liter. Calculate K_{sp} for this salt.

14.31 How many moles of Ag_2CrO_4 will dissolve in one liter of $0.10M$ $AgNO_3$?

14.32 Would a precipitate form in the following solutions?
(a) 5×10^{-2} mole of $AgNO_3$ and 1.0×10^{-3} mole of $NaC_2H_3O_2$ dissolved in 1.0 liter of solution
(b) 1.0×10^{-2} mole of $Ba(NO_3)_2$ and 2.0×10^{-2} mole NaF dissolved in 1.0 liter of solution
(c) 1.0 mole of $AgNO_3$ and 1.0 mole $HC_2H_3O_2$ in 1.0 liter of solution
(d) 500 ml of $1.4 \times 10^{-2}M$ $CaCl_2$ and 250 ml of $0.25M$ Na_2SO_4 mixed to give a final volume of 750 ml

14.33 A solution is prepared by mixing 100 ml of $0.20M$ $AgNO_3$ with 100 ml of $0.10M$ HCl. What are the concentrations of all species present in the solution when equilibrium is reached?

14.34 A solution is known to contain $0.010M$ Pb^{2+} and $0.010M$ Ni^{2+}. How much must the pH be adjusted to achieve the maximum separation when the solution is saturated with H_2S?

14.35 A solution containing $0.10M$ Zn^{2+} and $0.10M$ Fe^{2+} is saturated with H_2S. What must the H^+ concentration be to separate these ions by selectively precipitating ZnS? What is the smallest Zn^{2+} concentration that can be achieved without precipitating any of the Fe^{2+} as FeS?

14.36 What would the H^+ concentration have to be in order to prevent the precipitation of HgS when a $0.0010M$ $Hg(NO_3)_2$ solution is saturated with H_2S? Can you explain why HgS is insoluble in concentrated ($12M$) HCl?

14.37 Explain, on the basis of the equilibria involved, why ZnS is able to be dissolved in concentrated hydrochloric acid.

14.38 How many moles of HCl must be added to 1.0 liter of water to dissolve completely 0.20 mole of FeS? Remember that a saturated H_2S solution is $0.1M$.

14.39 Use the data in Table 14.5 to determine the molar solubility of AgI in 0.010M KCN solution.

14.40 The solubility of $Zn(OH)_2$ in $1M$ NH_3 is 5.7×10^{-3} mole/liter. Determine the value of the instability constant of the complex ion, $Zn(NH_3)_4^{2+}$. Ignore the ionization of NH_3.

14.41 How many moles of solid NH_4Cl must be added to 1.0 liter of water in order to dissolve 0.10 mole of solid $Mg(OH)_2$? *Hint.* Consider the simultaneous equilibria:

$$Mg(OH)_2 \rightleftharpoons Mg^{2+} + 2OH^-$$
$$NH_3 + H_2O \rightleftharpoons NH_4^+ + OH^-$$

14.42 Determine the pH of each of the following salt solutions:
(a) $1.0 \times 10^{-3}M$ $NaC_2H_3O_2$
(b) 0.125M NH_4Cl

(c) 0.10M Na_2CO_3
(d) 0.10M NaCN

14.43 What is the percent hydrolysis of a 0.10M solution of pyridinium chloride, C_5H_5NHCl?

14.44 A 0.10M solution of the sodium salt of a weak acid has a pH of 9.35. What is the K_a of the weak acid?

14.45 When 50 ml of 0.20M HF is titrated with 0.10M NaOH, what is the pH
(a) After 5.0 ml of base has been added?
(b) When half of the HF has been neutralized?
(c) At the equivalence point?

14.46 Using the data in Tables 14.1 and 14.3, choose an indicator that is suitable for the titration of
(a) Acetic acid with sodium hydroxide
(b) Aniline with hydrochloric acid

15
Electrochemistry

Electrochemistry is concerned with the conversion of electrical energy into chemical energy in **electrolytic cells,** as well as with the conversion of chemical energy into electrical energy in **galvanic** or **voltaic cells.** In an electrolytic cell a process called **electrolysis** takes place in which the passage of electricity through a solution provides sufficient energy to cause an otherwise nonspontaneous oxidation-reduction reaction to take place. A galvanic cell, on the other hand, provides a source of electricity that results from a spontaneous oxidation-reduction reaction taking place in solution.

Electrochemical processes have a practical importance in chemistry and in everyday life. Electrolytic cells can provide us with information concerning the chemical environment as well as the energy that is required for many important oxidation-reduction reactions to take place. For many years now, galvanic cells, such as the dry cell, have powered our flashlights, radios, and children's toys, while cells such as the lead storage battery have achieved widespread applications, especially in the automotive industry. More recently fuel cells, in which the energy available from the combustion of fuels is converted directly into electricity, are finding many uses, especially in space vehicles. Electrochemical know-how has aided scientists in producing modern equipment for pollution analysis and biomedical research. With the aid of tiny electrochemical probes scientists are beginning to study the chemical reactions taking place in living cells.

All of these processes will be discussed in this chapter. However, before we begin let us first understand, qualitatively, how electrolytic solutions conduct electricity.

15.1 Metallic and Electrolytic Conduction

In order for a substance to be classified as a conductor of electricity, it must be able to allow electrical charges within it to be moved from one point to another for the purpose of completing an electrical circuit. From our discussion of solids in Chapter 7 we know that most metals are conductors of electricity because of the relatively free movement of their *electrons* throughout the metallic lattice, as can be described by either the electron sea model or the band theory of solids. This conduction is simply called **metallic conduction.** We also know from Chapter 5 that solutions containing electrolytes also have the ability to conduct electricity. In this case, however, there are no "free" electrons to carry the current. How, then, do these solutions conduct?

We can determine whether or not a solution is a conductor of electricity by using an apparatus similar to that shown in Figure 5.2. When the two electrodes, which are connected to a source of electricity, are immersed into a solution, we then observe whether or not the bulb of the apparatus lights. The bulb will burn brightly with a strong electrolyte, somewhat dimmer with a weak electrolyte, and not at all with a nonelectrolyte. We find, therefore, that the presence of ions is necessary if a liquid is to conduct electricity; a situation that is fulfilled only by solutions of electrolytes and molten salts.

If the source of electricity to the two electrodes is a direct current, as shown in Figure 15.1, each ion in the liquid is then attracted to the electrode of opposite charge. Thus, when the voltage is applied, the positive ions migrate toward the negative electrode and the negative ions move toward the positive electrode. This movement of ionic charges through the liquid, brought about by the application of electricity, is called **electrolytic conduction.**

When the ions of the liquid come in contact with the electrodes, chemical reactions take place. At the positive electrode (where there is a deficiency of electrons) the negative ions deposit electrons and are therefore oxidized. At the negative electrode (which has an excess of electrons) the positive ions pick up electrons, and are reduced. Thus, during electrolytic conduction oxidation is occurring at the positive electrode and reduction is taking place at the negative electrode. The liquid will continue to conduct electricity only as long as the oxidation-reduction reactions occurring at the electrodes continue.

The electrons that are deposited during the oxidation reaction are immediately pumped out of the solution by the voltage source and a constant supply of electrons is always being pumped into the negative electrode. During electrolytic conduction, then, we have electrons flowing through the exterior wire and ions flowing through the solution. This situation is illustrated in Figure 15.2a.

The ionic movement, as well as the reactions at the electrodes, must take place so that electrical neutrality is maintained. This means that even in the most minute part of the liquid, whenever a negative ion moves away, a positive ion must also leave, or another negative ion must immediately take its place (Figure 15.2b). In this way every portion of the liquid is electrically neutral at all times. During the reactions at the electrodes, electrical neutrality is assured by having equal numbers of electrons deposited and picked up. Whenever an electron, for example, is deposited at the positive electrode,

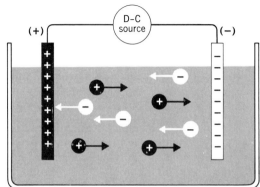

Figure 15.1
Ion flow in an electrolytic cell.

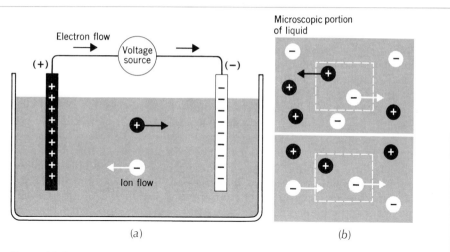

Electron flow

Voltage source

(+)   (−)

Ion flow

(a)   (b)

Figure 15.2

Electrolytic conduction. (a) Electrolytic cell. (b) Maintaining electrical neutrality on a microscopic Scale.

one electron must simultaneously be picked up at the negative electrode. It is the chemical consequences of these last two processes that we focus our attention on next.

15.2
Electrolysis

The chemical reactions that occur at the electrodes during electrolytic conduction constitute electrolysis. When liquid (molten) sodium chloride, for example, is electrolyzed, we find that the Na^+ ions move toward the negative electrode and the Cl^- ions move toward the positive electrode (Figure 15.3). The reactions that take place at the electrodes are:

Positive electrode $2Cl^- \longrightarrow Cl_2 + 2e^-$ oxidation
Negative electrode $Na^+ + e^- \longrightarrow Na$ reduction

Anode   D.C. source   Cathode
(+)   (−)

$2Cl^- \longrightarrow Cl_2 + 2e^-$ $Na^+ + e^- \longrightarrow Na$

Cl_2 formed   Na formed

Figure 15.3
Electrolysis of molten NaCl.

In electrochemistry we assign the terms **cathode** and **anode** according to the chemical reaction that is taking place at the electrode. *Reduction always takes place at the cathode and oxidation always at the anode.* Thus in our electrolysis reactions above we label the negative electrode the cathode and the positive electrode the anode. In galvanic cells, however, we shall find, following this procedure for the labeling of electrodes, that the positive electrode is the cathode and the negative electrode is the anode.

The net chemical change that takes place in the electrolytic cell (the **cell reaction**) is obtained by adding together the anode and cathode reactions in such a way that the same number of electrons are gained and lost. This is the same procedure that we used in the ion-electron method of balancing oxidation-reduction reactions in Chapter 5. Thus, in this case we must multiply the reduction half-reaction by two to get

$$2Cl^-(l) \longrightarrow Cl_2(g) + 2e^-$$
$$\underline{2Na^+(l) + 2e^- \longrightarrow 2Na(l)}$$
$$2Na^+(l) + 2Cl^-(l) \longrightarrow Cl_2(g) + 2Na(l)$$

In this electrolytic cell, then, sodium is formed at the cathode and chlorine gas at the anode. This is one of the major sources of pure sodium metal and chlorine gas in the United States.

The electrolysis of aqueous solutions of electrolytes is somewhat more complex because of the ability of water to be oxidized as well as reduced. The oxidation reaction for water is

$$2H_2O \longrightarrow O_2 + 4H^+ + 4e^- \tag{15.1}$$

and the reduction reaction takes the form

$$2H_2O + 2e^- \longrightarrow H_2 + 2OH^- \tag{15.2}$$

In acidic solutions another reaction that may take place is the reduction of H^+, which is

$$2H^+ + 2e^- \longrightarrow H_2 \tag{15.3}$$

Reaction 15.3 is not, however, a major reaction in most dilute aqueous solutions that we shall consider.

In aqueous solution, then, we must keep in mind the oxidation and reduction, respectively, of the anion and cation of the solute plus the oxidation and reduction of H_2O. Whether the solute anion or water is going to be oxidized, or whether the solute cation or water is going to be reduced, depends on the relative ease of the two competing reactions, as we shall see in the next few examples.

ELECTROLYSIS OF AQUEOUS NaCl. In the electrolysis of aqueous NaCl the following two anode (oxidation) reactions are possible:

(1) $2Cl^-(aq) \longrightarrow Cl_2 + 2e^-$
(2) $2H_2O \longrightarrow O_2 + 4H^+(aq) + 4e^-$

and the following two cathode (reduction) reactions are possible:

(3) $Na^+(aq) + e^- \longrightarrow Na$
(4) $2H_2O + 2e^- \longrightarrow H_2 + 2OH^-(aq)$

We can, of course, experimentally determine the outcome of this electrolysis by simply examining the products that are formed at the electrodes. Here we find that, in concentrated solutions, chlorine gas is produced at the anode and hydrogen gas at the cathode. Therefore, during the electrolysis of an aqueous solution of NaCl, the two half-reactions and the cell reaction would be

$$2Cl^-(aq) \longrightarrow Cl_2 + 2e^-$$
$$2H_2O + 2e^- \longrightarrow H_2 + 2OH^-(aq)$$
$$\overline{2H_2O + 2Cl^-(aq) \longrightarrow Cl_2(g) + H_2(g) + 2OH^-(aq)}$$

In this case, then, the Na^+ is more difficult to reduce than the H_2O and the Cl^- is more easily oxidized than H_2O. Hence we find the reduction of H_2O with the oxidation of Cl^-. It is interesting to note that in industry mercury has often been used as the electrode material for this reaction and, as a result, has been a source of mercury water pollution.

ELECTROLYSIS OF AQUEOUS $CuSO_4$. As in our last example we have two possible oxidation and two possible reduction reactions for the electrolysis of $CuSO_4$. These reactions in this case are

Oxidation   **(1)** $2SO_4^{2-}(aq) \longrightarrow S_2O_8^{2-}(aq) + 2e^-$
              (2) $2H_2O \longrightarrow O_2 + 4H^+(aq) + 4e^-$

and

Reduction   **(3)** $Cu^{2+}(aq) + 2e^- \longrightarrow Cu$
             (4) $2H_2O + 2e^- \longrightarrow H_2 + 2OH^-(aq)$

During this electrolysis we find that oxygen is produced at the anode and copper metal is deposited on the cathode. Therefore, we would write for the electrolysis of aqueous $CuSO_4$:

$$2H_2O \longrightarrow O_2 + 4H^+(aq) + 4e^-$$
$$2Cu^{2+}(aq) + 4e^- \longrightarrow 2Cu$$
$$\overline{2H_2O + 2Cu^{2+}(aq) \longrightarrow O_2 + 4H^+(aq) + 2Cu}$$

Note that we multiplied the equation for the reduction of Cu^{2+} by 2 in order to have equal numbers of electrons lost and gained.

In the electrolysis of aqueous $CuSO_4$ the H_2O is more easily oxidized than the SO_4^{2-} and the Cu^{2+} is more readily reduced than the H_2O.

ELECTROLYSIS OF AQUEOUS $CuCl_2$. We should be able to apply what we have learned about the electrolysis of aqueous solutions to the electrolysis of $CuCl_2$. We would expect that the two species that could be oxidized are water and chloride ion, and that the two species which could be reduced are water and Cu^{2+}. Since we already know that Cl^- is more easily oxidized than H_2O and that Cu^{2+} is more readily reduced than H_2O, we expect the reaction

$$2Cl^-(aq) \longrightarrow Cl_2 + 2e^-$$
$$Cu^{2+}(aq) + 2e^- \longrightarrow Cu$$
$$\overline{Cu^{2+}(aq) + 2Cl^-(aq) \longrightarrow Cl_2 + Cu}$$

This is exactly what is found experimentally.

ELECTROLYSIS OF AQUEOUS NA_2SO_4. Once again we call upon what we have learned previously in our discussion of electrolysis. We know that water is more easily oxidized than SO_4^{2-} and that water is more readily reduced than Na^+. Therefore, in this solution H_2O is both oxidized and reduced, giving us

$$2H_2O \longrightarrow O_2 + 4H^+(aq) + 4e^-$$
$$\underline{4H_2O + 4e^- \longrightarrow 2H_2 + 4OH^-(aq)}$$
$$6H_2O \longrightarrow O_2 + 2H_2 + 4OH^-(aq) + 4H^+(aq)$$

Note that we had to multiply the reduction reaction by 2 to achieve the same number of electrons as in the oxidation reaction.

The overall reaction, as written above, can be simplified further by recalling that H^+ and OH^- will react to give H_2O.

$$4OH^- + 4H^+ \longrightarrow 4H_2O$$

Thus the true overall reaction for the electrolysis of a stirred aqueous Na_2SO_4 reaction is

$$2H_2O \longrightarrow O_2 + 2H_2$$

which is simply the reaction for the electrolysis of H_2O. Sodium sulfate does not participate in this electrolysis in the sense that it is not consumed at the electrodes; yet we would find experimentally that it (or some other similar salt) is needed if electrolysis of H_2O is to occur. What then is the role of the Na_2SO_4? The Na_2SO_4 is needed in order to maintain electrical neutrality. During the oxidation of H_2O, H^+ ions are produced in the immediate vicinity of the anode. A negative ion must also be present in that vicinity to neutralize the positive charges. This is fulfilled by the SO_4^{2-} ion. Likewise at the cathode, where OH^- ions are produced, there must be a positive ion present to neutralize the charges on the OH^-, and keep the solution electrically neutral.

Let us now examine some practical applications of the electrolysis process. We have already seen one such application in the production of pure metallic sodium by the electrolysis of molten sodium chloride. The commercial production of other metals such as aluminum, magnesium, and copper also employs electrolysis.

ALUMINUM. As you are undoubtedly aware, aluminum finds many important uses as a structural metal because of its strength and light weight. Its commercial availability has been made possible through the application of electrochemical reduction.

If we were to electrolyze an aqueous solution of an aluminum salt, such as $AlCl_3$, we would find that H_2O is more easily reduced than the Al^{3+}. Therefore an aqueous solution of an aluminum salt cannot be used to produce the metal. A 22-year-old graduate of Oberlin College, Charles Hall, invented a process, in 1886, whereby molten Al_2O_3 is used. He prepared a mixture of Al_2O_3 with cryolite, Na_3AlF_6, and electrolyzed it in the molten state. The cryolite, he found, reduced the melting temperature from 2000° C for Al_2O_3 to 1000° C for the mixture. A diagram of the electrolysis cell is shown in Figure 15.4. The vessel holding the melted mixture is made of iron lined with carbon and serves as the cathode. Carbon rods that serve as the

Carbon electrodes

(+)
(−)

Molten cryolite—Al_2O_3 mixture

Molten Al

Figure 15.4

Production of aluminum by the Hall process.

anode are inserted into the melt. As the oxidation-reduction reactions proceed, pure aluminum is produced at the cathode and sinks to the bottom of the vessel. The reactions at the electrodes are

$$\text{Anode} \qquad 3O^{2-} \longrightarrow \tfrac{3}{2}O_2(g) + 6e^-$$
$$\text{Cathode} \quad \underline{2Al^{3+} + 6e^- \longrightarrow 2Al(l)}$$
$$2Al^{3+} + 3O^{2-} \longrightarrow \tfrac{3}{2}O_2(g) + 2Al(l)$$

Today other materials are used in place of the cryolite. These materials permit operation at still lower temperatures and are less dense than the cryolite used by Hall. This lower density of the electrolyte mix permits easier separation of the molten aluminum.

MAGNESIUM. Magnesium, another structural metal that is important because of its light weight, occurs to an appreciable extent in sea water. Magnesium ions are precipitated from sea water as the hydroxide and the $Mg(OH)_2$ is then converted to the chloride by treatment with hydrochloric acid. After evaporation of the water the $MgCl_2$ is melted and electrolyzed, magnesium being produced at the cathode and chlorine being evolved at the anode. The overall net reaction is simply

$$MgCl_2 \longrightarrow Mg + Cl_2$$

COPPER. An interesting application of electrolysis is the refining, or purification, of copper metal. When first separated from its ore, copper metal is about 99% pure, with iron, zinc, silver, gold, and platinum as major impurities. In the refining process the impure copper is used as the anode in an electrolytic cell containing aqueous copper sulfate as the electrolyte. The cathode of the cell is constructed of high purity copper (Figure 15.5).

When electrolysis is carried out, the voltage across the cell is adjusted so that only copper and other more active metals, such as iron or zinc, are able to dissolve at the anode. The silver, gold, and platinum do not dissolve and simply slough off and settle to the bottom of the electrolysis cell. At the cathode only the most easily reduced species, Cu^{2+}, is caused to pick up electrons; hence, only copper is deposited.

The net result of the operation of this cell is that copper is transferred from the anode to the cathode while the Fe and Zn impurities remain in solution as Fe^{2+} and Zn^{2+}. Afterwards the silver, gold, and platinum "sludge" is removed from the apparatus and sold for enough money to pay for the cost of

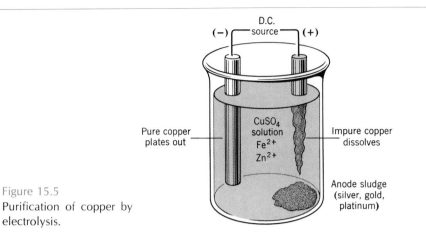

Figure 15.5
Purification of copper by electrolysis.

the electricity required in the electrolysis. As a result, the purification copper (about 99.95% pure) costs nearly nothing! However, the total production cost is still considerable since it includes the mining of the crude ore and its initial purification.

ELECTROPLATING. We have just seen how copper can be "plated out" on an electrode in an electrolysis cell. The plating out of a metal in this fashion is called **electroplating.** If we replaced the cathode in the cell in Figure 15.5 with another metal, the surface of that metal will also become covered with a layer of pure copper when the current is applied. Other metals can be electroplated as well as copper, which makes this process of great commercial importance. In the manufacturing of automobiles, for example, various parts, such as steel bumpers, are electroplated with chromium for beauty as well as for protection against corrosion.

Michael Faraday was the first to describe, in a quantitative fashion, the relationship that exists between the amount of current used and the extent of the chemical change which takes place at the electrodes during electrolysis. The extent of an electrolysis reaction is, of course, related to the number of moles of electrons lost or gained in the oxidation-reduction reactions, respectively. For example, in the reaction for the reduction of silver ion to silver metal,

$$Ag^+ + e^- \longrightarrow Ag(s)$$

one mole of electrons reacts with one mole of silver ions to give one mole, or 107.87 grams, of solid silver (the amount of silver deposited can be experimentally determined by weighing the cathode before and after the current is supplied to the cell). Thus, in this case, when 107.87 grams of silver are deposited on the cathode, we know that one mole of electrons must have passed through the cell.

The amount of electricity that must be supplied to a cell in order for *one mole* of electrons to undergo reaction, in the above case as well as in any other oxidation-reduction reaction, is called a **faraday** ($\mathscr{F}$) of electricity. In the above example, then, one faraday was supplied to produce the 107.87 grams of silver, and it would take two faradays to produce 215.74 grams of silver, and so on.

15.3
Faraday's Laws of Electrolysis

Another important unit of electricity is the **coulomb (coul)**. One coulomb of electricity is the amount of charge that moves past any given point in a circuit when a current of one ampere (amp) is supplied for one second. Thus

$$1 \text{ coul} = 1 \text{ amp sec}$$

Experimentally it is found that one faraday is equivalent to 96,487 coulombs, or 96,500 when rounded off to three significant figures. Thus

$$1 \mathscr{F} = 96,500 \text{ coul} = 1 \text{ mole of electrons}$$

The number of faradays of electricity (moles of electrons) passed through a cell may be found by dividing the number of coulombs by 96,500 (see Example 15.1 below).

In Chapter 5 it was pointed out that in electrochemistry the gram equivalent weight of any substance is the weight that reacts with or produces one mole of electrons. From our discussion above we can now write that one gram equivalent weight of a substance requires the passage of 96,500 coulombs, or 1 faraday, of electricity. This leads us to a clear understanding of **Faraday's law** which states that *the number of gram equivalent weights of a substance produced at either electrode during electrolysis is exactly equal to the number of faradays of electricity that is passed through the cell.*

With a knowledge of Faraday's law we can perform many simple, but chemically important, calculations. Some of these are illustrated in the following sample problems.

Example 15.1

Referring to the electrochemical cell shown in Figure 15.5
(a) How many grams of Cu would be deposited if a current of 1.5 amps is supplied for 2 hr?
(b) How many hours, at a current of 2 amp, will it take to produce 5.0 grams of Cu?

Solution

(a) From our discussion above we know that faradays of electricity can be converted into grams of material by applying Faraday's law. First, we must convert the 1.5 amps for 2 hr into coulombs, and then into faradays.

$$1.5 \text{ amp} \times 2 \text{ hr} \times \left(\frac{3600 \text{ sec}}{1 \text{ hr}}\right) = 10,800 \text{ coul}$$

$$10,800 \text{ coul} \times \left(\frac{1 \mathscr{F}}{96,500 \text{ coul}}\right) = 0.11 \mathscr{F}$$

From the reduction reaction for Cu^{2+}

$$Cu^{2+} + 2e^- \longrightarrow Cu(s)$$

we see that the gram equivalent weight of copper is its atomic weight divided by two (the number of electrons gained per copper ion).

$$1 \text{ equiv Cu} = \frac{63.54}{2} \text{ g Cu} = 31.77 \text{ g Cu}$$

Thus for copper, 31.77 g is equivalent to one faraday and 0.11 $\mathscr{F}$ corresponds to

$$0.11\,\mathscr{F} \times \left(\frac{31.77\ \text{g Cu}}{1\,\mathscr{F}}\right) \sim 3.49\ \text{g Cu}$$

(b) The answer to this question is found by simply reversing the procedure used in (a). We first convert the 5.0 grams into faradays, then into coulombs, and finally into hours.

Knowing that the gram equivalent weight of copper is 63.54/2 or 31.77 g, we can calculate the faraday equivalent of 5.0 grams of copper by

$$5.0\,\text{g Cu} \times \left(\frac{1\,\mathscr{F}}{31.77\ \text{g Cu}}\right) \sim 0.16\,\mathscr{F}$$

and converting faraday to coulombs we have

$$0.16\,\mathscr{F} \times \left(\frac{96{,}500\ \text{coul}}{1\,\mathscr{F}}\right) = 15{,}400\ \text{coul}$$

Since one coulomb is equal to one amp sec, we have

$$15{,}400\,\text{coul} \times \left(\frac{1\ \text{amp sec}}{1\ \text{coul}}\right) \times \left(\frac{1}{2\ \text{amp}}\right) \sim 7700\ \text{sec}$$

Finally, to convert to hours,

$$7{,}700\,\text{sec} \times \left(\frac{1\ \text{hr}}{3600\ \text{sec}}\right) = 2.1\ \text{hr}$$

It was pointed out earlier that we can experimentally determine the weight of a substance that has been deposited on an electrode during electrolysis by weighing the electrode before and after the current has been supplied. The apparatus used in experiments of this kind is called a **coulometer.** In Figure 15.6 we see two such coulometers connected in series so that the

Silver coulometer

$$\left[\begin{array}{c}1\,\mathscr{F}\ \text{deposits 107.87 g Ag} \\ \text{on a cathode}\end{array}\right]$$ 1 $\mathscr{F}$ passes through this cell when 107.87 g of Ag (1 mole Ag) is deposited in the other cell

Figure 15.6
The use of a coulometer.

same current, and thus the same number of faradays, passes through both cells. With the aid of this apparatus it is possible to use a known oxidation-reduction reaction in one cell to provide an experimental measure of the equivalent weight of an unknown in the other cell. This type of analysis is illustrated in the next example.

Example 15.2
In the left cell of Figure 15.6 we place a solution containing Ag^+ ions, and in the right cell a solution whose metal ion is unknown (X). The same current is passed through both cells for the same amount of time. When the current is turned off and the electrodes are rinsed, dried, and weighed, it is found that 3.50 g of silver was deposited during the same period of time that 2.50 g of element X was deposited. What is the equivalent weight of element X?

Solution
This rather long-winded problem has a relatively simple solution. Since the current and time are the same for both cells in the series circuit, the same number of faradays was passed through both. We can calculate the number of faradays supplied by using the information from the Ag^+ cell.

From our earlier discussion in this section we know that the reduction reaction for Ag^+ is

$$Ag^+ + e^- \longrightarrow Ag(s)$$

Thus the equivalent weight of silver is exactly equal to its atomic weight. According to Faraday's law, then, for silver

$$107.87 \text{ g} \sim 1 \mathscr{F}$$

and the faraday equivalent of 3.5 g of silver is then

$$3.50 \text{ g} \times \left(\frac{1 \mathscr{F}}{107.87 \text{ g}} \right) = 0.0324 \mathscr{F}$$

This means that 0.0324 $\mathscr{F}$ of electricity was passed through both cells. In the cell containing the unknown, then, 0.0324 faradays deposited 2.50 g of the substance. Thus, for the unknown

$$0.0324 \mathscr{F} \sim 2.50 \text{ g } X$$

and one faraday is equivalent to

$$1 \mathscr{F} \times \left(\frac{2.50 \text{ g } X}{0.0324 \mathscr{F}} \right) \sim 77.2 \text{ g } X$$

Hence the equivalent weight of element X is equal to 77.2 g.

**15.4
Galvanic Cells**

We now move into an area of electrochemistry where spontaneous oxidation-reduction reactions taking place in solution cause electrons to flow through an external circuit, and thus provide a source of electricity. An example of a spontaneous oxidation-reduction reaction taking place in a solution can be seen simply by placing a piece of metallic zinc into a solution of $CuSO_4$. A brownish spongelike layer begins to form on the piece of zinc and, at the same time, the blue color of the $CuSO_4$ begins to disappear. The brownish substance forming on the zinc is metallic copper, and we write the two half-reactions that occur as

$$Cu^{2+} + 2e^- \longrightarrow Cu$$
$$Zn \longrightarrow Zn^{2+} + 2e^-$$

We see from these reactions that the Cu^{2+} ions are spontaneously removed from the solution and are replaced by the colorless Zn^{2+} ions. Thus the blue color of the solution disappears as more and more Zn^{2+} ions are formed.

As long as these spontaneous reactions take place at the surface of the zinc, no useful flow of electrons can be obtained. We can, however, take advantage of the electron flow that accompanies these oxidation-reduction reactions if we separate the copper solution from the zinc by placing them in a cell similar to that shown in Figure 15.7a or 15.7b. In an apparatus of this design the electrons produced by the oxidation of the zinc must travel through the wire and into the electrode in the $CuSO_4$ solution. The electrons are then picked up by the Cu^{2+} ions and reduction takes place. The electrons flowing through the external wire provide a source of electricity.

Although the zinc and copper have to be separated to obtain a useful flow of electrons, complete isolation of the two species leads to an electrical imbalance at the electrodes, and the electron flow would soon cease. We can see how electrical imbalance would occur if we imagine that the two half-cells were completely isolated from each other and the oxidation-reduction reactions still continued to take place. On the left side of this hypothetical setup, where Zn^{2+} ions are produced, the solution would soon possess an excess of Zn^{2+} ions and, as a result, attain an overall positive charge. In the right compartment, where the Cu^{2+} ions are removed, the SO_4^{2-} ions left behind would soon give this solution a negative charge. We know, of course, that this situation could never take place, so that completely separating the two compartments does not lead to a continuous flow of electrons.

If the ions are allowed to travel from one compartment to the other, elec-

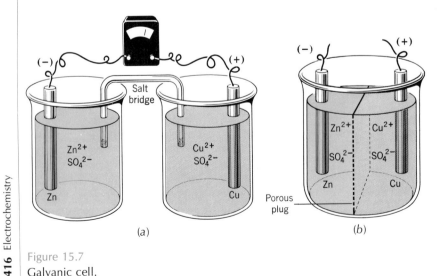

(a)

(b)

Figure 15.7
Galvanic cell.

trical neutrality can be maintained. The flow of ions, however, should not be so rapid that the half-reactions occur at the surface of the zinc electrode. The salt bridge in Figure 15.7a and the porous plug in Figure 15.7b allow for the slow mixing of the ions in the two solutions. A salt bridge is a U-shaped tube filled with an electrolyte such as KNO_3 or KCl in gelatin. Cations from the salt bridge can move into one compartment to compensate for the excess negative charge, while the anions from the salt bridge diffuse into the other compartment to neutralize the excess positive charge. The porous plug in Figure 15.7b serves the same purpose as the salt bridge. With either the salt bridge or the porous plug in place there is a continuous electron flow through the external wire and ion flow through the solution as a result of the spontaneous oxidation-reduction reactions taking place in the galvanic cell.

THE SIGNS OF THE ELECTRODES IN GALVANIC CELLS. Earlier we defined the anode in electrochemistry as the electrode where oxidation takes place and the cathode as the one where reduction occurs. In the galvanic cell just described, oxidation takes place in the zinc compartment, so that the zinc bar would be the anode and the copper electrode would be the cathode. Since electrons are released (oxidation reaction) at the zinc anode and removed (reduction reaction) at the copper cathode, the zinc bar possesses an excess of electrons compared to the copper. As a result the zinc bar is the negative electrode and the copper is the positive electrode. This means that in galvanic cells the anode is negative and the cathode is positive, quite the opposite of what we found to be true in electrolytic cells.[1]

15.5
Cell Potentials

The electric current obtained from a galvanic cell is a result of electrons being pushed or forced to flow from the negative electrode, through an external wire, to the positive electrode. The force with which these electrons move through the wire is called the **electromotive force,** or **emf,** of the cell and is measured in volts. The emf of any galvanic cell depends on the nature of the chemical reactions taking place within the cell, the concentrations of the species undergoing oxidation and reduction, and the temperature of the cell, which we shall take to be 25° C unless otherwise noted. The greater the tendency or potential of the two half-reactions to occur spontaneously, the greater will be the emf of the cell. Thus the emf of the cell can also be called the **cell potential,** designated by the symbol $\mathscr{E}_{cell}$.

Since the cell potential depends on the concentration of the species in the cell, we define a reference potential, which is called the standard potential of the cell, $\mathscr{E}^{\circ}_{cell}$, as that emf obtained when all species are at unit activity (see Section 12.4). Since activities and concentrations do not differ by very much for dilute solutions, for simplicity, we shall use concentrations in place of activities. Thus a solution of a substance at unit activity will be assumed to

[1] This labeling of electrodes in the galvanic cells is, however, consistent with electrolytic cells when we consider the movement of the ions within the solution. The Zn^{2+} ions produced at the anode and the SO_4^{2-} ions freed at the cathode must mingle with each other if electrical neutrality is to prevail in the solution. To accomplish this, some of the Zn^{2+} ions must move toward the cathode and some of the SO_4^{2-} ions must move toward the anode. Thus we have cations moving toward the cathode and anions moving toward the anode, which is precisely the same situation in electrolytic cells.

be one in which its concentration is $1M$. If a gas is involved in a half-reaction, a pressure of 1 atm best approximates unit activity. Whenever we write the half-reaction or the overall cell reaction for any galvanic cell, we must specify concentrations. Thus the half-reactions and the overall reaction for the Zn/Cu cell in the last section would be written[2]

$$Zn(s) \longrightarrow Zn^{2+}(1M) + 2e^-$$
$$\underline{Cu^{2+}(1M) + 2e^- \longrightarrow Cu(s)}$$
$$Zn(s) + Cu^{2+}(1M) \longrightarrow Zn^{2+}(1M) + Cu(s)$$

The cell potential, $\mathscr{E}_{cell}$, obtained from galvanic cells is measured with an instrument called a **potentiometer.** A potentiometer is a very accurate device for measuring electrical potentials, in which the voltage of the cell being measured is balanced by an equal potential in the opposing direction. When the voltage from the potentiometer exactly balances that from the galvanic cell, no current flows, and the emf of the cell is read directly from the potentiometer. With this technique the maximum emf or voltage of the cell is obtained.

Somewhat less than the maximum emf of a cell is obtained when the cell's potential is measured with a voltmeter. For this device to operate, current must be continually flowing out of and through the cell during the measurement. As a result, part of the cell's emf is used up in order to overcome the electrical resistance of the cell, wires, and voltmeter, and the rest of the emf is what is measured. Thus, whenever a voltmeter is employed to measure cell potential, we always obtain voltages that are less than the actual (or maximum) emf of the cell.

A very important and useful concept can be developed if we attempt to answer the question, what is the origin of the cell potential? To answer this question we shall use the Zn/Cu cell we just described above. In this cell we have a solution containing Zn^{2+} ions about one electrode and a solution containing Cu^{2+} ions around the other. Each of these ions have a certain tendency to acquire electrons from their respective electrodes and become reduced. In other words, there is associated with a reduction half-reaction, such as

$$Zn^{2+} + 2e^- \longrightarrow Zn(s)$$

or

$$Cu^{2+} + 2e^- \longrightarrow Cu(s)$$

in our cell, a certain intrinsic tendency to proceed from left to right that we can describe by its **reduction potential.** The larger the reduction potential for any half-reaction, the greater is its ability to undergo reduction.

15.6
Electrode Potentials

---

[2] Here is another type of notation sometimes used to describe galvanic cells:

$$Zn(s)/Zn^{2+}(1M)//Cu^{2+}(1M)/Cu(s)$$

The species involved in the half-reaction taking place at the anode is written first, where the slash line, /, represents a phase boundary [Zn(s) being the solid electrode and Zn^{2+} in the solution]. The double slash represents a salt bridge between the two half-reactions and, finally, the cathodic reaction is written with a slash separating the phases here as well.

When the cell reaction takes place, then, what we actually observe is a kind of "tug-of-war." Each of the species in solution attempts to pull electrons from its electrodes so as to become reduced. The species with the greatest ability to acquire electrons, that is, the substance with the largest reduction potential, wins the tug-of-war and does undergo reduction. The loser, on the other hand, must supply the electrons to the winner and that substance is therefore oxidized.

The potential measured for the cell represents the difference in the abilities of the two ions to become reduced, that is, the cell emf corresponds to the *difference* between the reduction potentials of the two half-reactions. If we represent the **standard reduction potentials** for Cu^{2+} and Zn^{2+} as $\mathscr{E}^{\circ}_{Cu}$ and $\mathscr{E}^{\circ}_{Zn}$ respectively, then the cell potential for the Zn/Cu cell can be written as

$$\mathscr{E}^{\circ}_{cell} = \mathscr{E}^{\circ}_{Cu} - \mathscr{E}^{\circ}_{Zn}$$

Experimentally, we can only measure *positive* cell potentials. Thus in order to obtain the overall standard cell potential for this cell, or for any other spontaneous cell reaction, we always subtract the smaller reduction potential from the larger one.

Since, experimentally, only overall cell potentials can be measured, we are only capable of obtaining differences between the reduction potentials for any two half-reactions. How then can we obtain the reduction potential for any specific half-reaction? Clearly, if the cell potential and the $\mathscr{E}^{\circ}$ for one of the half-reactions are known, the $\mathscr{E}^{\circ}$ for the other half-reaction can be calculated.

15.7
Standard Reduction Potentials

We have just seen that we cannot actually measure the $\mathscr{E}^{\circ}$ for a single electrode; instead we can only measure the $\mathscr{E}^{\circ}_{cell}$, which is the potential difference between the $\mathscr{E}^{\circ}$'s for the half-reactions. Therefore, in order to obtain the reduction potential for several single electrodes, we must choose one and arbitrarily assign an emf to it. If we assign the chosen electrode a value of zero (0.000) volts, then this electrode becomes the zero point on our potential scale, and all other single electrodes take on potentials compared to this standard.

The electrode chosen to be the standard is the hydrogen electrode, shown in Figure 15.8a. The hydrogen electrode consists of a platinum wire encased in a glass sleeve with hydrogen gas passing through it at a pressure of 1 atm. The platinum wire is attached to a platinum foil that is coated with a black velvet-looking layer of finely divided platinum that serves as a catalyst for the reaction,

$$2H^+(aq) + 2e^- \longrightarrow H_2(g)$$

This assembly is then immersed into an acid solution whose hydrogen ion concentration is $1M$. When the hydrogen electrode and the copper electrode, for example, are placed together in a cell, we have the galvanic cell shown in Figure 15.8b.

If the reduction potential of the species in the other half of the cell is greater than that for the hydrogen electrode (i.e., if it has a positive value), the hydrogen electrode is forced to undergo oxidation. The corresponding half-reaction for the oxidation of the hydrogen electrode is

$$H_2(g) \longrightarrow 2H^+ + 2e^-$$

Figure 15.8
The hydrogen electrode. (a) The hydrogen electrode. (b) The hydrogen electrode used in a galvanic cell.

If, on the other hand, the reduction potential of the other half-reaction is less than 0.000 volt (this species would have a negative reduction potential) the hydrogen electrode undergoes reduction,

$$2H^+ + 2e^- \longrightarrow H_2(g)$$

and causes the other species to become oxidized.

To illustrate this idea let us take a closer look at the Cu/H_2 cell shown in Figure 15.8b. The potential for this cell, as measured with a potentiometer, turns out to be +0.34 volt. We now must determine whether the copper is easier or more difficult to reduce than the hydrogen, in order to obtain the proper form for the $\mathscr{E}^\circ_{cell}$ equation. In this cell we find that copper is plated out on its electrode and, therefore the Cu^{2+} ion is being reduced. Thus the correct form for the $\mathscr{E}^\circ_{cell}$ for the Cu/H_2 cell is

$$\mathscr{E}^\circ_{cell} = \mathscr{E}^\circ_{Cu} - \mathscr{E}^\circ_{H_2}$$

Since $\mathscr{E}^\circ_{H_2}$ is 0.000 volt, then

$$\mathscr{E}^\circ_{cell} = \mathscr{E}^\circ_{Cu} = +0.34 \text{ volt}$$

Therefore, the reduction potential for the reduction of copper,

$$Cu^{2+} + 2e^- \longrightarrow Cu$$

is +0.34 volt.

Now let us examine a cell in which hydrogen is more easily reduced than the species at the other electrode. For this example we shall replace the copper electrode in Figure 15.8b with a zinc electrode. The $\mathscr{E}^\circ_{cell}$ for this Zn/H_2 cell is measured to be +0.76 volt.

Following our first example we must now determine the relative ease of reduction of the zinc ion compared to the H$^+$ ion. In this cell, by the gradual disappearance of the zinc electrode, we observe that the zinc is oxidized,

$$Zn(s) \longrightarrow Zn^{2+} + 2e^-$$

Therefore Zn^{2+} must be more difficult to reduce than H$^+$. The $\mathscr{E}^\circ_{cell}$ for this cell would then correspond to

$$\mathscr{E}^\circ_{cell} = \mathscr{E}^\circ_{H_2} - \mathscr{E}^\circ_{Zn}$$

Since $\mathscr{E}^\circ_{H_2}$ is zero, we have

$$\mathscr{E}^\circ_{cell} = 0 - \mathscr{E}^\circ_{Zn}$$

or

$$\mathscr{E}^\circ_{cell} = -\mathscr{E}^\circ_{Zn} = +0.76 \text{ volt}$$

Therefore,

$$\mathscr{E}^\circ_{Zn} = -0.76 \text{ volt}$$

We see that the reduction potential for the zinc electrode in this cell is a negative quantity, which is consistent with our discussion above on the sign convention of the electrode for those species more difficult to reduce than hydrogen.

With a knowledge of the reduction potentials for the zinc and copper electrodes, we can now *predict* the cell potential for the Zn/Cu cell. This can be done even if we had no previous knowledge of the species undergoing oxidation or reduction. We know that from the sizes of the reduction potentials of Cu and Zn, +0.34 and −0.76 volt, respectively, that the only way we can subtract these two values from each other and obtain a positive $\mathscr{E}^\circ_{cell}$ is if we subtract the −0.76 volt from the +0.34 volt. Thus

$$\mathscr{E}^\circ_{Cu-Zn} = \mathscr{E}^\circ_{Cu} - \mathscr{E}^\circ_{Zn} = +0.34 - (-0.76)$$

and we find that

$$\mathscr{E}^\circ_{Cu-Zn} = 1.10 \text{ volts}$$

This is precisely the value that is obtained experimentally when we measure the potential of this cell.

We are now in a position to determine the reduction potentials for many different half-reactions, since all we must do is construct galvanic cells in which the reduction potential of one half-cell is known, relative to the hydrogen electrode. Some reduction potentials determined in this manner are given in Table 15.1. In this table the reduction potential of the hydrogen electrode is placed in the middle, with the species more difficult to reduce than hydrogen listed above it and those more easily reduced placed below it. Such a table of reduction potentials serves many useful purposes.

1. From a table of reduction potentials we can, at a glance, pick out substances that are good oxidizing agents and those that are good reducing agents. Any species that appears on the left of the equal signs in these equations serves as an oxidizing agent, if it undergoes reduction during the course of a chemical reaction. Since the substances at the bottom of the table are more easily reduced than those

Table 15.1
Standard Reduction Potentials at 25° C

Half-Reaction	$\mathcal{E}°$ (volts)
$Li^+ + e^- \rightleftharpoons Li$	-3.05
$K^+ + e^- \rightleftharpoons K$	-2.92
$Ba^{2+} + 2e^- \rightleftharpoons Ba$	-2.90
$Ca^{2+} + 2e^- \rightleftharpoons Ca$	-2.76
$Na^+ + e^- \rightleftharpoons Na$	-2.71
$Mg^{2+} + 2e^- \rightleftharpoons Mg$	-2.38
$Al^{3+} + 3e^- \rightleftharpoons Al$	-1.71
$Mn^{2+} + 2e^- \rightleftharpoons Mn$	-1.03
$2H_2O + 2e^- \rightleftharpoons H_2 + 2OH^-$	-0.83
$Zn^{2+} + 2e^- \rightleftharpoons Zn$	-0.76
$Cr^{3+} + 3e^- \rightleftharpoons Cr$	-0.74
$Fe^{2+} + 2e^- \rightleftharpoons Fe$	-0.44
$PbSO_4 + 2e^- \rightleftharpoons Pb + SO_4^{2-}$	-0.36
$Ni^{2+} + 2e^- \rightleftharpoons Ni$	-0.25
$Sn^{2+} + 2e^- \rightleftharpoons Sn$	-0.14
$Pb^{2+} + 2e^- \rightleftharpoons Pb$	-0.13
$Fe^{3+} + 3e^- \rightleftharpoons Fe$	-0.04
$2H^+ + 2e^- \rightleftharpoons H_2$	0.00

at the top, their ability to serve as oxidizing agents increases as we proceed down the table. Thus we could conclude from their positions on this table, that H^+ is a better oxidizing agent than Zn^{2+}, and that F_2 is a better oxidizing agent than Cl_2. In brief, then, good oxidizing agents are those species on the left of the equal sign at the bottom of the table.

Each of the half-reactions listed in Table 15.1 is reversible. We saw, for example, in the last section, that H_2 is oxidized to H^+ when placed in a cell with copper, and that H^+ is reduced to H_2 when placed against zinc. When the reactions in Table 15.1 are forced to proceed from right to left, that is, when they are caused to be the oxidation step in the overall reaction, then the species appearing at the right in Table 15.1 are functioning as reducing agents by being oxidized. All of the substances appearing on the right side of the reactions in Table 15.1, then, could behave as reducing agents, with those species at the top right of the table, such as Li, being the best and those at the bottom right, such as F^-, being the poorest.

Table 15.1 (Continued)

Half-Reaction	$\mathscr{E}°$ (volts)
$AgCl + e^- \rightleftharpoons Ag + Cl^-$	0.22
$Hg_2Cl_2 + 2e^- \rightleftharpoons 2Hg + 2Cl^-$	0.27
$Cu^{2+} + 2e^- \rightleftharpoons Cu$	0.34
$Cu^+ + e^- \rightleftharpoons Cu$	0.52
$I_2(aq) + 2e^- \rightleftharpoons 2I^-$	0.54
$Fe^{3+} + e^- \rightleftharpoons Fe^{2+}$	0.77
$Ag^+ + e^- \rightleftharpoons Ag$	0.80
$Br_2(aq) + 2e^- \rightleftharpoons 2Br^-$	1.09
$O_2 + 4H^+ + 4e^- \rightleftharpoons 2H_2O$	1.23
$MnO_2 + 4H^+ + 2e^- \rightleftharpoons Mn^{2+} + 2H_2O$	1.28
$Cr_2O_7{}^{2-} + 14H^+ + 6e^- \rightleftharpoons 2Cr^{3+} + 7H_2O$	1.33
$Cl_2(g) + 2e^- \rightleftharpoons 2Cl^-$	1.36
$2ClO_3{}^- + 12H^+ + 10e^- \rightleftharpoons Cl_2 + 6H_2O$	1.47
$8H^+ + MnO_4{}^- + 5e^- \rightleftharpoons Mn^{2+} + 4H_2O$	1.49
$PbO_2 + SO_4{}^{2-} + 4H^+ + 2e^- \rightleftharpoons PbSO_4 + 2H_2O$	1.69
$H_2O_2 + 2H^+ + 2e^- \rightleftharpoons 2H_2O$	1.78
$S_2O_8{}^{2-} + 2e^- \rightleftharpoons 2SO_4{}^{2-}$	2.00
$F_2 + 2e^- \rightleftharpoons 2F^-$	2.87

2. From a table of this type we can rather quickly find which combination of reactants leads to a spontaneous oxidation-reduction reaction and which does not (when all reactants and products are in their standard states). It can also be determined whether or not a given reaction as written will proceed spontaneously in the forward direction. For example, consider the reaction

$$Cu + 2H^+ \longrightarrow Cu^{2+} + H_2$$

This reaction, as written can be taken to be the sum of the two half-reactions

$$Cu(s) \longrightarrow Cu^{2+} + 2e^-$$

and

$$2H^+ + 2e^- \longrightarrow H_2(g)$$

We recall from the last section that the $\mathscr{E}°_{cell}$ for a spontaneous reaction is always determined as the difference between the reduction potential of the species being reduced and the reduction potential of the substance that is oxidized. This is why, for example, we wrote that the $\mathscr{E}°_{cell}$ for the Zn/Cu cell is given as

$$\mathscr{E}°_{cell} = \mathscr{E}°_{Cu} - \mathscr{E}°_{Zn}$$

For our *proposed* reaction of Cu with H^+, the H^+ is reduced and the Cu(s) is oxidized; the $\mathscr{E}°_{cell}$ would be written as

$$\mathscr{E}°_{cell} = \mathscr{E}°_{H+} - \mathscr{E}°_{Cu}$$

Substituting the reduction potentials from Table 15.1 into the above equation, we have

$$\mathscr{E}°_{cell} = 0.00 - 0.34 \text{ volt}$$
$$\mathscr{E}°_{cell} = -0.34 \text{ volt}$$

Since $\mathscr{E}°_{cell}$ must be positive for a spontaneous reaction, we may conclude that the reaction between Cu(s) and H^+, *as written*, is non-spontaneous. The results of this calculation allow us to understand why copper does not dissolve in acids such as HCl. Another important point to note here is that the reverse of the reaction considered above, namely,

$$Cu^{2+} + H_2 \longrightarrow Cu(s) + 2H^+$$

is spontaneous. The $\mathscr{E}°_{cell}$ for the reaction in this direction would be

$$\mathscr{E}°_{cell} = \mathscr{E}°_{Cu} - \mathscr{E}°_{H_2}$$

for which we calculate that $\mathscr{E}°_{cell}$ is

$$\mathscr{E}°_{cell} = +0.34 \text{ volt}$$

Thus the $\mathscr{E}°_{cell}$ for the forward and reverse reactions are *equal* in magnitude but *opposite* in sign.

Based on the evidence for the Zn/H_2, H^+/Cu, and the Zn/Cu cells we can conclude that in order to obtain positive $\mathscr{E}°_{cell}$, we must always subtract an $\mathscr{E}°$, for a half-reaction that is in a higher position on the table from that which is in a lower one. The following example illustrates how this can be applied.

Example 15.3
How, under standard conditions, may we obtain a spontaneous reaction between the following sets of half-reactions?
(a) $Zn^{2+} + 2e^- \rightleftharpoons Zn(s)$
     $Mn^{2+} + 2e^- \rightleftharpoons Mn(s)$
(b) $Cr^{3+} + 3e^- \rightleftharpoons Cr(s)$
     $MnO_2(s) + 4H^+ + 2e^- \rightleftharpoons Mn^{2+} + 2H_2O$
Give the overall spontaneous reaction and calculate the $\mathscr{E}°_{cell}$ for each set.

Solution
In light of our previous discussion we can answer this question by first finding the relative positions of the two species in each set in Table 15.1.
(a) Since Mn^{2+} appears above Zn^{2+}, it must have a smaller reduction potential than Zn^{2+} and its $\mathscr{E}°$ must therefore be subtracted from $\mathscr{E}°_{Zn}$. The $\mathscr{E}°_{cell}$

for this pair of half-reactions would then take the form

$$\mathscr{E}^{\circ}_{cell} = \mathscr{E}^{\circ}_{Zn} - \mathscr{E}^{\circ}_{Mn}$$

Substituting the reduction potentials from Table 15.1 into this expression, we have

$$\mathscr{E}^{\circ}_{cell} = -0.76 - (-1.03)$$
$$\mathscr{E}^{\circ}_{cell} = +0.27 \text{ volt}$$

Since the Zn^{2+} has the higher reduction potential, it causes the Mn to undergo oxidation. The corresponding two half-reactions and the spontaneous overall reaction would be

$$Zn^{2+} + 2e^- \longrightarrow Zn(s)$$
$$\underline{\qquad Mn(s) \longrightarrow Mn^{2+} + 2e^-}$$
$$Zn^{2+} + Mn(s) \longrightarrow Zn(s) + Mn^{2+}$$

(b) Following the procedure outlined in part (a), we find that Cr(s) has a position higher in Table 15.1 than does MnO_2; therefore, $\mathscr{E}^{\circ}_{Cr}$ must be subtracted from $\mathscr{E}^{\circ}_{MnO2}$ to give a positive $\mathscr{E}^{\circ}_{cell}$.

$$\mathscr{E}^{\circ}_{cell} = \mathscr{E}^{\circ}_{MnO_2} - \mathscr{E}^{\circ}_{Cr}$$
$$\mathscr{E}^{\circ}_{cell} = 1.28 \text{ volt} - (-0.74 \text{ volt})$$
$$\mathscr{E}^{\circ}_{cell} = +2.02 \text{ volt}$$

Since MnO_2 has the higher reduction potential, it is reduced and causes the Cr(s) to undergo oxidation. To obtain the balanced equation for the spontaneous reaction, we must also make the number of electrons gained and lost be the same.

$$3[MnO_2(s) + 4H^+ + 2e^- \longrightarrow Mn^{2+} + 2H_2O]$$
$$\underline{2[Cr(s) \longrightarrow Cr^{3+} + 3e^-]}$$
$$3MnO_2(s) + 12H^+ + 2Cr(s) \longrightarrow 3Mn^{2+} + 6H_2O + 2Cr^{3+}$$

Note that even though we were forced to adjust the coefficients in order to balance the equation, the standard cell potential was still obtained in the same manner as before, namely as the difference between the $\mathscr{E}^{\circ}$ values in Table 15.1.

3. In the process of combining half-reactions from Table 15.1, we see that among the reactants in the spontaneous reaction there is a set of substances that appear on the left side of one half-reaction plus a set of substances that appear on the right side of another. In part (b) of Example 15.3, for instance, we have $MnO_2 + 4H^+$ from the left side of one half-reaction and Cr(s) from the right side of the other. The order of these reactions in Table 15.1 is

$$Cr^{3+} + 3e^- \rightleftharpoons Cr(s) \qquad\qquad \mathscr{E}^{\circ} = -0.74$$
$$MnO_2(s) + 4H^+ + 2e^- \rightleftharpoons Mn^{2+} + 2H_2O \qquad \mathscr{E}^{\circ} = 1.28$$

and we see that the reactants in the overall spontaneous reaction are those substances related by the diagonal line running from lower left to upper right. As a general statement, then, *we can say that when comparing reactants and products in their standard states any species on the left of a given half-reaction will react spontaneously with a substance that is found on the right of a half-reaction located above it in Table 15.1.* We could use this rule of thumb, for example, to tell us that Br_2 will react spontaneously with I^- to produce Br^- and I_2, while Br_2 will *not* react spontaneously with Cl^-. Our rule, therefore, permits us to determine the course of a reaction without having to worry about subtracting electrode potentials in the proper sequence.

4. A point worth noting is that a collection of half-reactions, such as that found in Table 15.1, enables us to predict the outcome of many chemical reactions when we know only a relatively few half-reactions and their corresponding reduction potentials. From the 36 half-reactions listed in Table 15.1, for example, we can predict the results of 630 different chemical reactions! A table of this type, therefore, provides us with a very compact way of storing chemical information.

5. With a knowledge of the reduction potentials listed in Table 15.1, we account for the course of electrolysis reactions. For example, we can produce copper by electrolyzing its aqueous solution, but we cannot obtain aluminum in this same fashion. From Table 15.1 we see that the reduction potential of copper is $+0.34$ volt and that for H_2O it is -0.83 volt. Thus copper ion is more readily reduced than H_2O and will plate out on the electrode according to the half-reaction

$$Cu^{2+} + 2e^- \longrightarrow Cu(s)$$

In the case of aluminum, however, we find that the reduction potential for Al^{3+} is -1.66 volt, which makes it more difficult to reduce than water. This means that when an aqueous solution containing Al^{3+} ions is electrolyzed, the H_2O will preferentially be reduced.

It was pointed out in Section 10.8 that the thermodynamic criterion for spontaneity of a chemical reaction is that the change in free energy, ΔG, for the reaction has to be a negative quantity. In Section 10.10 we saw that ΔG also represents the maximum amount of useful work obtainable from a chemical reaction. The relationship between ΔG and maximum work (W_{max}) for any system takes the form

**15.8
Spontaneity of
Oxidation-
Reduction
Reactions**

$$\Delta G = -W_{max}$$

What is W_{max} for an electrochemical cell?

The work derived from an electrochemical cell is, perhaps, not unlike that obtained from a waterwheel, shown in Figure 15.9. The amount of work that can be obtained from this waterwheel depends on two things: (1) the amount of water flowing over the wheel and (2) the force with which the water is flowing. Thus the expression for the work that can be derived from this waterwheel would be written as

$$\text{Work} = (\text{amount of } H_2O)(\text{force of } H_2O)$$

The work obtained from an electrochemical cell is also dependent on (1)

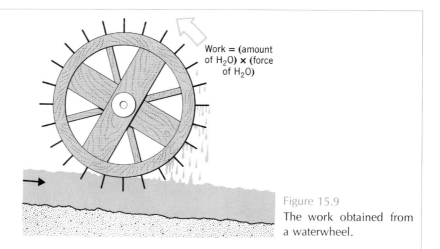

Work = (amount of H_2O) × (force of H_2O)

Figure 15.9

The work obtained from a waterwheel.

the amount of electricity flowing and (2) the force with which it moves. The amount of electricity generated by a cell (i.e., the number of moles of electrons) is equal to the number of electrons (n) involved in the oxidation-reduction reaction times the faraday,

$$\text{Amount of electricity} = n\mathscr{F}$$

The value of n depends on the nature of the half-reactions taking place in the cell and can be derived once the specific reactions are known. For example, in the Zn/Cu cell there are two electrons involved in both of the half-reactions; therefore, n for this cell would be 2. The symbol $\mathscr{F}$ in the above equation is the faraday, 96,500 coulombs, or 23,060 cal/volt, which is the caloric equivalent of the faraday.

The force with which the current moves in a cell corresponds to the emf of that cell. When the emf is a maximum, the work derived from the cell is also a maximum. In Section 5 of this chapter we saw that the maximum emf, $\mathscr{E}°_{cell}$, is that voltage obtained when no current is flowing through the cell. We can experimentally determine the maximum emf of any cell with the aid of a potentiometer.

Thus the equation for maximum work for an electrochemical cell is

$$W_{max} = \text{(amount of electricity)(maximum emf)}$$

or

$$W_{max} = n\mathscr{F}\mathscr{E}°_{cell}$$

Since $\Delta G = -W_{max}$, then

$$\Delta G = -n\mathscr{F}\mathscr{E}_{cell} \tag{15.4}$$

When all species are in their standard states, as identified by $\mathscr{E}°$, then ΔG becomes the standard free energy change for the reaction and is given the symbol $\Delta G°$. Thus, the above equation becomes

$$\Delta G° = -n\mathscr{F}\mathscr{E}°_{cell} \tag{15.5}$$

We can, with the aid of Equation 15.5 calculate the standard free energy change for an oxidation-reduction reaction from a knowledge of its standard cell potential. Consider, for example, the Zn/Cu cell.

$$Zn(s) \longrightarrow Zn^{2+} + 2e^-$$
$$\underline{Cu^{2+} + 2e^- \longrightarrow Cu(s)}$$
$$Zn(s) + Cu^{2+} \longrightarrow Zn^{2+} + Cu(s)$$

The n for this reaction is 2 since two electrons are transferred, $\mathscr{F} = 23{,}060$ cal/volt and $\mathscr{E}^\circ_{cell}$, either derived from Table 15.1 or determined experimentally, is $+1.10$ volts therefore

$$\Delta G^\circ = -2\left(23{,}060 \; \frac{cal}{volt}\right)(+1.10 \text{ volts})$$
$$\Delta G^\circ = -50.7 \text{ kcal}$$

This relationship between the standard cell potential and the standard free energy is an extremely important one because it ties together two different aspects of spontaneity, while at the same time giving us a readily accessible pathway for calculating standard free energy changes. Experimentally, we have only to measure the standard cell emf from which we can then compute the value of ΔG°. But still more important, perhaps, is that through this equation we can derive even more useful thermodynamic quantities.

In Chapter 12, Section 4, we obtained a useful equation for determining equilibrium constants. We see in Example 12.1 (page 325) that the equation showing the relationship between the standard free energy and the thermodynamic equilibrium constant can be written as

15.9
Thermodynamic Equilibrium Constants

$$\Delta G^\circ = -RT \ln K \qquad (15.6)$$

or, in terms of logarithms to the base 10,

$$\Delta G^\circ = -2.303 \; RT \log K \qquad (15.7)$$

Combining Equation 15.7 with Equation 15.5, we have

$$\Delta G^\circ = -n\mathscr{F}\mathscr{E}^\circ = -2.303 RT \log K$$

or simply

$$n\mathscr{F}\mathscr{E}^\circ = 2.303 RT \log K$$

This equation can be solved either for $\mathscr{E}^\circ$,

$$\mathscr{E}^\circ = \frac{2.303 RT}{n\mathscr{F}} \log K \qquad (15.8)$$

or for K,

$$\log K = \frac{n\mathscr{F}\mathscr{E}^\circ}{2.303 RT} \qquad (15.9)$$

If we choose to restrict ourselves to discussing reactions that take place at 25° C(298° K), the quantity $2.303 RT/\mathscr{F}$ becomes a constant at this temperature and has the value

$$\frac{2.303 RT}{\mathscr{F}} = \frac{2.303(1.987 \text{ cal/mole °K})(298° \text{ K})}{23{,}060 \text{ cal/volt}} = 0.0592$$

Thus at 25° C our equation for $\mathscr{E}°$ becomes

$$\mathscr{E}° = \frac{0.0592}{n} \log K \qquad (15.10)$$

By the same token, our equation for $\log K$ is

$$\log K = \frac{n\mathscr{E}°}{0.0592} \qquad (15.11)$$

We see, therefore, that from a knowledge of the standard cell potential, the equilibrium constant for the cell reaction can be calculated. For the Zn/Cu cell we have

$$\log K = \frac{n\mathscr{E}°}{0.0592} = \frac{2(+1.10)}{0.0592} = 37.1$$

hence

$$K \approx 1 \times 10^{37}$$

From the magnitude of this equilibrium constant, we would certainly say that the spontaneous Zn/Cu cell reaction would go very nearly to completion.

Example 15.4

Using Table 15.1, determine whether or not the overall oxidation-reduction reaction

$$Sn(s) + Ni^{2+} \longrightarrow Sn^{2+} + Ni(s)$$

is spontaneous and calculate its equilibrium constant.

Solution

The two half-reactions of this overall reaction would be

$$Sn(s) \longrightarrow Sn^{2+} + 2e^-$$

and

$$Ni^{2+} + 2e^- \longrightarrow Ni(s) \qquad \mathscr{E}° = -0.25$$

Writing the Sn as a reduction, we have

$$Sn^{2+} + 2e^- \longrightarrow Sn \qquad \mathscr{E}° = -0.14$$

From the way the reaction is written, the $\mathscr{E}°$ for the cell would be

$$\mathscr{E}°_{cell} = -0.25 - (-.14)$$

and

$$\mathscr{E}°_{cell} = -0.11 \text{ volt}$$

This means that under *standard conditions* (i.e., unit activity) the reaction in this direction would be nonspontaneous. We can still calculate the equilibrium constant in the same fashion as outlined above. Since two electrons are transferred during the reaction,

$$\log K = \frac{2(-0.11)}{0.0592} = -3.7$$

so that

$$K = 2 \times 10^{-4}$$

From the size of this equilibrium constant we would say that indeed this reaction would not occur to any appreciable extent in the forward direction.

Thus far we have limited our discussion to those cells containing reactants at unit activity. In the laboratory, however, chemists usually do not restrict themselves to only this one set of conditions, and it is found that the cell emf, and in fact the direction of the cell reaction, are controlled by the activities (concentrations) of the species taking part in the reaction. Let us examine this now from a quantitative point of view.

An equation that summarizes how the free energy of the reactants and products of a given reaction varies with temperature and concentration was developed in Section 12.4. For the generalized reaction

$$aA + bB \longrightarrow eE + fF$$

this equation takes the form

$$\Delta G = \Delta G^\circ + RT \ln \left(\frac{a_E{}^e a_F{}^f}{a_A{}^a a_B{}^b} \right)$$

where ΔG is the free energy change under conditions other than unit activity, ΔG° is the free energy change at unit activity, and the fraction

$$\frac{a_E{}^e a_F{}^f}{a_A{}^a a_B{}^b}$$

is the mass action expression in terms of activities. As was pointed out in Section 12.4, as well as at other times, in dilute solutions and mixtures of gases at low pressures, the activity is very nearly the same as molar concentration, so that we can write this same expression in terms of molar concentration. In doing this the mass action expression takes the form,

$$\frac{[E]^e [F]^f}{[A]^a [B]^b}$$

and the expression for the free energy change becomes

$$\Delta G = \Delta G^\circ + RT \ln \left(\frac{[E]^e [F]^f}{[A]^a [B]^b} \right)$$

Equations 15.4 and 15.5 (in the last section) show the relationship between ΔG and $\mathscr{E}$, and ΔG° and $\mathscr{E}^\circ$, respectively, for an oxidation-reduction reaction. Substituting these expressions for ΔG and ΔG° into the above equation, we have

$$-n\mathscr{F}\mathscr{E} = -n\mathscr{F}\mathscr{E}^\circ + RT \ln \left(\frac{[E]^e [F]^f}{[A]^a [B]^b} \right)$$

which can be rearranged to give

$$\mathscr{E} = \mathscr{E}^\circ - \frac{RT}{n\mathscr{F}} \ln \left(\frac{[E]^e [F]^f}{[A]^a [B]^b} \right)$$

Changing this equation from ln (logarithm to the base e) to log (logarithm to the base 10) we have,

$$\mathscr{E} = \mathscr{E}° - \frac{2.303RT}{n\mathscr{F}} \log \left(\frac{[E]^e[F]^f}{[A]^a[B]^b} \right) \tag{15.12}$$

This equation, first developed by Walter Nernst in 1889, now bears his name and is called the Nernst equation.

We can see from Equation 15.12 that when all ionic species are present at unit concentration, the log term becomes zero (log 1 = 0) and the emf of the cell becomes $\mathscr{E}°$, that is, at unit concentration $\mathscr{E} = \mathscr{E}°$. This, of course, must be true in light of our basic definition of $\mathscr{E}°$ in Section 15.6. When the species in a cell are not present at unit activity, $\mathscr{E}$ is generally not equal to $\mathscr{E}°$ and the Nernst equation must be employed in order to calculate $\mathscr{E}$. For example, in the case of the Zn/Cu cell, whose cell reaction is

$$Zn(s) + Cu^{2+} \longrightarrow Cu(s) + Zn^{2+}$$

the Nernst equation takes the form

$$\mathscr{E} = \mathscr{E}° - \frac{2.303RT}{n\mathscr{F}} \log \frac{[Zn^{2+}]}{[Cu^{2+}]}$$

Since the activity of any pure solid is equal to unity, Zn(s) and Cu(s) are omitted from the mass action expression. At 25° C we have seen that $2.303RT/\mathscr{F} = 0.0592$. For this reaction $n = 2$, and we have

$$\mathscr{E} = \mathscr{E}° - \frac{0.0592}{2} \log \frac{[Zn^{2+}]}{[Cu^{2+}]}$$

or

$$\mathscr{E} = \mathscr{E}° - 0.030 \log \frac{[Zn^{2+}]}{[Cu^{2+}]}$$

Thus we see that $\mathscr{E}$ can be calculated for any particular cell if the Zn^{2+} and Cu^{2+} concentrations are known. This use of the Nernst equation is illustrated in the next example.

Example 15.5
Calculate the emf of the Zn/Cu cell under the following conditions:

$$Zn(s) + Cu^{2+}(0.020M) \longrightarrow Cu(s) + Zn^{2+}(0.40M)$$

Solution
Using the Nernst equation for this cell we have

$$\mathscr{E} = \mathscr{E}° - \frac{2.303RT}{n\mathscr{F}} \log \frac{[Zn^{2+}]}{[Cu^{2+}]}$$

We have already seen that $\mathscr{E}°$ for this cell is +1.10 volts, which can be substituted into this equation along with the zinc and copper ion concentrations to give us

$$\mathscr{E} = +1.10 - \frac{0.0592}{n} \log \frac{(0.40)}{(0.020)}$$

Since two electrons are transferred in this process, $n = 2$ and we have

$$\mathscr{E} = 1.10 - \frac{0.0592}{2} \log \frac{(0.40)}{(0.020)}$$

$$\mathscr{E} = 1.10 - 0.030(1.30)$$

and

$$\mathscr{E} = 1.06 \text{ volts}$$

Thus we see that under these conditions of concentration the voltage obtained from this cell is slightly less than that obtained at unit activity.

CONCENTRATION CELLS. Just as the cell emf is dependent on the concentration of the ions involved in the half-reaction, so we find that the reduction potential of the individual half-reactions are also determined by the concentration of the ions involved. This effect of the concentration on the reduction potential can also be given by the Nernst equation. For example, if we consider the half-reaction,

$$Zn^{2+} + 2e^- \rightleftharpoons Zn(s)$$

the Nernst equation at 25° C would take the form

$$\mathscr{E} = \mathscr{E}° - \frac{0.0592}{2} \log \frac{1}{[Zn^{2+}]}$$

where we again take the activity of $Zn(s)$ to be equal to one, and that of Zn^{2+} to be equal to its concentration. This statement permits us to construct an electrochemical cell in which the same species are present in both the anode and cathode compartments, but where their concentrations are not the same. Such a cell is called a **concentration cell** and is illustrated in Figure 15.10.

In Figure 15.10 we have a cell composed of two zinc electrodes placed in separate solutions of $ZnSO_4$ whose Zn^{2+} concentrations are different and are separated by a salt bridge. The concentration of the Zn^{2+} on the left (1.0M) is 100 times greater than the Zn^{2+} concentration in the right compartment, and when the circuit is completed a spontaneous reaction takes place in a direction that tends to make the two Zn^{2+} concentrations become equal. Thus in the more concentrated side Zn^{2+} ions disappear forming $Zn(s)$, in order to decrease the Zn^{2+} concentration, and in the more dilute side, more Zn^{2+} will be produced. Thus we have, in the more concentrated compartment,

$$Zn^{2+}(1M) + 2e^- \longrightarrow Zn(s) \quad \text{(reduction)}$$

and on the more dilute side

$$Zn(s) \longrightarrow Zn^{2+}(0.01M) + 2e^- \quad \text{(oxidation)}$$

From our earlier discussions we know that the potential of a cell is found by subtracting the reduction potential of the half-cell in which oxidation occurs from the reduction potential of the half-cell that undergoes reduction. For this concentration cell, then,

$$\mathscr{E}_{cell} = \mathscr{E}_{conc} - \mathscr{E}_{dil}$$

Potentiometer

emf = 0.0592 volts

Zn^{2+}
SO_4^{2-}

Zn

Zn^{2+}
SO_4^{2-}

Zn

Figure 15.10

1.0 M Zn^{2+}

0.01 M Zn^{2+}

Concentration cell.

where $\mathscr{E}_{conc}$ and $\mathscr{E}_{dil}$ are the electrode potentials of the concentrated and dilute half cells, respectively. These are given as

$$\mathscr{E}_{conc} = \mathscr{E}_{Zn}^{\circ} - \frac{0.0592}{2} \log \frac{1}{[Zn^{2+}]_{conc}}$$

and

$$\mathscr{E}_{dil} = \mathscr{E}_{Zn}^{\circ} - \frac{0.0592}{2} \log \frac{1}{[Zn^{2+}]_{dil}}$$

Therefore

$$\mathscr{E}_{cell} = \underbrace{(\mathscr{E}_{Zn}^{\circ} - \mathscr{E}_{Zn}^{\circ})}_{zero} - \frac{0.0592}{2} \left(\log \frac{1}{[Zn^{2+}]_{conc}} - \log \frac{1}{[Zn^{2+}]_{dil}} \right)$$

or

$$\mathscr{E}_{cell} = -\frac{0.0592}{2} \log \frac{[Zn^{2+}]_{dil}}{[Zn^{2+}]_{conc}}$$

Substituting the concentrations of Zn^{2+} into this expression allows us to compute the cell potential.

$$\mathscr{E}_{cell} = -\frac{0.0592}{2} \log \frac{(0.01)}{(1)}$$

$$\mathscr{E}_{cell} = 0.0592 \text{ volts}$$

In general, for any concentration cell we could write

$$\mathscr{E}_{cell} = -\frac{0.0592}{n} \log \frac{[M^{n+}]_{dil}}{[M^{n+}]_{conc}}$$

The voltage obtained from this type of cell is usually small and will continually decrease as the concentrations in the two compartments approach each other. The voltage becomes zero when the concentration of the ions in each compartment of the concentration cell are the same.

SOLUBILITY PRODUCT CONSTANT. The Nernst equation can also be useful in determining the solubility product constant of an insoluble salt. To find the K_{sp} of $PbSO_4$, for example, a student might design his experiments in the following fashion: a galvanic cell is prepared consisting of a Pb/Pb^{2+} versus an Sn/Sn^{2+} electrode with a salt bridge connecting them. In the tin compartment the Sn^{2+} concentration is held constant at $1M$ while the Pb^{2+} concentration in the lead compartment is decreased to a minimum by adding excess SO_4^{2-}. The SO_4^{2-} concentration around the lead electrode is then adjusted to $1M$ and the emf of this cell is found to be $+0.22$ volt. It is also observed that the Pb electrode is negative with respect to the Sn electrode, thereby indicating that the Pb is undergoing oxidation while the Sn^{2+} is reduced. The cell reaction here is

$$Pb(s) + Sn^{2+}(1M) \longrightarrow Pb^{2+}(?) + Sn(s)$$

and the calculated $\mathscr{E}°$ would be

$$\mathscr{E}° = \mathscr{E}°_{Sn} - \mathscr{E}°_{Pb}$$
$$\mathscr{E}° = (-0.14 \text{ volt}) - (-0.13 \text{ volt})$$
$$\mathscr{E}° = -0.01 \text{ volt}$$

(Note that *if* the Sn^{2+} and Pb^{2+} concentrations were both $1M$, the reaction would be spontaneous from right to left, rather than from left to right.)

We can calculate the concentration of Pb^{2+} by using the Nernst equation, which takes the form,

$$\mathscr{E} = \mathscr{E}° - \frac{0.0592}{n} \log \frac{[Pb^{2+}]}{[Sn^{2+}]}$$

for this cell reaction. We know $\mathscr{E}, \mathscr{E}°$, and $[Sn^{2+}]$ and, since there are two electrons transferred in this reaction, the above equation becomes, after substitution,

$$0.22 \text{ volt} = -0.01 \text{ volt} - \frac{0.0592}{2} \log \frac{[Pb^{2+}]}{(1)}$$

Solving for log $[Pb^{2+}]$ gives us

$$-0.22 \text{ volt} = 0.01 \text{ volt} + 0.030 \log [Pb^{2+}]$$

or

$$\log [Pb^{2+}] = \frac{-0.22 \text{ volt} - 0.01 \text{ volt}}{0.030 \text{ volt}}$$

and

$$\log [Pb^{2+}] = -7.7$$

Taking the antilog we find then that the concentration of Pb^{2+} in this cell is

$$[Pb^{2+}] = 2 \times 10^{-8}M$$

The expression for K_{sp} of $PbSO_4$ is

$$K_{sp} = [Pb^{2+}][SO_4{}^{2-}]$$

Since

$$[Pb^{2+}] = 2 \times 10^{-8}$$
$$[SO_4{}^{2-}] = 1M$$

then

$$K_{sp} = (2 \times 10^{-8})(1) = 2 \times 10^{-8}$$

which is the value that was given in Table 14.4

DETERMINATION OF pH. An extremely important application of the Nernst equation is that it can be used to calculate the concentration of a single ionic species by experimentally measuring the potential of a carefully designed cell. We have already seen one example of this in the determination of K_{sp}. If we were to use the Cu/H_2 cell, discussed in the last section, we could determine the H^+ concentration of a solution and then calculate its pH. The cell reaction for the Cu/H_2 is

$$Cu^{2+} + H_2(g) \longrightarrow Cu(s) + 2H^+$$

and the corresponding form of the Nernst equation at 25° C is

$$\mathscr{E} = \mathscr{E}° - \frac{0.0592}{n} \log \frac{[H^+]^2}{[Cu^{2+}]p_{H_2}}$$

If the concentration of Cu^{2+} is $1M$ and the pressure of H_2 is 1 atm, this equation reduces to

$$\mathscr{E} = \mathscr{E}° - \frac{0.0592}{n} \log [H^+]^2$$

which is the same as

$$\mathscr{E} = \mathscr{E}° - \frac{0.0592(2)}{n} \log [H^+]$$

If we rewrite this equation to take the form

$$\mathscr{E} = \mathscr{E}° + \frac{0.0592(2)}{n} (-\log[H^+])$$

We see that since $\mathscr{E}°$ and the quantity $0.0592(2)/n$ are both constant for a specific reaction. Then

$$\mathscr{E} \propto - \log[H^+]$$

By definition, pH $= -\log[H^+]$ and, consequently, we have

$$\mathscr{E} \propto pH$$

Thus by measuring the emf of a galvanic cell containing a reference electrode (such as the Cu, Cu^{2+} electrode here) and the hydrogen electrode, the pH of a solution can be calculated. One such application is shown in the next example.

Example 15.6

A galvanic cell consisting of a Cu versus a hydrogen electrode was used to determine the pH of an unknown solution. The unknown was placed in the hydrogen electrode compartment and the pressure of the hydrogen gas was controlled at 1 atm. The concentration of Cu^{2+} was $1M$ and the emf of the cell at 25° C was determined to be +0.48 volt. Calculate the pH of this unknown solution.

Solution

The cell reaction for the Cu/H_2 cell is

$$Cu^{2+}(1M) + H_2(g)(1 \text{ atm}) \longrightarrow Cu(s) + 2H^+(?M)$$

for which we write the Nernst equation as

$$\mathscr{E} = \mathscr{E}^\circ - \frac{0.0592}{n} \log \frac{[H^+]^2}{[Cu^{2+}]}$$

or, since $[Cu^{2+}] = 1M$,

$$\mathscr{E} = \mathscr{E}^\circ - \frac{(0.0592)(2)}{n} \log[H^+]$$

The $\mathscr{E}^\circ$ for this cell, from the last section, is calculated to be +0.34 volt; the value of n is 2. Substituting these values as well as the measured value of $\mathscr{E}$ into the equation, we have

$$+0.48 = +0.34 - 0.0592 \log[H^+]$$

and hence

$$-\log[H^+] = \frac{0.48 - 0.34}{0.0592} = 2.2$$

Therefore the pH of this solution is 2.2

The last example represents but a single case where, with the proper choice of electrodes, the concentration of a single ionic species can be selectively measured. In this particular area of electrochemistry scientists through many years of research, have developed many practical electrodes whose emf depends on the concentration of only one species. Such electrodes are called ion-selective electrodes and are used in conjunction with a reference electrode, whose potential always remains constant and is of a known value. Thus when an ion-selective electrode is placed in a solution with a reference electrode, only the ion-selective electrode changes in emf, and the measured voltage can immediately be used to calculate the concentration of the species being determined. Such electrodes have been found to be of great importance in such areas as chemical analysis, pollution analysis, clinical measurements, oceanography, and geology. Some of these ion-selective electrodes and their practical applications are discussed in the next section.

<table>
<tr><td>15.11
Ion-Selective
Electrodes</td><td>

One type of ion-selective electrode, shown in Figure 15.11, consists of a very thin-walled membrane that is sealed onto one end of a hollow tube. Inside the tube, in contact with the membrane is a reference solution and immersed into this solution is a wire. The wire extends from the reference solution through the tube and out the top, to make electrical contact with the outside circuitry. The material used to make the membrane as well as the composition of the reference solution depends on the species that is to be measured. Some of the cations and anions whose concentration can be determined by these electrodes are listed in Table 15.2. Let us now discuss a few of these electrodes.

</td></tr>
</table>

The glass electrode is one such ion-selective electrode. The membrane in this electrode is made of an extremely thin piece of glass, and the reference solution inside is a dilute solution HCl, whose H^+ concentration is known and which remains constant. The wire electrode is a silver wire coated with silver chloride. The emf of the glass electrode is sensitive to the relative concentration of H^+ inside and outside across the thin glass membrane. Since the H^+ concentration inside is constant, the emf of the electrode, in effect, is determined by the concentration of H^+ in the solution in contact with the membrane on the outside.

The glass electrode can be made selective to various other monovalent cations such as Na^+, K^+, and NH_4^+ by suitable changes in the composition of the glass. Still other ions can be detected by electrodes if the glass membrane is replaced by a solid crystal. For example, when a solid crystal such as LaF_3 is used, fluoride ion concentrations can be detected, and when Ag_2S is used, silver and sulfide ion concentrations can be determined. Other ions such as chloride, cyanide, and lead can be determined using a membrane made of silver mixed with silver sulfide.

Ion-selective electrodes have also become very important and useful in the study of biological processes. One of these, called an enzyme-substrate electrode, employs a glass electrode sensitive to ammonium ion coated with

Reference solution

Thin-walled membrane

Figure 15.11
Ion selective electrode.

Table 15.2

**Ions Whose Concentration Can Be Determined
By An Ion-Selective Electrode**

Cations	Anions
Cadmium (Cd^{2+})	Bromide (Br^-)
Calcium (Ca^{2+})	Chloride (Cl^-)
Copper (Cu^{2+})	Cyanide (CN^-)
Hydrogen (H_3O^+)	Fluoride (F^-)
Lead (Pb^{2+})	Nitrate (NO_3^-)
Mercury (Hg^{2+})	Perchlorate (ClO_4^-)
Potassium (K^+)	Sulfide (S^{2-})
Silver (Ag^+)	Thiocyanate (SCN^-)
Sodium (Na^+)	

a thin layer of a gel containing an enzyme, as shown in Figure 15.12. Enzymes are large organic molecules that catalyze very specific chemical reactions in biochemical systems and if, for example, the enzyme in the gel is urease, the decomposition of urea to produce ammonium ion will occur when the solution around the electrode contains urea. This ammonium ion is detected by the electrode and, in effect, the electrode becomes sensitive to the presence of urea in the solution being tested.

The application of electrochemistry to biochemical research is still in its infancy. An illustration of how this field of research is progressing is the rela-

Glass electrode sensitive
to ammonium ion

Enzyme suspended in gel
that coats the thin-walled
membrane

Figure 15.12

Enzyme substrate electrode.

tively recent development of miniature ion-selective electrodes that can actually be placed into a living cell to monitor changes in the concentrations of ions during the life process.

15.12
Some Practical Galvanic Cells

As the final section in this chapter, let us discuss some galvanic cells that play an important part in our lives.

DRY CELL. This type of cell is used in flashlights, portable radios, toys, and the like. A cut-away diagram of a typical dry cell is shown in Figure 15.13. Dry cells have an exterior layer of either cardboard or metal to serve only as a seal against the atmosphere. Inside this outer shell is a zinc cup that serves as the anode. The zinc cup is filled with a moist paste consisting of ammonium chloride, manganese dioxide, and finely divided carbon. Immersed into this paste is a graphite rod, which serves as the cathode. The chemical reactions that take place when the circuit is completed are actually quite complex and, in fact, are not completely understood. The following, however, is perhaps a reasonable estimate of what occurs.

At the anode zinc is oxidized,

$$Zn \longrightarrow Zn^{2+} + 2e^- \qquad \text{(anode)}$$

while at the carbon cathode the MnO_2/NH_4Cl mixture undergoes reduction to give a complex mixture of products. *One* of these reactions appears to be

$$2MnO_2 + 2NH_4^+ + 2e^- \longrightarrow Mn_2O_3 + 2NH_3 + H_2O \qquad \text{(cathode)}$$

The ammonia produced at the cathode reacts with part of the Zn^{2+} formed at the anode to give the complex ion, $Zn(NH_3)_4^{2+}$. Because of the complex nature of the dry cell, no simple overall cell reaction can be written.

Dry cells cannot be effectively recharged and, therefore, have a relatively short lifetime (as compared to the rechargeable lead storage battery, for example).

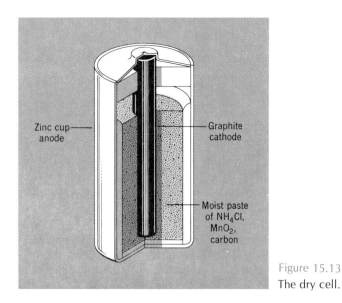

Zinc cup anode

Graphite cathode

Moist paste of NH_4Cl, MnO_2, carbon

Figure 15.13
The dry cell.

LEAD STORAGE BATTERY. The common automobile battery is a lead storage battery that usually delivers either six or 12 volts, depending on the number of cells used in its construction. The inside of the battery consists of a number of galvanic cells connected to each other in series (Figure 15.14).

To increase the current output, each of the individual cells contains a number of lead anodes connected together, plus a number of cathodes, composed of PbO_2, also joined together. These electrodes are immersed in an electrolyte composed of dilute sulfuric acid (actually about 30% by weight in a fully charged cell). A single lead storage cell delivers two volts so that a 12-volt battery contains six such cells connected in series.

When the external circuit is complete, and the battery is in operation the following oxidation-reduction reactions take place

(Anode) $\qquad Pb(s) + SO_4^{2-} \longrightarrow PbSO_4(s) + 2e^-$
(Cathode) $\qquad PbO_2(s) + 4H^+ + SO_4^{2-} + 2e^- \longrightarrow PbSO_4(s) + H_2O$

and the overall reaction is

$$Pb(s) + PbO_2(s) + 4H^+ + 2SO_4^{2-} \longrightarrow 2PbSO_4(s) + 2H_2O$$

These batteries are unique because the electrode reactions can be reversed by placing across the electrodes a voltage that is slightly larger than that which the battery can deliver. The recharging operation is performed in such a way that the negative external voltage is applied to the negative pole and the positive voltage to the positive pole. In doing this, some of the H_2SO_4 that is used up while the battery is in operation is restored. This is accomplished by the generator or alternator of the car, and in more serious cases with the aid of a battery charger.

A convenient method of estimating the degree to which the battery has been discharged is by checking the density (or specific gravity) of the electrolyte. If the battery is in a weakened state, the electrolyte will be mostly water (the product of our overall reaction) and have a density somewhere near

One cell of a
storage battery

(+)

(−)

H_2SO_4
electrolyte

Alternating
plates of
Pb and PbO_2

PbO_2
(cathode)

Pb
(anode)

Figure 15.14
Lead storage battery.

1 g/ml. If, however, the battery is in good operating order, with a full charge, the density of the electrolyte will be somewhat higher than 1 g/ml (the density of concentrated sulfuric acid is 1.8 g/ml). The mechanic in a garage can perform this test with the aid of a hydrometer, a device having a float that sinks to a depth that is a function of the density of the liquid in which it is immersed.

FUEL CELLS. Fuel cells are another means by which chemical energy may be converted into electrical energy. When gaseous fuels, such as H_2 and O_2, are allowed to undergo reaction in a carefully designed environment, electrical energy can be obtained. This type of cell finds great importance in space vehicles where the fuels used in such cells can also be the same as those used to power the rockets.

A diagram of a H_2/O_2 fuel cell is shown in Figure 15.15. In this cell there are three compartments separated from one another by porous plugs. The hydrogen gas is fed into one compartment and the oxygen gas is fed into another. These gases then diffuse (not bubble) slowly through the plugs and react with an electrolyte that is in the center compartment. The plugs are made of a conducting material, such as carbon, with a sprinkling of platinum to act as a catalyst, and the electrolyte is an aqueous solution of a base.

At the cathode the oxygen undergoes reduction producing OH^- ions, which can be expressed as

$$O_2(g) + 2H_2O(l) + 4e^- \longrightarrow 4OH^-$$

These OH^- ions travel to the anode where they undergo reaction with H_2:

$$H_2(g) + 2OH^- \longrightarrow 2H_2O + 2e^-$$

The net reaction in the cell is

$$2H_2 + O_2 \longrightarrow 2H_2O$$

The fuel cell is operated at a high temperature so that the water that is formed as a product of the cell reaction evaporates and may be condensed and used as drinking water for an astronaut. Several of these cells are usually

Figure 15.15
Hydrogen-oxygen fuel cell.

placed together so that several kilowatts of power can be continually produced.

Fuel cells offer several advantages over other sources of energy. Unlike the dry cell or storage battery, the cathode and anode reactants may be continually supplied so that, in principle, energy could be withdrawn indefinitely from a fuel cell, as long as the outside supply of fuel is maintained. Another advantage of the fuel cell is that the energy is extracted from the reactants under more nearly reversible conditions. Therefore, the thermodynamic efficiency of the reaction, in terms of producing useful work, is higher than when the reactants such as H_2 and O_2 are burned to produce heat that must be subsequently harnessed to produce work. These two advantages suggest that the development of fuel cells will undoubtedly continue at an accelerated pace in the future, particularly in light of recent energy shortages caused by greater demands for energy.

**Review
Questions
and Problems**

15.1 Distinguish between the following:
 (a) Electrolytic and galvanic cells
 (b) Metallic and electrolytic conduction
 (c) Anode and cathode in both electrolytic and galvanic cells
 (d) Oxidation and reduction
 (e) Half-reactions and cell reactions

15.2 Is it possible to obtain the pure metals listed below by electrolyzing an aqueous solution of these salts? Why or why not?
 (a) Mercury (d) Iron
 (b) Lithium (e) Barium
 (c) Silver

15.3 Give the half-reactions and the cell reaction for the electrolysis of aqueous solutions of the following salts: $CrCl_3$, $ZnBr_2$, $FeSO_4$, $MgSO_4$.

15.4 How many grams of Na and Cl_2 would be produced if a current of 25 amps was applied for 8 hours to the cell shown in Figure 15.3?

15.5 How many faradays would be required to reduce 1 mole of each of the following to the indicated product?
 (a) Cu^{2+} to Cu°
 (b) Fe^{3+} to Fe^{2+}
 (c) MnO_4^- to Mn^{2+}
 (d) F_2 to F^-
 (e) NO_3^- to NH_3

15.6 How many grams of O_2 and H_2 are produced in one hour when water is electrolyzed at a current of 0.5 amps? What would be the volume, at STP, of O_2 and H_2?

15.7 In an experiment two coulometers were connected in series, one containing $CuSO_4$, the other an unknown salt. It was found that 1.25 g of copper were plated out during the same period of time as 3.42 g of the unknown metal. What is the equivalent weight of the unknown?

15.8 How long would it take to plate out 5 g of copper from a solution of $CuSO_4$ at a current of 5 amps?

15.9 How many faradays of electricity are required to produce the following?
 (a) 10 ml of O_2 (at STP) from aqueous Na_2SO_4
 (b) 10 g of Al from molten Al_2O_3 (in cryolite)
 (c) 5 g of Na from molten NaCl
 (d) 5 g of Mg from molten $MgCl_2$

15.10 Convert each of the answers in problem 9 into (a) coulombs and (b) seconds if the current were 5 amps.

15.11 In the electrolysis of Na_2SO_4, only H_2 and O_2 are produced as products. What is the function of the Na_2SO_4 in this system?

15.12 Calculate the number of electrons that corresponds to one coulomb of charge.

15.13 Why is a potentiometer used to measure the voltage of a galvanic cell rather than a voltmeter?

15.14 Given the following sets of half-reactions, write the net cell reaction and calculate $\mathscr{E}°$ for the spontaneous changes that will occur.
  (a) $Hg_2Cl_2 + 2e^- \rightleftharpoons 2Hg + 2Cl^-$
      $PbSO_4 + 2e^- \rightleftharpoons Pb + SO_4^{2-}$
  (b) $AgCl + e^- \rightleftharpoons Ag + Cl^-$
      $Cu^{2+} + 2e^- \rightleftharpoons Cu$
  (c) $Mn^{2+} + 2e^- \rightleftharpoons Mn$
      $Cl_2(g) + 2e^- \rightleftharpoons 2Cl^-$
  (d) $Al^{3+} + 3e^- \rightleftharpoons Al$
      $Br_2(aq) + 2e^- \rightleftharpoons 2Br^-$

15.15 Predict whether or not, under standard conditions, the following overall cell reactions (not balanced) would occur in the direction written. Write the proper reaction for any incorrect ones.
  (a) $Fe^{3+} + Sn \rightarrow Fe^{2+} + Sn^{2+}$
  (b) $Mg^{2+} + Al \rightarrow Mg + Al^{3+}$
  (c) $Zn^{2+} + Cu \rightarrow Zn + Cu^+$
  (d) $PbO_2 + SO_4^{2-} + 4H^+ + 2Hg + 2Cl^- \rightarrow Hg_2Cl_2 + PbSO_4 + H_2O$

15.16 Write the Nernst equation, calculate $\mathscr{E}°$ and $\mathscr{E}$ for the following reactions:
  (a) $Cu^{2+}(0.1M) + Zn(s) \rightarrow Cu(s) + Zn^{2+}(1.0M)$
  (b) $Sn^{2+}(0.5M) + Ni(s) \rightarrow Sn(s) + Ni^{2+}(0.01M)$
  (c) $F_2(g) + 2Li(s) \rightarrow 2Li^+(1M) + 2F^-(0.5M)$
  (d) $Zn(s) + 2H^+(0.01M) \rightarrow Zn^{2+}(1M) + H_2(g)(1\ atm)$
  (e) $2H^+(1.0M) + Fe(s) \rightarrow H_2(1\ atm) + Fe^{2+}(0.2M)$

15.17 Use the shorthand notation described in the footnote on page 418 to illustrate the galvanic cells consisting of a salt bridge and the following electrode pairs:
  (a) H_2, $H^+(1M)$ versus AgCl, Ag
  (b) Sn, $Sn^{2+}(1M)$ versus Ni, $Ni^{2+}(1M)$
  (c) Al, $Al^{3+}(1M)$ versus Fe, $Fe^{3+}(1M)$
  (d) Hg_2Cl_2, Hg versus $Fe^{3+}(1M)$, $Fe^{2+}(1M)$
  (e) Hg_2Cl_2, Hg versus Mn, $Mn^{2+}(1M)$

15.18 Which is the better reducing agent?
  (a) Li^+ or Ca
  (b) Cl_2 or F_2
  (c) Al^{3+} or H_2O

15.19 Which is the better oxidizing agent?
  (a) Cl_2 or ClO_3^-
  (b) O_2 or $Cr_2O_7^{2-}$
  (c) $S_2O_8^{2-}$ or H_2O_2

15.20 Calculate the equilibrium constants for the following cell reactions:
  (a) $Ni(s) + Sn^{2+}(aq) \rightarrow Ni^{2+}(aq) + Sn(s)$
  (b) $Cl_2(g) + 2Br^-(aq) \rightarrow Br_2(l) + 2Cl^-(aq)$
  (c) $Fe^{2+}(aq) + Ag^+(aq) \rightarrow Ag(s) + Fe^{3+}(aq)$

15.21 A student set up a galvanic cell to measure the K_{sp} of CuS. On one side of the cell he had a copper electrode dipping into a $0.10M$ Cu^{2+} solution and on the other side a zinc electrode in a Zn^{2+} solution. The Zn^{2+} concentration was held constant at $1M$ and the Cu^{2+} brought to a minimum by saturating the Cu^{2+} solution with H_2S. The emf of the cell was read as $+0.67$, with the Cu electrode serving as the cathode. Calculate the Cu^{2+} concentration and the K_{sp} of CuS. Compare your answer to the K_{sp} reported in Table 14.4. In a saturated solution the concentration of H_2S is $0.10M$.

15.22 Calculate the concentrations of the ions labelled with a (?) in the following cells (the cathode in each case is underlined):
  (a) $Sn/Sn^{2+}(?)//\underline{Pt/H_2(1\ atm)/H^+(1\ M)}$
      $\mathscr{E} = +0.12$ volt
  (b) $Fe/Fe^{2+}(1M)//\underline{Pt/H_2(1\ atm)/H^+(?)}$
      $\mathscr{E} = +0.50$ volt
      What is the pH of this solution?
  (c) $Pt/H_2(1\ atm)/H^+(1M)//\underline{Cu/Cu^{2+}(?)}$
      $\mathscr{E} = +0.40$ volt

15.23 Calculate the $\mathscr{E}°$, $\mathscr{E}$, and ΔG for the following cell reactions:
  (a) $Al(s) + Ni^{2+}(0.80M) \rightarrow Al^{3+}(0.020M) + Ni(s)$
  (b) $Ni(s) + Sn^{2+}(1.10M) \rightarrow Sn(s) + Ni^{2+}(0.010M)$
  (c) $Cu^+(0.050M) + Zn(s) \rightarrow Cu(s) + Zn^{2+}(0.010M)$

15.24 The solubility product constant of AgCl is 1.7×10^{-10}. What will be the potential of a cell constructed using the H_2 electrode ($[H^+] = 1\ M$, $p_{H_2} = 1$ atm) versus a half-cell containing a silver wire coated with AgCl immersed in $0.010M$ HCl?

15.25 Calculate the potential generated by the concentration cell: $Fe/Fe^{2+}(0.10M)//Fe^{2+}(0.0010M)/Fe$.

15.26 (a) What is a reference electrode?
(b) What is an ion-selective electrode?

16
Covalent Bonding and Molecular Structure

In Chapter 4 we found that chemical bonds can be broadly classified into two main categories: ionic bonds and covalent bonds. The ionic bond arises as a purely electrostatic attraction between oppositely charged particles and is therefore nondirectional, in the sense that the arrangement of ions in a cluster is determined simply by the balancing of attractive and repulsive forces between the ions, and not by the electronic structures of the ions. It is for this reason that the structures of ionic compounds are determined almost exclusively by packing considerations (Chapter 7). The covalent bond, on the other hand, has very definite directional properties, and covalently bound substances, such as molecules or polyatomic ions, have characteristic shapes that are usually retained when these substances undergo physical changes such as melting or vaporization.

The simple picture of the covalent bond as a pair of dots shared between two atoms is clearly not sufficient to explain molecular structure. In this chapter we shall examine some of the theories that have been developed to account for covalent bonding and the shapes of molecules. It should be kept in mind throughout this discussion that each theory represents an attempt to describe the *same* physical phenomenon. None of the theories are perfect—otherwise we would only have to consider one of them. Each has its usefulness, and each has its weaknesses. The theory that a chemist applies in a particular circumstance depends largely on what aspect of the covalent bond he is attempting to explain and, to some extent, his own feelings about the validity of the various theories.

16.1 Valence Bond Theory

There are two important approaches to chemical bonding that are based on the results of quantum mechanics. One of these, called the **valence bond theory,** permits us to retain our picture of individual atoms coming together to form a covalent bond. The other, called **molecular orbital theory,** views a molecule as a set of positive nuclei with orbitals that extend over the entire molecule. The electrons that populate these **molecular orbitals** do not belong to any individual atoms but, instead, to the molecule as a whole. We will look at the molecular orbital theory in more detail in Section 16.5.

The basic postulate of the valence bond theory is that when two atoms come together to form a covalent bond, an atomic orbital of one atom overlaps with an atomic orbital of the other, and the pair of electrons that we have come to associate with a covalent bond is shared between the two atoms in the region where the orbitals overlap. It is also postulated that the strength of the covalent bond is proportional to the *extent* of overlap of the

Figure 16.1
Formation of H_2 by overlap of $1s$ orbitals.

atomic orbitals. As a consequence the atoms in a molecule tend to position themselves so that there is a maximum amount of orbital overlap.

Let's see how these postulates can be applied to some familiar compounds. The simplest of these is the hydrogen molecule that is formed from two hydrogen atoms, each having a single electron in a $1s$ orbital. According to valence bond theory, we would view the H—H bond as resulting from the overlap of the two $1s$ orbitals, as shown in Figure 16.1.[1]

In the HF molecule we have a somewhat different state of affairs. Fluorine has the valence shell electron configuration

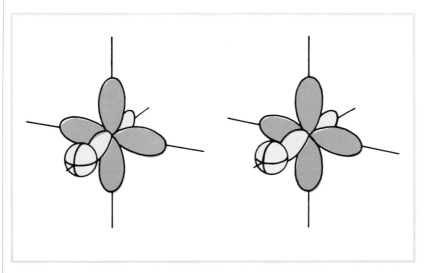

Figure 16.2
Formation of HF by overlap of partially filled fluorine $2p$ orbital with $1s$ orbital of hydrogen. Heavy shading indicates filled orbital, light shading indicates partially filled orbital.

---

[1] In this illustration, as well as in others throughout this chapter it is important to keep in mind that we are looking at schematic representations of orbital wave functions.

where we find one of the $2p$ orbitals occupied by a single electron. It is with this partially occupied $2p$ orbital that the hydrogen $1s$ orbital overlaps, as illustrated by Figure 16.2. In this case the hydrogen electron and the fluorine electron can pair up and be shared between the two nuclei. Note that the $2s$ orbital of the hydrogen atom does not overlap with an already filled atomic orbital on fluorine since there would then be three electrons in the bond (two from the fluorine $2p$ orbital and one from the hydrogen $1s$ orbital), a situation that is not permitted in this theory. *Only two electrons may be shared in one set of overlapping orbitals.*

Suppose we now consider the molecule, H_2O. Here we have two hydrogen atoms bound to a single oxygen atom. The outer-shell electronic structure of oxygen,

$$O \quad \underline{\uparrow\downarrow} \quad \underline{\uparrow\downarrow} \; \underline{\uparrow} \; \underline{\uparrow}$$
$$ 2s \phantom{\quad \underline{\uparrow\downarrow}} 2p$$

tells us that there are two unpaired electrons in p orbitals, and we predict that the two hydrogen atoms, with their electrons in $1s$ orbitals, will bond to the oxygen by means of overlap of their $1s$ orbitals with these partially filled oxygen p orbitals. This is shown in Figure 16.3. Since the p orbitals are oriented at 90° to one another, we expect the H—O—H bond angle in water also to be 90°. Actually this angle is 104.5°. One explanation for this discrepancy (we shall see another one later) is that since the O—H bonds are highly polar, the H atoms carry a substantial positive charge and hence repel one another. This factor tends to increase the H—O—H angle. However, since the best overlap between the hydrogen $1s$ orbitals and the oxygen $2p$ orbitals occurs at an angle of 90°, the H—O—H angle cannot increase too much without there being a considerable loss in bond strength. There are thus two factors

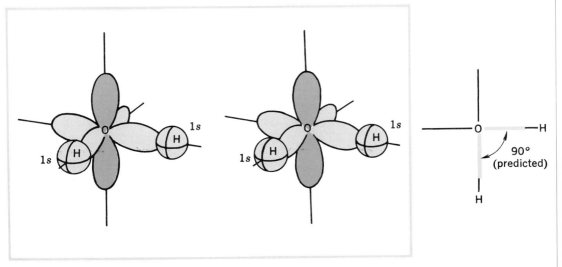

Figure 16.3
Bonding in H_2O. Overlap of two half-filled oxygen $2p$ orbitals with the hydrogen $1s$ orbitals.

working in opposition to each other, one tending to increase the bond angle and one tending to reduce it to 90°, and it appears that a balance is obtained when the angle is 104.5°. Qualitatively, the valence bond theory can account for the geometry of the water molecule. We can also apply the theory to the ammonia molecule with reasonable success. Nitrogen, being in Group VA, has three unpaired electrons in its p subshell. We would expect that the three hydrogen atoms in NH_3 would bond to the nitrogen such that they would tend to lie along the x, y, and z axes at 90° to one another (Figure 16.4a). As in the water molecule, the H—N—H angles are larger than 90°, having values in this case of 107°; as with H_2O, we might attempt to explain this angle in terms of repulsion between the protons. In any event, we obtain a picture of the NH_3 molecule that has a geometry referred to as pyramidal (i.e., pyramidlike), with the nitrogen atom at the apex of the pyramid and the three hydrogen atoms at the corners of the base, as shown in Figure 16.4b.

The very simple picture of the overlap of half-filled atomic orbitals we have just developed cannot be used to account for all molecular structures. For example, with carbon we would initially expect only two bonds to be formed with hydrogen since the valence shell of carbon contains only two unpaired electrons.

$$C \quad \underset{2s}{\uparrow\downarrow} \quad \underset{2p}{\uparrow \;\; \uparrow} \quad \underline{}$$

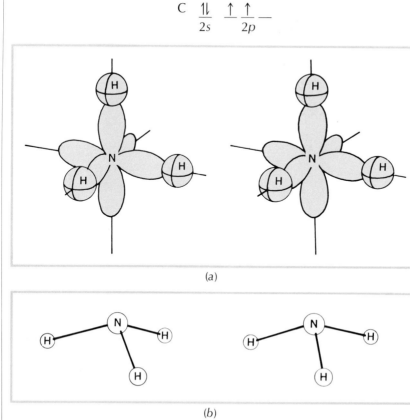

(a)

(b)

Figure 16.4
Bonding in NH_3 gives pyramidal molecule.

The species CH_2, however, does not exist as a stable molecule. Instead the simplest compound between carbon and hydrogen is methane, which has the formula CH_4. Attempting to explain the structure of this molecule by spreading the electrons out to give

$$\underline{\uparrow} \quad \underline{\uparrow} \; \underline{\uparrow} \; \underline{\uparrow}$$
$$2s \qquad 2p$$

suggests that three of the C—H bonds will be formed by overlap of hydrogen $1s$ orbitals with carbon $2p$ orbitals, while the remaining bond would be the result of the overlap of the carbon $2s$ orbital with a hydrogen $1s$ orbital. This fourth C—H bond should certainly be different from the other three bonds, since it is formed from different kinds of orbitals. It has been found experimentally, however, that *all* four C—H bonds are identical and that the molecule has a structure in which the carbon atom lies at the center of a tetrahedron with the hydrogen atoms located at the four corners (Figure 16.5). How does valence bond theory explain the structure of this molecule?

The solution to this apparent dilemma is found in the mathematics of quantum mechanics. In that theory, (see Sections 3.12 and 3.16), the solution of Schrödinger's wave equation provides us with a series of wave functions, ψ, each of which describes a different atomic orbital. It is the property of these mathematical functions that when they are squared, they enable us to calculate the probability of locating the electron at some point in space around the nucleus and, in fact, the spheres and figure-eights that we have been drawing roughly correspond to pictorial representations of the probability distributions predicted by the wave functions for s and p orbitals, respectively.

What is important to us here is that it is possible to combine these wave functions by appropriately adding or subtracting them to give new functions that are referred to as **hybrid orbitals.** In other words, two or more atomic orbitals are mixed together to produce a new set of orbitals and, invariably, these hybrid orbitals possess different directional properties than the pure atomic orbitals from which they are created. For example, Figure 16.6 illus-

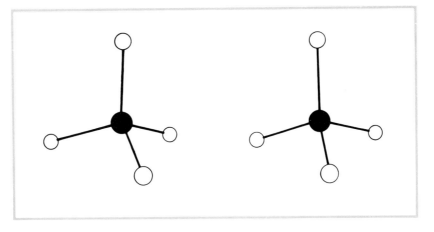

Figure 16.5
The structure of methane. Solid sphere-carbon.

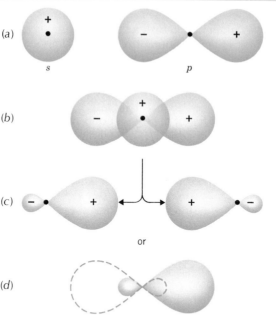

(a)

s p

Figure 16.6

Formation of two sp hy- (b)
brid orbitals from an s
and a p orbital. (a) s and p
orbitals drawn separately.
(b) s and p orbitals before
hybridization. (c) Two sp
hybrid orbitals are formed (c)
(drawn separately). (d) The
two sp hybrid orbitals
drawn together to show
their directional proper-
ties. Note that one orbital (d)
points to the left, the other
to the right.

or

trates the result of the combination of an s and a p orbital to provide a new
set of two **sp hybrid orbitals.** In this drawing you will notice that we have indicated that the wave function for a p orbital has positive numerical values in
some regions about the nucleus and negative values in others.[2] The s orbital,
on the other hand, has the same algebraic sign everywhere. Therefore, when
these wave functions are alternatively added and subtracted, the new orbitals
that result become larger in those regions where both functions have the same
sign and smaller in regions where they are of opposite sign.

Hybrid orbitals formed in the manner just described possess some interesting properties. We see for each orbital that one lobe is much larger than
the other and because of this a hybrid orbital can overlap well in only one
direction—the direction in which the orbital protrudes the most. A hybrid orbital is therefore very strongly directional in its ability to enter into covalent
bond formation. Furthermore, since hybrid orbitals extend out farther from the
nucleus than do unhybridized orbitals, they are able to overlap more effectively with orbitals of other atoms. Consequently, bonds formed from hybrid
orbitals tend to be stronger than those formed from ordinary atomic orbitals.

Thus far we have examined what occurs when one s and one p orbital
are mixed together. Other combinations of orbitals are also possible, with the
number of orbitals in a hybrid set, as well as their orientations, being determined by which atomic orbitals are combined. Table 16.1 contains a listing

---

[2] Certain wave functions (e.g., p orbital wave functions) can have either a positive or negative
algebraic sign, depending on which region around the nucleus is examined. The square of a
wave function, ψ^2, of course, must always be positive (a negative times a negative equals a positive).

Table 16.1
Hybrid Orbitals

Hybrid Orbitals	Number of Orbitals	Orientation
sp	2	Linear
sp^2	3	Planar triangle
sp^3	4	Tetrahedral
dsp^2	4	Square planar
dsp^3	5	Trigonal bipyramid
d^2sp^3	6	Octahedral

of the sets of hybrid orbitals that can be used to explain most of the molecular structures that we shall encounter in this book. Their directional properties are illustrated in Figure 16.7. Notice that the number of each kind of atomic orbital included in a combination is specified by an appropriate superscript on the atomic orbital type. Thus the d^2sp^3 hybrids are formed from two d orbitals, one s orbital, and three p orbitals.

Let us see now how we can use the information contained in Table 16.1 and Figure 16.7 to account for the structures of some typical molecules. We might begin with the substance BeH_2 in which there are two H atoms bound to the central beryllium. The electronic structure of Be is

$$\text{Be} \quad \underset{1s}{\underline{\uparrow\downarrow}} \quad \underset{2s}{\underline{\uparrow\downarrow}} \quad \underset{2p}{\underline{}\,\underline{}\,\underline{}}$$

In order to form two covalent bonds with H atoms, the Be atom must provide two half-filled (i.e., singly occupied) orbitals. This can be accomplished by creating a pair of sp hybrids and placing one electron in each of them.

$$\text{Be} \quad \underset{1s}{\underline{\uparrow\downarrow}} \quad \underset{sp}{\underbrace{\underline{\uparrow}\;\underline{\uparrow}}} \quad \underset{\text{unhybridized } 2p \text{ orbitals}}{\underbrace{\underline{}\;\underline{}}}$$

The two H atoms can then bond to the beryllium atom by overlap of their respective singly occupied s orbitals[3] with the singly occupied Be sp hybrids as shown in Figure 16.8. Because of the orientation of the sp hybrid orbitals the H atoms are forced to lie on opposite sides of the Be, and we predict that a linear H—Be—H molecule would result. It is interesting (and perhaps comforting) to find experimentally that BeH_2 is indeed a linear molecule.

Let us now return to our problem of the structure of CH_4. If we use hybrid orbitals on the carbon atom, we find that in order to provide four orbitals with which hydrogen $1s$ orbitals can overlap, we must use a set of sp^3 hybrids.

$$\text{C} \quad \underset{2s}{\underline{\uparrow\downarrow}} \quad \underset{2p}{\underline{\uparrow}\;\underline{\uparrow}\;\underline{}} \qquad \text{(unhybridized)}$$

---

[3] The electronic configuration of H is $1s^1$.

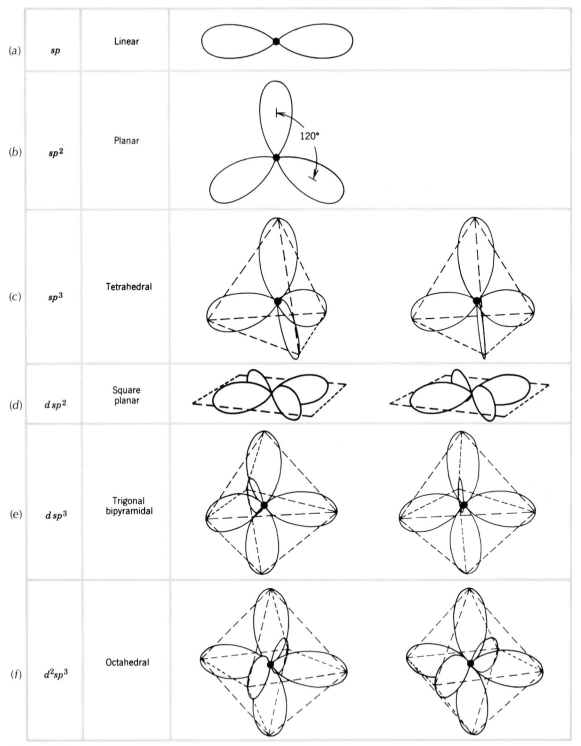

Figure 16.7
Directional properties of hybrid orbitals. The minor lobes have been omitted for the sake of clarity. (a) *sp*, linear. (b) *sp²*, planar. (c) *sp³*, tetrahedral. (d) *dsp²*, square planar. (e) *dsp³*, trigonal bipyramidal. (f) *d²sp³* octahedral.

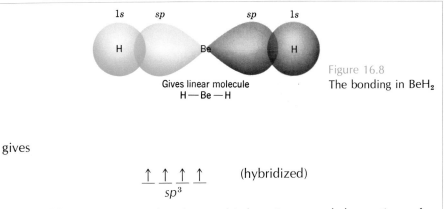

1s   sp   sp   1s

H   Be   H

Gives linear molecule
H—Be—H

Figure 16.8

The bonding in BeH$_2$

gives

$$\underline{\uparrow}\ \underline{\uparrow}\ \underline{\uparrow}\ \underline{\uparrow}$$
$$sp^3$$
         (hybridized)

From Table 16.1 we see that these orbitals point toward the vertices of a tetrahedron. Hence, when the four hydrogen atoms are attached to the carbon by orbital overlap with these sp^3 hybrids, a tetrahedral molecule results, as shown in Figure 16.9. Again we find the predicted structure in agreement with that found by experiment.

    In our earlier discussion we viewed the structure of H$_2$O and NH$_3$ as resulting from the use of partially filled p atomic orbitals on the oxygen and nitrogen atoms, respectively. An alternative view of the bonding in these molecules employs sp^3 hybrid orbitals on the central atom. In the tetrahedral set of hybrids the orbitals are oriented at angles of 109.3° to each other. The bond angles in water (104.5°) and ammonia (107°) are not too different from the tetrahedral angle and, using water as an example, we might consider the molecule to result from the overlap of hydrogen 1s orbitals with two partially occupied sp^3 orbitals on the oxygen atom.

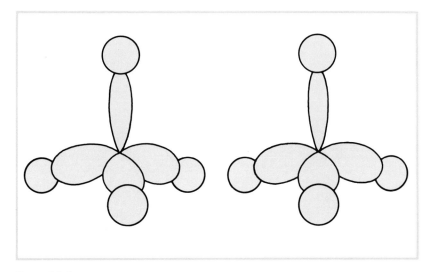

Figure 16.9
The formation of methane by overlap of hydrogen 1s orbitals with carbon sp^3 hybrids.

$$O \quad \underline{\uparrow\downarrow} \quad \underline{\uparrow\downarrow} \; \underline{\uparrow} \; \underline{\uparrow}$$
$$\underset{2s}{} \qquad \underset{2p}{}$$

unhybridized

$$\underline{\uparrow\downarrow} \; \underline{\uparrow\downarrow} \; \underline{\uparrow} \; \underline{\uparrow}$$
$$\underset{sp^3}{}$$

hybridized

$$\underline{\uparrow\downarrow} \; \underline{\uparrow\downarrow} \; \underline{\uparrow\cdot} \; \underline{\uparrow\cdot}$$
$$\underset{sp^3}{}$$

H_2O molecule
(dots = H electrons)

Notice that only two of the hybrid orbitals are involved in bond formation
while the other two contain nonbonded "lone pairs" of electrons. In the case
of ammonia, three of the sp^3 orbitals are used in bonding while the fourth orbital contains a lone pair of electrons (Figure 16.10). There is, in fact, rather
strong experimental evidence to indicate that this lone pair does indeed project out from the nitrogen atom as implied in this picture of the NH_3 molecule.
It is worth noting that in our previous description of NH_3 we found this lone
pair of electrons in an s orbital that would have spread the electron pair symmetrically about the nucleus.

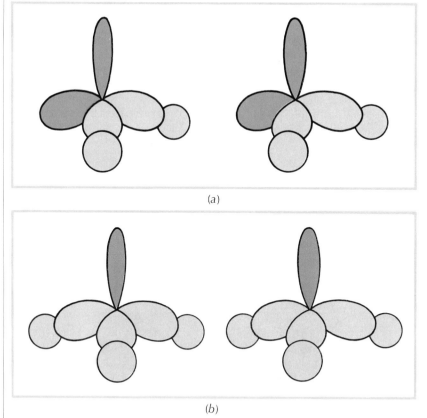

(a)

(b)

Figure 16.10

The use of sp^3 hybrids for bonding in H_2O and NH_3 (heavy shading indicates a filled
hybrid orbital; light shading indicates a half-filled orbital). Spheres represent hydrogen
$1s$ orbitals.

In the case of H_2O and NH_3 the H—X—H bond angles (104.5° and 107°, respectively) are less than the tetrahedral angle of 109° that is observed in the molecule, CH_4. One way to account for this is through the influence of the lone-pair electrons present in hybrid orbitals of the central atom. A pair of electrons in a bond is attracted to two nuclei and, therefore, might be expected to occupy a smaller effective volume than a pair of electrons in a non-bonded orbital, which experience the attraction of only one nucleus. The lone-pair electrons, then, because of their greater space requirement, tend to crowd together the electron pairs located in the bonds and hence reduce the bond angle to something less than 109°. On this basis we anticipate a greater reduction in bond angle for water than for ammonia, since the former has two lone pairs while ammonia has but one.

As a final example, consider the molecule SF_6. Sulfur, being in Group VIA, has six valence electrons distributed over the $3s$ and $3p$ subshells,

$$\text{S} \quad \underset{3s}{\underline{\uparrow\downarrow}} \quad \underset{3p}{\underline{\uparrow\downarrow}\ \underline{\uparrow}\ \underline{\uparrow}} \quad \underset{3d}{\underline{\ }\ \underline{\ }\ \underline{\ }\ \underline{\ }\ \underline{\ }}$$

Here we have shown the empty $3d$ subshell as well as the $3s$ and $3p$ subshells that contain electrons. In order for sulfur to form six covalent bonds to fluorine, six half-filled orbitals must be created. This can be accomplished by using two of the unoccupied $3d$ orbitals and a d^2sp^3 hybrid set is formed.

$$\text{S} \quad \underset{d^2sp^3}{\underline{\uparrow}\ \underline{\uparrow}\ \underline{\uparrow}\ \underline{\uparrow}\ \underline{\uparrow}\ \underline{\uparrow}} \quad \underset{3d \text{ (unhybridized)}}{\underline{\ }\ \underline{\ }\ \underline{\ }}$$

$$\text{SF}_6 \quad \underset{d^2sp^3}{\underline{\uparrow\cdot}\ \underline{\uparrow\cdot}\ \underline{\uparrow\cdot}\ \underline{\uparrow\cdot}\ \underline{\uparrow\cdot}\ \underline{\uparrow\cdot}} \quad \underset{3d}{\underline{\ }\ \underline{\ }\ \underline{\ }} \quad \text{(dots = F electrons)}$$

Since the d^2sp^3 orbitals point toward the corners of an octahedron, SF_6 is expected to have an octahedral geometry, which it does.

Before moving on a word should be said about the coordinate covalent bond. An example of this, you remember, is provided by the ammonium ion.

$$\left[\begin{array}{c} \text{H} \\ \ \ \overset{\times\times}{} \\ \text{H}\overset{\times}{\text{:}}\text{N}\overset{\times}{\text{:}}\text{H} \\ \ \ \overset{\times\times}{} \\ \text{H} \end{array}\right]^+ \quad \text{or} \quad \left[\begin{array}{c} \text{H} \\ | \\ \text{H—N}{\rightarrow}\text{H} \\ | \\ \text{H} \end{array}\right]^+$$

In terms of valence bond theory we can imagine the coordinate covalent bond in this ion to be formed by the overlap of an *empty* $1s$ orbital centered on a proton with the completely filled lone-pair orbital on the nitrogen of an ammonia molecule. The electron pair is then shared in the region of orbital overlap. Once the bond is formed it is a full-fledged covalent bond whose properties do not depend on its origin. Consequently the four N—H bonds in NH_4^+ are identical and the ion, although tetrahedral, is usually represented as simply

This same argument can be extended to other coordinate covalent bonds as well.

16.3
Multiple Bonds

Double and triple bonds occur when two and three pairs of electrons, respectively, are shared between two atoms. As examples, we have seen the molecules ethylene, C_2H_4, and acetylene, C_2H_2.

H—C≡C—H

ethylene   **acetylene**

The bonding in ethylene is usually interpreted in the following way. In order to form bonds to three other atoms (two hydrogens and one carbon), each carbon atom employs a set of sp^2 hybrids.

$$C \quad \underset{2s}{\uparrow\downarrow} \quad \underset{2p}{\uparrow \; \uparrow} \; \underline{}$$

gives

$$C \quad \underset{sp^2}{\uparrow \; \uparrow \; \uparrow} \quad \underset{p \text{ (unhybridized)}}{\uparrow}$$

Two of these hybrid orbitals are used for overlap with hydrogen $1s$ orbitals while the third sp^2 orbital overlaps with a similar orbital on the other carbon atom, as shown in Figure 16.11a. This, then, accounts for all of the C—H bonds in C_2H_4 as well as *one* of the electron pairs shared between the two carbons.

Because of the way the sp^2 orbitals are created, each carbon atom also has an unhybridized p orbital that is perpendicular to the plane of the sp^2 orbitals and that projects above and below the plane of these hybrids (Figure 16.11b). When the two carbon atoms are joined together, these p orbitals approach each other sideways and, in addition to the bond formed from the overlap of sp^2 orbitals, a second bond is formed in which the electron cloud is concentrated above and below the carbon-carbon axis. This we see illustrated in Figure 16.11c.

In terms of this interpretation, the double bond in ethylene consists of two distinctly different kinds of bonds, and to differentiate between them a specific notation is employed. A bond that concentrates electron density along the line joining the bound nuclei is called a **σ-bond** (**sigma bond**). The overlap of the sp^2 orbitals of adjacent carbons therefore gives rise to a σ-bond. The bond that is formed by the sideways overlap of two p orbitals, and that provides electron density above and below the line connecting the bound nuclei is called a **π-bond** (**pi bond**).[4] Thus in ethylene we find the double bond to consist of one σ-bond and one π-bond. Notice that in this double bond the two electron pairs manage to avoid one another by occupying different regions in space.

[4] In later chapters we refer to a π-bond formed by the sideways overlap of two p orbitals as a $p\pi$-$p\pi$ bond.

(a)

(b)

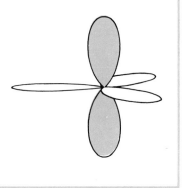

(c)

Figure 16.11

The bonding in ethylene, C_2H_4. (a) Overlap of hydrogen $1s$ orbitals with sp^2 hybrid orbitals on carbon. The carbon atoms are also bound by overlap of sp^2 hybrid orbitals. (b) Unhybridized p orbital is perpendicular to plane of sp^2 hybrid orbitals. (c) Formation of π bonds by sideways overlap of p orbitals.

Another point to note is that bonds formed by the overlap of the hydrogen 1s orbitals with carbon sp hybrid orbitals (Figure 16.11a) also concentrate electron density along a line joining bound atoms. Therefore, these C—H bonds would also be termed σ-bonds.

In acetylene each carbon is bound to only two other atoms, a hydrogen and a carbon atom. Two orbitals are needed for this purpose and a pair of sp hybrid orbitals are used.

$$C \quad \underline{\uparrow\downarrow} \quad \underline{\uparrow} \ \underline{\uparrow} \ \underline{}$$

$$\underline{\uparrow \ \uparrow} \quad \underline{\uparrow \ \uparrow}$$
$$\quad sp \qquad p$$

This leaves on each carbon atom two singly occupied unhybridized p orbitals that are mutually perpendicular as well as perpendicular to the sp hybrids. When the carbon atoms join by way of σ-bond formation between an sp hybrid orbital on each carbon, the p orbitals can also overlap to yield two π-bonds that surround the axis between the carbon nuclei (Figure 16.12). A triple bond therefore consists of one sigma and two pi-bonds. The two π-bonds in acetylene (or in any triple bond) give a total electron distribution that is cylindrical about the bond axis. This is shown in Figure 16.12b.

In Chapter 4 we saw that there are instances in which we cannot draw a single satisfactory Lewis dot formula for a molecule or ion. Some examples, you might remember, are SO_2, SO_3, and NO_2^-. Sulfur dioxide, for instance, was drawn as

$$:\ddot{O}\!-\!\ddot{S}\!=\!\ddot{O}: \longleftrightarrow :O\!=\!\ddot{S}\!-\!\ddot{O}:$$

and it was stated that the actual electronic structure of this molecule corresponds to a resonance hybrid of these two structures.

Electron-dot formulas, as we have drawn them, closely correspond to the valence bond pictures that were developed in the preceding sections. Each

**16.4
Resonance**

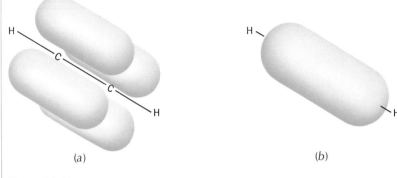

(a)

(b)

Figure 16.12
The triple bond in acetylene consists of one sigma bond and two π bonds. (a) Two π bonds. (b) Cylindrical electron distribution about bond axis.

pair of dots drawn between two atoms denote a pair of electrons shared in a region where atomic orbitals of the bonded atoms overlap. When we draw one of the resonance structures of SO_2, we are therefore referring to a bonding picture in which one S—O bond consists of a single σ-bond while the other is composed on one σ- and one π-bond.

When the valence bond theory was developed, it was recognized that there were numerous instances where a single valence bond structure was inadequate in accounting for the molecular structure. Consequently the concept of resonance evolved. The inability in these cases to draw a single picture to describe the electron density in the molecule is one drawback of valence bond theory. Nevertheless, the correspondence between the valence bond structures that are based on orbital overlap and the simple electron-dot formulas make the valence bond concept a very useful one.

<table>
<tr><td>

16.5
Molecular Orbital Theory

</td><td>

In our discussion of atomic structure in Chapter 3 we saw that around an atomic nucleus there exists a set of atomic orbitals. The electronic structure of a particular atom was derived by feeding the appropriate number of electrons into this set of atomic orbitals such that (1) no more than two electrons populated a single orbital and (2) each electron was placed into the lowest energy orbital available.

Molecular orbital theory proceeds in much this same way. According to this theory, there exists in a molecule a certain arrangement of atomic nuclei and, spread out over these nuclei, there is a set of **molecular orbitals.** The electronic structure of the molecule is obtained by feeding the appropriate number of electrons into these molecular orbitals following the same rules that apply to the filling of atomic orbitals.

No one is quite sure what shapes molecular orbitals have in any particular molecule or ion. What *appears* to be an approximately correct picture is obtained by combining the atomic orbitals that reside on the nuclei making up the molecule. These combinations are achieved by either adding or subtracting the wave functions corresponding to the atomic orbitals that overlap one another. This is shown in Figure 16.13 for the $1s$ orbitals on two identical nuclei. Notice that when the two wave functions are *added*, the resulting

</td></tr>
</table>

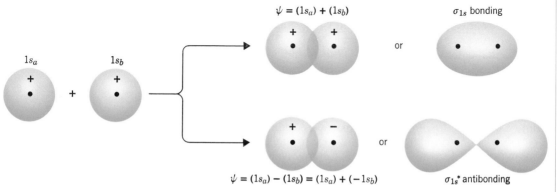

Figure 16.13
The combination of atomic 1s orbitals to give bonding and antibonding molecular orbitals.

molecular orbital has a shape that concentrates electron density between the two nuclei. Electrons placed in such a molecular orbital tend to hold the nuclei together and tend to yield a stable molecule. For that reason this orbital is called a **bonding molecular orbital.** Because the electron density in the orbital is centered along the line joining the atomic nuclei, it is a σ type of orbital; since it is derived in this case from two $1s$ atomic orbitals, we refer to it as the σ_{1s} **molecular orbital.**

You will also observe in Figure 16.13 that a second molecular orbital is obtained by subtracting one atomic orbital wave function from the other. In this instance a molecular orbital is produced that places the maximum electron density outside of the region between the two nuclei. If the electrons of a molecule are placed into this molecular orbital, they do not help to cement the nuclei together and, in fact, the unshielded nuclei repel one another. Consequently, electrons placed into this molecular orbital lead to a destabilization of the molecule and, as a result, the orbital is said to be **antibonding.** This antibonding orbital also has its greatest electron density along the line that passes through the two nuclei and is thus a σ-type orbital. Its antibonding character is denoted by an asterisk superscript; thus it is called the σ_{1s}^* **molecular orbital.** As you might expect, we can also draw similar pictures for the combination of any pair of s orbitals; therefore, in a diatomic molecule we also have σ_{2s}, σ_{2s}^*, σ_{3s}, σ_{3s}^*, etc. molecular orbitals.

In a molecule the p orbitals are also capable of interacting to produce bonding and antibonding molecular orbitals, as illustrated in Figure 16.14. Here we have arbitrarily chosen to denote the internuclear axis as the z axis of our coordinate system so that the p orbitals that point toward one another correspond to p_z orbitals. Again we find that one combination of orbitals gives a bonding molecular orbital, with electron density placed between the two nuclei, while the second combination places most of the electron density outside of the region between the nuclei. The p_z orbitals, like s orbitals, form σ-type molecular orbitals and for $2p_z$ orbitals, then, there would result σ_{2p_z} and $\sigma_{2p_z}^*$ molecular orbitals.

Having chosen the z axis as the internuclear axis, we find that the p_x and p_y orbitals on the two nuclei of our molecule are forced to overlap in a sideways fashion to produce π and π^* molecular orbitals (Figure 16.14). Also keep in mind that the π_{p_x} and $\pi_{p_x}^*$ orbitals are the same as the π_{p_y} and $\pi_{p_y}^*$ orbitals, respectively, with the exception that they are situated at 90° to each other when viewed down the molecular axis.

We have now examined for a diatomic molecule the shapes of the molecular orbitals that can be considered to arise as a consequence of the overlap of atomic orbitals. To discuss the electronic structure of a diatomic molecule, however, we must know the relative energies of these orbitals. Once this has been established, we can then proceed with filling the orbitals with electrons, following the rules that were mentioned earlier.

Let us first consider the σ_{1s} and σ_{1s}^* orbitals. Electrons placed into the bonding orbital lead to stable bond formation and, therefore, to an energy lower than that of two separate atoms. On the other hand, electrons placed into the antibonding orbital lead to a destabilization of the molecule and thus to a state higher in energy than the atoms from which the molecule is formed. We can represent this schematically as shown in Figure 16.15a where the energies of the atomic orbitals of the separate atoms appear on either side of

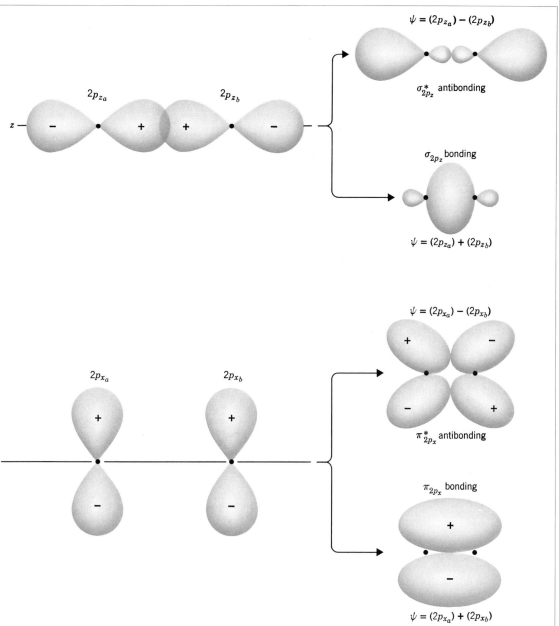

Figure 16.14
Formation of molecular orbital from atomic *p* orbitals.

Figure 16.15
(a) The energies of the bonding and antibonding σ_{1s} molecular orbitals. (b) Bonding in H_2.

the energy level diagram while the energies of the molecular orbitals appear in the center.

Using this simple diagram we can examine the bonding in the H_2 molecule. There are two electrons in H_2 that we place in the lowest energy molecular orbital, the σ_{1s}, (Figure 16.15b). The electron distribution in H_2 is therefore that described by the shape of the σ_{1s} orbital. Notice that this picture is the same as that developed in the valence bond view of H_2. This should not be too surprising since both theories are attempting to describe the same molecular species.

Before moving on, let us also see why the molecule He_2 does *not* exist. The species He_2 would have four electrons, two of which would be placed into the σ_{1s} orbital and a second pair that would be forced to occupy the σ_{1s}^* orbital. The pair of electrons in the antibonding orbital cancel out the stabilizing influence of the bonding pair. As a result, the net bond order, which we can define as

$$\left(\begin{array}{c}\text{Net bond}\\\text{order}\end{array}\right) = \frac{(\text{no. of } e^- \text{ in bonding MO's}) - (\text{no. of } e^- \text{ in antibonding MO's})}{2}$$

has a value of zero for He_2. Since the bond order in He_2 is zero, He_2 is not a stable molecule and is not observed to exist.

For diatomic molecules of second period elements we really only need to consider molecular orbitals that are derived from the interaction of the $2s$ and $2p$ orbitals. The $1s$ orbitals are essentially buried beneath the valence-shell orbitals and are therefore not involved to any appreciable extent in the bonding in these species, The energy level diagram for the molecular orbitals created from the $2s$ and $2p$ orbitals is shown in Figure 16.16a. Let's see how this energy level diagram can be used to account for the bonding in the molecules N_2, O_2, and F_2.

Nitrogen is in Group VA and, therefore, each nitrogen atom contributes from its valence shell five electrons to the N_2 molecule. This means that we must place 10 electrons into our set of molecular orbitals. As shown in Figure 16.16b, two electrons enter the σ_{2s}, two go into the σ_{2s}^*, two more into the σ_{2p_z} and, finally, two into each of the bonding π orbitals, π_{2p_x} and π_{2p_y}. As before, the two σ_{2s}^* antibonding electrons cancel the effect of the σ_{2s} bonding electrons, leaving us with a net total of six bonding electrons (two

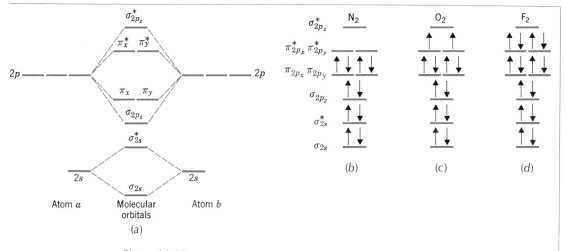

Figure 16.16

Molecular orbital electron configuration in N_2, O_2 and F_2.

each in the σ_{2p_z}, π_{2p_x}, and π_{2p_y} orbitals). If, as usual, we take two electrons to represent a "bond," we find that N_2 is held together by a triple bond that is composed of one σ- and two π-bonds. As with H_2, we arrive at the same resultant description of the bonding in N_2 with both the valence bond and molecular orbital theories.

The real mark of success for molecular orbital theory is seen in its description of the O_2 molecule. This species is found experimentally to be paramagnetic with two unpaired electrons. An attempt to derive a valence bond picture for O_2, however, gives us

$$\underset{\times\times}{\overset{\times\times}{O}}\colon\colon\underset{\cdot\cdot}{\overset{\cdot\cdot}{O}}$$

where, to satisfy the octet rule, all of the electrons appear in pairs.

The molecular orbital description of O_2 is seen in Figure 16.16c. The first 10 of the 12 valence electrons populate all of the same molecular orbitals as in N_2. The final two electrons must then be placed in the $\pi^*_{2p_x}$ and $\pi^*_{2p_y}$ antibonding orbitals; since these two orbitals are of the same energy, the electrons spread themselves out with their spins in the same direction (Hund's rule, see Section 3.14). These two antibonding π electrons also cancel one of the pair of π bonding electrons, so that in the final analysis we see O_2 to be held together by a *net* double bond (one σ- and one net π bond) and we further note that the molecule is predicted to have two unpaired electrons, in precise agreement with experiment.

Finally, with F_2 (which contains two more electrons than O_2) we find that the two π^* antibonding orbitals are filled (Figure 16.16d). This leaves one net single bond, and once again the valence bond and molecular orbital theories give the same result.

The success of molecular orbital theory is not restricted merely to diatomic molecules. In more complex molecules, however, the energy level diagrams are more difficult to predict, and we shall not attempt to extend the theory much further. One useful concept in molecular orbital theory that we can look at further, however, is the idea that molecular orbitals may extend over more than two nuclei. It is this aspect of molecular orbital theory that allows one to avoid the concept of resonance.

Consider, for example, the molecule SO_3. From experiment we know this to be a planar molecule (all four atoms lie in the same plane) with all three S—O bonds the same. This structure can be explained if we assume that the sulfur employs a set of sp^2 hybrid orbitals to form σ-bonds with the three oxygen atoms. This leaves one unhybridized p orbital on the sulfur that can overlap *simultaneously* with p orbitals on the three oxygen atoms, as shown in Figure 16.17. The result is the creation of a molecular orbital that extends over all four nuclei such that the electron densities in the S—O bonds are all the same. Obviously, there is no need to draw more than one bonding picture for the molecule; molecular orbital theory thus manages to avoid the problem that valence bond theory attempts to solve with resonance.

One of the primary goals of chemical bonding theory is to explain and, hopefully, to predict molecular structure. We have seen, for example, how valence bond theory rationalizes molecular structure through the use of hybrid orbitals. Another theory that is exceedingly simple in its concept and remarkably successful in its ability to predict molecular geometry accurately is called the **electron pair repulsion theory.** In applying it, it is not necessary to employ the notion of atomic orbitals at all. We shall see, instead, that if an

**16.6
Electron Pair
Repulsion Theory
of Molecular
Structure**

(a)

(b)

Figure 16.17
Simultaneous overlap of atomic p orbitals in the SO_3 molecule. (a) p orbitals on the sulfur and oxygen atoms. (b) Delocalized π molecular orbital.

electron-dot structure can be drawn for a molecule, its general shape can be predicted.

The electron pair repulsion theory proposes that the geometric arrangement of atoms, or groups of atoms (which we shall generally refer to as ligands) about some central atom is determined *solely* by the repulsion between the electron pairs present in the valence shell of the central atom. Consider, for example, the molecule BeH_2. The electron dot structure would be given as

$$H \overset{x}{\circ} Be \overset{x}{\circ} H$$

where the x's are Be electrons and the circles are hydrogen electrons. This particular molecule violates the octet rule, and there are only two pairs of electrons located in the valence shell of Be. These electron pairs would like to be as far apart as possible so that the repulsion between them is at a minimum. When there are two electron pairs in the valence shell, this minimal repulsion occurs when the electron pairs are located on opposite sides of the nucleus, so that we have

$$\overset{x}{\circ} - Be - \overset{x}{\circ}$$

The hydrogen atoms are attached to the Be through these electron pairs and the molecule should therefore have the *linear* structure,

$$H - Be - H$$

As we have seen before, this is the structure of BeH_2.

We can also extend this reasoning to situations involving double or triple bonds. For instance, the CO_2 molecule has the dot structure

$$: \overset{..}{O} :: C :: \overset{..}{O} :$$

where we see that there are double bonds between C and O. Both pairs of electrons in a double bond are confined to the same general region in the valence shell of an atom, so that in terms of their effect on determining molecular geometry, a group of four electrons in a double bond behaves much as a group of two electrons in a single bond. In the valence shell of carbon, then, we have *two* groups of four, and they will locate themselves on opposite sides of the C nucleus so that the repulsion between the two groups is at a minimum:

$$:: - C - ::$$

As before, the ligands (in this case, oxygen) are attached to the central atom through these electron pairs and we again have a *linear* structure.

$$O = C = O$$

MORE THAN TWO PAIRS (OR GROUPS OF PAIRS) OF ELECTRONS. When there are more than two pairs (or groups of pairs) of electrons in the valence shell we find other geometric arrangements as shown in Figure 16.18. Electron pairs arranged in the valence shell in this manner lead to minimum repulsions. Let us see how we can use these electron pair arrangements to predict molecular structure.

Number of Electron Pairs	Geometric Arrangement of Electron Pairs	
2	Linear	
3	Planar triangle	
4	Tetrahedral	
5	Trigonal bipyramidal	
6	Octahedral (eight sides– 6 corners)	

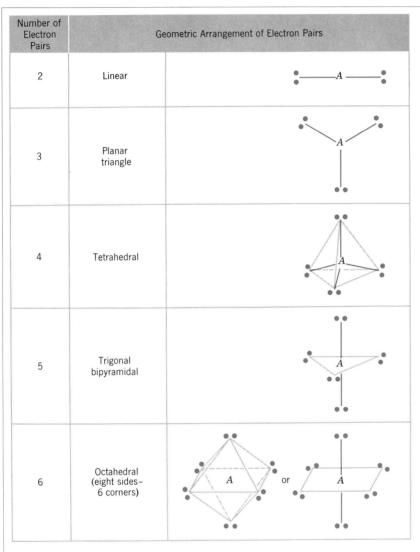

Figure 16.18
Positions of electron pairs that lead to minimum electrostatic repulsions.

THREE GROUPS OF ELECTRONS IN THE VALENCE SHELL. The molecule BCl_3 has the dot structure

$$:\overset{\displaystyle ..}{\underset{\displaystyle ..}{Cl}}:$$
$$:\overset{\displaystyle ..}{\underset{\displaystyle ..}{Cl}}\overset{\displaystyle \times}{\underset{\displaystyle .}{B}}\overset{\displaystyle ..}{\underset{\displaystyle ..}{Cl}}:$$

Thus there are three electron pairs around boron. We therefore expect the three chlorine atoms to be arranged about the B atom in a planar triangle. This is indeed the structure of BCl_3.

Let us now consider the molecule SO_2. The electron dot structure for one of the two resonance structures is

$$\ddot{\text{O}} : \ddot{\text{S}} :: \ddot{\text{O}} :$$

About the sulfur there are again three groups of electrons, two groups each with one pair, and one group with two pairs (the double bond). These groups of electrons are situated at the corners of a triangle with the sulfur in the center.

Attaching the oxygen atoms, one to a single pair and one to the double pair, we have

The SO_2 molecule is thus *not* linear, but instead has an angular, or bent, shape. The nonlinear shape is caused in this case by the presence of the non-bonded, *lone pair* of electrons on the sulfur. We see, therefore, that lone pairs of electrons in the valence shell of an atom influence the molecular geometry.

In summary, when there are three groups of electrons about an atom, they are arranged at the corners of a triangle. If all of them are bonded to ligands, we have a molecule, AX_3, having a planar triangular shape. If only two groups are bonded, leaving one lone pair, we have a species AX_2E (where we use E to represent the lone pair) in which the atomic nuclei are situated so as to give an angular structure.

FOUR GROUPS OF ELECTRONS IN THE VALENCE SHELL. When there are four groups of electrons in the valence shell of an atom, they will situate themselves at the vertices of a tetrahedron (Figure 16.18). As shown in Figure 16.19, there are three kinds of molecules of interest that can be formed with four groups of electrons: AX_4, AX_3E, and AX_2E_2 (note that there is *no* uncertainty about the geometry of any AXE_n molecule; A and X must be in a straight line).

The molecule, AX_4, of which CH_4 is an example, has a tetrahedral shape with ligands bonded through all four electron pairs. The species AX_3E, on the other hand, contains *one* lone pair (e.g., NH_3) and has a geometry that we refer to as pyramidal (shaped like a pyramid). Finally, the molecule AX_2E_2 (e.g., H_2O) has two lone pairs and has an angular structure. We see, then, that with four groups of electrons there are three different possible molecular shapes.

FIVE ELECTRON PAIRS. With five electron pairs in the valence shell of the central atom, there are four types of molecules to be considered: AX_5, AX_4E, AX_3E_2, and AX_2E_3. These are illustrated in Figure 16.20. Note that when there are five electron pairs about an atom, any lone pairs that are present occur in the triangular plane.

SIX ELECTRON PAIRS. Of the five different possibilities, only three molecule types are known: AX_6, AX_5E, and AX_4E_2. The molecular geometries that are associated with these formulas are shown in Figure 16.21.

Example

(a)	AX_4	CH_4
(b)	AX_3E	NH_3
(c)	AX_2E_2	H_2O

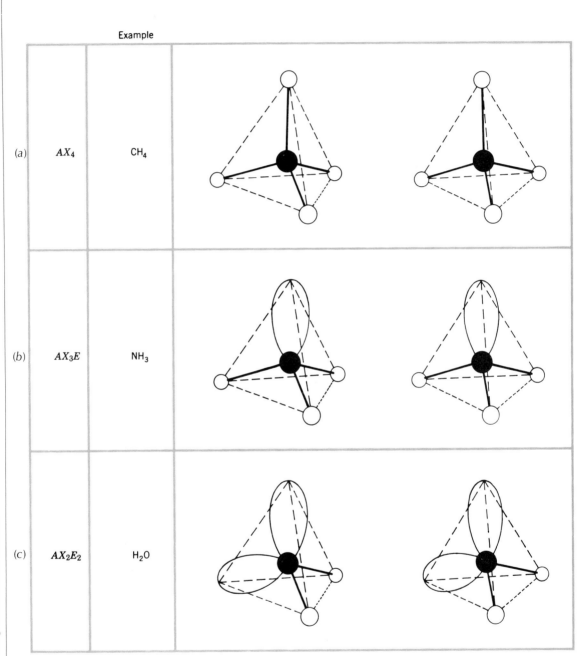

Figure 16.19
Geometries of molecules in which the central atom has four pairs of electrons. (a) AX_4 (for example, CH_4). (b) AX_3E (for example, NH_3). (c) AX_2E_2 (for example, H_2O).

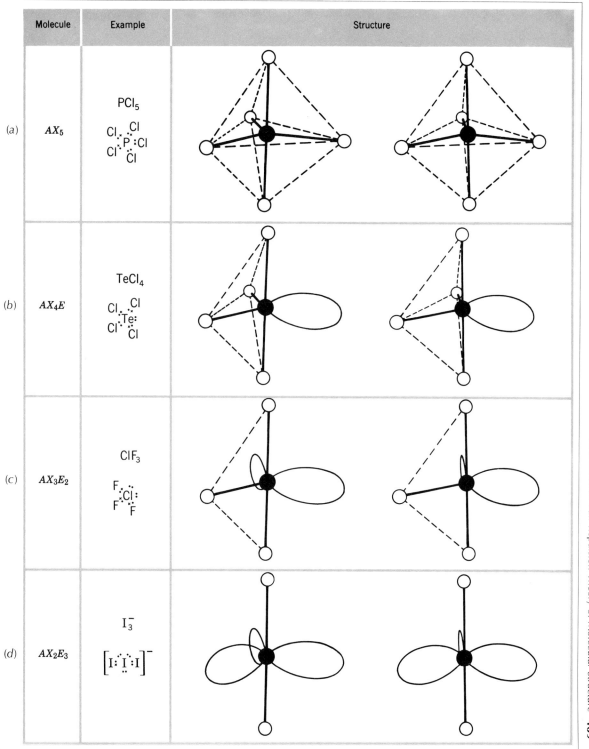

Figure 16.20
Molecular structures that result when the central atom has five electron-pair groups.

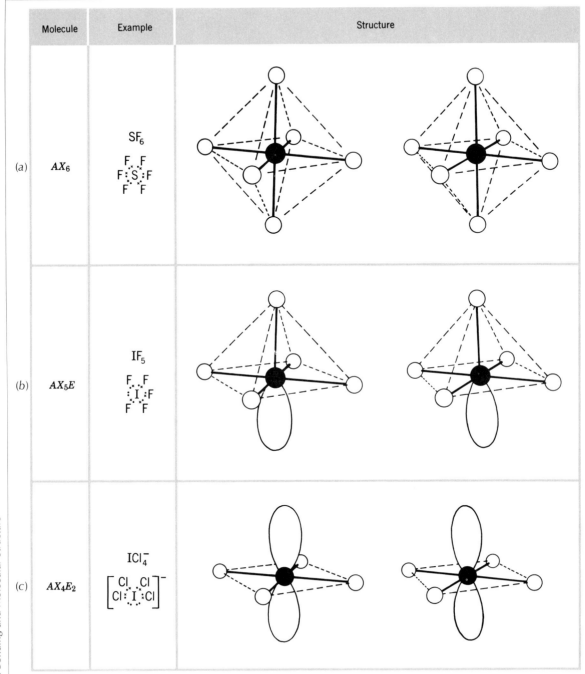

Molecule	Example	Structure

(a) AX_6 — SF_6

(b) AX_5E — IF_5

(c) AX_4E_2 — ICl_4^-

Figure 16.21

Molecular structures that result when the central atom has six electron-pair groups.

SUMMARY. The theory that has just been presented above can be applied with great success to a very large number of molecules and ions formed by the representative elements. The key to using this theory lies in your ability to write an electron-dot structure for the molecule. Once this has been done, you merely count up the number of groups of electrons (either single or multiple pairs) and decide what geometry the electron pairs will assume: linear for two groups, triangular for three, etc. The appropriate number of ligand atoms are then attached through the electron pairs, and you are now in a position to predict the structure of the molecule, that is, the structure assumed by the central atom and the ligands (when we speak of molecular structure, we refer to the positions of the atoms, *not* to the location of the electrons even though we use the latter to predict the former). A summary of the molecular shapes found for different molecule types is given in Table 16.2.

By now you have undoubtedly noticed that both the valence bond theory, through the use of hybrid orbitals, and the electron pair repulsion theory lead to very similar, if not identical, results in the prediction of molecular geometry. These two theories are actually very much the same, with the electron pair repulsion theory yielding structures that are the same as those we would obtain from the valence bond theory if we assumed complete hybridization in every case (sp^3 hybrids in H_2O, sp hybrids on Be in BeH_2, d^2sp^3 hybrids on S in SF_6, etc.).

While the electron repulsion pair theory is quite good for predicting molecular geometry, it tells us nothing about the relative stabilities of bonds or why a bond is formed at all. For these aspects of the chemical bond we must rely on valence bond and molecular orbital theory.

Table 16.2
Summary of Molecular Shapes

Molecule Type	Shape
AX_2	Linear
AX_3	Planar triangular
AX_2E	Angular (bent)
AX_4	Tetrahedral
AX_3E	Pyramidal
AX_2E_2	Angular (bent)
AX_5	Trigonal bipyramidal
AX_4E	Distorted tetrahedral
AX_3E_2	T-shaped
AX_2E_3	Linear
AX_6	Octahedral
AX_5E	Square pyramidal
AX_4E_2	Square planar

16.1 What is the basic concept upon which the valence bond theory is based?

16.2 Use valence bond theory to explain the bonding in the Cl_2 molecule.

16.3 How can you account for the fact that the H—S—H bond angle in H_2S is approximately 92°?

16.4 What is a hybrid orbital?

16.5 Why is it necessary to employ hybrid orbitals when attempting to account for the structure of methane, CH_4?

16.6 On the basis of the electronic structure of the central atom, suggest what kind of hybrid orbitals would be involved in the bonding in each of the following:
(a) BCl_3 (d) $AlCl_6^{3-}$ (g) PCl_3
(b) NH_4^+ (e) BeH_2 (h) SF_4
(c) PCl_5 (f) $SbCl_6^-$ (i) ClO_4^-

16.7 From a knowledge of the hybrid orbitals used for bonding, predict the structures of each of the species in Question 16.6.

16.8 Describe a σ-bond; a π-bond. What constitutes a double bond? A triple bond?

16.9 Which of the species in Question 16.6 have one or more bonds that would be considered to have been formed by way of coordinate covalent bonding? How, if at all, does a coordinate covalent bond differ from a normal covalent bond once it has been formed?

16.10 What angles exist between the orbitals in
(a) sp^3 hybrids   (c) sp hybrids
(b) sp^2 hybrids   (d) d^2sp^3 hybrids

16.11 Describe, in detail, the bonding in the N_2 molecule.

16.12 Draw the resonance structures for (a) NO_3^- and (b) NO_2^-.

16.13 How does molecular orbital theory view the formation of a molecule? How does molecular orbital theory differ from valence bond theory?

16.14 What is the difference between a bonding and an antibonding molecular orbital? How do their energies compare?

16.15 Use Figure 16.16a to discuss the bonding in the molecules Li_2, Be_2, B_2, and C_2. What can you predict about the stabilities of these species when an electron is removed from each?

16.16 How does molecular orbital theory avoid the concept of resonance?

16.17 What is the basic postulate of the electron pair repulsion theory?

16.18 Use the electron pair repulsion theory to predict the geometry of each of the following:
(a) NF_3 (g) BrF_5 (l) $SnCl_4$
(b) PH_4^+ (h) CCl_4 (m) SO_3
(c) CO_3^{2-} (i) $AlCl_6^{3-}$ (n) BF_4^-
(d) NO_2 (j) $SbCl_5$ (o) OF_2
(e) $SeCl_4$ (k) $SnCl_2$ (p) XeF_4
(f) ICl_2^-

16.19 Use your predictions from Question 16.18 to suggest the type of hybrid orbitals that would be used in valence bond theory to account for these geometries.

16.20 The species H_2^+ and He_2^+ have been observed. Use molecular orbital theory to account for their existence.

16.21 On page 455 it was suggested that lone electron pairs tend to repel rather strongly electron pairs between bonded atoms; in other words, lone pair/bond pair reuplsions are greater than bond pair/bond pair repulsions. This will lead to structural distortions of some of the idealized molecular geometries pictured in Figures 16.19, 16.20, and 16.21. Predict the nature of these distortions and sketch the shapes of the resulting molecules.

17
Chemistry of the Representative Elements: Part I

In previous chapters you learned many of the concepts the chemist has applied in developing his understanding of how the elements react with each other and the kinds of compounds they produce. These earlier discussions were focused primarily on the concepts themselves, with examples of chemical behavior being used to reinforce and justify them. With these concepts available to us now, we shall direct our emphasis in the opposite direction and begin to examine, in more or less a systematic way, some of the physical and chemical properties of the elements and their compounds. For example, we shall look at what kinds of reactions the elements and their compounds undergo, what types of compounds are formed when they react and, in many cases, the types of structures that result. Chemistry discussed from this point of view is often called **descriptive chemistry.**

In the following pages quite a lot of factual information will be presented. You are not expected to memorize all of it. Our discussion of descriptive chemistry is aimed at providing you with an awareness of certain important kinds of chemical reactions and compounds. Where they exist, trends in chemical and physical behavior will be pointed out and structural similarities or differences described. A knowledge of these trends in behavior will help you tie together some of the bits and pieces of factual knowledge that you may need at a later time.

17.1
Metals, Nonmetals, and Metalloids

It is typically human of us to attempt to classify and compartmentalize knowledge in an effort to make it easier to assimilate and understand. In the discussion of the chemistry of the elements, however, this is a somewhat risky undertaking, since many of the trends in chemical properties run across boundaries that we may attempt to set up. Nevertheless, it is sometimes useful to classify the elements according to at least some of their properties.

In Section 3.9 we stated that the elements could be divided into three main categories: metals, nonmetals, and metalloids, the latter having properties that lie between the other two. We shall follow this general classification in our present discussion, although it should be kept in mind that no sharp boundaries exist. In fact many aspects of behavior extend over more than one category.

Figure 17.1 illustrates the rough breakdown of the elements into our three classes. As you can see, most of the elements exhibit metallic properties and, furthermore, most of the metals belong to the transition and inner transition (lanthanide and actinide) series. In Chapters 17 and 18 our discussion shall be devoted primarily to the chemistry of the representative elements (i.e., the A-Group elements). The transition and inner transition elements will

Figure 17.1
Distribution of metals, metalloids and nonmetals in the periodic table.

be treated separately in Chapter 19. Again, as we might expect, some characteristics that are exhibited by the representative elements are also observed for the transition elements, and vice versa.

Before we examine the chemistry of the elements in any detail, let's first take a rather brief qualitative look at what kinds of properties will identify an element as a metal, nonmetal, or a metalloid and what sort of trends in metallic and nonmetallic character exist throughout the periodic table.

Metals have many physical properties with which we all are very familiar: for example, their electrical and thermal conductivities as well as their characteristic luster. Ductility (the ability of a substance to be drawn into wire) and malleability (its ability to be hammered into thin sheets) are two other physical properties that we associate with metals.

We can also recognize metals by their chemical properties. In earlier chapters we found that metals form oxides that are basic in nature, reacting with acids to give salts and producing hydroxide ion when they dissolve in water. Thus iron would be classed as a metal because its oxide, Fe_2O_3, reacts with HCl.

$$Fe_2O_3 + 6HCl \longrightarrow 2FeCl_3 + 3H_2O$$

Sodium and calcium could be identified as metals because of the reaction of their oxides with water.

$$Na_2O + H_2O \longrightarrow 2NaOH$$
$$CaO + H_2O \longrightarrow Ca(OH)_2$$

Of course their oxides also react directly with acids, for example,

$$CaO + 2HCl \longrightarrow CaCl_2 + H_2O$$

Another characteristic of metals, which you may have already noticed, is that in their compounds they generally exhibit positive oxidation states.[1]

The general properties of metals can be explained quite well in terms of their electronic structures. As a group, metals possess relatively few electrons in their valence shells and, consequently, the attainment of a noble gas electron configuration is often impossible by way of sharing of electrons in ordinary covalent bonds. In addition, the metals generally have low ionization energies (I.E.) and electron affinities (E.A.). They therefore do not react readily with each other because they are not inclined to form covalent bonds and also because they have little desire to pick up electrons from another metal atom. They do form compounds with nonmetals, however, in which they tend to lose electrons and acquire a positive charge and, as we have seen before, ionic bonding is an important factor to consider in the chemistry of metals.

In the free state metals exist in a unique type of lattice in which they tend to lose their valence electrons to the lattice as a whole. As described in Chapter 7, the result is an array of positive metal ions imbedded in a sea of very mobile electrons. Since these electrons are able to move easily and rapidly through the solid, they are able to transport thermal energy quickly and metals conduct heat well. Of course their electrical conductivity is also explained by this model. The important point here is that the metallic lattice is a consequence of the small number of valence electrons and the low ionization energies of the metal atoms.

By contrast, nonmetals are characteristically very poor conductors of electricity and, in fact, are classed as insulators. Many of them, such as H_2, N_2, O_2, and the Group 0 elements, are gases at room temperature. Those that are solids conduct heat poorly and tend to be brittle. Chemically we find that the oxides of the nonmetals are acidic. Many of them react with water to yield hydronium ion, and they react with bases to form salts containing oxoanions. The reaction of carbon dioxide with NaOH is a typical example.

$$2NaOH + CO_2 \longrightarrow Na_2CO_3 + H_2O$$

From their positions in the periodic table we see that nonmetals generally require only a small number of electrons to reach a noble gas structure. Consequently in their elemental state they combine with themselves to produce covalently bonded molecules containing two or more atoms. We shall examine these in more detail in Chapter 18. The important point is that in the process of achieving a noble gas structure all of the electrons usually become paired and are localized on individual molecules. As a result, no electrons remain available for electrical conduction and an insulator results. This absence of mobile electrons also means that the only mechanism for heat

---

[1] There are compounds of certain elements in which the metal is assigned a zero or even a negative oxidation number; Nickel carbonyl, $Ni(CO)_4$, for example, contains a nickel atom bonded to four neutral CO molecules and is an example of Ni in the zero oxidation state. In $NaCo(CO)_4$ the cobalt atom is assigned an oxidation number of -1.

transfer through a nonmetal is by the vibrational motions of the atoms. Since this is much slower than electronic motion, nonmetals conduct heat poorly.

Chemically, nonmetals combine with themselves, other nonmetals, and with metals. They are characterized by high ionization energies and electron affinities and, therefore, by high electronegativities. In their binary compounds with metals, for instance NaCl or Al_2O_3, they assume negative oxidation states. In those compounds containing more than one nonmetal, the usual rules for assigning oxidation numbers apply; the more electronegative element is assigned the negative oxidation state. Thus in CO_2, since oxygen is more electronegative than carbon, we say that the oxidation number of carbon is positive (by our rules we obtain +4 for the oxidation state of carbon).

The metalloids are elements with outer-electron shells that are approximately half-filled, and with electronegativities lying between the metals and the nonmetals. As a class, they have electrical conductivities that are much smaller than those of metals but still much larger than those of nonmetals, and these semiconductor properties (discussed in Section 7.9) have made them extremely useful in electronic devices (e.g., in diodes and transistors).

Chemically, they exhibit both positive and negative oxidation states. They combine with nonmetals and, in some cases, also with metals. Their oxides tend to be amphoteric, although their acid properties outweigh the basic ones. In fact, in many ways the metalloids behave chemically more as nonmetals than as metals; the chemistries of these two classes of elements will be discussed together in Chapter 18. The remainder of this chapter is devoted to the representative metals.

The trends in the metallic character of the elements can be examined by following the variations in electronegativity throughout the periodic table, since it is this property, which reflects both ionization energy and electron affinity, that is primarily responsible for metallic or nonmetallic behavior.

Within a period, ionization energy and electron affinity both increase with increasing atomic number; hence the electronegativity also increases. Consequently, as we proceed from left to right across a period, the metallic character of the elements gradually decreases and their nonmetallic character, of course, increases. In period 3, for example, Na, Mg, and Al are typical metals. In this sequence, however, the oxides become less basic (more acidic) so that we find Al_2O_3 to be soluble in strong bases. Na_2O and MgO, on the other hand, do not react with base. Continuing to the right, silicon is classed as a metalloid and exhibits typical semiconductor properties. The remaining elements in the period, P, S, Cl, and Ar are nonmetals that, except for Ar (which has no tendency to react) form characteristically acidic oxides.

As we descend a group, ionization energy, electron affinity, and electronegativity decrease. The metallic character of the elements therefore increase accordingly. The Group IVA elements clearly illustrate this trend. Carbon, at the head of the group, is nonmetallic in virtually every way.[2] Below carbon are the metalloids, silicon, and germanium, and below these

---

[2] One form of carbon, graphite, does conduct electricity. This, however, is more an accident of the bonding in graphite than evidence for metallic behavior. The structure of graphite and the mechanism of its electrical conductivity is discussed in Chapter 18.

the metals tin and lead. Of the latter two, tin exists at low temperatures in a crystalline form that is nonmetallic whereas lead, at the bottom of the group, demonstrates only metallic properties.

This trend toward increasing metallic character as we proceed down within a group appears in Group VA too. The pair of elements at the top, nitrogen and phosphorus, are nonmetals, arsenic and antimony behave as metalloids, and bismuth, at the bottom of the group, shows mainly metallic characteristics.

Since the metallic nature of the elements increases from right to left within a period, and from top to bottom in a group, the most metallic elements occur in the lower left corner of the periodic table. Similarly, we expect the most nometallic elements to be found in the upper right corner of the table. These trends in properties also permit us to understand why the metalloids occur as a band running diagonally from upper left to lower right. In fact, if we omit the transition elements and draw our periodic table with only the representative elements, we find the division rather even (Figure 17.2).

17.3
Preparation of Metals

Most metals, including those of both the representative elements and the transition elements, always occur in nature in a combined state, primarily as oxides, sulfides, carbonates, silicates, or halides. Regardless of the type of ore, however, metals almost always exist in positive oxidation states, and in order to produce the free element a chemical reduction must be brought about. The nature of this reduction process depends on the ease with which the metal can be reduced.

Some metals are so easily reduced that many of their compounds can be decomposed just by heating them at relatively low temperatures. Priestly, for example, in his experiments on oxygen, produced metallic mercury and oxygen from mercuric oxide by simply heating it with sunlight focused on the HgO by means of a magnifying glass. In this case HgO decomposes quite spontaneously at elevated temperatures according to the equation

$$2HgO(s) \longrightarrow 2Hg(g) + O_2(g)$$

IA	IIA	IIIA	IVA	VA	VIA	VIIA	0
						H	He
Li	Be	B	C	N	O	F	Ne
Na	Mg	Al	Si	P	S	Cl	Ar
K	Ca	Ga	Ge	As	Se	Br	Kr
Rb	Sr	In	Sn	Sb	Te	I	Xe
Cs	Ba	Tl	Pb	Bi	Po	At	Rn
Fr	Ra						

Figure 17.2

The representative elements. Metalloids are indicated by gray squares.

The practicality of using a thermal decomposition reaction of this type to produce a free metal depends on the extent to which the reaction proceeds to completion at a given temperature. In Chapter 10 we saw that at 25° C the position of equilibrium in a reaction is governed by $\Delta G°$. If we take $\Delta G'$ to be the equivalent of $\Delta G°$, but at some other temperature, we have the relationship,

$$\Delta G' = \Delta H' - T \Delta S'$$

where $\Delta H'$ and $\Delta S'$ are the heat and entropy changes that accompany the reaction. For most systems ΔH and ΔS do not change much with temperature so that $\Delta H'$ and $\Delta S'$ can reasonably be approximated by $\Delta H°$ and $\Delta S°$.

The thermal decomposition reaction will be feasible when $\Delta G'$ is negative, since under these conditions an appreciable amount of product will be formed. We must now look at the magnitudes of $\Delta H'$ and $\Delta S'$, since they control the sign and magnitude of $\Delta G'$. Since a gas (O_2) and sometimes the metal vapor is produced in the decomposition, the process occurs with a sizable increase in entropy; hence ΔS will be positive.

The enthalpy change for the decomposition, $\Delta H'$, is simply the negative of the heat of formation of the oxide and, since $\Delta H_f'$ is generally negative for metal oxides, $\Delta H'$ for the decomposition reaction will be positive. As a result, the sign of $\Delta G'$ is determined by the difference between two positive quantities, $\Delta H'$ and $T \Delta S'$.

If the metal oxide has a high negative heat of formation (i.e., if a great deal of energy is evolved when the oxide is formed), then $\Delta H'$ for the decomposition will have a large positive value. Consequently, the difference $\Delta H' - T\Delta S'$ will be negative *only* at very high temperatures where $T\Delta S'$ is larger than $\Delta H'$. We express this by saying that the metal oxide is very stable with respect to thermal decomposition. On the other hand, if the $\Delta H_f'$ of the metal oxide is relatively small, as with HgO and certain other oxides (e.g., AgO, CuO, and Au_2O_3), $\Delta H'$ for the decomposition reaction is a small positive quantity and $\Delta G'$ for the reaction becomes negative at relatively low temperatures. These oxides, therefore, are said to have relatively low thermal stabilities.

Except in a few cases, thermal decomposition is not a practical way of producing the free metals. Instead, their compounds are reacted with some substance that is a better reducing agent than the metal being sought. One of the most common agents used for the reduction of metal oxides is carbon. Tin and lead, for example, can be produced by heating their oxides with carbon.

$$2SnO + C \xrightarrow{\text{heat}} 2Sn + CO_2$$

$$2PbO + C \xrightarrow{\text{heat}} 2Pb + CO_2$$

Carbon is used in large quantities in commercial metallurgy because of its abundance and low cost. Its importance in the reduction of iron ore and in steel making will be examined in Chapter 19.

Hydrogen is another reducing agent that can be used to liberate metals of moderate chemical activity from their compounds. For instance, tin and lead oxides will also be reduced when heated under a stream of H_2.

$$SnO + H_2 \xrightarrow{\text{heat}} Sn + H_2O$$

$$PbO + H_2 \xrightarrow{\text{heat}} Pb + H_2O$$

The use of a more active metal to carry out the reduction is also possible. In Chapter 15 we saw that a galvanic cell could be established between two different metals, for example, Zn and Cu. In that cell the more active reducing agent, Zn, causes the Cu^{2+} to be reduced. Aluminum was first prepared in 1825 by the reaction of aluminum chloride with the more active metal, potassium.

$$AlCl_3 + 3K \longrightarrow 3KCl + Al$$

As a practical source of metals, the reduction of compounds with other elements that are better reducing agents suffers from a serious limitation, specifically, the availability (and cost) of the reducing agent. Each possible reducing agent must then be generated by reacting one of its compounds with a still better reducing agent. Ultimately, of course, there must be some "best" reducing agent. How could this substance be prepared if there were no better reducing agent available that could be used to reduce its compounds?

The solution to this dilemma is electrolysis where, by applying a suitable potential, virtually any oxidation-reduction process can be brought about. Consequently, metals that themselves are very powerful reducing agents are nearly always prepared by electrolysis. Among the representative elements these include the very active elements in Groups IA and IIA (in Chapter 15 the electrolysis of molten NaCl was described). Although less active than the alkali and alkaline earth metals, the metallic elements in Group IIIA are generally produced by electrolysis too. When a molten salt is used for the electrolysis, a halide is generally employed because of their usually lower melting points.

**17.4
Chemical
Properties
and Typical
Compounds**

In the following discussion of the chemistry of the A group metals we shall not attempt to provide an exhaustive review of all of the many kinds of compounds that they form. Instead, we shall look at some of the similarities and differences in their chemical behavior.

The elements within any given group are expected to exhibit similar properties and this is particularly evident among the elements in Groups IA and IIA. For example, the alkali metals in Group IA (we exclude hydrogen from this discussion because of its clearly nonmetallic behavior) are all extremely reactive elements that are capable of reducing water to produce hydrogen and the metal hydroxide

$$2M + 2H_2O \longrightarrow H_2 + 2MOH$$

They yield only compounds in which their oxidation state is $+1$, and many of their compounds have similar solubilities in water.

The reactivity of the alkali metals is high toward all of the nonmetals. Their behavior toward oxygen is especially interesting, however. Only lithium reacts directly with oxygen to produce the normal oxide, Li_2O. Sodium, when it reacts with O_2, gives a pale yellow peroxide

$$2Na + O_2 \longrightarrow Na_2O_2$$

This compound contains the peroxide ion, O_2^{2-}

$$\left[:\ddot{O}:\ddot{O}:\right]^{2-}$$

and reacts with water to produce H_2O_2 by hydrolysis

$$Na_2O_2 + 2H_2O \longrightarrow 2Na^+ + 2OH^- + H_2O_2$$

The remaining alkali metals, K, Rb, and Cs, combine with molecular oxygen to form rather deeply colored yellow-orange superoxides. For example,

$$K + O_2 \longrightarrow KO_2$$

The superoxide ion, O_2^-, can be considered an oxygen molecule that has gained a single electron. This electron pairs with one of the two unpaired electrons in O_2 so that the O_2^- ion still has one unpaired electron and is paramagnetic.

The superoxides also react with water to yield hydrogen peroxide; however, they also yield molecular oxygen as one of the products.

$$2MO_2 + 2H_2O \longrightarrow 2M^+ + 2OH^- + H_2O_2 + O_2$$

Like the alkali metals, the alkaline earths exhibit very marked group similarities. For example, they all show only a +2 oxidation state and are also quite reactive, although less so than their neighbors in Group IA. However, in both of these groups the greatest group similarities are exhibited by the elements in period 3 and below. Thus in Group IA the elements Na and K are more nearly alike than are Li and Na. Among the alkaline earths we again find a closer similarity between the elements Mg and Ca than between Be and Mg. In fact, in general the period 2 elements reveal a somewhat unique chemistry among both the metals and the nonmetals.

In looking for similarities in the chemistry of the A group metals, it is interesting to find that a *diagonal relationship exists between the chemistry of the first member of a group and that of the second member of the following group.* Thus lithium is, in some ways, more like magnesium than like sodium. For example, many lithium and magnesium salts have similar solubilities that differ markedly from those of the other Group IA elements. The salts MgF_2 and LiF, for instance, are insoluble in water while the corresponding salts NaF, KF, etc. are soluble. In addition, the lithium halides (LiCl, LiBr, LiI) and $LiClO_4$ are appreciably soluble in relatively nonpolar solvents such as alcohol, as are the corresponding magnesium salts. On the other hand sodium salts tend to be quite insoluble in alcohol.

Still another point of similarity between Li and Mg is their reactions with molecular nitrogen. Both react readily at elevated temperatures to produce nitrides.

$$6Li + N_2 \longrightarrow 2Li_3N$$
$$3Mg + N_2 \longrightarrow Mg_3N_2$$

The other Group IA metals are unreactive toward N_2.

The similarities in chemical behavior of lithium and magnesium are generally interpreted in terms of their ratios of ionic charge to ionic radius. This ratio is called the **ionic potential** and is usually symbolized by the Greek letter ϕ. Thus we have

$$\phi = \frac{q}{r}$$

where q is the charge and r is the ionic radius.

As it happens, both Li^+ and Mg^{2+} have similar ionic potentials. The Mg^{2+} ion is larger than Li^+, but it also has a higher charge. These two factors combine to make their ionic potentials nearly the same. Because of this each ion (i.e., both Li^+ and Mg^{2+}) behaves in very much the same fashion when placed into a chemical environment where ionic interactions are important.

The diagonal relationship between Li and Mg is repeated for the elements Be and Al and, in this latter case, it is even more striking. Once again we have a pair of ions[3] that have very similar ionic potentials. The ionic radii of Be^{2+} and Al^{3+} are 0.35 and 0.51Å, respectively, from which their ionic potentials are 5.88 and 5.71, quite close indeed. (For comparison, magnesium, which is found in Group IIA just below Be, has an ionic potential of 3.03). As a result the chemistries of Be and Al are remarkably alike. For example:

1. Both Be and Al react with oxygen to form an oxide coating that protects the metal beneath from further reaction. Consequently, even though they have highly negative reduction potentials suggesting that they should be readily oxidized ($\mathscr{E}^\circ_{Be} = -1.85$ volts, $\mathscr{E}^\circ_{Al} = -1.66$ volts), both dissolve only slowly in acids.

2. The oxides of Be and Al, BeO, and Al_2O_3, are extremely high melting (BeO, melting point = $2530°$ C; Al_2O_3, melting point = $2045°$ C) and very hard. Presumably the high concentration of positive charge on these very small cations leads to very strong lattice forces and hence to an unusually high lattice energy that must be overcome in order to melt the oxides. Aluminum oxide is interesting because the presence of trace impurities imparts brilliant colors to its gem quality crystals. When the impurity is Cr^{3+}, ruby results; when the impurities are Fe^{2+}, Fe^{3+}, and Ti^{4+}, the gem blue sapphire is produced. Both of these gems are currently manufactured synthetically in large quantities.

3. Beryllium and aluminum both dissolve in strong base with the evolution of hydrogen while magnesium, just below Be in Group IIA, does not dissolve in base. The chemical equations for these reactions can be represented as

$$Be + 2OH^- \longrightarrow BeO_2^{2-} + H_2$$
$$2Al + 2H_2O + 2OH^- \longrightarrow 2AlO_2^- + 3H_2$$

The formulas given for these ions, BeO_2^{2-} (beryllate) and AlO_2^- (aluminate), are really oversimplifications since the actual species that exist in solution are more complex, *probably* $Be(OH)_4^{2-}$ and $Al(OH)_4(H_2O)_2^-$. Note the stoichiometric equivalence,

$$Be(OH)_4^{2-} \qquad \text{and} \qquad BeO_2^{2-} + 2H_2O$$
$$Al(OH)_4(H_2O)_2^- \qquad \text{and} \qquad AlO_2^- + 4H_2O$$

---

[3] Actually, in all known cases, there is no evidence for the existence of a simple Be^{2+} ion. Instead, some degree of covalent bonding is involved in all Be compounds.

We see here that it is frequently very difficult to ascertain the actual chemical identity of species in solution because of possible interactions with the solvent.

4. Beryllium and aluminum halides are covalent. As we shall see later in this chapter, the high ionic potential of the Be^{2+} and Al^{3+} ions leads to a substantial degree of covalent bonding in their halides, for instance $BeCl_2$ and $AlCl_3$. Magnesium chloride and the other alkaline earth chlorides, on the other hand, are essentially ionic in character.

The structures of these covalent halides are also somewhat similar. The simple species, $BeCl_2$ and $AlCl_3$, are electron deficient, having only four and six electrons, respectively, in the valence shell of the metal atom. As a result both function as Lewis acids by combining with electron pair donors (bases). In the absence of a suitable base, however, their strong desire to achieve an octet is satisfied by forming coordinate covalent bonds to chlorine atoms of a neighboring metal halide molecule. In solid $BeCl_2$, for example, two such coordinate covalent bonds are formed to each Be atom and a long linear chainlike structure is formed. This is illustrated by the formula

The chlorine atoms in this structure serve as bridges between adjacent Be atoms by making their lone pairs of electrons available for coordinate covalent bonding. The chain is not planar, of course, since each Be in the structure is surrounded by four electron pairs. In Chapter 16 we found that this requires that the geometry around each Be atom be tetrahedral. In other words, each Be atom in the chain lies at the center of a tetrahedron with the bridging chlorine atoms at the apexes, as shown stereoscopically in Figure 17.3.

In the simple aluminum halide species the aluminum atom is covalently bonded to three halogen atoms, and in order to complete its octet the aluminum needs to form only one coordinate covalent bond. As a result **dimeric** species having the formula Al_2X_6 and containing two AlX_3 molecules are formed. This is illustrated for Al_2Cl_6 as

Once again the metal atom has four electron pairs in its valence shell and is surrounded tetrahedrally by halogen atoms as shown in Figure 17.4. As a final note on this discussion, it is interesting that despite the close similarities found between Be and Al in many of their chemical properties, the physiological effects of these metals are quite different. The body is able to tolerate rather large amounts of aluminum; however, even very small quantities of beryllium, particularly inhaled as its oxide, are quite lethal.

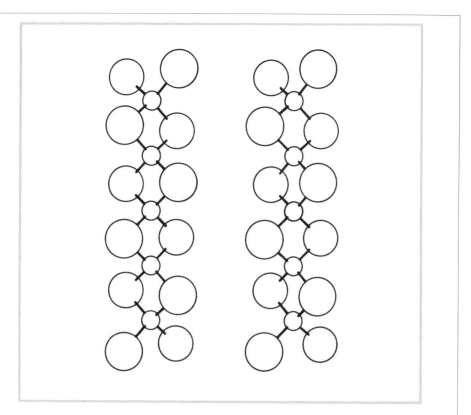

Figure 17.3

Structure of $(BeCl_2)_x$ (small spheres, Be; large spheres, Cl).

17.5
Oxidation States

As we know, the oxidation states that an element exhibits are governed by its electronic structure. Here we wish to examine the oxidation states that characterize a particular group and the trends in their ease of formation.

GROUP IA. The elements of this group are characterized by a single electron in an s orbital located outside a filled, noble gas core. Loss of this one electron gives the characteristic +1 oxidation state for the group. Since the single electron is loosely held, it is easily lost and the alkali metals are readily oxidized, as can be seen from their highly negative reduction potentials and low ionization energies in Table 17.1.

A close examination of these two quantities, however, reveals an apparent contradiction. Note that the ionization energy (I.E.) decreases as we proceed down within the group, suggesting that it becomes progressively easier to strip an electron from the atom as we go from Li to Cs. In Chapter 3 we saw that this is, in fact, expected. We would also anticipate that the reduction potentials should become more negative as the I.E. becomes smaller since the elements should become more easily oxidized. This trend is indeed followed from Na downward; however, Li has an $\mathscr{E}°$ that is more negative than Na (or any of the other alkali metals for that matter). Why is this so?

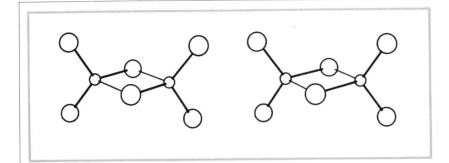

Figure 17.4
Structure of Al_2X_6 (small spheres, Al; large spheres, X).

The ionization energy, remember, is a measure of the ease with which a *gaseous* atom loses electrons to produce a gaseous cation. The reduction potential, on the other hand, is concerned with the loss of electrons by the *solid* metal to form the corresponding cation in *aqueous solution* where it is hydrated by the water molecules surrounding it. This latter process is more complex than simply removing an electron from the isolated metal atom. To understand the trends in the reduction potentials, we must break down the overall reaction into several steps. If we concentrate on the enthalpy changes involved in the reaction, we can construct the diagram in Figure 17.5. We see that the net enthalpy change is the sum of three energy terms. Two of these are endothermic, the sublimation energy, ΔH_{subl}, which is the energy needed to convert the solid into gaseous atoms, and the ionization energy that we have already examined. The third quantity, called the hydration energy, is strongly exothermic. It corresponds to the energy *released* when the cation is placed into the solvent cage where it is surrounded by the water dipoles oriented in such a way that their negative ends are directed at the positive ion (Figure 17.6).

Among the alkali metals the sublimation energy remains approximately constant as we descend the group while the ionization energy decreases. To

Table 17.1
Some Properties of the Alkali Metals

	Ionization Energies (kcal/mole)		$\mathscr{E}°$ (volts)	M^+ Radius (Å)
	First	Second		
Lithium	123.6	1735	−3.04	0.68
Sodium	118.0	1084	−2.71	0.98
Potassium	99.5	730	−2.92	1.33
Rubidium	95.9	631	−2.93	1.48
Cesium	89.2	539	−2.92	1.67

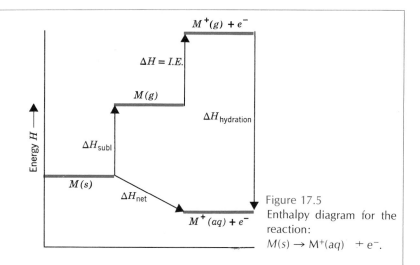

$M^+(g) + e^-$

$\Delta H = I.E.$

$M(g)$

$\Delta H_{hydration}$

Energy H →

ΔH_{subl}

$M(s)$

ΔH_{net}

$M^+(aq) + e^-$

Figure 17.5
Enthalpy diagram for the reaction:
$M(s) \rightarrow M^+(aq) + e^-$.

reach the peak on the energy diagram, then, we require the greatest amount of energy for Li and the least amount for Cs. However, because of its small size and high ionic potential, upon hydration Li$^+$ interacts much more strongly with the water dipoles than do any of the other Group IA ions. As a result the hydration energy of Li$^+$ is unusually large, being much greater than for the other M^+ ions in the group and, consequently, the net overall enthalpy change is most exothermic for Li. This in turn causes Li to be more easily oxidized than any other alkali metal. In other words, the extraordinarily high hydration energy of the small Li$^+$ ion more than compensates for its relatively high ionization energy and causes Li to have an unexpectedly high negative $\mathscr{E}°$.

Associated with the easy loss of electrons from the alkali metals is their interesting behavior in liquid ammonia. We have already seen that these metals are capable of reducing water to liberate H$_2$. Ammonia is not as easily reduced as water and, when placed into this solvent, alkali metals dissolve without reaction to form deep blue solutions. It is generally agreed that this color, which is identical for liquid ammonia solutions of all of the alkali metals (as well as for Ca, Sr, and Ba from Group IIA), is a result of the presence of free electrons that have become solvated by ammonia molecules. Ap-

Figure 17.6
Solvation of a cation by water dipoles.

Figure 17.7
Solvated electron in liquid ammonia.

parently, when the metal dissolves in NH_3, it loses its valence electron to become a cation. This electron becomes surrounded by NH_3 molecules arranged so that the positive ends of their dipoles are directed at the negatively charged electron, as shown in Figure 17.7, thereby stabilizing it through solvation. Solutions containing alkali metals in liquid ammonia are, as we would expect from the presence of readily available electrons, excellent reducing agents.

GROUP IIA. The elements in this group each have a filled s subshell outside a noble gas core. When they react they always lose both of these electrons to produce a +2 oxidation state (e.g., Mg^{2+}, Ca^{2+}). In general, the Group IIA elements are easily oxidized, and we see that the reduction potentials of the alkaline earths are all highly negative (Table 17.2), although they are generally more positive than those of their neighbors in Group IA. This is a reflection of the fact that, within a given period, the alkaline earth metals have larger ionization energies than the alkali metals and, hence, lose their electrons with more difficulty.

Within the group, Ca and the heavier elements are able to reduce water at room temperature according to the reaction

$$M + 2H_2O \longrightarrow M(OH)_2 + H_2$$

Table 17.2
Some Properties of the Alkaline Earth Metals

	Ionization Energies (kcal/mole)			$\mathscr{E}°$ (volts)	M^{2+} Radius (Å)
	First	Second	Third		
Beryllium	213.9	417.7	3529	−1.70	0.30
Magnesium	175.4	344.8	1838	−2.34	0.65
Calcium	140.9	273.8	1181	−2.87	0.94
Strontium	130.6	253.1	1005	−2.89	1.10
Barium	119.6	229.3	818.7	−2.90	1.29
Radium	122	234	—	−2.92	1.52

Magnesium, while unable to react with cold water, will liberate hydrogen from boiling water or steam. Beryllium, on the other hand, will not react with water at all.

An interesting question to consider is why the alkaline earth metals do not form compounds in which they exhibit a +1 state, even though much more energy must be expended to remove two electrons from the atom instead of only one. For example, why do we not observe the formation of CaCl but, instead, only $CaCl_2$?

To answer the question, we must once again look further than simply the ionization energies of the isolated Ca^0 atom and Ca^+ cation (140.9 and 273.8 kcal/mole, respectively). These quantities tell us that in the gas phase Ca^+ is, indeed, preferred energetically over Ca^{2+}. In the solid or in solution, however, there are additional factors to consider.

In aqueous solution the stabilizing influence of the hydration energy is dominant. Here we find that the +2 ion, because of its smaller size and higher charge, interacts much more strongly with the solvent than does the +1 ion. It therefore has a much larger hydration energy. This additional hydration energy for the Ca^{2+} ion, for example, more than compensates for the additional energy required to remove the second electron from the calcium atom. These energy changes are illustrated in Figure 17.8 where we see that even though it takes more energy to create the Ca^{2+} ion than to form the Ca^+ ion, the much larger hydration energy of Ca^{2+} causes the $Ca^{2+}(aq)$ species to be lower in energy (i.e., more stable) than the corresponding $Ca^+(aq)$ ion.

A similar argument can be made for the relative stabilities of solids containing Ca^+ or Ca^{2+}. In this case we find that it is the very large lattice energies for the M^{2+} salts, compared to M^+ compounds, that is able to compensate for the extra energy needed to remove the second electron from the alkaline earth atom.

A question we might ask at this point is, if the additional hydration energy for Ca^{2+} over Ca^+ can serve to stabilize the more highly charged ion, why do we not observe Ca^{3+} salts? To form Ca^{3+} compounds we would have

Figure 17.8
Relative enthalpy changes for the oxidation of $Ca°(s)$ to $Ca^+(aq)$ and $Ca^{2+}(aq)$.

to supply an additional amount of energy equal to the third ionization energy of calcium. The reaction

$$Ca^{2+} \longrightarrow Ca^{3+} + e^-$$

requires breaking into the closed-shell argon core of the Ca^{2+} ion, a process that requires the input of a huge amount of energy. For Ca, the third I.E. has a value of 1181 kcal/mole! This extremely large investment cannot be recovered by the hydration of the Ca^{3+}, even though ΔH_{hyd} is larger for Ca^{3+} than for Ca^{2+}. As a result the $Ca^{2+}(aq)$ ion is more stable (of lower energy) than the $Ca^{3+}(aq)$ species. These same arguments also apply to the other members of Group IIA, so that only a +2 state is observed.

GROUP IIIA. In this group we begin to see, for the first time, more than one stable oxidation state, at least among the heavier elements. While aluminum shows only a +3 oxidation state, corresponding to the loss of all three of its outer electrons, gallium, indium, and thallium exist in both a +1 state (brought about by the loss of the single outer-shell p electron) and a +3 state (which arises, as with Al, from the loss of all three valence electrons).

It is found that the relative stability of the lower oxidation state increases with increasing atomic number within the group. For instance, Ga^{3+} is more stable than Ga^+ while Tl^+ is more stable than Tl^{3+}. This trend in the stabilities of the high and low oxidation states persists in later groups, as we shall see, and results, it appears, from a decreasing stability of M—X bonds with increasing size of the metal atom. As a result, insufficient energy is recovered in bond formation to make up for the energy that must be expended to remove additional electrons.

GROUP IVA. Here we have only two elements that can be classed as true metals, Sn and Pb. Their outer-shell electron configurations correspond to s^2p^2 and two oxidation states, +2 and +4, are found. As in Group IIIA, the relative stability of the +2 state increases going down within the group. Thus, Sn^{2+} solutions are mild reducing agents owing to the fairly easy oxidation of Sn to the +4 state. By contrast, Pb^{2+} in aqueous media has virtually no tendency to become oxidized to the +4 state. In fact, Pb^{IV} compounds[4] tend to be rather good oxidizing agents in acid solution, thereby implying a strong tendency of Pb^{IV} to be converted to the more stable Pb^{II}. In Chapter 15 we saw that this is put to practical use in the lead storage battery where PbO_2 serves as the cathode while Pb serves as the anode. In the sulfuric acid electrolyte the net cell reaction is

$$Pb + PbO_2 + 2H_2SO_4 \longrightarrow 2PbSO_4 + 2H_2O$$

GROUP VA. Only one true metal, bismuth, is found in this group. It has an outer-shell configuration of $6s^26p^3$ and, like the elements at the bottom of the preceding two groups, it too forms compounds in two oxidation states. These consist of a +3 state, from the loss of the three p electrons, and a +5 state that

---

[4] We shall see later in this chapter that highly charged ions tend to become involved in a significant degree of covalent bonding. It is doubtful whether a true Pb^{4+} ion can actually exist in chemical compounds and therefore the formal oxidation state of the lead, in roman numerals, is used instead of an ionic charge.

corresponds to the loss (at least in principle) of all five valence electrons. In actual fact, no Bi^{5+} ion exists, and the assignment of a +5 oxidation state to Bi is really a result of the rules that we use in computing oxidation numbers.

As with other representative elements at the bottom of a group, the lower oxidation state of Bi is more stable than the higher one. For example, the oxide Bi_2O_3 forms readily, but the formation of the higher oxide, Bi_2O_5, requires severe oxidizing conditions. Because of the difficulty encountered in preparing them, few compounds containing Bi in the +5 state are known and those that do exist, for example $NaBiO_3$, have a strong tendency to acquire electrons and are very powerful oxidizing agents.

Frequently we tend to think of metal compounds with nonmetals as essentially ionic in character and, in many instances, this is indeed true. However, we should remember that no bond is purely ionic and, most important, there is a gradual transition between ionic and covalent bonding. What we wish to examine in this section are some of the factors that allow us to predict trends in ionic (or covalent) bonding within a series of compounds.

The basic concept that we shall expand upon is that the electron cloud surrounding an atom or ion is somewhat soft and "mushy" and can be distorted when in the presence of another ion. Consider, for example, the two ions in Figure 17.9a. The positive ion here is depicted as smaller than the negative one, reflecting the fact that cations are generally smaller than anions. Because of its positive charge, the cation tends to draw the electron cloud of the anion toward itself. This gives an electron distribution somewhat like that in Figure 17.9b in which the anion has been distorted so that in addition to being an ion, it is now something of a dipole as well. We say, therefore, that the electron cloud of the anion has been **polarized** by the cation.

When the anion becomes polarized, electron density is drawn into the region between the two nuclei. Since a covalent bond consists of electrons shared *between* nuclei, the polarization of the anion results in the partial formation of a covalent bond. The extent to which this polarization occurs determines the degree of covalent character in the ionic bond. The limit of this, of course, is where the electrons of the anion are pulled toward the cation so much that a nonpolar covalent bond is formed.

At this point we must ask, what factors influence the degree to which anion polarization occurs? As we might expect, one of these is the ionic potential of the cation, that is, the ratio of its charge to its radius. We have said before that this quantity is a measure of the "concentration" of charge, and cations in which the charge is highly concentrated are better able to distort a neighboring electron cloud than cations in which the positive charge is thinly spread and diffuse.

(a) Unpolarized anion and cation.

(b) Electric charge on anion distorted by cation.

$(\delta-)$ $(\delta+)$

Figure 17.9
Polarization of an anion by the positive charge on the cation. (a) Unpolarized anion and cation. (b) Electric charge on anion distorted by cation.

The ionic potential increases with decreasing size of the cation and, hence, *small ions are more effective at producing anion polarization than are large ones*. The ionic potential also increases as the charge on the cation gets larger, so that *for a given size, multiply charged ions produce a greater degree of polarization than do singly charged ions*. We can see how these ideas apply if we examine some examples.

As we proceed down within a group, the ions of a given charge become progressively larger and their ionic potentials therefore decrease. In Group IA, for instance, we find the lithium ion to be very small, Na^+ somewhat larger, K^+ larger still, and so on. Since Li^+, because of its size, has the largest ionic potential in this series of ions, it should also form compounds with the greatest degree of covalent character. As it happens, among the alkali metals Li is in fact the only element that forms compounds with organic molecules (called organolithium compounds) that have properties usually associated with covalently bonded substances, such as low melting points and appreciable solubilities in nonpolar solvents. The corresponding compounds of the other alkali metals, however, are predominantly ionic.

In Group IIA we would expect the degree of covalent character to be greater because of the greater charge on the ions of the group. It has already been mentioned that Be compounds are all covalent to at least some degree and that there is little evidence for the existence of a true Be^{2+} ion. Magnesium compounds, as pointed out earlier, are in many ways similar to Li compounds, and Mg forms organomagnesium compounds which, like organolithium compounds, exhibit properties characteristic of covalent substances. However, Mg differs from Be in that there are many compounds of Mg, such as the oxide for example, which are distinctly ionic. Finally, the metals below Mg in Group IIA, because of their larger size and smaller ionic potentials, form compounds that are essentially ionic. In Group IIA, as in the first group, we see that as we descend the group there is a trend toward decreasing covalent character in the bonds.

The effect of charge is also seen for compounds such as $SnCl_4$ and $PbCl_4$. These, at first glance, appear to be salts of the cations Sn^{4+} and Pb^{4+}; however, each is quite covalent. For example, both are *liquids* with low melting points (melting point $SnCl_4 = -33°$ C, $PbCl_4 = -15°$ C). In fact, we might compare them with the other Group IVA tetrahalides CCl_4 (melting point $= -23°$ C), $SiCl_4$ (melting point $= -70°$ C) and $GeCl_4$ (melting point $= -49.5°$ C). The very high ionic potential of a $+4$ ion thus appears to polarize an anion such as Cl^- to the extent that covalently bonded molecules result.

In addition to vertical trends in covalent character, we can also look at variations in the nature of the bonding as we move from left to right within a period. Here, as you know, the cations become progressively smaller and more highly charged, for example, Li^+, Be^{2+}, and B^{3+}, etc. Consequently the ionic potential increases rapidly and the degree of covalent bonding in their compounds does so too. Thus LiCl is predominantly ionic, $BeCl_2$ is covalent, and so are the remaining halides as we continue across period 2 (i.e., BCl_3, CCl_4, etc.).

In period three the cations are larger than those in the preceding period, with the result that the covalent character of the bonds shows up later in the period. For example, NaCl and $MgCl_2$ are both essentially ionic, and it is not until Group IIIA, with $AlCl_3$, that the halides become largely covalent. We might predict, however, that with a given anion the bonds to Mg^{2+} would be less ionic than those to Na^+. Recall, for instance, that covalent organomagnesium compounds exist while the corresponding compounds with sodium are ionic.

The type of outer-shell electron configuration possessed by the cation also affects its ability to polarize an anion. Cations with pseudonoble gas configurations of 18 electrons ($ns^2np^6nd^{10}$), such as those formed from the metals immediately following a transition series, appear to have higher ionic potentials than similarly sized ions having the same net charge but with only an octet of electrons. This is because the d electrons that are added to the atom as the transition series is crossed do not completely shield the nuclear charge that has also increased. The net effect is that a cation with a pseudonoble gas structure presents to a neighboring anion an *effective* positive charge that is actually somewhat greater than its net ionic charge. *As a result these cations behave as if their effective ionic potentials were higher than those calculated from their net ionic charges and, consequently, they are more effective at distorting the electron cloud of the anion.*

This effect is illustrated in Table 17.3 where the melting points of several metal chlorides are presented. We associate high melting points with a high degree of ionic bonding where the attractive forces between particles within the solid are very strong. On the other hand, low melting points are characteristic of substances that contain covalently bonded molecules in which the attractive forces between neighboring molecules are relatively weak. As a rough guide, then, we can use melting points of compounds of similar composition as a measure of their degree of covalent bonding and we expect their melting points to decrease as the bonding becomes progressively more covalent.[5]

Table 17.3
Melting Points of Some Metal Chlorides

Pre-transition Elements			Post-transition Elements		
M-Cl distance (Å)		Compound	Melting Point (°C)		
M-Cl distance (Å)		Compound	Melting Point (°C)	M-Cl distance (Å)	Compound	Melting Point (°C)

Pre-transition Elements				Post-transition Elements			
M-Cl distance (Å)		Compound	Melting Point (°C)	M-Cl distance (Å)		Compound	Melting Point (°C)
Na-Cl	2.81	NaCl	800	Ag-Cl	2.77	AgCl	455
Rb-Cl	3.29	RbCl	715	Tl-Cl	3.31	TlCl	430
Ca-Cl	2.74	$CaCl_2$	772	Cd-Cl	2.76	$CdCl_2$	568
Sr-Cl	3.02	$SrCl_2$	873	Pb-Cl	3.02	$PbCl_2$	501
				Hg-Cl	2.91	$HgCl_2$	276
Mg-Cl	2.46	$MgCl_2$	708	Zn-Cl	2.53	$ZnCl_2$	283
				Sn-Cl	2.42	$SnCl_2$	246

[5] This generalization breaks down for polymeric covalent structures such as those found in quartz or diamond. In this case, because of the size of the giant molecules, very high melting points are found even though the bonding is covalent.

In Table 17.3 we compare salts having similar metal-chlorine bond distances. The salts on the left of the table have cations with an octet in their outer shell while those on the right contain cations with either a pseudonoble gas $s^2p^6d^{10}$ structure (Ag^+, Zn^{2+}, Cd^{2+}, Hg^{2+}) or a pseudonoble gas structure plus two electrons (Tl^+, Pb^{2+}, Sn^{2+}). The effectiveness of these post-transition metal ions at polarizing the chloride ion is revealed by the consistently lower melting points of their salts. Since on each line we have cations of essentially the same effective size (as shown by nearly the same M-Cl distances), the difference in covalence must be related to their effective ionic potentials, being less for elements that precede a transition series than for elements at the end of the series or for those in Groups IIIA and IVA that follow the transition elements.

The covalent character of a metal-nonmetal bond is also influenced by the nature of the anion. For a given cation the bond becomes less ionic as the anion becomes more easily deformed; therefore we must examine trends in anion polarizability. In general, it is found that *for a given charge, anions become more readily distorted as they become larger.* In a large anion the outer electrons are further from the nucleus and are spread more thinly. Consequently they are able to be influenced to a greater degree by a cation. Thus among the halides F^-, Cl^-, Br^-, and I^- we expect to find the greatest amount of covalent character in compounds in which a metal is combined with I^-. This is indeed the case as demonstrated, for example, by the colors of the silver halides (Table 17.4).

Ionic compounds generally do not abosrb light in the visible region of the spectrum. Therefore, they appear white, or colorless, when bathed in white light (e.g., NaCl is a colorless crystalline substance). The absorption that does take place with ionic compounds occurs in the shorter wavelength (higher frequency, higher energy) ultraviolet region. The energy absorbed by the compound is used to shift an electron from the anion to the cation, for example,

The absorption of a band of wavelengths from the ultraviolet "rainbow" produces a **charge transfer absorption band.**

As the bond between the metal and nonmetal becomes more covalent

Table 17.4
Colors of Silver Halides

Compound	Color	Anion Radius (Å)
AgF	White	1.36
AgCl	White	1.81
AgBr	Cream	1.95
AgI	Yellow	2.16

(i.e., as electron density shifts away from the anion in the direction of the cation) less energy is required to produce the charge transfer. As a result, the absorption band shifts toward the visible region of the spectrum, where the removal of some colors from white light gives rise to reflected colors that represent the remainder of the visible spectrum.

The depth of color in compounds (particularly those of the nontransition elements) can often be taken to be a measure of the degree of covalent character in the metal-nonmetal bonds. Among the silver halides we see that as the anion grows in size, the compounds become progressively deeper in color indicating a progressive rise in covalent character in the Ag—X bonds.[6]

We also find a similar relationship between ionic character and anion size if we compare metal oxides ($r_{O^{2-}} = 1.40$ Å) and sulfides ($r_{S^{2-}} = 1.84$ Å). Both of these ions are colorless as evidenced by the fact that both Na_2O and Na_2S are colorless. However, with aluminum we find Al_2O_3 to be white while Al_2S_3 is yellow, suggesting a greater degree of ionic bonding in the oxide. This is further supported by comparing the melting points of these two compounds; Al_2O_3 melts at 2045° C while Al_2S_3 melts at a much lower temperature, 1100° C. Additional examples are provided in Table 17.5.

For anions of a given size, polarization also increases with an increase in the charge on the anion. We can compare, for example, compounds containing chloride ($r = 1.81$ Å) and sulfide ($r = 1.84$ Å) as shown in Table 17.6. In each case the sulfide has a deeper color than the corresponding chloride salt, indicating that the compounds containing the more highly charged sulfide ion are more covalent. We also see that in general most metal sulfides, except those of the alkali and alkaline earth metals, are deeply colored and possess a quite substantial degree of covalent bonding.

17.7

Hydrolysis

The concepts developed in the last section, you may have noticed, are very similar to those presented in Chapter 13, where we discussed the relative

Table 17.5

Properties of Some Metal Oxides and Sulfides

Oxides				Sulfides		
	Color	Melting Point (°C)			Color	Melting Point (°C)
Al_2O_3	White	2045		Al_2S_3	Yellow	1100
Ga_2O_3	White	1900		Ga_2S_3	Yellow	1255
Sb_2O_3	White	656		Sb_2S_3	Yellow-red	550
Bi_2O_3	Yellow	860		Bi_2S_3	Brown-black	685 decomp.
SnO_2	White	1127		SnS_2	Yellow	882
ZnO	White	1975		ZnS	White	1850

[6] The salts NaF, NaCl, NaBr, and NaI, which are predominantly ionic, are colorless. This indicates that the halide ions themselves are colorless. The color of the AgX compounds is therefore a reflection of covalent bonding.

Table 17.6
Colors of Metal Chlorides and Sulfides

Chloride ($r = 1.81$ Å)		Sulfide ($r = 1.84$ Å)	
AgCl	White	Ag_2S	Black
CuCl	White	Cu_2S	Black
AuCl	Yellow	Au_2S	Brown-black
$CdCl_2$	White	CdS	Yellow
$HgCl_2$	White	HgS	Black or red, depending on crystal structure
$PbCl_2$	White	PbS	Black
$SnCl_2$	White	SnS	Black
$AlCl_3$	White	Al_2S_3	Yellow
$GaCl_3$	White	Ga_2S_3	Yellow
$BiCl_3$	White	Bi_2S_3	Brown-black

degrees of acidities of Brønsted acids. In fact, they are essentially identical, and we can use them to understand trends in metal ion hydrolysis.

Let us consider, for example, the hydrolysis of a metal ion, M^{n+}. In water, as you may recall, the ion will be surrounded by water dipoles oriented with their negative ends toward the cation. The positive charge of the ion will distort the electron cloud surrounding the oxygen nucleus of a neighboring water molecule, inducing a certain degree of covalent bonding to the M^{n+} —OH_2 bond. The electron density drawn into the M—O bond, however, tends to come from the O—H bonds of the H_2O. As a result the hydrogen atoms become more positively charged and therefore become more easily removed as H^+. We see then that the extent of hydrolysis depends on the amount of charge removed from the O—H bonds and transferred to the M—O bonds. However, the degree of covalent character produced in the M—O bonds is determined, in turn, by the ionic potential of the cation. Consequently, we expect that the tendency for the $M(H_2O)_x^{n+}$ species to lose protons (i.e., the tendency for the cation to hydrolyze) will be greatest for ions having large ionic potentials.

This is, in fact, precisely what is found. The ions of Groups IA and IIA, for the most part, have virtually no tendency to hydrolyze. Beryllium is an exception because of its small size and high charge and, hence, high ionic potential. Aqueous solutions of Be salts are therefore acidic because of the reaction

$$Be(H_2O)_4^{2+} + H_2O \longrightarrow Be(H_2O)_3(OH)^+ + H_3O^+$$

Of the ions of the metals in Groups IIIA, IVA, and VA, almost all exhibit a substantial degree of hydrolysis. In our earlier discussions (Chapter 13) we saw the hydrolysis of Al^{3+}, which probably exists as $Al(H_2O)_6^{3+}$ in water. The elements below Al, in their +3 states, have a pseudonoble gas structure and, consequently, have high ionic potentials just as Al^{3+} does. As a result the Ga^{3+}, In^{3+} and Tl^{3+} ions tend to be rather extensively hydrolyzed too. Because of their much lower ionic potentials, the +1 states of these three metals are expected to hydrolyze to a much lesser degree. This is true for Tl^+, compared to Tl^{3+}; since the other two, Ga^+ and In^+, do not exist in aqueous solution we can only speculate about their behavior.

The aqueous chemistry of the Group IVA metals too, is influenced by hydrolysis. Tin(II) solutions, for example, are extensively hydrolyzed and Pb^{2+}, although larger in size and therefore of lower ionic potential than Sn^{2+}, is also hydrolyzed to a degree. In the higher oxidation state characteristic of the metals of this group we might expect even greater hydrolysis and, indeed, this occurs.

Moving to the right again, we find that Bi^{3+}, because of its high charge, is very extensively hydrolyzed. In this case, in fact, there is evidence for the species BiO^+ (called the bismuthyl ion) produced by the reaction

$$Bi^{3+} + H_2O \longrightarrow BiO^+ + 2H^+$$

Thus the BiO^+ ion can be looked upon as the product of very severe hydrolysis in which two protons of a water molecule are lost entirely. We shall encounter ions of this type again in our discussion of the transition elements where we meet ions of high charge (and hence high ionic potential).

Review Questions and Problems

17.1 On a piece of paper, sketch out the shape of the periodic table and roughly mark off those regions where we find the metals, nonmetals, and the metalloids.

17.2 Write the symbol for (a) a representative element, (b) an alkali metal, (c) an alkaline earth metal, (d) an inner transition element.

17.3 What are three physical properties that are generally characteristic of metals? What chemical characteristics do metals possess in common?

17.4 Give the outer shell electron configuration of each of the following: (a) Ca, (b) Tl, (c) Bi, (d) Cs, (e) Sn.

17.5 Why, under ordinary conditions, do metals not form simple molecular species with each other by sharing electrons in covalent bonds?

17.6 How does the band theory of solids (Chapter 7) differentiate between conductors, semiconductors, and insulators?

17.7 How does the metallic character of the elements depend on electronegativity? What vertical and horizontal trends in metallic character exist in the periodic table? Illustrate these trends for the elements in the second period and in Group IVA.

17.8 What is meant by amphoteric? Write chemical equations to illustrate the amphoteric behavior of beryllium and aluminum.

17.9 Using Table 10.1, predict which of the oxides, CuO or ZnO, should be most stable with respect to thermal decomposition to the free elements.

17.10 From the data in Tables 10.1 and 10.4, calculate the temperature at which the thermal decomposition of ZnO should become feasible.

17.11 Write chemical reactions illustrating the chemical reduction of a metal compound using (a) carbon, (b) hydrogen, (c) sodium, (d) electrolysis.

17.12 Write chemical equations for the reactions that occur when the alkali metals are exposed to molecular oxygen.

17.13 Write chemical equations for the hydrolysis of (a) an oxide, (b) a peroxide, (c) a superoxide.

17.14 List three ways that lithium and magnesium are similar to one another.

17.15 To what do we attribute the diagonal

relationships that exist between Li and Mg, and Be and Al?

17.16 Compare the structures of $BeCl_2$ and $AlCl_3$. Why doesn't $AlCl_3$ form a linear chain polymer as $BeCl_2$ does? What type of hybrid orbitals would be used by the Be and Al in these species?

17.17 To what do we attribute the unusually high negative reduction potential of lithium? Why do we not observe Li^{2+} in compounds?

17.18 Given the following thermodynamic data, calculate the hydration energy for the Na^+ ion

$$\Delta H_f^\circ \text{ of } Na^+(aq) = -57.28 \; \frac{kcal}{mole}$$

$$\Delta H_{atom}^\circ \text{ of } Na = 25.98 \; \frac{kcal}{mole}$$

I.E. of $Na =$

$$118.0 \; \frac{kcal}{mole} \text{ (from Table 17.1)}$$

17.19 Using the data for the atomization energy of Na (Problem 17.18) and the first and second ionization energies for Na in Table 17.1, compute the value of the hydration energy required to produce a negative ΔH_f for $Na^{2+}(aq)$.

17.20 What is the active reducing agent in solutions of alkali metals in liquid ammonia? What prevents the formation of stable solutions of alkali metals in water?

17.21 Why do the alkaline earth metals only form compounds in which they exhibit a +2 oxidation state? On this basis, and using the data in Table 3.8 on page 81 can you suggest why Al only exhibits a +3 oxidation state and not a +4 state?

17.22 Among the representative elements, why are the lower oxidation states preferred by elements at the bottom of a given group in the periodic table?

17.23 Predict the better oxidizing agent in each of the following pairs.
(a) Bi_2O_3 or Bi_2O_5
(b) SnO_2 or SnO
(c) PbO_2 or PbO
(d) $TlCl$ or $TlCl_3$

17.24 Predict which of the following exhibit the greatest degree of covalent bonding.
(a) Bi_2O_3 or Bi_2O_5
(b) PbO or PbS
(c) CaO $(r_{Ca^{2+}} = 0.99 \; Å)$ or SnO $(r_{Sn^{2+}} = 0.93 \; Å)$
(d) Na_2S or MgS
(e) $LiCl$ $(r_{Cl^-} = 1.81 \; Å)$ or Li_2S $(r_{S^{2-}} = 1.84 \; Å)$

17.25 Which of the following compounds should have the highest melting point?
(a) $CaCl_2$ or $SnCl_2$
(b) $BeCl_2$ or BeF_2
(c) $BeCl_2$ or $MgCl_2$
(d) $SnCl_2$ or $SnCl_4$

17.26 For each pair, predict which compound should be most deeply colored.
(a) $HgCl_2$ $(r_{Cl^-} = 1.81 \; Å)$ or HgS $(r_{S^{2-}} = 1.84 \; Å)$
(b) Ag_2O or Ag_2S
(c) SnO or PbO
(d) SrS $(r_{Sr^{2+}} = 1.12 \; Å)$ or PbS $(r_{Pb^{2+}} = 1.20 \; Å)$

17.27 For each of the pairs below, pick the ion that should undergo the greatest degree of hydrolysis. Explain your choice in each case.
(a) Li^+ or Be^{2+}
(b) Be^{2+} or Ca^{2+}
(c) B^{3+} or Al^{3+}
(d) Tl^+ or Tl^{3+}
(e) Sn^{4+} or Sn^{2+}

17.28 When a solution of $BiCl_3$ in concentrated HCl is diluted with water, a precipitate having the formula BiOCl is formed. Explain this reaction in terms of hydrolysis of Bi^{3+}.

17.29 Construct a table showing the oxidation states found for each of the representative metals. Look up the reduction potentials for each of these species (i.e., for the reaction $M^{n+} + ne^- \rightarrow M$). What vertical and horizontal trends exist in the standard reduction potentials? What atomic properties do these trends correlate with?

18
Chemistry of the Representative Elements: Part II

In this chapter we shall conclude our discussion of the representative elements by considering the chemistries of the metalloids and nonmetals. These two classes of elements are conveniently treated together because in many ways their chemical properties are similar, owing to the covalent nature of many (if not most) of their compounds.

As in Chapter 17 we shall continue to concentrate on points of similarity and trends in behavior which allow us to correlate more efficiently the factual descriptive information. In this vein we shall discuss, in addition to the free elements themselves, three classes of compounds formed by the metalloids and nonmetals, specifically their compounds with hydrogen, oxygen, and the halogens.

Among the various aspects of the chemistry of these elements that we shall encounter in this chapter, one of the most fascinating is the variety of structures exhibited by both the free elements and their compounds. This is particularly pronounced for these elements because of the predominance of covalence among their compounds. Covalent bonds, you recall, are strongly directional in nature, and we shall come to see that the basic structural units formed by atoms covalently bound to one another can be combined in various ways to give quite complex overall structures. What you should note, however, is that even the most complex structures can be understood by considering them in terms of the *simple* structural units of which they are composed.

18.1
The Free Elements

There are two aspects of the elemental forms of the metalloids and nonmetals that we shall examine. The first of these, discussed in this section, is the way in which these elements may be obtained in their free state. The second, which we shall attend to in the next section, is the variety of structural forms that they exhibit.

In the last chapter we made the rather broad generalization that metals, in their compounds, usually exist in only positive oxidation states. As it happens, this statement can be extended for the most part also to the metalloids, since most of their compounds contain the metalloid combined with a nonmetal, either in a molecular structure such as SiO_2 or in an oxoanion such as is found in the silicates. In these combinations the metalloids have a lower electronegativity than the nonmetal and, consequently, exist in positive oxidation states. As a result, the free elements are produced by the reduction of metalloid compounds, usually using either carbon or hydrogen as a chemical reducing agent. For example, boron is obtained by passing a mixture of BCl_3 vapor and hydrogen gas over a hot wire, upon which occurs the reaction

$$2BCl_3 + 3H_2 \longrightarrow 2B + 6HCl$$

On the other hand, elemental silicon is produced by heating SiO_2 with carbon in an electric furnace where the reaction

$$SiO_2 + C \longrightarrow Si + CO_2$$

takes place spontaneously once the temperature exceeds approximately 3000° C (below this temperature the reverse reaction is actually favored).

The remaining metalloids may be obtained from their oxides by heating them with either carbon or hydrogen; for example,

$$GeO_2 + C \longrightarrow Ge + CO_2$$
$$GeO_2 + 2H_2 \longrightarrow Ge + 2H_2O$$

Similarly, we also have

$$2As_2O_3 + 3C \longrightarrow 4As + 3CO_2$$
$$2Sb_2O_3 + 3C \longrightarrow 4Sb + 3CO_2$$

and

$$As_2O_3 + 3H_2 \longrightarrow 2As + 3H_2O$$
$$Sb_2O_3 + 3H_2 \longrightarrow 2Sb + 3H_2O$$

Unlike the metals and metalloids, it is difficult to make general statements about the preparation of the nonmetals. Some of them, such as the noble gases, are always found uncombined in nature. Others, while present in many naturally occurring compounds, also are found extensively in the free state as well. For instance, our atmosphere is composed primarily of elemental nitrogen, N_2 (about 80%) and oxygen, O_2 (about 20%). While both nitrogen and oxygen are found in a vast number of compounds, certainly their most economical source is simply the air itself. The atmosphere is also the major source of the noble gases, even though they are present in only very small quantities. Of the noble gases, only helium and radon are not obtained primarily from the atmosphere. Helium is found in gaseous deposits beneath the earth's crust where it has collected after being produced by the capture of electrons by α-particles (He nuclei) that are formed during the radioactive decay of elements such as uranium. Radon itself is radioactive and is produced by the radioactive decay of still heavier elements. Since radon spontaneously decomposes into other elements, it occurs in only minute quantities in nature.

Sulfur and carbon are two other elements that occur naturally in both the combined and free states. There are, for instance, many naturally occurring sulfates (e.g., $BaSO_4$, $CaSO_4 \cdot 2H_2O$) and sulfides (FeS_2, CuS, HgS, PbS, ZnS). In the free state sulfur has been found in large underground deposits from which it is mined using a rather clever method developed by an engineer, Herman Frasch in 1890. The process, which has come to bear his name, involves forcing superheated water under pressure into the sulfur deposit, causing the sulfur to melt. Once molten, the sulfur-water mixture is then foamed to the surface using compressed air. Huge quantities of this element, mined by the Frasch process, are used annually to produce such important industrial chemicals as sulfuric acid.

Turning to carbon, we find that most of the naturally occurring compounds are carbonates, for example, limestone ($CaCO_3$). In the free state carbon is found in two forms, diamond and graphite.

Since nonmetals combine with each other as well as with metals, no strict generalizations can be made concerning their recovery from compounds. When combined with a metal, the nonmetal is found in a negative oxidation state and, therefore, in order to generate the free element an oxidation must be brought about. For example, the halogens Cl_2, Br_2, and I_2 can be conveniently prepared in the laboratory in this manner by reacting one of their salts with an oxidizing agent such MnO_2 in acid solution, as in the equation

$$2X^- + MnO_2 + 4H^+ \longrightarrow X_2 + Mn^{2+} + 2H_2O$$

where $X = Cl$, Br, or I.

The halogens themselves can also serve as oxidizing agents in replacement reactions. Since the tendency to acquire electrons decreases as we proceed downward in a group, the ability of the halogen to serve as an oxidizing agent decreases too. This is seen in their reduction potentials (Table 18.1), which decrease from fluorine to iodine. As a result, a given halogen is a better oxidizing agent than the other halogens below it in Group VIIA and is able to displace them from their binary compounds with metals. Thus F_2 will displace Cl^-, Br^-, and I^-, while Cl_2 will displace only Br^- and I^- but not F^-, and so on. This is illustrated by these typical reactions.

$$F_2 + \begin{Bmatrix} 2NaCl \\ 2NaBr \\ 2NaI \end{Bmatrix} \longrightarrow 2NaF + \begin{Bmatrix} Cl_2 \\ Br_2 \\ I_2 \end{Bmatrix}$$

$$Cl_2 + \begin{Bmatrix} 2NaBr \\ 2NaI \end{Bmatrix} \longrightarrow 2NaCl + \begin{Bmatrix} Br_2 \\ I_2 \end{Bmatrix}$$

$$Cl_2 + NaF \longrightarrow \text{no reaction}$$

$$Br_2 + 2NaI \longrightarrow 2NaBr + I_2$$

$$Br_2 + \begin{Bmatrix} NaF \\ NaCl \end{Bmatrix} \longrightarrow \text{no reaction}$$

Table 18.1
Reduction Potentials of the Halogens

	$\mathscr{E}°$ (volts)
$F_2 + 2e^- \rightleftharpoons 2F^-$	2.87
$Cl_2 + 2e^- \rightleftharpoons 2Cl^-$	1.36
$Br_2 + 2e^- \rightleftharpoons 2Br^-$	1.09
$I_2 + 2e^- \rightleftharpoons 2I^-$	0.54

Fluorine, because of its position as the most powerful chemical oxidizing agent, can only be obtained by electrolytic oxidation. This process must be carried out in the absence of water, since water is more easily oxidized than the fluoride ion and, if H_2O is present, the reaction

$$2H_2O \longrightarrow O_2 + 4H^+ + 4e^- \qquad \mathscr{E}° = -1.23 \text{ volts}$$

will occur in preference to

$$2F^- \longrightarrow F_2 + 2e^- \qquad \mathscr{E}° = -2.87 \text{ volts}$$

In practice, a molten mixture of KF and HF, which has a lower melting point than KF alone, is electrolyzed, producing H_2 at the cathode and F_2 at the anode.

Nonmetals can also be extracted from their compounds by reduction if the nonmetal happens to exist in a positive oxidation state. We have seen how this applies to the metalloids, and the general procedure employed there can be used also for the nonmetals. For instance, elemental phosphorus is produced from a phosphate such as $Ca_3(PO_4)_2$ where it is found in the $+5$ state. In this case, the $Ca_3(PO_4)_2$ is heated with a mixture of carbon and SiO_2 (sand).

$$Ca_3(PO_4)_2 + 3SiO_2 + 5C \longrightarrow 3CaSiO_3 + 5CO + 2P$$

In this reaction the SiO_2 is present to combine with the calcium. Since the reduction of SiO_2 by carbon requires extremely high temperatures, only the phosphorus is reduced to the element.

We have seen that when atoms react they have a strong tendency to acquire a noble gas electron configuration, either by mutual sharing of electrons or by electron transfer. Except for the noble gases themselves, the nonmetals and metalloids have incomplete valence shells and, therefore, their individual atoms tend to combine until a noble gas structure is achieved. For example, hydrogen atoms unite to form H_2 so that each H atom acquires the He configuration

$$H\cdot + H\cdot \longrightarrow H:H$$

In a similar fashion, the atoms of the other nonmetals and metalloids combine to give structures containing two or more atoms.

One of the controlling factors in determining the complexity of the molecular structures that the nonmetals and metalloids exhibit is the ability of the second period elements, C, N, and O, to enter into multiple bonding by way of overlap of adjacent p orbitals ($p\pi$-$p\pi$ double and triple bonds) as described in Chapter 16. As it turns out, *there is very little tendency for elements in the third and succeeding periods to form multiple bonds of this type.* Presumably this is because their larger size prevents them from approaching each other too closely. Consequently, for these heavier elements, the sideways overlap of p orbitals, required for $p\pi$-$p\pi$ multiple bonding, is not very effective, and the formation of two (or three) separate single bonds tends to be preferred energetically over the formation of one double (or triple) bond. The net result, regardless of how we justify it, is that *elements of the second period are able to form multiple bonds fairly readily, while the elements below them in the following periods have a tendency to prefer single bonds.* This phenomenon is particularly striking when we examine the structures of the elements in their free state.

We have seen that some of the elements in the second period form stable diatomic molecules, for example, N_2, O_2, and F_2. In these three cases the valence shells of the atoms are completed by sharing three, two, and one electron, respectively, with nitrogen and oxygen participating in $p\pi$-$p\pi$ bonding. It is because these elements are capable of achieving a stable electron configuration by sharing electrons with a *single* neighbor that they are able to form simple diatomic molecules in their elemental state.

Oxygen, in addition to forming the stable species O_2, also can exist in another molecular form, O_3, called **ozone,** the structure of which may be represented as a resonance hybrid

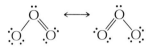

This unstable molecule (as evidenced by its endothermic heat of formation from O_2, $\Delta H_f = +34$ kcal/mole) can be generated by the passage of an electric discharge through ordinary O_2, and its pungent odor can often by detected in the vicinity of electrical equipment. It is also formed in limited quantities in the upper atmosphere by the action of ultraviolet radiation from the sun on O_2. It is blue in color and is exceedingly reactive.

Ozone is currently believed to be one of the major constituents of photochemical smog, a type of air pollution that has accompanied the increased use of the internal combustion engine.

When an automobile engine operates, it gives off a mixture of gases in its exhaust, including unburned hydrocarbons from the fuel and small amounts of NO produced from the oxidation of atmospheric N_2. The series of reactions that produce the smog are thought to include the following:

1. Oxidation of NO by atmospheric oxygen

$$2NO + O_2 \longrightarrow 2NO_2$$

2. A photochemical reaction in which a molecule of NO_2 absorbs a photon (energy $= h\nu$)

$$NO_2 + h\nu \longrightarrow NO + O$$

3. Reaction of atomic oxygen with molecular oxygen to give ozone

$$O + O_2 \longrightarrow O_3$$

4. A host of other reactions in which hydrocarbons react with O_3 and nitrogen oxides to produce a range of unpleasant products.

A characteristic property of photochemical smog is its oxidizing properties that result primarily from high concentrations of O_3.

The existence of an element in more than one form, either as the result of differences in molecular structure, as with O_2 and O_3, or as a consequence of differences in the packing of atoms or molecules in the solid, is a phenomenon called **allotropism.** Oxygen is only one of several nonmetals that exist in different allotropic forms, although the phenomenon is not limited to the nonmetals since tin, an element with mostly metallic properties, exhibits a metallic lattice at high temperature and a nonmetallic lattice at low temperature.

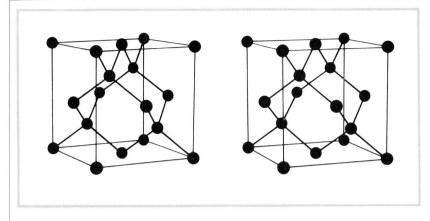

Figure 18.1
Structure of diamond.

If we next turn our attention to carbon, also a second period element, we see that it must share four electrons to complete its octet. Now there is just no way for carbon to form a "quadruple bond" and a simple, C_2 species is not stable under ordinary conditions. Instead, carbon tends to complete its octet in either of two ways so that two allotropic forms of elemental carbon are found. One of these is diamond. In diamond (see Chapter 7) each carbon atom is covalently bonded to four others located at the corners of a tetrahedron. Each of those atoms, in turn, are bonded to three more, and so on, as illustrated in Figure 18.1. In this fashion a three-dimensional network is created so that a diamond crystal (e.g., a gem quality diamond) consists of a huge number of carbon atoms covalently bonded together in one gigantic molecule. Since breaking a diamond crystal involves rupturing a very large number of covalent bonds, diamond is very hard.

In the second allotrope of carbon, graphite, the atoms are arranged in the form of hexagonal rings fused together in large planar sheets, perhaps somewhat reminiscent of chicken wire (Figure 18.2). Each carbon atom is surrounded by three nearest neighbors located at angles of 120° from one another, and the molecular framework is therefore based on σ-bonds produced by overlap of sp^2 hybrid orbitals on the carbon atoms. On each of the carbon atoms throughout the entire structure there remains an unhybridized p orbital, each containing one electron, and these p orbitals are situated ideally for $p\pi$-$p\pi$ overlap. The result, then, is a huge delocalized π electron cloud extending across the graphite sheet above and below the plane of the carbon

Figure 18.2
Sigma-bond framework of a graphite sheet.

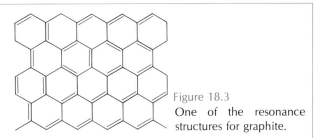

Figure 18.3

One of the resonance structures for graphite.

atoms.[1] The free movement of electrons in this π cloud accounts for the electrical conductivity of graphite. An electron can be pumped into the cloud at one end of the sheet and another removed from the other end to give a net transfer of electrons through the solid.

In the total graphite structure the planes of carbon atoms are stacked in layers so that each carbon atom lies above another in every second layer, as shown in Figure 18.4. Within any given layer, adjacent carbon atoms are fairly close together (1.41 Å) while the spacing between successive planes is much greater (3.35 Å). The different planes of carbon atoms are not held together by covalent bonds but instead, by very much weaker van der Waals forces. As a result, the layers are able to slide over one another with relative ease and, as you may know, graphite has many applications as a dry lubricant. As every school child is taught, the "lead" in a "lead pencil" is actually graphite and when one writes with a pencil, the layers of graphite slide off one another onto the paper.

In graphite carbon exhibits multiple bonding, as do nitrogen and oxygen in their molecular forms. When we move now to the third and successive periods, a different state of affairs exists. Here we have atoms that prefer to form only single bonds with other atoms and that have relatively little tendency to participate in $p\pi$-$p\pi$ multiple bonds. The molecular structures of the free elements reflect this.

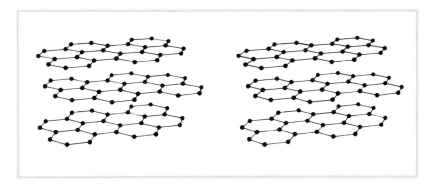

Figure 18.4

Total graphite structure.

[1] In the valence bond approach we imagine that adjacent carbon p orbitals overlap to form simple π-bonds and that the total bonding picture can be viewed as a composite of resonance structures such as that shown in Figure 18.3.

Chlorine, because it only needs to form a single covalent bond to complete its octet, exists as diatomic molecules. Bromine and iodine form diatomic Br_2 and I_2 for the same reason. The structures of the remaining non-metals, however, are considerably more complex. With sulfur, for instance, we find that each atom must share two electrons to fill its valence shell, and it does so by forming two single bonds to *two different* sulfur atoms. These, in turn, must also be bonded to two separate S atoms and a —S—S—S—S— sequence is produced. Actually, in its most stable form the sulfur atoms are arranged in puckered, eight-membered S_8 rings having a crownlike structure, illustrated in Figure 18.5. Selenium too forms Se_8 rings in one of its allotropic forms. Selenium and tellurium also exist in a gray form in which there are long Se_x and Te_x chains, respectively.

Elemental sulfur exhibits rather interesting behavior when it is heated. At room temperature the most stable allotrope of sulfur contains the S_8 rings packed into a rhombic crystal structure. If the sulfur is melted and allowed to cool and solidify slowly, a second allotrope is produced in which the S_8 rings are stacked in a monoclinic crystal structure. The monoclinic form, however, is stable only above $95.5°$ C and, when allowed to stand at room temperature, gradually reverts to the rhombic modification. The reverse transformation (rhombic $\rightarrow$ monoclinic) can be brought about too if the rhombic form is held at a temperature above $95.5°$ C, although the rate of transformation is slow. Usually when sulfur is heated, the rhombic form melts before it has an opportunity to revert to the monoclinic.

Liquid sulfur, at its melting point, is yellow, relatively nonviscous, and is composed primarily of S_8 rings in the random orientations that characterize a liquid. As the liquid is heated to higher temperatures, it begins to thicken and darken, becoming a dark red, molasseslike substance. At still higher temperatures it thins out again and becomes lighter in color and finally boils at $445°$ C. This rather unusual behavior is explained in the following way.

As the temperature of the liquid sulfur is raised, thermal motion, transmitted into molecular vibrations, begins to break S—S bonds, and the S_8 rings begin to open, producing S_8 chains in which the end S atoms each have one unpaired electron.

When one terminal S atom encounters another, a single bond can be created by the pairing of electrons, and the S_8 chains begin to couple together to produce first S_{16}, then S_{24}, S_{32}, S_{40}, etc. until extremely long chains possessing perhaps as many as several hundred thousand S atoms are produced. These chains become tangled and intertwined and slip past one another only with great difficulty, thereby causing a marked increase in the viscosity of the liquid. At still higher temperatures the more violent thermal motions cause the long chains to start to break down into smaller fragments, and the liquid becomes mobile again.

It's interesting to note that if the thickened liquid sulfur is cooled rapidly,

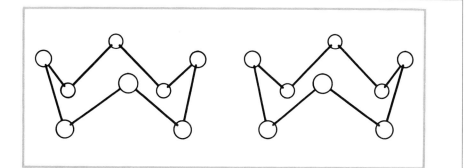

Figure 18.5
The structure of the S_8 ring.

by being poured into cold water, for example, the sulfur atoms do not have an opportunity to be transformed into S_8 rings and, as a result, a supercooled liquid called **amorphous sulfur** (also referred to as plastic sulfur) is produced that has many of the elastic properties of rubber. When allowed to stand, the S_x chains of the amorphous sulfur gradually revert to the more thermodynamically stable S_8 rings of the rhombic form.

Let's move now to the left and examine the element phosphorus. Being in Group VA, an atom of phosphorus has three unpaired electrons.

$$\cdot \overset{\textstyle \cdot}{\underset{\textstyle \cdot \cdot}{P}} \cdot$$

Therefore, to achieve a noble gas structure it must acquire three more. Since there is little tendency for this element to form multiple bonds, as nitrogen does when it forms N_2, the octet is completed by the formation of three single covalent bonds to three different phosphorus atoms.

The simplest elemental form of phosphorus is a waxy solid called white phosphorus consisting of P_4 molecules in which each P atom lies at the corner of a tetrahedron, as illustrated in Figure 18.6. In this structure we see that each P atom is nicely bound to three others. This particular allotrope of

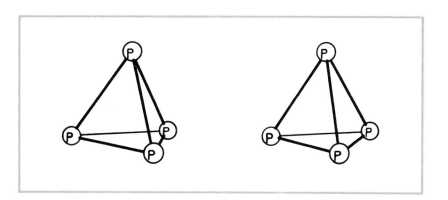

Figure 18.6
The structure of white phosphorus, P_4.

phosphorus is very reactive. It ignites spontaneously in air to produce the oxide and, for this reason, is used militarily in incendiary devices.

A second allotrope that is much less reactive is red phosphorus and, at the present time, its structure is unknown, although it has been suggested that it contains P_4 tetrahedra linked at the corners.

The third allotropic form is black phosphorus formed by heating the white variety at very high pressures. It has a layer structure in which each P atom in a layer is singly bonded to three others (Figure 18.7). This drawing represents a portion of one of these layers (each consisting of two planes of P atoms). In the solid the layers are stacked upon each other with only weak van der Waals forces between them. Consequently black phosphorus looks and behaves physically much like graphite. Like red phosphorus it is quite unreactive.

The elements below phosphorus, arsenic, and antimony are also able to form somewhat unstable yellow allotropic forms containing As_4 and Sb_4 molecules. The most stable modifications have a metallic appearance with structures similar to black phosphorus.

Finally, we look at the heavier elements in Group IVA, silicon and germanium. These elements each must acquire, through sharing, four electrons and they do so by the formation of single bonds. As a result Si and Ge (as well as the low temperature nonmetallic form of tin) have the diamond structure shown in Figure 18.1. Since silicon and germanium do not appear able to enter into $p\pi$-$p\pi$ bonding, graphitelike structures do not exist.

There is still one element whose structure we have not considered, namely boron. This period 2 element, found in Group IIIA, is quite unlike any of the others, since there is no simple way for it to complete its valence shell. Sharing electrons in such a way as to give three ordinary single bonds still would leave each boron atom with only six electrons around it; as a result there is no easy way to understand its structure.

Boron exists in several different crystalline forms, each of which is characterized by clusters of 12 boron atoms located at the vertices of an icosahedron (a 20-sided geometric figure) as shown in Figure 18.8a. Each boron atom within a given cluster is equidistant from five others and, in the solid, each of these is also joined to yet another boron atom outside the clus-

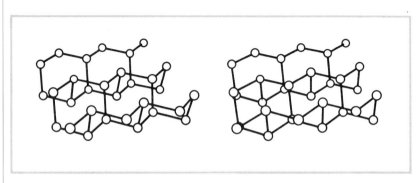

Figure 18.7
The structure of black phosphorus.

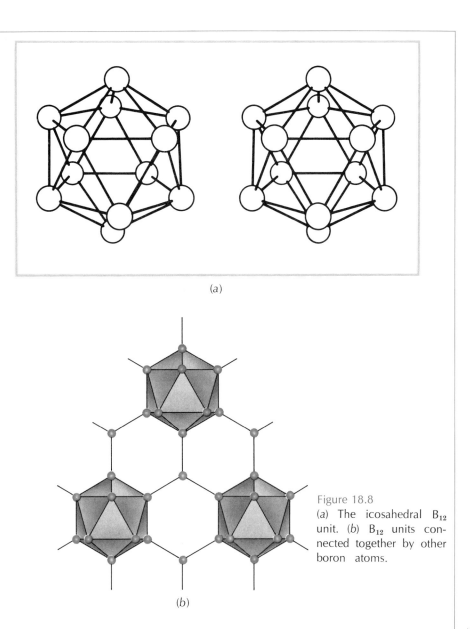

(a)

(b)

Figure 18.8
(a) The icosahedral B_{12} unit. (b) B_{12} units connected together by other boron atoms.

ter (Figure 18.8b). The electrons available for bonding are therefore delocalized to a large extent over many boron atoms.

The linking together of B_{12} units produces a large three-dimensional network solid that is very difficult to break down. Boron, therefore, is very hard (it is the second hardest element) and has a very high melting point (about 2200° C).

18.3 Oxidation Numbers

The oxidation number concept is useful in coordinating the descriptive chemistry of the metals because most compounds of the metals are appreciably ionic and the metal atom therefore carries a positive charge that is at least approximated by the oxidation number assigned to it. With many (if not most) of the compounds of the nonmetals and metalloids,[2] however, this is not the case, particularly in those instances where the element, because of the rules of assigning oxidation numbers, exists in a high positive oxidation state. Thus sulfur in SO_4^{2-} certainly does not carry a $+6$ charge, as its oxidation number would imply. As a result, oxidation numbers are only of somewhat limited usefulness in discussing and comparing the chemistries of the nonmetals and metalloids.[3] With these words of caution, let's take a very brief look at the oxidation numbers of these elements.

As we have seen, when a nonmetal occurs in a binary compound with a metal, such as NaCl or Mg_2Si, the oxidation number of the nonmetal will be negative. These compounds are largely ionic in character, and the oxidation number of the nonmetal corresponds to the charge on the anion. We saw in Chapter 3 that this anionic charge is determined by the number of electrons that must be acquired by the element in order to achieve a noble gas structure. As a result we find the halogens pick up one electron to form anions with a -1 charge (e.g., F^-, Cl^-, Br^-, I^-). In like manner the elements in Group VIA acquire two electrons and form anions with a charge of -2 (e.g., O^{2-}, S^{2-}), while those in Group VA gain three electrons to produce anions with a -3 charge (e.g., N^{3-} in Li_3N), and so forth.

When a nonmetal is combined with another nonmetal, its oxidation state is determined by whether the other nonmetal has a higher or lower electronegativity. The basic rules of assigning oxidation numbers thus apply and, because of the myriad of ways that the nonmetals combine with each other, it is a property of nonmetals that they exhibit multiple oxidation states. Nitrogen, for example, occurs in every integral oxidation state between -3 and $+5$, as seen in Table 18.2. Fractional oxidation numbers also occur among the nonmetals, as we saw in Chapter 3 where the $S_4O_6^{2-}$ ion, containing sulfur with an oxidation number of $+5/2$, was used as an example. In every case that we shall consider, the highest positive oxidation state equals the group number. With the halogens (Group VIIA) it is $+7$, as found in the ClO_4^- ion, while the Group VIA elements show a maximum oxidation number of $+6$ (S in SO_4^{2-}). In a similar fashion, for Group VA the maximum is $+5$ (N in NO_3^-) and in Group IVA it is $+4$ (Si in SiO_2).

Hydrogen is quite a unique element, possessing but one electron in its $1s$ orbital. It can achieve the noble gas structure of helium either by acquiring an additional electron to give the hydride ion, H^-, as it does when it reacts with the very active metals, or it can complete its valence shell through covalent bond formation. Because of its relatively high electronegativity, it forms covalent bonds with the nonmetals. As a class, binary compounds with hydrogen are called hydrides, whether or not the hydrogen exists in a positive or negative oxidation state.

Except for the very special case of boron, which we shall treat in some detail later, and the hydrogen difluoride ion HF_2^- in which a H atom lies

18.4
Nonmetal Hydrides

---

[2] For the remainder of this chapter we shall not make the formal distinction between metalloids and nonmetals. The term nonmetal will be used to refer to any element that is clearly not a metal.

[3] They still, of course, retain their usefulness when we wish to discuss redox reactions involving their compounds.

Table 18.2
Oxidation States of Nitrogen

Oxidation State	Example	Preparative Reaction
−3	NH_3 (ammonia)	$N_2 + 3H_2 \longrightarrow \mathbf{2NH_3}$
−2	N_2H_4 (hydrazine)	$2NH_3 + NaOCl \longrightarrow \mathbf{N_2H_4} + NaCl + H_2O$
−1	NH_2OH (hydroxylamine)	$NaNO_2 + NaHSO_3 + SO_2 + H_2O \longrightarrow 2NaHSO_4 + \mathbf{NH_2OH}$
0	N_2	$NH_4NO_2(s) \xrightarrow{\Delta} \mathbf{N_2}(g) + 2H_2O(g)$
+1	N_2O (nitrous oxide)	$NH_4NO_3(s) \xrightarrow{\Delta} \mathbf{N_2O}(g) + 2H_2O(g)$
+2	NO (nitric oxide)	$4NH_3 + 5O_2 \longrightarrow \mathbf{4NO} + 6H_2O$
+3	N_2O_3 (dinitrogen trioxide)	$NO + NO_2 \xrightarrow{-20°C} \mathbf{N_2O_3}$
+4	NO_2 (nitrogen dioxide)	$2NO + O_2 \longrightarrow \mathbf{2NO_2} \xrightarrow{Cool} \mathbf{N_2O_4}$
+5	HNO_3 (nitric acid)	$3NO_2 + H_2O \longrightarrow \mathbf{2HNO_3} + NO$

equidistant between two F atoms (i.e., [F—H—F]⁻), hydrogen is capable of binding to only one other atom by way of electron sharing, since it seeks only one additional electron. As a result the structural chemistry of the simple hydrides is rather straightforward and, perhaps, even somewhat mundane. With the halogens it forms compounds with the general formula HX (e.g., HF, HCl). Similarly we find, quite expectedly, that the Group VIA elements form molecules of general formula H_2X (e.g., H_2O, H_2S), the elements of Group VA form H_3X (usually written XH_3, e.g., NH_3, PH_3), and those in Group IVA form H_4X (or XH_4, e.g., CH_4, SiH_4). The geometries of these molecules are all readily predicted by the electron pair repulsion theory discussed in Section 16.6.

In addition to these simple hydrides containing a single atom of the nonmetal, there are others that possess two or more nonmetal atoms. Some examples are given in Table 18.3. All of these compounds are characterized by nonmetal atoms linked directly to one another, a phenomenon called **catenation.** Thus hydrogen peroxide, H_2O_2, has the Lewis structure

Similarly, we have others such as[4]

| hydrazine | disilane | propane |

[4] The halogens do not exhibit this property except in the simple elemental state such as F_2, Cl_2, etc.

Table 18.3
Catenation Among Nonmetal Hydrides

Group IVA	CH_4	SiH_4	GeH_4	SnH_4	
	C_2H_6	Si_2H_6	Ge_2H_6	Sn_2H_6	
	C_3H_8	Si_3H_8	Ge_3H_8		
	.	.			
	.	.			
	.	.			
	C_nH_{2n+2}	Si_6H_{14}			
	+				
	Many others				
Group VA	NH_3	PH_3	AsH_3	SbH_3	BiH_3
	N_2H_4	P_2H_4			
Group VIA	H_2O	H_2S	H_2Se	H_2Te	H_2Po
	H_2O_2	H_2S_2			
		$H_2S_n(n = 1 - 6)$			

The ability of nonmetals to form compounds in which they bond to other like atoms varies greatly. In Table 18.3, you will notice, for example, that in Group VIA only oxygen and sulfur form such compounds. In Group VA we find that both nitrogen and phosphorus catenate, but the chain length seems to be limited to two atoms. When we proceed to Group IVA, all of the elements, down to and including tin exhibit this property and here we find chains containing three, four, and even more atoms. We also see that the tendency toward catenation generally decreases downward in a group, as evidenced by the trend toward shorter chains demonstrated by the heavier elements in Group IVA, Ge and Sn.

Of all of the elements, carbon has the greatest capacity to form bonds to itself. In fact, the broad area of organic chemistry is concerned entirely with hydrocarbons and compounds that are derived from them by substituting other elements for hydrogen. Organic compounds, then, are compounds in which the molecular framework consists primarily of carbon-carbon chains. The unique ability of carbon to form such diverse compounds containing these long stable carbon chains is undoubtedly the reason why life has evolved around the element carbon instead of around another element such as silicon.

Catenation is a property that is not restricted to the nonmetal hydrides alone; it occurs in other compounds as well. Sulfur, for example, has a rather marked tendency to form bonds to other sulfur atoms, as we saw in our discussion of the free element. This carries over to its compounds too. For instance, if an aqueous solution containing S^{2-} is heated with elemental sulfur, a series of polysulfide ions, S_2^{2-}, S_3^{2-}, . . . , S_x^{2-}, are formed. We might illustrate this kind of reaction as

$$\left[:\ddot{S}:\right]^{2-} + \ddot{S}:\ddot{S}:\ddot{S}: \longrightarrow \left[:\ddot{S}:\ddot{S}:\ddot{S}:\ddot{S}:\right]^{2-}$$

Addition of strong acid to these solutions produces the corresponding hydrides (H_2S_2, H_2S_3, . . . , H_2S_x).

Another similar reaction occurs when an aqueous solution containing sulfite ion is heated with sulfur. This reaction produces **thiosulfate** ion,[5] $S_2O_3^{2-}$,

Oxidation of thiosulfate produces another catenated sulfur species, **tetrathionate** ion, $S_4O_6^{2-}$, having the structure

Other ions of similar structure are **dithionate** ($S_2O_6^{2-}$), **trithionate** ($S_3O_6^{2-}$), **pentathionate** ($S_5O_6^{2-}$), and **hexathionate** ($S_6O_6^{2-}$). Their structures are, respectively,

In addition to sulfur, some other common nonmetals also form catenated species that are not hydrides. Some examples are:

1. *Carbon.* For example, oxalic acid:

2. *Nitrogen.* For example, azides containing the ion N_3^- (derived from hydrazoic acid, HN_3):

$$\left[:\ddot{N}{=}N{=}\ddot{N}: \right]^-$$

[5] The prefix thio in this case implies substitution of sulfur for oxygen. Thus the thiosulfate ion is a sulfate ion in which one oxygen atom has been replaced by a sulfur atom.

3. *Oxygen.* For example, peroxides such as peroxydisulfate ion, $S_2O_8^{2-}$, which may be considered to be derived, at least in a formal way, from H_2O_2,

Nonmetal hydrides are produced as products of many different chemical reactions; however, we shall consider only two general methods of preparation here. One of these is the direct combination of the elements, as illustrated, for example, by the reaction of hydrogen with either chlorine

$$H_2 + Cl_2 \longrightarrow 2HCl$$

or with oxygen

$$2H_2 + O_2 \longrightarrow 2H_2O$$

18.5
Preparation of the Hydrides

However, this method is not applicable to all of the hydrides, as we can see by examining some of their thermodynamic properties shown in Table 18.4. Here we see that only the hydrides of the more active nonmetals possess negative free energies of formation. Those lying below the heavy line in the table have positive free energies of formation and from a practical standpoint cannot be prepared directly from the free elements. Instead an indirect procedure must be employed.

The rates of reaction toward hydrogen vary substantially among the nonmetals. In period 2, for instance, fluorine reacts immediately with hydrogen

Table 18.4
Standard Enthalpies and Free Energies of Formation of Nonmetal Hydrides

	XH_n ΔG_f° (kcal/mole) ΔH_f° (kcal/mole)			
BH_3 Not stable, simplest hydride is B_2H_6	CH_4 -17.9 -12.1	NH_3 -11.0 -3.9	H_2O -57.8 -54.6	HF -64.8 -65.3
	SiH_4 $+8.2$ $+13.6$	PH_3 $+1.3$ $+3.2$	H_2S -4.9 -8.0	HCl -22.1 -22.8
	GeH_4 (positive)	AsH_3 $+15.9$ $+16.5$	H_2Se $+7.1$ $+3.8$	HBr -8.7 -12.8
		SbH_3 (?)	H_2Te $+36.9$ $+33.1$	HI $+6.3$ $+0.4$

when they are placed in contact. On the other hand, H_2 and O_2 mixtures are stable virtually indefinitely, unless the reaction is initiated in some way, for example, by applying heat or introducing a catalyst.

Nitrogen is even less reactive than oxygen, not only toward hydrogen, but toward nearly all other chemical reagents as well. Presumably this is because of the high stability of the N_2 molecule that arises as a consequence of its strong triple bond (the bond energy of N_2 is 226 kcal/mole, compared to 120 and 38 kcal/mole for O_2 and F_2, respectively).

The production of ammonia by reaction of N_2 and H_2 is undoubtedly one of the most important industrial chemical reactions, since virtually all useful nitrogen compounds, such as chemical fertilizers, for example, may be prepared from NH_3 in one way or another.

The Haber process, developed in Germany during World War I, employs the reaction

$$N_2(g) + 3H_2(g) \rightleftharpoons 2NH_3(g) \qquad \Delta H^\circ = -22 \text{ kcal}$$

As we can see, the reaction is exothermic and we predict, on the basis of the arguments presented in Chapter 12, that the greatest yield of NH_3 would be achieved if the reaction were permitted to come to equilibrium at low temperature and high pressure. However, the reaction takes place very slowly at ordinary temperatures, even in the presence of a catalyst (iron containing a small amount of oxide serves as a heterogeneous catalyst). Consequently, a high temperature is used, even though the quantity of NH_3 produced is somewhat reduced. The actual conditions that are employed are a pressure of approximately 1000 atm and a temperature of 400 to 500° C.

The second method of preparation of nonmetal hydrides involves the addition of protons, from a Brønsted acid, to the conjugate base of a nonmetal hydride, a reaction that we might depict as

$$X^{n-} + nHA \longrightarrow H_nX + nA^-$$

where X^{n-} is the conjugate base of the hydride H_nX, and HA is the Brønsted acid. Let's look at some examples.

The hydrogen halides are commonly prepared in the laboratory by treating a halide salt with a nonvolatile acid such as sulfuric or phosphoric acid.

$$NaCl(s) + H_2SO_4(l) \longrightarrow HCl(g) + NaHSO_4(s)$$
$$NaCl(s) + H_3PO_4(l) \longrightarrow HCl(g) + NaH_2PO_4(s)$$

In these examples HCl is removed as a gas, which causes the reaction to proceed to completion.

With the heavier halogens, Br and I, sulfuric acid cannot be used since it is a sufficiently strong oxidizing agent to oxidize the halide ion to the free halogen. For example, when treated with H_2SO_4, I^- reacts as follows,

$$2I^- + HSO_4^- + 3H^+ \longrightarrow I_2 + SO_2 + 2H_2O$$

Phosphoric acid, being a much weaker oxidizing agent than H_2SO_4, simply supplies protons to I^-, and HI can therefore be produced in a reaction analogous to the production of HCl above, that is,

$$NaI(s) + H_3PO_4(l) \longrightarrow HI(g) + NaH_2PO_4(s)$$

As we proceed from right to left across a period (e.g., from fluorine toward carbon) we have seen that the acid strength of the H_nX compounds decrease. Thus HF is a stronger acid than H_2O which, in turn, is stronger than NH_3, and so forth. This means that the strengths of their corresponding conjugate bases *increase* from right to left ($C^{4-} > N^{3-} > O^{2-} > F^-$). As a result, the strength of the Brønsted acid required to react with the anion of the nonmetal to produce the hydride decreases. For example, the production of HF, whose conjugate base, F^-, is weak, requires a strong acid such as H_2SO_4. Oxide ion, on the other hand, is a much stronger base than F^- and when treated with even a relatively weak source of protons, oxide ion gobbles them up to produce water.

$$O^{2-} + 2H^+ \longrightarrow H_2O$$

This is a reaction that we have seen before in Chapters 5 and 13.

Nitride ion, N^{3-}, is expected to be even a stronger base than O^{2-}. Therefore it is not surprising to find that Mg_3N_2 reacts with the weak acid, H_2O, to produce NH_3 in a reaction that we can interpret as a hydrolysis of the N^{3-} ion.

$$Mg_3N_2 + 6H_2O \longrightarrow 3Mg(OH)_2 + 2NH_3$$

Metal carbides, which can be prepared by heating an active metal with carbon, also react with water in the same fashion. Aluminum carbide, for instance, which contains C^{4-} ions, hydrolyzes according to the reaction

$$Al_4C_3 + 12H_2O \longrightarrow 4Al(OH)_3 + 3CH_4$$

There are also carbides in which the carbon atoms exist in discrete pairs which we might write as C_2^{2-}, such as in Ca_2C_2. This ion has the structure

$$[:C:::C:]^{2-}$$

and upon hydrolysis yields acetylene, C_2H_2

$$CaC_2 + 2H_2O \longrightarrow Ca(OH)_2 + C_2H_2$$

This general method of preparation also extends to the third, fourth and fifth periods too, with the same trends in the strength of the Brønsted acid required to liberate the hydride. In period 3 we have these anions:

Group	IV	V	VI	VII
	Si^{4-}	P^{3-}	S^{2-}	Cl^-

We again expect the anions to become increasingly basic as we move from right to left (i.e., from Cl^- to Si^{4-}); therefore the strength of the Brønsted acid needed to protonate the anion decreases. To form HCl from NaCl a strong acid is required. Sulfide ion, on the other hand, is sufficiently basic to be highly hydrolyzed in aqueous solution (as we found in Chapter 14), and solutions containing a soluble sulfide such as Na_2S always have a strong odor of H_2S because of the reaction[6]

---

[6] Actually there is a two-step equilibrium involving both HS^- and H_2S. See Chapter 14.

$$S^{2-} + 2H_2O \rightleftharpoons H_2S + 2OH^-$$

Many insoluble metal sulfides dissolve in acids with the evolution of H_2S.

Phosphides, like sulfides, also hydrolyze on contact with water. However, because the P^{3-} ion is more basic than the S^{2-} ion, the hydrolysis proceeds essentially to completion. Thus aluminum phosphide, AlP, reacts with water to produce phosphine, PH_3.

$$AlP + 3H_2O \longrightarrow Al(OH)_3 + PH_3$$

Moving left to Group IVA we again find that a hydrolysis reaction serves to prepare silicon hydrides. A metal silicide such as Mg_2Si (which can be formed by simply heating Mg and Si together) reacts with water to generate a mixture of silanes; SiH_4, Si_2H_6, Si_3H_8, etc., up to Si_6H_{14}.

The heavier nonmetals behave in much the same fashion as those above them. Thus H_2Se and H_2Te, like H_2S, can be prepared by adding an acid to a metal selenide or telluride. Arsine, AsH_3, like phosphine, PH_3, is made by the hydrolysis of a metal arsenide such as Na_3As or AlAs, and the germanes, GeH_4, Ge_2H_6, and Ge_3H_8 are produced by the action of dilute HCl on Mg_2Ge.

18.6
Boron Hydrides

We now come to a very interesting and unique series of hydrogen compounds, the boron hydrides. From its electronic structure we might expect boron to have chemical properties somewhat similar to those of aluminum (a metal), which is just below it in Group IIIA. The extremely small size of the B^{3+} "ion," and its correspondingly high ionic potential, however, prevent it from forming ionic compounds and bonds formed to boron are always covalent.

Because of its three valence electrons, we might predict that the simplest boron hydride would be BH_3

$$\overset{\overset{\textstyle H}{\textstyle \cdots}}{H:\overset{\cdots}{B}:H}$$

While this compound has been observed, it is not stable. Instead, the simplest stable molecule has the formula B_2H_6, the structure of which has been determined to be that shown in Figure 18.9. If you look carefully at this structure, you will see that it appears to contain a total of eight B—H bonds. The B_2H_6

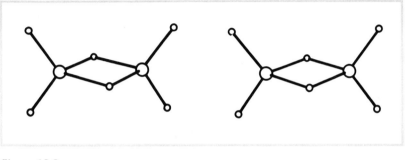

Figure 18.9
The diborane molecule, B_2H_6. Large atom, boron; small atom, H.

molecule, however, possesses only 12 valence electrons—sufficient to form only six ordinary single bonds.

In the diborane molecule, B_2H_6, there are two different kinds of bonds. Four of the bonds are ordinary B—H bonds, in the sense that they are formed from the overlap of an sp^3 orbital on boron with a $1s$ orbital on hydrogen. At two electrons apiece, these account for eight of the 12 valence electrons.

The other bonds correspond to those in which hydrogen atoms serve as bridges between the two boron atoms. We might view each of these bonds as arising from the simultaneous overlap of one sp^3 hybrid orbital from each B atom, with the $1s$ orbital of the bridging hydrogen atom, as shown in Figure 18.10. This picture leads to two bonds, called **three center bonds,** that each extend over three nuclei and that each contain a single pair of electrons. This produces a rather odd situation where three atoms in a three center bond are held together by only two electrons.

We might note that valence bond theory, which always considers pairs of electrons shared between two atoms, cannot easily handle this, and other, boron hydrides. In fact, boron compounds containing three center bonds are often spoken of as being electron deficient since there does not appear to be, on the basis of our usual view of bonding, sufficient electrons to account for all of the bonds. Molecular orbital theory, on the other hand, is not embarrassed by this molecule since we simply have a case of a delocalized molecular orbital extending over three nuclei, not too much different in principle than the delocallized π-bonds in the SO_3 molecule discussed in Chapter 16.

In general, we have seen that nonmetal hydrides are formed in such a way that the nonmetal achieves a noble gas configuration consisting of four electron pairs in its valence shell. We have seen before that these electron pairs tend to situate themselves at the vertices of a tetrahedron and therefore the molecular structures that we find for most of the hydrides are based primarily on the tetrahedron.

The boron hydrides offer us some interesting variations in geometry. With B_2H_6 there is an approximately tetrahedral arrangement of hydrogen atoms about each boron (Figure 18.9), and we can say that the total molecular structure is derived by the sharing of an edge between two tetrahedra.

18.7
Geometric Structures of the Nonmetal Hydrides

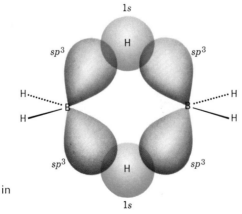

Figure 18.10
Three-center bonding in B_2H_6.

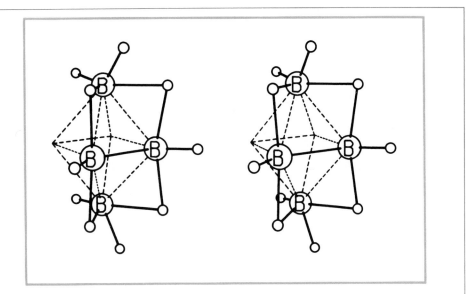

Figure 18.11
Structure of B_4H_{10}.

There are, in addition to B_2H_6, more complex boron hydrides. Two of these are B_4H_{10} and B_5H_9 whose structures are shown in Figures 18.11 and 18.12, respectively. In each case, notice that the boron atoms are found to occupy positions that correspond, at least roughly, to most (although not all) of the vertices of another familiar geometric figure, the octahedron.

Still other boron hydrides have structures in which the boron atoms sit at some of the vertices of an icosahedron. This geometric figure, you might recall, describes the structure of elemental boron (Figure 18.8). Examples of

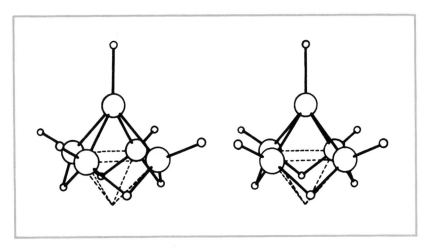

Figure 18.12
Structure of B_5H_9.

(a)

(b)

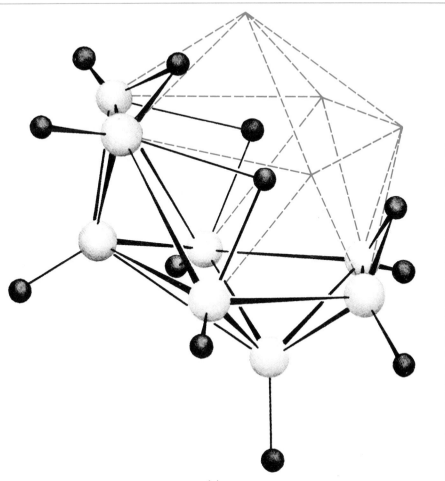

(c)

Figure 18.13
The structures of (a) $B_{12}H_{12}^{2-}$ (Large spheres = Boron), (b) $B_{10}H_{14}$, and (c) B_8H_{12}. Note the similarity to the B_{12} unit. (From *The Chemistry of Borons and Its Compounds,* Earl L. Muetterties. Copyright 1967, John Wiley and Sons, New York. Reprinted by permission.)

these more complex boron hydrides, including the $B_{12}H_{12}^{2-}$ ion, are shown in Figure 18.13.

18.8
Oxygen
Compounds
of the Nonmetals

The oxides and those compounds derived from them, the oxoacids (acids containing oxygen) and oxoanions (anions containing oxygen), are among the most important compounds of the nonmetals. Here we find many of the common reagents encountered in the laboratory, for example, acids such as H_2SO_4 and HNO_3 and their salts, the sulfates and nitrates. In addition, most minerals are either oxides, such as SiO_2 (silica) and Fe_2O_3 (hematite), or contain oxoanions of nonmetals, for example, the carbonates (limestone) and silicates (asbestos, mica, and others).

We have already seen that nonmetal oxides exhibit acidic properties and that a relatively simple relationship exists between an oxide and the acid derived from it by reaction with water. Thus we have many reactions in which water combines with an oxide; for example,

$$SO_2 + H_2O \longrightarrow H_2SO_3$$
$$CO_2 + H_2O \longrightarrow H_2CO_3$$

You also will recall that neutralization of these acids gives their corresponding anions.

This relationship exists for many oxides, oxoacids, and oxoanions; however, there are also many cases in which the relationship is more a formality than an actual fact. For example, some oxides such as SiO_2 are insoluble and therefore do not react to give acids. In other instances the oxide cannot react with water to give an acid in which the nonmetal retains the same oxidation state. Nitric oxide, NO, is unreactive toward water while nitrogen dioxide **disproportionates** (i.e., undergoes an oxidation-reduction in which the same reactant is both oxidized *and* reduced).

$$3NO_2 + H_2O \longrightarrow 2HNO_3 + NO$$

In still other cases the parent acid of an anion either is unknown or cannot be obtained in the pure state. Carbonic acid, H_2CO_3, for example, decomposes as its aqueous solution is concentrated by evaporation of the solvent.

Table 18.5 contains a list of many of the oxides of the nonmetals. Here, once again, we find that there is no single preparative procedure that can be employed in every case because of the rather wide variety of properties of the nonmetals. We can, however, find some generalizations that apply in a reasonably large number of instances.

**18.9
Preparation of
Nonmetal Oxides**

Table 18.5
Typical Oxides of the Nonmetals

Group III	B_2O_3			
Group IV	CO	SiO_2	GeO_2	
	CO_2			
Group V	N_2O	P_4O_6	As_4O_6	Sb_4O_6
	NO	P_4O_{10}	$As_2O_5{}^a$	$Sb_2O_5{}^a$
	N_2O_3			
	NO_2; (N_2O_4)			
	N_2O_5			
Group VI	O_2	SO_2	SeO_2	TeO_2
	O_3	SO_3	SeO_3	TeO_3
Group VII	OF_2	Cl_2O	Br_2O	I_2O_5
	O_2F_2	ClO_2	BrO_2	I_2O_7
		Cl_2O_7		

[a]Molecular structure unknown.

Table 18.6
Thermodynamic Properties of Some Nitrogen Oxides

Oxide	ΔH_f° (kcal/mole)	ΔG_f° (kcal/mole)
$N_2O(g)$	+19.5	+24.8
$NO(g)$	21.6	20.7
$NO_2(g)$	8.1	12.4
$N_2O_4(g)$	2.3	23.5
$N_2O_5(g)$	2.7	27.5

One method that can be employed in most cases is simply the direct union of the elements, as typified by the reactions:

$$S + O_2 \longrightarrow SO_2$$
$$C + O_2 \longrightarrow CO_2 \quad \text{(excess oxygen)}$$
$$C + \tfrac{1}{2}O_2 \longrightarrow CO \quad \text{(limited supply of oxygen)}$$
$$H_2 + \tfrac{1}{2}O_2 \longrightarrow H_2O$$

Not all oxides can be prepared in this manner, however. For example, in Table 18.6 we see that many of the oxides of nitrogen have positive free energies of formation and, therefore, from what we know of thermodynamics, they cannot be synthesized directly from the elements.[7] In these instances, and in others too, indirect methods of preparation are employed.

The indirect procedures are, expectedly, many in number. However, once more, a few generalizations can be made. In some cases an oxide can be prepared from a lower oxide by further reaction with oxygen. The synthesis of SO_3, for example, consists of catalytic oxidation of SO_2,

$$2SO_2 + O_2 \longrightarrow 2SO_3$$

This reaction is quite favorable thermodynamically ($\Delta G^\circ = -33.46$ kcal) but is very slow in the absence of a catalyst such as platinum or V_2O_5. We also know that carbon monoxide can be readily burned to produce the dioxide, a reaction that is important since CO is frequently used as an industrial fuel. This reaction can also be applied to the removal of CO from automotive exhausts.

Another technique that can serve to produce oxides is the combustion of nonmetal hydrides. Methane, the chief constituent of natural gas, burns to produce CO_2 and H_2O:

$$CH_4 + 2O_2 \longrightarrow CO_2 + 2H_2O$$

---

[7] Since they also have positive ΔS_f°, it should be possible to prepare them at very high temperatures where $\Delta H - T\,\Delta S$, and hence ΔG, is negative. This is the case with NO, which is formed in small amounts near 3000° C. The high temperature production of NO in motor vehicle engines is a major source of urban air pollution.

In fact, the combustion of any hydrocarbon in the presence of an ample oxygen supply yields these same two products. When insufficient O_2 is available, as in an automobile engine, CO may be produced instead of CO_2.

A reaction of this general type, which is of great commercial importance, is the oxidation of ammonia. In this case a platinum catalyst is used and the reaction is

$$4NH_3 + 5O_2 \longrightarrow 4NO + 6H_2O$$

The NO formed in this reaction is readily oxidized further to produce NO_2,

$$2NO + O_2 \longrightarrow 2NO_2$$

which, as noted earlier, disproportionates when dissolved in water to yield HNO_3 and NO:

$$3NO_2 + H_2O \longrightarrow 2HNO_3 + NO$$

The commercial application of this sequence of reactions accounts for the major source of nitric acid and nitrates that are used in the manufacture of explosives, fertilizers, plastics, and many other useful substances. In fact, the development of this process in Germany by Wilhelm Ostwald, accompanied by the successful preparation of NH_3 from N_2 and H_2 by Haber, are said to have prolonged World War I, since the Allied blockade of Germany was unable to halt the German manufacture of munitions that had depended, prior to these processes, on the importation of nitrates from other countries.

Finally, another indirect method of obtaining nonmetal oxides makes use of oxidation-reduction reactions. For instance, when nitric acid serves as an oxidizing agent, the nitrate ion is reduced and, depending on conditions, nitrogen in any oxidation state can be produced. When concentrated nitric acid is used, the reduction product is frequently NO_2 while dilute solutions of HNO_3 often yield NO as the reduction product.

$$4HNO_3 + Cu \longrightarrow Cu(NO_3)_2 + 2NO_2 + 2H_2O \qquad \text{(concentrated)}$$
$$8HNO_3 + 3Cu \longrightarrow 3Cu(NO_3)_2 + 2NO + 4H_2O \qquad \text{(dilute)}$$

Similarly, hot concentrated sulfuric acid is a fairly potent oxidizing agent, the reduction product usually being SO_2; for example,

$$Cu + 2H_2SO_4 \longrightarrow CuSO_4 + SO_2 + 2H_2O$$

The structures that are found for the nonmetal oxides are once again determined significantly by the ability of the nonmetal to form multiple bonds, in this case to oxygen atoms. In our earlier discussion of the structures of the free elements we saw that period 2 elements have a substantial tendency to enter into such bonding; therefore the oxides of the period 2 elements are simple monomeric species. Carbon, for example, forms CO and CO_2, which we can write as

$$:C\equiv O: \qquad \text{and} \qquad :\overset{..}{O}=C=\overset{..}{O}:$$

respectively.

With nitrogen there are NO and NO_2. Both of these have an odd number of electrons and therefore are not able to satisfy the octet rule, since at least one atom must be left with an odd number of electrons. The structure of the

NO molecule is usually represented as a resonance hybrid of the dot structures

$$\cdot \ddot{N}{=}\ddot{O}: \quad \text{and} \quad :\ddot{N}{=}\ddot{O}\cdot$$

Molecular orbital theory gives a better picture of this molecule, which we might compare to the species $O_2{}^+$. These two (NO and $O_2{}^+$) are isoelectronic, that is, they have the same number of electrons; hence we might expect that electronically they would not be too different. The M.O. energy level diagram for $O_2{}^+$, obtained by removing one of the antibonding electrons from O_2 (Figure 16.16) is shown in Figure 18.14a. The corresponding energy diagram for NO is found in Figure 18.14b.

The molecule NO_2 is another interesting species. Using electron-dot formulas we can indicate its structure by two resonance forms, neither of which are capable of fully satisfying the octet rule.

In each of these structures an unpaired electron resides on the nitrogen atom and in samples of NO_2, there is a dimerization equilibrium,

$$2NO_2 \rightleftharpoons N_2O_4$$
(brown)   **(colorless)**

in which the N_2O_4 (dinitrogen tetroxide) molecule has the structure

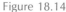

Figure 18.14
Molecular orbital energy level diagrams for $O_2{}^+$ and NO. (a) $O_2{}^+$. (b) NO. Note that the energies of the atomic orbitals on the isolated atoms ase not the same.

(one of four resonance forms)

In N_2O_4 the unpaired electrons of two NO_2 molecules have paired to form a covalent bond between the two nitrogen atoms.

In period 3 there is considerably less tendency for the nonmetals to enter into multiple bonding. Sulfur, though, does appear able to do so, as evidenced by the oxides SO_2 and SO_3. Recall that these can be represented by resonance structures such as

(two structures) (three structures)

With phosphorus there is even less of a tendency toward multiple bonding and when phosphorus reacts with oxygen two oxides are formed, depending upon reaction conditions

$$P_4 + 3O_2 \longrightarrow P_4O_6$$
$$P_4 + 5O_2 \longrightarrow P_4O_{10}$$

The structures of these oxides bear a remarkably simple relationship to the elemental P_4 unit which, as we saw in Figure 18.6, is tetrahedral in shape. In Figure 18.15a is shown the structure of P_4O_6 in which we find an oxygen placed along each of the six edges of the P_4 tetrahedron. In the P_4O_{10} molecule, Figure 18.15b, we find an additional oxygen atom attached to each of the P atoms at the vertices of the tetrahedron.

In the P_4O_6 molecule we can imagine that each of the three unpaired electrons of a phosphorus atom

$$\cdot \overset{\displaystyle \cdot}{\underset{\displaystyle \cdot \cdot}{P}} \cdot$$

is shared with one electron of an oxygen atom in a P-O-P bridge,

$$: \overset{\displaystyle \cdot}{\underset{\displaystyle \cdot}{P}} : \overset{\displaystyle \times \times}{\underset{\displaystyle \times \times}{O}} : \overset{\displaystyle \cdot}{\underset{\displaystyle \cdot}{P}} :$$

with each P atom participating in three such bridges. Each P atom in P_4O_6 therefore has an unshared lone pair of electrons. In P_4O_{10}, four additional oxygen atoms become attached to these P atoms by coordinate covalent bonds so that in P_4O_{10} each P atom is surrounded by four oxygen atoms in an approximately tetrahedral arrangement.

Silicon, in Group IVA, forms only one oxide, SiO_2. Silicon dioxide, while having the same empirical formula as carbon dioxide, is nevertheless quite different structurally since silicon has virtually no tendency to form Si-O multiple bonds. In quartz, and other forms of SiO_2, each Si atom is surrounded tetrahedrally by four oxygen atoms each of which is bound to another Si

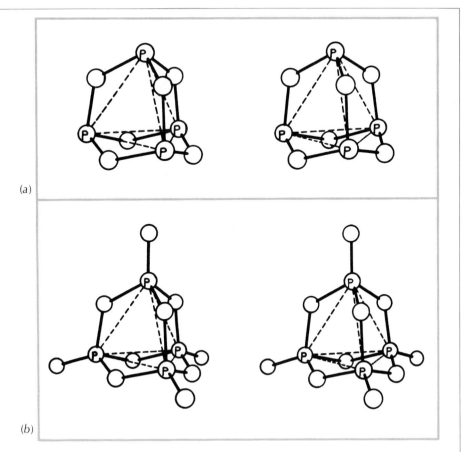

Figure 18.15

The oxides of phosphorus. (a) P_4O_6. (b) P_4O_{10}.

atom. The resulting structure, a portion of which is shown in Figure 18.16, is quite complex, extending in a network fashion in three dimensions.

A few features of this structure are worth noting. The overall hexagonal nature of quartz can be seen from Figure 18.17, which is the same as Figure 18.16 but with additional lines added for emphasis. Another interesting point is that the quartz structure is built up of —Si—O—Si—O—Si— spiral chains, one of which is outlined by the circle in Figure 18.17. In Figure 18.16 you are looking down the axis of these spiral chains; a side view of one of these chains containing three repetitions of the spiral is seen in Figure 18.18.

The existence of these spiral chains of SiO_4 tetrahedra in quartz impart an interesting property to the overall structure. Quartz crystals occur in two varieties that are similar to one another in the same way that your left and right hands are related, that is as mirror images of each other. In quartz this is because of the two possible directions of rotation of the spiral chains, in one case clockwise and in the other, counterclockwise, as shown in Figure 18.19.

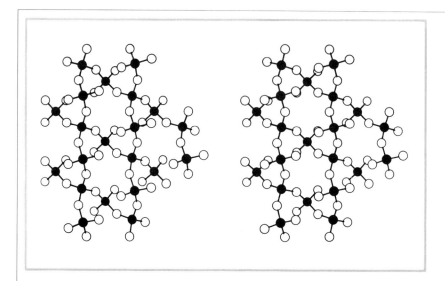

Figure 18.16
The structure of quartz. Solid spheres are silicon; open spheres are oxygen.

Spiral chain
rising up from
plane of paper

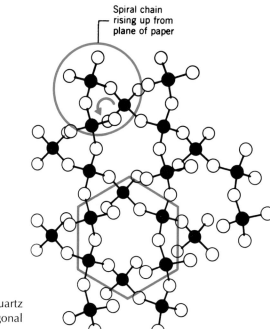

Figure 18.17
The structure of quartz showing the hexagonal nature of the crystal.

Chemistry of the Representative Elements: Part II

The oxoacids and their anions can be divided into two categories, simple monomeric acids and anions containing one atom of the nonmetal (Table 18.7), and complex polymeric acids and anions. We shall devote our attention in this section to the simple species.

In general, the oxoacids consist of an atom of a nonmetal to which is bonded one or more **hydroxyl groups,** —OH, plus, perhaps, additional oxygen atoms not bonded to hydrogen, to give a generalized formula

$$XO_m(OH)_n$$

Some typical examples are given below.

$PO(OH)_3$	$SO_2(OH)_2$	$NO_2(OH)$
H_3PO_4	H_2SO_4	HNO_3

Table 18.7
Monomeric Oxoacids and Oxoanions of the Nonmetals

Group IIIA	H_3BO_3			
	(no simple borates)			
Group IVA	H_2CO_3	$H_4SiO_4{}^a$	$H_4GeO_4{}^a$	
	$(CO_3{}^{2-})$	$(SiO_4{}^{4-})$	$(GeO_4{}^{4-})$	
Group VA	HNO_2	H_3PO_3		
	$(NO_2{}^-)$	$(HPO_3{}^{2-})$		
	HNO_3	H_3PO_4	H_3AsO_4	
	$(NO_3{}^-)$	$(PO_4{}^{3-})$	$(AsO_4{}^{3-})$	
Group VIA		H_2SO_3	H_2SeO_3	$H_2TeO_3{}^a$
		$(SO_3{}^{2-})$	$(SeO_3{}^{2-})$	$(TeO_3{}^{2-})$
		H_2SO_4	H_2SeO_4	$Te(OH)_6$
		$(SO_4{}^{2-})$	$(SeO_4{}^{2-})$	$(TeO(OH)_5{}^-)$
Group VIIA	HOF	HOCl	HOBr	HOI
		(OCl^-)	(OBr^-)	(OI^-)
		$HClO_2$	$HBrO_2$	HIO_3
		$(ClO_2{}^-)$	$(BrO_2{}^-)$	$(IO_3{}^-)$
		$HClO_3$	$HBrO_3$	H_5IO_6
		$(ClO_3{}^-)$	$(BrO_3{}^-)$	$(H_2IO_6{}^{3-})$
		$HClO_4{}^-$	$HBrO_4$	HIO_4
		$(ClO_4{}^-)$	$(BrO_4{}^-)$	$(IO_4{}^-)$

^{a}Not observed

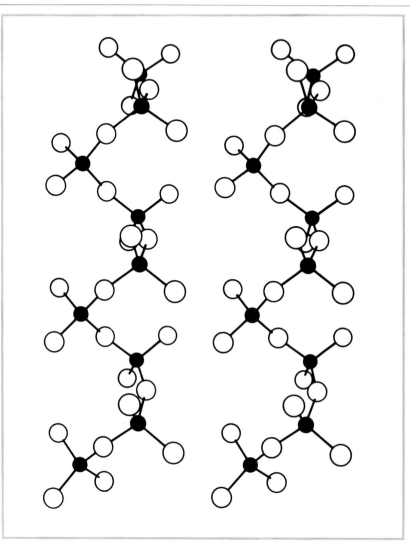

Figure 18.18
Spiral chains of SiO_4 tetrahedra.

In Chapter 13 we saw that polarization of the O—H bond of the hydroxyl group by the atom of the nonmetal permits the H atom to be removed, more or less readily, as an H^+ ion. To reflect this, the formulas of these acids are written with the hydrogens first, followed by the formula of the anion that remains when the protons are removed by neutralization.

The geometric structures of the monomeric acids and anions, like those of the simple oxides, are readily predicted using the electron pair repulsion theory. Rather than discuss them at length here, try Question 18.33 at the end of the chapter.

For the preparation of these substances, once again there are no specific reactions applicable to all cases. We have seen that the oxides often react

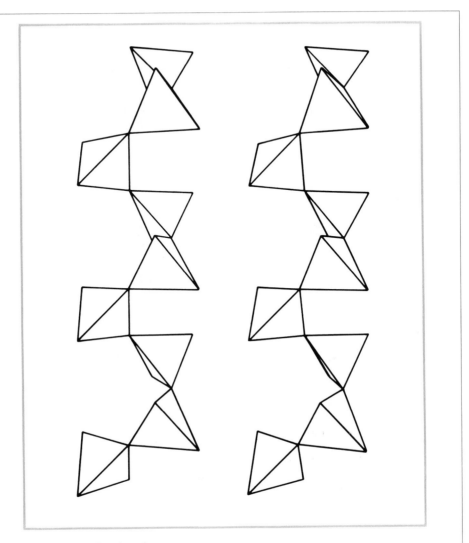

Figure 18.18 (Continued)

with water to form acids and that the oxoanions can be obtained from them by neutralization. In cases where more than one oxide is formed by the non-metal, different acids can be produced by hydrolysis. Thus we have seen that SO_2 and SO_3 produce H_2SO_3 and H_2SO_4, respectively.

With phosphorus, hydrolysis of the oxides P_4O_6 and P_4O_{10} yield H_3PO_3 and H_3PO_4. These latter two acids are interesting because only H_3PO_4 is able to provide three protons for neutralization and thus produce salts containing the PO_4^{3-} ion. Complete neutralization of H_3PO_3, on the other hand, gives the ion HPO_3^{2-}. The differences in behavior between these two acids can be traced to their structures.

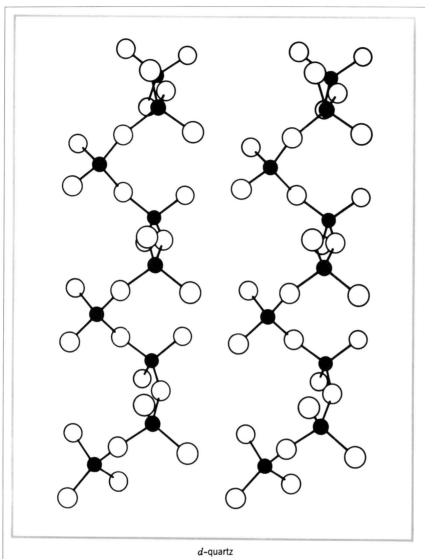

d-quartz

Figure 18.19
Left- and right-handed helices in quartz.

phosphoric acid   phosphorous acid
H_3PO_4 H_3PO_3

In each case phosphorus is bonded to four atoms. In H_3PO_3, however, one of the hydrogens is covalently bonded directly to the phosphorus atom and is

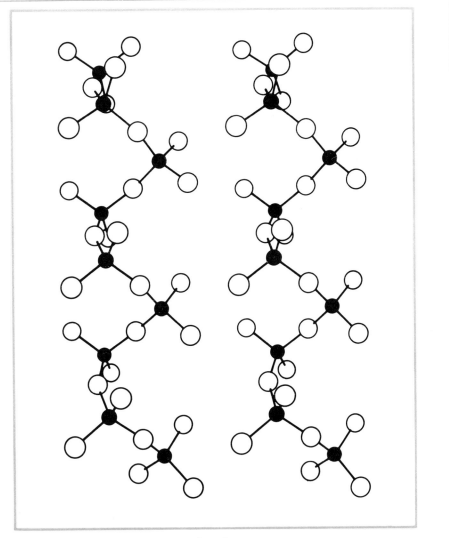

l-quartz

Figure 18.19 (Continued)

not able to be removed as a proton. Consequently H_3PO_3 is only a diprotic acid.

Oxoacids and their anions may also be prepared by oxidation-reduction reactions. We have seen, for instance, the reaction of NO_2 with water.

$$3NO_2 + H_2O \longrightarrow 2HNO_3 + NO$$

Another reaction that is at least somewhat similar occurs when an equimolar mixture of NO and NO_2 is dissolved in water.

$$NO + NO_2 + H_2O \longrightarrow 2HNO_2$$

Here we have nitrogen(II) oxide and nitrogen(IV) oxide which, taken together, are equivalent to N_2O_3, the acid anhydride of HNO_2. In fact, if this equimolar mixture of oxides is condensed to a liquid, or frozen to a solid, molecules having the formula N_2O_3 are created. A nitrite salt can also be prepared from this NO/NO_2 mixture by reacting it with a base.

$$NO + NO_2 + 2OH^- \longrightarrow 2NO_2^- + H_2O$$

Other oxoacids and anions that may be obtained by oxidation-reduction reactions are those containing the halogens. For example, with the exception of fluorine, the halogens react with water in a disproportionation reaction.

$$X_2 + H_2O \longrightarrow HOX + H^+ + X^-$$

Thus Cl_2 gives hypochlorous acid, $HOCl$, and hydrochloric acid, HCl; Br_2 yields $HOBr$ and HBr, etc. In the presence of a base we expect the corresponding acids to be neutralized and the oxoanions to be formed, thus driving these reactions toward completion. With chlorine, for example, we have

$$Cl_2 + 2OH^- \longrightarrow OCl^- + Cl^- + H_2O$$

and, in general, there is the reaction

$$X_2 + 2OH^- \longrightarrow OX^- + X^- + H_2O$$

However, here we have an interesting case of kinetics influencing the ultimate products of a chemical reaction.

The hypohalite ions themselves (OCl^-, OBr^-, OI^-) have a tendency to disproportionate to give the halate (XO_3^-) and halide (X^-).

$$3OX^- \rightleftharpoons XO_3^- + 2X^-$$

The equilibrium constants for this reaction with the different halogens are

$$
\begin{aligned}
3OCl^- &\rightleftharpoons ClO_3^- + 2Cl^- & K &\approx 10^{27} \\
3OBr^- &\rightleftharpoons BrO_3^- + 2Br^- & K &\approx 10^{15} \\
3OI^- &\rightleftharpoons IO_3^- + 2I^- & K &\approx 10^{20}
\end{aligned}
$$

The large values for these K's suggest that all of the OX^- ions should be transformed rather completely to the corresponding XO_3^- ions. Solutions of hypochlorite ion, however, are reasonably stable when kept cool although disproportionation to ClO_3^- does occur if a solution of hypochlorite is heated. The stability of cold solutions containing OCl^- is further illustrated by the fact that commercial liquid bleach consists of a 5% solution of $NaOCl$. Hypobromite ion, on the other hand, rapidly disproportionates at room temperature while OI^- reacts so rapidly that it is never even observed at all when I_2 is dissolved in base. We see then that it is slow kinetics that accounts for the "stability" of solutions of OCl^- rather than simply thermodynamics. If all of the OX^- ions were to disproportionate as rapidly as OI^-, the reaction of the elemental halogens with base would lead directly to the production of the XO_3^- ions.

There are other disproportionation reactions which yield oxoanions that can be caused to occur in the solid state by the application of heat. For example, strong heating of a nitrate such as $NaNO_3$ can drive off oxygen to give the nitrite

$$2NaNO_3 \longrightarrow 2NaNO_2 + O_2$$

Similarly, $KClO_3$ when heated at moderate temperatures, in the *absence* of a catalyst such as MnO_2, disproportionates to give the perchlorate

$$4KClO_3 \longrightarrow 3KClO_4 + KCl$$

(in the presence of a catalyst, or when heated to high temperatures, decomposition to KCl and O_2 occurs; $2KClO_3 \rightarrow 2KCl + 3O_2$).

It was mentioned earlier that there are some acids that do not exist in the pure liquid state but, instead, can only be obtained in solution. Nitrous acid, HNO_2, and chloric acid, $HClO_3$, are examples. Nitrous acid can be conveniently prepared by acidifying a nitrite salt.

$$NO_2^- + H^+ \rightleftharpoons HNO_2$$

In fact, on the basis of the equilibria involved we would expect that any weak acid can be produced in solution by adding a strong acid, such as HCl or H_2SO_4, to a salt of the weak acid.

Chloric acid, $HClO_3$, is strong and cannot be generated in the same manner as HNO_2. The solubility rules that you learned (hopefully) in Chapter 5 can, however, point to a preparative method, at least in this case. Mixing solutions of barium chlorate and sulfuric acid yields a precipitate of $BaSO_4$ and leaves $HClO_3$ in solution.

$$Ba(ClO_3)_2(aq) + H_2SO_4(aq) \longrightarrow BaSO_4(s) + 2HClO_3(aq)$$

As a general method of preparation of a strong acid, however, this obviously requires the use of a barium salt that is more soluble than $BaSO_4$.

18.12 Polymeric Oxoacids and Oxoanions

These substances consist of two or more nonmetal atoms linked together in some fashion via oxygen bridges. They occur primarily with the elements in the third and succeeding periods, although boron, a second period element, also forms complex borates (which we shall not discuss here).

In the polymeric species that we shall now examine the basic structural unit is the XO_4 tetrahedron. Polymeric acids or ions are constructed by linking XO_4 tetrahedra together through commonly shared corners, resulting in structures that range from simple to complex. However, we shall see that relatively simple, straightforward structural relationships exist between the polymeric acids and anions, and the simple oxoacids discussed in the last section.

The oxoacids formed by the nonmetals of period 3, in their highest oxidation states, can be represented by the formulas

orthosilicic acid (never isolated)	orthophosphoric acid	(ortho) sulfuric acid	(ortho) perchloric acid

These correspond to the "ortho" acids, although in the latter two the ortho prefix is omitted, as indicated by the parentheses.

As noted earlier, the hydrogen atoms of the —OH groups are acidic and may be removed by reaction with a base to yield the series of anions,

orthosilicate   orthophosphate   sulfate   perchlorate

Notice that the number of negative charges on the anion is equal, in each case, to the number of —OH groups attached to the central atom in the parent acid. In addition, in each of these anions the central atom is surrounded by four electron pairs (in the X—O bonds) and each therefore has a tetrahedral structure that we can depict as shown in Figure 18.20.

The polymeric oxoacids and anions are related to these simple species in a rather direct way, since they are formed, if not always in practice, at least in principle, by joining these simple tetrahedral units together into more complex structures. We may view this as accomplished by the removal of the constituents of a water molecule from two —OH groups on adjacent molecules of the acid. We can illustrate this in a general way as

where the resulting product contains an oxygen bridge that is characteristic of these polymeric species in general.

Another important point to notice is that this bridging oxygen atom had its origin in an —OH group bonded to the central atom. Therefore, the maximum complexity of the polymeric acid (or anion) is determined by the number of —OH groups that are bound to the nonmetal in the simple monomeric acid. Thus we expect that silicon can form a maximum of four oxygen bridges to other Si atoms, phosphorus can form three, sulfur two and chlorine, only one.

Let us now begin to examine some of these polymers. When one molecule of water is eliminated from between two molecules of acid, the species $H_nX_2O_7$ are formed (Table 18.8). These constitute the so called "pyro" acids, and their neutralization produces the corresponding oxoanions. In this table we have also included the oxide, Cl_2O_7, which is essentially structurally identical to the pyro-anions (Figure 18.21). In fact, Cl_2O_7 is prepared by reacting

Chemistry of the Representative Elements: Part II

Figure 18.20

Tetrahedral "ortho" anions, $XO_4{}^{n-}$)

Table 18.8

"Pyro" Acids and Anions

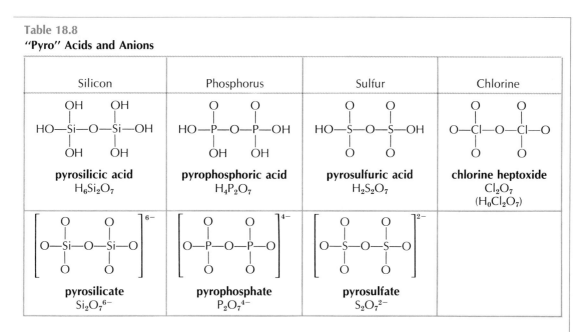

Silicon	Phosphorus	Sulfur	Chlorine
OH OH | | HO—Si—O—Si—OH | | OH OH **pyrosilicic acid** $H_6Si_2O_7$	O O | | HO—P—O—P—OH | | OH OH **pyrophosphoric acid** $H_4P_2O_7$	O O | | HO—S—O—S—OH | | O O **pyrosulfuric acid** $H_2S_2O_7$	O O | | O—Cl—O—Cl—O | | O O **chlorine heptoxide** Cl_2O_7 $(H_0Cl_2O_7)$
$\begin{bmatrix} O & & O \\ \| & & \| \\ O-Si-O-Si-O \\ \| & & \| \\ O & & O \end{bmatrix}^{6-}$ **pyrosilicate** $Si_2O_7^{6-}$	$\begin{bmatrix} O & & O \\ \| & & \| \\ O-P-O-P-O \\ \| & & \| \\ O & & O \end{bmatrix}^{4-}$ **pyrophosphate** $P_2O_7^{4-}$	$\begin{bmatrix} O & & O \\ \| & & \| \\ O-S-O-S-O \\ \| & & \| \\ O & & O \end{bmatrix}^{2-}$ **pyrosulfate** $S_2O_7^{2-}$	

HClO$_4$ with a powerful dehydrating agent such as P$_4$O$_{10}$. In addition, Cl$_2$O$_7$ is the acid anhydride of HClO$_4$ and reacts with water to form this acid.

The prefix "pyro" that is used with these acids and anions is derived from the fact that they may be formed by heating either the free acids, or one of their acid salts, to drive off the H$_2$O molecule from between two tetrahedral units. The pyrosulfate ion, for example, can be obtained by heating a salt such as KHSO$_4$.

$$2KHSO_4 \longrightarrow K_2S_2O_7 + H_2O$$

The acid, H$_2$S$_2$O$_7$, on the other hand, is formed by dissolving SO$_3$ in concentrated H$_2$SO$_4$,

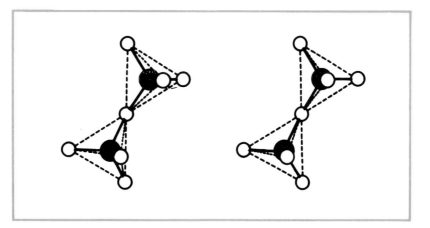

Figure 18.21
Structure of the "pyro" anions, $X_2O_7^{m-}$. Solid sphere = X.

$$SO_3 + H_2SO_4 \longrightarrow H_2S_2O_7$$

Pyrophosphoric acid is produced when H_3PO_4 is heated.

$$2H_3PO_4 \longrightarrow H_4P_2O_7 + H_2O$$

The pyrosilicate ion is observed only in certain minerals; for example, thortveitite, $Sc_2Si_2O_7$. The acid, $H_2Si_2O_7$, cannot be isolated.

The ease with which water may be removed from the acid to cause polymerization is not the same for all of the period 3 nonmetals but, instead, increases from Cl to Si. To prepare Cl_2O_7 from $HClO_4$ requires a very powerful dehydrating agent. As we move to the left, conditions become less severe until with silicon, the acidification of a solution containing a soluble silicate leads to the immediate precipitation of hydrated silica, $SiO_2(H_2O)_x$, the end result of large scale spontaneous dehydration—polymerization. With the exception of silicon, the dehydration process that leads to polymerization can be reversed by the addition of water to the polymeric species. Pyrosulfuric acid, for example, reacts with water to form H_2SO_4.

$$H_2S_2O_7 + H_2O \longrightarrow 2H_2SO_4$$

Except for chlorine, all of the dimeric "pyro" acids discussed above have additional —OH groups that can participate in further bridging, again by elimination of H_2O, to give the **"meta"** series of acids and anions. The sharing of two —OH groups of the original ortho acid thus can give a chainlike structure containing a very large number of repeating units as shown below.

The "meta" anions derived from the acids all consist of XO_4 tetrahedra linked through shared corners and possess the empirical formula $XO_3{}^{n-}$. The general structure is illustrated in Figure 18.22.

For silicon, the free acid is unknown while the metasilicate ion, $SiO_3{}^{2-}$, occurs in several minerals [e.g., spodumene, $LiAl(SiO_3)_2$, a major source of Li]. Not surprisingly, these are observed to have a fiberlike appearance.

metasilicic acid   **metasilicate ion**
(not isolated)

With phosphorus we have a similar situation. Heating NaH_2PO_4 drives off water to yield $(NaPO_3)_n$,

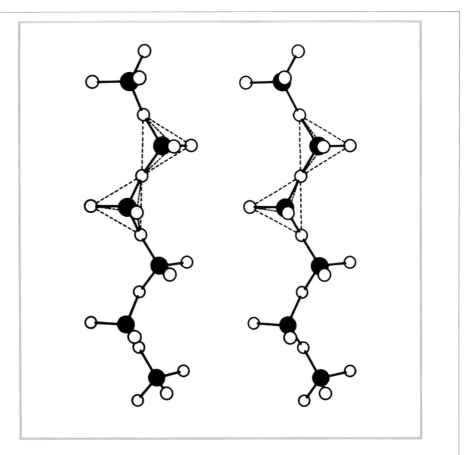

Figure 18.22
Simple linear $(XO_3{}^{n-})_x$ formed by sharing two corners on every XO_4 tetrahedron. (Solid spheres $= X$.)

$$nNaH_2PO_4 \longrightarrow (NaPO_3)_n + nH_2O$$

where we again have a linear polyanion

which we can consider to be derived from metaphosphoric acid.

Note that only one of the nonbridging oxygens in metaphosphoric acid belongs to an OH group while in metasilicic acid both do. As a result the metaphosphate ion carries a charge of -1 while the SiO_3 unit bears a -2 charge.

A metaphosphate that is a component of many solid detergent mixtures is sodium tripolyphosphate, $Na_5P_3O_{10}$, which contains the ion,

The presence of this ion in the wash water has several beneficial effects. For example, it forms soluble complexes with iron and manganese ions, which can stain clothing, and also helps to keep dirt particles suspended so that they can be readily washed away.

In liquid detergents sodium or potassium pyrophosphate ($P_2O_7^{4-}$) is used because it hydrolyzes to PO_4^{3-} less rapidly than the $P_3O_{10}^{5-}$ ion.

Earlier we saw that the removal of an H_2O molecule from a pair of $HClO_4$ molecules leads to the oxide Cl_2O_7. By the same token we might expect that the removal of two molecules of water from H_2SO_4 might lead to a linear chainlike SO_3 species.

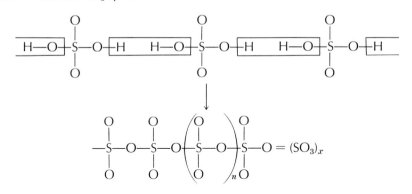

Such a substance does actually exist. We have spoken earlier of SO_3 as a simple molecule whose structure can be represented by resonance forms such as

However, when a trace of H_2O is added to pure SO_3 a polymerization takes place to produce a fibrous solid.

We can view this reaction in the following way. An H^+ from the H_2O adds itself to the oxygen of one SO_3 molecule.

This species then adds another SO_3 unit,

a process that is repeated over and over to produce a long chain of SO_3 molecules. Finally, the positive charge on the sulfur atom at the end of the chain is neutralized by the addition of the OH^- left over from the water molecule that had lost its proton. The end result is a species

If the chain is extremely long, its empirical formula reduces to simply SO_3 since the H^+ and OH^- make up an insignificant part of the whole molecule.

In addition to infinite chain anions, the formation of two bridges to one atom can also lead to cyclic structures. For example, heating $(NH_4)_2HPO_4$ leads to the reaction

$$3(NH_4)_2HPO_4 \longrightarrow H_3P_3O_9 + 6NH_3 + 3H_2O$$

in which $H_3P_3O_9$ is cyclotrimetaphosphoric acid (quite a mouthful, isn't it?)

This is a triprotic acid and yields salts such as $Na_3P_3O_9$ containing the cyclotrimetaphosphate ion

In a similar fashion, discrete polymeric anions are also found with silicon. For example, there is the $Si_6O_{18}^{12-}$ ion found in the mineral beryl,

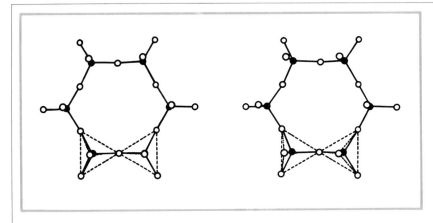

Figure 18.23
Cyclic structure of the $Si_6O_{18}^{12-}$ anion. Notice that there are 12 nonbridging oxygen atoms that each carry a single negative charge. (Solid spheres = X.)

$Be_3Al_2(Si_6O_{18})$. This ion is pictured in Figure 18.23. Note that once again we have a series of SiO_4 tetrahedra in which each share two corners and where the empirical formula of the *anion* is SiO_3^{2-}.

The formation of cyclic species even occurs with SO_3, which solidifies in a form containing rings composed of three SO_3 units,

entirely analogous to the $P_3O_9^{3-}$ ion.

The sharing of more than two corners of the tetrahedron, which produces either three or four oxygen bridges, leads to even more complex structures. Phosphorus has only three oxygen atoms that can form bridges and when they do so, we return to the oxide, P_4O_{10} (Figure 18.15). It was pointed out earlier that each of the phosphorus atoms in this molecule is tetrahedrally surrounded by oxygen atoms. The structure of P_4O_{10} is thus related to H_3PO_4 by the elimination of water from the three —OH groups, so that four PO_4 tetrahedra can each share three of their corners with other tetrahedra.

Silicon, with four oxygen atoms that can function as bridges, forms the most complex series of anions. In addition to the simple strands of SiO_4 tetrahedra in the infinite chain anion SiO_3^{2-} (Figure 18.22), double strands, formed by the sharing of three corners by *every other* SiO_4 unit, are also found. This time, an infinite *double* chain results, a small segment of which is illustrated in Figure 18.24. In this, as well as in the other silicate anions, each unshared oxygen atom carries a negative charge and the repeating unit along the chain is $Si_4O_{11}^{8-}$. This anion is found in asbestos, $CaMg_3(Si_4O_{11})$ and, as we might expect, asbestos has a fiberlike nature because of the presence of the long $(Si_4O_{11}^{8-})_x$ chains that line up more or less parallel to one another.

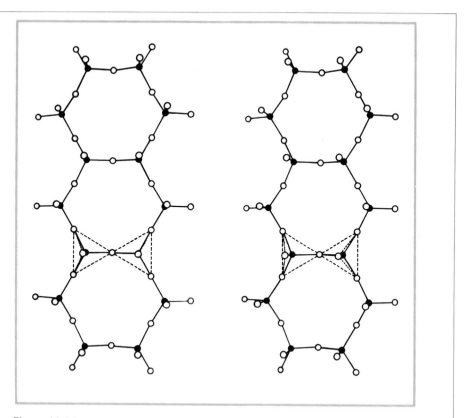

Figure 18.24
Linear double chain in which every other SiO$_4$ tetrahedron shares three corners. (Solid spheres = Si.)

Still another type of silicate is formed if *each* SiO$_4$ tetrahedron shares three of its corners so that each Si is attached to three others by oxygen bridges. When this occurs a planar sheet of SiO$_4$ units results. A portion of one of these is shown in Figure 18.25 with the repeating unit, Si$_2$O$_5^{2-}$, outlined by the rectangle.

A number of different minerals are known to contain these $(Si_2O_5^{2-})_x$ sheets. They differ in the way the silicate layers are stacked, and the nature of the cations and other anions that are also present in the structure; however, all have certain similarities to each other. Some examples are talc and soapstone in which the $(Si_2O_5^{2-})_x$ sheets are packed together with cations in such a way that there is a minimum of attractive forces between successive layers. These layers therefore slide over each other easily and both of these minerals feel slippery.

In other related minerals there is substitution of another element for Si. In mica, for instance, every fourth Si atom is replaced by an Al^{3+} ion. The properties of mica are therefore different from talc; however, the layer structure of mica is clearly apparent.

When Si finally shares all four of its oxygen atoms with other Si atoms, a

Figure 18.25
Planar sheet silicate formed by sharing three corners on every SiO_4 tetrahedron.

three-dimensional framework occurs that has the empirical formula SiO_2. This is found in the mineral quartz whose structure was discussed earlier (see Figures 18.16 to 18.19)

In concluding this section we note that there are great similarities of structure among the oxides, polymeric anions, and even the simple anions of the third period nonmetals. Some of these relationships are summarized in Table 18.9. In all of these species, as we hope you have noticed, the simple tetrahedron is found, and even the most complex structures are simply obtained by combining tetrahedra in many different ways.

Table 18.9
Summary of Oxoanions of the Nonmetals

	Silicon	Phosphorus	Sulfur	Chlorine
Simple anions (no X—O—X bridges)	SiO_4^{4-}	PO_4^{3-}	SO_4^{2-}	ClO_4^{2-}
"Pyro" anions (one X—O—X bridge)	$Si_2O_7^{6-}$	$P_2O_7^{4-}$	$S_2O_7^{2-}$	Cl_2O_7
"Meta" anions (two X—O—X bridges)	$(SiO_3)_x^{2x-}$	$(PO_3)_x^{x-}$	$(SO_3)_x$	
Two-dimensional sheets (three X—O—X bridges)	$(Si_2O_5)_2^{2x-}$	P_4O_{10}		
Three-dimensional network (four X—O—X bridges)	$(SiO_2)_x$			

The halogens (F, Cl, Br, and I) as a class form a large number of different compounds with the nonmetals. We shall look at some of their binary compounds in this section, with an eye toward those factors that control the kinds of compounds that are formed as well as one aspect of their reactivity, their tendency to react with water.

In Table 18.10 you will find a list (although not an exhaustive one) of many of the compounds that are formed between the nonmetals and the halogens. The structures of the substances found in this table can, without exception, be predicted on the basis of the electron pair repulsion theory.

In Table 18.10 we see that most of the nonmetals form more than one compound with a given halogen. The number of halogen atoms that may become bound to any particular nonmetal can be related to two factors. One of these is the electronic structures of the elements that are combined together; the other has to do with the sizes of the atoms.

Each halogen atom contains seven electrons in its valence shell and

Table 18.10

Halogen Compounds of the Nonmetals

Group IIIA	$BX_3(X=F,Cl,Br,I)$			
	BF_4^-			
Group IVA	$CX_4(X=F,Cl,Br,I)$	SiF_4	GeF_4	
		SiF_6^{2-}	GeF_6^{2-}	
		$SiCl_4$	$GeCl_4$	
Group VA	$NX_3(X=F,Cl,Br,I)$	$PX_3(X=F,Cl,Br,I)$	AsF_3	SbF_3
	N_2F_4	PF_5	AsF_5	SbF_5
		PCl_5		
		PBr_5		
Group VIA	OF_2	SF_2	SeF_4	TeF_4
	(O_2F_2)	SCl_2	SeF_6	TeF_6
	OCl_2	S_2F_2	$SeCl_2$	$TeCl_4$
	OBr_2	S_2Cl_2	$SeCl_4$	$TeBr_4$
		SF_4	$SeBr_4$	TeI_4
		SCl_4		
		SF_6		
Group VIIA	ICl	ClF_3	ClF_5	IF_7
	IBr	BrF_3	BrF_5	
	BrF	ICl_3	IF_5	
	$BrCl$	IF_3		
	ClF			

requires but one more to achieve the stable noble gas configuration. As a result there is little tendency for them to form $p\pi$-$p\pi$ multiple bonds with other nonmetals. Furthermore, the halogens ordinarily do not accept electrons in the formation of coordinate covalent bonds because this would mean the addition of two electrons to a valence shell that already contains seven, thereby exceeding the stable octet by one electron.[8]

On this basis, then, we can divide the halogen compounds into two groups, those which obey the octet rule and those which do not. The compositions of the compounds in the first category are determined by the number of electrons that a given nonmetal requires to reach an octet, since each atom bonded to a halogen atom furnishes one electron. For example, in Group VIIA (the halogens themselves) only one electron is needed and only one bond is formed. Thus the halogens are diatomic and substitution of one halogen for another is possible, as we see for substances such as ClF, BrF, BrCl, BrI and ICl.

$$:\overset{\displaystyle ..}{\underset{\displaystyle ..}{Cl}}:\overset{\displaystyle \times\times}{\underset{\displaystyle \times\times}{F}}\overset{\times}{}$$

In Group VIA each element requires two electrons to reach an octet and, hence, we find that compounds such as OF_2 and SCl_2 are formed. Similarly, in Group VA three electrons are given to the central atom by three halogens in molecules like NF_3, PF_3, AsF_3, etc. On the other hand the Group IVA elements pick up four electrons from four halogen atoms in compounds such as CCl_4 and $SiCl_4$.

Boron, in Group IIIA, is once again a special case. Since the boron atom has only three valence electrons, it forms only three ordinary covalent bonds to the halogens. With fluorine, however, BF_3 can add on an additional F^- ion to form the BF_4^-, tetrafluoroborate, anion.

$$:\overset{..}{\underset{..}{F}}: \quad\quad\quad\quad\quad\quad\quad :\overset{..}{\underset{..}{F}}:$$
$$:\overset{..}{\underset{..}{F}}\overset{\times}{\underset{\times}{B}} + \left[:\overset{..}{\underset{..}{F}}:\right]^- \longrightarrow \left[:\overset{..}{\underset{..}{F}}\overset{\times}{\underset{\times}{B}}:\overset{..}{\underset{..}{F}}:\right]^-$$
$$:\overset{..}{\underset{..}{F}}: \quad\quad\quad\quad\quad\quad\quad :\overset{..}{\underset{..}{F}}:$$

We see that the fourth B—F bond can be considered to be a coordinate covalent bond (although by now you know that we really cannot distinguish the source of the electrons once the bond has been formed). Since the boron halides are electron deficient, in the sense that there is less than an octet of electrons in the valence shell of boron, they are all powerful Lewis acids and the formation of the BF_4^- ion is a typical Lewis acid-base reaction.

In the second category of halogen compounds we have substances in which more than four pairs of electrons surround the central atom. These are limited to those nonmetals beyond the second period, since the second period elements have a valence shell that can contain a maximum of only eight electrons corresponding to the completion of the $2s$ and $2p$ subshells. The elements below the second period, however, also have in their valence

---

[8] The halide ions (e.g., Cl^-) do *furnish* electron pairs toward the formation of coordinate covalent bonds and are common ligands in complex ions.

shell a low energy set of vacant d orbitals as well as the s and p subshells. These d orbitals may be used, through hybridization, to make additional electrons available for bonding, as illustrated below for sulfur in the molecule SF_6.

In its ground state the electron configuration of sulfur can be represented as

$$S \quad [Ne] \quad \underset{3s}{\uparrow\downarrow} \quad \underset{3p}{\uparrow\downarrow \;\uparrow\;\uparrow} \quad \underline{\;}\;\underline{\;}\;\underset{3d}{\underline{\;}}\;\underline{\;}\;\underline{\;}$$

The use of two $3d$ orbitals permits the formation of a set of d^2sp^3 hybrid orbitals that are each singly occupied.

$$S \quad [Ne] \quad \underset{d^2sp^3}{\uparrow\;\uparrow\;\uparrow\;\uparrow\;\uparrow\;\uparrow} \quad \underline{\;}\;\underline{\;}\;\underline{\;} \quad \text{unhybridized } 3d \text{ orbitals}$$

This provides six electrons that may each pair with one electron of a fluorine atom to form SF_6.

$$S(\text{in } SF_6) \quad [Ne] \quad \underset{d^2sp^3}{\uparrow x\;\uparrow x\;\uparrow x\;\uparrow x\;\uparrow x\;\uparrow x} \quad \underline{\;}\;\underline{\;}\;\underline{\;} \quad \text{unhybridized } 3d$$

(x's represent fluorine electrons)

The structure of SF_6 is octahedral (Figure 18.26) as we would expect, both from the directional properties of the d^2sp^3 hybrid set and from the electron pair repulsion theory, which says that when there are six electron pairs in the valence shell of an atom, they will situate themselves at the vertices of an octahedron.

In summary, then, we can account for many of the halogen compounds with nonmetals below the second period on the basis of an expansion of the octet that occurs through the use of available d orbitals in hybrids that make additional valence electrons available for bonding.

The second factor that influences the number of halogen atoms that become bound to a nonmetal is the relative sizes of the different atoms. In Table 18.10 you will notice that the compounds that contain a large number of halogen atoms are formed from nonmetals found toward the bottom of the periodic table. This makes sense since these nonmetals are expected to be large and therefore able to accommodate a relatively large number of bonded atoms with a minimum of crowding. On the other hand, an element near the top of a group would be small and only a relatively few halogen atoms could be expected to be packed about it.

The Group VIA halides serve quite nicely to illustrate this point. Sulfur is able to form species such as SF_4 and SF_6 with the smallest halogen, fluorine. With chlorine, however, only SCl_4 exists while SBr_4 and SI_4 are not known. The much larger atom, Te, however, forms all of the tetrahalides, TeF_4, $TeCl_4$, $TeBr_4$, and TeI_4.

Figure 18.26
Octahedral SF_6.

In Group VIIA too we find that size influences the number of atoms that can be bonded to a given halogen. Both chlorine and bromine form compounds having the formulas XF_3 and XF_5; however, ClF_5 has only recently been synthesized and is difficult to prepare. Iodine, on the other hand, is able to accommodate up to seven fluorine atoms to give IF_7 whose structure is given in Figure 18.27.

Another interesting facet of the chemistry of the nonmetal halides is their reactivity toward compounds containing an —OH group, the most familiar of which is water. Here once again we find that both thermodynamics and kinetics are involved in determining the course of reactions.

The kind of reaction on which we shall focus our attention is the hydrolysis of the nonmetal halide to produce either the oxoacid, or an oxide, plus the corresponding hydrogen halide. Some examples are the reactions of PCl_5, $SiCl_4$, and SF_4 with water.

$$PCl_5 + 4H_2O \longrightarrow H_3PO_4 + 5HCl$$
$$SiCl_4 + 2H_2O \longrightarrow SiO_2 + 4HCl$$
$$SF_4 + 2H_2O \longrightarrow SO_2 + 4HF$$

These reactions occur very rapidly and proceed to completion with the evolution of considerable amounts of heat. In fact, it is quite common for many of the halogen compounds of the elements below period 2 to react very rapidly in this same way. For example, the tin(IV) and lead(IV) halides, which we saw were covalent in Chaptes 17, also hydrolyze in this manner with the formation of a mixture of species including complexes of Sn^{4+} and Pb^{4+} with the halide ion.

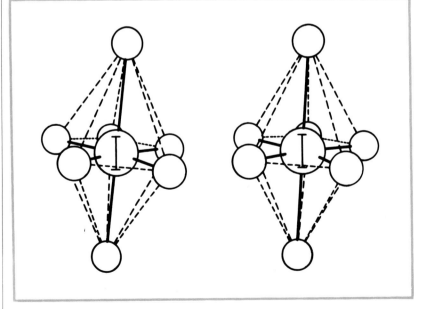

Figure 18.27
Structure of IF_7 (pentagonal bipyramid).

There are also some nonmetal halogen compounds that are quite *unreactive* toward water, for example, CCl_4, SF_6, and NF_3. In these cases it is unfavorable kinetics, rather than thermodynamics, that prevents the hydrolysis from taking place.

We can compare, for example, the potential hydrolysis reactions of CCl_4 and $SiCl_4$. Calculations based on thermodynamics imply that both should proceed very nearly to completion.

$$SiCl_4(l) + 2H_2O(l) \longrightarrow SiO_2(s) + 4HCl(aq) \qquad \Delta G° = -67.3 \text{ kcal}$$
$$CCl_4(l) + 2H_2O(l) \longrightarrow CO_2(g) + 4HCl(aq) \qquad \Delta G° = -90.1 \text{ kcal}$$

In fact, we see from the values of $\Delta G°$ that the hydrolysis of CCl_4 is even more "spontaneous" than $SiCl_4$. Kinetically, however, the hydrolysis of CCl_4 is essentially prohibited. This is attributed to the absence of a low energy path for the hydrolysis of CCl_4. Attack by water on the carbon atom of CCl_4 is prevented by the crowding of the Cl atoms. In $SiCl_4$, on the other hand, the larger Si atom provides a greater opportunity for attack and, in addition, the presence of low energy $3d$ orbitals in the valence shell of the Si atom permits a temporary bonding of the water molecule to the Si atom prior to the expulsion of a molecule of HCl. The mechanism of this hydrolysis is believed to be

Repetition of this process eventually yields $Si(OH)_4$ (orthosilicic acid), which loses water spontaneously to give the hydrated SiO_2.

The stability of SF_6 and NF_3 toward hydrolysis can also be attributed to the absence of a low energy reaction mechanism. Like CCl_4, SF_6 should also undergo hydrolysis quite spontaneously, the value of $\Delta G°$ for the reaction being tremendous.

$$SF_6(g) + 4H_2O(l) \longrightarrow H_2SO_4(aq) + 6HF(g) \qquad \Delta G° = -101 \text{ kcal}$$

However, the crowding of the fluorine atoms around the sulfur atom apparently prevents attack by water (even up to 500° C), as well as by most other reagents. This crowding is absent with SF_4, and hydrolysis by water is instantaneous.

The resistance of NF_3 toward attack by water cannot be attributed to interference by the fluorine atoms as in SF_6, since the NF_3 molecule is pyramidal, with the nitrogen atom being quite openly exposed to an attacking water molecule. We might compare NF_3 with NCl_3, which *does* hydrolyze (if it doesn't explode first—NCl_3 is extremely unstable). The mechanism for this reaction appears to involve the initial formation of a hydrogen bond from water to the lone pair on the nitrogen atom, followed by expulsion of hypochlorous acid.

The ultimate products of the hydrolysis are ammonia and HOCl. This mechanism is not favorable for NF_3 because of its very low basicity. In this case the highly electronegative fluorine atoms draw electron density from the nitrogen atom. As a result, it has been suggested that the lone pair of electrons on nitrogen, in NF_3, may not be available to serve as a point of attachment for the H_2O molecule which, as we see above, is a necessary step in the mechanism proposed for the hydrolysis of NCl_3.

In our discussion of the nonmetals we have not mentioned compounds of the noble gases. These are rather unusual substances since, on the basis of the electronic structure of the noble gases, we would perhaps not have predicted their existence. In fact, until 1962 chemists firmly believed that these elements were totally incapable of forming compounds (other than several **clathrates** in which the noble gas atoms are trapped in cagelike sites within a crystalline lattice). For this reason chemists had referred to them as the inert gases. Today they are spoken of as the noble gases in recognition of the fact that they do react but nevertheless possess a very low degree of chemical reactivity.

18.14 Noble Gas Compounds

The first real chemistry of the noble gases was discovered in 1962 by Neil Bartlett at the University of British Columbia. He had found that molecular oxygen, O_2, reacts with PtF_6 to form an orange-red compound, O_2PtF, containing the ion, O_2^+. Since the ionization energies of O_2 and Xe are nearly the same (12.5 and 12.1 eV, respectively), he reasoned that Xe should react in the same way that O_2 does; when he reacted Xe with PtF_6 he isolated a yellow compound, containing Xe, that was formulated as $XePtF_6$.

After the initial report by Bartlett, it was not long before chemists at Argonne National Laboratory found that Xe also reacts directly with fluorine at elevated temperatures. This reaction yields a series of fluorides, XeF_2, XeF_4, and XeF_6. Other reactions and compounds were soon discovered and a partial list of the known Xe compounds is given in Table 18.11. The oxides and oxofluorides result from the hydrolysis of the fluorides,

$$XeF_6 + 3H_2O \longrightarrow XeO_3 + 6HF$$
$$XeF_6 + H_2O \longrightarrow XeOF_4 + 2HF$$

Some of these compounds are quite unstable and tend to decompose. This is particularly true for the oxides XeO_3 and XeO_4 that explode (XeO_3 has $\Delta H_f^\circ = +96$ kcal/mole). Others, on the other hand, appear quite stable. For example, Cs_2XeF_8 does not decompose even when heated to 400° C and the fluorides have moderately high melting points, suggesting a modest degree of thermal stability.

Table 18.11
Some Compounds of Xenon

		Melting Point (°C)	Physical Form
Fluorides	XeF_2	140	Colorless crystals
	XeF_4	114	Colorless crystals
	XeF_6	47.7	Colorless crystals
Oxides	XeO_3	Explodes	Colorless crystals
	XeO_4	Explodes	Colorless gas
Salts	$XePtF_6{}^a$	—	Red-orange crystals
	$CsXeF_7$	Decomp > 50°	Colorless solid
	Cs_2XeF_8	Decomp > 400°	Yellow solid

^{a}Since shown to be more complex; $Xe(PtF_6)_x$ where x lies between 1 and 2.

The structure and bonding in these compounds is quite interesting. Since Xe has four pairs of electrons in its valence shell, corresponding to a completed 5s and 5p subshell, unpairing of electrons and expansion of the octet must occur to provide unpaired electrons for bonding. Let us consider XeF_2 and XeF_4.

The electronic structure of Xe can be represented as

$$Xe \quad \underset{5s}{\underline{\text{↿⇂}}} \quad \underset{5p}{\underline{\text{↿⇂}}\ \underline{\text{↿⇂}}\ \underline{\text{↿⇂}}} \quad \underset{5d}{\underline{\ }\ \underline{\ }\ \underline{\ }\ \underline{\ }\ \underline{\ }}$$

In order to form XeF_2 one electron must be promoted to the 5d subshell, followed by hybrid orbital formation. The smallest hybrid set that will accommodate all of our electrons is dsp^3.

$$Xe \quad \underset{5s}{\underline{\text{↿⇂}}} \quad \underset{5p}{\underline{\text{↿⇂}}\ \underline{\text{↿⇂}}\ \underline{\text{↿}}} \quad \underline{\text{↿}}\ \underset{5d}{\underline{\ }\ \underline{\ }\ \underline{\ }\ \underline{\ }}$$

gives

$$Xe \quad \underset{dsp^3}{\underline{\text{↿⇂}}\ \underline{\text{↿⇂}}\ \underline{\text{↿⇂}}\ \underline{\text{↿}}\ \underline{\text{↿}}} \quad \underset{\text{unhybridized }5d}{\underline{\ }\ \underline{\ }\ \underline{\ }\ \underline{\ }}$$

The two unpaired electrons can now be used in bonding to fluorine to give

$$Xe \quad \underset{dsp^3}{\underline{\text{↿⇂}}\ \underline{\text{↿⇂}}\ \underline{\text{↿⇂}}\ \underline{\text{↿x}}\ \underline{\text{↿x}}} \quad \underset{\text{unhybridized }5d}{\underline{\ }\ \underline{\ }\ \underline{\ }\ \underline{\ }}$$
(x's represent fluorine electrons)

In Chapter 16 we saw that the dsp^3 hybrids point to the vertices of a trigonal bipyramid. In terms of the electron pair repulsion theory these five electron pairs will also be situated in this fashion, and from our rules (presented on page 467), we expect that the three lone pairs will locate them-

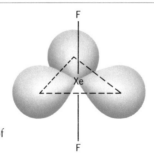

Figure 18.28
Molecular structure of XeF_2.

selves in the triangular plane with the fluorine atoms above and below (Figure 18.28). The XeF_2 molecule is therefore linear.

In the case of XeF_4 we must provide four unpaired electrons for bonding to fluorine. This requires promotion of two electrons to the $5d$ subshell and the formation of d^2sp^3 hybrid orbitals.

$$Xe \quad \underline{\uparrow\downarrow} \quad \underline{\uparrow\downarrow} \; \underline{\uparrow} \; \underline{\uparrow} \quad \underline{\uparrow} \; \underline{\uparrow} \; \underline{} \; \underline{} \; \underline{}$$
$$5s \qquad 5p \qquad\qquad 5d$$

gives

$$Xe \quad \underline{\uparrow\downarrow} \; \underline{\uparrow\downarrow} \; \underline{\uparrow} \; \underline{\uparrow} \; \underline{\uparrow} \; \underline{\uparrow} \qquad \underline{} \; \underline{} \; \underline{} \; \underline{}$$
$$d^2sp^3 \qquad\qquad\qquad \text{unhybridized } 5d$$

Finally, bonding with fluorine gives

$$Xe(\text{in } XeF_4) \quad \underline{\uparrow\downarrow} \; \underline{\uparrow\downarrow} \; \underline{\uparrow x} \; \underline{\uparrow x} \; \underline{\uparrow x} \; \underline{\uparrow x} \qquad \underline{} \; \underline{} \; \underline{} \; \underline{}$$
$$d^2sp^3 \qquad\qquad\qquad \text{unhybridized } 5d$$

In XeF_4 there are six electron pairs about the Xe and both valence bond theory, with its d^2sp^3 hybrids, as well as the electron pair repulsion theory predict that these are directed toward the corners of an octahedron. As we have seen before, in a similar situation, the two lone pairs occupy positions on opposite sides of a square plane containing the four ligand atoms, so that XeF_4 has a square planar structure (Figure 18.29).

Since the initial discovery of noble gas compounds by Bartlett, three of the noble gases have been demonstrated to form compounds, Rn, Xe, and Kr.

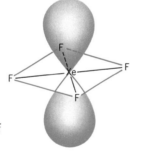

Figure 18.29
Molecular structure of XeF_4.

The lighter ones, because of their much higher ionization energies do not appear able to form chemical compounds.

Bartlett's work on Xe has taught chemists an important lesson. So firmly convinced were they that the noble gases were totally unreactive that after some initial experiments attempting to react Xe with fluorine had failed in the 1930s, no further efforts were made to explore the possibility that they were not inert. It is interesting that the noble gas compounds that have been obtained do not present any particular problem in bonding. In fact, many of the interhalogen compounds that had already been found (e.g., BrF_5, ICl_4^-, IF_7) have the same number of electrons in the valence shell of the central atom as do the noble gas compounds. The same concepts that we applied to other compounds in which the octet was exceeded, therefore, works with the noble gas compounds too. The failure by chemists to recognize the possibility that the noble gases might react reflects a "blind-spot" in their thinking that was probably founded in an overzealous acceptance of the stability and inertness of the ns^2np^6 octet of electrons.

Review Questions and Problems

18.1 Write chemical equations for
(a) The chemical reduction of BCl_3 with hydrogen.
(b) The production of Si from SiO_2 using carbon as a reducing agent.
(c) The reduction of As_2O_3 with hydrogen.

18.2 What is the major source of N_2, O_2? Why is He found in underground deposits?

18.3 What are the two allotropic forms of carbon? How do they differ from one another in terms of structure and physical properties?

18.4 What are the two allotropic forms of sulfur? How do they differ? Outline the physical changes that take place when sulfur is gradually heated to its boiling point and relate them to the structural changes that occur.

18.5 Write a balanced chemical equation for the production of Cl_2 from HCl by reaction with $KMnO_4$ (the manganese is recovered as $MnCl_2$). What is the net ionic equation for the reaction?

18.6 Complete the following chemical equations. If no reaction occurs write N.R.
(a) $Cl_2 + KI \rightarrow$
(b) $Br_2 + CaF_2 \rightarrow$

(c) $I_2 + MgCl_2 \rightarrow$
(d) $F_2 + SrCl_2 \rightarrow$

18.7 Why can't fluorine be produced by the electrolysis of aqueous NaF? What will be the electrolysis reaction in this case?

18.8 What appears to be the dominant factor in determining the complexity of the molecular structures of the elemental nonmetals?

18.9 Compare the structures of N_2 and P_4. What is the P—P—P bond angle in P_4? How does this compare to the normal bond angles that would occur with overlap of p orbitals? Can you suggest how this might account for the high reactivity of P_4?

18.10 What are the structural differences between white and black phosphorus?

18.11 How do we account for the electrical conductivity of graphite?

18.12 What is the structural unit that occurs in elemental boron? Practice sketching its shape.

18.13 What limitations are there on the usefulness of the oxidation number concept as a means of classifying compounds of the nonmetals? What is the oxidation number of sulfur in each of the following: SO_2, $S_2O_3^{2-}$, SO_4^{2-},

$S_2O_6^{2-}$, $S_5O_6^{2-}$, $S_6O_6^{2-}$? Draw the structure of each species. Among these compounds, what (if any) relationship exists between oxidation number and the number of covalent bonds formed by each sulfur atom?

18.14 What are the maximum oxidation numbers of the nonmetals in each of the A-Groups? Give an example for each of the nonmetals.

18.15 What are the molecular shapes of each of the following: PH_3, H_2S, CH_4, H_2Se, Si_2H_6?

18.16 What explanation can be offered for the fact that the nonmetals in the third and succeeding periods do not appear to form stable $p\pi$-$p\pi$ bonds?

18.17 What is meant by catenation? Which element exhibits this property in the largest number of compounds?

18.18 The structures of the polythionate ions ($S_2O_6^{2-}$, $S_3O_6^{2-}$, $S_4O_6^{2-}$, etc.) were written on page 511 with the sulfur atoms in a straight line. The S—S—S angle in these compounds is actually in the neighborhood of 100° (rather than 180°). Make a series of sketches that more accurately depict the structures of these ions.

18.19 The peroxydiphosphate ion has the same general composition as the peroxydisulfate ion, $S_2O_8^{2-}$, with P substituted for sulfur. Draw the electron-dot structure for the peroxydiphosphate ion and indicate the charge on the ion. (Hint. The charge is not −2.)

18.20 What two general methods for the preparation of nonmetal hydrides were presented in this chapter?

18.21 What is the Haber process? Why is the chemical reaction in the Haber process run at a high pressure and high temperature?

18.22 Under the proper conditions, HI will react with H_2SO_4 to generate I_2 and H_2S. Write a balanced chemical equation for this reaction.

18.23 Write balanced chemical equations for the hydrolysis of each of the following

(a) Mg_3N_2 (d) Ca_3P_2
(b) CaC_2 (e) Na_2S
(c) Mg_2Si (f) NaF

18.24 What is the structure of B_2H_6? What type of bonding exists in this molecule?

18.25 What is a disproportionation reaction? Give two examples.

18.26 List three methods of preparing non-metal oxides and write a chemical reaction to illustrate each one.

18.27 Hydrogen forms ionic hydrides with the Group IA elements and also with the heavier elements in Group IIA. Write chemical equations for the reactions of Na and Ca with elemental H_2.

18.28 Write a chemical equation for the combustion of C_8H_{18} (octane) in the presence of excess oxygen.

18.29 Write chemical equations for the reaction of the following with oxygen: (a) SiH_4 (b) PH_3 (c) AsH_3 (d) H_2S

18.30 Write chemical equations for the production of HNO_3 using N_2, H_2, O_2, and H_2O as starting materials.

18.31 Compare the molecular structures of P_4, P_4O_6, and P_4O_{10}.

18.32 What is the basic structural unit in the structure of quartz? Why do two kinds of quartz crystals exist? What is the relationship between them?

18.33 What are the geometric structures of the following:
(a) PO_4^{3-} (d) $S_2O_3^{2-}$
(b) ClO_2^{-} (e) SO_3^{2-}
(c) NO_3^{-} (f) ClO_3^{-}

18.34 Give examples of three methods of preparing oxoacids or their anions.

18.35 Why are solutions of OCl^- stable while OI^- disproportionates into I^- and IO_3^-?

18.36 Use electron-dot structures to show how $H_2S_2O_7$ results from the removal of one molecule of H_2O from two molecules of H_2SO_4.

18.37 Pyrophosphoric acid may be obtained from its sodium salt in the following way. The sodium salt is dissolved in water and treated with lead nitrate to precipitate lead pyrophosphate which, when treated with H_2S, liberates the free acid. Write chemical equations for all of the reactions that take place.

18.38 Write generalized formulas for the "ortho" anions, "pyro" anions and "meta" anions of the period 3 elements.

18.39 Illustrate the types of structures that are obtained by joining XO_4 tetrahedra through the sharing of the following:
(a) One corner
(b) Two corners
(c) Three corners
(d) Four corners

18.40 What are the structures of the following:
(a) PCl_3 (e) ClF_5 (h) SiF_4
(b) BBr_3 (f) BrF_4^- (i) SF_6
(c) SF_4 (g) SCl_2 (j) I_3^-
(d) ClF_3

18.41 In SF_6, d^2sp^3 hybrid orbitals are used to make all six valence electrons of sulfur available for bonding. What type of hybrid orbitals would be used in the bonding in SF_4? How does the resulting molecular structure compare with that predicted by the electron pair repulsion theory?

18.42 How does size influence the number of halogen atoms that can combine with a given nonmetal in (a) period 2 (b) period 5?

18.43 Sulfur hexafluoride is so stable toward reaction with other reagents and toward decomposition that it is used as a gaseous insulator in high voltage generators. What accounts for this inert behavior?

18.44 Why is $SiCl_4$ so susceptible to attack by water while CCl_4 is unreactive toward water?

18.45 Write electron-dot structures for XeO_3 and XeO_4. What geometric structures would you expect them to exhibit? What structure would you expect for $XeOF_2$ (Xe bonded to two fluorine atoms and one oxygen atom)?

19
The
Transition
Elements

In this chapter we consider the collection of elements, generally called the transition elements, which fit into the periodic table between Groups IIA and IIIA. We saw in Chapter 3 that these elements arise as a consequence of the gradual filling of *d* and *f* subshells and, as a rule, a transition element is usually considered to be one that possesses an incompletely filled *d* or *f* subshell in either the free state or in one of its compounds. We shall also include in this discussion the elements in Group IIB, zinc, cadmium, and mercury, that are found at the extreme right of the transition elements and that complete a transition series (a horizontal row) by having their outer *d* subshells filled.

To discuss their properties it is convenient to divide the transition elements into two categories. These are the **d-block elements** (or **main transition elements**), which in our condensed version of the periodic table are located in the main body of the table between Groups IIA and IIIA, and the **inner transition elements** that correspond to the two long rows of 14 elements each that *are placed just below the table*. The *d*-block elements themselves consist of three rows that are frequently referred to as the first, second, and third transition series.

    Like the representative elements, most of the *d*-block transition elements possess certain vertical similarities in chemical and physical properties and are therefore divided into groups, designated as B-groups. They begin with Group IIIB on the left and proceed through Group VIIB. Next there follows a set of nine elements collectively termed Group VIII, and finally, on the right, we find Groups IB and IIB. This numbering sequence of the B groups results because the group numbers are chosen to correspond to the highest positive oxidation state that their elements normally exhibit.

    The division of the periodic table into A and B groups (e.g., IIIA and IIIB) suggests that there may be certain parallels between the two, and to a limited degree this is true. The similarities, however, are primarily restricted to likenesses of composition, structure, and maximum positive oxidation state, instead of chemical reactivity. Some examples of these are found in Table 19.1.

    The Group VIII elements, which lie between Groups VIIB and IB, are classed differently from the other *d*-block elements because they have no counterparts among the representative elements. Within this group there are greater *horizontal similarities* than vertical ones and the description of the behavior of these elements is usually organized on the basis of horizontal groups of three elements each, called **triads.** Each triad is named after the

19.1
General
Properties

554

Table 19.1

Some Similarities of Chemical Composition Between *A* and *B* Group Compounds

Group	Compounds		Group	Compounds
IVA	CCl_4, $SnCl_4$, CO_2		IVB	$TiCl_4$, TiO_2
VA	PO_4^{3-}, $POCl_3$		VB	VO_4^{3-}, $VOCl_3$
VIA	SO_4^{2-}, $S_2O_7^{2-}$		VIB	CrO_4^{2-}, $Cr_2O_7^{2-}$
VIIA	ClO_4^-, Cl_2O_7		VIIB	MnO_4^-, Mn_2O_7
IA	NaCl		IB	CuCl, AgCl
IIA	$CaCl_2$		IIB	$ZnCl_2$

best-known element within it. Thus we have the iron triad, palladium triad, and the platinum triad.

As a class, the transition elements are all typical metals; they are good conductors of heat and electricity (silver has the highest electrical and thermal conductivity of any metal) and possess a characteristic metallic luster. Their chemical and physical properties, however, cover a wide range. Most are hard and melt at high temperatures. Tungsten (atomic symbol W), for example, has the highest melting point of any element, approximately 3400° C. In contrast, mercury melts at −39° C and, of course, is a liquid at room temperature.

The chemical reactivity of the free elements varies greatly too. Most react directly with nonmetals such as oxygen and the halogens to produce the corresponding oxides and halides. In fact, some of the transition elements are so easily oxidized that they react with water to liberate hydrogen. This is true for scandium (Sc), yttrium (Y), lanthanum (La), and the lanthanide elements (atomic numbers 58 to 71) that have very negative reduction potentials and react according to the equation

$$M + 3H_2O \longrightarrow \tfrac{3}{2}H_2 + M(OH)_2$$

Other transition elements, such as Pt and Au, are very resistant to oxidation and are insoluble in both protonic acids like HCl as well as oxidizing acids such as HNO_3. It is interesting, however, that these two metals do dissolve slowly in a 3:1 mixture of HCl and HNO_3 (called *aqua regia*).

Despite some rather marked differences in behavior, there are several characteristics that the transition elements have in common with each other:

(a) **Multiple oxidation states.** With only a few exceptions, the transition elements tend to exhibit more than one oxidation state.

(b) **Many of their compounds are paramagnetic.** Because the transition elements tend to have partially completed *d* or *f* subshells in both the free state and in their compounds, the metal atoms often possess unpaired electrons. These impart the property of paramagnetism.

(c) **Many (if not most) of their compounds are colored.** The origin of the colors of complex ions of the transition elements will be discussed in Section 19.11.

(d) **They have a strong tendency to form complex ions.** As a group, these elements form a huge number of complex ions of varying degrees of complexity. The last seven sections of this chapter are devoted to a discussion of their structures and bonding.

19.2
Electronic Structure and Oxidation States

In Chapter 3 we saw that as we proceed from left to right across a period through the main transition elements, there is a gradual filling of the d subshell that lies just below the outer shell. In period 4, for example, this corresponds to the 3d subshell, as seen for the electronic structures of the first row elements given in Table 19.2. Each of these elements possesses a completed argon core with additional electrons in the 3d and 4s subshells. Notice once again that chromium and copper are anomalous owing to the extra stability that is associated with half-filled and filled subshells. Similar irregularities are found in the second and third transition series (Table 19.3), although other factors in addition to those having to do with half-filled and filled subshells are apparently involved and, therefore, no simple correlations can be made.

Among the inner transition elements, the lanthanides and actinides (so named because lanthanum and actinium have properties more or less typical of their respective series), there is a gradual filling of an f subshell that lies *two* shells below the outer shell. Thus, in Table 19.4 we see that as we pass through the lanthanides in period 6 the 4f subshell is completed, whereas in the following period, as we pass through the actinides, the 5f subshell becomes populated.

The chemical and physical properties of the transition elements are controlled, of course, by their electronic structures. With the d-block elements the outer s and underlying d subshells are of nearly equal energy and, consequently, when these elements react, the d electrons are able to participate in bonding. The importance of the d electrons in determining the chemistry of these elements accounts for their varied chemical properties, including the multiplicity of oxidation states, illustrated for the first transition series in Table 19.5.

Examination of Table 19.5 reveals several interesting points. First, we see that *except for Sc all of the first row elements show a +2 oxidation state*. You might recall that, in general, when electrons are lost by an atom, those lost first come from the subshell having the highest principal quantum number. Consequently the 4s electrons are lost before the 3d; therefore the +2 state results simply from the loss of the two outer 4s electrons. Exceptions to this are Cr and Cu where a 3d electron is also lost.

A second point is that *elements at the left of a transition series prefer the highest oxidation state and, as we proceed to the right, the stabilities of the lower oxidation states increase relative to the higher ones*. Thus with scan-

Table 19.2
Electronic Structures for Elements in the First Transition Series

Sc[Ar]$3d^14s^2$	Fe[Ar]$3d^64s^2$
Ti[Ar]$3d^24s^2$	Co[Ar]$3d^74s^2$
V[Ar]$3d^34s^2$	Ni[Ar]$3d^84s^2$
Cr[Ar]$3d^54s^1$	Cu[Ar]$3d^{10}4s^1$
Mn[Ar]$3d^54s^2$	Zn[Ar]$3d^{10}4s^2$

Table 19.3
Electronic Structures of Elements of the Second and Third Transition Series

Period 5 Second Transition Series	Period 6 Third Transition Series
$Y[Kr]5s^24d^1$	$La[Xe]6s^25d^1$
$Zr[Kr]5s^24d^2$	$Hf[Xe,4f^{14}]5d^26s^2$
$Nb[Kr]5s^14d^4$	$Ta[Xe,4f^{14}]5d^36s^2$
$Mo[Kr]5s^14d^5$	$W[Xe,4f^{14}]5d^46s^2$
$Tc[Kr]5s^14d^6$	$Re[Xe,4f^{14}]5d^56s^2$
$Ru[Kr]5s^14d^7$	$Os[Xe,4f^{14}]5d^66s^2$
$Rh[Kr]5s^14d^8$	$Ir[Xe,4f^{14}]5d^76s^2$
$Pd[Kr]5s^04d^{10}$	$Pt[Xe,4f^{14}]5d^96s^1$
$Ag[Kr]5s^14d^{10}$	$Au[Xe,4f^{14}]5d^{10}6s^1$
$Cd[Kr]5s^24d^{10}$	$Hg[Xe,4f^{14}]5d^{10}6s^2$

dium all three electrons ($3d^14s^2$) are lost, and only the maximum +3 state is observed. With titanium, +2, +3, and +4 states are found; however, in aqueous solution the +4 state, corresponding to the "loss" of all four outer electrons, is most stable. In fact, the +2 state is so unstable that it reduces water to produce H_2 and Ti(IV), indicating that an oxidizing agent as weak as H_2O is capable of removing electrons from Ti^{2+}.

As we continue to move to the right through the first transition series, we find that up to manganese the maximum oxidation state that is observed corresponds to the group number. However, the most stable state gradually becomes the lower one, while the higher oxidation states tend to become good oxidizing agents. With chromium in Group VIA, for example, the most stable state is Cr^{3+}. The +2 state is very easily oxidized to +3 whereas the +6

Table 19.4
Electronic Structure of the Lanthanide and Actinide Elements

Lanthanides	Actinides
$La[Xe]5d^16s^2$	$Ac[Rn]6d^17s^2$
$Ce[Xe]4f^26s^2$	$Th[Rn]6d^27s^2$
$Pr[Xe]4f^36s^2$	$Pa[Rn]5f^26d^17s^2$
$Nd[Xe]4f^46s^2$	$U[Rn]5f^36d^17s^2$
$Pm[Xe]4f^56s^2$	$Np[Rn]5f^57s^2$
$Sm[Xe]4f^66s^2$	$Pu[Rn]5f^67s^2$
$Eu[Xe]4f^76s^2$	$Am[Rn]5f^77s^2$
$Gd[Xe]4f^75d^16s^2$	$Cm[Rn]5f^76d^17s^2$
$Td[Xe]4f^96s^2$	$Bk[Rn]5f^86d^17s^2$
$Dy[Xe]4f^{10}6s^2$	$Cf[Rn]5f^{10}7s^2$
$Ho[Xe]4f^{11}6s^2$	$Es[Rn]5f^{11}7s^2$
$Er[Xe]4f^{12}6s^2$	$Fm[Rn]5f^{12}7s^2$
$Tm[Xe]4f^{13}6s^2$	$Md[Rn]5f^{13}7s^2$
$Yb[Xe]4f^{14}6s^2$	$No[Rn]5f^{14}7s^2$
$Lu[Xe]4f^{14}5d^16s^2$	$Lr[Rn]5f^{14}6d^17s^2$

Table 19.5

Oxidation States of the First Transition Series Elements

Sc	(+2)	Not known
	+3	Only oxidation state; Sc reduces H_2O to $H_2(g)$
Ti	+2	Not stable in H_2O, reduces water
	+3	Prepared by reducing Ti(IV) with Zn
	+4	Most stable oxidation state
V	+1	
	+2	Easily oxidized
	+3	Stable
	+4	Most stable under ordinary conditions
	+5	V_2O_5 moderate oxidizing agent
Cr	+2	Very easily oxidized
	+3	Most stable
	+6	CrO_4^{2-}, $Cr_2O_7^{2-}$; good oxidizing agent
Mn	**+2**	Most stable
	+3	Stable in complex ions
	+4	MnO_2; good oxidizing agent
	+6	MnO_4^{2-} (manganate ion) stable in basic solution only
	+7	MnO_4^- (permanganate ion) very powerful oxidizing agent
Fe	+2	Stable but easily oxidized to +3
	+3	Most stable
	+4	
	+6	Rare
Co	**+2**	Most stable in water
	+3	Co^{3+} oxidizes water, stable in complex ions
Ni	**+2**	Most stable
	+3	Rare, powerful oxidizing agent
Cu	+1	Cu^+ disproportionates in water, $2Cu^+ \longrightarrow Cu^{2+} + Cu$. Stable in complex ions and insoluble CuCl
	+2	Most stable in water
Zn	**+2**	Only oxidation state for zinc

state, found in the CrO_4^{2-} (chromate) and $Cr_2O_7^{2-}$ (dichromate) ions, is a good oxidizing agent. For manganese, the next element to the right, the +2 oxidation state is most stable and higher states tend to be very readily reduced. Permanganate ion, MnO_4^-, containing Mn(VII) is a very powerful oxidizing agent.

$$MnO_4^- + 8H^+ + 5e^- = Mn^{2+} + 4H_2O \qquad \mathscr{E}° = +1.49$$

After manganese, the occurrence of high positive oxidation states is rare. With iron the +2 state is fairly easily oxidized to Fe^{3+}, although stable solutions (as well as compounds) containing Fe^{2+} are readily prepared. Moving to the right to cobalt, the +2 state tends to be most stable and, in fact, in the absence of complex ion-forming agents Co^{3+} oxidizes water.

$$2H_2O + 4Co^{3+} \longrightarrow O_2 + 4H^+ + 4Co^{2+}$$

For nickel the +3 state is very rare, and only the +2 state is observed under ordinary conditions. Following nickel we have copper, which forms compounds in both the +1 and +2 states, although the +1 state is easily oxidized. Finally we arrive at zinc, which can lose but two electrons to form Zn^{2+} having a pseudonoble gas configuration.

The trend toward increasing stability of the lower oxidation states as we proceed across a transition series occurs for the second and third series too, as seen in Table 19.6. In the second row, for instance, the maximum oxidation state is most stable for Zr while the lowest oxidation state is most stable for Ag. In the third transition series we again see that the lower oxidation states become more prevalent and increasingly more stable as we move to the right. Thus mercury, at the end of the series in Group IIB, forms compounds in both the +1 and +2 states. The +1 oxidation state of mercury seems surprising at first glance since it would appear to contain a single electron in an s orbital outside a pseudonoble gas core. However, experimental evidence indicates that the mercury(I) species is actually a dimer, Hg_2^{2+}. For example, mercurous compounds such as Hg_2Cl_2 are diamagnetic, indicating that no unpaired electrons are present. A simple Hg^+ ion would contain an odd number of electrons (79) and could not possibly have all of them paired. In the Hg_2^{2+} ion, two Hg^+ ions seem to be joined by a covalent bond, thus pairing the odd electrons on each mercury atom. Other evidence for the exis-

Table 19.6
Oxidation States of Second and Third Transition Series Elements
(common oxidation states in boldface type)

Y	+3		La	+3
Zr	+2, +3, **+4**		Hf	+3, **+4**
Nb	+2, +3, +4, **+5**		Ta	+2, +3, +4, **+5**
Mo	+2, +3, **+4, +5, +6**, +8		W	+2, +3, +4, **+5, +6**
Tc	+2, +3, **+4**, +5, +6, **+7**		Re	**+3, +4**, +5, +6, **+7**
Ru	**+2, +3, +4**, +5, +6, +7, +8		Os	+2, +3, **+4**, +5, **+6**, **+8**
Rh	+1, +2, **+3, +4**, +5, +6		Ir	+1, +2, **+3, +4**, +5, +6
Pd	**+2**, +3, **+4**		Pt	**+2**, +3, **+4**, +5, +6
Ag	**+1**, +2, +3		Au	**+1, +3**
Cd	**+2**		Hg	**+1, +2**

tence of the Hg_2^{2+} species relates to equilibria, as illustrated in Problem 19.9 at the end of the chapter.

Among the representative elements we found that as we proceed downward within a group the lower oxidation states become increasingly more stable compared to the higher ones. In the transition elements this trend is reversed and *in the second and third transition series it is found that the higher oxidation states are preferred.* For example, in Group VIB the CrO_4^{2-} ion, with Cr in the +6 oxidation state, is a potent oxidizing agent indicating that the chromate ion wants to pick up electrons to reduce the Cr to a lower oxidation state. In contrast to this, the MoO_4^{2-} ion and other complex polynuclear molybdates are poor oxidizing agents owing to the stability of the Mo(VI) species.

In a similar fashion, comparing the Group VIIB elements we find that the MnO_4^- ion is a very powerful oxidizing agent whereas the ReO_4^- ion has very little tendency to be reduced. In Group VIII, iron forms compounds in the +2 and +3 states, although there are unusual and relatively unstable compounds containing iron in oxidation states as high as +6. For example, the FeO_4^{2-} ion (containing Fe(VI)) is an even stronger oxidizing agent in acid solution than the permanganate ion. Below iron, however, we find ruthenium and osmium in oxidation states as high as +8 (e.g., RuO_4 and OsO_4).

In contrast to the wide range of chemical properties of the *d*-block elements, the lanthanides exhibit a remarkable sameness of properties. The 4*f* subshell, which is only partially filled for most of these elements, is buried beneath the outer 5*d* and 6*s* subshells and does not interact to an appreciable extent with the surrounding chemical environment. Consequently the chemistry of the lanthanides, like that of lanthanum itself, is predominantly that of the +3 ion and differences in behavior depend primarily on differences in ionic size.

The actinide elements exhibit a greater variation in oxidation numbers than do the lanthanides (e.g., uranium forms compounds in the +3, +4, +5, and +6 oxidation states). This is sometimes ascribed to the notion that the 5*f* orbitals of the actinides project out further toward the periphery of the atom than do the 4*f* orbitals of the lanthanides. As a result, the 5*f* orbitals are able to become involved to a greater degree in chemical bonding; hence more complex chemistry is observed.

We have seen before that many trends in properties can be correlated with variations that occur in atomic and ionic radii. This is true among the transition elements as well as the representative elements.

19.3 Atomic and Ionic Radii

In Chapter 3 the horizontal and vertical trends in atomic size were discussed. Let us briefly review them here. As we move across a given transition series there is only a gradual decrease in atomic radius. This is because the 3*d* electrons that are added to the atom shield the outer 4*s* electrons quite well from the increasing nuclear charge and, as a result, the effective nuclear charge experienced by the outer electrons rises only slowly. Consequently, only a small size decrease occurs (Table 19.7).

Vertically, we find a rather large increase in size among the *d*-block elements between periods 4 and 5 (from the first transition series to the second), just as we expect. However, between periods 5 and 6 there is only a very small size increase (and in some cases, none at all). This is a consequence of the **lanthanide contraction.** The gradual size decrease that

Table 19.7
Atomic Radii (Å)

Sc	Ti	V	Cr	Mn	Fe	Co	Ni	Cu	Zn
1.44	1.32	1.22	1.17	1.17	1.16	1.16	1.15	1.17	1.25

Y	Zr	Nb	Mo	Tc	Ru	Rh	Pd	Ag	Cd
1.62	1.45	1.34	1.29	1.27	1.24	1.25	1.28	1.44	1.41

La	Hf	Ta	W	Re	Os	Ir	Pt	Au	Hg
1.69	1.44	1.34	1.30	1.28	1.26	1.26	1.29	1.44	1.44

occurs across the lanthanide series (from atomic numbers 58 to 72) apparently just cancels the size increase that would be expected as we go from period 5 to period 6. Consequently, the period 6 elements that follow the lanthanides are essentially the same size as those above them in period 5.

These variations in size have some very pronounced chemical and physical consequences. They can be correlated, for example, with variations in ionization energies (I.E.), as shown in Table 19.8. Here we see that the gradual horizontal size decrease that we associated with an increase in effective nuclear charge is also accompanied by an increase in I.E. As we might expect, this increasing difficulty encountered in removing an outer electron from the isolated atoms is reflected in a gradual (although not altogether uniform) rise in their standard reduction potentials. In other words, as we proceed across the table from left to right, it generally becomes more difficult to oxidize the elements.

The effect of the lanthanide contraction is demonstrated in these properties as well. Among the representative elements the I.E. generally decreases as we proceed down a group, paralleling the increase in size. This phenomenon is also observed among the transition elements on going from period 4 to period 5. However, from periods 5 to 6 there is an increase in nuclear charge without an accompanying increase in size and, as a result, the I.E. increases. This in turn manifests itself in reduction potentials that tend to be quite high

Table 19.8
Ionization Energy (eV)

Sc	Ti	V	Cr	Mn	Fe	Co	Ni	Cu	Zn
6.54	6.83	6.74	6.76	7.43	7.90	7.86	7.63	7.72	9.39

Y	Zr	Nb	Mo	Tc	Ru	Rh	Pd	Ag	Cd
6.37	6.95	6.88	7.18	7.45	7.36	7.45	8.33	7.57	8.99

La	Hf	Ta	W	Re	Os	Ir	Pt	Au	Hg
5.59	6.98	7.89	7.98	7.87	8.71	8.98	8.96	9.22	10.43

for the third transition series elements, thereby accounting for their virtually inert behavior toward many oxidizing agents.

Still another consequence of the lanthanide contraction is the very high density of the third row elements. For example, from Rh to Ir we find an increase of 89 mass units (about an 87% increase in mass) with only a 0.01 Å increase in radius. Iridium therefore has 87% more mass packed into essentially the same size atom; it is not surprising, therefore, to find that Ir is about 81% more dense than Rh ($d_{Ir} = 22.5$ g/ml, $d_{Rh} = 12.4$ g/ml).

Metallurgy is the process whereby a metal is extracted from its ore and brought to the point where it can be put to practical use. In Chapter 17 we discussed one aspect of this, the reduction of metal compounds to produce the free element. The desirable physical properties of many of the transition metals, such as high strength, hardness, and high melting points, make them extremely important in modern technology. Some of the methods and procedures that are used to obtain them from their ores are worth examining.

An **ore** is a substance that contains a particular desirable constituent in a high enough concentration that its extraction from the ore is economically worthwhile. Thus, for example, many minerals may contain small amounts of iron but only those that are rich in iron are used to prepare this metal. As the earth's reserves of rich ores are consumed, it will become the job of the chemist to devise new ways to obtain metals such as iron from less rich ores.

Generally, we can divide metallurgical processes into three categories:

1. **Concentration.** Ores that contain substantial amounts of impurities, such as rock, must often be treated to concentrate the metal-bearing constituent. Pretreatment of an ore is also carried out to convert some metal compounds into substances that can be more easily reduced.

2. **Reduction.** This is the topic discussed in Chapter 17. The particular procedure employed for a given metal depends on its ease of reduction to the free state.

3. **Refining.** Often, during reduction, substantial amounts of impurities become introduced into the metal. Refining is the process whereby these impurities are removed and the composition of the metal adjusted (alloys formed) to meet specific applications.

Let us now take a brief look at each of these steps as they apply to some important metals.

CONCENTRATION. Not all ores have to be subjected to a pretreatment step prior to reduction, although most of them must. These pretreatment procedures involve the separation of the metal-bearing component of the ore from unwanted or interfering impurities. This is particularly important for low grade ores in which the desired metal is present only in small amounts.

As expected, different methods are applied to different ores, depending on the specific properties of the impurities and the metal compounds. We can divide these procedures into two classes: *physical separations,* in which the chemical compositions of the constituents are not altered, and *chemical separations,* which make use of the chemical properties of the different substances in the ore. Some metals, such as silver and gold, are found in deposits as the free element, and their recovery simply involves removing them from the rock and sand with which they are mixed.

One of the earliest forms of physical separation was used by the "forty-niners" in panning for gold. A mixture of sand containing (hopefully!) particles of metallic gold was placed in a shallow pan with water. The mixture was swirled about and the sand was washed away, leaving the gold dust in the bottom of the pan. The success of this procedure was based on the fact that gold is about nine times as dense as the sand and gravel impurities. As a result, the lighter impurities are more easily washed away than the more dense metal.

Another way of removing metallic gold (and silver) from its ore is to treat the mixture with metallic mercury (a liquid) in which silver and gold dissolve to form an alloy called an **amalgum.** The silver and gold are later recovered by distilling away the mercury that is reclaimed and used again.

A physical separation technique that can be applied to the sulfide ores of zinc, copper, and lead is called **flotation.** In this process, illustrated in Figure 19.1, the ore is finely ground and added to a mixture of water and oil, containing suitable additives. The metal-bearing component of the ore becomes coated by the oil while the unwanted material, called the **gangue,** is wetted by the water. A stream of air is then blown through the mixture and the oil-covered mineral is carried to the surface by bubbles where it is trapped in a froth that can be removed to recover the metal compound. The gangue, on the other hand, simply settles to the bottom of the apparatus and is later discarded.

Chemical methods of concentrating the metal-bearing component of an ore vary considerably because of the variety of chemical properties exhibited by the metals and their compounds. For example, aluminum, whose electrolytic reduction was described in Chapter 15, occurs in deposits of bauxite, a form of Al_2O_3. In this case the ore is concentrated by taking advantage of the

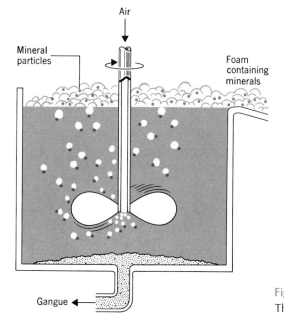

Figure 19.1
The flotation process.

amphoteric behavior of aluminum. The bauxite is treated with concentrated base that dissolves the Al_2O_3 to produce aluminate ion, AlO_2^-.[1]

$$Al_2O_3 + OH^- \longrightarrow 2AlO_2^- + H_2O$$

After it is removed from the gangue, the solution is acidified, precipitating $Al(OH)_3$, which yields pure Al_2O_3 when heated.

$$2Al(OH)_3 \longrightarrow Al_2O_3 + 3H_2O$$

This purified aluminum oxide serves as the charge in the Hall process discussed previously.

Another chemical pretreatment that is often given to a sulfide ore is called **roasting.** Here the ore is heated in air, converting the metal sulfide to an oxide that is more conveniently reduced.

$$2PbS + 3\ O_2 \longrightarrow 2PbO + 2SO_2$$
$$2ZnS + 3\ O_2 \longrightarrow 2ZnO + 2SO_2$$

This industrial process is a severe source of air pollution unless the SO_2 is recovered.

REDUCTION. In Chapter 17 we saw that there are several ways of reducing a metal compound to produce the free element. All of these methods have found application in the production of metals commercially. Titanium, for example, is prepared by reacting $TiCl_4$ with the more active metal, magnesium. The $TiCl_4$ is produced from rutile, a fairly pure source of TiO_2, by reaction with chlorine gas and carbon.

$$TiO_2 + 2C + 2Cl_2 \longrightarrow TiCl_4 + 2CO$$

The titanium tetrachloride is a volatile liquid (boiling point $= 136°\,C$) and can be separated from impurities by distillation. The final reduction to the metal follows the equation

$$TiCl_4 + 2Mg \longrightarrow Ti + 2MgCl_2$$

Titanium is a very useful metal because it is considerably lighter than steel ($d_{Fe} = 7.86$ g/ml, $d_{Ti} = 4.51$ g/ml), yet does not lose its strength at high temperature as does aluminum. It is used in large quantities in jet engines and replaces aluminum and steel in other aircraft applications.

Active metals such as magnesium are very expensive reducing agents because they themselves are difficult and costly to prepare. As a result, less expensive reducing agents are employed whenever possible. One of the cheapest of all is carbon, in the form of coke, which is produced from coal by heating it at high temperatures in the absence of air. This treatment drives off the volatile components of the coal (from which other important chemicals are derived) leaving nearly pure carbon behind.

Some typical reductions of metal oxides with carbon were illustrated in Section 17.3, and many of the transition metals (e.g., Zn, Cd, Fe, Co, Ni, Mo) can be prepared in this way. Undoubtedly the most important chemical reduction brought about with carbon is that of iron oxide, Fe_2O_3, to iron. This

[1] The exact nature of this species is not known with certainty. See page 481.

$$3Fe_2O_3 + CO \longrightarrow 2Fe_3O_4 + CO_2$$

$$Fe_3O_4 + CO \longrightarrow 3FeO + CO_2$$

$$FeO + CO \longrightarrow Fe + CO_2$$

$$CO_2 + O \longrightarrow 2CO$$

$$C + O_2 \longrightarrow CO_2$$

Charge added at top

500°C

850°C

1000°C

Compressed air

1300°C

1900°C

Slag

Molten iron

Figure 19.2
The blast furnace.

is accomplished in the **blast furnace,** developed in about 1300 A.D. and that in modern times takes the form shown in Figure 19.2.

The blast furnace is charged with a mixture of limestone, coke, and iron ore. The ore is generally composed primarily of Fe_2O_3 with impurities of SiO_2 ($\sim 10\%$) and smaller amounts of compounds containing sulfur, phosphorus, aluminum and manganese. Heated air is forced in at the bottom of the furnace where it reacts with carbon in a very exothermic reaction to produce carbon dioxide.

$$C + O_2 \longrightarrow CO_2 \qquad \Delta H = -94 \text{ kcal}$$

The large amount of heat generated in this region of the furnace raises the temperature to nearly 1900° C. As the hot gases rise, the CO_2 reacts with additional carbon in an endothermic reaction to form carbon monoxide, the active reducing agent in the furnace.

$$CO_2 + C \longrightarrow 2CO \qquad \Delta H = 41 \text{ kcal}$$

The reduction of the iron oxide takes place in a series of steps. Near the top of the furnace, Fe_2O_3 is reduced to Fe_3O_4.

$$3Fe_2O_3 + CO \longrightarrow 2Fe_3O_4 + CO_2$$

Farther down, in a hotter region of the furnace, this is reduced to FeO.

$$Fe_3O_4 + CO \longrightarrow 3FeO + CO_2$$

Finally, still farther down the FeO is reduced to the metal which, at these high temperatures, is a liquid and trickles down to form a pool of molten metal at the base of the tower.

$$FeO + CO \longrightarrow Fe + CO_2$$

The function of the limestone in the furnace is to provide a basic medium with which acidic oxides, such as SiO_2 and P_2O_5, can react. At elevated temperatures limestone, $CaCO_3$, decomposes to form lime (CaO) and CO_2 according to the equation,

$$CaCO_3 \longrightarrow CaO + CO_2$$

The lime then reacts with the acidic oxides,

$$CaO + SiO_2 \longrightarrow CaSiO_3$$
$$3CaO + P_2O_5 \longrightarrow Ca_3(PO_4)_2$$
$$CaO + Al_2O_3 \longrightarrow Ca(AlO_2)_2$$

The products of these reactions have relatively low melting points and are liquids when they are formed. The mixture, called **slag,** also runs to the base of the furnace where it floats atop the molten iron. As these two layers are formed, the charge in the furnace settles and additional limestone-coke-ore mixture is added at the top. In this way the blast furnace operates continuously, with fresh charge being added at the top and molten iron and slag being tapped off at the bottom. These furnaces are often run for months at a time before being shut down for routine maintenance.

The liquid iron, when it is withdrawn from the blast furnace, is called **pig iron** and consists of about 95% Fe and approximately 4% carbon, with small amounts of silicon, manganese, phosphorus, and sulfur. This somewhat impure iron is very hard and can be poured into molds as **cast iron.** The slag that comes from the furnace can be used in making cement.

REFINING. In the process of separating a metal from its ore, impurities are often introduced that impart undesirable properties to the final product. Consequently, it is generally necessary to purify the metal before it can be put to practical use. This purification process is called **refining.**

The specific procedure employed for refining a given metal depends on the chemical and physical properties of the metal as well as the properties of the impurities. As a result there is no single method applicable to a very large number of different metals. We saw in Chapter 15 that copper can be economically refined electrolytically. This occurs, however, primarily because the silver and other precious metals recovered from the electrolytic cell offset the generally high cost of electricity.

An interesting process for refining nickel, called the Mond process, makes use of the relative ease of formation of a compound formed between nickel and carbon monoxide.

$$Ni + 4CO \longrightarrow Ni(CO)_4$$

Compounds of this general type are called **carbonyls:** nickel carbonyl, besides being easily formed, is also very volatile (and very poisonous). The impure nickel is therefore treated with CO at a moderately low temperature of 60° C where the $Ni(CO)_4$ that is formed exists as a gas. This is circulated to another portion of the apparatus where it is heated to about 200° C and decomposes to give pure nickel plus CO, which can be recycled through the process.

The most important commercial refining process involves the conversion of pig iron into steel. This requires that impurities such as silicon, sulfur, and phosphorus be removed and the carbon content lowered significantly from the approximately 4% introduced into the pig iron in the blast furnace.

Modern steelmaking began with the introduction of the **Bessemer converter** in England in 1856. A batch of molten pig iron from the blast furnace, weighing about 25 tons, is transferred to a tapered cylindrical vessel containing a refractory lining (Figure 19.3). The composition of the lining is determined in part by the nature of the impurities in the iron. Since these impurities are usually silicon, phosphorus, and sulfur, whose oxides are acidic, a basic lining of dolomite (a $MgCO_3$, $CaCO_3$ mineral) is generally used. A blast of air (or oxygen) is blown through the melt from a set of small holes at the bottom of the vessel. The oxygen passing through the molten metal converts the silicon, phosphorus, and sulfur to oxides that then react with the lining to form a slag. The carbon in the pig iron is also oxidized to CO and its concentration is also reduced. The conversion of the pig iron to steel by this process is rapid, requiring about 15 minutes, and gives rise to a spectacular display of

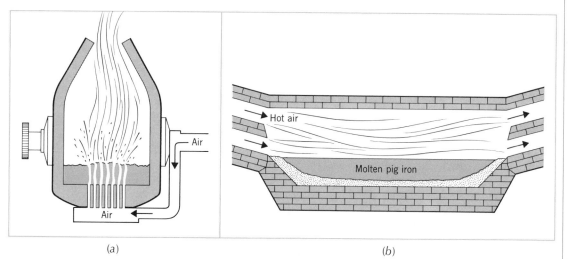

(a)　　　　　　　　　　　　　　　　(b)

Figure 19.3
(a) Bessemer converter. (b) Open hearth furnace.

fire and showers of sparks. The reaction is difficult to control, however, and the quality of the steel produced in the Bessemer converter can be quite variable.

A somewhat newer method that has almost replaced the Bessemer process employs an **open hearth furnace,** a large shallow hearth usually lined with a basic oxide refractory (e.g., MgO, CaO). The furnace is charged with a mixture of pig iron, Fe_2O_3, scrap iron, and limestone. A mixture of burning gases and hot air is played over the surface of the charge to maintain it in a molten state while a series of chemical reactions take place. Impurities in the steel are oxidized by the Fe_2O_3 and air. Carbon dioxide bubbles out of the mixture, keeping it stirred, while the SiO_2 and other acidic oxides combine with CaO (from the limestone) and the refractory lining to form a slag. This entire process takes much longer than the Bessemer process, requiring 8 to 10 hours to complete. However, the quality of the steel is much more easily controlled since chemical analyses can be constantly carried out on samples of the mixture. The increased length of time required to process a batch of steel is also offset by the fact that much larger quantities (about 200 tons) can be handled at one time. In addition, prior to being poured from the furnace, other metals (e.g., cobalt, chromium, nickel, vanadium, and tungsten) can be added to the steel to form alloys having special properties. A typical stainless steel, for instance is composed of approximately 72% iron, 19% chromium and 9% nickel.

Modern methods of chemical analysis, making use of high speed computers, have enabled a return to a modified form of the Bessemer process. This newer procedure, which is replacing the open hearth furnace because of its speed, involves forcing a mixture of powdered $CaCO_3$ and oxygen gas into the molten pig iron. This rapidly burns away the impurities, which form a slag. The characteristic emission spectra of the elements in the steel permit rapid chemical analysis, and additives can be incorporated into the steel in the proper proportions to give a product with the desired properties. This **basic oxygen process** takes only about 20 to 25 minutes to complete, thereby giving a very substantial savings in time (and, of course, *money*) over the open hearth process.

In Chapter 3 we saw that the presence of unpaired electrons in an atom or molecule impart to the substance the property called paramagnetism. The tiny electron magnets cause the atom or molecule as a whole to behave as a small magnet. When these are placed into a magnetic field, the microscopic magnets tend to align themselves with and be attracted toward the field. However, thermal motion operates to randomize the orientations of the little magnets with the net result that only a relatively small number of tiny magnets are aligned with the field at any particular instant. Consequently an ordinary paramagnetic substance is drawn only weakly into an external magnetic field.

It is characteristic of the transition elements that they and their compounds possess a partially filled d or f subshell; as a result, many of these species exhibit the phenomenon of paramagnetism. The prediction of magnetic properties of transition metal compounds is therefore one of the requisites of a theory of bonding that is applicable to these substances. We shall explore this a little further when we discuss complex ions later in this chapter.

19.5
Magnetism

Related to the property of paramagnetism is the phenomenon called **ferromagnetism,** observed for the three pure elements, iron, cobalt, and nickel. Ferromagnetic materials, like paramagnetic ones, are also attracted to a magnetic field; however, the magnitude of the interaction for a ferromagnetic substance is approximately a million times stronger than with paramagnetic materials. How does this occur?

The origin of ferromagnetism is the same as paramagnetism—that is, the existence of unpaired electrons in the ferromagnetic material. In these substances it is believed that there exist regions, called **domains,** containing very large numbers of paramagnetic atoms that have their atomic magnets all lined up in the same direction, as illustrated in Figure 19.4. Ordinarily these domains are randomly oriented in the ferromagnetic solids so that even though each domain behaves as a relatively large magnet, their effects cancel. When placed in a magnetic field, the domains tend to become oriented much the same as the atomic magnets of a paramagnetic substance. In this case, however, each time one domain becomes aligned with the field, millions of tiny atomic magnets become aligned. As a result the interaction between the ferromagnetic solid and the magnetic field is very much larger than that experienced by paramagnetic substances.

When the magnetic field is removed from a paramagnetic substance, its atomic magnets very quickly become randomly oriented and no permanent magnetism is induced. For a ferromagnetic substance, however, the domains tend to remain in the orientation that they found themselves when the external magnetic field was present. This alignment of domains in the absence of an external field causes the substance to possess a residual magnetism; we

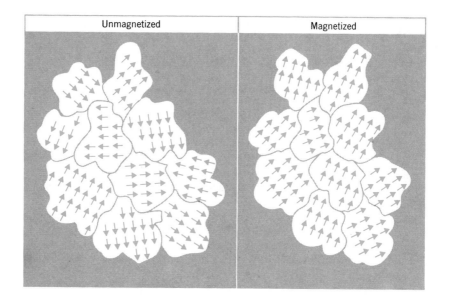

| Unmagnetized | Magnetized |

Figure 19.4
Domains in a ferromagnetic solid.

say that it has become *permanently magnetized*. Any piece of iron (e.g., a pin) can be magnetized simply by stroking it with another permanent magnet.

A permanent magnet is really not permanent, since the magnetism may be destroyed by either heating the solid or by pounding it. In the first case the increased thermal motion causes the domains to become randomly oriented whereas in the second instance violent vibrational motions cause the domains to twist and turn and become disoriented.

The phenomenon of ferromagnetism is associated only with the solid state. Iron, for example, is no longer ferromagnetic when it is melted. Instead it exhibits only paramagnetism. Melting of the solid thus appears to destroy the domains, and each individual atom in the liquid behaves more or less independently of the others nearby.

Even in the solid state not all elements containing unpaired electrons are ferromagnetic. Manganese, for example, possesses five unpaired electrons compared to only four for iron; yet pure iron is ferromagnetic but pure manganese is not. Apparently a requirement for ferromagnetism is that the spacings between paramagnetic ions be just right so that they may lock onto each other to form a domain. Nonferromagnetic metals, in which the ions are too close together, can sometimes be made ferromagnetic by forming an alloy. Such is the case with manganese where the addition of the proper amount of copper permits the Mn^{2+} ions in the metallic lattice to interact strongly and form domains, thereby producing a ferromagnetic alloy.

The transition elements are known for their ability to form many complex compounds (e.g., complex ions) in which the metal cation is surrounded by two or more ions or molecules, generally referred to as ligands. These complexes are also called **coordination compounds** because, from the point of view of the valence bond theory (the first bonding theory to be applied to them), they are considered to be held together by *coordinate* covalent bonds between the ligands and the metal. Because of the extremely large number of these coordination compounds, as well as their unusual colors, magnetic properties, structures, and chemical reactions, the study of them has become one of the major areas of inorganic chemical research.

As the study of biochemistry has progressed, it has also become evident that many biologically important molecules owe their biological activity to a metal ion held in a "complex ion" within the molecule. Hemoglobin, containing iron(II) atoms, is a well-known example. The importance of metal ions in biosystems (not to mention the increased availability of research funding in biochemistry), has recently turned many inorganic chemists into bioinorganic chemists. We shall take a closer look at metal-containing biomolecules in Chapter 21.

The father of modern coordination chemistry was Alfred Werner, who received the Nobel Prize in Chemistry in 1913 for his work on these compounds. Werner was the first to recognize that metal ions could combine with other molecules or ions through more than one type of "valence" to produce relatively stable complex species. He was also the first to propose structures of complex ions that were consistent with their properties.

Metal complexes are formed with many kinds of ligands, including ions such as Cl^-, CN^-, and NO_2^-, as well as neutral molecules like H_2O or NH_3. Nearly all ligands, however, have one thing in common; they possess a lone pair of electrons that may be shared with the metal cation in coordinate

19.6
Coordination Compounds

covalent bonds. In this sense the formation of a complex can be viewed as a Lewis acid-base reaction; in general, we can expect that the ligands in coordination compounds will all be Lewis bases with few exceptions.

In a complex, the ligands that are attached to the metal are considered to be in a **first coordination sphere,** and in solution they are held tightly by the metal ion, compared to other ions and molecules that might also be present nearby in the mixture. When the formula of a metal complex is written, it is usually the practice to indicate the species that are bonded to the metal in the first coordination sphere by enclosing them and the metal ion within square brackets. An example is the ion

$$[CoCl_6]^{3-}$$

These brackets are *not* to be confused with those that we used earlier when we wished to denote molar concentration.[2] Note that the charge on the complex is indicated *outside* the brackets, showing that the entire complex ion carries, in this example, a charge of -3.

Ligands such as Cl^- or NH_3, which have one atom that can bond to a metal cation, are said to be **monodentate** (one "tooth") ligands. There are also many molecules and ions that are able to attach themselves to a metal ion through more than one donor atom to produce a cyclic *ring* type of arrangement. Two very common examples that have been much studied are oxalate ion, C_2O_4,

and ethylenediamine, $H_2N—CH_2—CH_2—NH_2$,

They bond to a metal ion as shown below.[3]

With cobalt(III), for example, these two ligands form complexes such as $[Co(C_2O_4)_3]^{3-}$ and $[Co(H_2NCH_2CH_2NH_2)_3]^{3+}$, whose structures are illustrated

---

[2] In our earlier discussions of complex ion equilibria, we avoided this notation specifically to prevent such confusion.

[3] The origin of chemical terminology is sometimes rather colorful. Complexes of this general type are often called **chelates,** from the Greek *chele* meaning claw. The ligand in this case bites the metal with two claws (donor atoms) much like a crab.

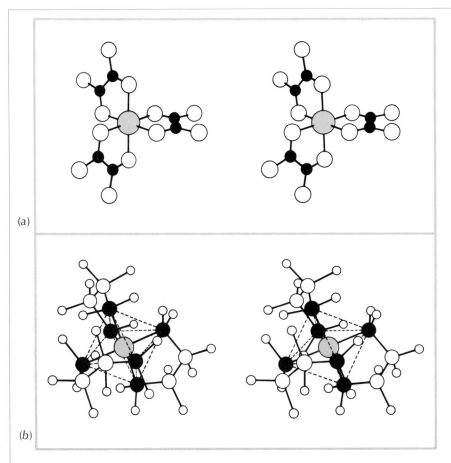

(a)

(b)

Figure 19.5

Structures of $[Co(C_2O_4)_3]^{3-}$ and $[Co(en)_3]^{3+}$. (a) $[Co(C_2O_4)_3]^{3-}$. The blue atom is cobalt, solid black atoms are carbon, and white spheres are oxygen atoms. (b) $[Co(en)_3]^{3+}$. The blue atom is cobalt, solid black atoms are nitrogen, the large white atoms are carbon, and the small white atoms are hydrogen.

in Figure 19.5. Ethylenediamine is such a common ligand in coordination chemistry that in writing formulas containing it the abbreviation, **en,** usually is used. As a result, the latter ion is generally written as simply $[Co(en)_3]^{3+}$.

Molecules or ions, such as $C_2O_4{}^{2-}$, or en, that have two atoms that may coordinate to a metal ion are said to be **bidentate** ligands. There are also more complex **polydentate** ligands containing three, four or more donor atoms. Table 19.9 contains a list of some common mono- and bidentate ligands, as well as a few examples of polydentate ligands.

The term **coordination number** (C.N.) refers to the total number of ligand atoms that are bound to a given metal ion in a complex. These atoms may be supplied by either monodentate or polydentate ligands, or both. Thus, in the three complexes, $[CoCl_6]^{3-}$, $[Co(en)_2Cl_2]^+$, and $[Co(en)_3]^{3+}$, the C.N. of cobalt

**19.7
Coordination
Number**

Table 19.9
Common Ligands Found in Complex Ions

Monodentate	H_2O	Water	Br^-	Bromide
	NH_3	Ammonia	I^-	Iodide
	CN^-	Cyanide	NO_2^-	Nitrite
	OH^-	Hydroxide	SCN^-	Thiocyanate
	F^-	Fluoride	$S_2O_3^{2-}$	Thiosulfate
	Cl^-	Chloride		

Bidentate

oxalate

ethylenediamine

o-phenanthroline

or

dipyridyl

or

Polydentate

(coordinating atoms indicated with asterisks)

Diethylenetriamine (three coordinating atoms)

$$H_2\overset{*}{N}—CH_2—CH_2—\overset{*}{N}H—CH_2—CH_2—\overset{*}{N}H_2$$

Ethylenediaminetetraacetate (six coordinating atoms)

also called EDTA

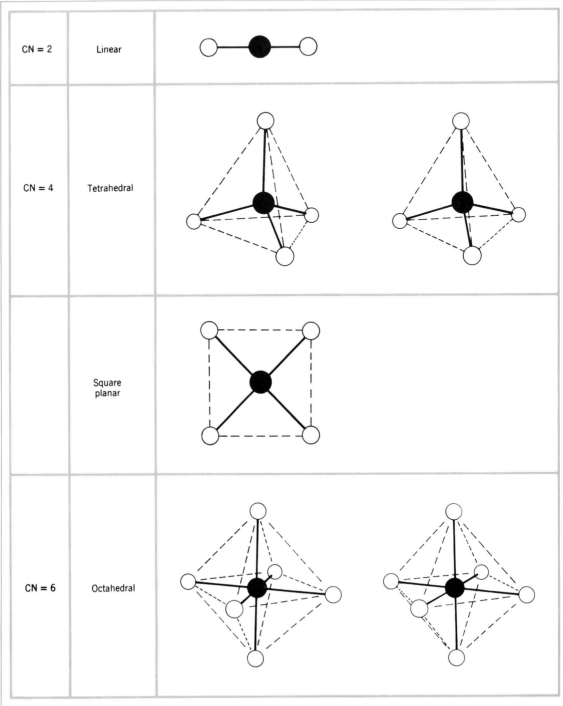

CN = 2	Linear	
CN = 4	Tetrahedral	
	Square planar	
CN = 6	Octahedral	

Figure 19.6
Structure types for coordination 2, 4, and 6.

Figure 19.7
Two-dimensional repre-
sentation of octahedral
coordinate.

is the same, since in each case there are *six* donor atoms about the Co^{3+} ion.

Coordination numbers ranging from two to more than eight are observed in various coordination compounds, the C.N. in any given instance being determined by the nature of the metal ion, its oxidation state, and to some extent, the ligands and the environment surrounding the complex. The most common coordination numbers are observed to be two, four, and six. The basic structural types that are found for these are shown in Figure 19.6.

By far the most frequently occurring coordination number in transition metal complexes is six, and the geometry that is observed in nearly all instances is octahedral. A simple two-dimensional way of representing the octahedral geometry is shown in Figure 19.7. The dotted rectangle represents the square plane (viewed in perspective) that joins the upper and lower pyramids in the octahedron. The six solid lines connect the center of the metal cation to the coordinated ligand atoms. This arrangement is illustrated in Figure 19.7 for the complex ion, $[CoCl_6]^{3-}$.

19.8
Nomenclature

In an effort to communicate with each other, chemists attempt to devise systematic approaches to the naming of chemical compounds. On a periodic basis, the International Union of Pure and Applied Chemistry (I.U.P.A.C.), composed of a group of chemists drawn from all over the world, meets to discuss current problems in nomenclature. In this way the systematic nomenclature of compounds continually evolves to meet the chemist's needs as new compounds and structures are discovered.

Below are some of the rules that have been developed by the I.U.P.A.C. to name coordination complexes. Some of the names assigned to complexes following these rules may sound odd, and even funny. Remember, however, that the chemist is primarily interested in formulating a name that is able to transmit the maximum amount of information in the shortest possible name. The end result is therefore often difficult to pronounce.

RULES OF NOMENCLATURE OF COORDINATION COMPOUNDS

1. **Cationic species are named before anionic species.** This is just like other cases of ionic compounds, such as NaCl, which is named as sodium chloride (cation, anion).
2. **Within a complex ion, the ligands are named first, followed by the metal ion.** This is opposite to the sequence in which they appear in the formula. For example, the complex $[Co(NH_3)_6]^{3+}$ is named by specifying the ammonia first, then the cobalt.
3. **The names of anionic ligands end in the suffix -o.**
   (a) Ligands whose names end in *ide* have this suffix replaced by -o

Anion		Ligand
Chloride	Cl^-	Chloro
Bromide	Br^-	Bromo
Cyanide	CN^-	Cyano
Oxide	O^{2-}	Oxo

(b) Ligands whose names end in *-ite* or *-ate* become *-ito* and *-ato*, respectively.

Anion		Ligand
Carbonate	CO_3^{2-}	Carbonato
Thiosulfate	$S_2O_3^{2-}$	Thiosulfato
Thiocyanate	SCN^-	Thiocyanato (bonded through sulfur) Isothiocyanato (bonded through nitrogen)
Oxalate	$C_2O_4^{2-}$	Oxalato
Nitrite (bonded through oxygen, ONO^-)[a]	NO_2^-	Nitrito

[a] An exception to this is NO_2^- when bonded through nitrogen, in which case it is named as *nitro*.

4. **Neutral ligands are given the same names as the neutral molecule.** Thus ethylenediamine as a ligand is called ethylenediamine in the name of the complex. Two very important exceptions to this, however, are:

H_2O Aquo
NH_3 Ammine (note double *m*)

5. **When there are more than one of a particular ligand, the number is specified by di = 2, tri = 3, tetra = 4, penta = 5, hexa = 6, etc. When confusion might result, the prefixes bis = 2, tris = 3, tetrakis = 4, etc. are employed.** Thus the presence of two chloride ions is specified as *dichloro*. However, since ethylenediamine already contains the term *di*, two of these molecules are indicated by placing the name of the ligand in parentheses preceded by the term bis; that is, *bis(ethylenediamine)*.

6. **Negative (anionic) complex ions always end in the suffix -ate.** This suffix is appended to the English name of the metal atom in most cases.

Element	Metal as Named in Anionic Complex
Aluminum	Aluminate
Chromium	Chromate
Manganese	Manganate
Nickel	Nickelate
Cobalt	Cobaltate
Zinc	Zincate
Molybdenum	Molybdate
Tungsten	Tungstate

For some metals, the *ate* is appended to the Latin stem.

Element	Stem	Metal as Named in Anionic Complex
Iron	Ferr-	Ferrate
Copper	Cupr-	Cuprate
Lead	Plumb-	Plumbate
Silver	Argent-	Argentate
Gold	Aur-	Aurate
Tin	Stann-	Stannate

In neutral or positively charged complexes the metal *always* appears with the common English name for the element.

7. **The oxidation number of the metal in the complex is written in Roman numerals within parentheses following the name of the metal.** For example,

$[Co(H_2O)_6]^{3+}$ is the hexaaquocobalt(III) ion.
$[CoCl_6]^{3-}$ is the hexachlorocobaltate(III) ion.

Note that the charge on the complex is obtained as the *algebraic sum* of the oxidation number of the metal and the charges on the ligands.

Some additional examples illustrate these rules.

$[Ni(CN)_4]^{2-}$	Tetracyanonickelate(II) ion
$[Co(NH_3)_4Cl_2]^+$	Dichlorotetraamminecobalt(III) ion
$Na_3[Cr(NO_2)_6]$	Sodium hexanitrochromate(III)
$[Ag(NH_3)_2]^+$	Diamminesilver(I) ion
$[Ag(CN)_2]^-$	Dicyanoargentate(I) ion
$[Co(en)_3]Cl_3$	Tris(ethylenediamine)cobalt(III) chloride
$[Cr(NH_3)_3Cl_3]$	Trichlorotriamminechromium(III)

When two different compounds have the same molecular formula, but differ in the way that their atoms are arranged, they are said to be **isomers** of one another. For example, there are two compounds having the general formula:

$$Cr(NH_3)_5SO_4Br$$

One of these we should formulate as

$$[Cr(NH_3)_5SO_4]Br$$

because it yields a precipitate of AgBr when treated in aqueous solution with $AgNO_3$ but does not give a precipitate of $BaSO_4$ when treated with $Ba(NO_3)_2$. This latter observation means that the SO_4^{2-} is not free in the solution and, hence, must be bound to the chromium.

The second compound is written as

$$[Cr(NH_3)_5Br]SO_4$$

and produces $BaSO_4$ when treated with $Ba(NO_3)_2$. On the other hand, addition of $AgNO_3$ to a solution of the compound does not yield AgBr.

The two compounds just described have different chemical properties and are clearly different chemical substances, even though they are composed of the same number of the same kinds of atoms. This particular type of isomerism is not uncommon among coordination compounds and is called **ionization isomerism.**

Another type of isomerism that is very important is called **stereoisomerism,** and results when a given molecule or ion can exist in more than one structural form in which the same atoms are bound to one another but find themselves oriented differently in space. To illustrate this, we shall focus our attention on octahedral complexes because they represent the most common structural type.

The simplest form of stereoisomerism results when a complex has the general formula Ma_4b_2 in which a and b represent monodentate ligands. An example would be the ion, $[Co(NH_3)_4Cl_2]^+$. (How would you name it?) This complex can exist in two different isomeric forms, called **geometrical isomers,** as shown in Figure 19.8. As you can see, in one of these isomers the two b ligands are located across from one another on opposite sides of the metal ion. Such an isomer is given the designation **trans** (Latin *trans* means "across"). The other isomer has the two b ligands adjacent to one another and is referred to as the **cis** isomer (L. *cis* = on the same side). Thus the two

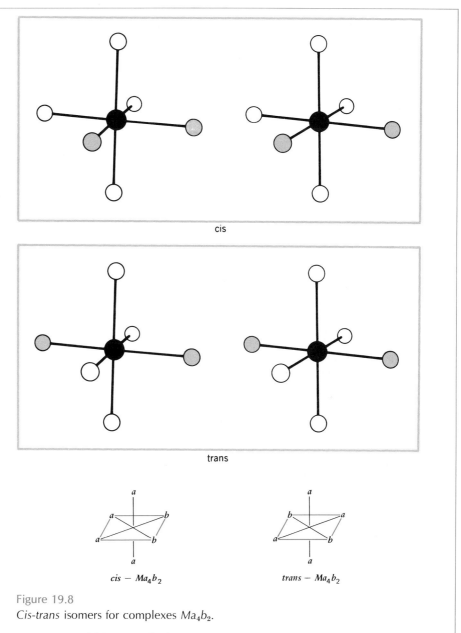

$cis - Ma_4b_2$ $trans - Ma_4b_2$

Figure 19.8
Cis-trans isomers for complexes Ma_4b_2.

isomers would be specified as

$$trans\text{-}[Co(NH_3)_4Cl_2]^+$$

and

$$cis\text{-}[Co(NH_3)_4Cl_2]^+$$

    Because *cis*- and *trans*- isomers possess different structures, they are different chemical species, each with its own set of chemical and physical prop-

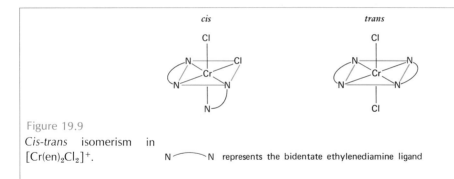

Figure 19.9

Cis-trans isomerism in $[Cr(en)_2Cl_2]^+$.   N⌒N represents the bidentate ethylenediamine ligand

erties. While these properties may often be similar, the fact that they are different clearly tells us that the two structures represent truly different compounds.

Geometrical isomers also occur when there are bidentate ligands in a complex, as illustrated by the *cis* and *trans* forms of the ion $[Cr(en)_2Cl_2]^+$ shown in Figure 19.9. Once again, in the *trans* form we see that the chloride ligands are on opposite sides of the metal while in the *cis* form they are alongside one another.

A second form of stereoisomerism is called **optical isomerism.** Optical isomers, which, as we shall see, affect polarized light differently, bear the same relationship to each other as do your left and right hands; that is, they are **nonsuperimposable mirror images** of one another. To see what this means, try this simple experiment. Place your right hand in front of a mirror, with the palm toward the mirror, and hold your left hand alongside with the palm facing you (Figure 19.10). Notice that the image of your *right* hand in the mirror looks the same as your left hand and hence we can say that your

Left hand   Mirror image of right hand

Figure 19.10

Illustration of nonsuperimposable mirror images.

left and right hands are mirror images of each other. The nonsuperimposable aspect arises because your left and right hands, while similar in appearance, do not match exactly when one is placed over the other, both with palms down; the thumbs point in different directions. This difference is perhaps even more clearly seen if you attempt to place your right hand into a left-hand glove; it doesn't fit properly. Thus your left and right hand, and optical isomers too, cannot be superimposed on each other.

An example of a complex that exists as two nonidentical mirror image isomers is the $[Co(en)_3]^{3+}$ ion. The mirror image relationship between the two isomers shown in Figure 19.11 is demonstrated in Figure 19.12 (where the hydrogen atoms have been omitted for simplicity). A pair of isomers related in this manner are called **enantiomers.**

In such complexes it is the arrangement of the chelate rings that gives rise to the optical isomerism and, in general, any octahedral complex con-

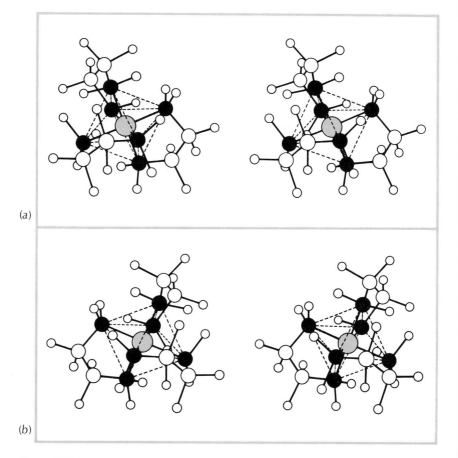

(a)

(b)

Figure 19.11

The complex $[Co(en)_3]^{3+}$. The two structures shown are nonsuperimposable mirror images of one another.

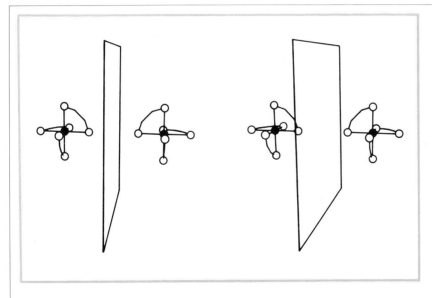

Figure 19.12
Mirror image relationship between optical isomers of $[Co(en)_3]^{3+}$.

taining three bidentate ligands will exist as two optical isomers. In two dimensions these are usually represented as shown in Figure 19.13.

Optical isomerism is also important for the *cis* form of complex ions containing two bidentate ligands and two monodentate ligands; for example, *cis*-$[Co(en)_2Cl_2]^+$ (Figure 19.14a). The *trans* form of this complex does not exhibit optical isomerism, however, because it and its mirror image are identical. In Figure 19.14b the *trans* isomer is drawn with a plane cutting through it. Notice that all of the atoms on one side of this plane are oriented, relative to the plane, exactly as the atoms on the other side. Such an imaginary plane is called a mirror plane, and, it is fact that any molecule or ion that has a structure such that it possesses a mirror plane will not exist as two distinct optical isomers.

In general, the properties of optical isomers are identical except for the way in which they interact with outside influences that are able to distinguish between left and right handedness. The situation here is analogous to having a group of baseball players, some of whom are left-handed and some of whom are right-handed. Since they are all able to toss a baseball with equal ease, a baseball will not differentiate between left- and right-handed players. The same applies to a bat, since there are no left- or right-handed baseball bats. A fielder's glove, however, will fit only one hand. A glove designed to

The Transition Elements

Figure 19.13
Two-dimensional representation of optical isomerism for the complex ion $[Co(en)_3]^{3+}$.

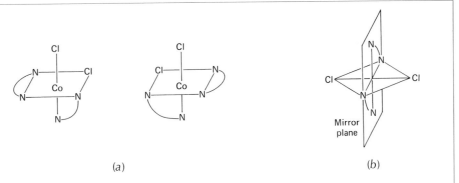

Figure 19.14

(a) Optical isomers for cis-[Co(en)$_2$Cl$_2$]$^+$. (b) trans-[Co(en)$_2$Cl$_2$]$^+$.

be worn on the left hand cannot be used by a player who catches the ball in his right hand. In this case, the glove differentiates between these two kinds of players because it too has a left or right handedness to it. In this same fashion, optical isomers interact in an identical way with most chemical reagents and physical probes. They do differ, however, in the way in which they react toward polarized light.

Light, in general, is composed of electromagnetic radiation that possesses both electric and magnetic vectors. These vectors oscillate in a sinusoidal fashion perpendicular to the direction from which the light wave is propagated (Figure 19.15). If we examine the electric vectors, all different orientations are observed in an unpolarized beam. However, when such a beam is passed through a polarizing medium, only the vibrations in one direction (plane) remain. The result is called plane polarized light. A unique feature of optical isomers is that when plane polarized light is passed through them (or their solutions) the plane of polarization is rotated through some angle, θ, as shown in Figure 19.16. Substances that exhibit this property are also said to be **optically active.** One enantiomer (optical isomer) causes the light to be rotated to the right (clockwise when viewed down the axis of the oncoming light beam) and is said to be **dextro-rotatory,** whereas the other enantiomer causes the polarized beam to be rotated to the left and is described as **levo-rotatory.** The two isomers are therefore designated as d or l depending on the direction of rotation of the polarized light.

An equal mixture of two enantiomers tends to rotate polarized light to both the left and right simultaneously. These effects therefore cancel each other, and such a mixture shows no optical activity; it is said to be **racemic.** In almost all cases, when enantiomers are produced in a chemical reaction, they are formed in equal numbers so that a racemic mixture is formed. One of the arts in chemistry is the separation of optical isomers from one another.

19.10 Bonding in Coordination Compounds — Valence Bond Theory

There are three important properties of transition metal complexes that must be explained by a bonding theory: (1) structure, (2) magnetic properties, and (3) color (nearly all coordination complexes of the transition elements are colored).

In the earliest theories of coordination complexes the metal was considered to be attached to the ligands by way of coordinate covalent bonds,

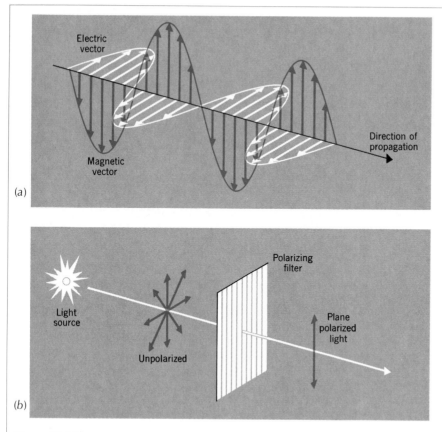

Figure 19.15
Polarized light. (a) Electromagnetic radiation composed of electric and magnetic vectors. (b) Orientation of electric vectors in unpolarized and polarized light.

and the first serious attempt to explain the structures and magnetic properties of complexes was made by applying the concepts of valence bond theory. Much of what follows is simply an extension of some of the ideas that were developed in Chapter 16.

Valence bond theory, you recall, views a bond formed by the overlap of two orbitals and the subsequent sharing of a pair of electrons between the two atoms in the region of overlap. Ligands, as a rule, do not possess unpaired electrons and, hence, the bonding in the complex must result from the overlap of ligand orbitals, containing lone pairs of electrons, with vacant orbitals on the metal ion, thereby giving rise to the coordinate covalent bond. To see how this occurs, let's consider an octahedral ion such as the blue-violet $[Cr(H_2O)_6]^{3+}$, characteristic of simple Cr(III) salts in aqueous solution. The electronic structure of a free chromium atom is

$$\text{Cr} \quad [\text{Ar}] \quad \underset{3d}{\uparrow \; \uparrow \; \uparrow \; \uparrow \; \uparrow} \quad \underset{4s}{\uparrow} \quad \underset{4p}{\_\_ \; \_\_ \; \_\_} \quad \underset{4d}{\_\_ \; \_\_ \; \_\_ \; \_\_ \; \_\_}$$

Figure 19.16
Rotation of plane polarized light by an optical isomer in solution. Plane of polarization is rotated by an angle θ as it passes through the solution containing the optically active compound. In this example the light is rotated to the left; the substance in the solution is said to be levo-rotatory.

where we have shown the empty $4p$ and $4d$ subshells as well as those which are occupied by electrons. The central Cr^{3+} ion in our complex results from the loss of three electrons (starting with the $4s$) to give

$$Cr \quad [Ar] \quad \underset{3d}{\uparrow\ \uparrow\ \uparrow \_\_\_} \quad \underset{4s}{\_} \quad \underset{4p}{\_\_\_} \quad \underset{4d}{\_\_\_\_\_}$$

Now, in order to obtain an octahedral geometry, we saw in Chapter 16 that a d^2sp^3 set of hybrid orbitals must be employed. In addition, these orbitals must be empty so that the electron pairs from the ligands can be placed into them when the bonds are formed. In this case, suitable hybrids can be constructed using the two vacant $3d$ orbitals. The $3d$ orbitals are preferred here over the $4d$ because they are lower in energy; therefore stronger bonds are formed when the $3d$ orbitals are used instead of the $4d$. If we use dots to represent ligand electrons, we can now write the electronic structure for the complex as

$$[Cr(H_2O)_6]^{3+} \quad \underset{3d}{\uparrow\ \uparrow\ \uparrow} \quad \overbrace{\underset{4s}{..\ ..} \quad \underset{4p}{..} \quad \underset{}{..\ ..\ ..}}^{d^2sp^3} \quad \underset{4d}{\_\_\_\_\_}$$

Experimental measurements demonstrate the presence of three unpaired electrons in this ion, as suggested by our bonding picture.

Let's consider next the emerald-green complex ion, $[Ni(H_2O)_6]^{2+}$. The electron configuration of the Ni^{2+} ion (obtained in the same manner as Cr^{3+} above) would be

$$Ni^{2+} \quad [Ar] \quad \underset{3d}{\uparrow\downarrow\ \uparrow\downarrow\ \uparrow\downarrow\ \uparrow\ \uparrow} \quad \underset{4s}{\_} \quad \underset{4p}{\_\_\_} \quad \underset{4d}{\_\_\_\_\_}$$

Once again we must have a set of vacant d^2sp^3 hybrids in order to obtain the octahedral geometry. However, this time we cannot use a pair of $3d$ orbitals to form the hybrid set. At best, we could only obtain one empty $3d$ orbital if we paired all of the electrons in the $3d$ subshell. When the hybrids are created, both d orbitals must come from the *same* subshell and, therefore, two $4d$ orbitals must be used. The electronic structure of the complex then becomes

$$[Ni(H_2O)_6]^{2+} \quad \underbrace{\uparrow\downarrow \ \uparrow\downarrow \ \uparrow\downarrow \ \uparrow \ \uparrow}_{3d} \quad \overbrace{\underset{4s}{\cdot\cdot} \quad \underset{4p}{\cdot\cdot \ \cdot\cdot \ \cdot\cdot} \quad \underset{4d}{\cdot\cdot \ \cdot\cdot} \ \_\_ \ \_\_ \ \_\_}^{d^2sp^3}$$

Note that the complex contains two unpaired electrons, again in agreement with experiment.

We have now seen two complex ions that can be considered to employ d^2sp^3 hybrids for bonding. In valence bond language, when $3d$ orbitals are used to form the hybrids, an **inner orbital complex** is said to result. On the other hand, when the $4d$ orbitals are used an **outer orbital complex** is formed.

In these last two examples there really was no choice as to which type of bonding (inner or outer orbital) would occur. Let's look at a situation, now, where we do have a choice. An example is Co(III). The electronic structure of the Co^{3+} ion is

$$Co^{3+} \quad [Ar] \quad \underset{3d}{\uparrow\downarrow \ \uparrow \ \uparrow \ \uparrow \ \uparrow} \quad \underset{4s}{\_} \quad \underset{4p}{\_ \ \_ \ \_} \quad \underset{4d}{\_ \ \_ \ \_ \ \_ \ \_}$$

In this case the d^2sp^3 hybrid can be formed in either of two ways. One is to make use of two $4d$ orbitals, thereby giving an outer orbital complex. The second is to pair the electrons together to produce two vacant $3d$ orbitals that can be used in the hybrids. This gives rise to an inner orbital complex.

Outer orbital[4]

$$Co^{III}X_6 \quad \underset{3d}{\uparrow\downarrow \ \uparrow \ \uparrow \ \uparrow \ \uparrow} \quad \overbrace{\underset{4s}{\cdot\cdot} \quad \underset{4p}{\cdot\cdot \ \cdot\cdot \ \cdot\cdot} \quad \underset{4d}{\cdot\cdot \ \cdot\cdot} \ \_\_ \ \_\_ \ \_\_}^{d^2sp^3}$$

Inner orbital

$$Co^{III}X_6 \quad \underset{3d}{\uparrow\downarrow \ \uparrow\downarrow \ \uparrow\downarrow} \ \cdot\cdot \ \cdot\cdot \quad \overbrace{\underset{4s}{\cdot\cdot} \quad \underset{4p}{\cdot\cdot \ \cdot\cdot \ \cdot\cdot}}^{d^2sp^3} \quad \underset{4d}{\_ \ \_ \ \_ \ \_ \ \_}$$

Notice that we can distinguish experimentally between these two possibilities by examining the number of unpaired electrons in the complex. The outer orbital complex has four unpaired electrons; the inner orbital complex has none. The latter complex is diamagnetic.

Both inner and outer orbital complexes are possible whenever the metal ion contains either four, five, or six electrons. In each case, two empty $3d$ orbitals can be created by the pairing of electrons. Failure to pair them, however, leads to the formation of outer orbital complexes.

---

[4] Recall that the Roman numeral superscript here is meant to indicate the oxidation state of the metal.

How do we make a choice between inner and outer orbital bonding? When there is a choice, what determines which of the two possibilities will occur? To answer these questions we must consider two opposing factors:

1. As stated earlier, when 3d orbitals are used to form the hybrid orbitals, stronger metal-ligand bonds result than when 4d orbitals are used. This favors the pairing of electrons and the production of inner orbital complexes.

2. The pairing of electrons required to produce the necessary vacant 3d orbitals for inner orbital bonding also requires an input of energy. Since this energy doesn't have to be invested if the 4d orbitals are used to form the hybrids, this factor favors the production of outer orbital complexes.

The way in which these two factors come into play is illustrated in Figure 19.17. In the first drawing we see that the energy released when the bonds are formed using hybrids composed of 3d orbitals is so great that it more than compensates for the pairing energy, and the resulting inner orbital complex is of lower energy (i.e., more stable) than the outer orbital complex. In this case the preferred complex would be the inner orbital one. On the right in Figure 19.17 we find the other situation where the energy released upon inner orbital bond formation does not lead to an overall lower energy than that achieved with the formation of an outer orbital complex. In this case outer orbital bonding would occur in preference to inner orbital bonding.

As a general rule, for first row transition metal ions with either a d^4 or d^6 electron configuration, most ligands tend to lead to the production of inner orbital complexes. Exceptions are the ligands H_2O and F^-, which usually produce outer orbital complexes.

With a $3d^5$ configuration the subshell is half-filled, and we have noted earlier that a half-filled subshell in which all electrons have the same spin possesses a certain extra stability. As a result, these electron configurations are difficult to disturb and electron pairing is difficult to accomplish. Consequently, metal ions with a d^5 structure tend to keep their electrons spread out with parallel spins and thus tend to form outer orbital complexes with most ligands. An exception to this occurs with cyanide ion. This particular ligand forms very stable metal-ligand bonds,[5] sufficiently stable that electron

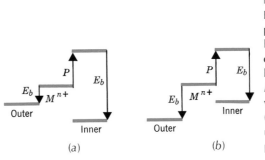

(a)   (b)

19.10 Bonding in Coordination Compounds—Valence Bond Theory

Figure 19.17

Energy changes in the production of inner orbital and outer orbital complexes. (a) inner orbital favored; $E_{outer} > E_{inner}$. (b) Outer orbital favored; $E_{outer} < E_{inner}$. ($E_b =$ Energy released upon bond formation. $P =$ pairing energy.)

[5] Cyanides, such as KCN or HCN, are very poisonous because of the ability of CN^- to irreversibly bind to iron atoms in hemoglobin and because CN^- is able to inactivate certain enzymes.

pairing of the d^5 configuration can occur. This is illustrated by the hexa-cyanoferrate(III) ion (also called ferricyanide ion), $[Fe(CN)_6]^{3-}$. For iron(III) we have

Fe³⁺   [Ar]   ↑ ↑ ↑ ↑ ↑   —   — — —   — — — — —

                            3d   4s   4p   4d

and for the $[Fe(CN)_6]^{3-}$ inner orbital complex,

$$d^2sp^3$$

$[Fe(CN)_6]^{3-}$ ↑↓ ↑↓ ↑   ‥ ‥   ‥   ‥ ‥ ‥   — — — — —

                              3d   4s   4p   4d

OTHER GEOMETRIES. Valence bond theory can also be applied to geometries other than octahedral. For example, the complex ion, $[Ni(CN)_4]^{2-}$, is found to have a square planar shape and is diamagnetic. In Table 16.1 we find that in order to have a square planar geometry, a dsp^2 set of hybrid orbitals must be used by the nickel ion. Once again we have nickel(II) and hence a d^8 electron configuration.

Ni²⁺   [Ar]   ↑↓ ↑↓ ↑↓ ↑ ↑   —   — — — —

                            3d   4s   4p

We can create the one empty d orbital needed for the hybrids by pairing the electrons in the 3d subshell, so that for the complex we have

$$dsp^2$$

$[Ni(CN)_4]^{2-}$ ↑↓ ↑↓ ↑↓ ↑↓ ‥   ‥   ‥ ‥   —

                          3d   4s   4p

Since we have paired all of the electrons the complex should be diamagnetic, which it is.

Tetrahedral complexes such as $[CoCl_4]^{2-}$ can also be accounted for. This ion, containing cobalt(II) with a d^7 configuration, makes use of tetrahedral sp^3 hybrids and has the electronic structure

$$sp^3$$

$[CoCl_4]^{2-}$ ↑↓ ↑↓ ↑ ↑ ↑   ‥   ‥ ‥ ‥

                        3d   4s   4p

From our discussion, valence bond theory appears to be quite effective at accounting for the structures and magnetic properties of complex ions. However, there are some problems with the theory. One very serious drawback is that it does not allow us to explain why complex ions exist in such a wide profusion of colors, even where they contain the same metal ion in the same oxidation state. Table 19.10, for example, contains a list of some typical cobalt(III) complexes and their colors. There are also certain complex ions that are quite difficult to account for in a reasonable and satisfying way. The ion $[Co(NO_2)_6]^{4-}$ is one of these. This ion contains Co(II), a d^7 ion,

Co²⁺   [Ar]   ↑↓ ↑↓ ↑ ↑ ↑   —   — — —   — — — — —

                          3d   4s   4p   4d

Table 19.10

Colors of Some Octahedral Cobalt(III) Complexes

Complex	Color
$[Co(H_2O)_6]^{3+}$	Pink
$[Co(NH_3)_6]^{3+}$	Yellow-red
$[Co(NO_2)_6]^{3-}$	Yellow
$[Co(en)_3]^{3+}$	Yellow-orange
$[Co(C_2O_4)_3]^{3-}$	Green
$[Co(en)(C_2O_4)_2]^-$	Red-violet
$[Co(EDTA)]^-$	Violet
$[Co(CN)_6]^{3-}$	Colorless

With water we saw that cobalt(II) forms an outer orbital complex containing three unpaired electrons. The $[Co(NO_2)_6]^{4-}$ ion, however, contains only one unpaired electron; therefore, it must be postulated that two of the three unpaired electrons in Co^{2+} become paired and that the third is promoted to the $4d$ subshell in order to make two $3d$ orbitals available for inner orbital complex formation. The final result, then, looks like this:

$$[Co(NO_2)_6]^{4-} \quad \underset{3d}{\underline{\uparrow\downarrow}\ \underline{\uparrow\downarrow}\ \underline{\uparrow\downarrow}\ \underline{..}\ \underline{..}} \quad \overset{d^2sp^3}{\overbrace{\underset{4s}{\underline{..}}\ \underline{..}\ \underset{4p}{\underline{..}\ \underline{..}\ \underline{..}}}} \quad \underset{4d}{\underline{\uparrow}\ \underline{\ }\ \underline{\ }\ \underline{\ }\ \underline{\ }}$$

Thus with valence bond theory we can explain the magnetic properties of this ion, but they certainly are not what we would have predicted. Let us now look at another bonding theory that manages to avoid some of the pitfalls of valence bond theory as applied to these transition metal complexes.

19.11 Crystal Field Theory

A second theory of bonding in transition metal complexes, that has been extensively applied over the past 20 years, is called **crystal field theory** (CFT). It was developed by physicists in the early 1930s to deal with metal ions trapped in crystalline lattices, and it was not until the early 1950s that chemists realized that it could be applied to coordination complexes in general. It differs from valence bond theory in that it views the complex as held together by purely electrostatic attractions, that is, *in its simplest form, CFT ignores covalent bonding*. The most significant aspect of the theory, however, is its concern with the effect that the ligands have on the energies of the *d* orbitals of the metal.

Generally, the ligands in a transition metal complex are either anions, or they are polar molecules. In the latter case the negative ends of the ligand dipoles point in the direction of the metal cation. Let us examine how these ligands affect the *d* orbitals. The simplest complex ion that we can consider for this purpose is the $[Ti(H_2O)_6]^{3+}$ cation, consisting of a Ti^{3+} ion surrounded octahedrally by six water molecules. Titanium(III) has a single $3d$ electron,

$$Ti^{3+} \quad \uparrow\!\!-\!\!-\!\!-\!\!-\!\!-\ -\ -\ -\!\!-\ -\ -$$

$3d$ $4s$ $4p$

Which of the five $3d$ orbitals will this electron prefer to occupy?

In the free gaseous Ti^{3+} ion all of the $3d$ orbitals have exactly the same energy; we say that they are **degenerate.** In the presence of the ligands, however, some of this degeneracy is removed and we find that not all of the $3d$ orbitals have the same energy. To understand why, we must consider the spatial arrangement of the five $3d$ orbitals.

In Chapter 3 we discussed the shapes of the p orbitals; each one consists of a pair of lobes directed along a coordinate axis. The d orbitals are somewhat more complex, as we can see in Figure 19.18. Four of them, labeled d_{xy}, d_{xz}, d_{yz}, and $d_{x^2-y^2}$, have the same shape, being composed of four lobes each. The fifth, the d_{z^2}, consists of two large lobes directed along the positive and negative z axis plus a donut of charge in the xy plane. For our purposes here, it is important to notice that two of these d orbitals have lobes that are pointed along the coordinate axes (the $d_{x^2-y^2}$ and d_{z^2} orbitals) while the other three (the d_{xy}, d_{xz}, and d_{yz}) have lobes that point between the axes at 45° angles to them.

d_{xy}

d_{xz}

d_{z^2}

$d_{x^2 - y^2}$

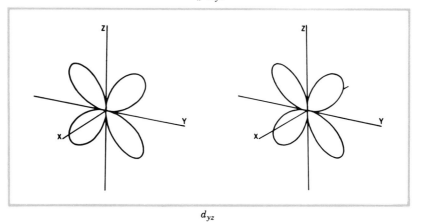

d_{yz}

Figure 19.18
Directional properties of the d orbitals.

We can construct an octahedral complex ion by placing the six ligands along the coordinate axes as shown in Figure 19.19. If we now look at the d orbitals of the central ion in the complex, Figure 19.20, we see that the $d_{x^2-y^2}$ and d_{z^2} orbitals have their lobes pointing directly at the ligands. The d_{xy}, d_{xz}, and d_{yz} orbitals, on the other hand, point between the ligands. Let's imagine, now, placing an electron into one of these d orbitals. If we place it into either the $d_{x^2-y^2}$ or d_{z^2} orbital, it will be forced to spend much of its time in the vicinity of the ligands and, hence, it will tend to be strongly repelled by the negative charge of the ligands (remember that pointing toward the central ion we have either a negative charge because of an anionic ligand or the negative end of a ligand dipole). On the other hand, this repulsion will be much less if the electron is placed into either the d_{xy}, d_{xz}, or d_{yz} orbitals where it can avoid the ligands.

In the $[Ti(H_2O)_6]^{3+}$ ion, then, the single d electron of the Ti^{3+} ion will experience the least repulsion and, hence, be of lowest energy, when it is located in the d_{xy}, d_{xz}, or d_{yz} orbitals. In effect, then, the energies of the d orbitals are no longer all the same. Three of them are lower in energy while the other two are higher in energy. This splitting of the energies of the d orbitals of a metal ion, that occurs when the ion is placed into the electrostatic field of the ligands, is indicated graphically in Figure 19.21. For reasons beyond the scope of this book, for an octahedral field the upper level containing two orbitals is labeled $\mathbf{e}_g$ and the lower level, consisting of three orbitals, is referred to as the $\mathbf{t}_{2g}$ level. The energy difference between the e_g and t_{2g} levels is denoted as Δ.

In the lowest energy state of the $[Ti(H_2O)_6]^{3+}$ ion, the single $3d$ electron of titanium will be located in one of the orbitals in the low energy t_{2g} level. When the ion absorbs light, the energy absorbed is able to promote this electron to the e_g level, as shown in Figure 19.22. The color of the absorbed light depends on the magnitude of Δ.

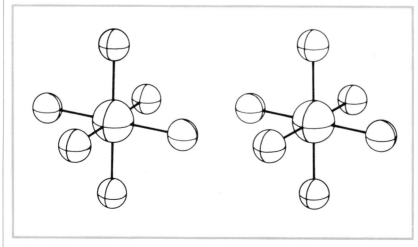

Figure 19.19
Octahedral arrangement of ligands about a central metal ion.

Recall that the energy of a photon is related to the frequency of the light, ν, by the equation

$$E = h\nu \qquad (h \text{ is Planck's constant})$$

and that the wavelength is, in turn, related to frequency by

$$\lambda = \frac{c}{\nu} \qquad (c \text{ is the speed of light})$$

Thus high frequency is associated with high energy and short wavelenth. As the magnitude of Δ increases, more energy is required to raise the electron from the t_{2g} to the e_g level. Hence, light of higher frequency (shorter wavelength) must be used. Generally, for most complex ions the wavelength of the absorbed light required to cause the promotion lies in the visible region of the spectrum. Since in the absorption process, a portion of the visible spectrum is removed from white light as it passes through (or is reflected from) the complex, the observed color of the complex is due to the wavelengths that remain. For example, the $[Ti(H_2O)_6]^{3+}$ complex has a Δ that corresponds to energies associated with light in the green region of the spectrum. When white light is passed through a solution of this complex, green is removed and the solution appears violet.

For a given metal ion, different ligands have different effects on the splitting of the d orbitals, that is, on Δ. By examining the absorption spectra of various complexes, we can arrange the ligands in order of their ability to produce a large Δ. This series is called the **spectrochemical series** and can be given in abbreviated form as

$$I^- < Br^- < Cl^- < F^- < OH^- < H_2O < NH_3 < en < NO_2^- < CN^-$$

Thus I^- is poorest at splitting the energies of the t_{2g} and e_g orbitals, and CN^- is best. What is particularly interesting is that this same series applies for essentially any metal in any oxidation state. However, whereas the order is the same, the actual magnitude of Δ for a given complex in a given geometry depends on the ligand, the metal and its oxidation state.

As with nearly any generalization that we might make in attempting to describe chemical properties, there are exceptions. This is true here with the order of ligands in the spectrochemical series where, in some instances, the relative positions of neighboring ligands is reversed. With cobalt(III), for example, Cl^- appears to produce a greater crystal field splitting than F^-. Nevertheless, the spectrochemical series often serves as a useful guide in understanding (and sometimes even predicting) properties of complexes. For example, we have just seen how CFT accounts for the colors of complexes. We can explain their magnetic properties as well. Consider, for example, the cobalt(III) complexes of F^- and Cl^-. The metal ion here contains six d electrons and, in a weak crystal field (one that produces a small Δ), they will be unpaired as much as possible, as shown in Figure 19.23a, to give a complex with four unpaired electrons. This is what occurs with F^- in the $[CoF_6]^{3-}$ ion.

When the ligand produces a large crystal field (and hence a large Δ) we have the possibility of pairing all of the d electrons in the t_{2g} level (Figure 19.23b) to produce a diamagnetic complex. This will occur if the magnitude of Δ is greater than the energy needed to pair the electrons in a given orbital. In other words, when the pairing energy (let's call it P) is less than Δ, more

(a)

(b)

(c)

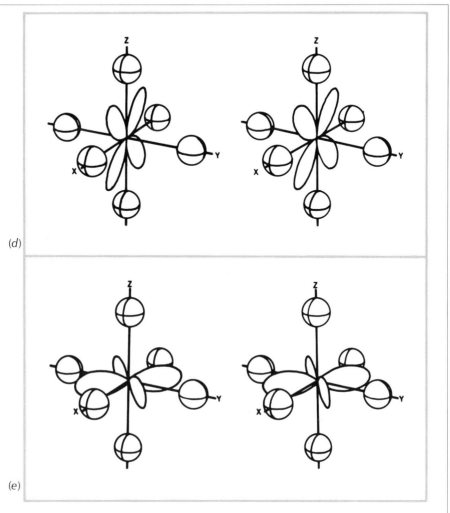

(d)

(e)

Figure 19.20
Interaction of ligands with the d orbitals of the metal. (a) d_{z^2}. (b) $d_{x^2-y^2}$. (c) d_{yz}. (d) d_{xz}. (e) d_{xy}.

Figure 19.21
Splitting of energies of the d orbitals in an octahedral crystal field.

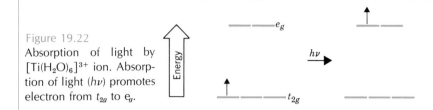

Figure 19.22
Absorption of light by
$[Ti(H_2O)_6]^{3+}$ ion. Absorption of light ($h\nu$) promotes
electron from t_{2g} to e_g.

energy is required to place the electron in the e_g orbital than is required to pair them and place them in the t_{2g} level. This happens when the ligand is Cl^-.

For Co(III) complexes in general (or for that matter, any d^6 system) a paramagnetic complex with four unpaired electrons will occur whenever $\Delta < P$; diamagnetic complexes will be formed when $\Delta > P$. In general, we speak of these two possibilities as **high spin** complexes (minimum pairing of electrons) and **low spin** complexes (maximum pairing of electrons from the e_g into the t_{2g}). Comparing them to the valence bond treatment, we find that low spin complexes correspond to inner orbital complexes whereas high spin complexes correspond to outer orbital complexes.

The possibility of both low spin and high spin complexes exists when the central metal ion contains 4, 5, 6 or 7 d electrons. The electron configurations of the t_{2g} and e_g levels in these species are left to you as an exercise (Problem 19.34). For d^1, d^2, and d^3 systems the electrons will naturally prefer the three low energy t_{2g} orbitals since no pairing is required; therefore, only one type of electron configuration will be found for them. Likewise, with a d^8 or d^9 configuration six electrons will be forced to occupy the t_{2g} (thereby filling it) and the e_g level will contain either two or three electrons, respectively. Once again we see that d^8 and d^9 ions will each have only one type of electron configuration in an octahedral complex.

With this as background, we note that the magnetic properties of the $[Co(NO_2)_6]^{4-}$ ion, which presented such a problem with the valence bond theory, are easily explained in terms of the CFT. Recall that this complex con-

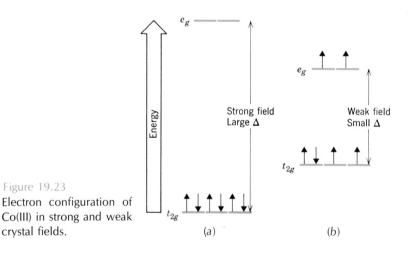

Figure 19.23
Electron configuration of Co(III) in strong and weak crystal fields.

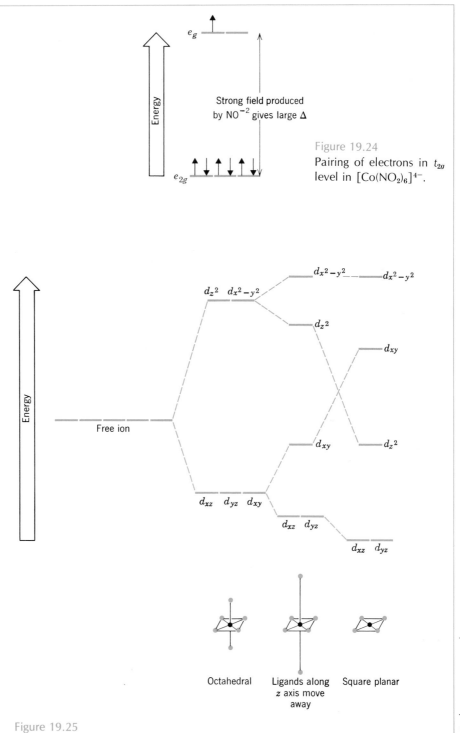

e_g

Energy

Strong field produced
by NO^{-2} gives large Δ

e_{2g}

Figure 19.24
Pairing of electrons in t_{2g}
level in $[Co(NO_2)_6]^{4-}$.

Energy

Free ion

d_{z^2} $d_{x^2-y^2}$

$d_{x^2-y^2}$ ------- $d_{x^2-y^2}$

d_{z^2}

d_{xy}

d_{xy}

d_{z^2}

d_{xz} d_{yz} d_{xy}

d_{xz} d_{yz}

d_{xz} d_{yz}

Octahedral   Ligands along   Square planar
                z axis move
                away

Figure 19.25
Splitting pattern of the *d* orbitals changes as the geometry of the complex changes.

All electrons are paired

Free ion Ni^{2+}

Square planar complex

$d_{x^2-y^2}$

d_{xy}

d_{y^2}

d_{xz} d_{xy}

Figure 19.26
Diamagnetic [Ni(CN)$_4$]$^{2-}$,
a square planar complex.

tains the Co^{2+} ion, a d^7 system. From the spectrochemical series we also note that NO$_2^-$ produces a very strong crystal field; therefore, we would expect that Δ is probably quite large, larger in fact than the pairing energy. Under these circumstances there will be pairing of electrons in the t_{2g} level, as we see in Figure 19.24. Since the t_{2g} level can accommodate only six electrons (three pairs), the seventh electron is forced to occupy the e_g level. The complex is therefore low spin (analogous to inner orbital of the valence bond theory) and will contain a single unpaired electron, in agreement with experiment. Thus we see that there is really nothing unusual about the [Co(NO$_2$)$_6$]$^{4-}$ ion, quite opposite to what we would have concluded based on the valence bond theory.

Crystal field theory can be extended to other geometries besides octahedral, the difference being that other splitting patterns are observed. For example, a square planar complex can be thought of as being derived from an octahedral complex by removing the ligands that lie along the z axis. As shown in Figure 19.25, when this occurs the energies of the d_{z^2}, d_{xz} and d_{yz} orbitals decrease because an electron placed into them experiences less repulsion than in an octahedral complex. Also, by removing the ligands along the z axis, those along the x and y axes can move in slightly and therefore the energies of the $d_{x^2-y^2}$ and d_{xy} orbitals rise somewhat. In the [Ni(CN)$_4$]$^{2-}$ ion, the energy separation between the d_{xy} and $d_{x^2-y^2}$ is large enough that the eight d electrons of the Ni^{2+} ion can exist as four pairs (Figure 19.26).

In tetrahedral complexes the splitting pattern of the d orbital is that shown in Figure 19.27. Notice that the order of the energy levels is exactly opposite to that found in octahedral complexes. The magnitude of Δ is also considerably smaller (actually $\Delta_{\text{tetr}} \approx 4/9\ \Delta_{\text{oct}}$ for the same ligands and metal ion). The small Δ observed for tetrahedral complexes is always less than the pairing energy, and tetrahedral complexes are always of the high spin variety. Note that this agrees with the valence bond theory in which sp^3 hybrids were used by the metal, thereby making all of the d orbitals available for spreading out the d electrons.

d_{xz} d_{yz} d_{xy}

Free ion

d_{z^2} $d_{x^2-y^2}$

Tetrahedral field

Figure 19.27
Splitting pattern for a
tetrahedral field. $\Delta_{\text{tet}} \approx \frac{4}{9}\ \Delta_{\text{oct}}$.

Review Questions and Problems

19.1 What is a transition element? What is an inner transition element?

19.2 What accounts for the occurrence of a +2 oxidation state for a large number of transition elements?

19.3 What relationship (if any) exists between the A and B groups in the periodic table?

19.4 What is *aqua regia*?

19.5 Give four general properties of the transition elements.

19.6 Review the electronic structures of the first row transition elements.

19.7 How do the relative stabilities of high and low oxidation states of the transition elements vary as we move from left to right across a given period?

19.8 Compare the relative stabilities of high and low oxidation states for the elements in Group VIB.

19.9 Saturated solutions of mercurous chloride were prepared by adding the solid to solutions having various chloride ion concentrations. The total concentration of mercury in each solution was then determined. Here are the results.

Chloride Ion Concentration	Moles of Mercury/Liter
1.0M	2.2×10^{-18}
0.5M	8.8×10^{-18}
0.2M	5.5×10^{-17}
0.1M	2.2×10^{-16}

Use the concepts that you learned having to do with equilibrium and solubility product to show that these data are only consistent with mercury(I) having the formula Hg_2^{++} (and hence mercury(I) chloride being Hg_2Cl_2), and not with Hg^+ (and therefore HgCl).

19.10 What is the lanthanide contraction? How does it affect the properties of the *d*-block elements in period 6?

19.11 What are the three steps that are involved in extracting a metal from its ore and making it ready for practical use?

19.12 Describe the process called flotation. What is meant by *roasting* as applied to metallurgy?

19.13 Write chemical reactions for (a) the reduction of Fe_2O_3 in the blast furnace and (b) the production of slag from SiO_2 and $CaCO_3$.

19.14 What is the Mond process?

19.15 Compare the properties of paramagnetism and ferromagnetism. Why can a ferromagnetic material become permanently magnetized?

19.16 Define ligand, first coordination sphere, coordination compound, monodentate ligand, polydentate ligand, chelate, and coordination number.

19.17 Sketch the common structures found for complex ions with coordination number 4 and coordination number 6.

19.18 Give I.U.P.A.C. names for each of the following:

(a) $[Ni(NH_3)_6]^{2+}$
(b) $[CrCl_3(NH_3)_3]^0$
(c) $[Co(NO_2)_6]^{3-}$
(d) $[Mn(C_2O_4)_3]^{3-}$
(e) MnO_4^-

19.19 What are the I.U.P.A.C. names for the following:

(a) $[AgI_2]^-$
(b) $[Cr(NH_3)_5Cl]^{2+}$
(c) $[Co(H_2O)_4(NH_3)_2]Cl_2$
(d) $[Co(en)_2(H_2O)_2]_2(SO_4)_3$
(e) $[Cr(NH_3)_4Cl_2]Cl$

19.20 Write chemical formulas for the following:

(a) Dicyanotetraaquoiron(III) ion
(b) Oxalatotetraamminenickel(II)
(c) Potassium hexacyanomanganate(III)
(d) Tetrachlorocuprate(II)
(e) Tetraoxochromate(VI) ion
(f) Tetrachloroaurate(III) ion

(g) Dinitrobis(ethylenediamine)-iron(III) sulfate

(h) Carbonatotetraamminecobalt(III) nitrate

(i) Ethylenediaminetetraacetatoferrate(II) ion

(j) Dithiosulfatoargentate(I) ion

19.21 What is meant by *isomer*? What are stereoisomers?

19.22 Sketch the isomers of $[Co(NH_3)_2Cl_4]^-$. Identify *cis* and *trans* isomers. How many isomers are there for the complex, $[Co(NH_3)_3Cl_3]$? Sketch them.

19.23 Sketch the isomers of $[Cr(en)_2Cl_2]^+$. Identify *cis* and *trans* isomers and indicate any isomers that exhibit optical isomerism.

19.24 What are enantiomers? What is meant by *racemic*?

19.25 What is the difference between an inner orbital complex and an outer orbital complex?

19.26 Use valence bond theory to predict the electron configuration, the type of bonding (inner orbital or outer orbital), and the number of unpaired electrons for each of the following:
(a) $[VCl_6]^{3-}$
(b) $[Ni(NH_3)_6]^{2+}$
(c) $[Fe(NH_3)_6]^{3+}$
(d) $[Co(CN)_6]^{3-}$

19.27 What magnetic properties would you predict for the square planar complex $[Cu(NH_3)_4]^{2+}$?

19.28 Sketch on appropriate coordinate axes the shapes of the five *d* orbitals.

19.29 Diagram the crystal field splitting of the *d* orbitals, and indicate the electron population of each energy level, in the paramagnetic complex $[Mn(H_2O)_6]^{3+}$. Label the energy levels.

19.30 How does crystal field theory account for the colors of complex ions?

19.31 What relationship exists between Δ (the crystal field splitting) and the pairing energy in determining whether a given complex will be paramagnetic or diamagnetic?

19.32 What are meant by high spin and low spin complexes? How do these compare with inner orbital and outer orbital complexes in the valence bond theory?

19.33 Sketch the CFT splitting patterns of the *d* orbitals for (a) square planar and (b) tetrahedral complexes.

19.34 Using the CFT splitting pattern for an octahedral complex, indicate the high spin and low spin distribution of electrons among the t_{2g} and e_g levels for the configurations: (a) d^4, (b) d^5, (c) d^6, (d) d^7.

20
Organic Chemistry

As far back as the eighteenth century chemists were able to distinguish between two types of compounds, those derived from plants and animals and those from the mineral constituents of the earth. The latter type, called inorganic substances, have received much of our attention thus far in this text. This chapter and the next is devoted to a discussion of organic compounds, many of which find their origin in the nature of living things but most of which have been synthesized in the laboratory.

One does not have to delve very far into the chemistry of life before observing that the element carbon is present in *all* of the molecules in life's makeup. Of all of the 100 odd elements, carbon is the only one found universally in these substances and, thus, organic chemistry has become known as the study of carbon and its compounds. Besides carbon, organic molecules contain relatively few other elements; among the most prevalent are hydrogen, oxygen, nitrogen, and to a lesser extent phosphorus and sulfur.

Up until 1828 chemists believed that the only source of these organic compounds was from nature itself. It was thought that man was not able to synthesize them in the laboratory because nature's "vital force" was missing. In 1828 Friedrich Wöhler first synthesized urea (a compound found in urine) from inorganic materials. Wöhler evaporated an aqueous solution of the inorganic salt ammonium cyanate which resulted in the production of urea:

$$NH_4OCN(aq) \longrightarrow CO(NH_2)_2$$

ammonium cyanate   **urea**

As a result of this first synthesis many scientists began to attempt to prepare other organic materials with a good deal of success. Gradually the vital force theory was abandoned, and during the next several years a great profusion of organic compounds were made. Today there are well over a million known organic compounds, compared to about 100,000 inorganic substances, with new ones being discovered every day.

The great abundance of organic compounds is a result of carbon's ability to bond to itself to form long chain, branched-chain, ring and branched-ring molecules, while maintaining its normal valence of four. In comparison, silicon forms only relatively short chains when it is bonded to itself. Silicon does not have the ability of forming chains containing thousands of atoms bonded together as carbon does as, for example, in rubber, plastics, and synthetic fibers.

In organic chemistry we are fortunate to be able to categorize systematically the prodigious number of compounds into a relatively small number of groups quite successfully. We begin this classification with the **hydrocarbons,** which are compounds containing only carbon and hydrogen. All of the remaining types can then be looked on as derived from these hydrocarbons.

Hydrocarbons, which are primarily found in natural gas, petroleum, and coal can be divided into two main categories: **aliphatic hydrocarbons,** which are the straight-chain, branched-chain, and cyclic compounds, and **aromatic hydrocarbons,** which are the benzene (C_6H_6)-like compounds. The aliphatic hydrocarbons can be further subdivided into two groups based on the multiplicity of the carbon-carbon bond: **saturated hydrocarbons** that contain only carbon-carbon single bonds; and **unsaturated hydrocarbons** that possess at least one carbon-carbon double bond or triple bond.

SATURATED HYDROCARBONS. There is only one type of noncyclic saturated hydrocarbon while, as we shall soon see, there are two types of noncyclic unsaturated hydrocarbons. The compounds that constitute the saturated hydrocarbons are collectively called the **alkanes** or **parafins.** The first 10 members of the straight-chain (normal) alkanes are listed in Table 20.1 in order of an increasing number of carbon atoms in the chain.

The names of each of the members of this group are composed of two parts. The first part, *meth, eth, prop,* and so on in Table 20.1, reflects the number of carbon atoms in the chain. The second part, which is the same for all the members, is "ane" after the parent name alk*ane.* Thus we have methane, an alkane with one carbon, ethane having two carbons, propane consisting of three carbons, and so on. You should become familiar with the names of all 10 of these simpler alkanes for they serve as the basis for naming many of the remaining organic compounds.

In listing the alkanes by increasing number of carbon atoms, as in Table 20.1, two things become apparent. First, the molecular formula for each of the members of this series can be represented by a single general formula,

Table 20.1
First Ten Members of the Straight-Chain Alkanes

Molecular Formula	Name	Boiling Point (°C)
CH_4	methane	−161
CH_3—CH_3	ethane	−89
CH_3—CH_2—CH_3	propane	−44
CH_3—CH_2—CH_2—CH_3	butane	−0.5
CH_3—CH_2—CH_2—CH_2—CH_3	pentane	36
CH_3—CH_2—CH_2—CH_2—CH_2—CH_3	hexane	68
CH_3—CH_2—CH_2—CH_2—CH_2—CH_2—CH_3	heptane	98
CH_3—CH_2—CH_2—CH_2—CH_2—CH_2—CH_2—CH_3	octane	125
CH_3—CH_2—CH_2—CH_2—CH_2—CH_2—CH_2—CH_2—CH_3	nonane	151
CH_3—CH_2—CH_2—CH_2—CH_2—CH_2—CH_2—CH_2—CH_2—CH_3	decane	174

C_nH_{2n+2}, where n is the number of carbons in the molecular chain. For example, the formula for the alkane with four carbon atoms would be C_4H_{10} and, according to Table 20.1, would be called butane. Second, we see that any two successive members of the series differ from each other by a single CH_2 group (this becomes quite apparent after ethane). Such a series, where one member differs from the next by the same repeating cluster of atoms is called a **homologous series.** The alkanes form such a series.

In organic chemistry it is important for us to be able to write molecular formulas, which indicate the number of each of the various atoms present in a molecule, as well as structural formulas, which show the relative positions of each of the atoms in a molecule. Since carbon forms the backbone of all organic compounds, it is mainly the shape of the carbon skeleton that is responsible for the overall shape of the various molecules. As pointed out in the introduction to this chapter, the normal valence for carbon in organic compounds is four. Thus, in order to accommodate four single bonds, each of the carbons in a noncyclic alkane must be sp^3 hybridized.

$$\underset{1s}{\underline{\uparrow\downarrow}}\quad \underset{2s}{\underline{\uparrow\downarrow}}\quad \underset{2p}{\underline{\uparrow\;\uparrow\;\_}} \qquad\qquad \underset{1s}{\underline{\uparrow\downarrow}}\quad \underset{sp^3\ \text{hybrid}}{\underline{\uparrow\;\uparrow\;\uparrow\;\uparrow}}$$

unhybridized carbon

Thus in the straight- and branched-chain alkanes carbon is tetrahedrally surrounded by bonded atoms. There are several ways of representing tetrahedral carbon in two dimensions, as illustrated in Figure 20.1 for methane. The most common, and the simplest, two-dimensional representation used for most organic compounds is the structural formula shown as Figure 20.1a. For comparison a three-dimensional representation of methane showing the correct H—C—H bond angle of 109° is drawn in Figure 20.2. The remaining members of the straight-chain alkanes are drawn simply by connecting together several tetrahedral carbons along with their respective hydrogens. In Figure 20.3 we see the two- and three-dimensional representations of a four carbon alkane, butane.

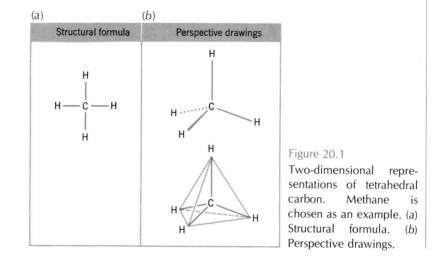

(a) Structural formula (b) Perspective drawings

Figure 20.1

Two-dimensional representations of tetrahedral carbon. Methane is chosen as an example. (a) Structural formula. (b) Perspective drawings.

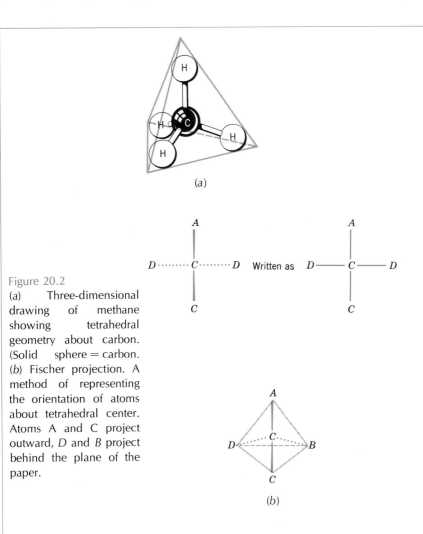

(a)

$$A$$
$$D \cdots\cdots C \cdots\cdots D \quad \text{Written as} \quad D \text{———} C \text{———} D$$
$$C \qquad\qquad C$$

(b)

Figure 20.2
(a) Three-dimensional drawing of methane showing tetrahedral geometry about carbon. (Solid sphere = carbon. (b) Fischer projection. A method of representing the orientation of atoms about tetrahedral center. Atoms A and C project outward, D and B project behind the plane of the paper.

Although the two-dimensional structural formulas drawn in Figure 20.3 show a straight carbon chain, we know that as a result of the tetrahedral nature of the bonding about carbon atoms, these atoms cannot be in a straight line. They are drawn this way only for simplicity. Since we are ordinarily limited to drawing such structural formulas in two dimensions, we should practice by drawing most of the formulas for the first 10 straight-chain alkanes listed in Table 20.1. Besides the straight-chain alkanes, there also exist branched-chain alkanes and cycloalkanes. We shall discuss these further in a later section.

UNSATURATED HYDROCARBONS. The unsaturated hydrocarbons can be divided into two groups: **alkenes** or **olefins,** which contain at least one carbon-carbon double bond and **alkynes,** which contain a carbon-carbon triple bond.

Following a procedure similar to that used for the alkanes, we can introduce the alkenes by considering a series of compounds with an increasing

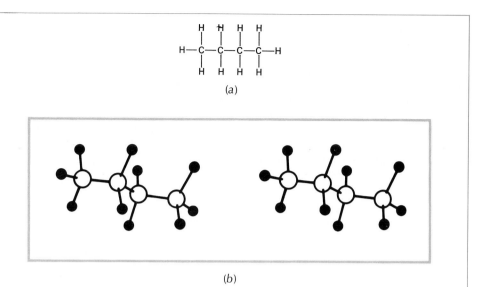

(a)

(b)

Figure 20.3

Two- and three-dimensional illustrations of the structure of butane. (a) Two-dimensional drawing. (b) Three-dimensional drawing. Note that carbon atoms (white spheres) are not actually in a straight line.

number of carbon atoms in the chain. The simplest alkene contains two carbons:

(C_2H_4)

which is then followed by the three-carbon alkene

(C_3H_6)

Next come the alkenes with the formulas C_4H_8, C_5H_{10}, C_6H_{12}, and so on. Thus we see that the alkenes, like the alkanes, form a homologous series, with a CH_2 group being the difference between any two successive members. We also see that, like the alkanes, the alkenes can collectively be represented by one general formula. The formula that is the same for all alkenes is C_nH_{2n}, where n is the number of carbons in the molecule.

The names of the straight-chain alkenes can also be viewed as consisting of two parts. The first, which is the same used in the naming of the alkanes, indicates the number of carbons in the chain: eth-, two; prop-, three; but-, four; and so on. To this stem we add "ene," which tells us that there is a carbon-carbon double bond present. Thus the first few members of this group would be called ethene (C_2H_4), propene (C_3H_6), butene (C_4H_8), etc. These compounds are also commonly called ethylene, propylene, and butylene, respectively. We shall discuss the nomenclature of these and the other hydrocarbons in Section 20.3.

The bonding in ethene was discussed earlier in Section 16.3 and will, therefore, only be reviewed briefly here. The carbons atoms participating in the double bond are each sp^2 hybridized. The double bond then is formed by the sp^2-sp^2 (σ) overlap of the hybrids on the two carbon atoms, plus the overlap of their remaining pure p orbitals giving rise to a π-bond. Each carbon is also bound to two other atoms and, because of the geometry of the sp^2 hybrids and the restriction that the unhybridized p orbitals must overlap in the π-bond, a planar configuration of atoms with approximately 120° bond angles is formed.

Besides the straight-chain alkenes there also exist alkenes with more than one double bond, alkenes that have other groups attached to the straight chain, and alkenes that are cyclic in structure.

The second type of unsaturated hydrocarbon that still remains to be discussed here is the **alkynes,** a series of compounds containing a carbon-carbon triple bond. The simplest, but truly one of the most important alkynes is HC≡CH, which is called ethyne or more commonly acetylene. The next two members of this group are CH_3—C≡CH, propyne, and CH_3—CH_2—C≡CH, butyne. In naming these compounds we once again use the stems *eth-, prop-,* and *but-* to mean 2, 3, and 4 carbons, respectively. To this stem is added "yne" to denote the existence of the carbon-carbon triple bond.

The bonding in the carbon-carbon triple bond was also discussed earlier (Section 16.3). The hybridization used by the two carbons in the triple bond is sp. Thus one of the bonds is formed by an sp-sp overlap. The remaining two bonds are both π-bonds that are formed by the pure p-p overlap of the carbons, p_x-p_x and p_y-p_y for example.

Beginning with hydrocarbons containing four carbon atoms, we find that besides the normal straight-chain structures, there also exist branched-chain alkanes bearing the same molecular formula. For example, we find that there are two compounds with the formula C_4H_{10}.

20.2
Isomers in Organic Chemistry

<table>
<tr><td>(1)</td><td>(2)</td></tr>
<tr><td>straight chain</td><td>branched chain</td></tr>
</table>

Two or more compounds that have the same molecular formula but differ in the sequence in which the atoms are joined together are said to be **structural isomers.** Thus there are two structural isomers of butane, each with its own chemical and physical properties.

Each of the remaining members of the alkane series show an even greater number of isomers. With C_5H_{12}, for example, we can write three structures:

(1)   (2)   (3)

In the case of C_6H_{14} we find five isomers, which are listed in Table 20.2. In this table we have left room for you to add the names of each of the isomers following our discussion of nomanclature in the next section.

Structural isomerism also exists in the alkenes. Butene (C_4H_8), for example, can be written as a straight chain with the double bond between the first and second carbon,

$(CH_3—CH_2—CH=CH_2)$

isomer 1

or with the double bond between the middle two carbons,

$(CH_3—CH=CH—CH_3)$

isomer 2

or as a branched isomer in which two CH_3 groups are attached to one of the carbon atoms participating in the carbon-carbon double bond.

$((CH_3)_2C=CH_2)$

isomer 3

Table 20.2
The Five Isomers of Hexane

Isomer	Name

These are indeed structural isomers of butene, as we can see from their structural and molecular formulas. The arrangement of atoms in the molecules are different in all three cases while the molecular formula, C_4H_8, is the same. Thus in alkenes we have isomers that result from branching of the carbon chains as well as isomers that result from a difference in the relative position of the double bond.

Another type of isomerism exists in organic compounds in which the sequence of atoms in the molecule is the same but in which the relative positions of the atoms or groups of atoms are different. In general this type of isomerism is called **stereoisomerism,** and is found in two forms: geometric isomerism and optical isomerism (see Section 19.10).

GEOMETRICAL ISOMERISM. In organic compounds *cis* and *trans* geometrical isomers can occur in molecules that possess one or more carbon-carbon double bonds. This is illustrated by one of the isomers of butene,

cis **isomer**   *trans* **isomer**

These are not identical molecules because there is not free rotation of the atoms about the double bond. If there were, one of the —$CH(CH_3)$ groups could rotate 180° to convert *cis* into *trans* or vice versa. However, such a rotation would cause the unhybridized *p* orbitals that overlap to give the bond to become misaligned and thus destroy the π-bond. In effect, then, conversion of *cis* to *trans* in this manner would require the breaking of a bond. Hence this rotation does not occur readily. As a consequence, *cis* and *trans* isomers of this type can be isolated.

OPTICAL ISOMERISM. Optical isomers, as we saw in Section 19.9, are molecules that have the same formula but possess structures that are non-superimposable mirror images of each other. In organic compounds the presence of an **asymmetric carbon atom** is responsible for optical isomerism. An asymmetric carbon is bonded to four *different* atoms or groups of atoms; for example,

where *A*, *B*, *D*, and *E* represent different groups that are bonded to the carbon. For instance, the isomer of heptane with the structural formula

has an asymmetric carbon atom, the one marked with an asterisk. This can be seen more readily if we rewrite the structural formula for this isomer in a slightly more condensed fashion.

By using the two structures above as our examples, let us see how compounds with an asymmetric carbon atom give rise to optical isomers and why molecules without one cannot. To show this, we compare the structures of two different isomers of a compound to see if one is the nonsuperimposable mirror image of the other. If this is true, then they are optical isomers.

In the case of our general molecule the two structures shown in Figure 20.4 are mirror images of each other with nonsuperimposable structures. Note that when one is placed over the other, the D's and B's line up, but the E's and A's do not. Furthermore, we would find that, regardless of the amount of manipulation of the two structures, we could never bring about the situation where one structure would be exactly superimposable upon the other.

If, in each structure there are two groups that are alike, we destroy the asymmetry of the central carbon and the two structures no longer represent optical isomers; they are, in fact, identical. For example, by replacing the E in our general formula with another A, we would then have the two structures, shown in Figure 20.5, that are directly superimposable on each other. Thus we see that an asymmetric carbon atom must be present in order for optical isomerism to occur.

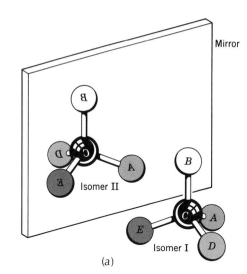

(a)

Figure 20.4

Optical isomers. (a) Isomer II is reflection of isomer I in mirror, that is it is the mirror image of isomer I. (b) Isomer II is rotated about B—C bond so *that atoms D in both I and* II are in same relative posistion. Isomers I and II are not superimposable. Atoms B, C, and D match but A and E do not.

(b)

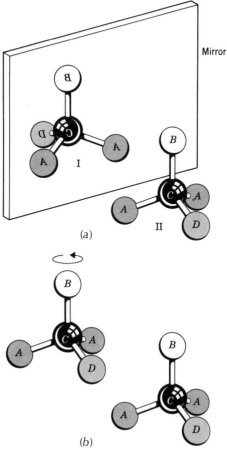

Figure 20.5
(a) Lack of optical isomers when two groups attached to carbon are the same. (b) "Isomer" I rotated about B—C bond so that atoms D are matched. These two are identical and are *not* isomers of each other.

(a)

(b)

With the existence of structural, geometrical, and optical isomers, the molecular formulas given in Table 20.1 really represent the formulas for well over 100 different hydrocarbons. In order to name these alkanes, as well as the thousands of other organic compounds in the remaining groups, chemists from all over the world have been meeting periodically in an attempt to formulate an acceptable systematic way of naming all of the organic compounds. The system of nomenclature, in its present modification, will now be discussed for the alkanes and will then be added to as new groups of compounds are introduced.

20.3
Nomenclature

The systematic nomenclature of organic compounds, like that of coordination compounds, has also been established by the I.U.P.A.C. The application of the I.U.P.A.C. system to the saturated and unsaturated hydrocarbons is defined by the following set of rules.

**RULES OF
NOMENCLATURE
OF ORGANIC
COMPOUNDS**

1. **The longest unbroken chain of carbon atoms in a molecule serves as the parent name for any hydrocarbon or its derivative.** With the alkenes this longest chain must contain the double bond; in the al-

kynes the triple bond must be included. For example, each of the following compounds would be named as a derivative of pentane since in each case the longest carbon chain consists of five carbon atoms.

Sometimes, when a structural formula is drawn, the longest carbon chain is not written in a straight line. For instance, the compound below has an eight carbon chain (can you find it?) and would be named as a derivative of octene (note the double bond).

2. **The stem of the parent name indicates the number of carbon atoms in the chain and has the following endings: ane** when only single bonds are present, **ene** when there is a carbon-carbon double bond, and **yne** when a triple bond is present. When two double bonds are present the ending is **diene** and when three are present the ending shall be **triene.** Thus we would have

$$CH_3\text{---}CH_2\text{---}CH_2\text{---}CH_3 \qquad \text{butane}$$
$$CH_3\text{---}CH_2\text{---}CH\text{=}CH_2 \qquad \text{butene}$$
$$CH_3\text{---}CH_2\text{---}C\text{≡}CH \qquad \text{butyne}$$
$$CH_2\text{=}CH\text{---}CH\text{=}CH_2 \qquad \text{butadiene}$$

3. **Branched isomers are named as derivatives of straight-chain hydrocarbons in which one or more hydrogen atoms are replaced by hydrocarbon fragments.** For example, isomer 2 of butane (see Section 20.2) can be considered to be derived from propane where a CH_3 group has replaced one of the middle hydrogen atoms.

Any hydrocarbon fragment, such as the CH_3 group in the example above, which is not in, but is attached to, the longest unbroken chain of carbon atoms is said to be a **radical.** The names and locations of these radicals always precede the name of the parent chain.

In naming the radicals the alkane stem indicating the number of carbon atoms present is used, and to this stem is added the ending "yl." The formulas and names of some of the more common radicals are listed in Table 20.3.

Thus isomer 2 of butane has in it a three carbon chain as its longest and a CH_3 group attached to the middle carbon; therefore it would be called

$$CH_3—CH—CH_3 \qquad \text{methylpropane}$$
$$| \atop CH_3$$

Table 20.3
The Names of Some Common Organic Radicals

Radical	Formula	Name
H | H—C— | H	$CH_3—$	methyl
H H | | H—C—C— | | H H	$CH_3—CH_2—$	ethyl
H H H | | | H—C—C—C— | | | H H H	$CH_3—CH_2—CH_2—$	propyl
H H H H | | | | H—C—C—C—C— | | | | H H H H	$CH_3—CH_2—CH_2—CH_2—$	butyl

Likewise we would have

$$CH_3\text{—}CH_2\text{—}CH\text{—}CH_3 \qquad \text{methylbutane}$$
$$\underset{\displaystyle CH_3}{\vert}$$

and

$$CH_3\text{—}C\text{=}CH_2 \qquad \text{methylpropene}$$
$$\underset{\displaystyle CH_3}{\vert}$$

4. **In order to denote the positions of the radicals attached to the parent chain, as well as the double and triple bonds, a numbering system is used.** The numbering starts from the end carbon that gives the lowest number to the radical or multiple bond. In the case of a radical, the number identifying its position immediately precedes its name, whereas the numbers identifying a double bond precede the name of the parent chain. Examples of this would include

$$CH_3\text{—}CH_2\text{—}CH_2\text{—}CH\text{—}CH_2\text{—}CH_3 \qquad \text{3-methylhexane}$$
$$\underset{\displaystyle CH_3}{\vert}$$

$$CH_3\text{—}CH\text{=}CH\text{—}CH_3 \qquad \text{2-butene}$$

$$CH_3\text{—}CH\text{=}CH\text{—}CH_2\text{—}CH\text{=}CH_2 \qquad \text{1,4-hexadiene}$$

$$CH_3\text{—}CH_2\text{—}C\text{=}CH\text{—}CH_3 \qquad \text{3-methyl-2-pentene}$$
$$\underset{\displaystyle CH_3}{\vert}$$

5. **In compounds where more than one radical is attached to a carbon chain then: (1) when they are identical the prefixes di(2), tri(3), tetra(4) and so on, immediately precede its name and (2) if they are different, they are listed in alphabetical order along with their respective locations.** Some examples of this rule are

$$\overset{\displaystyle CH_3 \quad CH_3}{\overset{\vert \qquad \vert}{CH_3\text{—}CH\text{—}CH\text{—}CH_3}} \qquad \text{2,3-dimethylbutane}$$

$$\overset{\displaystyle CH_3 \quad CH_3}{\overset{\vert \qquad \vert}{CH_3\text{—}CH\text{—}C\text{—}CH_3}} \qquad \text{2,2,3-trimethylbutane}$$
$$\underset{\displaystyle CH_3}{\vert}$$

$$CH_3\text{—}CH_2\text{—}CH\text{—}CH_2\text{—}CH_2\text{—}CH_2\text{—}CH_2\text{—}CH_3 \qquad \text{6-ethyl-2-methyloctane}$$
$$\underset{\displaystyle CH_3}{\underset{\vert}{\underset{CH_2}{\vert}}} \qquad\qquad\qquad \underset{\displaystyle CH_3}{\vert}$$

As an exercise, you should now go back and fill in the names of all the isomers of hexane in Table 20.2.

COMMON NAMES. Quite a large number of organic compounds were known prior to the introduction of the I.U.P.A.C. rules. As a result, many of them had already been named using other systems of nomenclature, and some of these names are carried over into current usage. For example, methylpropane

$$CH_3—CH—CH_3$$
$$\quad\quad\; |$$
$$\quad\quad CH_3$$

is more commonly referred to as isobutane. This name indicates that there are four carbons present (butane), but that they are not in a straight chain and that the molecule is an isomer of butane, hence iso. In general, alkanes that possess this arrangement

are called *iso*-compounds. Thus we have

| isobutane | isopentane | isohexane |

In the case of the alkynes, the first member, acetylene, is so important that its name is used in the common nomenclature of all the remaining members. Thus we have

Alkyne	Common	I.U.P.A.C.
$CH_3—C{\equiv}CH$	methylacetylene	propyne
$CH_3—CH_2—C{\equiv}CH$	ethylacetylene	1-butyne
$CH_3—CH_2—CH_2—C{\equiv}CH$	propylacetylene	1-pentyne

Example 20.1
(a) Give the I.U.P.A.C. name for the following compounds:

(1)

(2)

(3) H₃C   CH₃
        C=C
   H₃C   CH₃

(4) CH₃—CH—CH=CH—CH=CH₂
            |
           CH₃

(5) CH₃—CH₂—C≡C—CH₂—CH₃
(b) Write structural formulas for the following:
    (1) 2,3-dimethylbutane
    (2) 2-pentyne
    (3) 2-ethyl-1-butene
    (4) 1,5-octadiene
    (5) 2-ethyl-3-methyl-1-pentene

Solution

(a) (1) 2-methylpentane (common: isohexane)
    (2) 4-ethyl-2,2,6-trimethylheptane
    (3) 2,3-dimethyl-2-butene
    (4) 5-methyl-1,3-hexadiene
    (5) 3-hexyne
(b) (1)   CH₃—CH—CH—CH₃
              | |
             CH₃ CH₃

    (2)   CH₃—CH₂—C≡C—CH₃

    (3)   CH₂—CH₃
                      |
          CH₃—CH₂—C=CH₂

    (4)   CH₃—CH₂—CH=CH—CH₂—CH₂—CH=CH₂

    (5)   CH₂—CH₃
                  |
          CH₂=C—CH—CH₂—CH₃
                  |
                 CH₃

CYCLIC ALKANES. With alkanes of three carbons and higher it is possible for one end of the molecule to become attached to the other end, forming a ring. An example of a three carbon hydrocarbon that exists in a cyclic or ring structure is

**20.4
Cyclic
Hydrocarbons**

Table 20.4
Some Cycloalkanes

Structural Formula	Molecular Formula	Name
	C_3H_6	cyclopropane
	C_5H_{10}	cyclopentane
	C_6H_{12}	cyclohexane
	C_7H_{14}	cycloheptane
	$CH_3C_6H_{11}$	methylcyclohexane

This compound, as well as all the other members of this group, contains only carbon-carbon single bonds and is, therefore, saturated. These hydrocarbons are named, then, in the same manner as the straight-chain alkanes, but with the prefix cyclo- being added. The compound above would therefore be named cyclopropane (*cyclo* meaning ringed, *prop* for three carbon atoms, and *ane* because it is a saturated alkane). Some of the other cycloalkanes, with their molecular and structural formulas, are listed in Table 20.4.

In cyclopropane, shown above, the C—C—C bond angle is 60° while in cyclobutane the C—C—C bond angle is about 90°. In both cases the bond angle is much less than the stable tetrahedral bond angle of 109° that is exhibited by carbon in the straight-chain alkanes. These two species then are said to be quite strained and, as a result, are very unstable. This manifests itself in their high degree of reactivity. Cyclopropane, for example, has a tendency to explode spontaneously when mixed with air, whereas the straight-chain propane, C_3H_8, will only react with oxygen when the mixture is ignited.

The next two members that follow cyclobutane, that is, cyclopentane and cyclohexane, are quite stable. In cyclopentane, which has a nonplanar pentagonal structure, the C—C—C bond angle is very nearly 109° and, as a result, we expect it to be stable. In cyclohexane, on the other hand, we find that in order to achieve more closely the tetrahedral angle of 109°, the ring is warped or puckered. In a planar or flat hexagon

the C—C—C bond angle would have to be 120° and, as a result, the structure would be strained. Two of the strain free structures of cyclohexane are shown in Figure 20.6.

CYCLIC ALKENES. Cyclic alkenes are formed by alkenes of three carbons and higher and are named in a fashion similar to that for the cycloalkanes. For example, cyclopentene would be

Other examples include

1,3-cyclopentadiene

1,3-cyclohexadiene

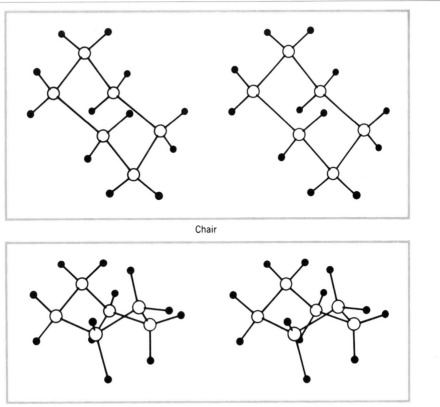

Chair

Boat

Figure 20.6
Two strain free structures of cyclohexane.

In these compounds the numbering begins at the carbon with the first double bond and continues in the direction that leads to the smallest numbers for the remaining double bonds. Following this idea the compound

would be called 1,3,5-cyclohexatriene. This is more commonly called **benzene.**

20.5
Aromatic
Hydrocarbons

Since benzene has properties so unlike those of the other cycloalkenes, it is placed in a separate class. Benzene, and the host of benzene-like compounds (those containing a "benzene ring"), are collectively called the *aromatic compounds.* In general these compounds have a pleasant odor, but many of them are extremely harmful. However, it is interesting to note that the benzene ring is found in many biologically important compounds such as vitamins, proteins, and hormones. In order to make these compounds, the human body must synthesize the aromatic benzene ring which, in many cases, proves to be an impossible task. We must, therefore, obtain them from an outside source. Plants have the ability to synthesize some of the needed aromatics and therefore are essential to our diets.

The main structural feature in the aromatic compounds that is responsible for their distinctive chemical properties is the benzene ring. We saw in the last section that the stoichiometry of benzene corresponds to that of the cyclotriene,

Physical and chemical evidence, however, reveals that this does not give an accurate representation of the molecule. For instance, it is found experimentally that all of the C—C bond distances in benzene are the same. If there were, in fact, alternating double and single bonds, some bond distances (C—C double bonds) would be shorter than others. In addition, the benzene molecule does not readily undergo chemical reactions typical of molecules containing double bonds.

The uniform bond distances in benzene can be accounted for by resonance. The two resonance structures (also called Kekule structures[1]) that are usually written are

These are usually represented simply as

and

---

[1] These structures are named after the German chemist August Kekulé who first proposed them in 1865.

It is understood that at each vertex of the hexagon there is a carbon atom bound to a hydrogen atom.

On an atomic orbital level we see that in order to achieve a C—C—C bond angle of 120°, the carbon atoms must sp^2 hybridize. Thus on each of the carbons in the benzene ring we have three of the valence electrons in an sp^2 hybrid and a fourth in a pure p orbital. The carbon skeleton showing only the hybridization can be seen in Figure 20.7. This same skeleton showing the usual dash for the sigma bonds as well as the electrons in the pure p orbitals is illustrated in Figure 20.8.

According to *valence bond theory* the two resonance structures of benzene would be formed by the overlap of pairs of adjacent p orbitals, as shown in Figure 20.9.

In Section 16.5, however, it was pointed out that the *molecular orbital theory* can quite successfully explain the bonding in polyatomic molecules

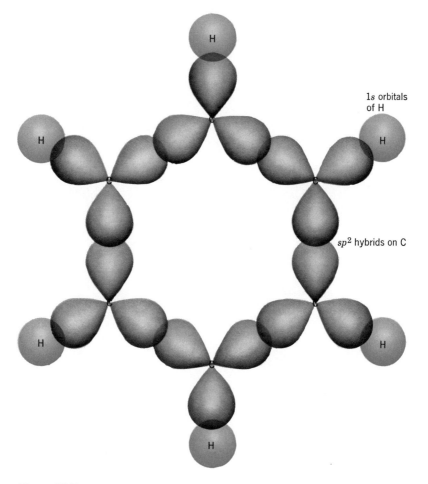

Figure 20.7

Sigma bond framework in benzene. Each carbon uses sp^2 hybrid orbitals.

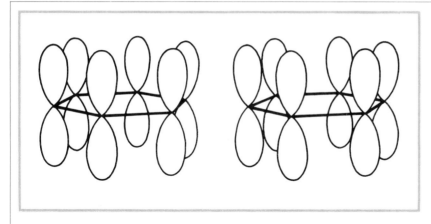

Figure 20.8
Electrons in unhybridized *p* orbitals on C atoms.

without resorting to resonance. According to this theory the six unhybridized atomic orbitals, which have the proper alignment, overlap with each other and form a molecular orbital that extends over the entire molecule. In this respect the electron density, and of course the length, of those bonds involved in the production of the molecular orbital are identical. In benzene the unhybridized *p* orbitals on each of the six carbons are in perfect position to form such a molecular orbital. This situation is illustrated in Figure 20.10 for benzene, where we see the sigma bonds as the solid lines between the

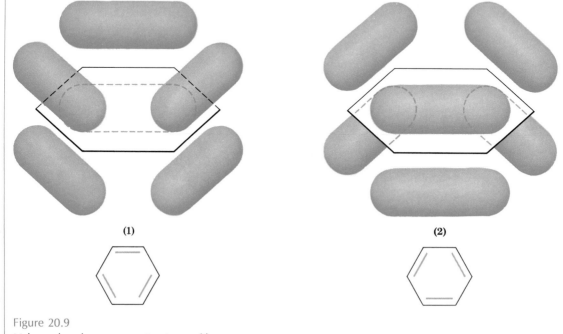

(1)   (2)

Figure 20.9
Valence bond resonance structures of benzene.

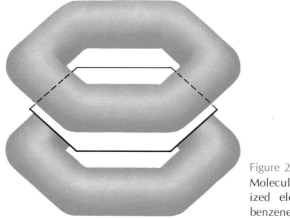

Figure 20.10

Molecular orbital delocalized electron cloud for benzene.

carbons and the π-bonds as a dispersed cloud with electron density above and below the carbon ring. Electrons belonging to a molecular orbital that extends over several nuclei (such as the six π-electrons in benzene) rather than to any specific bond are said to be delocalized. The benzene ring is frequently drawn as

to emphasize the delocalized nature of the π-electrons.

This delocalization of the π-electron cloud in benzene tends to stabilize the molecule by some 36 kcal/mole above that which is calculated for 1,3,5-cyclohexatriene from thermochemical data. Thus we say that the *delocalization energy* or *resonance stabilization energy* of benzene is 36 kcal/mole. A similar high resonance stabilization energy is found in all members of the aromatic family.

NOMENCLATURE OF SOME BENZENE DERIVATIVES. The nomenclature of the aromatic hydrocarbons follows much the same pattern as was outlined for the alkanes and alkenes. Here, *benzene is generally taken to be the parent and the branched groups are named as before.* For example, the compound

is called methylbenzene. Its common name is *toluene.* Another example is ethylbenzene,

In a compound where there is more than one substituent attached to the ring, there are two systems that may be used to identify their positions. In one of these the carbon atoms in the ring are numbered 1 to 6, beginning with the carbon that is bonded to the first group and continuing in such a direction as to lead to the lowest numbers for the remaining radicals. For example, the compound

would be called 1,2-dimethylbenzene, whereas the compound

would be named 1,3-dimethylbenzene.

When the positions of the substituent groups are named, rather than numbered, the terms **ortho, meta,** and **para** are used, and the common name of the parent compound is usually employed. The terms ortho, meta, and para, respectively, label the next three consecutive positions that are adjacent to the first group. For example, the ortho, meta, and para positions on toluene are

Thus the compounds

would be called ortho-methyltoluene and meta-methyltoluene, respectively. These compounds also have common names, which are ortho-xylene and meta-xylene (or simply o-xylene and m-xylene) respectively.

In some of the larger, more complex, compounds *it is sometimes easier to consider the benzene radical (C_6H_5-) as the attached group and the longest carbon chain as the parent.* In this case the benzene radical is called **phenyl,** and the same rules developed earlier for the alkanes and alkenes are employed. For example, we have which is called phenylethene. Its more common name is *styrene* and will be seen again in the last section of this chapter.

Another example, but with two C_6H_5 radicals is

named 1,2-diphenylethene. Note that this compound can exist in either the *cis* or *trans* configuration.

Finally, consider the compound

This compound is called diphenyl.

20.6 Derivatives of the Hydrocarbons

There still remains a whole host of organic compounds yet to be discussed, which are not hydrocarbons. These compounds are looked upon as derived from a hydrocarbon (thus the term derivative) where an atom or a group of atoms has replaced a hydrogen in the parent chain or ring. Usually these attached atoms or groups bestow some characteristic property to the molecule so that any molecule with the same grouping will react chemically in a similar fashion. Groups bestowing such properties on organic compounds are called **functional groups** and aid us in further classifying the remaining organic compounds into groups.

We have already seen two such functional groups within the hydrocarbons: the carbon-carbon double bond and the carbon-carbon triple bond. Compounds containing these functional groups were called alkenes and alkynes, respectively. Some typical reactions of these compounds will be discussed below. Some of the other important functional groups, such as the halogens (X), hydroxyl group (—OH, in alcohols) and so on are listed in Table 20.5.

For convenience, all organic compounds can be viewed as containing two parts: a hydrocarbon fragment that is generally denoted as R (aliphatic) or Ar (aromatic) plus one or more functional groups, such as those listed in Table 20.5. Thus we have R—X or Ar—X, an alkyl halide, or aryl halide, respectively, R—OH an alcohol, etc. Let's now take a brief look at some of these functional groups and some of the properties they impart to organic compounds.

ALKENES AND ALKYNES. The chemical properties of the unsaturated hydrocarbons are due mainly to the high electron density of their double or triple bonds. These compounds are considered to be Lewis bases because they are

Table 20.5
Functional Groups in Organic Compounds

Functional Group	Name	Compound Class	Example
C=C	double bond	alkenes	
C≡C	triple bond	alkynes	H—C≡C—H
F,Cl,Br,I	halo	halides	
—OH	hydroxyl	alcohols	
	carbonyl	aldehydes	
	carbonyl	ketones	
	carboxyl	carboxylic acids	
	amino	amines	
	amido	amides	
—NO$_2$	nitro	—	(TNT)
NH$_2$ | —CH$_2$COOH	—	amino acids	NH$_2$ | CH$_3$CHCOOH (alanine)

rich in electrons and will readily undergo reaction with compounds poor in electrons (Lewis acids). As you might recall from Section 13.5, any species that seeks a nucleus with which it can share a pair of electrons is called a nucleophile (Lewis base) and those species to which the nucleophile becomes bound is called an electrophile. Thus both alkenes and alkynes are nucleophiles that readily undergo reactions with electrophilic agents. During this type of reaction it often happens that the electrophile is simply added to the reacting alkene or alkyne. This type of reaction is termed an **addition reaction.** Some examples of electrophilic addition reactions of the alkenes and alkynes are

$$CH_3—CH{=}CH—CH_3 + HBr \longrightarrow CH_3—CH_2—\underset{\underset{Br}{|}}{CH}—CH_3$$

$$CH_3—CH{=}CH_2 + Br_2 \longrightarrow CH_3—\underset{\underset{Br}{|}}{CH}—CH_2Br$$

$$HC{\equiv}CH + Cl_2 \xrightarrow{FeCl_3} \underset{\underset{Cl}{|}}{HC}{=}\underset{\underset{Cl}{|}}{CH}$$

$$HC{\equiv}CH + HCN \xrightarrow{CuCl} CH_2{=}CHCN$$

In the latter two reactions the $FeCl_3$ and CuCl are acting as catalysts.

ALKYL HALIDES. In this group of compounds a halogen has replaced a hydrogen in the parent hydrocarbon. Thus the halogen is the attached group and the hydrocarbon is the parent. In naming these compounds the same rules developed for the alkanes and alkenes apply, to which, we add that the name of the halogen ends in "o," that is, fluoro, chloro, bromo, and iodo. A few examples of alkyl halides are:

chloromethane   dichloromethane   chloroethane

tetrachloromethane
(carbon tetrachloride)   trichloroethene   dichlorodifluoromethane

Carbon tetrachloride and trichloroethene are used as dry cleaning fluids whereas dichlorodifluoromethane is called "Freon-12" and is used as a coolant in refrigerators and air conditioners.

Alkyl halides can be prepared by electrophilic addition to alkenes as we have just seen. Another method that can be used involves the substitution of a halogen for a hydrogen on an alkane.

Here the Br replaces the hydrogen that is bonded least strongly to the carbon in the molecule.

Perhaps the most important method of preparing alkyl halides is from alcohols. During this reaction the OH of the alcohol is displaced by a halide ion (nucleophile). Examples of this are

$$CH_3—CH_2—CH_2—CH_2—OH + HBr \longrightarrow$$
$$CH_3—CH_2—CH_2—CH_2—Br + H_2O$$

and

Because of the polar nature of the C—X bond the alkyl halides can undergo either a *nucleophilic substitution* (S_N), or an *elimination* (E) reaction when they are combined with basic reagents. In general these reactions can be written as

S_N $R—X + Z^- \longrightarrow R—Z + X^-$

where Z^- is an appropriate nucleophile.

In nucleophilic substitution (S_N) reactions a nucleophile actually replaces the halogen on the alkyl halide. Some nucleophiles that can be used in this type of reaction include OH^-, H_2O, and NH_3. For example, we have

$$CH_3—Br + OH^- \longrightarrow CH_3—OH + Br^-$$

$$CH_3—CH_2—Cl + H_2O \longrightarrow CH_3—CH_2—OH + HCl$$

$$CH_3—Br + NH_3 \longrightarrow CH_3—NH_2 + HBr$$
(amine)

An elimination reaction is just the reverse of an addition reaction. In this case a hydrogen and a halogen on two adjacent carbons of the alkyl halide are eliminated, and a double bond between these two carbon atoms is created.

removed as HCl

The OH⁻ of the strong base removes an H⁺ from the carbon next to the halogen, with a simultaneous formation of a double bond and the expulsion of the halide ion. If the base is moderately weak, there is a competition between the substitution and the elimination paths. Substitution is favored when extremely weak bases are used, whereas a very strong base favors the elimination pathway.

**20.7
Alcohols**

Essential to any discussion of the preparation and reactions of the alcohols are compounds belonging to the aldehydes, ketones, carboxylic acids, ethers, and esters. This discussion therefore will also serve as an introduction to the chemical and physical properties of these important organic groups.

Members of the group of organic compounds known as alcohols all contain the functional group —OH. Some typical alcohols with their I.U.P.A.C. as well as their common names are:

Formula	I.U.P.A.C. Name	Common Name
CH₃—OH	methanol	methyl alcohol
CH₃—CH₂—OH	ethanol	ethyl alcohol
CH₃—CH₂—CH₂—OH	1-propanol	n-propyl alcohol

As you can see, the name of an alcohol is obtained from the name of the parent alkane by replacing the e by *ol*. Also, when necessary the position of the OH group is identified by number. Thus we have

2-propanol is commonly known as isopropyl alcohol and is used widely as rubbing alcohol.

The number of groups attached to the carbon to which the OH is bonded, aids us in further classifying the alcohols into groups. the —C—OH grouping is called the **carbinol group** and the carbon of this group is referred to as the **carbinol carbon.** Compounds in which there is one hydrocarbon group (R) attached to the carbinol carbon are known as **primary alcohols.** Alcohols which contain two such R groups attached to the carbinol carbon are known as **secondary alcohols. Tertiary alcohols,** on the other hand, have three R groups bonded to the carbinol carbon.

primary alcohol   secondary alcohol   tertiary alcohol

Some specific examples are.

a primary alcohol   a secondary alcohol   a tertiary alcohol

Following the I.U.P.A.C. nomenclature, these alcohols are called ethanol, 2-butanol, and 2-methyl-2-propanol. Their common names are ethyl alcohol, sec-butyl alcohol, and tert-butyl alcohol.

Of all of the alcohols, ethanol is the least poisonous to the human body. When ethanol is consumed, however, even in small quantities it causes the blood vessels to dilate, resulting in a lowering of the blood pressure followed by a general feeling of relaxation. In larger amounts ethanol causes intoxication and excessively prolonged use can permanently damage the liver and eventually lead to death.

Methanol, also known as wood alcohol, is a deadly poison whereas the remaining alcohols, other than ethanol, are somewhat "milder poisons." Methanol can cause blindness and eventually total loss of motor control and death.

Methanol is prepared in large quantities from the reaction between carbon monoxide and hydrogen in the presence of a metal oxide catalyst at high temperature and pressure. The balanced equation for this preparation is

$$CO + 2H_2 \xrightarrow[\substack{3000 \text{ psi} \\ 350°–400°C}]{ZnO/Cr_2O_3} CH_3OH$$

The common method of preparing ethanol is through the fermentation of carbohydrates (sugars or starch). Sugars are converted into ethanol and carbon dioxide by the action of yeast in the absence of oxygen.

$$C_6H_{12}O_6 \xrightarrow{yeast} 2C_2H_5OH + 2CO_2$$
carbohydrate   ethanol

Some other preparations of alcohols, which can be carried out in the laboratory **include hydration of alkenes, hydrolysis of esters,** and **reduction of aldehydes and ketones.** The latter two preparations involve reactants yet to be discussed but are still recognizable by their functional groups using Table 20.5.

In the *hydration of an alkene* a water molecule is added directly to the alkene when an acid catalyst is present. The water molecule adds directly across the double bond at elevated temperatures and pressures. For example, consider the reactions

In the first reaction a primary alcohol is produced and in the second a secondary alcohol is formed. The product of the second reaction is predicted by *Markovnikov's*[2] *rule,* which states that **during an electrophilic addition to ordinary alkenes the hydrogen of the reactant becomes attached to that carbon of the alkene that is already bonded to the most hydrogen atoms.** In propene, the end carbon has two hydrogen atoms while the middle carbon has only one. The H of the H_2O then is attached to the end carbon, and the OH is attached to the middle carbon. Other reagents that follow this rule during their reaction with alkenes are the hydrogen halide acids, H*X*. Using Markovnikov's rule we can also see how a tertiary alcohol can be produced from 2-methylpropene.

During the *hydrolysis of an ester* an alcohol is produced along with a carboxylic acid. The functional group for this latter type of compound is

$-C\begin{smallmatrix}\diagup O \\ \diagdown OH\end{smallmatrix}$. The reaction can be catalyzed by either an acid or a base. An example of an acid catalyzed hydrolysis is

        methyl acetate   **methanol**   **acetic acid**
    I.U.P.A.C.: methyl ethanoate   **ethanoic acid**

[2] The publication stating Vladimir W. Markovnikov's rule was published in 1905, one year after his death.

This equilibrium is established rapidly, and in order to drive the reaction to the right, the alcohol can be removed by distillation. Since this is a reversible reaction, an ester can also be produced by reaction of the appropriate acid and alcohol.

In the base catalyzed hydrolysis, also called **saponification,** the acid that is produced in the forward reaction is neutralized, thereby shifting the position of equilibrium to the right. For example,

$$CH_3-CH_2-CH_2-O-\overset{\overset{\displaystyle O}{\|}}{C}-CH_3 + NaOH \overset{H_2O}{\rightleftharpoons}$$

propyl ethanoate

$$CH_3-CH_2-CH_2-OH + NaO-\overset{\overset{\displaystyle O}{\|}}{C}-CH_3$$

Aldehydes (functional groups $-\overset{\overset{\displaystyle O}{\diagup\hspace{-1mm}\diagdown}}{C}\diagdown H$) and ketones (functional group $-\overset{\overset{\displaystyle O}{\diagup\hspace{-1mm}\diagdown}}{C}-$) can produce alcohols by their reduction with H_2. The hydrogen in both cases is added across the $-\overset{\overset{\displaystyle O}{\diagup\hspace{-1mm}\diagdown}}{C}-$ group, which is called the **carbonyl group.** A metal catalyst, such as pure nickel, is needed and the reaction must be performed at high pressure. For example, we have the reaction for the conversion of ethanal to ethanol

ethanal   **ethanol**
common: acetaldehyde

and the conversion of propanone to 2-propanol.

$$CH_3-\overset{\overset{\displaystyle O}{\|}}{C}-CH_3 \xrightarrow[\text{1-5 atm}]{H_2,\ Ni} CH_3-\overset{\overset{\displaystyle OH}{|}}{CH}-CH_3$$

propanone   **2-propanol**
common: acetone

We see that *when an aldehyde is reduced, a primary alcohol is produced and when a ketone undergoes reduction, a secondary alcohol is formed.*

Aldehydes and ketones are used mainly as organic solvents and as starting materials for the preparation of many other organic materials. In the I.U.P.A.C. method for naming aldehydes and ketones the longest carbon chain containing the functional group is used as the parent. The **e** of the corresponding alkane is dropped and the suffix **-al** is added for aldehydes and **-one** for ketones.

REACTIONS OF ALCOHOLS. In the laboratory alcohols can be used to produce many useful organic materials such as alkyl halides, alkenes, ethers, esters, aldehydes, and ketones.

Alkyl halides can be produced by the reaction of alcohols with either a hydrogen halide or a phosphorus halide. In both cases the C—O bond is broken upon forming the products.

$$CH_3—CH_2—CH_2—OH + PCl_5 \longrightarrow CH_3—CH_2—CH_2—Cl + POCl_3 + HCl$$

At elevated temperatures alcohols can be dehydrated with sulfuric acid to produce **alkenes.** For example,

When more powerful dehydrants, such as phosphorus(V) oxide (P_4O_{10}) or alumina (Al_2O_3) are employed, a lower temperature can be used. This reaction can be seen as

When alcohols undergo reaction with an alkyl iodide, an **ether** is produced at a rather slow rate. During this reaction the O—H bond of the carbinol group is broken.

$$CH_3—CH_2—I + CH_3CH_2OH \rightleftharpoons CH_3—CH_2—O—CH_2—CH_3 + HI$$

   ethyl iodide   **ethanol**   **diethyl ether**

Diethyl ether is used mainly as an anesthetic.

We saw earlier that **organic esters** can be produced by the reaction between an alcohol and a carboxylic acid. Because of their pleasant odors many esters are used in artificial flavors and fragrances. Some of the fragrances of esters include apple, apricot, banana, orange, pineapple, and wintergreen. For example, the ester that smells like essence of bananas, *n*-amyl acetate, is prepared by the reaction of *n*-amyl alcohol (*n*-pentanol) with acetic acid in the presence of an acid catalyst.

$$CH_3—C\overset{O}{\underset{OH}{\diagup}} + CH_3—CH_2—CH_2—CH_2—CH_2—OH \overset{H^+}{\longrightarrow}$$

  acetic acid   **n-amyl alcohol**

$$CH_3—\overset{O}{\overset{||}{C}}—O—CH_2—CH_2—CH_2—CH_2—CH_3$$

                  n-amyl acetate

The other organic esters can similarly be formed by choosing the correct alcohol to react with the proper carboxylic acid.

When an alcohol is treated with an inorganic acid an **inorganic ester** is produced. For example, when nitric acid is *cautiously* added to ethyl alcohol,

the product is ethyl nitrate (quite explosive). Organic phosphate esters of the type

$$R-O-\overset{\overset{\displaystyle O}{|}}{\underset{\underset{\displaystyle OH}{|}}{P}}-OH$$

are very important in biological systems. More will be said about them in the next chapter.

An **aldehyde,** and eventually a **carboxylic acid,** is produced when a *primary alcohol is oxidized.* In the body these oxidation products partially account for the toxicity of ethyl alcohol. In the laboratory this oxidation is often carried out with dichromate ion, $Cr_2O_7^{2-}$, in acid solution. As an example of this type of oxidation we have

methanol	methanal	formic acid
	(formaldehyde)	(methanoic acid)

and

ethanol	ethanal	acetic acid
	(acetaldehyde)	(ethanoic acid)

When secondary alcohols are similarly oxidized the product is a ketone.

propanone
(acetone)

Tertiary alcohols, on the other hand, resist oxidation and usually no reaction is seen with many oxidizing agent.

Organic chemists take advantage of the various products that are produced by the different types of alcohols. As it turns out, this is a very convenient way of differentiating between the various alcohols and can be used whenever a test is needed for such a differentiation. The reason for the various products seems to be the number of hydrogens that are attached to the carbinol carbon. During the oxidation reaction one hydrogen is eliminated from the carbinol carbon as well as the one on the OH. With a primary alcohol, therefore, one hydrogen remains on the carbon after oxidation, whereas with a secondary alcohol none remain. Since no hydrogens are attached to the carbinol carbon in a tertiary alcohol, no reaction occurs when H_2CrO_4 or when other similar oxidizing materials are added.

POLYHYDRIC ALCOHOLS. Polyhydric alcohols are compounds that contain more than one hydroxyl group. Examples of these are ethylene glycol with two OH groups

$$H_2C-OH$$
$$|$$
$$H_2C-OH$$

(1,2-dihydroxyethane)

and glycerol

$$H_2C-OH$$
$$|$$
$$HC-OH$$
$$|$$
$$H_2C-OH$$

(1,2,3-trihydroxypropane)

which contains three.

Ethylene glycol is soluble in water and is the main ingredient in commercial antifreeze for automobiles. Glycerol, on the other hand, is important biologically, as we shall see in the next chapter.

Both these compounds can be nitrated by their cautious addition to a mixture of nitric and sulfuric acid.

glycol dinitrate

glyceryl trinitrate

Both of these are inorganic esters and must be handled with extreme caution. Glyceryl trinitrate is also called nitroglycerine and is used with glycol dinitrate in the production of dynamite.

20.8
Amines
and Amides

The amines are a group of compounds identified by the functional group —NH_2. As a result they are commonly viewed as derivatives of ammonia. Below are some typical amines.

NH₃	Ammonia	I.U.P.A.C.
CH₃—NH₂	methyl amine	aminomethane
CH₃—CH₂—NH₂	ethyl amine	aminoethane
CH₃—CH₂—NH—CH₂—CH₃	diethyl amine	ethylaminoethane
(CH₃)₃—C—NH₂	t-butyl amine	2-amino-2-methylpropane
H₂N—CH₂—CH₂—NH₂	ethylene diamine	1,2-diaminoethane
	aniline	aminobenzene

Amines can be classified as being either primary, secondary, or tertiary, depending on the number of R groups attached to the nitrogen.

primary amine   **secondary amine**   **tertiary amine**

All the amines in our list above are primary amines except diethyl amine, which is a secondary amine.

Amines also exist in which the nitrogen is a member of a ring. Examples of this type are

pyridine

piperidine

These compounds are referred to as **heterocyclics** because not all the atoms in the ring are identical.

Amides are identified by the functional group $-\overset{\overset{\text{O}}{\|}}{\text{C}}-\text{NH}_2$ Some examples of this type of compound are

$CH_3-\overset{\underset{\|}{\text{O}}}{\text{C}}-NH_2$ acetamide   nicotinamide

benzamide $H_2N-\overset{\underset{\|}{\text{O}}}{\text{C}}-NH_2$ urea

The functional groups of the amines and amides are found in many important biological compounds that will be discussed in the next chapter. These compounds include the nucleic acids, the amino acids, thiamin, riboflavin, and biotin. The heterocyclic amines are also found as a basic unit of a group of compounds known as **alkaloids.** Alkaloids are rather complex compounds containing nitrogen that are found in plants. Compounds such as nicotine, codeine, morphine, and lysergic acid diethylamide (LSD) are all alkaloids.

20.9
Polymers

Polymers are very large molecules that are made by bonding together many smaller molecules that we call **monomers.** For example, polyethylene (plastic food wrap) is prepared by linking together a large number of $CH_2=CH_2$ monomer units to give a hydrocarbon having the general formula, $(-CH_2-CH_2-)_n$. Many naturally occurring substances are polymers; for example, rubber, starch, proteins, and the nucleic acids. Manmade polymers include such familiar materials as Bakelite, Melmac, Nylon, Dacron, Plexiglass, Teflon, and polyvinyl chloride (PVC).

Polymers can be made either by direct addition of their monomeric units (to make an **addition polymer**) or by condensation in which a small molecule is lost when the monomers are linked together (forming a **condensation polymer**). Vinyl chloride

$$CH_2=CH$$
$$\overset{|}{Cl}$$

can be made to polymerize in the presence of a peroxide initiator to form polyvinyl chloride. This reaction can be seen as

$$n\ CH_2=CH \xrightarrow{\text{peroxide}} \left(CH_2-CH-CH_2-CH\right)_n$$
$$\qquad\quad \overset{|}{Cl} \qquad\qquad\qquad \overset{|}{Cl}\quad\ \overset{|}{Cl}$$

In general the formula for polyvinyl chloride is $\left(CH_2-CH\right)_n$. This material

$$\qquad\qquad\qquad\qquad\qquad\qquad\qquad\qquad\qquad \overset{|}{Cl}$$

finds many uses, for example, in phonograph records, plastic pipe and raincoats.

Condensation polymers are prepared by the reaction between two species each of which have more than one functional group. The resulting polymer is said to be a **copolymer** because it consists of two different monomers. For example, nylon is formed when a **dicarboxylic acid** (i.e., a carboxylic aicd that contains two —COOH groups) reacts with a **diamine** (an amine with two —NH$_2$ groups). The overall reaction for the production of nylon is

$$n\text{HOOC}(CH_2)_4\text{COOH} + n\text{H}_2\text{N}(CH_2)_6\text{NH}_2 \xrightarrow[280°\ C]{-H_2O} \left(-\overset{}{\underset{\overset{\|}{O}}{C}}(CH_2)_4\overset{}{\underset{\overset{\|}{O}}{C}}-\overset{\overset{H}{|}}{N}(CH_2)_6\overset{\overset{H}{|}}{N}-\right)_n + H_2O$$

adipic acid   hexamethylenediamine

Nylon

Dacron, a polyester, is prepared by the reaction of methyl terephthalate (a diester) with ehylene glycol in the presence of an acid or base.

$$n\text{CH}_3\text{OOC}\langle\bigcirc\rangle\text{COOCH}_3 + n\text{HOCH}_2\text{CH}_2\text{OH} \xrightarrow[\text{base}]{\text{acid or}}$$

methyl terephthalate   **ethylene glycol**

$$\left(-\underset{\underset{\text{O}}{\|}}{\text{C}}\langle\bigcirc\rangle\underset{\underset{\text{O}}{\|}}{\text{C}}-\text{OCH}_2\text{CH}_2\text{O}-\right)_n + \text{CH}_3\text{OH}$$

Dacron   **methanol**

(a)

(b)

Figure 20.11
Bakelite. (a) Polymerization of salicyl alcohol can give a linear polymer. (b) Continued polymerization, with cross linking, gives a rigid three-dimensional structure called bakelite.

In the production of nylon H_2O is eliminated and with dacron, methanol is eliminated. Table 20.6 contains a number of important polymers, the reactants needed for their production, and whether they are addition or condensation polymers.

The formation of bonds between adjacent polymer molecules can also be brought about. In this case, the greater the degree of this **cross-linking** between parallel rows of polymer molecules, the stronger will be the material. Bakelite, for example, owes its strength and hardness to the three-dimensional network of covalent bonds throughout the entire polymer, as illustrated in Figure 20.11.

Table 20.6
Compositions of Some Common Polymers

Monomer	Polymer	Type
$CH_2{=}CH_2$ ethylene	polyethylene	addition
$CH_2{=}CHCl$ vinyl chloride	polyvinyl chloride	addition
$F_2C{=}CF_2$ tetrafluoroethylene	Teflon	addition
⬡—CH=CH$_2$ styrene	polystyrene	addition
$CH_2{=}C(CH_3)COOCH_3$ methyl methacrylate	Plexiglass	addition
$HOOC{-}(CH_2)_4{-}COOH + NH_2{-}(CH_2)_6NH_2$ adipic acid hexamethylenediamine	Nylon	condensation
$HOOC{-}C_6H_4{-}COOH + HO{-}CH_2CH_2{-}OH$ terephthalic acid ethylene glycol	Dacron	condensation
$C_6H_5OH + HCHO$ phenol formaldehyde	Bakelite	condensation
$C_6H_5OH + HC{-}\!\!-\!\!-CH$ ... $CH{=}CHO$ phenol furfural	Durite	condensation
melamine $+ HCHO$	Melmac	condensation

Natural rubber, too, can be made harder and stronger by a process known as **vulcanization.** In this reaction sulfur bridges between different chains create cross links that lead to a tougher material.

20.1 What is meant by aliphatic hydrocarbon, aromatic hydrocarbon, hydrocarbon derivative, structural isomer, olefin, alkane, alkene and alkyne, and asymmetric carbon atom.

20.2 Draw the structural formulas for and name the nine isomers of heptane. Which of these isomers would give rise to optical isomerism?

20.3 Draw all the possible isomers of hexene and show geometric isomers wherever possible.

20.4 Name the following compounds:

(a) H_3C
$\quad$ CH—CH$_2$—CH—CH$_2$—CH$_3$
H_3C $\qquad\qquad$ CH$_3$

(b)   CH$_3$—CH—CH$_2$—CH—CH$_2$—CH$_3$
$\qquad\quad$ CH$_2$ $\qquad$ CH$_3$
$\qquad\quad$ CH$_3$

(c) $\qquad\qquad$ CH$_3$
CH$_3$—CH$_2$—CH—CH$_2$—CH—CH$_2$—CH$_3$
$\qquad\qquad\qquad\qquad\qquad$ CH$_2$—CH$_2$—CH$_3$

(d)   CH$_3$—CH$_2$—CH=CH—CH—CH$_3$
$\qquad\qquad\qquad\qquad\quad$ CH$_2$—CH$_3$

(e) $\qquad\qquad\qquad$ CH$_3$
CH$_3$—CH$_2$—CH—CH$_2$—CH—CH$_3$
$\qquad\qquad\quad$ CH$_3$

20.5 Starting with 1-propanol how could one prepare the following:
(a) propene
(b) 1-bromopropane
(c) propanoic acid
(d) propyl acetate

20.6 Predict the results of the following reactions:

(a)
$$CH_3—CH_2OH + CH_3—\overset{\overset{\displaystyle O}{\|}}{C}—OH \xrightarrow{H^+}$$

(b)   CH$_3$—CH$_2$—CH$_2$—OH $\xrightarrow[\Delta]{KMnO_4}$

(c) $\qquad\qquad$ OH
CH$_3$—CH$_2$—CH—CH$_3$ $\xrightarrow{KMnO_4}$

(d)   CH$_3$—CH$_2$—CH$_2$—CH$_2$—OH $\xrightarrow{H_2SO_4}$

20.7 Write molecular and structural formulas for the following compounds:
(a) 2-methyl-1-butene
(b) 2,3-dimethyl-2-butanol
(c) 2-bromo-1-phenylpropane
(d) 3-methyl-2-pentanone

20.8 What chemical reactions discussed in this chapter could be used to distinguish between the following
(a) 2-propanol and 2-methyl-2-propanol
(b) 1-butanol and 2-butanol
(c) *n*-butane and 1-butene
(d) ethanal and 2-propanone

20.9 Complete the following reactions and classify each as either substitution or elimination.

(a)
$$CH_3—O—\overset{\overset{\displaystyle O}{\|}}{C}—CH_3 + H_2O \xrightarrow{H^+}$$

(b)   CH$_3$—CH$_2$—Cl + NaOH $\longrightarrow$

(c)   CH$_3$—CH$_2$—OH + HBr $\longrightarrow$

(d) $\qquad\qquad$ Cl
CH$_3$—C—CH$_3$ + H$_2$O $\xrightarrow{OH^-}$
$\qquad\quad$ CH$_3$

20.10 Name the following compounds using I.U.P.A.C. rules.

(a)

(b)

(c)

$$CH_3\!-\!CH_2\!-\!\overset{\displaystyle O}{\overset{\|}{C}}\!-\!CH_3$$

(d)

$$CH_3\!-\!\overset{\displaystyle CH_3}{\overset{|}{C}H}\!-\!CH_2\!-\!NH_2$$

20.11 Using any organic and inorganic reagents, how would one prepare the following:
(a) tert-butylchloride
(b) 2-butene
(c) 1-butanol
(d) 1-bromopentane

20.12 Explain what is meant by the following:
(a) Homologous series
(b) Functional group
(c) Nucleophilic substitution
(d) Elimination reaction
(e) Diene
(f) Polymer
(g) Monomer

20.13 Give a structural formula for each of the following:
(a) An aldehyde
(b) A ketone
(c) A carboxylic acid
(d) An amine
(e) An alcohol
(f) An ester
(g) An ether
(h) A heterocyclic amine

20.14 Knowing that polystyrene is an addition polymer, write its structural for-

mula showing at least three monomer units. The structure of styrene is shown on page 36.

20.15 Using Markovnikov's rule complete the following addition reactions. Name the reactants and products.

(a) $CH_3\!-\!CH_2\!-\!CH\!=\!CH_2 + HI \longrightarrow$

(b) $CH_3\!-\!CH\!=\!CH_2 + H_2O \xrightarrow[H_2SO_4]{}$

(c)

20.16 Write a chemical equation showing the saponification of propyl acetate. What drives this reaction toward completion?

20.17 Compare the products obtained by reduction of an aldehyde and a ketone with hydrogen.

20.18 Can you suggest why it is difficult to limit the oxidation of a primary alcohol to the production of an aldehyde?

20.19 What products (if any) are obtained by the mild oxidation, using $K_2Cr_2O_7$ in acid solution, of each of the following:
(a) $CH_3\!-\!CH_2\!-\!OH$

(b) $CH_3\!-\!\overset{\displaystyle CH_3}{\overset{|}{C}H}\!-\!CH_2\!-\!OH$

(c) $CH_3\!-\!\overset{\displaystyle CH_3}{\underset{\displaystyle OH}{\overset{|}{\underset{|}{C}}}}\!-\!CH_3$

(d) $CH_3\!-\!\overset{\displaystyle OH}{\overset{|}{C}H}\!-\!CH_3$

(e) $CH_3\!-\!CH_2\!-\!CHO$

(f) $CH_3\!-\!COOH$

(g) $CH_3\!-\!\overset{\displaystyle O}{\overset{\|}{C}}\!-\!CH_3$

20.20 Draw *all* of the structural isomers of $C_3H_4Cl_2$.

20.21 Geometrical isomers can be isolated for 1,2-dichloroethene but not for 1,2-dichloroethane. Why?

20.22 What is the difference between addition polymerization and condensation polymerization?

20.23 If ethylene glycol and dimethylmalonate were to form a condensation polymer by the elimination of methanol, what would be the structure of the repeating unit in the polymer chain?

20.23(cont'd)

ethylene glycol

dimethylmalonate

20.24 Diethylamine gives a basic solution in water. Why?

21
Biochemistry

Without question, one of the most active areas of chemical research today is the field of biochemistry which, as its name implies, is concerned with the chemistry that takes place in living systems. The modern biochemist views a living organism as a collection of organic molecules that interact with each other and with their environment in a very unique and special way. When isolated from a living system these biomolecules are themselves lifeless. They obey all of the laws of chemistry and thermodynamics that we have examined up to now, and it is the goal of the biochemist to understand the functions and intricate interactions of these molecules that give rise to the phenomenon that we call life.

The field of biochemistry is very large and quite complex and we certainly cannot hope to explore it fully in a single chapter. Instead, we shall be content to examine some of the types of biomolecules and their apparent functions in the operations of a living cell.

Nearly all of the compounds found in a living system are composed primarily of carbon, hydrogen, oxygen, nitrogen, and some sulfur, and for the most part, they depend on carbon for their molecular backbone. In general these molecules are very large, having molecular weights ranging up to a million or more. We shall see, however, that in many cases these **macromolecules** are constructed using a relatively small number of different simple molecules. For the purpose of discussing the various kinds of biomolecules we can place nearly all of them into one or another of four classes: **proteins, carbohydrates, lipids,** and **nucleic acids.**

Before we proceed remember that you are not expected to memorize the names and formulas of all of the different compounds that we shall discuss in this chapter. Instead, concentrate your attention on the types of molecules that are involved, the way that they combine with each other, and the general features of the structures that result.

21.1
Proteins

Proteins are very large molecules having molecular weights ranging from about 6000 to approximately 1,000,000. They constitute nearly 50 percent of the dry weight of cells and, depending on the individual protein, serve a variety of different functions within the cell. Many of them, such as the hormone insulin, for example, act as enzymes that catalyze biochemical reactions. We shall discuss these further in the next section. Others serve to transport substances through the organism. Hemoglobin, for instance, carries oxygen in the bloodstream and delivers it to different parts of the body. There are long fibrous proteins, such as *actin* and *myosin,* that are found in muscle. Another fibrous protein, *α-keratin,* serves as the major constituent of hair, nails, and skin while *collagen* is the prime constituent of tendons. Proteins are also found in toxins (poisonous materials) such as botulinus toxin as well as in antibodies and hormones.

Despite their wide range of functions, all proteins have something in common with one another. They are polymers made up by linking together in various combinations, a number of different simple monomeric units called **α-amino acids.**

An amino acid is a bifunctional organic molecule that contains both a carboxyl group, —COOH, as well as an amine group, —NH₂. In an α-amino acid the amine group is located on the carbon atom adjacent to the carboxyl group (i.e., the α-carbon atom). This gives a structure that we can generalize as

Since the —NH₂ group (like ammonia) is basic and the —COOH group acidic, in neutral solution the amino acid exists in an internal ionic form called a *zwitterion* where the proton of the —COOH group is transferred to the NH₂ to give

A very interesting and important fact is that in all organisms nearly all proteins are constructed using as building blocks a set of only 20 α-amino acids. These are shown in Figure 21.1. Note that except for proline, all fit the general formula above. Another point of interest is that except for glycine, all of these amino acids have four different groups attached to the α-carbon atom, which is therefore an asymmetric carbon atom. Each of these amino acids can therefore exist in two different isomeric forms (optical isomers). In proteins, however, only one isomer of each is commonly observed to occur. Apparently substitution of one isomer for another destroys the biological activity of the protein molecule.

In a protein molecule, amino acids are linked together to form a long chain. This can be viewed as the result of the elimination of a water molecule from between the —NH₂ group of one amino acid molecule and the —COOH group of another,

peptide bond or peptide linkage

The molecule that results is called a **peptide** and the group of atoms within the dotted line constitutes a **peptide bond** (amide bond), or **peptide linkage.** In the particular example above, the peptide is composed of two amino acids and is said to be a **dipeptide.** Since one end of this molecule contains a carboxyl group and the other a free —NH_2 group, additional amino acids may be joined to give ultimately a **polypeptide,** a long chain composed of many amino acid molecules linked by peptide bonds. A segment of such a chain could be indicated as

The backbone of the chain is thus the same series of atoms repeated over and over again, with only the R groups changing as we move along the chain.

Before we continue, we should note that in the formation of a protein the linking together of the different amino acids is not a random process. Each molecule of a given protein has the same sequence of amino acids along its polypeptide chain. In fact, it is this very sequence that imparts to a protein its own specific properties.

The amino acid sequence that exists in a polypeptide is called its **primary structure.** In addition to this, the polypeptide chain twists and turns and assumes a **secondary structure** that is determined by interactions such as hydrogen bonding that occur between different groups along the chain. An example of this is found in the fibrous protein, α-keratin (the major component of hair) in which the polypeptide chains coil themselves into the **α-helix,** shown in Figure 21.2. The hydrogen bonding, in this case, takes place between the oxygen atom in a carbonyl groups ($>C{=}O$) and a hydrogen atom attached to a nitrogen atom that lies in an adjacent loop of the helix (Figure 21.3). This therefore serves to hold the chain in its coiled shape.

In globular proteins, so named because of their overall shape, coiled polypeptide chains are also folded to give a complex three-dimensional structure, referred to as its **tertiary structure.** This is shown in Figure 21.4 for the protein myoglobin, a substance that stores oxygen in muscle tissue until it is needed in metabolic oxidation.

The tertiary structure of a protein is controlled by several different kinds of interactions that serve to hold the folded segments of the chain in place. For example, besides hydrogen bonding there are also ionic attractions that occur between a negatively charged deprotonated carboxyl group (like that found in the R group of glutamic acid) and a positively charged protonated amine group (like that found in lysine). This is shown in Figure 21.5.

The solvent is also important in determining the shape of the protein molecule. In the presence of the polar solvent water, nonpolar R groups such as the phenyl ring in phenylalanine (see Figure 21.1) are forced toward the center of the folded polypeptide chain, away from the solvent. This is the same phenomenon, you may remember, that leads to the low solubility of nonpolar substances in polar solvents. It too helps determine the tertiary

Figure 21.1
The 20 amino acids found in most proteins.

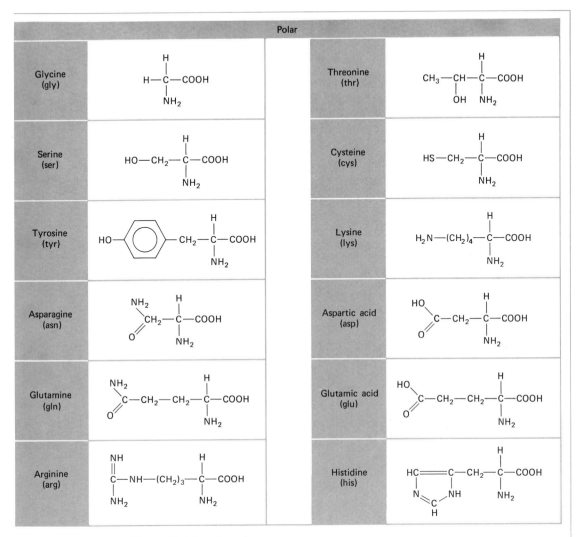

Figure 21.1 (continued).

structure of proteins because the polypeptide chain tends to fold in such a way that nonpolar groups do not contact the solvent.

Still another type of interaction that maintains the folded conformation of the protein molecule is the formation of covalent bonds between cysteine molecules located at different points along the chain. This occurs by partial oxidation of the —SH (thiol) group,

$$R\text{—}SH + HS\text{—}R \xrightarrow{\text{oxidation}} H_2O + R\text{—}S\text{—}S\text{—}R$$

The resulting linkage is called a **disulfide bridge.**

Disulfide bridges not only help to keep the polypeptide chain folded but

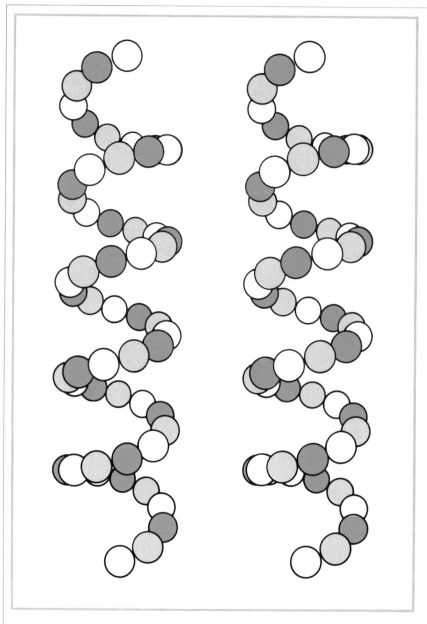

Figure 21.2

The alpha helix, composed of a repetition of amino acid units.

In the illustration N is white, C is pale blue and R is dark blue.

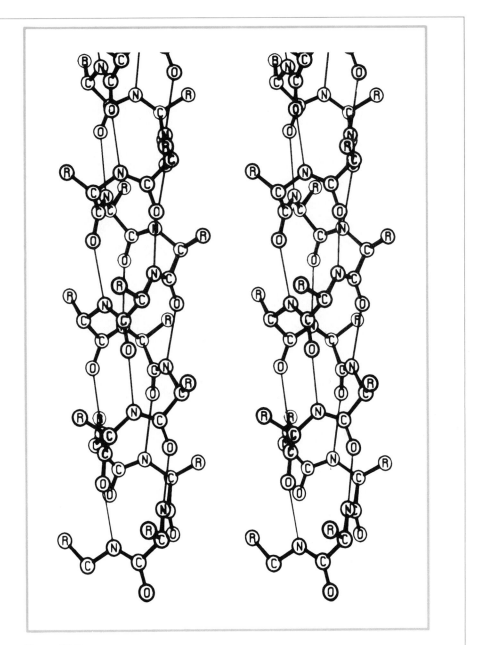

Figure 21.3
Hydrogen bonding (light lines) in the α-helix. Courtesy of Carroll K. Johnson, Oak Ridge National Laboratory, Oak Ridge, Tennessee.

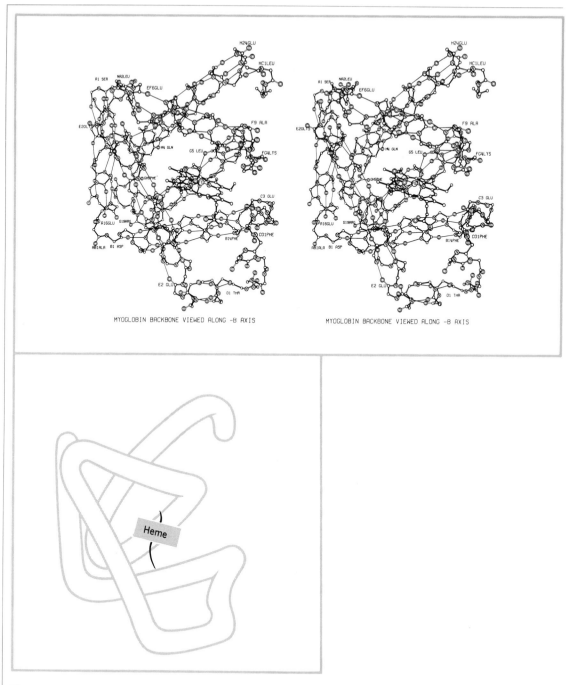

MYOGLOBIN BACKBONE VIEWED ALONG -B AXIS

MYOGLOBIN BACKBONE VIEWED ALONG -B AXIS

Heme

Figure 21.4
The tertiary structure of myoglobin. Courtesy of Carroll K. Johnson Oak Ridge National Laboratory, Oak Ridge, Tennessee.

can also bind two such chains together. For example, beef insulin (Figure 21.6) consists of two polypeptide chains that are cross-linked at two points by these disulfide bridges.[1]

There is an interesting sidelight to the subject of the disulfide bridge in protein chemistry. The curl (or lack of curl) in hair is determined by protein conformation locked in place by disulfide bridges. The "permanent wave" treatment that women (and, recently, men too) use to produce curly hair involves two chemical reactions. The hair is first treated with a chemical able to break the disulfide bridges so that the protein chains in the hair are free to twist into any desired shape, as determined by the curlers. A second solution is then applied that produces a mild oxidation, thereby causing the disulfide bridges to be reestablished. These newly formed disulfide bridges "set" the hair in the newly curled shape. The only reason a permanent wave isn't really "permanent" is because new hair grows that hasn't received the setting treatment.

Proteins that contain more than one independent polypeptide chain exhibit still another degree of structural sophistication, called **quaternary structure.** This is determined by the way in which the folded chains orient themselves with respect to one another. A good example of this occurs in hemoglobin, Figure 21.7. This protein consists of four polypeptide chains; two α-chains each containing 141 amino acids and two β-chains each with 146 amino acids.

In addition to the polypeptide chains this protein also contains groupings of atoms, called **heme groups,** that serve to bind oxygen so that it can be transported through the blood stream and be deposited at oxygen-poor cells. As depicted in Figure 21.7, the four folded polypeptide chains with their heme groups are packed together in a roughly tetrahedral fashion.

The heme group in hemoglobin is also found in myoglobin and accounts for the red color of blood and muscle tissue. The structure of heme is

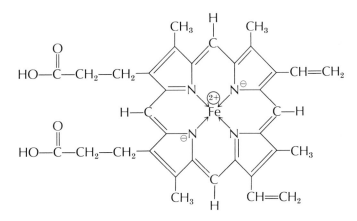

and is, in fact, a complex ion containing iron(II) enclosed within a square planar grouping of nitrogen atoms. In hemoglobin each heme group is at-

[1] The elucidation of the primary structure of this protein by Frederick Sanger won him the Nobel Prize in 1958.

Figure 21.5
Ionic interactions in determining the tertiary structure of proteins.

tached to its polypeptide by additional coordination to the nitrogen atom of a histidine, as shown in Figure 21.8. The sixth coordination site about the iron(II) is empty and is used to bind an oxygen molecule.

The basic square planar ligand structure in heme is called a **porphyrin** and forms very stable complexes with several different metal ions. For example, a structure very similar to heme containing Mg^{2+} instead of Fe^{2+} is found in chlorophyll, the green pigment found in plants that is used in photosynthesis. Still another porphyrin structure, containing Co^{2+} in the center, exists in a substance called vitamin B_{12} coenzyme (Figure 21.9). Thus, even though most of the structures of biomolecules are made up of carbon, hydrogen, nitrogen, and oxygen, some metals are also of critical importance to the well-being of a living organism.

21.2 Enzymes

Enzymes are globular proteins that serve to catalyze specific biochemical reactions with what can only be judged as amazing effectiveness. In some cases reactions are speeded up, with respect to their uncatalyzed paths, by factors ranging from 10^9 to 10^{20}! The result is that the catalyzed biochemical reactions are able to proceed with very nearly 100% yield of products, since competing side reactions that give undesirable products are extremely slow by comparison. In this way a buildup of by-products that would cause a waste disposal problem for the organism is avoided. Enzymes also provide the organism with a way of controlling the rates of the reactions that take place because biochemical reactions do not occur at appreciable rates in the absence of the catalyst. Removal, or at least temporary blockage, of a critically important enzyme "turns off" the chemistry which that particular enzyme catalyzes. Thus, in a very real sense enzymes direct the chemical reactions that take place in a living cell.

Some enzymes contain only one or two polypeptide chains. In the last section, for example, we saw the structure of beef insulin that is composed of two chains. Other enzymes require an additional substance called a **coenzyme** in order to function. Many of the vitamins that we must ingest to maintain good health are precursors of coenzymes. An example is vitamin B_{12}, the absence of which from the diet leads to a deficiency disease known as pernicious anemia. In the body vitamin B_{12} is converted to its coenzyme, the structure of which we saw in Figure 21.9. Many metals, like cobalt, are needed in small amounts to promote enzyme activity.

The mechanism by which an enzyme acts has long been the subject of intense research. It appears to depend on the ability of the enzyme to very selectively bind to a reactant molecule (called the **substrate** of the enzyme).

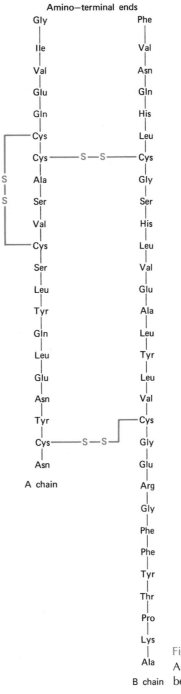

Figure 21.6

Amino acid sequence in beef insulin.

Figure 21.7

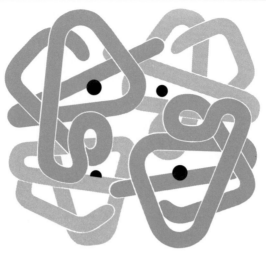

Figure 21.7
Quaternary structure of hemoglobin. Four globular protein molecules containing heme groups are packed into hemoglobin structure. Adapted from R. E. Dickerson and I. Geis, *The Structure and Action of Proteins,* W. A. Benjamin Co., Menlo Park, California, 1969. Original illustration copyright 1969 by R. E. Dickerson and I. Geis.

There thus seems to be a "lock and key" relationship between an enzyme and its substrate, where the substrate molecule just precisely fits into (or onto) the folded globular protein. There is evidence that when this occurs, there is a slight alteration in the shape of the enzyme that strains certain key bonds in the substrate, thereby making them more susceptible to chemical attack.

Some enzymes are very specific in their activity, affecting the rate of reaction of a single compound. Others are less choosy and simply promote a certain kind of chemical reaction on a whole class of compounds having similar structures. This behavior, too, is the direct result of the lock-and-key relationship between the enzyme and its substrate. An example is the enzyme *chymotrypsin* that accelerates the hydrolysis of the dashed bond in the compounds

Figure 21.8
Attachment of heme group to hemoglobin protein by coordination of histidine to the iron atom of heme. (a) Sixth coordination site is vacant. (b) When hemoglobin carries oxygen the O_2 is bound to Fe of heme at the sixth coordination site.

(a)   (b)

Figure 21.9

Vitamin B-12 coenzyme. Courtesy of Carroll K. Johnson, Oak Ridge National Laboratory, Oak Ridge, Tennessee.

where $X = N$ or O

An example is

In this case it is thought that the hydrophobic benzene ring serves to position the substrate molecule on the enzyme (Figure 21.10), which then interacts with the $>C{=}O$ group in a way which makes the dashed bond more susceptible to hydrolysis. Since the function of the enzyme depends on the hydrophobic tail and the proper location of the $>C{=}O$ group, a family of similar compounds are affected by the enzyme.

ENZYME INHIBITION. When a substance other than the enzyme substrate becomes bound to the active site of an enzyme, the catalytic activity is lost and the enzyme is said to be inhibited. In some cases this inhibition is irreversible. This occurs when the inhibitor becomes permanently bound to the en-

Figure 21.10
Enzyme action of chymotrypsin.

zyme, by covalent bond formation, and cannot be displaced. An example is diisopropylfluorophosphate,

a highly toxic nerve poison and an ingredient in some nerve gases. This molecule reacts with and poisons (inhibits) an enzyme called *acetylcholine esterase*, which is required for the transport of impulses along nerve tissue.

A second type of enzyme inhibition is called **competitive inhibition** in which there is a competition between the inhibitor and the substrate for the enzyme active site. This is a system in which there are two simultaneous equilibria, one between the enzyme (*E*) and substrate (*S*),

$$E + S \rightleftharpoons ES$$

and one between the enzyme and inhibitor (*I*),

$$E + I \rightleftharpoons EI$$

As we would predict from Le Chatelier's principle, increasing the substrate concentration displaces the inhibitor.

An example of competitive inhibition is the action of the sulfa drug, sulfanilamide. This molecule, shown below, bears a very close similarity to *p*-aminobenzoic acid, which is acted upon by an enzyme to produce an important coenzyme that is required by bacteria.

sulfanilamide   *p*-aminobenzoic acid

The sulfanilamide, by occupying the active site of the enzyme that works on the *p*-aminobenzoic acid, prevents the production of the required coenzyme and hence leads to the demise of the bacterium.

As a final note, enzyme activity is also affected by temperature and pH. These factors alter the conformation and shape of the globular protein structure; therefore, changes in temperature or pH can destroy the precise fit that must exist between the enzyme and its substrate in order to obtain the desired catalytic activity.

21.3
Carbohydrates

This important class of compounds is used by living organisms in a variety of ways: as a source of energy, as a source of carbon to be used in the synthesis of other biomolecules, and as a structural element in cells and tissues. Historically the name carbohydrate arose as a consequence of the empirical formula exhibited by many of them, $C_n(H_2O)_m$, which suggested that they were *hydrates of carbon*. Examples are glucose, $C_6H_{12}O_6$, having the empirical formula CH_2O, and sucrose (ordinary cane sugar) $C_{12}H_{22}O_{11}$, with the empirical formula $C_{12}(H_2O)_{11}$. The name carbohydrate has stuck with these substances even though it is now known that they do not contain intact water molecules.

Most carbohydrates, such as starch and cellulose, are very large molecules having enormous molecular weights. However, like the proteins, they are composed of many relatively simple units arranged in long polymeric chains. The simplest such units are called **monosaccharides,** and constitute the **simple sugars.** As a class, the monosaccharides are polyhydroxy aldehydes or ketones, the simplest of which is glyceraldehyde,

Glyceraldehyde is a **triose;** *tri* denoting three carbon atoms and *ose,* the characteristic ending used in naming the sugars (e.g., glu*cose,* su*crose,* and fru*ctose).*

Glyceraldehyde, like all of the other saccharides, contains an asymmetric carbon atom and exhibits optical isomerism. As with the amino acids, however, only one of the optical isomers of a monosaccharide is generally found to be formed in living systems. This specificity for a single optical isomer is illustrated by glucose, for example, which possesses four asymmetric carbon atoms (indicated by asterisks).

There are 16 possible optical isomers of glucose, but only the one above, D-glucose, is biologically significant.

By far the most important monosaccharides are those containing five and six carbon atoms, the pentoses and hexoses, respectively. Some of the more prominent ones are found in Table 21.1. The most common hexose is glucose, whose structure is shown above. However, glucose, like most of the other pentoses and hexoses, exists predominantly in a cyclic structure in which the molecule turns on itself as shown in Figure 21.11. When the ring is closed, the —OH group that is created from the aldehyde functional group can either point up or down (this is the —OH group on the right-most carbon atom in the structures drawn in Figure 21.11). Two isomers are thus created, α-D-glucose and β-D-glucose. As we shall see, the orientation of this —OH group is quite significant in the polysaccharides, starch, and cellulose.

Another important six carbon sugar is fructose. In its open-chain structure the molecule is a ketone.

Like glucose, however, it too prefers a cyclic structure, as shown in Figure 21.12. In this case a five-membered ring is formed.

The five-membered ring also occurs in two very important pentoses,

Table 21.1
Some Important Monosaccharides

Pentoses

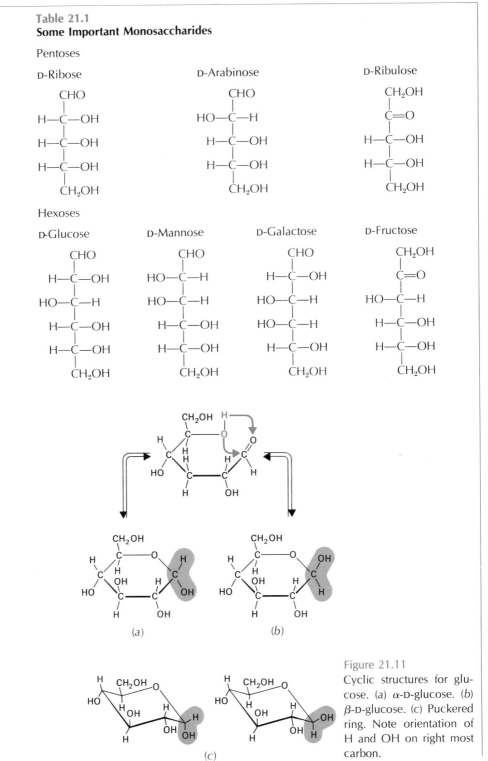

D-Ribose

$$CHO$$
$$H-C-OH$$
$$H-C-OH$$
$$H-C-OH$$
$$CH_2OH$$

D-Arabinose

$$CHO$$
$$HO-C-H$$
$$H-C-OH$$
$$H-C-OH$$
$$CH_2OH$$

D-Ribulose

$$CH_2OH$$
$$C=O$$
$$H-C-OH$$
$$H-C-OH$$
$$CH_2OH$$

Hexoses

D-Glucose

$$CHO$$
$$H-C-OH$$
$$HO-C-H$$
$$H-C-OH$$
$$H-C-OH$$
$$CH_2OH$$

D-Mannose

$$CHO$$
$$HO-C-H$$
$$HO-C-H$$
$$H-C-OH$$
$$H-C-OH$$
$$CH_2OH$$

D-Galactose

$$CHO$$
$$H-C-OH$$
$$HO-C-H$$
$$HO-C-H$$
$$H-C-OH$$
$$CH_2OH$$

D-Fructose

$$CH_2OH$$
$$C=O$$
$$HO-C-H$$
$$H-C-OH$$
$$H-C-OH$$
$$CH_2OH$$

(a)

(b)

(c)

Figure 21.11
Cyclic structures for glucose. (a) α-D-glucose. (b) β-D-glucose. (c) Puckered ring. Note orientation of H and OH on right most carbon.

Figure 21.12
Cyclic structure for fructose.

ribose and **deoxyribose** (Figure 21.13), sugars that are part of the backbone of RNA and DNA, respectively. We shall examine the structures of these in Section 21.5.

In the more complex sugars and the polysaccharides, monosaccharide units are condensed together by way of C—O—C bridges called **glycoside linkages.** Sucrose, for example, is a disaccharide consisting of a glucose and a fructose unit joined by eliminating H_2O from an —OH group on the glucose and an —OH group on the fructose. This is illustrated in Figure 21.14. As indicated, addition of H_2O to the glycoside linkage (i.e., hydrolysis) splits the sucrose molecule into the simple monosaccharides from which it is formed. This hydrolysis reaction is accelerated by the presence of dilute acid and, in general, polysaccharides can be broken down into their simple sugars by this reaction. Special enzymes in saliva start "digesting" carbohydrates in the mouth by this hydrolysis reaction.

The formation of two glycoside linkages by a single monosaccharide unit permits the formation of long polymeric chains that are called polysaccharides, the two most important of which are starch and cellulose. Starch (amylose) is composed of α-D-glucose units strung together whereas cellulose is composed of β-D-glucose units, as shown in Figure 21.15.

The difference between these two structures is rather subtle, but nevertheless has very profound effects. In starch the polysaccharide chains tend to coil in a helical structure with the polar —OH groups pointing outward. When placed into water these —OH groups on the starch molecule interact strongly with the polar solvent and cause the starch to be slightly water-soluble. Cellulose, on the other hand, forms linear chains that interact with

Figure 21.13
(a) Ribose and (b) deoxyribose.

(a)   (b)

Figure 21.14
Sucrose.

each other via hydrogen bonding. This phenomenon gives wood, which is composed of approximately 50% cellulose, its structural strength.

The relatively minor structural differences between starch and cellulose also account for the fact that starch can be digested by humans but cellulose cannot. In the digestive tract the starch molecule is hydrolyzed enzymatically, which requires a certain fit between the carbohydrate molecule and the enzyme. With cellulose this necessary fit is not achieved and, hence, cellulose is unaffected. In termites and cows, however, cellulose is hydrolyzed and digested.

Figure 21.15
Structures of the polysaccharides, starch and cellulose (a) Amylose (starch). (b) Cellulose.

A third class of biomolecules is made up of the **lipids,** water insoluble substances that can be extracted from other cell components by nonpolar organic solvents (e.g., hydrocarbon solvents, carbon tetrachloride, etc.). Lipids serve mainly as storage of energy-rich fuel for use in metabolism (e.g., in fats) and as a major structural element in cell membranes.

As was true with the proteins and carbohydrates, most lipids are composed of simpler substances. The primary building blocks of the lipids are called **fatty acids,** long unbranched hydrocarbon chains, from 12 to 28 carbon atoms long, terminated at one end with the carboxyl group characteristic of organic acids. Nearly all of the naturally occurring fatty acids have an even number of carbon atoms and occur with both saturated and unsaturated chains. Some typical examples are shown in Table 21.2.

Most lipids can be classed as either **neutral lipids** or **polar lipids.** Fats, for example, are neutral lipids and are esters of the fatty acids with the alcohol, glycerol,

The resulting triester is called a **triglyceride.**

As you might expect, many different triglycerides are found to occur, as determined by the nature and location of the fatty acids attached to the glycerol molecule. Lipids containing saturated fatty acids, such as *tristerin* (glycerol esterified with three steric acid molecules)

Table 21.2
Some Naturally Occurring Fatty Acids

Saturated		Melting Point (°C)
Lauric acid (coconut or palm kernel oil)	$CH_3(CH_2)_{10}COOH$	44
Myristic acid (nutmeg fat)	$CH_3(CH_2)_{12}COOH$	54
Palmitic acid (palm oil, animal fats)	$CH_3(CH_2)_{14}COOH$	63
Stearic acid (animal fats)	$CH_3(CH_2)_{16}COOH$	70

Unsaturated		Melting Point (°C)
Palmitoleic (butter fat)	$CH_3(CH_2)_5CH{=}CH(CH_2)_7COOH$	−1
Oleic (olive oil, animal fats)	$CH_3(CH_2)_7CH{=}CH(CH_2)_7COOH$	13.4
Linoleic (linseed oil)	$CH_3(CH_2)_4CH{=}CHCH_2CH{=}CH(CH_2)_7COOH$	−5
Linolenic (linseed oil)	$CH_3(CH_2)CH{=}CHCH_2CH{=}CHCH_2CH{=}CH(CH_2)_7COOH$	−11

are solids, whereas those containing three unsaturated fatty acids are liquids at room temperature. An example of this latter type is *triolein*, in which glycerol is esterified with three oleic acid molecules. This substance is the major constituent of olive oil. The liquid unsaturated triglycerides that are found in vegetable oils, such as olive oil and corn oil, serve as the basis of oleomargarine. Addition of hydrogen to the double bonds of unsaturated vegetable oils produces saturated chains and hence solid fats.

Fats, like other esters, can be **saponified** upon treatment with aqueous base. The products of this reaction are glycerol plus the anions of the fatty acids that were bound to the glycerol in the fat.

$$\begin{array}{c} H_2COOCC_{17}H_{35} \\ | \\ HCOOCC_{17}H_{35} \\ | \\ H_2COOCC_{17}H_{35} \end{array} \xrightarrow[H_2O]{OH^-} \begin{array}{c} H_2COH \\ | \\ HCOH \\ | \\ H_2COH \end{array} + 3C_{17}H_{35}COO^-$$

These anions constitute a soap and have rather peculiar properties that result from having a polar hydrophilic "head" and a nonpolar hydrophobic "tail."

$$\text{CH}_3\text{—CH}_2\text{—} \cdots \text{—CH}_2\text{—CH}_2\text{—CH}_2\text{—C} \overset{\displaystyle O}{\underset{\displaystyle O^-}{}}$$

$$\underbrace{\hspace{5cm}}_{\text{tail}} \quad \underbrace{\hspace{2cm}}_{\text{head}}$$

The polar end of the anion tends to be water-soluble while the other end, the nonpolar hydrocarbon tail, tends to be insoluble in water but soluble in nonpolar solvents. As a result, in water these anions group themselves together into small globules called **micells** in which the nonpolar tails "dissolve" in each other, leaving the polar heads facing outward toward the aqueous surroundings. This is illustrated in Figure 21.16. The same properties that lead to micell formation are also responsible for the ability of soap to dissolve grease. In this case the nonpolar tails dissolve in the grease particle and the polar heads dissolve in water (Figure 21.17). This keeps the grease particle suspended in water so that it can be rinsed away.

Another very important class of lipids are the **phospholipids.** These are polar lipids and, like the fats, are esters of glycerol. In this case, however, only two fatty acid molecules are esterified to glycerol, at the first and second carbon atom. The remaining end position of the glycerol is esterified to a molecule of phosphoric acid, which in turn is also esterified to another alcohol. This gives a general structure,

$$
\begin{array}{c}
O \\
\| \\
O\text{—P—O—}R'' \\
| \\
O \\
| \\
\text{H}_2\text{C—CH—CH}_2 \\
| \quad | \\
O \quad O \\
| \quad | \\
O{=}C \quad C{=}O \\
| \quad | \\
R \quad R'
\end{array}
$$

Figure 21.16
Micell formation with soap.

Figure 21.17

The dissolving of grease globule by soap.

An example is the phospholipid *phosphatidyl ethanolamine,*

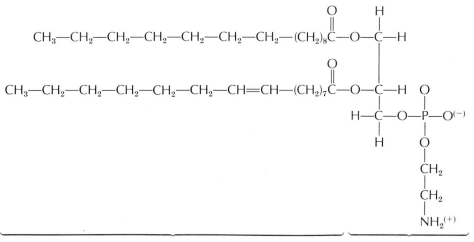

nonpolar tail   polar head

As indicated, this type of lipid contains a polar head and nonpolar tail, much the same as the anions of the fatty acids.

Cell membranes are composed of phospholipids and proteins in about equal proportion. The phospholipids in the membrane appear to be arranged in a double layer, or **bilayer** (Figure 21.18) in which the nonpolar tails face each other, thereby exposing the polar heads to the aqueous environment on either side of the membrane. As shown in Figure 21.19, the proteins found in the membrane are embedded in the mosaic formed by the lipids. Much research today is centered on the mechanism of transport of matter and energy across such membranes.

Finally, there is a third class of lipids that do not contain fatty acids and glycerol, and that do not undergo saponification when treated with a base. Included in this group are the **steroids,** complex substances that possess unusually high biological activity.

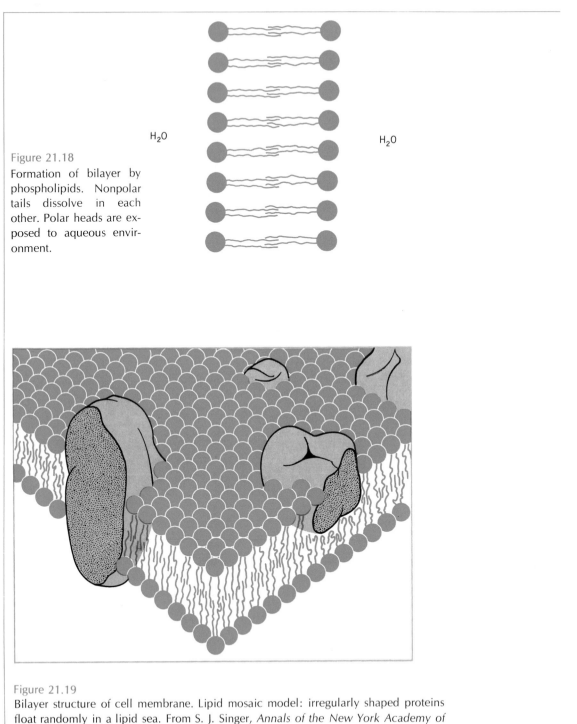

H_2O H_2O

Figure 21.18

Formation of bilayer by phospholipids. Nonpolar tails dissolve in each other. Polar heads are exposed to aqueous environment.

Figure 21.19

Bilayer structure of cell membrane. Lipid mosaic model: irregularly shaped proteins float randomly in a lipid sea. From S. J. Singer, *Annals of the New York Academy of Sciences,* Vol. 195, p. 21, 1962. Used by permission of publisher and author.

The steroids have in common a fused ring structure.

Examples of some important steroids, whose names you may have come across before, are shown below.

cholesterol

cortisone
(affects protein metabolism)

estrone

estradiol

(female sex hormones: note similarity in structures)

norethindrone: an
oral contraceptive
(the pill)

testosterone
(male sex hormone)

One of the most intriguing aspects of biochemistry has been, from its very beginnings, the mechanism whereby an organism transmits its genetic information from one generation to the next during cell division. It is now believed, with a good deal of confidence, that this process is controlled by a substance found in the nucleus of the cell, the **nucleic acid, DNA** (deoxyribunucleic acid). Furthermore, DNA in conjunction with **RNA** (ribonucleic acid, another type of nucleic acid) is responsible for the synthesis of the proteins that are characteristic of a given organism.

Nucleic acids, like the proteins and carbohydrates that we looked at earlier, are polymers. The simpler units that make up the nucleic acid are called **nucleotides,** and are themselves composed of three even simpler molecules. They include the following.

1. **A nitrogenous base.** These are heterocyclic organic compounds having two or more nitrogen atoms in the ring skeleton. They are called bases because the lone pairs of electrons on the nitrogen atoms make them Lewis bases. These substances are shown in Figure 21.20.
2. **A five-carbon sugar (pentose)** In RNA this sugar is ribose whereas in DNA the sugar is deoxyribose. These two are shown in Figure 21.21. Notice that they differ only at carbon atom number 2 in the ring.
3. **Phosphoric acid.** H_3PO_4, as we shall see, forms esters to —OH groups of the sugar to bind nucleotide segments together.

A molecule called a **nucleoside** is formed from these components by condensing a molecule of the base with the appropriate pentose. For example, adenine combines with ribose and deoxyribose at carbon no. 1 to give the compounds shown in Figure 21.22a. Adenosine, structure I, is an important constituent of ATP (adenosine triphosphate) and ADP (adenosine diphos-

Figure 21.20
Nitrogenous bases found in DNA and RNA. (a) Pyrimidine derivatives. (b) Purine derivatives.

Figure 21.21
(a) Ribose and (b) deoxy-ribose.

phate), both of which are involved in energy transfer processes in a cell. Finally, linkage of phosphoric acid to carbon atom no. 5 (Figure 21.22b) produces a **nucleotide,** the basic building block of both DNA and RNA.

The nucleic acids are condensation polymers of the nucleotide monomers and are formed by the creation of an ester linkage from the phosphoric acid residue on one nucleotide to the hydroxy group on carbon no. 3 in the pentose of the second nucleotide, as illustrated in Figure 21.23. The result is a very long polymeric chain, possessing up to a billion or so nucleotide units in DNA!

In Figure 21.20 it was indicated that the base compositions of DNA and RNA are not the same. In DNA the organic bases adenine (A), guanine (G),

Figure 21.22
Nucleosides and nucleotides (a) Adenine combines with ribose to form a nucleoside (b) Linkage of phosphoric acid to carbon 5 gives a nucleotide.

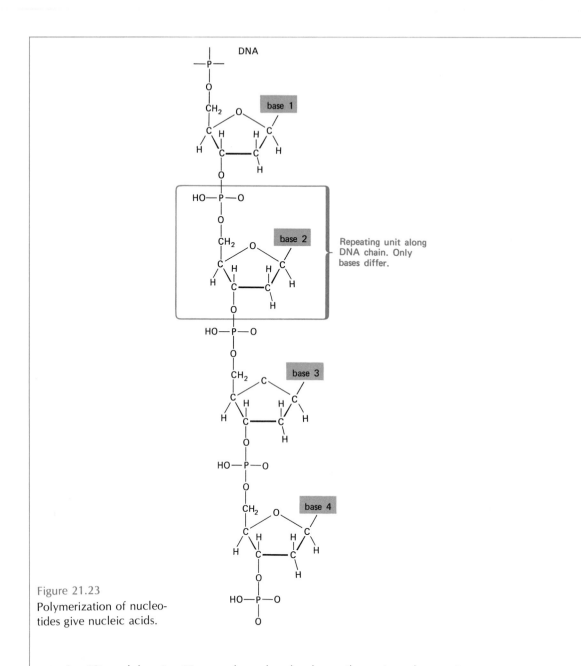

DNA

base 1

base 2

Repeating unit along
DNA chain. Only
bases differ.

base 3

base 4

Figure 21.23
Polymerization of nucleo-
tides give nucleic acids.

cytosine (C), and thymine (T) occur bound to the deoxyribose ring whereas, in
RNA the bases adenine, guanine, cytosine, and uracil (U) are found. As in the
proteins, the sequence of bases along the DNA or RNA chain establishes its
primary structure, which controls the specific properties of the nucleic acid.
For instance, the base sequence in DNA contains coded information that the
cell utilizes in the synthesis of its own characteristic proteins. We shall look at
this more closely in the next section.

The truly remarkable properties of DNA that account for its ability to reproduce itself exactly are dictated by its secondary structure in which two strands of DNA intertwine into the now much celebrated **double-helix** proposed by Watson and Crick, for which they received the Nobel Prize in 1962.[2] The key to the formation of the double-helix, as well as the function of DNA and RNA in protein synthesis, lies in the interaction of the nitrogenous bases by way of hydrogen bonding.

Consider, for example, the bases cytosine and guanine, situated across from one another on separate DNA strands. The structure of these bases, as shown below, is such that they are ideally suited to interact with each other through the formation of hydrogen bonds.

A similar relationship holds for the bases thymine and adenine.

Thus cytosine and guanine fit together like a hand in a glove, as do thymine and adenine. Now, in DNA there is always the same amount of C as G. In addition, the quantities of T and A are the same. Furthermore, the total

[2] M. F. H. Wilkins, whose X-ray data was used by Watson and Crick, also shared in this prize.

amount of C and T together is always equal to the combined total of G and A. If you think of this for a while, you will see that this suggests that C and G are paired together, as are T and A.

These observations, in conjunction with X-ray diffraction data led Watson and Crick to propose the double-helical structure of DNA, shown schematically in Figure 21.24. In this structure we see that for each G on one strand there is a C opposite it, across the axis of the helix, on the other strand. A similar relationship also holds for T and A. The two DNA strands are not identical, but rather complement one another, and it is this property that accounts for the replication of the DNA upon cell division.

It is believed that during cell division the two DNA strands begin to unravel, as shown in Figure 21.25, giving the two complementary chains that serve as templates for the construction of two new daughter chains. The restrictions on the base pairing (i.e., T with A and C with G) causes the newly formed strands to be identical to the departing complementary parent chain and, as a result, a pair of DNA double-helices are produced that are exact copies of the original.

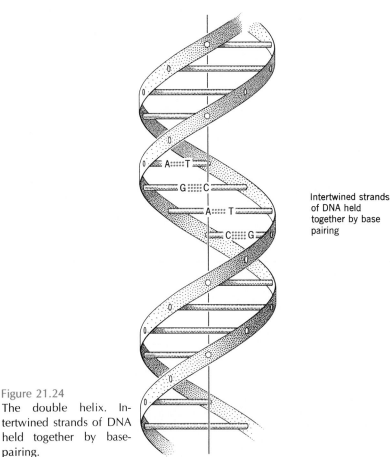

Intertwined strands of DNA held together by base pairing

Figure 21.24
The double helix. Intertwined strands of DNA held together by base-pairing.

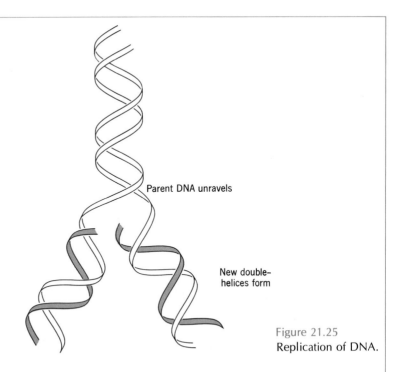

Parent DNA unravels

New double-helices form

Figure 21.25
Replication of DNA.

21.6
Protein Synthesis

The DNA found in the nucleus of a cell serves indirectly to determine the makeup of the proteins that are synthesized at *ribosomes* located outside the nucleus. The genetic information that determines the amino acid sequence in each of the enzymes is stored in the DNA in a genetic code that is made up of the sequence of bases along a DNA strand. The transcription of this code to the site of protein synthesis, as well as the decoding and construction of the polypeptide chain of the protein, is accomplished by the other nucleic acids, the RNA.

Unlike DNA, RNA occurs only in single strands. In addition, there are several types of RNA. One of these is called **messenger RNA, mRNA,** and serves to carry the genetic code from the DNA template within the nucleus to the ribosomes outside. Another type of RNA molecule, called **transfer RNA, tRNA,** is much smaller than mRNA. It acts as an amino acid carrier and, through a decoding mechanism that we shall examine momentarily, adds the amino acid to the growing polypeptide chain at just the right place at just the right time.

The mechanism by which this process is believed to occur is not really too different from that involved in the duplication of DNA itself. It is known, for instance, that the production of mRNA takes place within the nucleus on an untwisted segment of a DNA chain. This segment corresponds to the gene that is characteristic of the particular protein to be synthesized. The mRNA strand produced contains a sequence of bases that is determined, through base-pairing, by the sequence of bases in the DNA; however, the pairing scheme in this case is

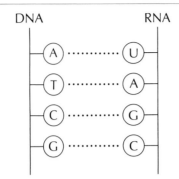

DNA   RNA

Thus uracil occurs in RNA instead of thymine. Once formed the mRNA is transported from the nucleus to the active site of protein synthesis.

It is now known that the genetic code that directs the insertion of amino acids in the proper sequence in the growing polypeptide chain consists of sets of three bases (called **codons**). The amino acids are brought to the proper place by tRNA, which is able to decipher the code.

A molecule of tRNA contains from 75 to 90 nucleotide units containing the bases U, A, G, C and, in addition, many minor bases. What is interesting is that although the tRNA's corresponding to different amino acids have different minor bases and base sequences, they can all be brought into a characteristic cloverleaf shape if maximum base pairing is assumed. For example, the yeast tRNA for the amino acid alanine has the structure shown in Figure 21.26. The molecule is folded in such a way as to give maximum base pairing (dotted lines). On the stem of this RNA molecule, alanine becomes attached by the action of the proper enzyme. The two arms to either side of the site of attachment of the amino acid appear to be important in the interaction of the tRNA with this enzyme. The bottom loop contains the three unit anticodon that attaches itself to the complementary codon on the mRNA chain.

The sequence of operations that leads to the synthesis of a polypeptide is summarized in Figure 21.27. In the bacterium, E. coli, for example, it is known that the amino acid sequence is initiated by a derivative of methionine. The tRNA containing this derivative becomes attached at the head of the mRNA through its codon-anticodon pairing. Once this is in place the next amino acid in the sequence is brought into place by its tRNA, which becomes bound to the mRNA at the second codon site. This is followed by an enzyme-induced formation of the peptide linkage and the subsequent departure of the tRNA from the first codon. Next, the third amino acid is delivered by its tRNA, which attached itself to the third codon. Again a peptide linkage is formed and the second tRNA leaves. This process is repeated over and over along the mRNA as the polypeptide chain grows in size. The chain is finally terminated when a "nonsense" codon (one which cannot be recognized by any tRNA molecule) is encountered.

THE GENETIC CODE. Through a series of very clever experiments, which earned Nirenberg, Holley, and Khorana a Nobel Prize in 1968, the base sequences in the genetic code triplets were determined. These are given in Table 21.3. Notice that most of the amino acids are specified by more than one code "word." Apparently for these the attachment of the tRNA to the mRNA is primarily determined by the first two bases in the sequence (which

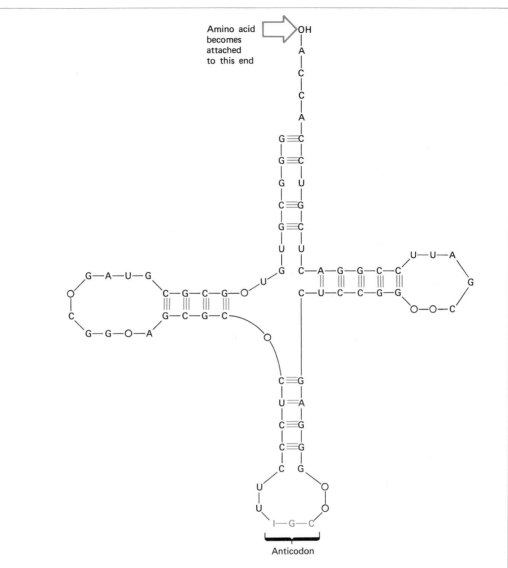

Anticodon

Figure 21.26
The base sequence in a yeast alanine tRNA.

are usually the same for a given amino acid). The interaction between the third base in the mRNA cocon and the tRNA anticodon thus does not appear to be as critical as the first two in determining specificity.

With this code we can now see how the amino acid sequence in a polypeptide is fixed by the primary structure of the DNA. Consider, for example, a segment of DNA having the base sequence,

(C) (G) (T) (C) (T) (A) (A) (A) (A) (A) (G) (G)

Figure 21.27
Synthesis of polypeptide.

The mRNA formed from it will have the base sequence,

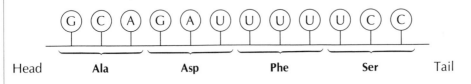

| Head | Ala | Asp | Phe | Ser | Tail |

which, as you can see, is composed of the code words for the polypeptide

Ala—Asp—Phe—Ser

The triplet code presented in Table 21.3 has been shown to be applicable to such diverse species as the bacterium, E. coli, the tobacco plant, the guinea pig, and even man. It is widely believed that this code is universal for all species.

MUTATIONS. Any factor that will have the effect of altering the base sequence in a DNA molecule will, in effect, alter the mRNA transcribed from it and thereby cause a change in the amino acid sequence that is produced

Table 21.3

The Genetic Code

		Second Base in Codon			
		U	C	A	G
First Base in Codon	U	UUU } phe UUC } UUA } leu UUG }	UCU } UCC } ser UCA } UCG }	UAU } tyr UAC } UAA[b] UAG[b]	UGU } cys UGC } UGA[b] UGG trp
	C	CUU } CUC } leu CUA } CUG }	CCU } CCC } pro CCA } CCG }	CAU } his CAC } CAA } gln CAG }	CGU } CGC } CGA } arg CGG }
	A	AUU } AUC } ile AUA } AUG met[a]	ACU } ACC } ACA } thr ACG }	AAU } asn AAC } AAA } lys AAG }	AGU } ser AGC } AGA } arg AGG }
	G	GUU } GUC } GUA } val GUG }	GCU } GCC } ala GCA } GCG }	GAU } asp GAC } GAA } glu GAG }	GGU } GGC } GGA } gly GGG }

[a]AUG appears to initiate an amino acid sequence with methionine.
[b]nonsense codons: these do not code for any amino acid. They serve to terminate peptide chains.

using that mRNA as a template. Providing that this does not prove fatal to the cell a mutation will have occurred that will be transmitted from one generation to another as the DNA reproduces itself upon cell division.

Often these mutations prove harmful, although not immediately lethal to the organism in which they occur and give rise to symptoms that cause them to be referred to as diseases. Since the origin of the diseases is in the genetic material of the cell, they are said to be genetic diseases. Many such diseases are recognized. Some of the more well known are cystic fibrosis, hemophelia, and sickle-cell anemia. Recently evidence has suggested that even schizophrenia is of genetic origin.

In sickle-cell anemia, for example, the red blood cells assume a crescent shape instead of the flat disklike shape of normal cells. This is caused by a mutated DNA in the gene that is responsible for the synthesis of hemoglobin. It has been found that in sickle-cell hemoglobin a *glutamic acid* residue in one of the hemoglobin chains is replaced by a *valine* residue. This alters the secondary and tertiary structure of the protein and reduces its ability to carry oxygen.

Sickle-cell anemia is thus a disease with its origin in the DNA of the cell nucleus and is therefore passed from one generation to the next because of its genetic nature.

21.1 What is an α-amino acid? Indicate how amino acids are linked together to give a dipeptide; a polypeptide.

21.2 Describe what is meant by the *primary structure of a protein*. What is meant by secondary structure, tertiary structure, and quaternary structure?

21.3 Indicate three functions served by proteins in a living organism.

21.4 What is a zwitterion? Why do amino acids form zwitterions?

21.5 Describe the α-helix structure found in many polypeptides. What holds the polypeptide chain in this helical conformation?

21.6 How do the properties of the solvent affect the tertiary structures of proteins? What interactions in addition to hydrogen bonding determine tertiary structure?

21.7 What is a porphyrin? Name two biologically important porphyrin structures.

21.8 In what sense do enzymes guide the chemistry that takes place in living organisms? What problems would a living cell encounter if it were not for the existence of enzymes?

21.9 Describe in qualitative terms how an enzyme operates. What is meant by *enzyme substrate*? What is enzyme inhibition? How do the sulfa drugs function?

21.10 Only one of the basic set of 20 amino acids does not exhibit optical isomerism. Which one is it? Why is it not optically active?

21.11 What is a monosaccharide? Give an example of (a) a pentose and (b) a hexose.

21.12 How many optical isomers of fructose are there?

21.13 Compare the structures of starch and cellulose. What is the major difference between them?

21.14 What is a lipid? What functions do they serve in living systems?

21.15 What is a fatty acid? Give an example of a triglyceride. What difference in physical properties exists between the saturated and unsaturated triglycerides?

21.16 Give the structure of triolein. What product is formed upon addition of H_2 to the double bonds in triolein?

21.17 What is a soap? How does it form micells? What is responsible for the ability of soap to dissolve grease?

21.18 How does a phospholipid differ from a triglyceride?

21.19 How are phospholipids believed to contribute to the structure of cellular membranes?

21.20 What characteristic structural feature is found among the steroids?

21.21 Describe the structure of a *nucleotide*. What is the difference between the nucleotide units in DNA and RNA?

21.22 Imagine that a single DNA strand contained a segment with the composition,

A C T C G A T C A T G

What would be the base sequence in the complementary strand?

21.23 What holds the two DNA strands together in the double helix? Illustrate the base pairing between C and G; between T and A.

21.24 What occurs with DNA during cell division? How is the DNA replicated?

21.25 What is meant by the *genetic code*? Describe the functions of mRNA and tRNA.

21.26 If a DNA strand contains the base sequence,

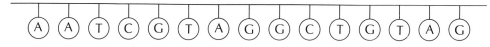

what will be the base sequence transcribed onto the mRNA?

21.27 What amino acid sequence is specified by the base sequence (from left to right) in the following mRNA strand?

21.28 What amino acid sequence is specified by the base sequence (starting at the left) in the DNA strand in Question 21.26?

21.29 Suppose that through some mutational change the first base was removed from the sequence in the DNA strand depicted in Question 21.26. What amino acid sequence would result in the polypeptide synthesized using this modified (mutant) DNA template?

22
Nuclear Chemistry

In our discussion of chemistry up to now we have paid very little attention to the atomic nuclei, other than that their charge determines the number of electrons in the neutral atom. There are, however, nuclear phenomena that have applications to chemistry and we shall explore them in this chapter. At the same time, we shall also explore the nuclear changes that occur in radioactive elements as well as those that can cause an otherwise stable element to become radioactive.

Henry Becquerel, a French chemist, was the first to discover that some elements, especially the heavier ones, spontaneously emit radiation and are said to be radioactive. He found, somewhat to his surprise, that when a photographic film was exposed to the salt potassium uranyl sulfate, $K_2UO_2(SO_4)_2$, the film was blackened. He reasoned that in order to accomplish this the uranium salt must emit a powerful radiation that upon striking the film blackened it. In later experiments he also observed that the rate at which radiation was emitted from this salt was directly dependent on the amount of sample present.

Two colleagues of Becquerel, Marie and Pierre Curie, were successful in isolating two other radioactive isotopes from uranium ores. These isotopes, which are more intensely radioactive than the uranium, were named polonium, Po, and radium, Ra. For their work in this field Becquerel and the two Curies were awarded the Nobel Prize for Physics in 1903.

We now know that the radiation emitted from radioactive materials is of three main types (as was first pointed out in Section 3.4): **alpha particles** (α), **beta particles** (β) and **gamma rays** (γ). Experimentally it is found that the α-particle carries a positive charge, the β-particle is negatively charged, and the γ-rays carry no charge. The actual charges and masses of these and other particles that we shall be concerned with are listed in Table 22.1.

When a substance spontaneously emits one or more of these types of radiation, it is said to be radioactive. The uranium in the potassium uranyl sulfate of Becquerel is radioactive and spontaneously emits alpha particles. The most abundant isotope of uranium is $^{238}_{92}U$, which upon emitting alphas becomes $^{234}_{90}Th$. This can be written as

$$^{238}_{92}U \longrightarrow ^{4}_{2}He + ^{234}_{90}Th$$

Thus we say that ^{238}U decays to ^{234}Th by the emission of alpha particles. Isotopic reactants and products in these decay reactions are also called **parent** and **daughter** isotopes, or **nuclides,** respectively. The term nuclide is applied

Table 22.1
Basic Types of Particles Emitted by Radioisotopes

Particle	Approximate Mass (amu)	Charge	Symbol	Type
Alpha	4	+2	^4_2He	Particle
Beta	0	−1	$^0_{-1}e$	Particle
Gamma			γ	Electromagnetic radiation
Neutron	1	0	^1_0n	Particle
Proton	1	+1	^1_1p (^1_1H)	Particle
Positron	0	+1	^0_1e	Particle

in a general way to a nucleus with a specified number of protons and neutrons. Thus the daughter isotope of nuclide $^{238}_{92}\text{U}$ is $^{234}_{90}\text{Th}$.

The thorium isotope produced in the above decay reaction is itself radioactive and spontaneously emits beta particles. As a result, when the thorium is formed, it then decays to an isotope of protactinium, the reaction being

$$^{234}_{90}\text{Th} \longrightarrow {}^0_{-1}e + {}^{234}_{91}\text{Pa}$$

$$\textbf{(parent)} \qquad \textbf{(}\beta\textbf{)} \qquad \textbf{(daughter)}$$

Highly energetic gamma rays are emitted by nearly all radioactive materials, a process that occurs without a change in the mass or the charge of the isotope. As a result they are often omitted when writing decay reactions like those shown above.

In balancing radioactive decay reactions, it is important to remember that the total mass and charge of all of the species on the left of the arrow must exactly balance with the total mass and charge of those species on the right (check the two reactions above to see if we wrote them correctly). Thus, whenever an isotope decays by alpha emissions, the daughter isotope always has an atomic mass number that is four units lower than the parent and an atomic number two units lower. In the case of a beta emitter, on the other hand, the daughter has an atomic mass number that is the same as the parent and an atomic number one unit *higher* than its parent.

Let's now try to answer the question proposed in Example 22.1.

Example 22.1
(a) ^{234}U decays by alpha emissions. What is its daughter isotope?
(b) ^{214}Pb decays to ^{214}Bi. By what type of radiation is this accomplished?

Solution

(a) $$^{234}_{92}\text{U} \longrightarrow {}^4_2\text{He} + {}^{230}_{90}\text{X}$$

The equation can first be written in this form; and then we simply identify the element with $Z = 90$. The complete equation is then written as

$$^{234}_{92}\text{U} \longrightarrow {}^4_2\text{He} + {}^{230}_{90}\text{Th}$$

(b) The statement of the problem means

$$^{214}_{82}Pb \longrightarrow ^{214}_{83}Bi + ^{0}_{-1}X$$

Thus the decay must be by beta emissions.

We have seen above that ^{238}U decays to ^{234}Th, which is also radioactive and decays to ^{234}Pa. This isotope is also unstable and decays to ^{234}U (how?) which is also radioactive and so on. This decay process will continue until a stable (nonradioactive) isotope of an element is formed. This entire scheme, where one isotope decays to another, etc. is called a **radioactive series** or **decay series.** ^{238}U, for example, decays by some 14 steps to stable ^{206}Pb.

KINETICS OF RADIOACTIVE DECAY. In Chapter 11 we saw that the rate of a chemical reaction is given by the change in the concentration of a reactant over a certain small time interval. For a radioactive substance, its concentration is directly proportional to the number of particles or rays emitted by the substance per minute. These emissions can be detected and counted by a suitable device such as the Geiger-Müller counter shown schematically in Figure 22.1.

If, for a particular radioactive isotope, we construct a graph in which the number of counts per minute, **cpm,** of particles emitted is plotted against time, we obtain a curve similar to that shown in Figure 22.2a. If the log of the cpm versus time is graphed, a straight line is obtained as shown in Figure 22.2b. Both of these plots are indicative of a first order reaction and, as it turns out, all radioactive decay reactions obey first order kinetics. The generalized rate law for any radioactive decay process would then take the form,

$$Rate = k(\text{concentration of isotope})^1$$

Figure 22.1
The Geiger-Müller counter. An alpha or beta particle enters the Geiger tube through the thin window shown at the left of the apparatus. As the particle passes through the gas inside the tube, it ionizes argon atoms along its path. These ions cause an electrical breakdown (discharge) between the wire and the wall of the tube, thereby producing a current pulse. This current pulse is readily amplified and counted electronically.

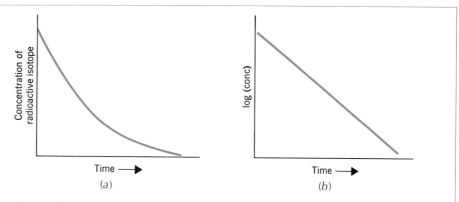

Figure 22.2
Kinetics of radioactive decay. (a) A graph of concentration versus time. (b) A logarithmic plot of concentration versus time. Straight line indicates first order process.

By applying what we have learned in Chapter 11 to the decay process, it is possible to derive an equation that is helpful in determining rate constants for decay reactions. According to Section 11.1, for any first order reaction, such as

$$A \longrightarrow B$$

we can express its rate law as

$$-\frac{\Delta[A]}{\Delta t} = k[A]$$

Rearranging this equation to take the form

$$\frac{\Delta[A]}{[A]} = k(\Delta t)$$

and using calculus, we can derive the equation,

$$\log \frac{[A]_0}{[A]} = \frac{kt}{2.30} \tag{22.1}$$

In this equation $[A]_0$ is the initial concentration of the isotope (concentration at time zero) and $[A]$ is its concentration at any time t. Equation 22.1 can be employed to calculate rate constants as shown in the following example.

Example 22.2
A chemist determined that after one week his initial 10.0 μg (1 μg = 10^{-6}g) of ^{222}Rn had decayed, and he now had 2.82μg of the radon. What is the rate constant for the alpha decay of ^{222}Rn?

Solution
The rate constant can easily be obtained by using Equation 22.1, where $[A]_0 = 10.0$ g, $[A] = 2.82$ g, and $t = 7$ days. Solving Equation 22.1 for k, we have

$$k = \frac{2.30}{t} \log \frac{[A]_0}{[A]}$$

and substituting the values given above into this equation, we have

$$k = \left(\frac{2.30}{7 \text{ days}}\right) \log \left(\frac{10.0}{2.82}\right)$$

or

$$k = 0.18 \text{ days}^{-1}$$

Another very important consequence of the kinetic studies described above is that the time required for half of any radioactive isotope to decay, called its **half-life,** $t_{1/2}$, is constant and is independent of the initial concentration of the isotope. Thus, if we were to begin with 10.0 grams of a particular isotope, after one half-life only 5.0 grams would remain. In a time equal to another half-life this would decay to 2.5 grams, and so on. This is represented graphically in Figure 22.3. In this figure we see, in general, how an isotope decays through many half-lives. We can show that the half-life of any substance is inversely proportional to the rate constant for its decay. The actual relationship as derived takes the form

$$t_{1/2} = \frac{0.693}{k} \tag{22.2}$$

From this equation we see that if the rate constant for decay is large, its half-life will be small and vice versa. Also we see that if either $t_{1/2}$, or k is known the other can be calculated.

Example 22.3
(a) What is the half-life of ^{222}Rn and (b) What is the fraction of a sample of Rn that decays in 1 week?

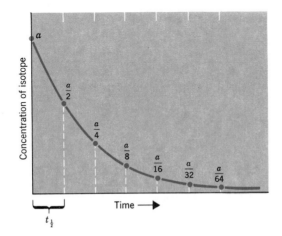

Figure 22.3
A graph of a first order radioactive decay illustrating the concept of half-life.

Solution

(a) We can calculate the half-life of Rn directly from Equation 22.2 by substituting in the value of k calculated in Example 22.2

$$t_{1/2} = \frac{0.693}{k}$$

and $k = 0.18$ days^{-1} so that

$$t_{1/2} = \frac{0.693}{0.18 \text{ days}^{-1}}$$

or

$$t_{1/2} = 3.85 \text{ days}$$

(b) We first have to calculate the fraction left after decay. By this we mean the ratio $[A]/[A]_0$. Therefore we employ Equation 22.1 for this purpose and solve it for $[A]_0/[A]$.

$$\log \frac{[A]_0}{[A]} = \frac{kt}{2.3}$$

Substituting in $k = 0.18$ days^{-1} and $t = 7$ days, we have

$$\log \frac{[A]_0}{[A]} = \frac{(0.18 \text{ days}^{-1})(7 \text{ days})}{2.3}$$

or

$$\log \frac{[A]_0}{[A]} = 0.548$$

and

$$\frac{[A]_0}{[A]} = 3.53$$

so that

$$\frac{[A]}{[A]_0} = 0.28$$

This is the fraction left after 1 week and the fraction decayed is simply $1 - 0.28$ or 0.72.

Half-lives are important for two reasons. First, they indicate the stability of a particular isotope; the longer the half-life the more stable is the material. Second, the half-life can be effectively used to measure the age of such things as rocks, bones, ancient pieces of art, and the like.

USES OF RADIOACTIVE DECAY. The age of rocks containing uranium can be dated by determining the ^{238}U to ^{206}Pb ratio. Remember that ^{206}Pb is the stable isotope to which ^{238}U eventually decays. Using a ratio of $^{238}U/^{206}Pb$ of 1 to 0 as zero time and a ratio of $^{238}U/^{206}Pb$ of 1 to 1 as the time for one half-life, that is 4.5×10^9 years, the age of these rocks can be closely approximated. The oldest rocks that have been found on earth have an age of 4.55×10^9 years using this method.

Rocks that do not contain uranium are presently being dated using a potassium-argon method. This method makes use of the reaction,

$$^{40}_{19}K + ^{0}_{-1}e \longrightarrow ^{40}_{18}Ar \qquad t_{1/2} = 1.3 \times 10^9 \text{ years}$$

The electron in this case comes from the $1s$ orbital of the potassium and is captured by the unstable $^{40}_{19}K$ nucleus. In the dating procedure the ratio of ^{40}K to argon is measured, and the age of the rock is determined in the same manner as the uranium dating described above.

The age of materials that were once living, such as bones and wood, can be quite accurately estimated by measuring their ratio of ^{14}C to ^{12}C. Carbon-14 is radioactive and is constantly being produced in the upper atmosphere by the bombardment of cosmic neutrons upon $^{14}_{7}N$, which is present there in large amounts. The equation for this reaction is

$$^{14}_{7}N + ^{1}_{0}n \longrightarrow ^{14}_{6}C + ^{1}_{1}p$$

The carbon-14 thus produced immediately begins to decay

$$^{14}_{6}C \longrightarrow ^{14}_{7}N + ^{0}_{-1}e \qquad t_{1/2} = 5770 \text{ years}$$

so that a steady state concentration (15 cpm) of this radioactive nuclide is maintained in the atmosphere. The ^{14}C becomes incorporated into the carbon dioxide ($^{14}CO_2$) in the air where it can be taken in by plants through the process of photosynthesis. The intake of ^{14}C into animals is by the consumption of such plants or by the consumption of plant-eating animals. While they are alive, plants and animals consume and excrete carbon so that they also maintain a steady state concentration of ^{14}C and are thus in equilibrium with their surroundings. Once they die, however, the ^{14}C that they possess is not replaced as it decays and hence the ^{14}C concentration begins to decrease. The half-life of the ^{14}C is 5770 years; therefore, if we find that the carbon-14 concentration in an object that had once been living has dropped to half its initial value, we could conclude that the object is 5770 years old. However, atmospheric nuclear testing has made it impossible for our current history to be dated this way by future archaeologists.

Example 22.4
A piece of charcoal from the ruins of a settlement in Japan was found to have a $^{14}C/^{12}C$ ratio that was 0.617 times that found in living organisms. How old is this piece of charcoal?

Solution
The answer to this problem can be obtained once again by employing Equations 22.2 and 22.1. From Equation 22.2 we can obtain the k for the decay of ^{14}C.

$$k = \frac{0.693}{5770 \text{ years}} = 1.20 \times 10^{-4} \text{ years}^{-1}$$

Substituting this value and that for the ratio of $^{14}C/^{12}C$ into Equation 22.1, we have

$$\log \frac{[A]_0}{[A]} = \frac{kt}{2.30}$$

$$\log \left(\frac{1.000}{0.617} \right) = \log 1.62 = 0.210 = \frac{(1.20 \times 10^{-4} \text{ years}^{-1})t}{2.30}$$

or

$$t = 4030 \text{ years}$$

22.2
Nuclear
Transformations

Lord Rutherford, in 1919, performed the first known **nuclear transformation** of one element into another. Rutherford found that by bombarding $^{14}_{7}N$ with alpha particles he produced the isotope of oxygen, $^{17}_{8}O$. The equation for this process is

$$^{14}_{7}N + ^{4}_{2}He \longrightarrow \left[^{18}_{9}F \right] \longrightarrow ^{17}_{8}O + ^{1}_{1}H$$

The $^{18}_{9}F$ isotope of fluorine is a very unstable intermediate that rapidly decays to the products above.

In 1933 Irene Curie[1] and her husband Frederick Joliot showed that other light elements could similarly be transformed by bombardment with alphas. For example, $^{27}_{13}Al$ can be transformed into $^{30}_{15}P$ by this process.

$$^{27}_{13}Al + ^{4}_{2}He \longrightarrow ^{30}_{15}P + ^{1}_{0}n$$

Reactions of this type, where an alpha is used for bombardment and a neutron is one of the products, is known as an alpha, neutron reaction symbolized by (α, n). A shorthand notation for the above reaction is $^{27}_{13}Al(\alpha, n)^{30}_{15}P$.

Since 1933 many isotopes have been produced by bombardment reactions, some where particles other than alphas have been used. One of the main problems with such experiments is that a positively charged nucleus is being bombarded with positively charged particles. Heavy elements, with their very highly positive nuclei, will repel particles with positive charges like the alpha. One way to circumvent this problem is to use neutrons as the bombarding particles.

Neutrons, which have no charge, are not repelled by the nucleus and, therefore, are excellent materials for bombardment reactions. The supply of neutrons for these reactions can be obtained from either of two sources; from a transformation reaction in which neutrons are produced or from a nuclear reactor in which fission reactions, which will be examined in Section 22.6, occur at a controlled rate. Examples of neutron producing reactions are the $^{27}_{13}Al(\alpha, n)^{30}_{15}P$ reaction mentioned above and

$$^{9}_{4}Be + ^{4}_{2}He \longrightarrow ^{12}_{6}C + ^{1}_{0}n$$

where the beryllium isotope undergoes an α, n reaction.

A second way that nuclear transformations can be brought about is through the use of particle accelerators. Particle accelerators, such as the cyclotron described in Figure 22.4, speed up particles to extremely high velocities and then direct them at target nuclei. At these speeds the positive

---

[1] Irene Joliot-Curie is the daughter of Marie and Pierre, the discoverers of polonium and radium.

Figure 22.4
Diagram of a cyclotron.

particles are able to overcome the coulombic repulsion of the nucleus and collide with it. Accelerators have been used extensively by Dr. Glenn Seaborg and his colleagues at the University of California in producing many of the transuranium elements, that is, elements 93 to 105. In these accelerators positive ions of such isotopes as ^2_1H (deuterium), $^{12}_6C$, $^{13}_6C$, $^{16}_8O$, $^{14}_7N$, $^{10}_5B$, as well as ^4_2He have been used in producing new, man-made, elements. A listing of these bombardment reactions and the elements they produce is shown in Table 22.2. Many of these elements are extremely short-lived and, as a result, only a few atoms, especially of the high atomic numbered isotopes, have ever been formed.

Experimentally it is observed that all of the elements with atomic numbers greater than 83 (bismuth) are radioactive and possess no known stable isotopes. On the other hand, all of the lighter elements, with the exception of

**22.3
Nuclear Stability**

Table 22.2
Elements Produced by Particle Accelerators

$$^{238}_{92}U + ^4_2He \longrightarrow ^{239}_{94}Pu + 3\ ^1_0n$$

$$^{239}_{94}Pu + ^4_2He \longrightarrow ^{240}_{95}Am + ^1_1H + 2\ ^1_0n$$

$$^{239}_{94}Pu + ^4_2He \longrightarrow ^{242}_{96}Cm + ^1_0n$$

$$^{244}_{96}Cm + ^4_2He \longrightarrow ^{245}_{97}Bk + ^1_1H + 2\ ^1_0n$$

$$^{238}_{92}U + ^{12}_6C \longrightarrow ^{246}_{98}Cf + 4\ ^1_0n$$

$$^{238}_{92}U + ^{14}_7N \longrightarrow ^{247}_{99}Es + 5\ ^1_0n$$

$$^{238}_{92}U + ^{16}_8O \longrightarrow ^{249}_{100}Fm + 5\ ^1_0n$$

$$^{253}_{99}Es + ^4_2He \longrightarrow ^{256}_{101}Md + ^1_0n$$

$$^{246}_{96}Cm + ^{13}_6C \longrightarrow ^{254}_{102}No + 5\ ^1_0n$$

$$^{252}_{98}Cf + ^{10}_5B \longrightarrow ^{257}_{103}Lw + 5\ ^1_0n$$

$$^{249}_{98}Cf + ^{12}_6C \longrightarrow ^{257}_{104}Ku + 4\ ^1_0n$$

technetium ($Z = 43$) and promethium ($Z = 61$), have one or more stable, nonradioactive isotopes. In addition, radioactive isotopes undergo nuclear transformations that lead ultimately to stable nuclei. Sometimes this is accomplished by a simple one-step process, while in other cases a series of nuclear reactions occur before a stable isotope is reached. A question that naturally arises from these observations is, what factors give rise to stable or unstable nuclei?

Little is known about the nature of the forces that hold a nucleus together. Some interesting facts concerning nuclear stability emerge, however, if we examine the numbers of protons and neutrons found in stable nuclei. For example, if we make a graph of the number of neutrons versus the number of protons in different nuclei, we find that all of the stable isotopes fall in a narrow band, which we might call a **band of stability,** as shown in Figure 22.5.

In this illustration we see that at low atomic numbers stable nuclei possess approximately equal numbers of protons and neutrons. Above about $Z = 20$, however, the number of neutrons always exceeds the number of protons and the neutron-to-proton ratio gradually increases to about 1.5 at the upper end of the band of stability. Apparently, as the number of protons in the nucleus increases there must be more and more neutrons present to help overcome the strong repulsion forces between the protons. It also seems that there is an upper limit to the number of protons that can exist in a stable nucleus, that number being reached at bismuth.

Nuclei that lie outside the band of stability are unstable and decay in a manner that tends to give them a stable neutron-to-proton (n/p) ratio. On this basis, then, we can understand why certain nuclei undergo the type of radioactive decay that they do. For instance, a nucleus that lies above the band of stability must either lose neutrons or gain protons in order to achieve stability. Thus we can understand why elements such as ^{14}C (which lies above the band) decay by beta emission since this process converts a neutron into a proton ($^{1}_{1}p$).

$$^{1}_{0}n \longrightarrow {}^{1}_{1}p + {}^{0}_{-1}e$$

For ^{14}C we have

$$^{14}_{6}C \longrightarrow {}^{14}_{7}N + {}^{0}_{-1}e$$

z = 20

z = 83

Number of protons

Figure 22.5
Band of stability.

22.3 Nuclear Stability **689**

Another way that an element located above the band can achieve a stable n/p ratio is by emitting a neutron, although this particular mode of decay is rare. An example is the decay of ^{137}I.

$$^{137}_{53}I \longrightarrow {}^{136}_{53}I + {}^{1}_{0}n$$

Elements located *below* the band of stability must increase their n/p ratio in order to achieve stability. This is accomplished generally in one of two ways. One of these involves the emission of a positron, a particle having the same mass as the electron but with a unit positive charge. The positron is symbolized as $^{0}_{1}e$. The ejection of a positron by an unstable nucleus converts a proton into a neutron,

$$^{1}_{1}p \longrightarrow {}^{1}_{0}n + {}^{0}_{1}e$$

An example is the decay of ^{11}C.

$$^{11}_{6}C \longrightarrow {}^{11}_{5}B + {}^{0}_{1}e$$

The second mode of decay that results in an increased n/p ratio is called **electron capture.** In this case the unstable nucleus captures an electron, usually from its own $1s$ orbital. Since the captured electron most often originates in the K shell the process is also called **K capture.** The addition of this electron to the nucleus transforms a proton into a neutron.

$$^{1}_{1}p + {}^{0}_{-1}e \longrightarrow {}^{1}_{0}n$$

Two examples of decay by K capture are

$$^{7}_{4}Be + {}^{0}_{-1}e \xrightarrow{\text{K capture}} {}^{7}_{3}Li$$

and

$$^{40}_{19}K + {}^{0}_{-1}e \xrightarrow{\text{K capture}} {}^{40}_{18}Ar$$

The vacancy created in the $1s$ subshell as a result of K capture is only temporary, and electrons from higher energy levels quickly drop to fill the $1s$ orbital. Since electrons are falling from higher energy levels to lower ones, energy is emitted in the form of electromagnetic radiation (light), in this instance in the X-ray region of the spectrum.

Elements having atomic numbers higher than 83, that is those beyond the end of the band of stability, cannot find their way to a stable n/p ratio by any of the decay modes that we just discussed. In these cases the unstable nuclei must lose both protons *and* neutrons. As a result their decay usually involves emission of alpha particles, since each alpha emission removes two protons and two neutrons simultaneously. Earlier, for example, we saw this type of decay process for uranium, that is,

$$^{238}_{92}U \longrightarrow {}^{4}_{2}He + {}^{234}_{90}Th$$

Another type of nuclear transformation that is available to the heavy elements is **fission** in which a heavy nucleus splits into several much lighter fragments, many of which may also lie outside the band of stability and hence may be radioactive. The smaller nuclei that are produced by fission, if they are unstable, are able to undergo the simpler types of decay in order to produce a stable nucleus. We shall take a closer look at nuclear fission in Section 22.6.

We may also observe that nuclei with even numbers of protons and neutrons are apparently more stable than those containing an odd number of these particles. For example, there are 157 stable isotopes in which there are an even number of both protons and neutrons, 52 isotopes having an even number of protons and an odd number of neutrons, and 50 with an even number of neutrons but an odd number of protons. By contrast, there are only five stable nuclides in which there are an odd number of both protons and neutrons.

Protons Neutrons	Even Even	Even Odd	Odd Even	Odd Odd
Stable nuclei	157	52	50	5

This phenomenon suggests that in stable nuclei protons and neutrons each tend to be paired, in much the same way that electrons become paired in the outer region of the atom, and that extra stability, as evidenced by the number of stable nuclides, results when pairing takes place with both protons and neutrons. On the other hand, when pairing cannot occur, as must be true when the numbers of protons and neutrons are both odd, very few stable isotopes occur (i.e., most isotopes having odd numbers of both protons and neutrons are radioactive).

A final observation on nuclear stability is that nuclei that contain certain specific numbers of protons and neutrons possess a degree of extra stability. These so-called *"magic numbers"* for protons and neutrons are 2, 8, 20, 28, 50 and 82, with an additional magic number of 126 for neutrons. When nuclei contain a magic number of both protons and neutrons, they are said to be *"doubly magic"* and are extremely stable. Examples are $^4_2$He, $^{16}_8$O, $^{40}_{20}$Ca, and $^{208}_{82}$Pb.

The occurrence of these magic numbers suggests a shell structure for the nucleus somewhat akin to the shell structure exhibited by electrons. For example, we have seen that very stable (unreactive) electron configurations occur when an atom contains magic numbers of 2, 8, 18, 36, or 54 electrons, corresponding to the noble gases He through Kr. In the nucleus, then, it seems that nuclear shells of either protons or neutrons become completed when these magic numbers are reached and that a particularly stable nucleus occurs whenever there is a completed shell of either neutrons or protons. Exceptionally stable nuclei result when the nucleus contains filled shells of protons and neutrons simultaneously.

22.4
Extension of the Periodic Table

The heaviest naturally occurring element is uranium and, as we saw earlier in this chapter, elements beyond $Z = 92$ are all artificially prepared by bombarding lighter nuclei with protons, alpha particles, and the positive ions of some of the second period elements. The discovery of these new elements quite expectedly prompted chemists to begin to think about a whole host of new elements with new and interesting properties to be studied. However, it soon became apparent that as the atomic number of the artificial element became higher, its half-life became shorter and, hence, the prospects for stable elements of very high atomic numbers became dim.

Recent calculations by many nuclear physicists, based on the nuclear shell model, now suggest that a closed nuclear shell for protons exists at $Z = 114$ and that one for neutrons occurs at 184. As a result, chemists have once again begun to speculate about the possibilities of new stable elements.

One proposed extension of the periodic table to include these heavy and superheavy elements is shown in Figure 22.6. Recall that the actinide series, which occurs as the result of the filling of the $5f$ subshell, ends at lawrencium, $Z = 103$. The next element, 104, therefore lies under hafnium if we follow the scheme for the filling of subshells developed in Chapter 3. Elements 104 to 112, therefore, would correspond to the filling of the $6d$ subshell. Next we have six elements in the p-block (113 to 118) that would have their $7p$ subshell gradually filled. Elements 119 and 120, in period 8, correspond to the completion of the $8s$ subshell and following element 121 there occurs a sequence of 32 inner transition elements, from $Z = 122$ to 153. These "superactinides" would be accounted for by the completion of first the $6f$ subshell (14 elements) followed by the filling of a $5g$ subshell (a g subshell

H 1																	He 2
Li 3	Be 4											B 5	C 6	N 7	O 8	F 9	Ne 10
Na 11	Mg 12											Al 13	Si 14	P 15	S 16	Cl 17	Ar 18
K 19	Ca 20	Sc 21	Ti 22	V 23	Cr 24	Mn 25	Fe 26	Co 27	Ni 28	Cu 29	Zn 30	Ga 31	Ge 32	As 33	Se 34	Br 35	Kr 36
Rb 37	Sr 38	Y 39	Zr 40	Nb 41	Mo 42	Tc 43	Ru 44	Rh 45	Pd 46	Ag 47	Cd 48	In 49	Sn 50	Sb 51	Te 52	I 53	Xe 54
Cs 55	Ba 56	La 57	Hf 72	Ta 73	W 74	Re 75	Os 76	Ir 77	Pt 78	Au 79	Hg 80	Tl 81	Pb 82	Bi 83	Po 84	At 85	Rn 86
Fr 87	Ra 88	Ac 89	(104)	(105)	(106)	(107)	(108)	(109)	(110)	(111)	(112)	(113)	(114)	(115)	(116)	(117)	(118)
(119)	(120)	(121)	(154)	(155)	(156)	(157)	(158)	(159)	(160)	(161)	(162)	(163)	(164)	(165)	(166)	(167)	(168)

Lanthanides	Ce 58	Pr 59	Nd 60	Pm 61	Sm 62	Eu 63	Gd 64	Tb 65	Dy 66	Ho 67	Er 68	Tm 69	Yb 70	Lu 71
Actinides	Th 90	Pa 91	U 92	Np 93	Pu 94	Am 95	Cm 96	Bk 97	Cf 98	Es 99	Fm 100	Md 101	No 102	Lr 103
Super-actinides	(122)	(123)	(124)											(153)

Figure 22.6

Extended periodic table. From G. Seaborg, Journal of Chemical Education, Vol. 46 p. 626, October, 1969. Used by permission.

would contain 9 orbitals that can accommodate 18 electrons; therefore we would have 18 more elements, 136 to 153). After the superactinides we would fill the 7d subshell (elements 154 to 162) and then the 8p subshell (163 to 168).

In the search for new elements it is expected that nuclides that differ much from $Z = 114$ would be extremely unstable and would decompose by fission with very short half-lives. However, in the vicinity of $Z = 114$ it has been suggested that fission should not occur and that the half-lives of these elements with respect to alpha and beta decay should be long enough so that it should be possible to detect them and perhaps even investigate their chemical properties. There is even the possibility that these superheavy elements will not be radioactive at all.

The relative stabilities of nuclides containing differing numbers of protons and neutrons have been dramatized in a drawing (Figure 22.7) published by Dr. Glenn T. Seaborg, formerly head of the U. S. Atomic Energy Commission. Here the stable nuclei of our band of stability are shown as a long peninsular extending out into a sea of instability. Stable nuclei correspond to points above sea level whereas submerged regions constitute unstable nuclei. Notice that nuclei with magic numbers of either protons or neutrons are shown as higher, more stable ridges while doubly magic nuclei are shown as mountains of stability.

The superheavy elements with approximately a magic number of 114 protons and either 184 or 196 neutrons are depicted as an island of stability separated from the peninsular by a region of high nuclear instability. As a result, in order to reach this island we cannot bombard stable (or relatively stable) nuclei with light particles such as $_2^4$He since this simply places the product nuclei into the sea of instability where they decompose rapidly

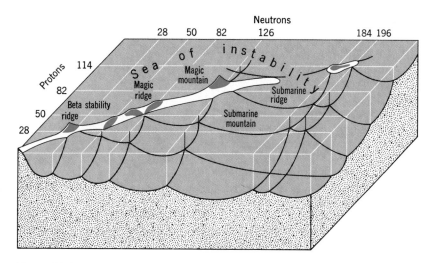

Figure 22.7
Known and predicted regions of nuclear stability, surrounded by a sea of instability. From G. Seaborg, Journal of Chemical Education, Vol. 46, p. 626, October, 1969. Used by permission.

before any additional mass and charge can be added. Consequently the jump to the island must be made in one step. At the present time this presents substantial problems because bombarding nuclei must contain a n/p ratio of at least 1.6 (184/114 = 1.61). However, light nuclei such as $^{40}_{20}Ar$, while possessing sufficient protons to give the desired atomic number by a reaction such as,

$$^{248}_{96}Cm + {}^{40}_{18}Ar \longrightarrow {}^{284}_{114}X + 4\,{}^{1}_{0}n$$

do not contain enough neutrons to place the product isotope within the island of stability. For example $^{284}_{114}X$ contains only $284 - 114 = 170$ neutrons, 14 less than the 184 that we would want in order to achieve a doubly magic nucleus. Research today is directed at obtaining suitable target and projectile nuclei that will give not only $Z = 114$ but also a sufficient number of neutrons to place the product nucleus within the bounds of the predicted island of stability.

Ordinarily, nuclear changes such as those involved in radioactive decay have very little direct effect on chemical reactions, although in some cases the high energy radiation emitted by radioactive nuclei can influence the products of reaction. This radiation is generally capable of disrupting chemical bonds. If the cleavage of a chemical bond occurs in DNA, for example, mutations of the DNA strand can be brought about. Mutational changes can also take place by reactions between DNA and the products of other cleavage reactions. For instance, free radicals are produced by the splitting of an H—O bond in water. This takes place through a series of steps, the first of which is the absorption of radiation by the water molecule and the subsequent ejection of an electron.

22.5 Chemical Applications

$$H_2O + h\nu \underset{\substack{\text{energy from}\\\text{radiation}}}{\longrightarrow} H_2O^+ + e^-$$

The final products of the reaction are a hydrogen atom, H· (the dot indicates the unpaired electron in the radical) and a *hydroxyl radical*, ·OH. The overall change is

$$H_2O + h\nu \longrightarrow H\cdot + \cdot OH$$

Free radicals are extremely reactive because of the presence of their unpaired electrons, and their interactions with DNA can cause mutations or otherwise disrupt the replication of the DNA strands.

It is only in rare cases that the chemist makes direct use of the energy emitted in nuclear transformations. Most chemical applications of radioactive nuclides stem from their ease of identification and detection, even when they are present in very small amounts. Hence radioactive isotopes are usually employed in **tracer studies** where they may be added in very small amounts and used to follow, or trace the course of a chemical reaction. The range of applications of these tracer techniques is limited only by the imagination and ingenuity of the experimenter. Let us take a brief look at some examples that demonstrate the scope of these applications.

ANALYTICAL CHEMISTRY. There are many examples of analytical uses for radioactive isotopes. One of these techniques, called **isotope dilution,** can be used when it is impossible to separate completely a desired substance from a mixture. In this case, a small measured amount of the substance containing a known quantity of a radioactive isotope is *added* to the mixture. After making sure that complete mixing has occurred, a small amount of the *pure* desired substance is separated from the mixture. This sample will contain some of the added radioactive isotope, and from the proportion of the labeled isotope present in the sample the total quantity of the substance in the original mixture can be computed.

Consider, for instance, a mixture of salts of similar solubilities, such as the mixture of KNO_3 and $NaCl$. By fractional crystallization only a portion of the KNO_3 can be separated from the mixture. As a result we cannot determine, in a simple fashion, how much of this salt is in the mixture.

Suppose, now, that 1.0 g of KNO_3 containing a small amount of radioactive ^{40}K is added to the salt mixture and then some KNO_3 (now containing K from the original mixture as well as from the added tagged KNO_3) is separated by fractional crystallization. If the specific activity[2] of this KNO_3 has dropped to 1% of the specific activity of the added KNO_3, then we know that only 1% of the added solid has been recovered in our KNO_3 sample and that the other 99% of the KNO_3 must have been present in the original mixture. In other words, after we had added the 1 g of labeled KNO_3 there was a 99-to-1 ratio of unlabeled to labeled salt. Therefore the original mixture must have contained 99 grams of KNO_3.

Another technique that is applicable to analytical chemistry is called **neutron activation analysis.** When nonradioactive isotopes are bombarded by neutrons, heavy isotopes of these elements can be produced. The product of this reaction may lie outside the band of stability and hence be radioactive. Even if another nonradioactive nucleus is produced, however, the absorption of these neutrons generally gives nuclei that are excited and that emit gamma radiation in much the same way that an excited atom emits light when it returns to the ground state.

$$^A_Z X + ^1_0 n \longrightarrow ^{A+1}_Z X^* \quad \text{(the asterisk indicates an excited nucleus)}$$
$$^{A+1}_Z X^* \longrightarrow ^{A+1}_Z X + h\nu \quad \text{(gamma photon)}$$

Since each element has its own characteristic γ emission spectrum, an analysis of the energies of the γ emissions from the activated sample allows its composition to be determined. In addition, from the intensity of the emitted γ radiation the concentration of each element can be computed.

This technique has some very unique advantages. First, it is nondestructive. Since the number of nuclei that must be activated to perform the analysis is small, most of the sample is unaffected. Second, as implied in the preceding sentence, the method is very sensitive and is, therefore, well suited to the analysis of trace amounts of impurities. In some cases, sensitivities of the order of 10^{-12} g can be achieved.

DESCRIPTIVE CHEMISTRY. Many of the elements having atomic numbers greater than $Z = 83$ (bismuth) have short half-lives and, hence, are not observed to occur naturally. Instead, they must be synthesized in particle accelerators; consequently only extremely small quantities of these elements have

---

[2] Specific activity is defined as the number of counts per minute per gram of sample.

ever been prepared. A question then arises; How can we study their chemistry if we cannot even obtain enough to be able to see them?

To arrive at the solution to this problem let us consider the element astatine. Astatine was first produced in the cyclotron by the reaction,

$$^{209}_{83}Bi + ^{4}_{2}He \longrightarrow ^{211}_{85}At + 2\,^{1}_{0}n$$

in which the ^{211}At produced has a half-life of only about 7.5 hours. The most stable isotope, ^{210}At, has a half-life of only 8.3 hours so that large quantities of the element cannot be accumulated.

Since astatine occurs in Group VIIA, we expect the element to be similar in some of its properties to iodine. To verify this the astatine is added as a tracer in reactions involving iodine, and the fate of the At is followed as the iodine undergoes reactions. If in a given reaction the At occurs in the products along with the iodine, we conclude that, in this reaction, At behaves just as I does. Hence, we have discovered something about the chemical behavior of an element that we cannot even see! For instance, it is observed that like iodine, elemental astatine is rather volatile since it is carried with the iodine when I_2 is sublimed. In solution At^- is carried from solution along with I^- upon the addition of Ag^+. Thus we conclude that AgAt is insoluble just as is AgI.

REACTION MECHANISMS. In Chapter 11 we saw that a study of the effect of the concentrations of the reactants on the rate of a chemical reaction can often give some insight into the mechanism of the reaction. Such studies, however, seldom answer all of the questions that we might ask about the reaction mechanism. Consider, for example, the reaction of an alcohol and an organic acid to produce an ester and water,

$$R\text{—}OH + HO\text{—}\overset{\displaystyle O}{\overset{\|}{C}}\text{—}R' \longrightarrow R\text{—}O\text{—}\overset{\displaystyle O}{\overset{\|}{C}}\text{—}R' + H_2O$$

Upon the formation of the ester molecule, two hydrogen atoms and one oxygen atom are removed from the alcohol and acid to become a molecule of water. There seems little doubt about the origin of the two hydrogen atoms; however, there is a question about which one of the —OH oxygen atoms is removed and finds its way into the H_2O molecule.

This question can be resolved by carrying out the reaction with a labeled oxygen (e.g., ^{18}O) incorporated into the OH group of either the alcohol or the acid. For instance, if the alcohol is labeled with ^{18}O, it is found that all of the labeled oxygen becomes incorporated into the ester. On the other hand, if the acid contains ^{18}O in the OH group, all of the labeled oxygen ends up in the water with none in the ester. It is clear, therefore, that the reaction involves the removal of the OH from the acid and the H from the alcohol.

Reaction using labeled alcohol

$$R\text{—}O^*\text{—}H + H\text{—}O\text{—}\overset{\displaystyle O}{\overset{\|}{C}}\text{—}R' \longrightarrow R\text{—}O^*\text{—}\overset{\displaystyle O}{\overset{\|}{C}}\text{—}R' + H_2O$$

Reaction using labeled acid

$$R\!-\!O\!-\!H + H\!-\!O^*\!-\!\overset{\displaystyle O}{\overset{\displaystyle \|}{C}}\!-\!R' \longrightarrow R\!-\!O\!-\!\overset{\displaystyle O}{\overset{\displaystyle \|}{C}}\!-\!R' + H_2O^*$$

Many similar experiments using tagged atoms have been employed to aid in the elucidation of a large number of reaction mechanisms, including biological and biochemical processes. For instance, labeled water can be added to the root system of a plant and its progression into the stem, and ultimately into the leaves, can be traced. Experiments using ^{14}C labeled CO_2 have been used to follow the course of carbon in photosynthesis in plants. In this case plants are exposed to CO_2 and, at various intervals, are killed and their cellular components separated to determine which compounds have had ^{14}C built into them. In this way the sequence of reactions in photosynthesis can be unraveled.

22.6 Nuclear Fission and Fusion

In the late 1930s a group of German chemists, Otto Hahn, Fritz Strassman, and Lise Meitner, found that when ^{235}U was bombarded with neutrons the unexpected products were the isotopes ^{139}Ba and ^{94}Kr, as well as three neutrons:

$$^{235}_{92}U + ^{1}_{0}n \longrightarrow ^{139}_{56}Ba + ^{94}_{36}Kr + 3\ ^{1}_{0}n$$

The significance of this accidental discovery was explained by Meitner and her cousin Otto Frisch who pointed out that fragmentation (splitting) of the ^{235}U was taking place, and that a very large amount of energy was emitted during the process. The splitting of an atom into two approximately equal parts is known as **fission.** The results of the fission of ^{235}U were tested and substantiated by Enrico Fermi at Columbia University in New York City and by physicists at Berkeley in California. Unfortunately, scientists saw military applications of the fission process, and it was Albert Einstein who alerted President Roosevelt to its possibilities. Roosevelt responded by establishing the "Manhattan Project," the research efforts of which led to the two bombs that were dropped on Hiroshima and Nagasaki, and ultimately to the production of energy by controlled nuclear fission.

Since three neutrons are produced during each fission of ^{235}U and since neutrons are required as a reactant for each fission process, then, potentially at least, the initial reaction is capable of triggering several additional reactions, as we can see in Figure 22.8. These reactions can in turn trigger many more and so on, permitting a nuclear chain reaction to take place. This is indeed what takes place if enough pure ^{235}U is present. Each fission reaction causes several others to take place, with the evolution of a tremendous amount of energy.

We mentioned earlier that the most abundant isotope of uranium found in its naturally occurring ores is ^{238}U. This isotope is nonfissionable and can, in fact, prohibit the fission chain reaction of the ^{235}U from occurring. The ^{238}U absorbs the neutrons emitted during the fission reaction, thus preventing the chain from continuing. There is then a minimum quantity of ^{235}U that must be present in order for the fission reaction to sustain itself. This minimum quantity needed for fission is called the isotope's **critical mass.**

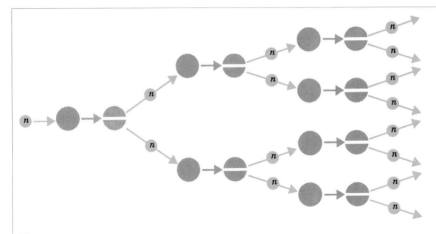

Figure 22.8
Fission chain reaction.

NUCLEAR REACTORS. In a nuclear reactor, fission reactions take place, but at a controlled rate; slow enough to avoid a chain explosion but fast enough to produce usable heat. The fission reactions are controlled by the use of control rods made of such materials as cadmium, which absorb neutrons and thus prohibit the chain from occurring too rapidly. When these rods are extended all the way into the reactor (or pile) fission occurs very slowly. The rate can be increased by withdrawing the rods. When they are pulled out, they absorb fewer and fewer neutrons and the reaction occurs faster and faster. The ideal position of the rods is at that point where the fission reaction is just able to sustain itself at a desired level.

The large amount of energy generated in the controlled nuclear fission reaction appears primarily as heat and hence nuclear reactors must be cooled. There are, of course, many current applications of this released thermal energy in the production of electrical power. The heat removed from the reactor is used to convert water to steam that is then used to drive turbines that generate electricity (Figure 22.9).

As with our supply of fossil fuels, there is only a limited supply of the fissionable ^{235}U, and nuclear reactors would face a somewhat uncertain future were it not possible to generate other fissionable isotopes. In a **breeder reactor** some of the control rods are replaced with rods containing ^{238}U. Some of the neutrons produced in the fission reaction are absorbed by the ^{238}U and give the reaction,

$$^{238}_{92}U + ^{1}_{0}n \longrightarrow ^{239}_{92}U$$

The $^{239}_{92}U$ decays rapidly to yield ultimately $^{239}_{94}Pu$ which, like $^{235}_{92}U$, is fissionable.

$$^{239}_{92}U \longrightarrow ^{239}_{93}Np + ^{0}_{-1}e$$
$$^{239}_{93}Np \longrightarrow ^{239}_{94}Pu + ^{0}_{-1}e$$

Breeder reactors thus have the useful property that they produce as much or *more* fissionable isotopes as they consume.

698 Nuclear Chemistry

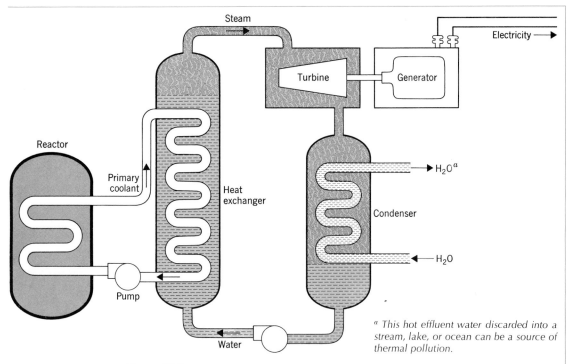

Steam

Electricity →

Turbine

Generator

Reactor

Primary coolant

Heat exchanger

H₂Oᵃ

Condenser

H₂O

Pump

Water

ᵃ This hot effluent water discarded into a stream, lake, or ocean can be a source of thermal pollution.

Figure 22.9

Application of Nuclear fission to the production of electricity.

Nuclear reactors, in general, have the undesirable effect of producing highly radioactive by-products, some with extremely long half-lives, that are nearly impossible to dispose of safely. Nevertheless, the energy crisis in recent years has placed increasing pressure on the need to develop additional sources of electrical power, including the exploitation of nuclear energy.

NUCLEAR FUSION. Quite the opposite of nuclear fission (fragmentation) is nuclear fusion. In this process two isotopes, usually very light ones, are brought together to form a new product. In so doing a very large amount of energy is released. The fusion reaction known as the hydrogen bomb is the reaction

$$\,^2_1H + \,^3_1H \longrightarrow \,^4_2He + \,^1_0n + energy$$

This is also one of the reactions taking place on the sun and accounts for the production of a good deal of its energy.

Fusion reactions, not surprisingly, possess a high energy of activation mainly because of the electrostatic repulsion between the two nuclei that are being joined together. As a result, they occur only at extremely high temperatures where their kinetic energies are sufficient to overcome this repulsion. In fact it is estimated that temperatures of approximately 200 million degrees Celsius are needed for fusion to occur (by comparison, the average temperature of the sun is 4×10^6 °C). The temperatures required to initiate such fusion reactions can be supplied by using an atomic (fission) bomb as a sort of "nuclear match." The energy obtained from one fusion reaction is sufficient to cause other reactions to occur; thus a chain reaction is set up resulting in a thermonuclear explosion.

As a potential source of commercial electrical power, the fusion process has several advantages over the fission reaction. First, the quantity of energy liberated in nuclear fusion is much greater than in fission. Another important advantage is that fusion reactions are relatively "clean" in the sense that the products of the fusion reaction are generally not radioactive. In fission reactions, on the other hand, many of the products and by-products are unstable radioactive nuclei. Fission reactors, therefore, pose a waste disposal problem not anticipated for potential fusion reactors. As a result scientists are attempting to establish controlled fusion reactors. One of the main obstacles to this, however, is the lack of a container that is able to hold a reaction mass that is at a temperature of 2×10^8 °C! Current research is aimed at maintaining the reacting mass of ions (called a **plasma**) suspended and enclosed within a powerful magnetic field.

Review Questions and Problems

22.1 What are the three main types of radiation emitted by radioactive nuclei?

22.2 Complete and balance the following nuclear equations:

(a) $^{81}_{36}Kr + ^{0}_{-1}e \rightarrow$?
(b) $^{104}_{47}Ag \rightarrow ^{0}_{1}e +$?
(c) $^{73}_{31}Ga \rightarrow ^{0}_{-1}e +$?
(d) $^{104}_{48}Cd \rightarrow ^{104}_{47}Ag +$?
(e) ? $+ ^{0}_{-1}e \rightarrow ^{54}_{24}Cr$
(f) $^{135}_{53}I \rightarrow ^{135}_{54}Xe +$?
(g) $^{245}_{97}Bk \rightarrow ^{4}_{2}He +$?
(h) $^{238}_{92}U + ^{12}_{6}C \rightarrow ^{246}_{98}Cf +$?
(i) $^{96}_{42}Mo + ^{2}_{1}H \rightarrow ^{1}_{0}n +$?
(j) $^{20}_{8}O \rightarrow ^{20}_{9}F +$?

22.3 Write balanced equations for the nuclear decay reactions below.

(a) Alpha emission by $^{11}_{5}B$
(b) Beta emission by $^{90}_{38}Sr$
(c) Neutron absorption by $^{107}_{47}Ag$
(d) Neutron emission by $^{88}_{35}Br$
(e) Electron absorption by $^{116}_{51}Sb$
(f) Positron emission by $^{70}_{33}As$
(g) Proton emission by $^{41}_{19}K$

22.4 Write nuclear equations for the following processes:

(a) $^{27}_{13}Al(\alpha,n)^{30}_{15}P$
(b) $^{209}_{83}Bi(d,n)^{210}_{84}Po$; $(d = $ deuteron, $^{2}_{1}H)$

(c) $^{15}_{7}N(p,\alpha)^{12}_{6}C$
(d) $^{12}_{6}C(p,\gamma)^{13}_{7}N$
(e) $^{14}_{7}N(\alpha,p)^{17}_{8}O$

22.5 Show that Equation 22.1 reduces to Equation 22.2 if we take $[A] = \frac{1}{2}[A]_0$.

22.6 The half-life for the decay of ^{109}Cd is 470 days. What is the value for the rate constant for this decay?

22.7 A sample of rock was found to contain 2.07×10^{-5} moles of ^{40}K and 1.15×10^{-5} moles of ^{40}Ar. If we assume that all of the ^{40}Ar came from the decay of ^{40}K, what is the age of the rock ($t_{1/2} = 1.3 \times 10^9$ years for ^{40}K)?

22.8 What is the significance of the *band of stability*? What decay processes are likely to occur for nuclides that have n/p ratios that place them above the band of stability?

22.9 Elements with atomic numbers greater than 83 generally decay by either α emission or fission. Why are the other forms of decay less likely for these nuclides?

22.10 What is a *magic number*? What magic numbers occur for protons? For neutrons? What magic numbers do we observe for orbital electrons?

22.11 In the absence of any specific information about their actual stability, rank the following nuclides in their expected order of decreasing stability.

$^{4}_{2}He$ $^{39}_{20}Ca$ $^{10}_{5}B$ $^{71}_{32}Ge$ $^{58}_{28}Ni$

22.12 Technetium and promethium do not possess any stable isotopes. In light of the discussion in Section 22.3, can you comment on this observation?

22.13 What chemical and physical properties would you predict for element number 114? If any of this element were formed at the time the universe came into being, where would be a likely place to search for it?

22.14 The element $^{34}_{17}Cl$ emits gamma radiation with energies of 0.14, 1.15, 2.27, 3.22, and 4.80 MeV (million electron volts). How does this observation support the nuclear shell theory?

22.15 Explain the difficulties that must be overcome if a stable element with $Z = 114$ is to be made by nuclear bombardment.

22.16 A large complex piece of apparatus has built into it a cooling system containing an unknown volume of cooling liquid. It is desired to measure the volume of the coolant without draining the lines. To the coolant was added 10 ml of methanol labeled with ^{14}C and having a specific activity of 580 cpm per gram. The coolant was permitted to circulate to assure complete mixing before a sample was withdrawn that was found to have a specific activity of 29 cpm per gram. Calculate the volume of coolant in the system. The den-sity of methanol is 0.792 g/ml and the density of the coolant is 0.884 g/ml.

22.17 Dinitrogen trioxide, N_2O_3, is largely dissociated into NO and NO_2 in the gas phase where there exists the equilibrium, $N_2O_3 \rightleftharpoons NO + NO_2$. In an effort to determine the structure of N_2O_3, a mixture of NO and N^*O_2 was prepared containing isotopically labeled N in the NO_2. After a period of time the mixture was analyzed and found to contain substantial amounts of both N^*O and N^*O_2. Explain how this is consistent with the structure for N_2O_3 being ONONO.

22.18 *Racemization* is a chemical reaction in which one optical isomer of a compound is converted into its mirror image. One possible mechanism for the racemization of octahedral complex ions containing three bidentate ligands involves the temporary loss of one of the ligands,

$$d\text{-}[M(AA)_3] \longrightarrow \begin{pmatrix} [M(AA)_2] \\ + \\ AA \end{pmatrix} \longrightarrow l\text{-}[M(AA)_3]$$

This can be pictured as shown in Figure 22.10. Can you suggest a simple experiment, making use of radioisotopes, that would be able to confirm whether or not this mechanism is operative in the racemization of the $[Co(C_2O_4)_3]^{3-}$ ion?

22.19 A complex ion of chromium(III) with oxalate ion was prepared from ^{51}Cr labeled $K_2Cr_2O_7$, having a specific activity of 843 cpm/gram, and ^{14}C la-

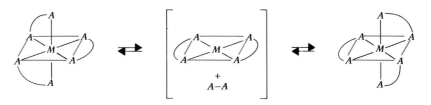

Figure 22.10
A possible mechanism for the racemization of an octahedral $[M(AA)_3]$ complex (AA = bidentate ligand).

beled oxalic acid, $H_2C_2O_4$, having a specific activity of 345 cpm/gram. Chromium-51 decays by electron capture with the emission of a gamma, whereas ^{14}C is a pure beta emitter. Because of the characteristics of the α and γ detectors, each of these isotopes may be counted independently. A sample of the complex ion was observed to give a gamma count of 165 cpm and an alpha count of 84 cpm. From these data determine the number of oxalate ions bound to each Cr(III) in the complex ion. (Hint. For the starting materials calculate the cpm per mole of Cr and oxalate, respectively.)

Appendix A
Some
Commonly
Encountered
Geometrical
Shapes

A subject of vital concern in chemistry is the geometry, or shapes, of complex molecules and ions. For example, knowledge of molecular geometry permits the testing of theories of bonding. In some instances knowledge of molecular geometry can also account for the products of chemical reaction, and it seems almost certain that the shapes of biologically important molecules significantly control their biological functions.

Many molecular structures can be considered to be derived from a small number of simple geometric figures. In fact, when speaking of the shapes of molecules and ions, the names of these simple geometric forms are often used. The major features of three of the solid shapes are described below.

Figure A.1
The tetrahedron.

Figure A.2
The tetrahedral methane molecule. The H—C—H angle is 109.5'.

A symmetrical tetrahedron, Figure A.1, is constructed of four equilateral triangles. There are four vertices. Molecular structures derived from the tetrahedron usually have one atom in the center that is bonded to four others at the vertices, as illustrated in the methane molecule, CH_4, in Figure A.2. Each H—C—H angle is 109.5 degrees. The molecule is said to be *tetrahedral*.

A trigonal bipyramid, Figure A.3, is composed of two trigonal pyramids that share a common base. The apex of one trigonal pyramid is above the base; the apex of the other is below the base (a trigonal pyramid is a pyramid that has a triangular base—a tetrahedron is a special case of a trigonal pyramid). Notice that the trigonal bipyramid has five vertices. Molecular shapes derived from this figure usually possess one atom in the center that is bonded to five others at the vertices. The molecule PCl_5 is illustrated in Figure A.4.

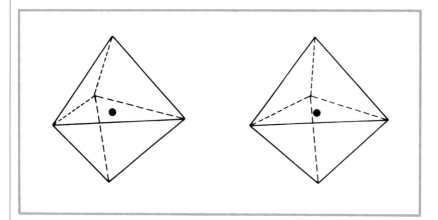

Figure A.3
The trigonal bipyramid.

Figure A.4
The trigonal bipyramidal
PCl_5 molecule.

Not all bond angles are equivalent in the trigonal bipyramid. Notice that the Cl—P—Cl angles in the horizontal *equatorial* triangular plane are 120 degrees, while those formed between an *axial* Cl atom (one that lies directly above or below the triangular plane) and an equatorial Cl atom are only 90 degrees.

An octahedron, Figure A.5 is an eight-sided geometric figure that might also be called a square bipyramid (two square pyramids that share a common square base). The octahedron possesses six vertices. Molecules that have this geometry generally have an atom in the center that is bonded to six others. An example is SF_6 (Figure A.6). Note that all of the F—S—F bond angles are the same, 90 degrees.

Figure A.5
The octahedron.

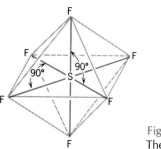

Figure A.6
The SF_6 molecule.

Appendix B
Naming
Inorganic
Compounds

This appendix gives the rules of nomenclature that apply to the naming of most simple inorganic compounds. The naming of complex inorganic substances and organic compounds will be found in Chapters 19 and 20, respectively.

A **binary compound** is composed of atoms of only two different elements. In naming these substances the less electronegative (more metallic) element is specified first by giving its ordinary English name. The name of the second element (almost always a nonmetal) is obtained by adding the suffix *ide* to its stem, as shown in Table B.1. Some typical examples are

NaCl	sodium chloride
SrO	strontium oxide
Al_2S_3	aluminum sulfide
Mg_3P_2	magnesium phosphide

Table B.1
Names of Anions Derived from Nonmetals

Group IV A	Group V A	Group VI A	Group VII A
C^{4-}; carbide[a]	N^{3-} nitride	O^{2-} oxide	F^- fluoride
Si^{4-}; silicide	P^{3-} phosphide	S^{2-} sulfide	Cl^- chloride
	As^{3-} arsenide	Se^{2-} selenide	Br^- bromide
		Te^{2-} telluride	I^- iodide

[a]Carbon also forms a number of complex carbides, for example, C_2^{2-} in CaC_2.

Many elements are commonly found to exist in more than one positive oxidation state. When the element is a metal, there are two methods that may be used to indicate its oxidation state. In the older of these methods the suffixes *-ic* and *-ous* are used to differentiate between high and low oxidation states. Thus the +3 and +2 oxidation states of chromium would be specified as

Cr^{3+}	chromic;	$CrCl_3$ chromic chloride
Cr^{2+}	chromous;	$CrCl_2$ chromous chloride

When the metal has a symbol derived from the Latin name for the element, its Latin stem is generally used. For example, with iron there are two common oxidation states, Fe^{3+} (ferric) and Fe^{2+} (ferrous). Other common examples are found in Table B.2. Notice that this system *only* differentiates between high and low oxidation states; it does not specify what the oxidation state of the metal is.

Table B.2
Metals Commonly Found in Two Oxidation States

Chromium	Manganese	Iron	Cobalt	Lead
Cr^{2+} chromous Cr^{3+} chromic	Mn^{2+} manganous Mn^{3+} manganic	Fe^{2+} ferrous Fe^{3+} ferric	Co^{2+} cobaltous Co^{3+} cobaltic	Pb^{2+} plumbous Pb^{4+} plumbic

Copper	Tin	Mercury
Cu^+ cuprous Cu^{2+} cupric	Sn^{2+} stannous Sn^{4+} stannic	Hg_2^{2+} mercurous (note that there are two Hg atoms) Hg^{2+} mercuric

The second and preferred method of indicating the oxidation state of the metal involves placing a Roman numeral, equal to the positive oxidation number, in parenthesis following the English name for the element. Thus Fe^{2+} and Fe^{3+} would be iron(II) and iron(III), respectively. The alternate names for the compounds $FeCl_2$ and $FeCl_3$ are therefore

$FeCl_2$ ferrous chloride   or   iron(II) chloride
$FeCl_3$ ferric   chloride   or   iron(III) chloride

When naming binary covalent compounds formed between two non-metals, a third system of nomenclature is often applied in which the numbers of each atom in a molecule of the compound is specified by a Greek prefix: *di* (2), *tri* (3), *tetra* (4), *penta* (5), *hexa* (6), *hepta* (7), *octa* (8), etc.

NO_2 nitrogen(IV) oxide   nitrogen dioxide
N_2O_4 nitrogen(IV) oxide   dinitrogen tetroxide
N_2O_5 nitrogen(V) oxide   dinitrogen pentoxide
PCl_3 phosphorus(III) chloride   phosphorus trichloride
PCl_5 phosphorus(V) chloride   phosphorus pentachloride

In some instances the prefix *mono* (1) is also used to avoid ambiguity.

CO_2 carbon dioxide
CO carbon monoxide

**B.2
Compounds
Containing
Polyatomic Ions**

Many ions are found to contain more than one atom and are therefore referred to in general as polyatomic ions. These species enter into ionic compounds as discrete units and generally stay intact in most chemical reactions. A list of these is given in Table 4.3 (page 92). As with binary compounds, salts that contain these ions are always named with the positive ion first. Some examples are:

Na_2CO_3 sodium carbonate $Ba(OH)_2$ barium hydroxide
$Ca(C_2H_3O_2)_2$ calcium acetate $(NH_4)_2SO_4$ ammonium sulfate

Acids can be described as substances which release H^+ when dissolved in water. A binary acid (also sometimes called a *hydro* acid) is a compound containing hydrogen and a nonmetal, having a general formula H_nX (e.g., HCl, H_2S). They are named as *hydro . . . ic acid,* where the stem of the name of the nonmetal is inserted in place of the dotted line. Examples are:

B.3
Binary Acids

HF	hydro*fluor*ic acid
HCl	hydro*chlor*ic acid
HBr	hydro*brom*ic acid
HI	hydr*iod*ic acid
H_2S	hydro*sulfur*ic acid

When these acids are reacted with hydroxide ion (a reaction called **neutralization**), an anion is formed, for example,

$$HCl + OH^- \longrightarrow H_2O + Cl^-$$
*hydro***chlor***ic acid*   **chlor***ide*

Note that *hydro . . . ic acids* give *. . . ide* salts, for example, sodium chloride.

Oxoacids are acids that contain hydrogen, oxygen, and at least one other element. Sulfuric acid, H_2SO_4, is an example. Where the third element (S in the case of H_2SO_4) can exist in more than one oxidation state, more than one oxoacid is possible. For example, two common oxoacids of sulfur are H_2SO_4 and H_2SO_3, containing sulfur in the +6 and +4 oxidation states, respectively. The acid containing the element in the higher oxidation state is given the suffix *ic,* whereas the acid having the element in the lower oxidation state is given the ending *ous.* Thus we have

B.4
Oxoacids

H_2SO_4	sulfuric acid
H_2SO_3	sulfurous acid

Salts produced by neutralization of these acids contain polyatomic anions (Table 4.3). The anion derived from the "ic" acid ends in *ate,* whereas the anion from the "ous" acid ends in *ite:*

H_2SO_4	sulfuric acid	SO_4^{2-}	sulfate
H_2SO_3	sulfurous acid	SO_3^{2-}	sulfite
HNO_3	nitric acid	NO_3^-	nitrate
HNO_2	nitrous acid	NO_2^-	nitrite
$HClO_3$	chloric acid	ClO_3^-	chlorate
$HClO_2$	chlorous acid	ClO_2^-	chlorite

Some elements form oxoacids in more than two oxidation states. In this case the prefixes *hypo* and *per* are used to designate a lower or higher oxidation state, respectively. A good example occurs among the oxoacids of the halogens.

hypochlorous	HClO	ClO^-	hypochlorite
chlorous	$HClO_2$	ClO_2^-	chlorite
chloric	$HClO_3$	ClO_3^-	chlorate
perchloric	$HClO_4$	ClO_4^-	perchlorate

Appendix C
Mathematics
for General
Chemistry

For many students, solving numerical problems is often the most difficult part of any chemistry course. In this appendix we review some of the mathematical concepts that you will find useful in your study of chemistry.

C.1
The Factor-Label
Method of Problem
Solving

Even after learning the principles of chemistry, students sometimes have difficulty in setting up the arithmetic correctly to give the proper numerical answer to a problem. The "factor-label" method uses the units associated with numbers as a guide in working out the arithmetic. The method is based on the idea that *units cancel from numerator and denominator in a fraction, just as numbers do*. For example, if the units *feet* appear in both numerator and denominator, they may be cancelled

$$\frac{3 \; \cancel{ft}}{2 \; \cancel{ft}} = \frac{3}{2}$$

Numerical problem solving uses this idea by employing valid relationships between units to create **conversion factors.** For instance, to convert 4 *yards* into *feet* the relationship

$$3 \text{ feet} = 1 \text{ yard}$$

is used to create a conversion factor (a fraction) with which the 4 yards is multiplied. Thus

$$4 \; \cancel{\text{yards}} \left(\frac{3 \text{ feet}}{1 \; \cancel{\text{yard}}} \right) = 12 \text{ feet}$$

Note that the factor was constructed deliberately so that *yards* cancelled and only the desired units, *feet*, remained. Had we inverted this conversion factor, we would obtain the wrong numerical answer *and* the wrong units,

$$4 \text{ yard} \left(\frac{1 \text{ yard}}{3 \text{ feet}} \right) = \frac{4}{3} \frac{\text{yard}^2}{\text{feet}}$$

Creation of a conversion factor can be accomplished from any *valid* relationship between a set of units. This can be an equality, as in the relationship between feet and yards (i.e., 3 feet equal 1 yard). It can also be an equivalency. For instance, for a student who earns 5 dollars per hour, there is an equivalence between dollars and time

$$5 \text{ dollars are equivalent to } 1 \text{ hour}$$

We shall use the symbol $\sim$ to stand for "are equivalent to." Thus, in the example just cited

$$5 \text{ dollars} \sim 1 \text{ hr}$$

If this student works 12 hours, we can use the dollar-hour relationship to construct a conversion factor that allows us to calculate his pay (before taxes!)

$$12 \text{ hours} \left(\frac{5 \text{ dollars}}{1 \text{ hour}} \right) = 60 \text{ dollars}$$

Note that hours cancel. You will see many examples in the text in which numerical problems are solved using this "factor-label" technique.

Quite often in science it is necessary to deal with numbers that are very large, such as Avogadro's number

$$602{,}000{,}000{,}000{,}000{,}000{,}000{,}000$$

or numbers that very small, such as the mass of a single molecule of water,

$$0.000\ 000\ 000\ 000\ 000\ 000\ 000\ 03 \text{ g}$$

These numbers are very cumbersome and difficult to work with without making mistakes in arithmetic computations. To aid us in handling these large and small numbers, a system called either **exponential notation** or **scientific notation** is employed. In this system, a number is expressed as a decimal part multiplied by 10 raised to an appropriate power. Thus

$$200 = 2 \times 10 \times 10 = 2 \times 10^2$$
$$205{,}000 = 2.05 \times 100{,}000 = 2.05 \times (10 \times 10 \times 10 \times 10 \times 10) = 2.05 \times 10^5$$

To determine the exponent on the 10, we can also count the number of places the decimal must be moved to produce the number that precedes the 10 when the number is expressed in the scientific notation

$$205\quad 000_{\circ} = 2.05 \times 10^5$$

5 places

Note that the exponent on the 10 is positive when the decimal is moved to the left. When it is moved to the right, the exponent is negative.

$$0_{\circ} 000000315 = 3.15 \times 10^{-7}$$

7 places

In carrying out arithmetic operations with numbers expressed in scientific notation, the following rules apply.

MULTIPLICATION. In multiplication, the decimal portions of the number are multiplied and the exponents on the 10 are *added* algebraically.

$$(2.0 \times 10^4) \times (3.0 \times 10^3) = (2.0 \times 3.0) \times 10^{(4+3)} = 6.0 \times 10^7$$
$$(4.0 \times 10^8) \times (-2.0 \times 10^{-5}) = (4.0 \times (-2.0)) \times 10^{(8+(-5))} = -8.0 \times 10^3$$

DIVISION. The decimal portions are divided, and the exponent on 10 in the denominator is *subtracted algebraically* from the exponent on 10 in the numerator.

$$\frac{8.0 \times 10^7}{4.0 \times 10^3} = \left(\frac{8.0}{4.0}\right) \times 10^{(7-3)} = 2.0 \times 10^4$$

$$\frac{6.0 \times 10^5}{2.0 \times 10^3} = \left(\frac{6.0}{2.0}\right) \times 10^{(5-(-3))} = 3.0 \times 10^8$$

$$\frac{9.0 \times 10^{-4}}{3.0 \times 10^{-6}} = \left(\frac{9.0}{3.0}\right) \times 10^{(-4-(-6))} = 3.0 \times 10^2$$

You have probably noticed that the usual practice is to express a number with the decimal point located between the first and second digit. There are, of course, other ways that these numbers can be written that are all equivalent, and you will undoubtedly find occasions where it is convenient to use a number in other than its standard form. An example of a few equivalent expressions of the same number are

$$3.15 \times 10^{-7} = 315 \times 10^{-9} = 0.0315 \times 10^{-5}$$

Notice that in converting from one to another, one part of the number is increased while the other is decreased. For instance, to change 8.25×10^6 to 825×10^4 multiply *and* divide by 100 (or 10^2)

$$8.25 \times 10^6 \left(\frac{100}{100}\right) = (8.25 \times 100) \times \left(\frac{10^6}{10^2}\right) = 825 \times 10^4$$

ADDITION AND SUBTRACTION. When carrying out addition and subtraction, each quantity must first be written with the same power of 10. Then addition or subtraction is performed on the decimal parts; the power of 10 remains the same. For example,

$$(2.17 \times 10^5) + (3.0 \times 10^4) = ?$$

If we express both numbers with the same power of 10, we have

$$
\begin{array}{cc}
2.17 \times 10^5 & 21.7 \times 10^4 \\
+0.30 \times 10^5 \quad \text{or} & +3.0 \times 10^4 \\
\hline
2.47 \times 10^5 & 24.7 \times 10^4
\end{array}
$$

TAKING A ROOT. To extract a root (e.g., the square root) the exponent on the 10 is made to be divisible by the desired root. For instance, to take the square root of 3.7×10^7, we first change the number so that the power of 10 is divisible by 2. Then we take the square root of the decimal part and divide the exponent by 2.

$$\sqrt{3.7 \times 10^7} = \sqrt{37 \times 10^6} = \sqrt{37} \times 10^3 = 6.1 \times 10^3$$

C.3
Logarithms

A logarithm is an exponent! **Common logarithms** are exponents to which 10 must be raised to give a specified number. For instance, the log $(100) = 2$ because $10^2 = 100$. Similarly, log $(1000) = \log(10^3) = 3$.

Since logarithms are exponents, when we perform mathematical operations the same rules that apply to exponents also apply to logarithms. Thus we have

Multiplication $\begin{cases} \text{add exponents} \\ \text{add logarithms} \end{cases}$

Division $\begin{cases} \text{subtract exponents} \\ \text{subtract logarithms} \end{cases}$

For example,

$$\boxed{10^3 \times 10^4} = 10^{3+4} = \boxed{10^7}$$

$$\log (10^3 \times 10^4) = \log (10^3) + \log (10^4) = 3 + 4 = 7 = \log (10^7)$$

Similarly, for division

$$\boxed{\frac{10^8}{10^6}} = 10^{8-6} = \boxed{10^2}$$

$$\log \left(\frac{10^8}{10^6}\right) = \log (10^8) - \log (10^6) = 8 - 6 = 2 = \log (10^2)$$

For decimal numbers between 1 and 10 their logarithms lie between 0 and 1, since

$$\begin{aligned} \log(1) &= 0 & (1 = 10^0) \\ \log(10) &= 1 & (10 = 10^1) \end{aligned}$$

For example, $\log 2 = 0.3010$ or

$$10^{0.3010} = 2$$

The logarithm of 2 and other numbers between 1 and 10 can be obtained from the table of logarithms in Appendix D.

To use this table to find the logarithm of a number, we use the extreme left column to locate the first two digits of the number, and the top horizontal row to locate the third digit. The value in the table corresponding to these is the logarithm of our number. For example, if we want $\log (4.61)$, we would locate 46 in the left column and proceed to the right until we were in the column headed by 1.

The answer is

$$\log(4.61) = 0.6637$$

This table is extremely easy to use as long as our numbers are expressed in this fashion, that is, as a decimal number between 1 and 10. If the number whose logarithm we seek does not appear this way, we can first express the number in exponential notation and then take its logarithm. For example, what is $\log (728)$?

$$\begin{aligned} \log (728) &= \log (7.28 \times 10^2) \\ \log (7.28 \times 10^2) &= \log (7.28) + \log (10^2) \end{aligned}$$

$$\begin{aligned} \log (7.28) &= 0.8621 \text{ (from Table)} \\ \log (10^2) &= \underline{2.0000} \\ \text{therefore,} \quad \log (728) &= 2.8621 \end{aligned}$$

What would be the value of $\log (0.00583)$? Once again we first express the number in exponential notation:

$$\begin{aligned} \log (0.00583) &= \log (5.83 \times 10^{-3}) \\ \log (5.83 \times 10^{-3}) &= \log (5.83) + \log (10^{-3}) \end{aligned}$$

$$\begin{aligned}\log 5.83 &= +0.7649 \\ \log 10^{-3} &= \underline{-3.0000}\end{aligned}$$

Adding these algebraically, we get

$$\log (0.00583) = -2.2351$$

Sometimes it is necessary to obtain the number whose logarithm is known. This is called taking the **antilogarithm.** The procedure is simply the reverse of that given above. For example, suppose we wish to find the number whose logarithm is 3.253.

$$\log x = 3.253$$

First, we divide the number into two parts, a positive integer and a decimal.

$$\begin{aligned}3.253 &= 3 + 0.253 = 0.253 + 3 \\ \log x &= (0.253 + 3)\end{aligned}$$

We locate 0.253 in the body of the log table and find that it is the log of 1.79; we also know that 3 is the log of 10^3. Therefore,

$$\begin{aligned}\log x &= \log (1.79) + \log (10^3) = \log (1.79 \times 10^3) \\ x &= 1.79 \times 10^3\end{aligned}$$

NATURAL LOGARITHMS. A system of logarithms encountered frequently in the sciences, known as natural logarithms, has as its base $e = 2.71828 \ldots$. In other words, natural logarithms are exponents to which e must be raised to give a number. The relationship between common logs and natural logs is seen below

$$\begin{aligned}\log_{10} (10) &= 1 \quad &\text{or} \quad 10^1 &= 10 \\ \log_e (10) &= 2.303 \quad &\text{or} \quad e^{2.303} &= 10\end{aligned}$$

With common logarithms we usually omit the base and write simply, $\log 10 = 1$. With natural logarithms the base e is omitted, and they are written

$$\ln 10 = 2.303$$

The conversion from base e to base 10 logarithm is accomplished by the equation

$$\ln x = 2.303 \log x$$

C.4 The Quadratic Equation

When an equation can be written in the form

$$ax^2 + bx + c = 0$$

in which the coefficients a, b, and c are known, two values (called roots) of the variable x can be obtained by substituting the values of a, b, and c into the expression

$$x = \frac{-b \pm \sqrt{b^2 - 4ac}}{2a}$$

For example, given the equation

$$x^2 - 5x + 4 = 0$$

what is the value of x? In this equation $a = 1$, $b = -5$, and $c = 4$. Thus

$$x = \frac{-(-5) \pm \sqrt{(-5)^2 - 4\,(1)(4)}}{2\,(1)} = \frac{5 \pm \sqrt{25 - 16}}{2}$$

$$x = \frac{5 \pm \sqrt{9}}{2} = \frac{5 \pm 3}{2}$$

Therefore,

$$x = \frac{2}{2} = 1 \quad \text{and} \quad x = \frac{8}{2} = 4$$

Both values of x are mathematically correct. Usually when a quadratic equation is encountered in a chemical problem only one of the roots has any real significance. Generally, the other root will be clearly meaningless; for instance, a negative concentration, which is impossible (you can't have less than nothing).

Appendix D
Logarithms

	0	1	2	3	4	5	6	7	8	9
10	0000	0043	0086	0128	0170	0212	0253	0294	0334	0374
11	0414	0453	0492	0531	0569	0607	0645	0682	0719	0755
12	0792	0828	0864	0899	0934	0969	1004	1038	1072	1106
13	1139	1173	1206	1239	1271	1303	1335	1367	1399	1430
14	1461	1492	1523	1553	1584	1614	1644	1673	1703	1732
15	1761	1790	1818	1847	1875	1903	1931	1959	1987	2014
16	2041	2068	2095	2122	2148	2175	2201	2227	2253	2279
17	2304	2330	2355	2380	2405	2430	2455	2480	2504	2529
18	2553	2577	2601	2625	2648	2672	2695	2718	2742	2765
19	2788	2810	2833	2856	2878	2900	2923	2945	2967	2989
20	3010	3032	3054	3075	3096	3118	3139	3160	3181	3201
21	3222	3243	3263	3284	3304	3324	3345	3365	3385	3404
22	3424	3444	3464	3483	3502	3522	3541	3560	3579	3598
23	3617	3636	3655	3674	3692	3711	3729	3747	3766	3784
24	3802	3820	3838	3856	3874	3892	3909	3927	3945	3962
25	3979	3997	4014	4031	4048	4065	4082	4099	4116	4133
26	4150	4166	4183	4200	4216	4232	4249	4265	4281	4298
27	4314	4330	4346	4362	4378	4393	4409	4425	4440	4456
28	4472	4487	4502	4518	4533	4548	4564	4579	4594	4609
29	4624	4639	4654	4669	4683	4698	4713	4728	4742	4757
30	4771	4786	4800	4814	4829	4843	4857	4871	4886	4900
31	4914	4928	4942	4955	4969	4983	4997	5011	5024	5038
32	5051	5065	5079	5092	5105	5119	5132	5145	5159	5172
33	5185	5198	5211	5224	5237	5250	5263	5276	5289	5302
34	5315	5328	5340	5353	5366	5378	5391	5403	5416	5428
35	5441	5453	5465	5478	5490	5502	5514	5527	5539	5551
36	5563	5575	5587	5599	5611	5623	5635	5647	5658	5670
37	5682	5694	5705	5717	5729	5740	5752	5763	5775	5786
38	5798	5809	5821	5832	5843	5855	5866	5877	5888	5899
39	5911	5922	5933	5944	5955	5966	5977	5988	5999	6010
40	6021	6031	6042	6053	6064	6075	6085	6096	6107	6117
41	6128	6138	6149	6160	6170	6180	6191	6201	6212	6222
42	6232	6243	6253	6263	6274	6284	6294	6304	6314	6325
43	6335	6345	6355	6365	6375	6385	6395	6405	6415	6425
44	6435	6444	6454	6464	6474	6484	6493	6503	6513	6522
45	6532	6542	6551	6561	6571	6580	6590	6599	6609	6618
46	6628	6637	6646	6656	6665	6675	6684	6693	6702	6712
47	6721	6730	6739	6749	6758	6767	6776	6785	6794	6803
48	6812	6821	6830	6839	6848	6857	6866	6875	6884	6893
49	6902	6911	6920	6928	6937	6946	6955	6964	6972	6981
50	6990	6998	7007	7016	7024	7033	7042	7050	7059	7067
51	7076	7084	7093	7101	7110	7118	7126	7135	7143	7152
52	7160	7168	7177	7185	7193	7202	7210	7218	7226	7235
53	7243	7251	7259	7267	7275	7284	7292	7300	7308	7316
54	7324	7332	7340	7348	7356	7364	7372	7380	7388	7396

	0	1	2	3	4	5	6	7	8	9
55	7404	7412	7419	7427	7435	7443	7451	7459	7466	7474
56	7482	7490	7497	7505	7513	7520	7528	7536	7543	7551
57	7559	7566	7574	7582	7589	7597	7604	7612	7619	7627
58	7634	7642	7649	7657	7664	7672	7679	7686	7694	7701
59	7709	7716	7723	7731	7738	7745	7752	7760	7767	7774
60	7782	7789	7796	7803	7810	7818	7825	7832	7839	7846
61	7853	7860	7868	7875	7882	7889	7896	7903	7910	7917
62	7924	7931	7938	7945	7952	7959	7966	7973	7980	7987
63	7993	8000	8007	8014	8021	8028	8035	8041	8048	8055
64	8062	8069	8075	8082	8089	8096	8102	8109	8116	8122
65	8129	8136	8142	8149	8156	8162	8169	8176	8182	8189
66	8195	8202	8209	8215	8222	8228	8235	8241	8248	8254
67	8261	8267	8274	8280	8287	8293	8299	8306	8312	8319
68	8325	8331	8338	8344	8351	8357	8363	8370	8376	8382
69	8388	8395	8401	8407	8414	8420	8426	8432	8439	8445
70	8451	8457	8463	8470	8476	8482	8488	8494	8500	8506
71	8513	8519	8525	8531	8537	8543	8549	8555	8561	8567
72	8573	8579	8585	8591	8597	8603	8609	8615	8621	8627
73	8633	8639	8645	8651	8657	8663	8669	8675	8681	8686
74	8692	8698	8704	8710	8716	8722	8727	8733	8739	8745
75	8751	8756	8762	8768	8774	8779	8785	8791	8797	8802
76	8808	8814	8820	8825	8831	8837	8842	8848	8854	8859
77	8865	8871	8876	8882	8887	8893	8899	8904	8910	8915
78	8921	8927	8932	8938	8943	8949	8954	8960	8965	8971
79	8976	8982	8987	8993	8998	9004	9009	9015	9020	9025
80	9031	9036	9042	9047	9053	9058	9063	9069	9074	9079
81	9085	9090	9096	9101	9106	9112	9117	9122	9128	9133
82	9138	9143	9149	9154	9159	9165	9170	9175	9180	9186
83	9191	9196	9201	9206	9212	9217	9222	9227	9232	9238
84	9243	9248	9253	9258	9263	9269	9274	9279	9284	9289
85	9294	9299	9304	9309	9315	9320	9325	9330	9335	9340
86	9345	9350	9355	9360	9365	9370	9375	9380	9385	9390
87	9395	9400	9405	9410	9415	9420	9425	9430	9435	9440
88	9445	9450	9455	9460	9465	9469	9474	9479	9484	9489
89	9494	9499	9504	9509	9513	9518	9523	9528	9533	9538
90	9542	9547	9552	9557	9562	9566	9571	9576	9581	9586
91	9590	9595	9600	9605	9609	9614	9619	9624	9628	9633
92	9638	9643	9647	9652	9657	9661	9666	9671	9675	9680
93	9685	9689	9694	9699	9703	9708	9713	9717	9722	9727
94	9731	9736	9741	9745	9750	9754	9759	9763	9768	9773
95	9777	9782	9786	9791	9795	9800	9805	9809	9814	9818
96	9823	9827	9832	9836	9841	9845	9850	9854	9859	9863
97	9868	9872	9877	9881	9886	9890	9894	9899	9903	9908
98	9912	9917	9921	9926	9930	9934	9939	9943	9948	9952
99	9956	9961	9965	9969	9974	9978	9983	9987	9991	9996

Appendix E
Solutions
to Selected
Numerical
Problems

Chapter 1
1.3 1.0370, 4.17×10^{-4}, 3.09×10^{-3}, 1.001×10^2, 9.0010
1.5 3900 kg, 4.2×10^9 m^3, $-40°$ C, 1800 cm/sec, 1.0×10^5 dm^3
1.9 The X to Y ratio is $1 : 1.78$
1.13 $C = 0.632$ and $H = 0.0526$
1.15 2K, 1S; 2Na, 1C, 3O; 3N, 12H, 1P, 4O; 4K, 1Fe, 6C, 6N; 3Na, 1Ag, 4S, 6O
1.19 56 km/hr
1.21 $37°$ C, $102.2°$ F

Chapter 2
2.3 (a) 12.5 g of N_2 (b) 12.5 g N_2 (0.89 g-atom)
2.5 8.5×10^{-1} g
2.7 (a) 5.68×10^{-22} g (b) 342 g
2.9 $(NH_4)SO_4$, $AlCl_3$, CH, $C_3H_8O_3$, CH_2O
2.11 (a) 2,3,1,6 (b) 2,1,1,2,2 (c) 3,2,1,6 (d) 1,3,1,3,3 (e) 1,1,1,1,2
2.13 $NaBH_4$
2.15 CH_3NO
2.17 (a) 2.50 moles (b) 13.0 g (c) 6.40 moles (d) 80.0 g
2.19 1574 kg
2.21 (a) 30.2 g (b) 0.83 g (c) 28.5 g (d) 94.4%
2.23 256 g
2.25 $C_8H_8O_3$

Chapter 3
3.3 8.0×10^{-20} coulombs
3.5 151.9 amu
3.7 8.08×10^{-2} mi
3.9 74571 Å, 46533 Å
3.11 1.64×10^{-11} erg
3.13 2.12 Å, 4.76 Å
3.15 3.0×10^{19} sec
3.21 $s = 2$, $p = 6$, $d = 10$, $f = 14$, $g = 18$, $h = 22$
3.33 13.6 eV

Chapter 4
4.15 LiF $\Delta = 4.58$, LiCl $\Delta = 3.03$, LiBr $\Delta = 2.78$, LiI $\Delta = 2.32$

4.17 $KCl\,O_2$, $Ba\,Mn\,O_4$, $Fe_3\,O_4$, $O_2\,F_2$, $I\,F_5$, $H\,O\,Cl$, $Ca\,S\,O_4$, $Cr_2\,(S\,O_4)_3$,

$\quad$ 1+ 3+ 2−$\quad$ 2+ 6+ 2−$\quad$ 8/3+ 2−$\quad$ 1+ 1−$\quad$ 5+ 1−$\quad$ 1+ 2− 1+$\quad$ 2+ 6+ 2−$\quad$ 3+ 6+ 2−

$Na_2S_2O_8$, O_3, Hg_2Cl_2

$\quad$ 1+ 6+ 2−$\quad$ 0$\quad$ 1+ 1−
$\qquad$ 1−

Chapter 5
5.17 1.24 g/ml
5.19 0.0602 M
5.21 (a) 150.9 g$\quad$ (b) 75.47 g$\quad$ (c) 30.19 g$\quad$ (d) 30.19 g
5.23 10.5 N
5.25 (a) 0.452 $N$$\quad$ (b) 0.500 $N$$\quad$ (c) 0.208 N
$\qquad$ (d) 1.42 $N$$\quad$ (e) 0.269 $N$$\quad$ (f) 0.338 N
5.27 34.3% $CuBr_2$
5.29 $TiCl_2$

Chapter 6
6.1 (a) 496 torr$\quad$ (b) 513 torr
6.3 1.54 g/liter
6.5 680 torr
6.7 26.2 ml
6.9 81.5 ml
6.11 He effuses 2.3 times faster than Ne.
6.13 2.00
6.15 8.3 atm
6.17 53.9 atm; 40,900 torr
6.19 (a) $CH_3$$\quad$ (b) 29.9$\quad$ (c) C_2H_6
6.21 528 ml

Chapter 7
7.3 $r_{Br^-} = 1.94$ Å $r_{Li^+} = 0.81$ Å
7.5 20.5°, 44.4°
7.7 $4.76 \times 10^{-25} cm^3$; 4.93×10^{-25}; 7.35×10^{-25}
7.9 (a) Al, Mg$\quad$ (b) P,S

Chapter 9
9.1 (a) 36% C_6H_6, 64% $C_7H_8$$\quad$ (b) 0.399 C_6H_6, 0.601 $C_7H_8$$\quad$ (c) 7.21 m
9.3 $X_{CuCl_2} = 7.4 \times 10^{-3}$, $X_{H_2O} = 0.993$, 0.417 m, 5.3% $CuCl_2$, 94.7% H_2O
9.7 KBr at 70° C, KNO_3 at 20°
9.9 53 g KNO_3
9.11 92.0 torr
9.13 $fp = -4.45°$ C, $bp = 101.2°C$
9.15 $C_8H_8O_4$
9.17 144 g/mole
9.19 $-0.2°$
9.21 271 torr
9.23 2000 g/mole
9.27 1.03
9.29 74.9 g, 41.8 g, 5.0 g, 10 g, 1.6 g, 1.6 g, 0.5 g; 135.4 g

Chapter 10
10.3 Step 1, $q = w = 75$ liter-atm; Step 2, $q = w = 130$ liter-atm; $\Delta E = 0$
10.5 $\Delta H = -77.2$ kcal/mole, $\Delta E = -77.8$ kcal/mole
10.7 $\Delta E = -942.7$ kcal/mole C_7H_8
10.9 (a) -204.0 kcal/mole (b) -341.5 kcal/mole (c) -95.9 kcal/mole
(d) -20.7 kcal/mole (e) -32.6 kcal/mole
10.11 -63.7 kcal
10.13 $q = 10.5$ kcal, $w = 0.59$ kcal, $\Delta E = 9.91$ kcal
10.17 -31.6 kcal
10.21 (b) $\Delta G° = -13.8$ kcal (c) $\Delta G° = -123.7$ kcal (e) $\Delta G° = -16.6$ kcal
10.23 (a) $\Delta S = -22.5°$ cal/deg
10.25 20,000 atm
10.27 331.8°K

Chapter 11
11.3 1.02×10^{-2} 2.04×10^{-2}, 4.08×10^{-2}
11.7 (a) rate $= k$ $[NOCl]^2$ (b) $k = 4 \times 10^{-8}$ liter/mole sec (c) 2.25
11.9 (a) rate $= k$ $[NO][Br_2]$ (b) rate $= k$ $[NO]^2[Br_2]$
11.15 $E_a = 38.7$ kcal, $A = 9.64 \times 10^{-17}$
11.17 3.89×10^{-1}

Chapter 12
12.3 a, d, c, b
12.5 2.81×10^{-2} liter/mole
12.7 5.25×10^{-6}
12.9 3×10^{61}
12.11 (a) 0.140 (b) 2.36×10^{-4} (c) $+4.95$ kcal
12.15 $[CO] = [Cl_2] = 2.08 \times 10^{-6}$ mole/liter $[COCl_2] = 2 \times 10^{-2}$ mole/liter
12.17 $[H_2] = [I_2] = 1.53 \times 10^{-2}M$, $[HI] = 1.13 \times 10^{-1}M$

Chapter 14
14.1 (a) 1.0×10^{-3}, 1.0×10^{-11}, 3.0   (e) 2.1×10^{-4}, 4.8×10^{-11}, 3.7
(b) 1.25×10^{-1}, 8.0×10^{-14}, 0.90   (f) 1.3×10^{-5}, 7.7×10^{-10}, 4.9
(c) 3.2×10^{-12}, 3.1×10^{-3}, 11.5   (g) 1.2×10^{-12}, 8.4×10^{-3}, 11.9
(d) 4.2×10^{-13}, 2.4×10^{-2}, 12.4   (h) 2.1×10^{-13}, 4.8×10^{-2}, 12.7
14.3 (a) $[H^+] = [C_2H_3O_2^-] = 4.2 \times 10^{-4}M$, $[HC_2H_3O_2] = 1 \times 10^{-2}M$
(b) $(NH_4^+) = [OH^-] = 4.2 \times 10_{-4}M$, $[NH_3] = 1 \times 10^{-2}M$
(c) $[H^+] = [CHO_2^-] = 2.1 \times 10^{-2}M$, $[HCHO_2] = 2.5M$
(d) $[H^+] = [F^-] = 2.8 \times 10^{-2}M$, $[HF] = 1.2M$
(e) $[OH^-] = [N_2H_5^+] = 4.4 \times 10^{-4}M$, $[N_2H_4] = 0.20M$
(f) $[OH^-] = [C_6H_5NH_3^+] = 3.5 \times 10^{-6}M$, $[C_6H_5NH_2] = 3.0 \times 10^{-2}M$
(g) $[H^+] = [OCl^-] = 3.7 \times 10^{-5}M$, $[HOCl] = 4.5 \times 10^{-2}M$
(h) $[H^+] = [CN^-] = 3.1 \times 10^{-6}M$, $[HCN] = 2.0 \times 10^{-2}M$
14.5 (a) 4.2% (b) 4.2% (c) 0.84% (d) 2.3% (e) 0.22% (f) 0.01%
(g) 0.082% (h) 0.02%
14.7 $0.55M$
14.9 7.8×10^{-8}
14.11 1.1×10^{-4}
14.13 6.6×10^{-4}

14.15 (a) 4.74 (b) 5.05 (c) 8.15 (d) 0.70
14.17 5.5
14.19 0.17 pH units
14.21 (a) $[Ag^+]^2[S^{2-}]$, (b) $[Ca^{2+}][F^-]^2$, (c) $[Fe^{3+}][OH^-]^3$,
    (d) $[Mg^{2+}][C_2O_4{}^{2-}]$, (e) $[Bi^{3+}]^2\ [S^{2-}]^3$, (f) $[Ba^{2+}][CO_3{}^{2-}]$
14.23 1.10×10^{-8}
14.25 (a) 8.4×10^{-14},   (b) 7.9×10^{-6},   (c) 3.9×10^{-5},   (d) 7.9×10^{-6},
    (e) 2.9×10^{-9}, (f) 9.3×10^{-3}
14.27 1.7×10^{-6}
14.29 1.31×10^3 liter
14.31 1.9×10^{-10}
14.33 $[H^+] = 0.05$, $[NO_3^-] = 0.10$, $[Ag^+] = 0.05$, $[Cl^-] = 3.4 \times 10^{-9}$
14.35 $3.0 \times 10^{-5} \leq [H^+] < 0.96M$; 3.2×10^{-5}
14.39 $5 \times 10^{-3} M$
14.41 0.52 moles
14.43 0.75%
14.45 (a) 2.18 (b) 3.19 (c) 8.01

Chapter 15

15.3 (a) $2Cl^- \leftrightharpoons Cl_2(g) + 2e^-$ ($\times 3$)
    $\dfrac{Cr^{3+} + 3e^- \leftrightharpoons Cr(s)\ (\times 2)}{6Cl^- + 2Cr^{3+} \rightleftharpoons 3Cl_2(g) + 2Cr(s)}$

    (b) $Zn^{2+} + 2e^- \leftrightharpoons Zn(s)$ ($\times 2$)
    $\dfrac{2H_2O \leftrightharpoons O_2 + 4H^+ + 4e^-}{2Zn^{2+} + 2H_2O \leftrightharpoons 2Zn(s) + O_2 + 4H^+}$

    (c) $Fe^{2+} + 2e^- \leftrightharpoons Fe(s)$ ($\times 2$)
    $\dfrac{2H_2O \leftrightharpoons O_2 + 4H^+ + 4e^-}{2Fe^{2+} + 2H_2O \leftrightharpoons 2Fe(s) + O_2 + 4H^+}$

    (d) $2H_2O + 2e^- \leftrightharpoons H_2 + 2OH^-$ ($\times 2$)
    $\dfrac{2H_2O \leftrightharpoons O_2 + 4H^+ + 4e^-}{6H_2O \leftrightharpoons 2H_2 + 4OH^- + O_2 + 4H^+}$
15.5 (a) $2\mathscr{F}$ (b) $1\mathscr{F}$ (c) $5\mathscr{F}$ (d) $2\mathscr{F}$ (e) $8\mathscr{F}$
15.7 87.02
15.9 (a) 1.78×10^{-3} (b) 1.11 (c) 0.22 (d) 0.41
15.15 (a) yes (b) no, the reverse reaction (c) no, reverse (d) yes
15.21 $3 \times 10^{-15} M$, 9×10^{-36}
15.23 (a) $\mathscr{E}° = 1.46$ volts, $\mathscr{E} = 1.49$ volts, $\Delta G = -206$ kcal
    (b) $\mathscr{E}° = 0.11$ volt, $\mathscr{E} = 0.17$ volt, $\Delta G = -7.84$ kcal
    (c) $\mathscr{E}° = 1.28$ volts, $\mathscr{E} = 1.26$ volts, $\Delta G = -58.1$ kcal
15.25 0.0592 volt

Chapter 17
17.19 > 1200 kcal/mole

Chapter 22
22.7 8.32×10^8 years
22.19 2

Index